Springer Series in Materials Science

Volume 318

The Springer Series in Materials Science covers the complete spectrum of materials research and technology, including fundamental principles, physical properties, materials theory and design. Recognizing the increasing importance of materials science in future device technologies, the book titles in this series reflect the state-of-the-art in understanding and controlling the structure and properties of all important classes of materials.

More information about this series at https://link.springer.com/bookseries/856

Alain Diebold · Tino Hofmann

Optical and Electrical Properties of Nanoscale Materials

Springer

Alain Diebold
SUNY Polytechnic Institute
Albany, NY, USA

Tino Hofmann
Department of Physics and Optical Science
University of North Carolina at Charlotte
Charlotte, NC, USA

ISSN 0933-033X ISSN 2196-2812 (electronic)
Springer Series in Materials Science
ISBN 978-3-030-80322-3 ISBN 978-3-030-80323-0 (eBook)
https://doi.org/10.1007/978-3-030-80323-0

This Springer imprint is published by the registered company Springer Nature Switzerland AG
The registered company address is: Gewerbestrasse 11, 6330 Cham, Switzerland

To my loving wife Annalisa, and our wonderful and inquisitive children Laura and Gregory and son in law Austin.

Alain Diebold

To my parents Rosieta and Eberhard and my wonderful wife Juliane and our curious children Anneliese and Oswin.

Tino Hofmann

Preface

The writing of this book is an endeavour of love for the subject. One of the goals of this book is to provide the ellipsometry and optical physics community an overview of 2D materials physics. One might say that the physics community work on topological materials speaks a different language from the materials science community. This book provides an introduction to both the physics of 2D materials and topological physics aimed at the materials science community whose research is in the area of optical characterization of materials.

It is clear from teaching optical physics at the graduate level that there is a need for providing advanced undergraduate and introductory graduate students with a text on optical physics. It is also clear that many students can be described as "math allergic". In order to address that need, this book provides additional details about derivations.

ACD and TH both gratefully acknowledge the expert and professional artwork of Yangzi Isabel Tian. Isabel endured roughly two years of weekly reviews of the figures and chapters. Her work brings the text to life. Many of the figures have been adapted for use in this book.

As with most books, a number of people provided critical assistance that made this book possible. We thank Zach Evenson who is our editor at Springer. He has patiently worked through the countless issues associated with the authorship and publication. We also thank Robert Hull of RPI who very kindly put us in contact with Zach making the book possible. ACD very gratefully acknowledges many discussions with Ed Seebauer whose guidance greatly influenced not only the text but the philosophy behind it. We note that Paul van der Heide and Brennan Peterson provided many useful comments for most of the chapters of the book.

We note the following read specific chapters and provided detailed review and commentary as follows:

Chapter 1. Gerald Jellison very kindly provided detailed edits and suggestions. Tony Heinz for discussions about characterizing nanoscale thick films.
Chapter 3. Stefan Schöche for help with the optical path of ellipsometers.
Chapters 4 and 5. Edmund Seebauer very kindly provided detailed edits and suggestions.

Chapter 6. Ji Ung Lee for several critically important comments including the need for the theory behind the integer quantum Hall effect for a 2D electron gas to explain the 1 part in 10^9 uncertainty.
Chapter 9. Shun-Qing Shen for discussions about Thoules, Kohmoto, Nightingale, and den Nijs quantization; Shiyuan Liu for discussions about $ZrTe_5$; Ana Akrap for discussions about $ZrTe_5$ and Cd_2As_3; and Ken Burch for comments on the optical characterization of tetradymites, especially Bi_2Se_3.
Appendix C. Emilio Cobanera for many discussions about topological classification and the topological periodic table and Eyal Cornfeld for guidance on interpreting the topological classification and topological periodic table that includes spatial symmetry.

Others have knowingly or unknowingly provided inspiration that also deserves recognition. Over the past several decades, the ellipsometry community has held one of the most enjoyable and scientifically useful conferences: the International Conference on Spectroscopic Ellipsometry (ICSE). Discussions at ICSE8 about nanoscale materials provided considerable inspiration. ICSE 8 was held in Barcelona in 2019 and was chaired by M. Isabel Alonso. Oriol Arteaga and Miquel Garriga were the other key organizers. Shiyuan Liu's invited talk and discussions clearly pointed out the challenges associated with determining the optical properties of anisotropic topological materials such as zirconium pentatelluride. ACD acknowledges the great fortune of being invited to present at ICSE 5 in Stockholm, Sweden. This trip provided an introduction to the ellipsometry community. At that time, Dave Aspnes and Jay Jellison were kind enough to discuss the topics that became inspiration. Both of them are so enthusiastic that one cannot help but follow their lead. I will never forget the time Dave derived a useful electromagnetic expression at an AVS meeting on the back of a napkin for ACD. We also acknowledge countless discussions with Stefan Zollner. The generous guidance of John Woollam and his company are specially mentioned as both inspiration and assistance. James Hilfiker, Tom Tiwald, Gerry Cooney, Blaine Johs and Harland Tompkins have been so generous with their time over the past countless years. We also acknowledge those at Woollam who help keep the equipment operational including Mark Brayton. The many scientists at NIST including Dave Seiler need special acknowledgement.

ACD acknowledges the students also provided insight into the challenges of teaching optical physics through their questions and comments. Vimal Kamineni, Manasa Medikonda, Raja Muthinti, Florence Nelson, Lay Wai Kong, Dhairya Dixit, Sam O'Mullane, Avery Green and Madhulika Korde all took the optical physics class that evolved into this text. Vimal Kamineni designed and built a photoluminescence system. The experience of mentoring undergraduate student Colin Wadsworth in ellipsometry and scatterometry has been very useful. Gert Leusink's group at TEL Technology Center America deserves special acknowledgement for funding research that included optical characterization of high k dielectrical films such as hafnium zirconium oxides. His group including Steve Consiglio, Dina Triyoso, Kanda Tapily and Rob Clark provided the students with materials that are essential to the semiconductor industry. We also acknowledge Tom Adams and Alex Reznicek who provided nanoscale films of $Si_{1-x}Ge_x$ alloys for optical characterization. The group postdocs especially Eric Berch and Sonal Dey also provided these students practical examples of the application of ellipsometry and critical mentoring.

The authors also hope that the book will be useful to the materials science community that uses ellipsometry, photoluminescence and Raman spectroscopy. The book is intended to address the need for increased communication between the users of these methods, the community of practitioners and the physics community.

The authors also recognize that many topics cannot be covered in this introductory text especially the mathematical foundations of topological physics. These topics might be a useful addition to a second edition of this book.

Niskayuna, NY, USA Alain Diebold
Charlotte, NC, USA Tino Hofmann

Acknowledgements

Yangzi Isabel Tian for all the artwork including the adaptation of published figures.

Zach Evenson for editing and guiding us through the publication process.

Robert Hull for connecting us to Springer.

General comments on the book—Paul van der Heide and Brennan Peterson. Review of the Preface—Howard Huff.

Chapter 1. Gerald Jellison.

Chapters 4 and 5. Edmund Seebauer.

Chapter 6. Ji Ung Lee for several critically important comments including the need for the theory behind the IQHE for a 2DEG to explain the 1 part in 10^9 uncertainty.

Shun-Qing Shen for discussions about TKNN quantization.

Chapter 9. Ana Akrap ($ZrTe_5$ and Cd_2As_3), Emilio Cobanera (topological classification and periodic table), Eyal Cornfeld (topological classification and periodic table), Ken Burch (tetradymites).

Niskayuna, NY, USA — Alain Diebold
Charlotte, NC, USA — Tino Hofmann

Contents

Chapter 1
The Interaction of Light with Solids: An Overview of Optical Characterization

Abstract In this chapter, we cover the basic principles involved in the interaction of light with crystals, thin films, and nanoscale materials necessary for discussing optical characterization. We discuss Fresnel's equations for bulk materials and thin films on substrates. The Fresnel equations for isotropic, uniaxial, and biaxial materials are all presented in terms of the complex refractive index. This chapter introduces ellipsometric characterization of the dielectric function of nanoscale materials, and it also discusses Raman spectroscopy and photoluminescence of 2D materials.

Light provides one of the most interesting means of characterizing new materials. As we know from observation of the world around us, materials are transparent, translucent, or opaque. We also know that these characteristics depend on the wavelength of the light. We can also observe that transparency can depend on thickness of some thin films present on a substrate. We can see that the thickness of some transparent films on an opaque substrate determines the perceived color of the film. One example of this is the thickness of silicon dioxide or silicon nitride on silicon. Even materials that we know are opaque for thick layers or bulk samples such as metal films can be transparent when they are thin enough. Here, we explore the application of these properties to the characterization of nanoscale materials.

The flow of this chapter is as follows: First the electromagnetic wave nature of light and is presented. Next, the fundamental concept of how light interacts with a solid, the dielectric function and the relationship between the complex refractive index, the absorption coefficient, and optical conductivity are introduced. Since the dielectric function is more completely described as a tensor, the tensor nature of optically uniaxial and biaxial crystals and films is discussed. The Fresnel equations that describe the reflection of the two polarization states of light are then presented. Following the discussion about the Fresnel equations, the particle picture of light is introduced. At the end of the chapter, Raman spectroscopy and photoluminescence are described.

A. Diebold and T. Hofmann, *Optical and Electrical Properties of Nanoscale Materials*, Springer Series in Materials Science 318,
https://doi.org/10.1007/978-3-030-80323-0_1

1.1 The Wave Nature of Light

Light is an electromagnetic wave with an electric field component and a magnetic field component which is perpendicular to the vector representing the electric field. Before discussing the wave equation for light, it is useful to show how it can be readily derived from Maxwell's equations for electromagnetism. We also note that historically, two different unit systems are found throughout the literature: Gaussian and the international system of units, SI. The reader can refer to undergraduate physics textbooks for further reference. Maxwell's equations are reviewed here so that the electromagnetic nature of the light wave is more easily understood. First, definitions of the electric displacement $\vec{D}$ in terms of the electric field $\vec{E}$, and magnetic field strength $\vec{H}$ in terms of the magnetic flux density $\vec{B}$ are stated.

$$\vec{D} = \varepsilon_0 \varepsilon_r \vec{E}; \; \vec{B} = \mu_0 \mu_r \vec{H} \tag{1.1}$$

where ε_0 is the permittivity of free space, ε_r is the relative dielectric constant which is a function of frequency and is a tensor for anisotropic materials, μ_0 is the magnetic permeability of vacuum, and μ_r is the relative magnetic permeability. The dielectric tensor is a function of the frequency of the light, so it is typically referred to as a dielectric function. Since many materials discussed in this book have anisotropic optical properties, the dielectric properties are expressed in tensor form as discussed in Sect. 1.2. For non-magnetic materials, the magnetic permeability is 1 for non-magnetic materials. The magnetic permeability for anisotropic magnetic materials is stated as a tensor. Maxwell's equations are expressed in terms of the electric displacement, electric field, magnetic flux density, and magnetic field strength. The charge current $\vec{J}$ and the free charge density $\rho = -\varepsilon_0 \varepsilon_r \vec{\nabla} \cdot \vec{E}$ appear in these equations. Here we state the differential form of *Maxwell's equations in free space.* The equations are stated in SI units in black text and cgs units in parenthesis using c for the speed of light [1–3].

Gauss' Law (1.2a) states that the electric field flux out of a closed surface is proportional to the charge enclosed in that surface. The electric field lines go from the positive charge to the negative charge. Gauss' law is more apparent when considering the integral form where the integral over a closed surface of the electric flux is equal to the enclosed charge Q, $\oint \vec{E} \cdot d\vec{A} = Q/\varepsilon_0$. It is more convenient to use the differential form of Gauss' law:

$$\vec{\nabla} \cdot \vec{D} = \rho \quad \left(\vec{\nabla} \cdot \vec{D} = 4\pi\rho\right) \tag{1.2a}$$

Gauss' Law for Magnetism is a statement that there are no sources or sinks for the magnetic field lines. In other words there are no magnetic charges as there are for electric charges. Instead, there are magnetic dipoles. The integral form of Gauss' Law for magnetism states that the total magnetic flux through a closed surface is zero as follows: $\oint \vec{B} \cdot d\vec{A} = 0$. The differential form of this law is:

$$\vec{\nabla} \cdot \vec{B} = 0 \quad \left(\vec{\nabla} \cdot \vec{B} = 0\right) \tag{1.2b}$$

The *Maxwell–Faraday Law* states that a time varying magnetic field creates an electric field.

$$\vec{\nabla} \times \vec{E} = -\frac{\partial \vec{B}}{\partial t} \quad \left(\vec{\nabla} \times \vec{E} = -\frac{1}{c}\frac{\partial \vec{B}}{\partial t}\right) \tag{1.2c}$$

The *Maxwell–Ampere Law* states that the magnetic field is produced by both current flow plus a changing electric field.

$$\vec{\nabla} \times \vec{H} = \vec{J} + \frac{\partial \vec{D}}{\partial t} \quad \left(\vec{\nabla} \times \vec{H} = \frac{4\pi}{c}\vec{J} + \frac{1}{c}\frac{\partial \vec{D}}{\partial t}\right) \tag{1.2d}$$

The complete statement of Maxwell's laws includes the effect of the polarizability and magnetization of the medium in which the electric and magnetic fields are present. The concept of a vector potential $\vec{A}$ for defining the magnetic flux density and the electric field leads to two very useful equations after substituting $\vec{B} = \vec{\nabla} \times \vec{A}$ into (1.2c):

$$\vec{B} = \vec{\nabla} \times \vec{A} \tag{1.3a}$$

$$\vec{E} = -\frac{\partial \vec{A}}{\partial t} \quad \left(\vec{E} = -\frac{1}{c}\frac{\partial \vec{A}}{\partial t}\right) \tag{1.3b}$$

For light moving in free space, no charge current flows, there is no free charge ρ, and (1.1) can be substituted into (1.2c) and (1.2d) to give:

$$\vec{\nabla} \times \vec{E} = -\mu_0 \mu_r \frac{\partial \vec{H}}{\partial t} \tag{1.4a}$$

$$\vec{\nabla} \times \vec{H} = \varepsilon_0 \varepsilon_r \frac{\partial \vec{E}}{\partial t} \tag{1.4b}$$

The curl in (1.4a) and (1.4b) indicates that the electric and magnetic fields will be perpendicular to each other. These two equations can be combined by taking the curl of (1.4a) and then inserting (1.4b):

$$\vec{\nabla} \times \vec{\nabla} \times \vec{E} = -\mu_0 \mu_r \frac{\partial \vec{\nabla} \times \vec{H}}{\partial t} = -\mu_0 \mu_r \varepsilon_0 \varepsilon_r \frac{\partial^2 \vec{E}}{\partial t^2} \tag{1.5}$$

The quantity $\vec{\nabla} \times \vec{\nabla} \times \vec{E}$ can be restated using the vector identity curl of the curl of a vector equals the gradient of the divergence minus the Laplacian of the vector:

$$\vec{\nabla} \times \vec{\nabla} \times \vec{E} = \vec{\nabla}\left(\vec{\nabla} \cdot \vec{E}\right) - \vec{\nabla}^2 \vec{E}$$

Since $\vec{\nabla} \cdot \vec{D} = 0$ since there is no free charge in vacuum,

$$\vec{\nabla} \times \vec{\nabla} \times \vec{E} = -\vec{\nabla}^2 \vec{E}$$

This equation can be inserted into (1.4) to give the wave equation. The wave equation for light describes the propagation of the electric field $\vec{E}$ in free space in SI units is:

$$\nabla^2 \vec{E} = \mu_0 \mu_r \varepsilon_0 \varepsilon_r \frac{\partial^2 \vec{E}}{\partial t^2} \tag{1.6}$$

This equation for light takes the form of a wave equation which illustrates the wave nature of light. In order for (1.5) to be a wave equation, the constants in front of the second derivative must be related to the velocity of the wave v as follows: $\mu_0 \mu_r \varepsilon_0 \varepsilon_r = \frac{1}{v^2}$. In free space, $\mu_r = 1$, and $\varepsilon_r = 1$ which led Maxwell in 1864 to the conclusion that the speed of light, $v = c = \frac{1}{\sqrt{\mu_0 \varepsilon_0}}$. Later, the speed of light was measured to be 2.998×10^8 ms^{-1}. Thus ***the wave equation for light in free space*** is:

$$\nabla^2 \vec{E} = \frac{1}{c^2} \frac{\partial^2 \vec{E}}{\partial t^2} \tag{1.7}$$

A linear polarized light wave is shown in Fig. 1.1, and the difference between linearly, circularly, and elliptically polarized light waves is shown in Fig. 1.2. Elliptically polarized light is a combination two perpendicular components of light with

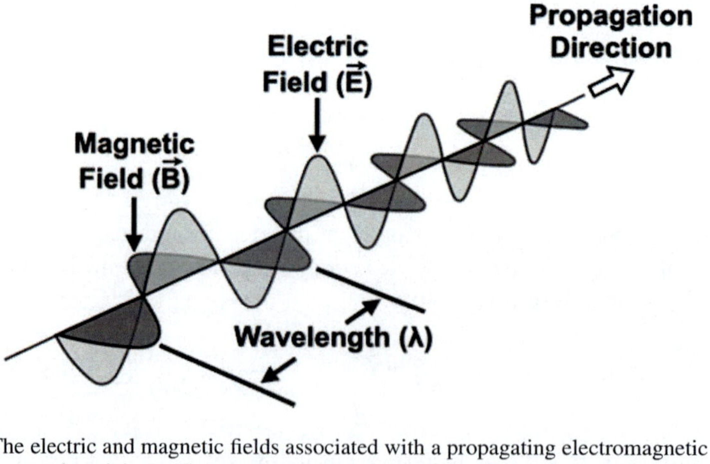

Fig. 1.1 The electric and magnetic fields associated with a propagating electromagnetic wave are shown. The wavelength is the distance between maxima of the electric (magnetic) field

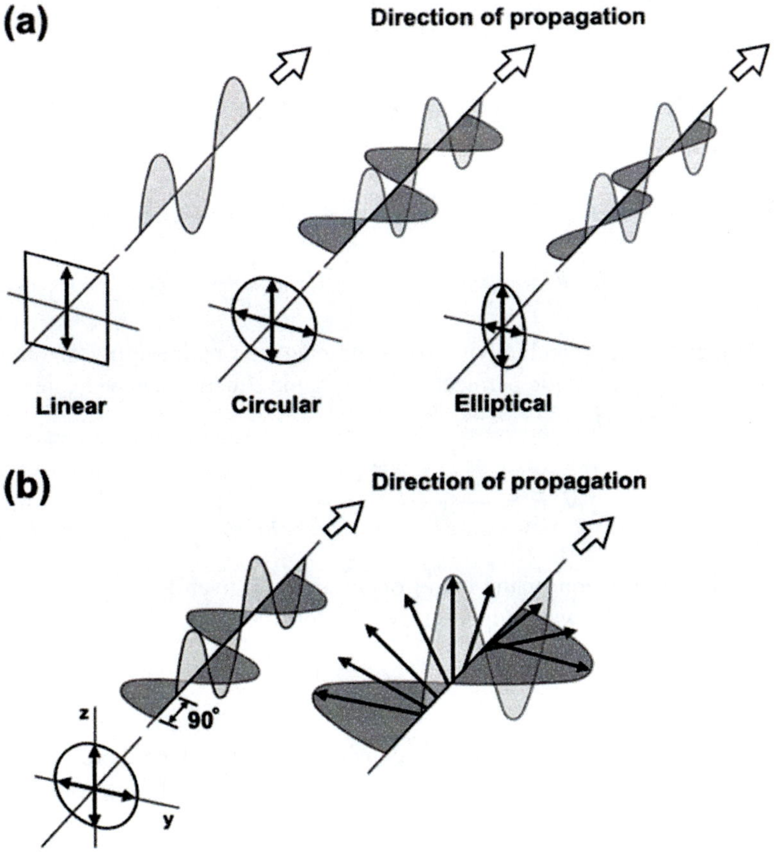

Fig. 1.2 The propagating electric field vector of linearly, circularly and elliptically polarized light is shown in (**a**), and **b** the electric field vector of right circularly polarized light will appear to be constantly rotating counterclockwise

a phase difference Δ between the waves. When light is traveling in a *non-absorbing medium* the refractive index is a real number, and the velocity is reduced by $v = \frac{c}{n}$ with $\mu_r = 1$ at optical frequencies as one can see from:

$$v = \frac{1}{\sqrt{\mu_0 \mu_r \varepsilon_0 \varepsilon_r}} = c \frac{1}{\sqrt{\varepsilon_r}} \text{ with } n = \sqrt{\varepsilon_r} \tag{1.8}$$

When light interacts with an absorbing medium such as a metal, the current density is not zero, and the current density and the electric field are connected by the complex electrical conductivity σ so that the current $J = \sigma \vec{E}$ [1]. Inserting this expression for J into (1.2d), (1.2d) and (1.2d) can be rewritten:

$$\vec{\nabla} \times \vec{E} = -\mu_0 \mu_r \frac{\partial \vec{H}}{\partial t} \tag{1.9a}$$

$$\vec{\nabla} \times \vec{H} = \sigma \vec{E} + \varepsilon_0 \varepsilon_r \frac{\partial \vec{E}}{\partial t} \tag{1.9b}$$

Once again, taking the curl of (1.9a) leads to:

$$\nabla^2 \vec{E} = \sigma \mu_0 \mu_r \frac{\partial \vec{E}}{\partial t} + \mu_0 \mu_r \varepsilon_0 \varepsilon_r \frac{\partial^2 \vec{E}}{\partial t^2} \tag{1.10a}$$

where σ is the electrical conductivity, μ_0 is the magnetic permeability of vacuum, and μ_r is the relative magnetic permeability. The same equation can be expressed in Gaussian (CGS) units [1, 3]:

$$\left(\nabla^2 \vec{E} = \frac{4\pi \sigma \mu_r}{c^2} \frac{\partial \vec{E}}{\partial t} + \frac{\varepsilon \mu_r}{c^2} \frac{\partial^2 \vec{E}}{\partial t^2} \right) \tag{1.10b}$$

In general, the direction of the travel of the light is given by a complex wave vector $\vec{k}$. Solutions to the wave equation take the form:

$$\vec{E} = \vec{E}_0 e^{-i(\vec{k} \cdot \vec{r} - \omega t)}, \tag{1.11}$$

where $\vec{E}_0$ is perpendicular to $\vec{k}$. First we consider the case where there is no light absorption by substituting (1.11) into (1.7) which results in a real wavevector where $k^2 = \left(\frac{n\omega}{c}\right)^2$ while substituting (1.11) into the wave equation when there is light absorption, (1.10a), results in a complex wavevector where $k^2 = \mu_0 \mu \varepsilon_0 \varepsilon_r \omega^2 + i\sigma \mu_0 \mu \omega$.

Light can be described in terms of its particle nature, photons, and its wave nature. The wavelength of light λ refers to the distance between the maximums of the electric field of the traveling light wave as shown in Fig. 1.1. Since we found that in general the wavevector k is a complex number, in general a material must have a *complex refractive index* $N = n + i\kappa$ where the extinction coefficient κ accounts for the absorption of light. The imaginary part of the refractive index κ *is not to* be confused with the wavevector $\vec{k}$. Frequently in the literature and thus in this book, the symbol k is used instead of κ for the imaginary part of the refractive index. The relationship between the wavelength λ, angular frequency ω, the k magnitude of the wavevector $\vec{k}$, and the velocity of light v in a medium with *complex refractive index* $N = n + i\kappa$ and the absorption coefficient α are given by:

$$\lambda = \frac{2\pi}{k}; \quad \frac{2\pi}{\lambda} = k = \frac{N\omega}{c}; \quad v = N/c \text{ and } \alpha \equiv \frac{4\pi\kappa}{\lambda} \tag{1.12}$$

Although it may be obvious, it is useful to state that the absorption of light means that the incident light is losing energy to the material it is traversing. Later in the chapter, the particle nature of light will be further discussed. The relationship between the photon energy, wavelength, angular frequency, and frequency v ($\omega = 2\pi v$) of light is given in terms of Plank's constant h by:

$$E = \hbar\omega = hv \tag{1.13}$$

In free space $n = 1$ and $E = \hbar\omega = hc/\lambda$. The relative dielectric constant ε_r used in Maxwell's equations above is a frequency dependent complex dielectric function $\varepsilon = \varepsilon_1 + i\varepsilon_2$. The relationship between the wavelength complex refractive index $N = n + i\kappa$ and complex dielectric function $\varepsilon = \varepsilon_1 + i\varepsilon_2$ and the complex optical conductivity σ in SI units (conductivity at optical frequencies) is:

$$\varepsilon = \varepsilon_1 + i\varepsilon_2 = N^2 = (n^2 - \kappa^2) + i2n\kappa = 1 + i\frac{\sigma(\omega)}{\varepsilon_0\omega} \tag{1.14a}$$

The dielectric function is relative to ε_0 the electrical permittivity of free space. In this book, we will refer to the real and imaginary parts of the dielectric function using ε_1 and ε_2. In CGS units, the relationship between the dielectric function and optical conductivity is:

$$\varepsilon = 1 + i\frac{4\pi\sigma(\omega)}{\omega} \tag{1.14b}$$

1.2 Dielectric Tensor of Bulk Crystals

For some crystals the interaction of light with the crystal depends on the direction of light in the crystal lattice, and thus the dielectric function is not isotropic. Well known examples include calcite and quartz crystals. In this section we discuss the tensor representation of the dielectric function. The complete dielectric tensor is:

$$\boldsymbol{\varepsilon} = \begin{pmatrix} \varepsilon_{xx} & \varepsilon_{xy} & \varepsilon_{xz} \\ \varepsilon_{yx} & \varepsilon_{yy} & \varepsilon_{yz} \\ \varepsilon_{zx} & \varepsilon_{zy} & \varepsilon_{zz} \end{pmatrix} \tag{1.15}$$

This form for the dielectric function is useful for a randomly oriented crystal. However, in most instances the sample is not randomly oriented. Let us assume that the optical axes lie along the *x*, *y*, *z* coordinate system axes. When the crystal is oriented so that the optical axes are not parallel to the *x*, *y*, *z* coordinate system axes, rotation matrices are used to rotate the coordinate system along the Euler angles as described in [4]. Many 2D materials are stacked layers that have van der Waals bonding between layers. These materials often stack along an optical axis. Exfoliated and deposited single layer or multilayer samples will also have the optical axis perpendicular to the surface. Thus, in many cases the optical axis of 2D films and bulk crystals will lie along the coordinate axis, and it is convenient to orient the coordinate system so that the z axis is parallel to the optical axis. Below, we relate the dielectric tensor to the optical axes of bulk crystal samples and thin films.

For optically isotropic crystals such as cubic, face center cubic, body center cubic, and amorphous crystals, the dielectric function is the same along all directions so that:

$$\boldsymbol{\varepsilon} = \begin{pmatrix} \varepsilon & 0 & 0 \\ 0 & \varepsilon & 0 \\ 0 & 0 & \varepsilon \end{pmatrix} \tag{1.16}$$

The complex dielectric function of several semiconductors such as Si, Ge, and GaAs is presented in Chap. 4.

For uniaxial crystals such as quartz [4, 5], rutile (TiO_2) [6], $AgGaS_2$, wurtzite (hexagonal) crystals such as α-GaN and α-AlN, and ZnO, when the optical axis lies along the z direction, the dielectric tensor is [4, 5]:

$$\boldsymbol{\varepsilon} = \begin{pmatrix} \varepsilon_x & 0 & 0 \\ 0 & \varepsilon_x & 0 \\ 0 & 0 & \varepsilon_z \end{pmatrix} = \begin{pmatrix} \varepsilon_o & 0 & 0 \\ 0 & \varepsilon_o & 0 \\ 0 & 0 & \varepsilon_e \end{pmatrix} \tag{1.17}$$

Here, ε_o is the ordinary dielectric function and ε_e is the extraordinary dielectric function. This leads to an ordinary and extraordinary refractive index. Both the ordinary and extraordinary refractive index of rutile TiO_2 are shown in Fig. 1.3. Because the refractive index is different along the ordinary and extraordinary directions, light travels at different speeds along these directions. The origin of the terminology ordinary and extraordinary and birefringence can be understood in terms of light interacting with a uniaxial crystal. If a uniaxial crystal is oriented so that the optical

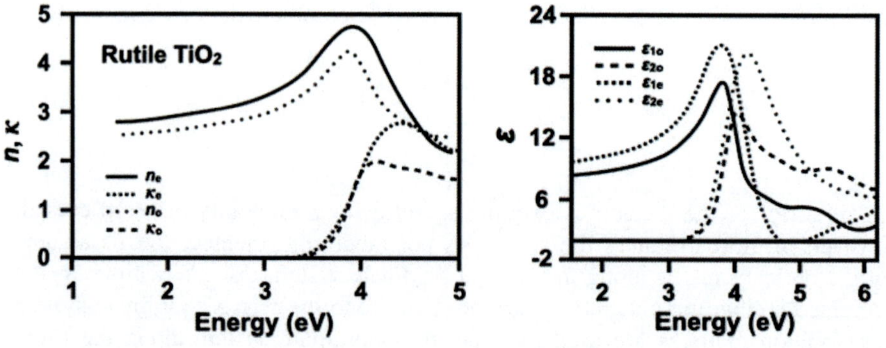

Fig. 1.3 The ordinary and extraordinary complex refractive index and complex dielectric function of the rutile phase of TiO_2 are shown from 1 to 5 eV [6]. Figure courtesy G. Jellison private communication

axis lies along the z direction (normal to the surface), and the incident light is traveling parallel to the optical axis (normal to the surface), then all polarization states of the light interact with the ordinary refractive index. When light is incident on this uniaxial crystal at an angle other than normal, the part of the light polarized perpendicular to the optical axis will interact with the ordinary part of the refractive index producing an ordinary ray, and the part of the light polarized perpendicular to the ordinary ray will interact with a refractive index that is an angular dependent mix of the ordinary and extraordinary refractive index. This is the extraordinary ray.

The unambiguous determination of the dielectric function requires measurement and analysis of datasets obtained for different crystal orientations and angles of incidence. However, some 2D samples are too thin for measurement of different crystal orientations. Determination of the dielectric function of uniaxial crystals can be done using reflection and transmission or through a prism coupler. Crystals with a hexagonal or tetragonal crystal symmetry are uniaxial. Graphite [7], graphene, and many transition metal dichalcogenides (TMD) are all hexagonal van der Waals materials with a single optical axis. Thin films of these TMD crystals will also have the optical axis normal to the surface.

Quartz and $AgGaS_2$ are optically active which means that they rotate the polarization of light [5]. Thus right and left circularly polarized light travels at different phase velocities. The gyration tensor $\boldsymbol{\alpha}$ of these materials is non-zero, and thus electric displacement $\vec{D}$ for optically active includes the impact of the magnetic field:

$$\vec{D} = \boldsymbol{\varepsilon}\vec{E} - i\boldsymbol{\alpha}\vec{H} \tag{1.18a}$$

And the magnetic displacement of optically active crystals includes the effect of the electric field where $\boldsymbol{\alpha}^T$ is the transpose of the matrix $\boldsymbol{\alpha}$:

$$\vec{B} = \mu\vec{H} + i\boldsymbol{\alpha}^T\vec{E} \tag{1.18b}$$

For optically active, uniaxial crystals, the gyration tensor is:

$$\boldsymbol{\alpha} = \begin{pmatrix} \alpha_x & 0 & 0 \\ 0 & \alpha_x & 0 \\ 0 & 0 & \alpha_z \end{pmatrix} \tag{1.18c}$$

Orthorhombic, monoclinic, and triclinic crystals are biaxial, i.e., they have two optical axes [6]. Let us again assume that the optical axes lie along the x, y, z coordinate system axes. For oriented, biaxial crystals such as c-$GeSe_2$, the dielectric tensor is:

$$\boldsymbol{\varepsilon} = \begin{pmatrix} \varepsilon_x & 0 & 0 \\ 0 & \varepsilon_y & 0 \\ 0 & 0 & \varepsilon_z \end{pmatrix} \tag{1.19}$$

For crystals with 4 optical axes such as monoclinic $CdWO_4$, and Ga_2O_3, the dielectric tensor is [8]:

$$\boldsymbol{\varepsilon} = \begin{pmatrix} \varepsilon_{xx} & \varepsilon_{xy} & 0 \\ \varepsilon_{xy} & \varepsilon_{yy} & 0 \\ 0 & 0 & \varepsilon_{zz} \end{pmatrix} \tag{1.20}$$

Determination of the wavelength (energy) dependent dielectric functions of 4 axis crystals requires measuring multiple crystal faces at multiple angles of incidence and multiple azimuthal angles.

1.3 Spectroscopic Ellipsometry

Spectroscopic ellipsometry is the wavelength dependent reflection or transmission of polarized light through a material [4, 9]. Spectroscopic ellipsometry is understood in terms of the wave nature of light discussed above. First, the case of a crystal or amorphous materials with isotropic optical properties [the dielectric tensor of (1.16)] is considered. The ellipsometric angles ψ and Δ are defined by the ratio of the complex Fresnel reflection coefficient r_p of the component of the incident light that is polarized parallel to the plane of incidence (see Fig. 1.4) to the complex reflection coefficient of the component of the light that is polarized perpendicular to the plane of incidence r_s as follows:

$$\tan\psi e^{i\Delta} \equiv \frac{r_p}{r_s} \text{ with } r_s(\boldsymbol{\varepsilon}_i,\ \theta,\ d_i) = \frac{E_{rs}}{E_{is}};\ r_p(\boldsymbol{\varepsilon}_i,\ \theta,\ d_i) = \frac{E_{rp}}{E_{ip}} \tag{1.21}$$

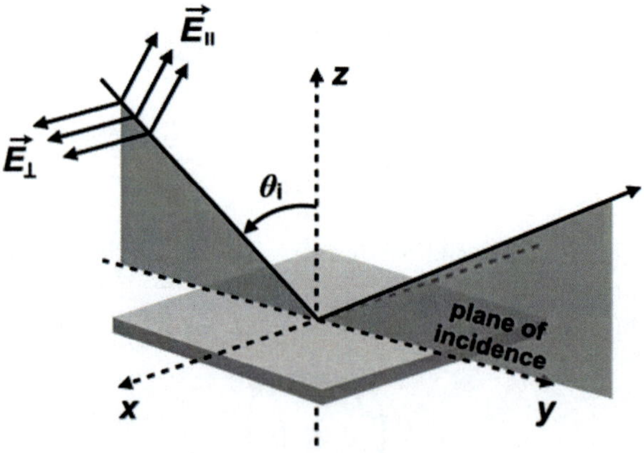

Fig. 1.4 Reflection of p $\left(\vec{E}_{\parallel}\right)$ and s $\left(\vec{E}_{\perp}\right)$ polarized light is shown with respect to the plane of incidence defined by the sample normal and the propagation direction of the light

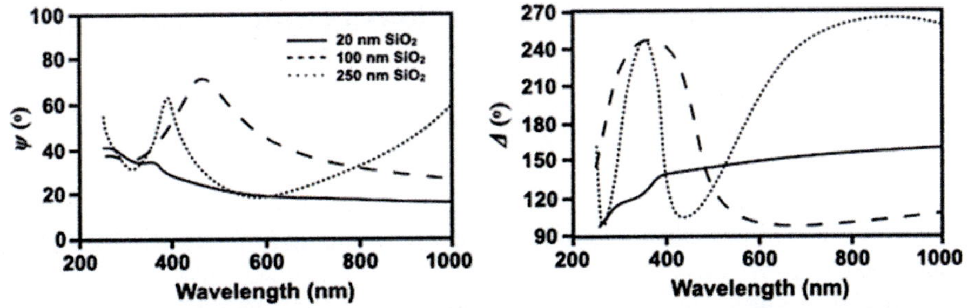

Fig. 1.5 Ellipsometric parameters ψ and Δ for silicon dioxide on silicon. Both materials have isotropic optical properties

Here, ε_i is the complex dielectric function(s) of material(s) i, θ is the angle of incidence and d_i is thickness of a film(s) i for samples composed of layer(s) of material(s). The electric field polarized s and p is shown in Fig. 1.4. As indicated in (1.15) the reflectivities are a function of the dielectric tensor of the material that the light is reflected from or transmitted through. Equation (1.21) describes the mathematical relationship for the ellipsometric angles ψ and Δ when there is no cross-polarized light scattering from s to p or from p to s polarization. In Fig. 1.5, we show an example of how the ellipsometric parameters ψ and Δ change as the thickness of silicon dioxide on silicon increases. Above the thin film limit, ellipsometry allows the determination of both the film thickness and the dielectric function. Below the thin film limit, N and d are correlated due to the small differences in path length even when different angles of incidence are used. If the film thickness of a very thin film is known then the dielectric function can be determined.

As mentioned above, the Jones matrix for generalized ellipsometry when cross-polarized light scattering occurs modifies the above discussion. First, there are reflection coefficients for cross-polarized specular reflection, i.e., incident p polarized light that reflects with both p and s polarization components and incident s polarized light that reflects with both s and p polarizations components [4, 9–12]. Thus the ellipsometric equations for ψ and Δ become:

$$\begin{aligned} \rho_{pp} &= {r_{pp}}/{r_{ss}} = \tan\psi_{pp} e^{-i\Delta_{pp}} \\ \rho_{ps} &= {r_{ps}}/{r_{ss}} = \tan\psi_{ps} e^{-i\Delta_{ps}} \\ \rho_{sp} &= {r_{sp}}/{r_{ss}} = \tan\psi_{sp} e^{-i\Delta_{sp}} \end{aligned} \tag{1.22}$$

where the reflection coefficients r_{pp}, r_{ps}, and r_{sp} are expressed in terms of the matrix elements of a transfer matrix which relates the incident and reflected to the transmitted p and s polarization components of the light. In other words, the light that is not reflected is transmitted.

1.4 Fresnel Equations for the Reflection of Light

In this section, we explore the reflection of light from bulk and thin film samples with the goal of describing the impact of 2D materials with anisotropic optical properties on reflectivity and ellipsometric measurements. Many 2D materials are one or few layer pieces of van der Waals materials where each layer is bonded by van der Waals forces to the layers above and below. Examples include graphene and few layer graphene, the trilayer transition metal dichalcogenides (TMD), and many topological insulators. For example, the tetradymite compounds (Bi_2Se_3, Bi_2Te_3, $Bi_2Se_xTe_{3-x}$, Sb_2Te_3 and alloys) have a quintuple layer (QL) structure with van der Waals bonding between the QLs. These materials are described in subsequent chapters.

Graphene, few layer graphene, and graphite have a single optical axis that is normal to the plane of the graphene layers. The single and few layer graphene samples lying on substrates such as a glass slide or on a silicon wafer with a thin silicon dioxide surface film represent a class of optically uniaxial materials with the optical axis normal to the film-substrate surface. This symmetry simplifies the Fresnel equations for the reflection of polarized light. Other examples that have the same symmetry include single and few trilayers of TMD on an isotropic substrate. It is useful to mention that graphene grown on most polymorphs of silicon carbide has a surface topography that results in the graphene not lying flat on the surface. The symmetry of the reflection of light from 2D materials is further discussed below.

In order to provide the necessary background for discussing ellipsometric measurement of 2D materials, we progress through a discussion of the Fresnel reflection coefficients for isotropic bulk and isotropic thin film on isotropic substrate before discussing the Fresnel equations for key examples of anisotropic thin films such as Graphene and TMD trilayers on a substrate. Azzam and Bashara as well as Fujiwara continue to provide an excellent references for ellipsometric measurement of both isotropic and anisotropic film-substrate combinations [4, 9].

1.4.1 Fresnel Description of the Reflection of Light from an Isotropic Material

The origin of the Fresnel equations for the amplitude of the reflection of polarized light is the boundary conditions for electromagnetic waves at an interface [4, 9]. The components of the electric field $\vec{E}$ and magnetic field $\vec{B}$ that are parallel to an interface must be continuous across that interface. Here, light specularly reflects from a smooth surface of an optically isotropic sample at an incident angle θ_i and the component of the light polarized along the plan of incidence is designated p and the component of the light perpendicular to the plane of incidence is designated s. The refractive index of the ambient (air) is labeled 0 and the material is labeled 1. This reflection configuration is used for all the examples in this section.

In Fig. 1.4, we show the reflection of s polarized light, $\overrightarrow{E_s}$ and $\overrightarrow{B_s}$, from an interface. The angle of incidence from the sample normal is θ_i as shown in Fig. 1.4. The electric field is parallel to the interface with incident magnitude E_{is}. It is useful to remember the cross-product relationship of (1.2c) which means that the electric field and magnetic field in the light wave are perpendicular to each other. Thus, the incident component of the magnetic field of magnitude B_s parallel to the surface is $-B_{is}\cos\theta_i$ and the magnitude of the reflected component that is parallel to the surface is $B_{ir}\cos\theta_r$. The transmitted electric field and magnetic field and the associated angle of transmission are E_{ts}, B_{ts} and θ_t. For specular reflection, $\theta_i = \theta_r$, so $B_{ir}\cos\theta_r = B_{ir}\cos\theta_i$. The boundary conditions for the s polarized light result in:

$$E_{is} + E_{rs} = E_{ts} \text{ and } -B_{is}\cos\theta_i + B_{rs}\cos\theta_r = -B_{ts}\cos\theta_t \quad (1.23)$$

Using the fact that the magnitude of the fields are related by $E = \frac{c}{N}B$ where N is the complex refractive index. The complex refractive index of the ambient is N_0, and the complex refractive index for the optically isotropic sample is N_1. The ambient (typically air) is assumed to be optically isotropic with $N_0 = 1$. We note that the electric field of the light is interacting with the refractive index parallel to the surface.

$$-B_{is}\cos\theta_i + B_{rs}\cos\theta_r = (-N_0E_{is}\cos\theta_i + N_0E_{rs}\cos\theta_i)/c = -N_1E_{ts}\cos\theta_t/c$$

$$-N_0E_{is}\cos\theta_i + N_0E_{rs}\cos\theta_i = -N_1E_{ts}\cos\theta_t = -N_1(E_{is} + E_{rs})\cos\theta_t$$

Rearranging we get:

$$-E_{is}(N_0\cos\theta_i - N_1\cos\theta_t) = -E_{rs}(N_0\cos\theta_i + N_1\cos\theta_t)$$

The amplitude of the reflection coefficient of the s polarized light is $r_s = \frac{E_{rs}}{E_{is}}$. We get:

$$r_s = \frac{E_{rs}}{E_{is}} = \frac{N_0\cos\theta_i - N_1\cos\theta_t}{N_0\cos\theta_i + N_1\cos\theta_t} \quad (1.24)$$

In a similar fashion, the boundary conditions for the reflection of p polarized light from the surface of an optically isotropic material are:

$$B_{ip} + B_{rp} = B_{tp} \; and \; E_{ip}\cos\theta_i - E_{rp}\cos\theta_r = E_{tp}\cos\theta_t \quad (1.25)$$

Using a similar analysis we get:

$$r_p = \frac{E_{rp}}{E_{ip}} = \frac{N_1\cos\theta_i - N_0\cos\theta_t}{N_1\cos\theta_i + N_0\cos\theta_t} \quad (1.26)$$

1.4.2 Isotropic Bulk Materials

Using the above analysis, we present the Fresnel equations for polarized light reflecting from a solid with a flat smooth surface with minimal roughness that is perpendicular to the solid normal [4, 9]. We will be consistent with the previous discussion and the solid has a complex refractive index N_1. Now we use Snell's law, $N_0 \sin\theta_i = N_1 \sin\theta_t$ where $\theta_i = \theta_0$ and $\theta_t = \theta_1$ to rewrite the Fresnel equations. A key point of comparison between the Fresnel equations for isotropic solids and many optically anisotropic 2D materials is that the refractive index is the same in the plane of the surface and perpendicular to the surface for an isotropic solid. Again, the ambient (air) is labeled 0 and the solid is labeled 1. The amplitude reflection coefficient for p polarized light from an optically isotropic material can be rewritten using Snell's law as:

$$r_{01p} = \frac{(N_1/N_0)^2\cos\theta_0 - \left((N_1/N_0)^2 - \sin^2\theta_0\right)^{1/2}}{(N_1/N_0)^2\cos\theta_0 + \left((N_1/N_0)^2 - \sin^2\theta_0\right)^{1/2}} \tag{1.27}$$

For $N_0 = 1$, where we use a to represent refection from a material in the ambient.

$$r_{01p} = \frac{N_1^2\cos\theta_0 - \left(N_1^2 - \sin^2\theta_0\right)^{1/2}}{N_1^2\cos\theta_0 + \left(N_1^2 - \sin^2\theta_0\right)^{1/2}} \tag{1.28}$$

For s polarized light, the light interacts with the in plane (parallel to the surface) part of the refractive index, and for the s polarized light the Fresnel equation is:

$$r_{01s} = \frac{\cos\theta_0 - \left((N_1/N_0)^2 - \sin^2\theta_0\right)^{1/2}}{\cos\theta_0 + \left((N_1/N_0)^2 - \sin^2\theta_0\right)^{1/2}} = \frac{N_0\cos\theta_0 - \left(N_1^2 - N_0^2\sin^2\theta_0\right)^{1/2}}{N_0\cos\theta_0 + \left(N_1^2 - N_0^2\sin^2\theta_0\right)^{1/2}} \tag{1.29a}$$

For $N_0 = 1$,

$$r_{01s} = \frac{\cos\theta_0 - \left(N_1^2 - \sin^2\theta_0\right)^{1/2}}{\cos\theta_0 + \left(N_1^2 - \sin^2\theta_0\right)^{1/2}} \tag{1.29b}$$

The *pseudodielectric function* is an important concept that is often used to analyze ellipsometric data [4]. The ellipsometric data for a bulk solid can be interpreted assuming that the sample has no surface roughness and no surface contamination or surface oxide layer. The resulting dielectric function (complex refractive index) is known as the pseudodielectric function. The effect of surface roughness as well as surface oxide and contamination must be accounted or in the determination of the real dielectric function [4]. The optical response of surface roughness can be included using an effective medium approximation which is discussed below. The optical response of a surface oxide layer is that of a very thin isotropic film on the bulk substrate.

1.4.3 Isotropic Thin Film on Isotropic Bulk Substrate

The derivation of the Fresnel amplitude reflection coefficients for an optically isotropic thin film on an isotropic substrate starts with a general expression for the amplitude reflection coefficients for light incident on media i which is on media j where $i, j = 0, 1$ or $i, j = 1, 2$ as shown in Fig. 1.1 [4, 9]:

$$r_{ijp} = \frac{N_j \cos\theta_i - N_i \cos\theta_j}{N_j \cos\theta_i + N_i \cos\theta_j}\ ;\ r_{ijs} = \frac{N_i \cos\theta_i - N_j \cos\theta_j}{N_i \cos\theta_i + N_j \cos\theta_j} \tag{1.30}$$

When multiple reflections inside the film are accounted for, the Fresnel amplitude reflection coefficients for the optically isotropic thin film with complex refractive index N_1 and thickness d on an isotropic substrate with complex refractive index N_2 for light of wavelength λ are [3, 6]:

$$r_{012p} = \frac{r_{01p} + r_{12p}e^{-i2\beta}}{1 + r_{01p}r_{12p}e^{-i2\beta}} \tag{1.31a}$$

$$r_{012s} = \frac{r_{01s} + r_{12s}e^{-i2\beta}}{1 + r_{01s}r_{12s}e^{-i2\beta}} \tag{1.31b}$$

$$\beta = \frac{2\pi d}{\lambda} N_1 \cos\theta_1 = \frac{2\pi d}{\lambda}\left(N_1^2 - N_0^2 \sin^2\theta_0\right)^{1/2} \tag{1.31c}$$

The ellipsometric angles ψ and Δ are defined the amplitude reflection coefficients for p and s polarized light (1.24 and 1.26 or 1.31a and 1.31b) by:

$$\tan\psi e^{i\Delta} \equiv \frac{r_p}{r_s} \tag{1.31d}$$

1.4.4 Ultra-Thin Dielectric Film Ellipsometry

The Drude Relationship: The thickness of dielectric films less than 5 nm in thickness can be estimated from the value of the ellipsometric parameter Δ at a single wavelength. The film thickness parameter β, (1.31c), is a function of the film thickness d, wavelength λ, and the complex refractive index of the film N_1. For films that are non-absorbing in the wavelength range of the measurement, N_1 is real. The real part of the refractive index for amorphous silicon dioxide, SiO_2, varies between 1.4 and 1.55

from 160 to 3000 nm. and other transparent dielectric films such as silicon nitride (Si_3N_4) varies between 2.1 at 350 nm and 1.97 at 2500 nm. Thus, β is small, and the s and p reflection coefficients and consequently ψ do not change appreciably with film thickness. However, the phase of the p and s components of the incident light are changed by reflection from the dielectric thin film-substrate sample. Changing the refractive index changes the slope of relationship between Δ and the thickness of a very thin film. Thus the thickness and refractive index are correlated for ultra-thin films.

1.4.5 Thin 2D Film on Transparent Solid

The optical response of 2D materials on transparent substrates has been analyzed using a number of models including that of an isotropic film or slab on a transparent, semi-infinite substrate with real refractive index N_2. The origin of this model comes from the study of antireflective coatings [13]. This model has been applied to monolayer and few layer 2D materials [14]. This model is sometimes called the 3D Slab model. It is useful to note that the non-absorbing nature of the substrate depends on the wavelength range. When the light is incident ***normal*** to the surface through an ambient with refractive index $N_0 = 1$, the reflectance of a thin film with complex refractive index N_1 of thickness d can be written as [13]:

$$R = |r|^2 \text{ where } r = \frac{N_1(1 - N_2)\cos\varphi + i\left(N_2 - N_1^2\right)\sin\varphi}{N_1(1 + N_2)\cos\varphi + i\left(N_2 + N_1^2\right)\sin\varphi} \tag{1.32a}$$

The complex phase shift of the light of wavelength λ after a single pass through the film is $\varphi = 2\pi N_1 d/\lambda$. Li and Heinz restated the reflectance in terms of the phase shift $\varphi_0 = 2\pi d/\lambda$ in vacuum [13, 14]:

$$R = |r|^2 \text{ where } r = \frac{(1 - N_2)\cos\varphi - i\left(N_2 - N_1^2\right)\varphi_0\left(\frac{\sin\varphi}{\varphi}\right)}{(1 + N_2)\cos\varphi - i\left(N_2 + N_1^2\right)\varphi_0\left(\frac{\sin\varphi}{\varphi}\right)} e^{-2i\varphi_0} \tag{1.32b}$$

Here, we used the notation of Li and Heinz [14], and R in (1.32a) is equivalent to equation 53 in [13]. This notation will be convenient when the 2D film approximation is discussed below. Since the light is normal to the surface, (1.32) applies to a thin film of uniaxial material with the axis normal to the surface on a semi-infinite substrate. In that case, the refractive index determined from normal incidence reflectivity is the in-plane (in this case ordinary refractive index). Fused silica is non-absorbing over a wide wavelength range, and sapphire is considered to be non-absorbing between 400 and 900 nm and to have a refractive index of ~1.77. At normal incidence, the uniaxial nature of sapphire does not alter this analysis. An incident angle dependent formula for the total reflectivity of a isotropic thin film is provided (53) in [13].

1.4.6 Effective Medium Approximation for Surface Roughness

As mentioned above, most solids have some surface roughness. The amount of surface roughness depends on the preparation of the sample before measurement. When surface roughness is present, it can be modeled using an effective medium approximation (EMA). The EMA model describes how the dielectric function of the EMA layer used to model surface roughness is calculated from a mix of the dielectric function of air with the dielectric function of the bulk material where the volume fraction of each material (air f_a and the bulk material $f_b = 1 - f_a$) is a function of the roughness. The thickness of the EMA layer is determined by the amplitude of the surface roughness. The Bruggeman EMA dielectric function $\varepsilon(EMA)$ is given by the solution to [4]:

$$f_a \frac{\varepsilon_a - \varepsilon(EMA)}{\varepsilon_a + 2\varepsilon(EMA)} + (1 - f_a)\frac{\varepsilon_b - \varepsilon(EMA)}{\varepsilon_b + 2\varepsilon(EMA)} = 0$$

In the case of surface roughness where air is mixed with the bulk material, the volume fraction is referred to as the void fraction. Other EMA models have been used [4].

1.4.7 Anisotropic Uniaxial Solid with Uniaxial Optical Axis Normal to the Surface

An example of this configuration is graphite or a bulk crystal of Gallium Nitride with the C-axis normal to the surface. The refractive index in the surface horizontal direction is referred to as the ordinary refractive index $\boldsymbol{N_o}$, and the refractive index along the C axis normal to surface is referred to as extraordinary $\boldsymbol{N_e}$.[4, 9] This symmetry illustrates the important principle that there is no cross-polarized light scattering for this configuration and thus the Jones Matrix is diagonal. The *Fresnel amplitude reflection coefficient for the component of the light that is s polarized light is not altered for this symmetry*, and the reflected light beam is referred to as the ordinary ray [4, 9, 13, 15–19]. This can be verified by comparing (1.29a)–(1.33). The component of the light that is p polarized is altered by reflection and is referred to as the extraordinary ray. The Fresnel amplitude reflection coefficients can be derived in a similar manner as above and for an anisotropic substrate with uniaxial axis normal to the surface are:

$$r_{01s} = \frac{N_0 \cos\theta_0 - \left(\boldsymbol{N_o^2} - N_0^2 \sin^2\theta_0\right)^{1/2}}{N_0 \cos\theta_0 + \left(\boldsymbol{N_o^2} - N_0^2 \sin^2\theta_0\right)^{1/2}} \tag{1.33}$$

The Fresnel amplitude reflection coefficient for p polarized light is different from the isotropic case due to the different index of refraction normal to the surface, $\boldsymbol{N_e}$.

$$r_{01p} = \frac{\boldsymbol{N_o N_e}\cos\theta_0 - N_0\left(\boldsymbol{N_e^2} - N_0^2\sin^2\theta_0\right)^{1/2}}{\boldsymbol{N_o N_e}\cos\theta_0 + N_0\left(\boldsymbol{N_e^2} - N_0^2\sin^2\theta_0\right)^{1/2}} \tag{1.34}$$

Equations (1.33) and (1.34) are identical to (1.1) and (1.4) in [16]. For the p polarized light, the index of refraction experienced by the extraordinary light ray, n_1, is dependent on the angle of incidence and for non-absorbing solids is given by [17]:

$$n_1^2 = \boldsymbol{N_o^2}\left(1 + \frac{N_0^2\sin^2\theta_0}{\boldsymbol{N_o^2}} - \frac{N_0^2\sin^2\theta_0}{\boldsymbol{N_e^2}}\right) \approx \boldsymbol{N_o^2}\left(1 + \frac{\sin^2\theta_0}{\boldsymbol{N_o^2}} - \frac{\sin^2\theta_0}{\boldsymbol{N_e^2}}\right) \tag{1.35}$$

The refractive index of the ambient is again $N_0 = 1$, which needs to be distinguished from the ordinary refractive index of the uniaxial solid, $\boldsymbol{N_o}$. Using only reflection ellipsometry for determination of the ordinary and extraordinary complex refractive index of anisotropic solids is difficult if the optical axis is perpendicular to the surface. Aspnes [18] found that the effective dielectric properties are denominated by the ordinary part of the refractive index when a uniaxial solid has its optical axis normal to the surface for cases where the magnitude of the dielectric function is larger than 1. This is can be seen in (1.6) of [18]. De Smet [17] concluded that for uniaxial solids with the optical axis normal to the surface, the ordinary and extraordinary refractive indices cannot be determined by reflectance or ellipsometry by changing the angle of incidence. The case of graphite provides an interesting example of why the ellipsometric analysis of uniaxial solids often seems insensitive to the optical properties normal to the c-axis. It is convenient to follow [20] and refer to the in-plane complex refractive index of graphene and graphite oriented with the c-axis normal to the surface as n_x, k_x, and the normal, out-of-plane complex refractive index as n_z, k_z. n_z of graphite varies linearly between 2 at ~250 nm and ~1.5 at 1000 nm, and $k_z = \boldsymbol{k_e} = 0$ between 250 and 1000 nm [20]. $k_x = \boldsymbol{k_o}$ of graphite is not a linear function of wavelength between 250 and 1000 nm. If we apply (1.35) and ignore the in plane absorption, one can estimate for incident angles between 55° and 70° that (ambient, $N_0 = 1$):

$$\theta_0 = 55^\circ \text{ for graphite}$$

$$\text{at 250 nm } n_z = 2 : n_1^2 = \boldsymbol{N_o^2}\left(1 + \frac{0.67}{\boldsymbol{N_o^2}} - 0.17\right)$$

$$\text{at 1000 nm for } n_z = 1.5 : n_1^2 = \boldsymbol{N_o^2}\left(1 + \frac{0.67}{\boldsymbol{N_o^2}} - 0.3\right)$$

$$\theta_0 = 65^\circ \text{ for graphite}$$

$$at\ 250\ \text{nm}\ n_z = 2 : n_1^2 = N_o^2\left(1 + \frac{0.82}{N_o^2} - 0.21\right)$$

$$\text{at}\ \ 1000\ \text{nm for}\ n_z = 1.5 : n_1^2 = N_o^2\left(1 + \frac{0.82}{N_o^2} - 0.37\right)$$

The ordinary part of the refractive index of graphite is complex. Due to variation of k_o with wavelength, the extraordinary ray is strongly influenced by the ordinary refractive index for graphite at typical angles of incidence. Grazing incidence can increase the influence of the extraordinary refractive index.

The ordinary and extraordinary parts of the complex refractive index of uniaxial solids can be determined using spectroscopic ellipsometry by measuring the solid along the optical axis and on a crystal face perpendicular to the optical axis as shown by Jellison [18]. Alonso, et al., determined the ordinary and extraordinary complex refractive index of $Nd_{1.85}Ce_{0.15}CuO_{4\text{-}\delta}$ by polishing a face of the crystal which would contain the optical axis [20]. Since the optical axis was perpendicular to surface normal, spectroscopic ellipsometry could be used to measure the sample at different azimuthal orientations and extract the ordinary and extraordinary complex refractive index [20]. In this case, (1.33) and (1.34) do not apply. A Jones vector approach was used [21]. These approaches are not feasible for ultra-thin films.

1.4.8 Anisotropic Uniaxial Solid with Uniaxial Optical Axis Parallel to the Surface

In this case, cross-polarized light scattering with a significant dependence on the orientation of the optical axis (azimuthal angle ϕ) with respect to the plane of reflection of the light [6]. The Jones matrix approach discussed in association with (1.22) must be used to describe light scattering. When the plane of reflection of the light is parallel to the in-plane optical axis ($\phi = 0°$) or perpendicular ($\phi = 90°$), there is no cross-polarized light scattering. At other azimuthal angles, cross-polarized light scattering occurs. In that case, $\rho_{ps} = \tan\psi_{ps}e^{-i\Delta_{ps}}$ and $\rho_{sp} = \tan\psi_{sp}e^{-i\Delta_{sp}}$ will have both real and imaginary components. The following relationships hold: $\rho_{ps}(\phi) = -\rho_{sp}(\phi)$ and $\rho_{ps}(\phi) = -\rho_{ps}(-\phi)$; $\rho_{sp}(\phi) = -\rho_{sp}(-\phi)$ [6].

1.4.9 Anisotropic Uniaxial Thin Film with the Optical Axis Normal to the Surface of an Isotropic Substrate (or Anisotropic Uniaxial Thin Film with the Optical Axis Normal to a Uniaxial Substrate with Optical Axis also Normal to the Surface)

An example of this configuration is a single or few layer graphene with the c-axis normal to the surface lying on a glass slide. The complex refractive index of the thin film is $\boldsymbol{N_o}$ in the plane of the film and $\boldsymbol{N_e}$ normal to the plane of the film. As with the uniaxial solid with the optical axis normal to the surface, the s-polarized light reflection is not affected by the anisotropy. Snell's law is $N_0 \sin\theta_0 = N_2 \sin\theta_2 = \boldsymbol{N_o} \sin\boldsymbol{\theta_o}$. The first reflection between ambient and the anisotropic thin film for the s polarized light is given by the same equation used for the anisotropic bulk solid [4, 9]:

$$r_{01s} = \frac{N_0 \cos\theta_0 - \left(\boldsymbol{N_o^2} - N_0^2 \sin^2\theta_0\right)^{1/2}}{N_0 \cos\theta_0 + \left(\boldsymbol{N_o^2} - N_0^2 \sin^2\theta_0\right)^{1/2}} \tag{1.36}$$

The reflection of the s-polarized light between the anisotropic layer and the isotropic substrate is given by:

$$r_{12s} = \frac{\left(\boldsymbol{N_o^2} - N_2^2 \sin^2\theta_2\right)^{1/2} - N_2 \cos\theta_2}{\left(\boldsymbol{N_o^2} - N_2^2 \sin^2\theta_2\right)^{1/2} + N_2 \cos\theta_2} \tag{1.37}$$

As a check on (1.37), one can show that it gives the same expression as (1.30) when the material is isotropic and not uniaxial. If $\boldsymbol{N_o} = N_1$ and $N_2 \sin\theta_2 = \boldsymbol{N_o} \sin\boldsymbol{\theta_o} = N_1 \sin\theta_1$. Then, one gets for this symmetry:

$$\begin{aligned} r_{12s} &= \frac{\left(N_o^2 - N_2^2 \sin^2\theta_2\right)^{1/2} - N_2 \cos\theta_2}{\left(N_o^2 - N_2^2 \sin^2\theta_2\right)^{1/2} + N_2 \cos\theta_2} = \frac{\left(N_1^2 - N_2^2 \sin^2\theta_2\right)^{1/2} - N_2 \cos\theta_2}{\left(N_1^2 - N_2^2 \sin^2\theta_2\right)^{1/2} + N_2 \cos\theta_2} \\ &= \frac{\left(N_1^2 - N_1^2 \sin^2\theta_1\right)^{1/2} - N_2 \cos\theta_2}{\left(N_1^2 - N_1^2 \sin^2\theta_1\right)^{1/2} + N_2 \cos\theta_2} = \frac{N_1 \cos\theta_1 - N_2 \cos\theta_2}{N_1 \cos\theta_1 + N_2 \cos\theta_2} \end{aligned}$$

When multiple reflections inside the film are accounted for, the total reflection coefficient for s polarized light from the uniaxial film on an isotropic substrate, one gets:

$$r_{012s} = \frac{r_{01s} + r_{12s} e^{-i2\beta_s}}{1 + r_{01s} r_{12s} e^{-i2\beta_s}} \quad \text{with } \beta_s = \frac{2\pi d}{\lambda}\left(\boldsymbol{N_o}^2 - N_0^2 \sin^2\theta_0\right) \tag{1.38}$$

For this symmetry $N_0 \sin\theta_0 = \boldsymbol{N_o} \sin\theta_{ord} = N_1 \sin\theta_1$,

$$\beta_s = \frac{2\pi d}{\lambda}\left(N_o^2 - N_0^2 \sin^2\theta_0\right) = \frac{2\pi d}{\lambda}\left(N_1^2 - N_1^2 \sin^2\theta_1\right) = \frac{2\pi d}{\lambda} N_1 \cos\theta_1 \quad (1.39)$$

Thus for uniaxial thin films on an isotropic substrate, the r_{012s} does not change from the reflectivity of an isotropic film. The p-polarized light does reflect differently and again the first reflection from ambient to the anisotropic film is given by:

$$r_{01p} = \frac{\boldsymbol{N}_o \boldsymbol{N}_e \cos\theta_0 - N_0\left(\boldsymbol{N}_e^2 - N_0^2 \sin^2\theta_0\right)^{1/2}}{\boldsymbol{N}_o \boldsymbol{N}_e \cos\theta_0 + N_0\left(\boldsymbol{N}_e^2 - N_0^2 \sin^2\theta_0\right)^{1/2}} \quad (1.40)$$

The reflection coefficient between the anisotropic film and the isotropic substrate is given by:

$$r_{12p} = \frac{N_2\left(\boldsymbol{N}_e^2 - N_2^2 \sin^2\theta_2\right)^{1/2} - \boldsymbol{N}_o \boldsymbol{N}_e \cos\theta_2}{N_2\left(\boldsymbol{N}_e^2 - N_2^2 \sin^2\theta_2\right)^{1/2} + \boldsymbol{N}_o \boldsymbol{N}_e \cos\theta_2} \quad (1.41a)$$

Using Snell's law $N_0 \sin\theta_0 = N_2 \sin\theta_2$ this can also be expressed as:

$$r_{12p} = \frac{N_2\left(\boldsymbol{N}_e^2 - N_0^2 \sin^2\theta_0\right)^{1/2} - \boldsymbol{N}_o \boldsymbol{N}_e \cos\theta_2}{N_2\left(\boldsymbol{N}_e^2 - N_0^2 \sin^2\theta_0\right)^{1/2} + \boldsymbol{N}_o \boldsymbol{N}_e \cos\theta_2} \quad (1.41b)$$

When multiple reflections inside the film are accounted for, the total reflection coefficient of p-polarized light from the uniaxial thin films on an isotropic substrate is given by:

$$r_{012p} = \frac{r_{01p} + r_{12p} e^{-i2\beta_p}}{1 + r_{01p} r_{12p} e^{-i2\beta_p}} \quad \text{with } \beta_p = \frac{2\pi d}{\lambda}\left(\frac{\boldsymbol{N}_o}{\boldsymbol{N}_e}\right)\left(\boldsymbol{N}_e^2 - N_0^2 \sin^2\theta_0\right) \quad (1.42)$$

For uniaxial substrates with the optical axis of the substrate normal to the surface, the first reflections result in the same coefficients r_{01s} and r_{01p} (1.36) and (1.40). In this case the film refractive index is $\boldsymbol{N}_{1o}$, $\boldsymbol{N}_{1e}$ and the substrate refractive index is $\boldsymbol{N}_{2o}$, $\boldsymbol{N}_{2e}$ and the ambient refractive index is N_0. The film-substrate reflections are [9]:

$$r_{12s} = \frac{\left(\boldsymbol{N}_{1o}^2 - N_0^2 \sin^2\theta_0\right)^{1/2} - \left(\boldsymbol{N}_{2o}^2 - N_0^2 \sin^2\theta_0\right)^{1/2}}{\left(\boldsymbol{N}_{1o}^2 - N_0^2 \sin^2\theta_0\right)^{1/2} + \left(\boldsymbol{N}_{2o}^2 - N_0^2 \sin^2\theta_0\right)^{1/2}} \quad (1.43a)$$

$$r_{12p} = \frac{\boldsymbol{N}_{1o}\boldsymbol{N}_{2e} \sin^2\theta_{1e}\left(\boldsymbol{N}_{1e}^2 - N_0^2 \sin^2\theta_0\right)^{1/2} - \boldsymbol{N}_{2o}\boldsymbol{N}_{1e} \sin^2\theta_{2e}\left(\boldsymbol{N}_{2e}^2 - N_0^2 \sin^2\theta_0\right)^{1/2}}{\boldsymbol{N}_{1o}\boldsymbol{N}_{2e} \sin^2\theta_{1e}\left(\boldsymbol{N}_{1e}^2 - N_0^2 \sin^2\theta_0\right)^{1/2} + \boldsymbol{N}_{2o}\boldsymbol{N}_{1e} \sin^2\theta_{2e}\left(\boldsymbol{N}_{2e}^2 - N_0^2 \sin^2\theta_0\right)^{1/2}} \quad (1.43b)$$

For uniaxial substrate with the substrate's optical axis normal to the surface (1.43a) and (1.43b) apply and the film refractive index is $\boldsymbol{N}_{1o}$, $\boldsymbol{N}_{1e}$ and the substrate refractive index is N_{2o}, N_{2e} and the ambient refractive index is N_0. The phase angles β_s and β_p remain unchanged [9].

For thin films with a strong optical axis, the difference between the extraordinary ray and ordinary ray increases with film thickness and r_{012p} will differ from that of an isotropic thin film. We also note that both β_s and β_p are very small over the visible wavelength range for single and few layer graphene and single trilayer and few trilayer transition metal dichalcogenides. For graphene with $d = 0.335$ nm, $\frac{2\pi d}{\lambda}$ varies from ~ 0.01 at 200 nm to ~ 0.002 at 1000 nm. The $\frac{2\pi d}{\lambda}$ factor will also be small for typical trilayer TMD materials where the thickness varies between ~ 0.32 and 0.36 nm.

When the film is thick enough, the ordinary and extraordinary parts of the complex refractive index can be determined by ellipsometry by using a combination of reflection and transmission measurements including measurements at different angles of incidence and different sample orientations [4, 9]. Another approach is measure a thin film with its optical axis perpendicular to surface normal. One can achieve this by depositing a thin film with uniaxial optical properties on an off-axis substrate so that the optical axis of the thin film is not normal to the surface of the sample. For example, wurtzite InN was grown on a GaN buffer layer on a $(1,\ \bar{1},\ 0,\ 2)$ r-plane sapphire substrate and subsequently characterized by spectroscopic ellipsometry [22]. Ellipsometric characterization of optically anisotropic samples where the optical axis is not normal to the surface requires using the transfer matrix approach is described later in this chapter.

1.4.10 Anisotropic Biaxial Solid with One Optical Axis Normal to the Surface and the Second Normal to the Plane of Incidence

The optical path is defined as the z-direction being surface normal and the plane of incidence of the light being the *xz*-plane so that the direction of the light is along the x-axis. The first optical axis is normal to the surface and the second optical axis is perpendicular (along the y axis) to the *xz*-plane. The complex refractive index for the biaxial solid labeled 1 along the *x*, *y* and *z* direction is given by N_{1x}, N_{1y}, N_{1z}. The complex reflectance ratio is given by [9, 18]:

$$r_{01p} = \frac{N_{1x}N_{1z}\cos\theta_0 - N_0\left(N_{1z}^2 - N_0^2\sin^2\theta_0\right)^{1/2}}{N_{1x}N_{1z}\cos\theta_0 + N_0\left(N_{1z}^2 - N_0^2\sin^2\theta_0\right)^{1/2}};$$

$$r_{01s} = \frac{N_0\cos\theta_0 - \left(N_{1y}^2 - N_0^2\sin^2\theta_0\right)^{1/2}}{N_0\cos\theta_0 + \left(N_{1y}^2 - N_0^2\sin^2\theta_0\right)^{1/2}} \tag{1.44}$$

Aspnes has provided a method for estimating the complex dielectric function along all three directions [18]. His method results in no cross-polarized light scattering allowing measurement using standard ellipsometric methods.

1.4.11 Anisotropic Biaxial Film on an Isotropic Substrate with One Optical Axis Normal to the Surface and the Second Normal to the Plane of Incidence

The complex refractive index for the biaxial film along the x-, y- and z-direction is given by N_{1x}, N_{1y}, N_{1z}.The first reflection between ambient and the anisotropic thin film for the s polarized light is given by the same equation used for the anisotropic bulk solid [4, 9]:

$$r_{01s} = \frac{N_0 \cos\theta_0 - \left(N_{1y}^2 - N_0^2 \sin^2\theta_0\right)^{1/2}}{N_0 \cos\theta_0 + \left(N_{1y}^2 - N_0^2 \sin^2\theta_0\right)^{1/2}} \tag{1.45}$$

The reflection coefficient of the s polarized light between the anisotropic layer and the isotropic substrate is given by:

$$r_{12s} = \frac{\left(N_{1y}^2 - N_2^2 \sin^2\theta_2\right)^{1/2} - N_2 \cos\theta_2}{\left(N_{1y}^2 - N_2^2 \sin^2\theta_2\right)^{1/2} + N_2 \cos\theta_2} \tag{1.46}$$

When multiple reflections inside the biaxial film are accounted for,

$$r_{012s} = \frac{r_{01s} + r_{12s} e^{-i2\beta_s}}{1 + r_{01s} r_{12s} e^{-i2\beta_s}} \quad \text{with } \beta_s = \frac{2\pi d}{\lambda}\left(N_{1y}^2 - N_0^2 \sin^2\theta_0\right) \tag{1.47a}$$

The first reflection between ambient and the anisotropic thin film for the p polarized light is given by the same equation used for the anisotropic bulk solid:

$$r_{01p} = \frac{\boldsymbol{N}_{1x}\boldsymbol{N}_{1z} \cos\theta_0 - N_0\left(\boldsymbol{N}_{1z}^2 - N_0^2 \sin^2\theta_0\right)^{1/2}}{\boldsymbol{N}_{1x}\boldsymbol{N}_{1z} \cos\theta_0 + N_0\left(\boldsymbol{N}_{1z}^2 - N_0^2 \sin^2\theta_0\right)^{1/2}} \tag{1.47b}$$

The reflection coefficient between the anisotropic film and the isotropic substrate is given by:

$$r_{12p} = \frac{N_2\left(\boldsymbol{N}_{1z}^2 - N_2^2 \sin^2\theta_2\right)^{1/2} - \boldsymbol{N}_{1x}\boldsymbol{N}_{1z} \cos\theta_2}{N_2\left(\boldsymbol{N}_{1z}^2 - N_2^2 \sin^2\theta_2\right)^{1/2} + \boldsymbol{N}_{1x}\boldsymbol{N}_{1z} \cos\theta_2} \tag{1.47c}$$

When multiple reflections inside the film are accounted for, the total reflection coefficient of p-polarized light from the uniaxial thin films on an isotropic substrate is given by:

$$r_{012p} = \frac{r_{01p} + r_{12p}e^{-i2\beta_p}}{1 + r_{01p}r_{12p}e^{-i2\beta_p}} \quad \text{with } \beta_p = \frac{2\pi d}{\lambda}\left(\frac{N_{1x}}{N_{1z}}\right)\left(N_{1z}^2 - N_0^2 \sin^2\theta_0\right) \quad (1.48)$$

1.5 Examples of Reflectance and Ellipsometry of 2D Films

An important aspect of measuring 2D films is how to model the optical reflection and transmission. The optical response of ultra-thin 2D materials in ellipsometry measurements is dominated by the in plane dielectric function and not by the dielectric response normal to the surface. The complicated nature of the optical response of 2D films is discussed in this section. There are several possible types of thin films:

- A thin film with isotropic optical properties on a substrate with isotropic properties.
  - The Fresnel Equations for reflection and transmission from an isotropic thin film on an isotropic substrate apply.
  - Example: Thin single crystal layer of an optically isotropic semiconductor that is thick enough to have bulk like electric band energies and a thin polycrystalline film with random grain orientation.
  - Example: Polycrystalline 2D materials with random grain orientation.
  - Often, single crystal 2D films such as single and few trilayer TMD films are modeled as isotropic.
- A thin film with that has one or more optical axes on an isotropic substrate.
  - The Fresnel equations for reflection and transmission need to be altered to include the ordinary and extraordinary light paths.
  - For uniaxial films the optical axis normal to the surface, the Fresnel Equations (1.38) and (1.42) for reflection and transmission from uniaxial thin films with the optical axis normal to the surface on an isotropic substrate apply.
  - Fresnel Equations for optically biaxial films depend on the orientation of the optical axis.
  - Examples of uniaxial films include a semiconductor with an optical axis such as hexagonal GaN on an isotropic or anisotropic substrate.
- A monolayer or few-layer film with one or more optical axes on an isotropic substrate.
  - The Fresnel Equations (1.38) and (1.42) for reflection and transmission from uniaxial thin films with the optical axis normal to the surface on an isotropic substrate apply.

- Uniaxial films on isotropic substrates with the optical axis normal to the surface are frequently observed for van der Waals materials. The normal incidence reflectance is discussed below and is approximated by (1.32).
- Examples include hexagonal materials such as graphene, few layer graphene on Si/SiO_2 or on glass.
- The sub 1 nm film thickness must be considered. The optical path length through single layer graphene and even few layer graphene as well as single trilayer TMD materials is so short that ordinary and extraordinary light paths do not significantly separate.
- The surface roughness of the substrate also plays a role in the reflection of light. Examples include graphene grown on 3C and 4h polytypes of SiC.

1.5.1 Graphene

Graphene is a layer from the uniaxial graphite crystal with the optical axis normal to the substrate surface. Graphite is a negative uniaxial material based on the values of n and k $\left(n_x = n_y = n_o > n_z = n_o\right)$ between 300 and 1500 nm [20, 23] $n_x \sim 2.8$; $k_x \sim 1.4$ and the value of n_z has been assumed to be 1 and in [20] to be between 1.9 to ~ 1.5 and $k_z = 0$ between 300 and 1000 nm. The interaction between graphene layers has a big impact on the band structure for the π bands so that the value of n_z may well be layer dependent [24]. The electron band structure of graphene will be reviewed in Chap. 7 along with a wide ranging discussion of the optical and electrical properties of graphene and few layer graphene. Here, we describe the properties of graphene required for discussing the reflection of un-polarized and polarized light. In graphene and TMD materials the optical transitions between the valence and conduction band are between states in a hexagonal lattice structure that have only in plane wavevectors k_x and k_y. The π to π^* transitions for graphene dominate the optical response for the wavelength range from near IR (1700 nm) to UV (at least 150 nm). Transitions from the sigma bonding are observed in EELS spectra at high energies ($\pi + \sigma$ at 15.4 eV) [25]. For transition metal dichalcogenides, the valence to conduction band transitions have only in plane wavevectors k_x and k_y especially at the K and K' points are between bands that are dominated (at the K and K' points) by the bands formed from the planar layer of transition metal atoms. *Thus, the dielectric response of graphene **in the optical wavelength range from these films is from the in-plane optical transitions.*** Merano [26] states that either a very small or no macroscopic dipole moment arises when an electric field is applied normal to the plane of graphene. This would be consistent with a refractive index of 1 normal to the surface. However, bilayer and multi-layer graphene should be polarizable normal to the plane of graphene. A film of graphene lying flat on a substrate has the optical axis normal to the surface. The surface roughness of graphene on SiO_2/Si and mica was found to have a RMS value of ~ 0.15 nm and ~ 0.025 nm respectively [27]. Graphene is known to conform to the surface of hexagonal boron nitride which is considered have a similar surface roughness to HOPG graphite of ~ 0.07 nm RMS for BN flakes thicker than 5 nm [28].

All of these values indicate a smooth surface. However, because the analysis area of an optical method can cover hundreds of microns, the roughness is not well characterized by a single RMS value. For example the RMS roughness measured by an atomic force microscope can vary with the analysis area, e.g., 1 μm × 1 μm versus 10 μm × 10 μm. Since the scanned area is divided into a set number of pixels often (~ 512 × 512), the measurement does not sample the same smallest and largest spatial wavelengths when the scanned area is changed. Thus, the power spectral density of surface roughness should be used to characterize the surface roughness [29]. The analysis area of spectroscopic ellipsometry depends on the specific measurement system. Since silicon wafers are often used as substrates, it is worth mentioning that the local and global surface roughness is well characterized and that information is available [30]. Thus, careful consideration of the true experimental geometry should be considered when interpreting measurements of 2D films in contrast to measurements of cleaved surfaces of optically anisotropic bulk crystals. Graphene formed on SiC surfaces will be conformal to the undulating topography of the surface [31, 32]. The 3C and 4 h polytypes of SiC have a surface morphology that results in the optical axis of graphene pointing away from surface normal as shown in Fig. 1.6 [32]. The surfaces also depend on whether the surface is the silicon or carbon face of the polymorph [32]. In this case, the dielectric response could be approximated as isotropic.

When light is incident on a graphene at an angle away from normal, the reflection of polarized light from graphene on an optically isotropic substrate is therefore given by the Fresnel equations for reflection from uniaxial thin films with the optical axis normal to the surface on an isotropic substrate, (1.36)–(1.42). Based on the

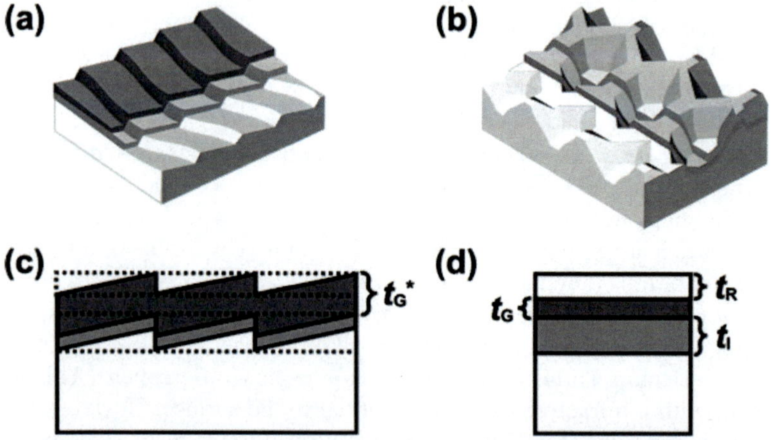

Fig. 1.6 Graphene grown on the 4H morph of silicon carbide (SiC). **a** Graphene on the silicon face of 4H SiC with the interface layer shown in light gray. **b** Graphene on the carbon face of 4H SiC. **c** Graphene on off-axis SiC. **d** The film stack can be approximated by a roughness layer of thickness t_R, graphene layer of thickness t_G, and interface layer of thickness t_I. Figure adapted and reprinted with permission from [32]. © 2012 American Chemical Society

discussion above, the reflection of the s polarized component is unaltered by the optical anisotropy of graphene and (1.28) is the appropriate Fresnel equation for this reflection. As described above, the p polarized light is altered by the optical anisotropy and p polarized light reflection is given by (1.42). As mentioned previously, the optical properties of graphene observed by ellipsometry are often considered to be dominated by the in-plane refractive index. In order to further justify that consideration for graphene, we examine the p polarized light reflection. It is interesting to note that if the extraordinary refractive index of graphene is 1 because there are no optical transitions in the visible wavelength range associated with an electric field normal to the surface (see Chapter 7), and we use the refractive index of air of 1, the Fresnel reflection coefficient for ambient–graphene would be:

$$r_{01p} = \frac{\boldsymbol{N_o N_e} \cos\theta_0 - N_0\left(\boldsymbol{N_e^2} - N_0^2 \sin^2\theta_0\right)^{1/2}}{\boldsymbol{N_o N_e} \cos\theta_0 + N_0\left(\boldsymbol{N_e^2} - N_0^2 \sin^2\theta_0\right)^{1/2}} \sim \frac{\boldsymbol{N_o} \cos\theta_0 - \left(1 - \sin^2\theta_0\right)^{1/2}}{\boldsymbol{N_o} \cos\theta_0 + \left(1 - \sin^2\theta_0\right)^{1/2}}$$
$$= \frac{\boldsymbol{N_o} \cos\theta_0 - \cos\theta_0}{\boldsymbol{N_o} \cos\theta_0 + \cos\theta_0} \tag{1.49}$$

If N_e is approximated by that of graphite [20] and thus real and linearly constant over much of the visible wavelength range, then r_{01p} would still be a function of the in-plane complex refractive index, $\boldsymbol{N_o}$. Using the same approach, the Fresnel reflection coefficient between the graphene and the isotropic substrate is also a function of $\boldsymbol{N_o}$.:

$$r_{12p} = \frac{N_2\left(\boldsymbol{N_e^2} - N_0^2 \sin^2\theta_0\right)^{1/2} - \boldsymbol{N_o N_e} \cos\theta_2}{N_2\left(\boldsymbol{N_e^2} - N_0^2 \sin^2\theta_0\right)^{1/2} + \boldsymbol{N_o N_e} \cos\theta_2} \sim \frac{N_2 \cos\theta_0 - \boldsymbol{N_o} \cos\theta_2}{N_2 \cos\theta_0 + \boldsymbol{N_o} \cos\theta_2} \tag{1.50}$$

Thus the wavelength dependent sensitivity of spectroscopic ellipsometry is due to the in-plane complex refractive index of graphene, and not a constant or near constant surface normal component of the complex refractive index. When graphene is lying flat on a transparent substrate and the light is at normal incidence, (1.32) applies since the measurement is sensitive to the isotropic, in-plane complex refractive index of graphene.

We note that in the complex refractive index of graphene has been obtained from spectroscopic reflectance and spectroscopic ellipsometry data using a variety of models from the isotropic thin film on a substrate model [23] to the anisotropic model where the optical properties of graphite are used for the refractive index normal to the surface [20]. The isotropic film model could also be referred to as an effective dielectric function or pseudo-dielectric function model, and this model can be very useful when using reflection ellipsometry for measuring the number of graphene layers in an area after deposition. The resulting complex refractive index for all these methods is very similar over a wide wavelength range (typically 200–1500 nm) as discuss in Chap. 7.

Graphene and few layer graphene fabricated using chemical vapor deposition has randomly oriented grains that are all in the same plane. Thus the optical properties of CVD graphene should be very close to that of exfoliated single crystal graphene [20, 23]. The symmetry of optical measurements for CVD graphene on a substrate will be the same as for single crystal graphene.

Merano [26] derived Fresnel reflection coefficients for a conducting 2D layer such as graphene. The conductivity associated with the high mobility, low energy states in electronic band structure near the K and K' points in the Brillouin zone which arise after low energy transitions from the π band energy to the π^* band touch is added onto the susceptibility of the 2D layer across all optical wavelengths. The optical response at visible wavelengths occur away from that part of the Brillouin zone.

1.5.2 Monolayer TMD's (Trilayers of Chalcogenide—Transition Metal—Chalcogenide)

Many monolayer Transition Metal Dichalcogenides are direct gap materials. The electronic band structure and dielectric functions of TMD materials will be described in a subsequent chapter, and here we discuss aspects of the band structure relevant to this discussion about their ellipsometric characterization. The optically active part of the electronic band structure of trilayers can be approximated by 3 bands formed from the d_{z^2}, $d_{x^2-y^2}$ and d_{xy} orbitals which describes the highest lying valence and lowest lying conduction bands. These bands have only in-plane wavevectors. This description holds for electronic states with wavevectors close to the K and K' points in the Brillouin zone where direct gap transitions occur. Away from the K points, the metal—chalcogenide bonds mix with the d orbitals and bond directions are no longer planar.

A variety of different analysis approaches from isotropic models to models that reflect the uniaxial nature of TMD films and crystals can be found in the literature. Most TMD crystals are optically uniaxial due to their hexagonal symmetry [33]. The anisotropy of the optical properties of bulk WS_2, WSe_2, α-$MoTe_2$, NbS_2, and $NbSe_2$ was demonstrated using reflectivity at 300 and 78 K [33]. However, determining the ordinary and extraordinary refractive index of single tri-layer of TMD materials is difficult. Consequently, it is not surprising that the complex refractive index/dielectric function of TMD trilayers has been modeled as an isotropic layer [34, 35]. Another challenge is obtaining a single crystal monolayer (single trilayer) samples. One study was done using monolayers deposited on sapphire [34], and the other study was done on exfoliated samples [35]. In both cases, the analysis area likely covers multiple nanocrystals of random in-plane orientation. The in-plane dielectric function (ordinary refractive index) of exfoliated monolayer MoS_2, $MoSe_2$, WS_2, and WSe_2 on fused silica substrates with the optical axis normal to the surface was studied using normal incidence reflectometry over the energy range 1.5–3 eV [36] This approach provides unambiguous measurement of the in-plane dielectric function. The dielectric functions were obtained using (1.32) and are shown in Fig. 1.7. The monolayer (trilayer) thickness values were obtained from the interlayer spacing in the bulk crystal. The free carrier contribution to the dielectric functions was estimated using

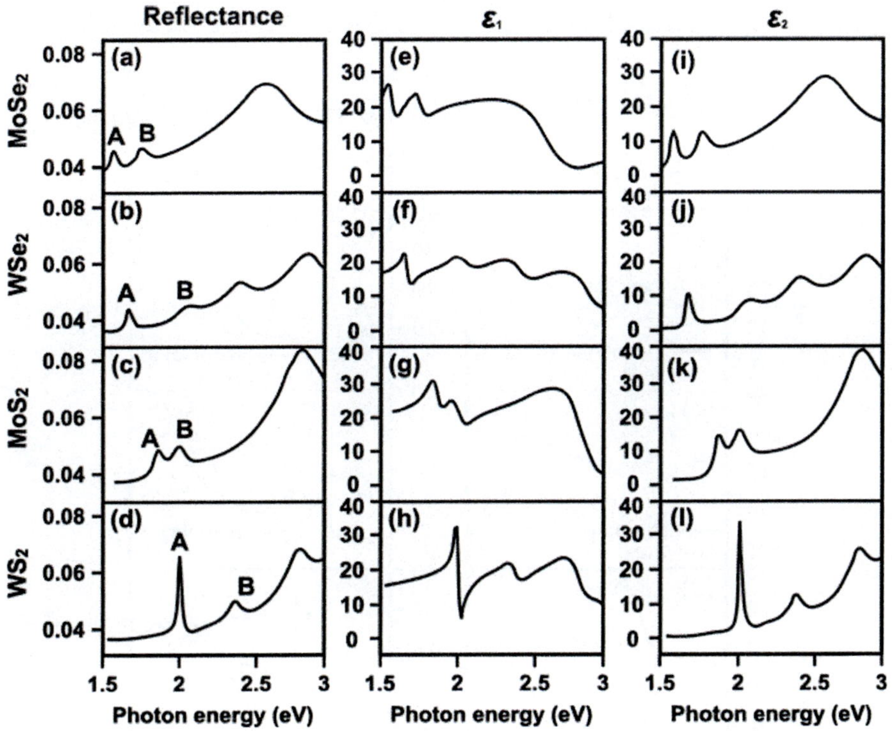

Fig. 1.7 The normal incidence reflectance and in-plane dielectric function of exfoliated monolayer MoS_2, $MoSe_2$, WS_2, and WSe_2 on fused silica substrates. The A and B shown in **a** through **d** are the spin-orbit split exciton transitions at the K point discussed in Chap. 8. The layer thickness values used to determine the dielectric function were $d_{MoS_2} = 0.615$ nm, $d_{MoSe_2} = 0.646$ nm, $d_{WS_2} = 0.618$ nm, $d_{WSe_2} = 0.649$ nm. Figure adapted and reprinted with permission from [36]. © 2014 American Physical Society

the Drude conductivity model discussed below to be less than 0.1 based on a carrier density of 10^{12} cm^{-2} [36]. Li et al., also show reflectance data that indicates that CVD MoS_2 has a nearly identical optical properties to exfoliated MoS_2 [36]. The in-plane dielectric functions of monolayer and bulk MoS_2, $MoSe_2$, WS_2, and WSe_2 show clear differences between 1.5 and 3 eV as shown in Fig. 1.8 [36]. The bulk dielectric functions in this comparison were also obtained using normal incidence reflectivity [37]. The same analysis was done for normal incidence reflectivity of monolayer (single trilayer) and few layer MoS_2 on sapphire where the refractive index was fixed a 1.77 over the wavelength range from 400 to 900 nm [38]. In this study, the optical conductivity was reported. Another interesting study showed that in-plane anisotropy of optically biaxial, triclinic ReS_2 can be measured using polarized light in FT-IR [39].

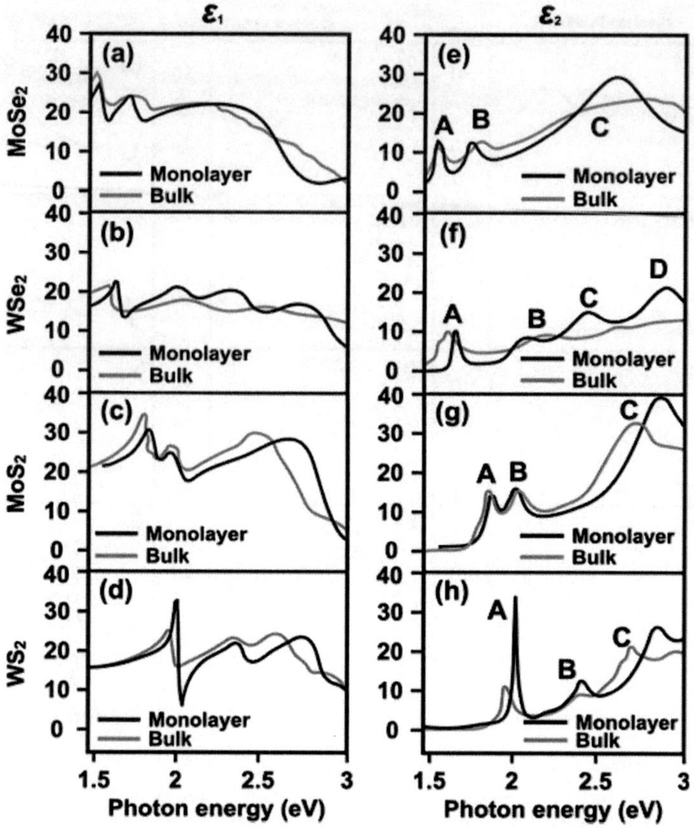

Fig. 1.8 A comparison of in-plane dielectric functions of monolayer MoS_2, $MoSe_2$, WS_2, and WSe_2 with bulk samples [36] and [37]. Figure adapted and reprinted with permission from [36]. © 2014 American Physical Society

1.5.3 Topological Insulators

Topological insulators have 2D carriers at their surface. The optical data from the near IR to the UV in topological insulators comes from an isotropic bulk response from the material. The topologically protected states are low energy. Only the Weyl metals seem to have a 3D nature to their topologically protected carriers. Since the topologically protected carriers are low energy, the Weyl metal optical response from the near IR to the UV is largely or entirely from the bulk material.

1.5.4 2D Slab and Surface Current Models for the Optical Conductivity of 2D Films

The challenges associated with the determination of the anisotropic optical properties of 2D materials can be avoided by using normal incidence reflectivity and determining the in-plane optical properties of 2D materials using reflectivity [14, 38, 39]. Li and Heinz have presented a 2D model for the optical conductivity based on (1.32) [14]. When the film (slab) with complex refractive index N_1 is very thin, the thickness d and thus the phase shifts $\varphi = 2\pi N_1 d/\lambda$ and $\varphi_0 = 2\pi d/\lambda$ are also small. Thus, the model is referred to as the Surface Current Model [40]. The initial analysis of the model makes some key assumptions (see supplemental [14]). The refractive index of the ambient is assumed to be 1, and the substrate is transparent so the refractive index of the substrate is N_2 is real and close to 1. As mentioned above, sapphire is a good example of a transparent substrate since it is non-aborbing between 400 and 900 nm, and $N_2 \sim 1.77$. Later we mention that this initial limitation on N_2 can be lifted allowing a complex value for N_2. Thus, $\sin\varphi/\varphi \to 1$, $\cos\varphi \to 1$, and $e^{-i2\varphi_0} \approx 1 - i2\varphi_0$. (1.32) can be approximated by the 2D Surface Current Model to first order in φ_0 as follows:

$$
\begin{aligned}
r &= \frac{(1-N_2)\cos\varphi - i\left(N_2 - N_1^2\right)\varphi_0\left(\frac{\sin\varphi}{\varphi}\right)}{(1+N_2)\cos\varphi - i\left(N_2 + N_1^2\right)\varphi_0\left(\frac{\sin\varphi}{\varphi}\right)} e^{-2i\varphi_0} \\
&\approx \frac{(1-N_2) - i\left(N_2 - N_1^2\right)\varphi_0}{(1+N_2) - i\left(N_2 + N_1^2\right)\varphi_0}(1 - i\,2\varphi_0) \\
&= \frac{(1-N_2) - i\left(N_2 - N_1^2\right)\varphi_0 - i\,2(1-N_2)\varphi_0 - 2\left(N_2 - N_1^2\right)\varphi_0^2}{(1+N_2) - i\left(N_2 + N_1^2\right)\varphi_0} \\
&\approx \frac{(1-N_2) - i\left(1 - N_1^2\right)\varphi_0 + i\,3(N_2 - 1)\varphi_0}{(1+N_2) - i\left(1 - N_1^2\right)\varphi_0 - i(N_2 - 1)\varphi_0}
\end{aligned}
$$

With $(N_2 - 1)\varphi_0 \sim 0$ and using the relationship between the dielectric function and the optical conductivity $\varepsilon = 1 + \frac{i\sigma}{\varepsilon_0\omega}$ **(SI units)**, the sheet optical conductivity σ^s of the thin film times the impedance of free space $Z_0 = 1/(\varepsilon_0 c)$ is $Z_0\sigma^s = -i\left(N_1^2 - 1\right)\varphi_0 \; with \; \sigma^s = \sigma d$ [14]:

$$
r = \frac{(1-N_2) - i\left(1 - N_1^2\right)\varphi_0}{(1+N_2) + i\left(1 - N_1^2\right)\varphi_0} = \frac{(1-N_2) - Z_0\sigma^s}{(1+N_2) + Z_0\sigma^s} \tag{1.51}
$$

Again, the reflectance $R = |r|^2$. The formula for the transmittance is $T = |t|^2$. This approach can be linearized when $\left|\frac{Z_0\sigma^s}{1+N_2}\right| \ll 1$ or $\left|\frac{Z_0\sigma^s}{1-N_2}\right| \ll 1$ resulting in [14]:

$$r = r_0\left(1 - \frac{2}{1 - N_2^2} Z_0\sigma^s\right) \tag{1.52a}$$

$$t = t_0\left(1 - \frac{2}{1 + N_2} Z_0\sigma^s\right) \tag{1.52b}$$

and the transmittance $T = |t|^2$. For cases where N_2 is complex, this model should be valid when $|\Delta T/T| \ll 1$ or $|\Delta R/R| \ll 1$ (see supplemental [14]).

Both models have been applied to the optical conductivity of graphene and monolayer WS_2 as shown in Fig. 1.9. Li and Heinz also presented criteria for the applicability of the 2D and linearized 2D models for obtaining conductivity from reflectance data which are summarized in Table 1.1 [14].

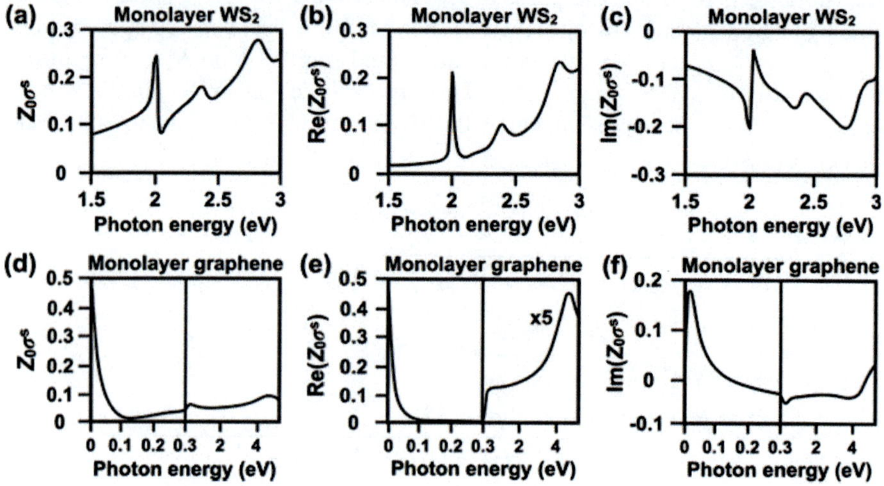

Fig. 1.9 The absolute, real, and imaginary optical conductivity obtained using the linearized model of (1.52) plotted as $Z_0\sigma^s$. Note that $Z_0 = 1/(\varepsilon_0 c)$. Figure adapted from [14].

Table 1.1 Criteria for use of 2D and linearized 2D surface current models for optical conductivity [14]

Criteria	2D Model	Linearized 2D Model
Bulk	$\lvert\varphi\rvert < 1$ or $\lvert N_1\rvert \ll \varphi^{-1}$	$\lvert\varphi\rvert < \lvert N_1\rvert^{-1}$ or $\lvert N_1\rvert \ll \varphi_0^{-1/2}$
Sheet	$\lvert Z_0\sigma^s\rvert \ll \lvert N_1\rvert$	$\lvert Z_0\sigma^s\rvert \ll 1$
Experimental[a]	$\lvert(\Delta t - \Delta r)/t\rvert \ll 1$	$\frac{\Delta T}{T} \ll 1$ or $\Delta R/R \ll 1$

[a]In terms of reflection coefficient r and transition coefficient t, or Reflectance R and Transmission T

1.6 Generalized Ellipsometry: Optical Transition Matrix Approach for Crystals and Thin Films with Arbitrarily Oriented Optical Axes

In this section, we describe the transfer matrix approach that is used for arbitrarily oriented materials [41]. The Jones equations for the reflection coefficients was discussed above and shown in (1.22). Here, the reflection coefficients r_{pp}, r_{ps}, and r_{sp} are expressed in terms of the matrix elements of a transfer matrix which relates the incident and reflected to the transmitted p and s polarization components of the light. In other words, the light that is not reflected is transmitted. The Transfer matrix T for a single layer of thickness d is obtained from multiplication of an incident matrix L_i, a partial transfer matrix T_p, and an exit matrix L_t [4, 10, 11]:

$$T \equiv L_i^{-1} T_p(-d) L_t \tag{1.53}$$

For a multilayer sample with N layers of thickness d_j the transfer matrix approach requires multiplication of the partial transfer matrix T_{pj} for each layer j as follows:

$$T \equiv L_i^{-1} \prod_{j=1}^{\mathrm{N}} T_{pj} L_t \tag{1.54}$$

Below the method used to obtain the transfer matrix is shown. First we relate the incoming, reflected, and transmitted electric fields to each other using the transfer matrix [4, 10, 11].

$$\begin{bmatrix} E_{is} \\ E_{rs} \\ E_{ip} \\ E_{rp} \end{bmatrix} = \begin{bmatrix} T_{11} & T_{12} & T_{13} & T_{14} \\ T_{21} & T_{22} & T_{23} & T_{24} \\ T_{31} & T_{32} & T_{33} & T_{34} \\ T_{41} & T_{42} & T_{43} & T_{44} \end{bmatrix} \begin{bmatrix} E_{ts} \\ 0 \\ E_{tp} \\ 0 \end{bmatrix} \tag{1.55}$$

The resulting cross-polarized reflection coefficients are:

$$\begin{aligned} r_{pp} &= \left(\frac{E_{rp}}{E_{ip}}\right)_{E_{is}=0} = \frac{T_{11}T_{43} - T_{13}T_{41}}{T_{11}T_{33} - T_{13}T_{31}} \\ r_{sp} &= \left(\frac{E_{rs}}{E_{ip}}\right)_{E_{is}=0} = \frac{T_{11}T_{23} - T_{13}T_{21}}{T_{11}T_{33} - T_{13}T_{31}} \\ r_{ss} &= \left(\frac{E_{rs}}{E_{is}}\right)_{E_{ip}=0} = \frac{T_{21}T_{33} - T_{23}T_{31}}{T_{11}T_{33} - T_{13}T_{31}} \\ r_{ps} &= \left(\frac{E_{rp}}{E_{is}}\right)_{E_{ip}=0} = \frac{T_{33}T_{41} - T_{31}T_{43}}{T_{11}T_{33} - T_{13}T_{31}} \end{aligned} \tag{1.56}$$

In order to implement the transition matrix approach, it is necessary to know the dielectric function along the experimental coordinate axes, x, y, and z. This dielectric function in the experimental coordinate axes can be obtained from the crystal's dielectric function by coordinate rotation. The dielectric function can be rotated to laboratory coordinates using the rotation matrix R for the Euler angels ϕ_E, θ_E, ψ_E following method [4, 10, 11, 41]:

$$R = \begin{bmatrix} \cos\phi_E \cos\psi_E - \sin\phi_E \cos\theta_E \sin\psi_E & -\cos\phi_E \sin\psi_E - \sin\phi_E \cos\theta_E \cos\psi_E & \sin\phi_E \sin\theta_E \\ \sin\phi_E \cos\psi_E + \cos\phi_E \cos\theta_E \sin\psi_E & -\sin\phi_E \sin\psi_E + \cos\phi_E \cos\theta_E \cos\psi_E & -\cos\phi_E \sin\theta_E \\ \sin\theta_E \sin\psi_E & \sin\theta_E \cos\psi_E & \cos\theta_E \end{bmatrix}$$

The rotation of an optically biaxial dielectric function to laboratory coordinates is:

$$\begin{pmatrix} \varepsilon_{xx} & \varepsilon_{xy} & \varepsilon_{xz} \\ \varepsilon_{yx} & \varepsilon_{yy} & \varepsilon_{yz} \\ \varepsilon_{zx} & \varepsilon_{zy} & \varepsilon_{zz} \end{pmatrix} = R^{-1} \begin{pmatrix} \varepsilon_x & 0 & 0 \\ 0 & \varepsilon_y & 0 \\ 0 & 0 & \varepsilon_z \end{pmatrix} R \tag{1.57}$$

The same approach would be used for uniaxial materials where $\varepsilon_x = \varepsilon_y$. An example of a uniaxial thin film where the Euler angles of the material are know is that of a wurtzite material grown on off-axis sapphire [16]. Otherwise, another method such as X-ray diffraction is required for determining the Euler angles.

The next step is to determine the transfer matrix T. This procedure is described elsewhere [4, 10, 11, 41] and here we show the results for the exit matrix for an isotropic substrate.

Assuming that the index of refraction of the ambient is n_i, the inverse incident matrix L_i^{-1} for an angle of incidence θ_i is given by:

$$L_i^{-1} = \frac{1}{2} \begin{bmatrix} 0 & 1 & -1/n_i \cos\theta_i & 0 \\ 0 & 1 & 1/n_i \cos\theta_i & 0 \\ 1/\cos\theta_i & 0 & 0 & 1/n_i \\ -1/\cos\theta_i & 0 & 0 & 1/n_i \end{bmatrix} \tag{1.58}$$

The exit matrix for an isotropic substrate is:

$$L_t = \begin{bmatrix} 0 & 0 & \cos\theta_i & 0 \\ 1 & 0 & 0 & 0 \\ -n_t \cos\theta_i & 0 & 0 & 0 \\ 0 & 0 & 0 & 0 \end{bmatrix} \tag{1.59}$$

The partial transfer matrix for a single film is approximated by a series expansion in the 4 by 4 matrix for Δ_B which Berreman's first order differential equation form of Maxwell's equations where z is the depth into the sample in a direction normal to the surface:

$$\frac{\partial \Psi}{\partial z} \equiv i\frac{\omega}{c}\Delta_B \qquad \Psi = \begin{bmatrix} E_x \\ E_y \\ H_x \\ H_y \end{bmatrix} \tag{1.60}$$

where c is the speed of light in vacuum, ω is the angular frequency of the light, and Ψ is the vector representing the electric field E and magnetic field H. The expression for Δ_B requires knowledge of the components of the dielectric tensor $\boldsymbol{\varepsilon}$ and $K_{xx} = n_i \sin\theta_i$ [4, 10, 11, 41]:

$$\Delta_B = \begin{bmatrix} -K_{xx}\frac{\varepsilon_{zx}}{\varepsilon_{zz}} & -K_{xx}\frac{\varepsilon_{zy}}{\varepsilon_{zz}} & 0 & 1-\frac{K_{xx}^2}{\varepsilon_{zz}} \\ 0 & 0 & -1 & 0 \\ \varepsilon_{yz}\frac{\varepsilon_{zx}}{\varepsilon_{zz}} - \varepsilon_{yx} & K_{xx}^2 - \varepsilon_{yy} - \varepsilon_{yz}\frac{\varepsilon_{zy}}{\varepsilon_{zz}} & 0 & K_{xx}\frac{\varepsilon_{yz}}{\varepsilon_{zz}} \\ \varepsilon_{xx} - \varepsilon_{xz}\frac{\varepsilon_{zx}}{\varepsilon_{zz}} & \varepsilon_{xy} - \varepsilon_{xz}\frac{\varepsilon_{zy}}{\varepsilon_{zz}} & 0 & -K_{xx}\frac{\varepsilon_{xz}}{\varepsilon_{zz}} \end{bmatrix} \tag{1.61}$$

The partial transfer matrix for a single film can be approximated by a series expansion in Δ_B as follows:

$$T_p(-d) = \beta_0 I + \beta_1\Delta_B + \beta_2\Delta_B^2 + \beta_3\Delta_B^3 \tag{1.62}$$

Here, I is the identity matrix and the series expansion coefficients β_i are expressed in terms of the Eigen values q_i of the determinant $\det(\Delta_B - qI)$ as follows [4]:

$$\beta_0 = -\sum_{j=1}^{4} q_k q_l q_m \frac{e^{-i\omega q_j d/c}}{(q_j - q_k)(q_j - q_l)(q_j - q_m)} \tag{1.63a}$$

$$\beta_1 = \sum_{j=1}^{4} (q_k q_l + q_k q_m + q_l q_m) \frac{e^{-i\omega q_j d/c}}{(q_j - q_k)(q_j - q_l)(q_j - q_m)} \tag{1.63b}$$

$$\beta_2 = -\sum_{j=1}^{4} (q_k + q_l + q_m) \frac{e^{-i\omega q_j d/c}}{(q_j - q_k)(q_j - q_l)(q_j - q_m)} \tag{1.63c}$$

$$\beta_3 = \sum_{j=1}^{4} \frac{e^{-i\omega q_j d/c}}{(q_j - q_k)(q_j - q_l)(q_j - q_m)} \tag{1.63d}$$

where the indices $k,\ l,\ m$ depend on j as follows [4]:

$$j = 1 \quad (k, l,\ m) = (2,\ 3,\ 4)$$

$$j = 2 \quad (k, l,\ m) = (1,\ 3,\ 4)$$

$$j = 3 \quad (k, l,\ m) = (1,\ 2,\ 4)$$

$$j = 4 \quad (k, l,\ m) = (1,\ 2,\ 3)$$

Schubert provides solutions for uniaxial and biaxial layers [10–12].

1.7 Optical Properties of Materials (Dielectric Function/Complex Refractive Index)

The complex refractive index/dielectric function of many stable materials is known and well documented. For example, the complex refractive index of many bulk semiconductors such as silicon, germanium, and gallium arsenide are tabulated and there are several well-known sets of data for silicon [42–44]. The complex refractive index of many dielectric materials such as silicon dioxide and silicon nitride are also well-known and tabulated [45]. These complex refractive indices are for undoped semiconductors at room temperature and stoichiometric dielectrics. It is important to note that thermally grown silicon dioxide will have a slightly different complex refractive index than deposited silicon dioxide. The relationship between the dielectric function (complex refractive index) and the electronic band structure is described in Chap. 2. In Chap. 2, the wave vector dependence of the electronic band energies for bulk crystals for the wave vectors inside the first Brillouin zone.

There are also model dielectric functions that can be fit when measuring the thickness of a thin film. Examples of this include the Lorentz oscillator, Tauc-Lorentz model, and Cody-Lorentz model. The derivation of these models is readily available and the reader can refer to a variety of [46, 47]. The imaginary part of the dielectric function for the *Lorentz model* is:

$$\varepsilon_2(E) = \frac{A\Gamma E}{\left(E_0^2 - E^2\right)^2 - \Gamma^2 E^2} \tag{1.64a}$$

$$\varepsilon_1(E) = 1 + \frac{A\left(E_0^2 - E^2\right)^2}{\left(E_0^2 - E^2\right)^2 - \Gamma^2 E^2} \tag{1.64b}$$

where ω_0 is the central frequency of the absorption is converted to an energy E_0, and Γ is the half width of the absorption at half of the maximum value. The Lorentz model is stated in terms of the photon energy instead of the frequency. The fitting parameter *A* is varied in order to obtain the best match to experimental data for the sample.

The Drude model for the dielectric function of metals provides a useful transition into the discussion about band structure in Chap. 2. First we discuss the Drude model using classical physics. The premise of the Drude model is that the electrons in a metal are not bound and are thus move through the metal freely. The electrons respond to the electric field of the light according to the frequency of the light. One can estimate the average thermal velocity v_e of the electrons at a temperature T using the equipartition theorem: $\frac{3}{2}K_B T = \frac{1}{2}m_e v_e^2$. This gives $v_e \sim 10^7$ cm s^{-1}. The electrons will accelerate in the electric field until they collide with another electron where the average time between collisions is τ. If we assume that the average $V_d = 0$ before the electric field of the light interacts with the electrons of charge e, then $\frac{\Delta V_d}{\Delta t} = \frac{V_d}{\tau} = a$ and using $F = e\vec{E} = m_e a$ the drift velocity is given by $V_d = e\vec{E}\tau/m_e$. Thus, the total force on the electrons of mass m_e with drift velocity V_d from light acting on the metal over a period of time much longer than a single frequency so that there is a frequency dependence $V_d(\omega)$ due to $\vec{E}(\omega)$ is:

$$F = e\vec{E}(\omega) = m_e a = m_e \frac{dV_d(\omega)}{dt} + m_e \frac{V_d(\omega)}{\tau} \tag{1.65}$$

The solution to this equation is:

$$V_d(\omega) = -\frac{e\tau}{m_e} \frac{\vec{E}(\omega)}{(1 - i\omega\tau)}$$

The dielectric function can be obtained using the current density J, the complex optical conductivity σ, and the frequency dependent electron velocity $V_d(\omega)$ for electron density n as follows:

$$\vec{J}(\omega) = \sigma(\omega)\vec{E}(\omega) = enV_d(\omega) = \frac{e^2 n\tau}{m_e} \frac{\vec{E}(\omega)}{(1 - i\omega\tau)}$$

Defining the static optical conductivity for the Drude model as $\sigma_0 = \frac{e^2 n\tau}{m_e}$ we can estimate the average time between collisions (relaxation time) using the resistivity ρ of a typical metal $\sigma_0 = 1/\rho$. For copper with $\rho = 1.6$ μΩ cm, for an electron density of $\sim 10^{23}$ cm^{-3} we have $\tau = 10^{-14}$ s. We can also estimate the average drift velocity due to the electric field: $V_d \sim 5 \times 10^3$ cm s^{-1} which is much less that the average thermal velocity. The optical conductivity can then be written in CGS units as:

$$\sigma(\omega) = \frac{e^2 n\tau}{m_e} \frac{1}{(1 - i\omega\tau)} = \sigma_0 \frac{1}{(1 - i\omega\tau)} \tag{1.66}$$

This allows us to state the Drude dielectric function for a free electron system (CGS units):

$$\varepsilon = \varepsilon_1 + i\varepsilon_2 = 1 + i\,\frac{4\pi\sigma(\omega)}{\omega}$$

The dielectric function can be written in terms of the plasmon frequency $\omega_P = \sqrt{\frac{4\pi e^2 n}{m_e}}$ in CGS units where it is important to remember that n is the electron density:

$$\varepsilon = 1 - \frac{\omega_P^2}{\left(\omega^2 + i\omega\tau^{-1}\right)}$$

Or in terms of the real and imaginary part of the Drude dielectric function:

$$\varepsilon_1 = 1 - \frac{\omega^2\omega_p^2\tau^2}{\omega^4\tau^2 + \omega^2} \text{ and } \varepsilon_2 = \frac{\omega\tau\omega_p^2}{\omega^4\tau^2 + \omega^2} \quad (1.67)$$

This leads to the well know 100% Reflectivity below the plasmon frequency. Based on the continuity requirement at the boundary between the ambient and the metal, the normal incidence reflectivity R is:

$$R = \left|\frac{1-\varepsilon}{1+\varepsilon}\right|^2 \approx 1 \quad (1.68)$$

For $\tau = 2.7 \times 10^{-14}$ s and $\omega \sim 10^{12}\ \text{s}^{-1}$, $\omega\tau = 2.7 \times 10^{-2}$, $ie\ \omega\tau < 1$ thus $\varepsilon \sim i\varepsilon_2$ and $\varepsilon_2 > 1$, thus $R = 1$ at low frequency. The real reflectivity of aluminum is approximated by the Drude model as shown in Fig. 1.10.

The optical properties of some metals can be approximated by a linear combination of the Drude and Lorentz dielectric functions. For nanoscale, polycrystalline metal films, the thickness dependence of the optical properties has been associated with the change in the relaxation time τ of the Drude component of the Drude–Lorentz model [49]. The grain size and the shape of the grains in polycrystalline metal films changes with thickness especially when the film thickness is at or below the grain

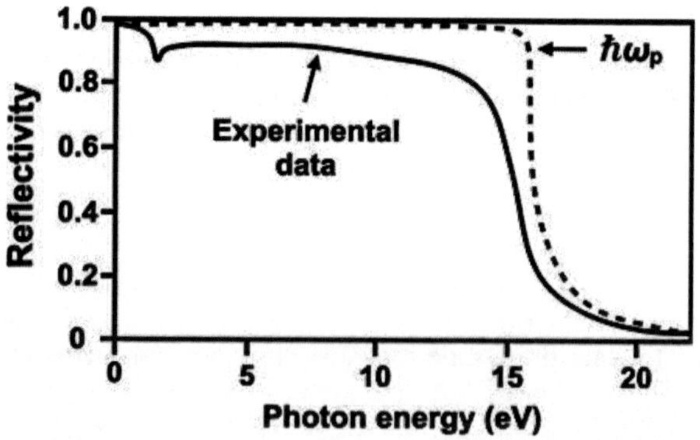

Fig. 1.10 The Drude model for the normal incidence reflectivity of Aluminum is shown along with the experimental reflectivity from [48]. Figure adapted and reprinted with permission from [48]. © 1963 American Physical Society

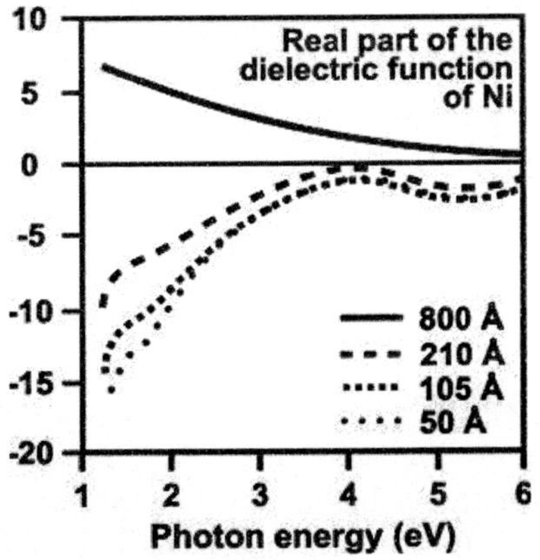

Fig. 1.11 The real part of the dielectric function of Ni containing 5% Pt is shown as a function of thickness. Figure adapted and reprinted from [49] with the permission of AIP Publishing

size found in thick films. The thickness dependence of the real part of the dielectric function of Ni alloy films is shown in Fig. 1.11.

1.8 The Particle Nature of Light

The wave particle duality of light is now well known. There is a long history of conjecture about the nature of light which is only briefly mentioned here, and a more complete history can be found in the historical introduction of Born and Wolf [2]. Many early theories considered light to be particle in nature. In the eleventh century in what has been described as the first comprehensive book on optics which was written the, Ibn al-Haytham considered light to be a particle [50]. Another proponent of the particle picture was Sir Isaac Newton. Based on the straight lines of reflection, he thought that only particles would exhibit this behavior, and he developed a corpuscular theory for light in the late 1600's. As discussed at the start of this chapter, James Clerk Maxwell first showed that light was a self-propagating electromagnetic wave using his theory of electromagnetism. The diffraction of light proves its wave nature, and the Young's double slit experiment demonstrates wave interference. In 1901, Max Planck explained Black Body radiation by relating the frequency of the emitted light to the frequency of an electrical oscillator in the black body that emitted the light. The energy from the oscillating system was transferred in fixed amounts or a quantum of energy. He proposed that these quanta were a linear function of the frequency through a constant h, Plank's constant, $E = hv$. His theory required that to be an integer number of oscillators emitting the light. In 1905, Albert Einstein explained the photoelectric effect using quanta of light, photons. Thus, the particle nature of light was firmly established. Light exhibits both particle and wave properties. The subsequent introduction of quantum mechanics relied on the wave–particle duality of matter. Although it is well known, light which has no rest mass displays particle properties. Atom, electrons, proton, and neutrons which also show both wave and particle properties have a resting mass. In the next section, we discuss Raman Spectroscopy which the scattering of light by vibrational modes in a crystal can be understood in both a classical wave picture and the particle picture of phonon scattering.

1.9 Raman Spectroscopy

Raman spectroscopy is widely applied to the study of the vibrational properties of bulk and nanoscale materials [51–53]. Raman can be used to characterize the stress in bulk and nanoscale semiconductor materials, and it is a critical means of determining the number of layers of van der Waals bonded 2D materials. Vibrational modes of solids are either Raman active or Infra-Red active depending on the polarization that occurs during vibration. First order Raman refers to light scattering from a single phonon, and second order Raman refers to two phonon scattering. The Raman spectra for the phonons of frequency ω will be observed as side peaks that are close in energy to the reflected light: $\omega_s = \omega_i \pm \omega$ where ω_i is the incident light and ω_s is the scattered light as shown in Fig. 1.12. Once laser became widely available, laser based excitation has been used to obtain Raman spectra. Due to wave vector (momentum) conservation, Raman spectra show the phonon modes that are close the center of the Brillouin zone. The maximum wave vector is at the end of the Brillouin zone, $\frac{\pi}{d}$ where d is the lattice constant. The lattice constant of silicon is 0.537 nm which gives a maximum wave vector of $\frac{\pi}{0.537 \times 10^{-7}\ \text{cm}} = 5.85 \times 10^7\ \text{cm}^{-1}$. Since the wave vector of 532 nm laser light is ~ $1.9 \times 10^4\ \text{cm}^{-1}$, only the phonon wave vectors close to 0 will be observed in first order Raman spectra. A simple Raman spectrum

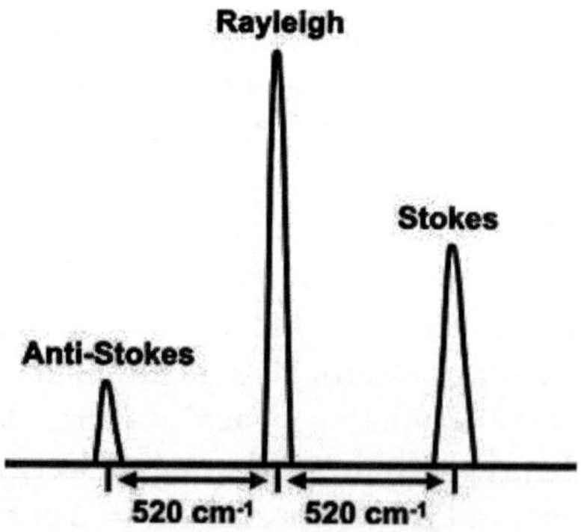

Fig. 1.12 Raman spectra of silicon (001) showing the Stokes, anti-Stokes, and Rayleigh peaks. The Rayleigh peak comes from the reflected laser light. Separating the Brillouin scattering peaks associated with the acoustic phonons requires a special experimental apparatus

for silicon is shown in Fig. 1.12. When a phonon is created by the scattering, $\omega_i - \omega$, the scattering is known as the Stokes process. When a phonon is annihilated by the light scattering, the process is known as anti-Stokes and $\omega_i + \omega$. Stokes lines are an excitation from the ground state to an excited state and are more intense that anti-Stokes lines which are an excitation from an excited state to the ground state [54].

Often, phonons are either Raman active, which is observable by Raman spectroscopy, or Infra-Red active. In Infra-Red spectroscopy, all the light absorption is through the phonons. Below, we describe Raman scattering by phonons. In Raman scattering, the initial step is light absorption by an electron. A phonon will be both Raman and Infra-Red active for crystals that have no center of inversion. Piezoelectric crystals are one example. The relative intensity of the Stokes and anti-Stokes lines $\frac{I_{anti-Stokes}}{I_{Stokes}} = e^{-\hbar\omega_0/K_BT}$ is a function of temperature, because the phonon occupation can be determined using Bose–Einstein Statistics for a phonon of frequency ω_0: $N = \left[e^{-\hbar\omega_0/K_BT} - 1\right]^{-1}$ [55]. K_BT is Boltzman's constant multiplied by the temperature. First order Raman phonon scattering is theoretically described below [54, 56].

We note that both acoustic and optical phonons can be observed in laser scattering. Adjacent atoms move together in acoustic phonons while in optical phonons the adjacent atoms move in opposition to each other. Acoustic phonons are lower in energy than the optical phonons. The acoustic phonons are given their name by the relationship between these phonons and speed of sound in the crystal. The reader is referred to textbooks on lattice dynamics for a complete description of phonons [57, 58]. Here we present the background information on phonons necessary for understanding the Raman spectra of bulk and nanoscale solids. The distinction between longitudinal phonons (L) and transverse phonons (T) is important especially since the use of a polarized light source can selectively excite specific phonon branches. Optical phonons are, by their nature, higher energy phonons. Longitudinal optical phonons are labeled (LO), transverse optical phonons (TO), longitudinal acoustic phonons (LA), and transverse acoustic phonons (TA). For example, the phonon dispersion curves (phonon vs wave vector) for diamond lattice crystals such as Si, Ge, and the diamond phase of carbon are different from the uniaxial wurtzite phase of GaN. The phonon modes for van der Waals materials depend on both the crystal symmetry and the number of layers. The symmetry of many of these materials changes with layer number. For example, inversion symmetry is not present in a single tri-layer of TMD materials while it is present in bi-layers. The symmetry of the crystal determines the phonon modes and which one is Raman or Infra-Red active. We emphasize as quantum mechanics requires, even in the ground state, phonon have vibrational motion and thus energy.

For a crystal having a unit cell with p atoms, there are $3p$ phonon branches as follows:

atoms in the primitive unit cell = p.
Acoustic phonon branches (1LA, 2TA).
$3p - 3$ optical phonons $(p - 1)$ LO and $2(p - 1)$ TO.

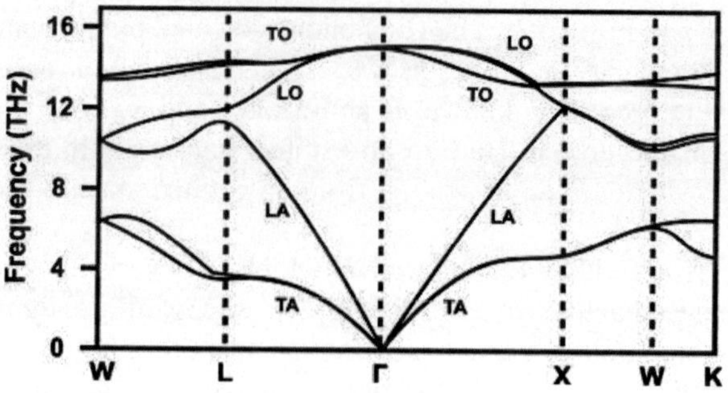

Fig. 1.13 Phonon dispersion curve for silicon showing the longitudinal and transverse acoustic and optical phonons. The two transvers phonons for both acoustic and optical phonons are degenerate. Figure adapted and reproduced with permission from [59].

The phonon dispersion curve for silicon is shown in Fig. 1.13 [59]. Silicon has two atoms per unit cell and thus 6 phonon modes. The LA, 2TA, LO, and 2TO phonon dispersions along the different directions in the Brillouin zone are highlighted by this figure which clearly shows that the optical phonons are at a higher energy than the acoustic phonons. Since phonons are quantum oscillations, they have vibrational motion and thus energy at the Brillouin zone center even at 0 K.

1.9.1 *Theory of Raman Scattering*

Both the wave and particle pictures of light can be used to describe Raman scattering [53]. A classical wave approach can be used to show Stokes and anti-Stokes scattering from phonons in a crystal. The induced dipole moment of crystal $\vec{P}$ is related to the polarizability tensor χ of the crystal and oscillating electric field $\vec{E}_0 \cos(2\pi \upsilon_0 t)$ of the incident light as follows:

$$\vec{P} = \chi \vec{E}_0 \cos(2\pi \upsilon_0 t) \tag{1.69}$$

Since the amplitude of the vibration $\vec{u}$ around the lattice positions in a crystal is small, the polarizability can be expanded as follows:

$$\chi = \chi_0 + \vec{u}\left(\frac{\partial \chi}{\partial t}\right)_{\vec{u}=0} + \cdots, \tag{1.70}$$

where χ_0 is the polarizability of the crystal when the atoms are at their equilibrium position. Using the harmonic approximation for the real part of the time dependent displacement for a monatomic crystal, $\vec{u} = u_0 \cos(2\pi \upsilon_c t)$ where υ_c is the vibrational frequency of the phonon mode. Then using (1.69), we have

$$\vec{P} = \chi_0 \vec{E}_0 \cos(2\pi v_0 t) + u_0 \cos(2\pi v_c t) \left(\frac{\partial \chi}{\partial t} \right)_{\vec{u}=0} \vec{E}_0 \cos(2\pi v_0 t) \tag{1.71}$$

The term $u_0 \cos(2\pi v_c t) \left(\frac{\partial \chi}{\partial t} \right)_{\vec{u}=0} \vec{E}_0 \cos(2\pi v_0 t)$ can be rewritten to show the Stokes and anti-Stokes contributions:

$$u_0 \left(\frac{\partial \alpha}{\partial t} \right)_{\vec{u}=0} \vec{E}_0 \{\cos(2\pi [v_0 - v_c]t) + \cos(2\pi [v_0 + v_c]t)\} \tag{1.72}$$

The time dependent polarization produces light.

A quantum description of first order Raman scattering from a phonon in a crystal requires third order perturbation theory [54]. When light scatters from a small molecule, the atomic positions can change. In a crystal, the atomic positions do not change when one electron is excited to a higher state. A second order perturbation theory for this process results in Raleigh (elastic) light scattering [54]. In Raman scattering, the initial and final electronic state of the crystal are the same.

1.9.1.1 First Order Raman Scattering

The incident light is absorbed and the electron goes to an excited state m. In Chap. 4, the theory behind the imaginary part of the dielectric function and thus the absorption of light by an electron causing a transition for the valence to the conduction band is discussed. In the polarization direction of light is a key part of the interaction between light and an electron. Thus the polarization direction of the light plays a role in determining what electronic transitions are allowed. The excited electron perturbs the system and creates a phonon through electron–phonon interaction. The electron drops in energy by an amount equal to the phonon energy to a state m'. Subsequently, the excited electron drops back to the same energy as the initial state emitting a photon. This is Stokes scattering, and it can be described using third order perturbation theory [54]. This process can produce polarized light since the polarization vector appears in the perturbation Hamiltonian for light emission. When the m state is real, the absorption process is resonant. The state m can also be virtual. It is also possible that the state m is virtual and the state m' is real. In that case, the emission process is resonant. The states m and m' can both be virtual.

The three steps in first order Raman scattering are represented by the matrix elements in the third order perturbation theory for the Stokes scattering intensity as follows [54]: $\langle m | \mathrm{H}_{i \to m} | i \rangle$ the matix element for the the absorption of the incident light taking the electron from the initial state to the m state (see Chap. 4); $\langle m' | H_{el-ph} | m \rangle$ the matrix element with the perturbation Hamiltonian H_{el-ph} for the electron–phonon interaction taking the m state to the m' state; and $\langle f | \mathrm{H}_{m \to f} | m' \rangle$ the transition matrix element for photon emission when the electron transitions from the m' excited state to the final state. As mentioned above, the final and initial state are the same. Using $\Delta E_{m'i} \equiv (E_{m'} - E_i) - i\gamma_\tau$ and $\Delta E_{mi} \equiv (E_m - E_i) - i\gamma_\tau$ where γ_τ is the broadening term for the two resonant transitions $E_{m'} - E_i$ and $E_m - E_i$, and ω_{ph} is the frequency of the phonon, E_L is the energy of the incident light (typically from a laser), the third order Raman intensity is:

$$I\left(\omega_{ph},\ E_L\right) = \sum_f \left| \sum_{m\ m'} \frac{\langle f|H_{m\to f}|m'\rangle\langle m'|H_{el-ph}|m\rangle\langle m|H_{i\to m}|i\rangle}{(E_L - \Delta E_{mi})\left(E_L - \hbar\omega_{ph} - \Delta E_{m'i}\right)} \right|^2 \tag{1.73}$$

The sum over m and m' is over all intermediate states, and the sum over f is for all the possible final (initial) states that can contribute.

1.9.1.2 Second Order Raman Scattering

Fourth order perturbation theory for 2nd order scattering processes can be used to describe both Stokes and anti-Stokes scattering [54]. Just as for first order scattering, the initial step is absorption of the incident light transitioning the electron to an excited state m. In the next step, both phonon emission and absorption are possible. In a single phonon emission, second order process, one transition is inelastic and one is elastic. The order of the inelastic-elastic scattering is interchangeable, and the elastic scattering could occur when the electron is scattered by a lattice defect or an impurity in the crystal. Four different scattering sequences are possible: phonon emission followed by elastic scattering to the m'; or elastic scattering followed by phonon absorption transitioning the electron to the m′ state. The next step is photon emission from the m′ state which can be either resonant when the m′ state is a real state, or it can be non-resonant if m′ is a virual state. In that way, both Stokes and anti-Stokes single phonon scattering can occur. A two phonon second order process is also possible, and below we show two phonon, second order scattering for GaAs in Fig. 1.15. In this case there are two inelastic scattering events where one phonon is emitted and one is absorbed. The sequence is interchangeable.

The key question for Raman spectroscopy of phonons is: which phonons will be observed by a specific experimental setup. In order to determine that, we describe how the Raman intensity is related to the light polarization direction and angle of incidence. Here we concentrate on first order Raman scattering. The Raman intensity can be expressed in terms of the polarization vectors of the incoming $\hat{e}_i$ and scattered $\hat{e}_s$ light and a Raman tensor R_T [60–62]. The incoming and scattered light vectors will be polarized along $x,\ y,\ z$ and the vectors are (1, 0, 0), (0, 1, 0), and (0, 0, 1). The Raman tensor for first order Raman scattering from phonons is related to the electric susceptibility (polarizability) tensor χ of the material being measured and the vibrational amplitude Q_j of a phonon of frequency ω_0. χ can be expanded into a Taylor series where the first term is associated with Rayleigh scattering, the second with first order (single phonon) scattering, and the third term is associated with second order (two phonon) scattering [56, 60]:

$$\chi_j = \chi_0 + \left(\frac{\partial \chi_j}{\partial Q_j}\right)_0 Q_j(\omega_0) + \left(\frac{\partial^2 \chi_j}{\partial Q_j \partial Q_i}\right)_0 Q_j(\omega_0) Q_i(\omega_0)$$
$$+ \text{higher order terms}$$

The Raman tensor is given by the second term in a Taylor series expansion of χ_j as follows [55, 62]:

$$R_{T\ j} = \left(\frac{\partial \chi_j}{\partial Q_j}\right)_0 Q_j(\omega_0) \tag{1.74a}$$

Loudon provided a table of the Raman tensors which we use here [51, 62]. The considerable impact of the laser on Raman is evident from reading [62]. The formalism allows determination of which Raman active modes can be observed in a specific experimental setup. The Raman intensity is given by the sum of the individual contributions to the scattering differential cross section $\frac{d\sigma}{d\Omega}$ for each Raman tensor for a specific vibration $R_{T\ j}$:

$$\frac{d\sigma}{d\Omega} \propto \sum_j \left|\hat{e}_s \cdot R_{T\ j} \cdot \hat{e}_i\right|^2 \tag{1.74b}$$

The polarization direction and crystal orientation determine which modes are observed. The Porto notation used to describe the measurement setup is A(BC)D where A and D are the propagation directions of the incoming and outgoing light respectively, and B and C are the polarization directions of the incoming and outgoing light respectively [52, 60, 61]. This is further described in Fig. 1.14. When a crystal is not oriented along the $x,\ y,\ z$ directions, the tensor elements need to be rotated to match the experimental setup. Some examples of how the Raman tensor and measurement setup terminology are provided below.

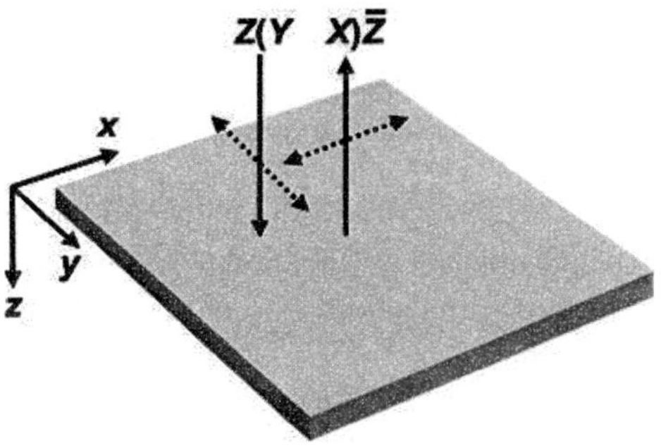

Fig. 1.14 Raman scattering is described in terms of the direction of the incoming light, the polarization direction of that light, the polarization of the detected light and its outgoing direction. Here, $Z(YX)\bar{Z}$ is shown

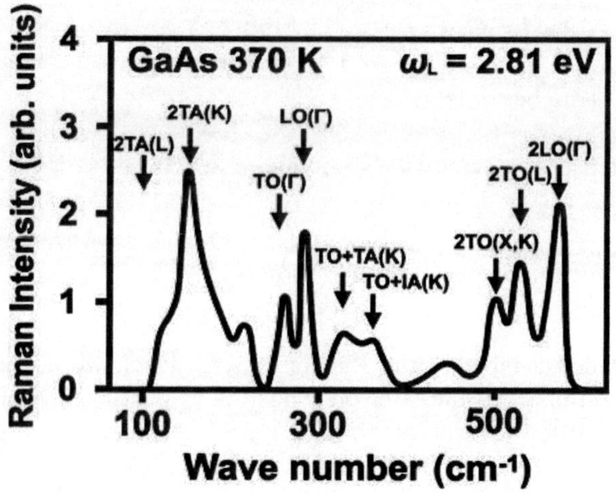

Fig. 1.15 Resonant Raman spectroscopy of GaAs showing the second order Raman scattering using a 2.81 eV light source. Figure adapted and reprinted with permission from [63]. © 1978 American Physical Society

While most Raman studies describe the use of one or perhaps a few laser wavelengths, resonant Raman spectroscopy has also been described [63]. Using tunable lasers, the laser light can be varied from below the band gap to above. In Chap. 4, direct gap transitions are described and the term critical point defined. Here, it is useful to mention that tuning the energy of the laser light enables second order Raman scattering. For example, a Raman peak for two TA phonons or a LO + TA can be observed for GaAs as shown in Fig. 1.15.

1.9.2 Diamond and Zinc Blende Crystals

For the diamond lattice there are 2 atoms per unit cell and thus there are 6 phonon bands 3 acoustic (1 LA and 2 TA) and 3 optical bands (1LO and 2 TO). At the Zone center Γ, the acoustic bands are all degenerate as are the optical bands. Both transverse phonon bands are also degenerate along the Δ direction from Γ (0, 0, 0) to the X point $\frac{2\pi}{a}(0, 1, 0)$ and along the Λ direction from Γ to the L point $\frac{\pi}{a}(1, 1, 1)$. Zinc Blende GaAs also has the same phonon structure at the zone center and along the Δ and Λ directions. Silicon has O_h crystal class and at the zone center the allowed phonon symmetries are F_{1u} and F_{2g} [60]. We note that the Raman tensors shown here are due to the susceptibility that arises from the phonon deformation potential. Only the optical F_{2g} modes are Raman active, and the F_{2g} modes have three T_{2g} Raman tensors for the optical phonons in terms of constants determined by either theory or experiment for each mode [51, 60]:

$$T_{2g}(x) = \begin{pmatrix} 0 & 0 & 0 \\ 0 & 0 & d \\ 0 & d & 0 \end{pmatrix}; \ T_{2g}(y) = \begin{pmatrix} 0 & 0 & d \\ 0 & 0 & 0 \\ d & 0 & 0 \end{pmatrix}; \ T_{2g}(z) = \begin{pmatrix} 0 & d & 0 \\ d & 0 & 0 \\ 0 & 0 & 0 \end{pmatrix}$$

The F_{1u} are acoustic modes. For Raman spectroscopy of the (001) surface, $T_{2g}(x)$ and $T_{2g}(y)$ correspond to TO modes polarized along the x- or y-direction respectively. The $T_{2g}(z)$ is the Raman tensor for the LO mode polarized along the z-axis. At the Γ, the 3 optical phonons at 520 cm^{-1} are degenerate. The $Z(XX)\bar{Z}$ results in $\frac{d\sigma}{d\Omega} = 0$, and for the $Z(XY)\bar{Z}$ measurement configuration we have:

$$\frac{d\sigma}{d\Omega} \propto \left| (1\,0\,0) \begin{pmatrix} 0 & 0 & 0 \\ 0 & 0 & d \\ 0 & d & 0 \end{pmatrix} \begin{pmatrix} 0 \\ 1 \\ 0 \end{pmatrix} \right|^2 + \left| (1\,0\,0) \begin{pmatrix} 0 & 0 & d \\ 0 & 0 & 0 \\ d & 0 & 0 \end{pmatrix} \begin{pmatrix} 0 \\ 1 \\ 0 \end{pmatrix} \right|^2$$
$$+ \left| (1\,0\,0) \begin{pmatrix} 0 & d & 0 \\ d & 0 & 0 \\ 0 & 0 & 0 \end{pmatrix} \begin{pmatrix} 0 \\ 1 \\ 0 \end{pmatrix} \right|^2$$

$$\frac{d\sigma}{d\Omega} \propto 0 + 0 + |d|^2$$

The zinc blende structure of III-V materials is not centrosymmetric. Experimental data for first order Raman scattering from III-V materials often shows both the LO and TO phonon modes. This is due to experimental setup since the optical phonons for the zinc blende structure have the T_{2g} Raman tensors stated above. The $Z(XY)\bar{Z}$ and $Z(YX)\bar{Z}$ optical configurations will both select the LO phonons. When the polarization direction is set along the appropriate [0, 1, 1] axes, the TO phonon can be observed [60].

1.9.3 Wurtzite and other Uniaxial Crystals

Here we describe Raman characterization of wurtzite crystals such as GaN whose symmetry is classified by the C_{6v}^4 space group. We note that group III nitrides can also have a zinc blende structure classified by the T_d^2 space group. The anisotropy of the wurtzite structure also results in an anisotropy of the electric field caused by polar phonons [61, 64]. Thus the optical phonons will be split into axial (A_1) or planar type (E_1) modes [61]. There are eight sets of phonons at the Γ point as follows: $2A_1$; $2E_1$; $2B_1$; and $2E_2$. Of these modes, the acoustic modes are A_1 and E_1, and the A_1; E_1; $2B_1$; and $2E_2$ are optical modes. The atomic vibrations for the optical modes are shown in Fig. 1.16. The superscript L refers to the low frequency branch and the superscript H refers to the high frequency branch of the E and B modes. The A and B modes vibrate along the c-axis while the E_1 and E_2 modes vibrate perpendicular the c-axis. The A_1 and E_1 modes are both Raman and Infra-Red active, the two B modes

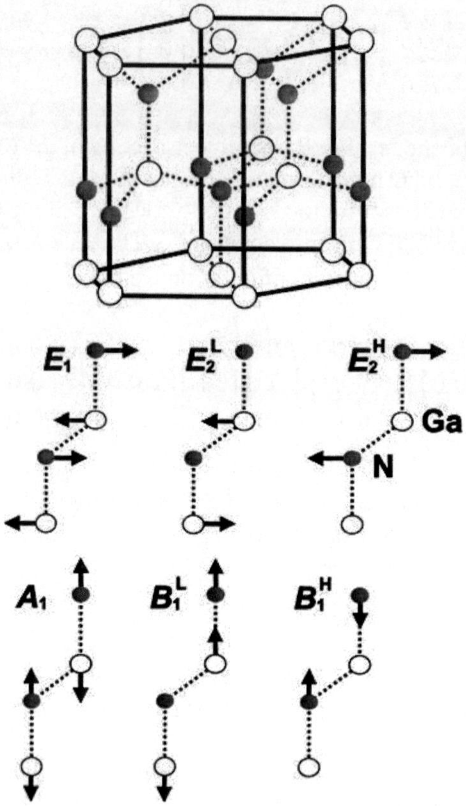

Fig. 1.16 The optical phonon modes for the crystals that have the wurtzite structure are shown. The *A* and *B* modes vibrate along the c-axis. Figure adapted from [61].

are Raman active, and the *B* modes are neither Raman nor Infra-Red active [61]. We note that the Raman tensors shown here are due to the susceptibility that arises from the phonon deformation potential. The Raman tensor for the various phonon modes are taken from Loudin [51]:

$$A_1(z) = \begin{pmatrix} a & 0 & 0 \\ 0 & a & 0 \\ 0 & 0 & b \end{pmatrix}; \qquad E_1(y) = \begin{pmatrix} 0 & 0 & 0 \\ 0 & 0 & c \\ 0 & c & 0 \end{pmatrix};$$

$$E_1(-x) = \begin{pmatrix} 0 & 0 & -c \\ 0 & 0 & 0 \\ -c & 0 & 0 \end{pmatrix}; \qquad E_2 = \begin{pmatrix} 0 & d & 0 \\ d & 0 & 0 \\ 0 & 0 & 0 \end{pmatrix}$$

It is convenient to align the *z*-axis with the c-axis of the crystal. The A_1 LO mode can be observed using the $Z(XX)\bar{Z}$ configuration, which we can verify using:

Table 1.2 The measurement configurations required for observing some of the phonon modes by Raman for a wurtzite structure [61]

Configuration	Observed phonon mode
$Z(XX)\bar{Z}$	A_1(LO)
$Z(XY)\bar{Z}$	E_2
$Z(YY)\bar{Z}$	A_1(LO)
$X(YY)\bar{X}$	A_1(TO),E_2
$X(ZZ)\bar{X}$	A_1(TO)
$X(ZY)\bar{X}$	A_1(TO)

$$\frac{d\sigma}{d\Omega} \propto \left| (1\ 0\ 0) \begin{pmatrix} a & 0 & 0 \\ 0 & a & 0 \\ 0 & 0 & b \end{pmatrix} \begin{pmatrix} 1 \\ 0 \\ 0 \end{pmatrix} \right|^2 = |a|^2$$

In Table 1.2, we show the Raman configurations which are required for observation of specific phonon modes. In Fig. 1.17, the Raman spectra of GaN is shown for different scattering configurations.

1.9.4 Van Der Waals (Layered) Materials

1.9.4.1 Transition Metal Dichalcogenides

The structure of transition metal dichalcogenides (TMD) is discussed in Chap. 8. Here, we provide a summary in order to provide information required for discussing the Raman tensors. TMD crystals having metals from groups 4 to 7 generally form layered structures. The crystals are formed from trilayers of chalcogenide-metal-chalcogenide which can have two different symmetries around the metal atom: the trigonal prismatic structure which has D_{3h} symmetry; and octahedral structure which has O_h symmetry. The symmetry of the bulk crystal formed from van der Waals binding of the trilayers depends on the stacking of the layers, and several polymorphs are possible. A number of TMD materials such as MoS_2, $MoSe_2$, $MoTe_2$, WS_2, WSe_2, and WTe_2 all have the D_{3h} symmetry. There are 2 metal atoms and 4 chalcogenide atoms in the unit cell of bulk TMD such as MoS_2 which results in 18 phonon modes at the Γ point for D_{3h} TMD crystals [52] which have symmetries as follows: $A_{1g} + 2A_{2u} + 2B_{2g} + B_{1u} + E_{1g} + 2E_{1u} + 2E_{2g} + E_{2u}$. The three acoustic modes are A_{2u} and E_{1u}, and the fifteen optical modes are $A_{1g} + A_{2u} + 2B_{2g} + B_{1u} + E_{1g} + E_{1u} + 2E_{2g} + E_{2u}$. Of these 18 modes, A_{2u} and E_{1u} are Infra-Red active and $A_{1g} + E_{1g} + 2E_{2g}$ are Raman active [52]. The other phonon modes cannot be observed by either Raman or Infra-Red. We note that the Raman tensors shown here are due to the susceptibility that arises from the phonon deformation potential. The Raman tensors for bulk TMD crystals having D_{3h} symmetry are as follows [52]:

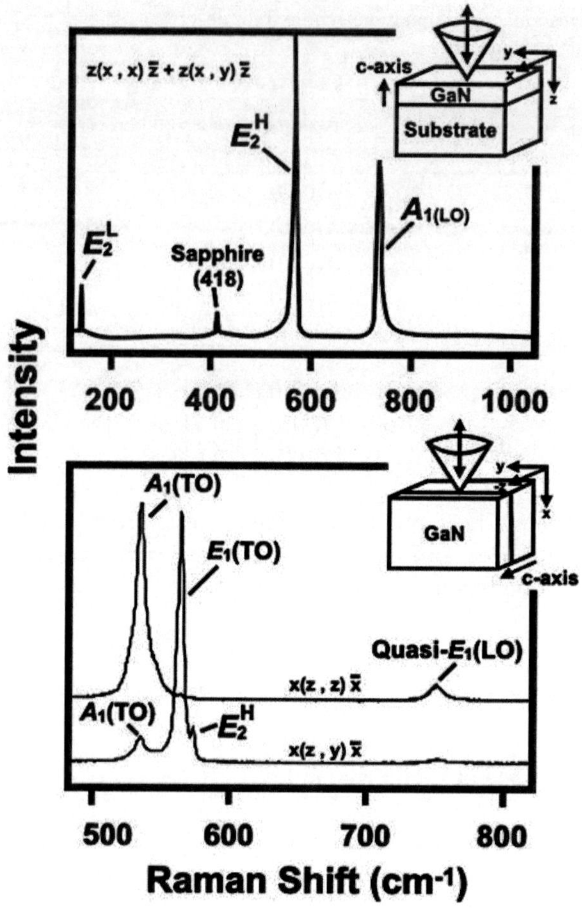

Fig. 1.17 Raman spectra of GaN for several measurement configurations. Figure adapted from [61].

$$A_{1g}(D_{6h}) = \begin{pmatrix} a & 0 & 0 \\ 0 & a & 0 \\ 0 & 0 & b \end{pmatrix}; \; E_{1g}(D_{6h}) = \begin{pmatrix} 0 & 0 & 0 \\ 0 & 0 & c \\ 0 & c & 0 \end{pmatrix}; \; E_{1g}(D_{6h}) = \begin{pmatrix} 0 & 0 & -c \\ 0 & 0 & 0 \\ -c & 0 & 0 \end{pmatrix};$$

$$E_{2g}(D_{6h}) = \begin{pmatrix} 0 & d & 0 \\ d & 0 & 0 \\ 0 & 0 & 0 \end{pmatrix}; \; E_{2g}(D_{6h}) = \begin{pmatrix} d & 0 & 0 \\ 0 & -d & 0 \\ 0 & 0 & 0 \end{pmatrix}$$

For the spectra shown in Fig. 1.18, the optical axis of the hexagonal TMD, the c-axis, is oriented along the experimental z axis, and the data was taken using the $Z(XX)\bar{Z}$ configuration so that only the A_{1g} and E_{2g} modes are observed [52]. The A_{1g} mode is not observed for the $Z(XY)\bar{Z}$ configuration. This spectrum was obtained

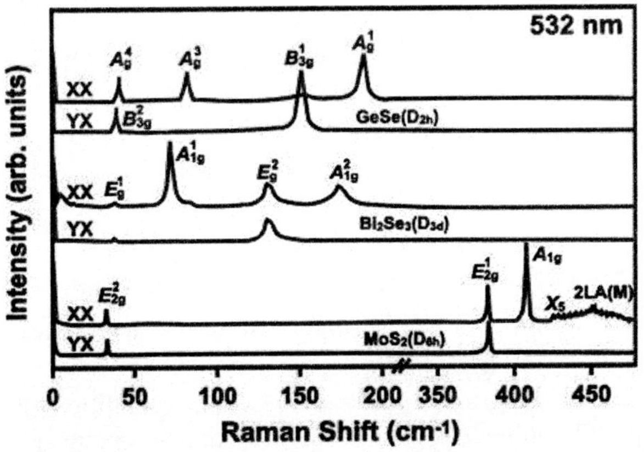

Fig. 1.18 Raman spectra of bulk GeSe, Bi_2Se_3 and MoS_2 for the $Z(XX)\ \bar{Z}$ and $Z(XY)\ \bar{Z}$ configurations. Figure adapted from [52].

at 532 nm. Again, the energy of the laser light is an important factor in determining the selection criteria. When 633 nm laser light is used to obtain the spectra, the second order 2 LA mode is observed which comes from the M point in the Brillouin zone [52]. The layers structure of bulk TMD crystals results in a layer dependent Raman shift for several peaks as well as shearing modes where the layers move relative to each other. The Raman spectra of TMD materials are further discussed in Chap. 8.

TMD materials have polar bonding. Fröhlich interactions (FI) arise from a coupling between longitudinal optical phonons and electrons due to the polar nature of the phonon [65]. This coupling is exceptionally strong for monolayer TMD when compared to electron–phonon coupling in non-polar materials [65, 66]. The relative contribution of the Fröhlich interactions to the phonon spectra of single trilayers (monolayers) of the 2H polymorph of MoS_2 has been compared to that of the deformation potential (d) [65]. Frequently in the literature, the phonon modes for single trilayers (monolayers) of TMD materials are designated by the equivalent phonon mode for bulk TMD materials [67, 68]. Thus the $\left(E'\right)_{(x)}$ mode is designated as the E_{2g} mode [65–67]. The Raman tensors for the different polymorphs of TMD materials is discussed in Chap. 8 and listed in Chap. 8 Table 8.8. The Raman tensor for the Fröhlich interaction for the $\left(E'\right)_{(x)}\ \{E_{2g}\}$ LO phonon for is [65, 66]:

$$R\Big(\left(E'\right)_{(x)}\ \{E_{2g}\}(LO:FI)\Big)=\begin{pmatrix} FI & 0 & 0\\ 0 & FI & 0\\ 0 & 0 & FI\end{pmatrix}$$

and the combined Raman tensor for the deformation potential (DP) and Fröhlich interaction (FI) is:

$$R\left((E')_{(x)}\{E_{2g}\}LO\right) = \begin{pmatrix} FI & d & 0 \\ d & FI & 0 \\ 0 & 0 & FI \end{pmatrix}; \; R\left((E')_{(y)}\{E_{2g}\}\,TO\right) = \begin{pmatrix} d & 0 & 0 \\ 0 & -d & 0 \\ 0 & 0 & 0 \end{pmatrix}$$

Left handed circularly polarized light $\sigma^+ = \frac{1}{\sqrt{2}}\begin{pmatrix} 1 \\ i \\ 0 \end{pmatrix}$ and right had circularly polarized light $\sigma^- = \frac{1}{\sqrt{2}}\begin{pmatrix} 1 \\ -i \\ 0 \end{pmatrix}$ can be used to determine the relative contribution of FI and DP [66]. Raman spectra obtained under resonance condition for exciton formation found that the FI dominates the on-resonance Raman spectra at low temperature [66]. When using circularly polarized light, the scattering intensity is:

$$I = |\sigma^*_{scattered} \cdot R(LO) \cdot \sigma_{incident}|^2 + |\sigma^*_{scattered} \cdot R(TO) \cdot \sigma_{incident}|^2$$

One can show that when left hand circularly polarized light is incident and one measures the scattered σ^+ light, then $I_{\sigma^+\sigma^+} = |AF|^2$. This is in contrast to the configuration where the scattered right hand light, σ^-, is measured: $I_{\sigma^+\sigma^-} = 2|d|^2$ [66].

1.9.4.2 Tetradymite Compounds (Bi_2Se_3, Bi_2Te_3, $Bi_2Se_xTe_{3\text{-}x}$, Sb_2Te_3 and Alloys)

The structure of tetradymite compounds is discussed in Chap. 9. Here, we discuss how the symmetry of these materials impacts their Raman spectra. The tetradymite compounds have a repeating five (quintuple) layer structure of chalcogenide-metal-chalcoginide-metal-chalcoginide with a hexagonal shape that can be observed when looking down the optical axis (layer stacking axis). The crystal has a rhombohedral crystal structure with a space group of $D^5_{3D}(R32/m)$. There are 15 phonon modes with the following symmetries at the Γ point: $2A_{1g} + 3A_{2u} + 2E_g + 3E_u$. The $2A_{1g}$ and $2E_g$ are Raman active, and the $3A_{2u} + 3E_u$ are Infra-Red active [52]. The Raman tensors for the Raman active modes are:

$$A_{1g}(D_{3d}) = \begin{pmatrix} a & 0 & 0 \\ 0 & a & 0 \\ 0 & 0 & b \end{pmatrix}; \; E_g(D_{3d}) = \begin{pmatrix} c & 0 & 0 \\ 0 & -c & d \\ 0 & d & 0 \end{pmatrix}; \; E_g(D_{3d}) = \begin{pmatrix} 0 & -c & -d \\ -c & 0 & 0 \\ -d & 0 & 0 \end{pmatrix}$$

The $A_{1g}(D_{3d})$ mode can be observed by both the $Z(XX)\bar{Z}$ and $Z(XY)\bar{Z}$ configurations while the $E_{1g}(D_{3d})$ modes can be observed only using the $Z(XX)\bar{Z}$ configuration.

1.9.4.3 Graphene, Few Layer Graphene, and Graphite

Graphene is a single layer of carbon atoms bonded by sp^2 orbitals. The interlayer bonding for few layer graphene and graphite is van der Waals bonding. The electronic band structure of graphene is discussed in Chap. 7 and the spatial symmetry classification for Bernal stacking is discussed in Chap. 2. The π and π^* valence and conduction bands formed from the p orbitals that lie above and below the graphene plane play a key role in Raman spectroscopy. The G (E_{2g}) and G' Raman bands appear at $\sim$ 1585 cm^{-1} and between 2500 and 2800 cm^{-1} for graphene and related systems, and are considered to a "signatures" for sp^2 carbons [69]. The G' can be used to determine the number of graphene layers [69], and the frequency depends on the energy of the laser excitation. It is important to note that the stacking pattern influences peak shape [69, 70]. One of the key uses of Raman spectroscopy in graphene research is determination of the number of layers [70]. A D band appears in sp^2 bonded carbons between 1250 – 1400 cm^{-1} when disorder (defects) is present [69]. The G (E_{2g}) transition is a result of a first order resonant Raman process involving the π and π^* bands [69, 70], and the G' band is due to a double resonant scattering process involving the π and π^* bands.

In monolayer graphene, there are vibrational modes that are Raman inactive including A_{2u}; B_{2g}; and E_{1u} modes. In both graphene and graphite, the G (E_{2g}) has been reported to appear at 1585 cm^{-1}. Trilayer graphene provides an interesting illustration of the influence on the layer stacking configuration on the Raman spectra, see Chap. 7 for a further discussion about layer stacking. In Bernal stacking, the stacking pattern repeats every other layer so that it is AB–AB–AB. Trilayer graphene with ABA (Bernal) and ABC stacking has been characterized by Raman [71, 72]. The spectral shape of the G' Raman band, which is observed between 2550 and 2570 cm^{-1}, of trilayer graphene has been shown to be dependent on the stacking sequence and laser excitation energy [72]. In that study, laser energies of 1.96, 2.33, and 2.94 eV resulted in observable differences in peak shape [72]. Interlayer sheer modes and layer breathing modes appear at low energies in trilayer graphene [71]. The sheer modes are labeled C and C'. For ABC stacking, the high (energy) sheer mode C has E' symmetry and is observable at normal incidence (light polarized in-plane) where the polarization of the light at 37 cm^{-1} [71]. For ABC stacking the C' sheer mode with E'' symmetry is not observable. For ABA trilayer graphene, the low (energy) C' sheer mode with E_g symmetry is observable for normal incidence excitation at 22 cm^{-1} [71]. The Raman of bilayer graphene is discussed in Chap. 7.

First we discuss the Raman tensor for the G band. It involves in-plane stretching of the C–C bonds. Degenerate, in-plane transverse optical phonon and longitudinal optical phonons are both excited [70]. It is the only Raman band in graphene that is due to normal 1st order Raman scattering [70]. Using α for the relative azimuthal angle for the polarization direction of the incident light relative to the graphene lattice and β for the relative azimuthal angle for the polarization direction of the scattered light, one can write the unit polarization vectors $e_L = (\cos(\alpha), \ \sin(\alpha), \ 0)$ and $e_S = (\cos(\beta), \ \sin(\beta), \ 0)$. The Raman tensors for the degenerate E_{2g} mode are

[73]:

$$E_{2g}(D_{6h}) = \begin{pmatrix} 0 & d & 0 \\ d & 0 & 0 \\ 0 & 0 & 0 \end{pmatrix}; \; E_{2g}(D_{6h}) = \begin{pmatrix} d & 0 & 0 \\ 0 & -d & 0 \\ 0 & 0 & 0 \end{pmatrix}$$

For normal incidence laser excitation, this gives an azimuthal angle independent intensity for the G (E_{2g}) band $\frac{d\sigma}{d\Omega} \propto \sum_j \left|\hat{e}_s \; R_{T\;j} \; \hat{e}_i\right|^2 = |d|^2$ [73].

The Raman tensor for the C and C' bands for the ABC stacking are [71]:

$$E' = \begin{pmatrix} a & b & 0 \\ b & -a & 0 \\ 0 & 0 & 0 \end{pmatrix}; \; E'' = \begin{pmatrix} 0 & 0 & c \\ 0 & 0 & d \\ c & d & 0 \end{pmatrix}$$

The Raman tensor for the C' band for the ABA stacking is[71]:

$$E_g = \begin{pmatrix} e & f & 0 \\ f & -e & 0 \\ 0 & 0 & 0 \end{pmatrix}.$$

Raman characterization of nanoscale materials has the advantage of a small analysis area due to the focusing ability of a laser relative to a traditional light source. The D band has proven to be a useful means of characterizing graphene after it has been processed for use in an electronic device [74]. A double resonant transition is responsible for the D band, and thus the peak shape is sensitive to changes in the π and π^* electronic band structure caused by defects or other changes to the electronic structure [69]. Lithographic and other processes are used to fabricate nanoribbons which have a width and edge dependent band structure [75–77]. The D and G bands both show significant broadening for the narrowest nanoribbons. The ratio of the D and G band intensity is often used to characterize damage to graphene [74, 75]. It is known that nanoribbons of graphene less than 100 nm in width develop a band gap with a width W dependence of $E_{BG} = \frac{\alpha}{(W - W_0)}$ where $\alpha = 0.2\frac{\text{eV}}{\text{nm}}$ and $W_0 = 16$ nm [76]. Thus, D band is also sensitive to the width of graphene nanoribbons [77]. The ratio of the D and G band intensities increases as nanoribbon width decreases [77]. A defect induced $2D$ band is observed at 2640 cm^{-1} and it is considered to be a reliable means of determining the number of graphene layers for nanoribbons [77]. Peak broadening due to changes in nanoribbon width should not interfere with use of the $2D$ band for ribbon width [77].

1.10 Photoluminescence

Photoluminescence (PL) can be used to characterize the quality of a crystal, the quality of an interface between a semiconductor and dielectric, the impurities present in a crystal, and the quality of semiconductor quantum wells. Electron–hole pairs (e–h) can be excited by light with greater energy than the band gap of a material and the photoluminescence comes from the recombination of these e–h pairs. Often, PL data is obtained using laser excitation at a few K to increase the lifetime of the excitons. We also note that e–h pairs can bind together and even form e–h liquids or gases. Excitons are discussed in Chap. 5. Photoluminescence is widely used to characterize TMD materials, and the lineshape of TMD materials is discussed below. Photoluminescence characterization of TMD materials is further discussed in Chap. 8. A low temperature PL spectra of boron doped silicon is shown in Fig. 1.19 as an example [78, 79].

The line shape observed in photoluminescence spectra depends on the dimensionality and crystal structure of the material. Spectra due to a single optical transition can be modeled by a Lorentzian function. Here, the Heisenberg uncertainty principle plays a key role in understanding the line shape. The Heisenberg uncertainty in the energy ΔE is limited by the uncertainty in the transition lifetime $\Delta\tau$ so that $\Delta E \Delta\tau \geq \hbar$. Thus, a short lifetime for the optical transition results in broader peak widths. Another contribution to the Lorentzian line shape is phonon-exciton

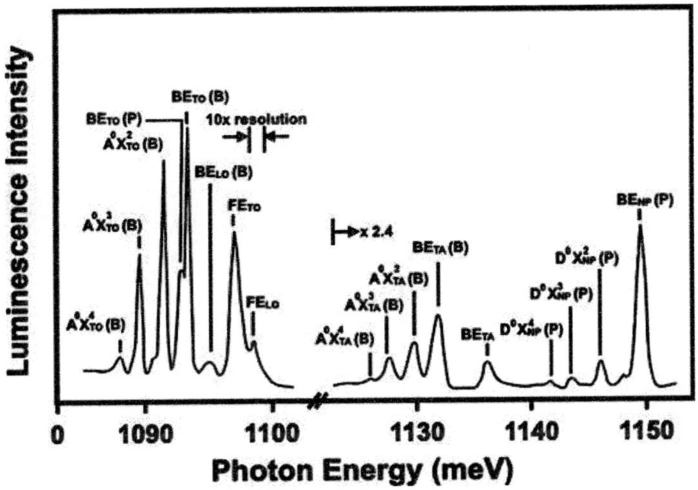

Fig. 1.19 Low temperature 4.2 K photoluminescence spectra of boron doped silicon. The figure is labeled as follows: boron (B), phosphorous (P), bound exciton (BE), free exciton (FE), transverse optical phonon (TO), longitudinal optical phonon (LO), bound multi-exciton complexes labeled A^0X^4, etc. For example, BE_{TO}(B) is an exciton bound to boron that scatters from a TO phonon. The data was obtained using light from a Hg arc lamp that was filtered to remove the Infra-Red at 1.5 W cm^{-2} [79]. The Figure adapted and reprinted from [78]. © 1980 The Physical Society of Japan and The Japan Society of Applied Physics

scattering. This results in phonon emission or absorption for some of the scattering events. Spectra from multiple transitions can be modeled by a Gaussian function. Often, the convolution of Lorentzian and Gaussian forms known as the Voigt function is used. The Lorentzian $L(\omega)$, Gaussian $G(\omega)$, and Voight $V(\omega)$ lineshapes are stated in terms of the angular frequency ω as follows:

$$L(\omega) = \frac{\Gamma_L}{2\pi}\left(\frac{1}{\hbar^2(\omega-\omega_0)^2-(\Gamma_L/2)^2}\right) \tag{1.75a}$$

$$G(\omega) = \frac{\Gamma_G}{\sqrt{2\pi}}e^{-\hbar^2(\omega-\omega_0)^2/2\Gamma_G^2} \tag{1.75b}$$

$$V(\omega) = \int_{-\infty}^{\infty} G(\omega')L(\omega-\omega')d\omega' \tag{1.75c}$$

with the frequency of the optical transition ω_0 and the full width at half maximums $\Gamma_L = \hbar/2\tau_L and\ \Gamma_G = \hbar/2\tau_G$ where the Gaussian and Lorentzian transition lifetimes τ_G and τ_L.

Photoluminescence spectra of 2D materials are presented in Chaps. 5, 7, and 8. Here, we discuss the line shape of TMD materials to illustrate the Lorentzian, Gaussian, and Voigt functions. Excitons in monolayer TMD materials such as MoS_2, $MoSe_2$, WS_2, and WSe_2 have large binding energies (~ 500 meV). The radiative lifetimes of these TMD materials is very short ($\tau_R \sim 1$ ps) when compared to direct gap III-V semiconductors such as GaAs and InGaAs which have orders of magnitude larger radiative lifetimes [80]. This short radiative lifetime broadens the PL peaks by 1 to 2 meV [80]. The lineshape of photoluminescence from monolayer TMD materials can be modeled using a Voigt function which is a convolution of a Lorentzian line shape due to lifetime broadening and phonon-exciton scattering with a Gaussian lineshape due to thermal motion. In Fig. 1.20, we show the A_{1s}^0 spectra of $MoSe_2$ between 1.64 and 1.67 eV at 3.2 K. The total linewidth was 5.47 meV and after deconvolution, the total Lorentzian broadening was 1.94 meV. The total Lorentzian broadening was mainly due to the fundamental radiative linewidth of 1.54 meV [80].

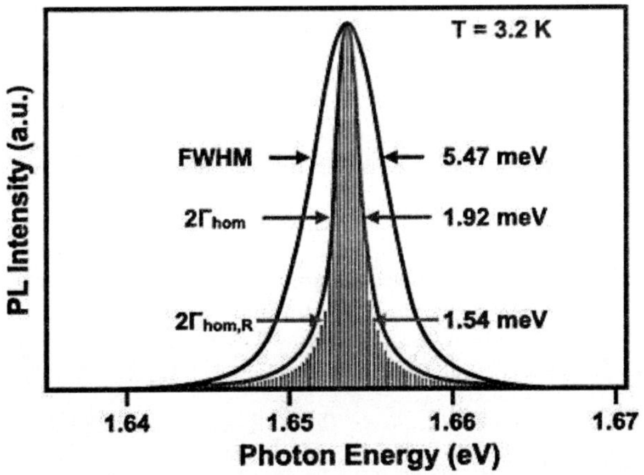

Fig. 1.20 The Photoluminescence Spectra of the A^0_{1s} transition which occurs at the K points due to the recombination of the exciton located in the Brillouin zone of $MoSe_2$. The spectrum was fit using a Voigt function. The Lorentzian broadening due to the short exciton lifetime ($\Gamma_{\mathrm{hom},\ R}$) is 1.54 eV. The spectra was obtained from an exfoliated $MoSe_2$ sample using 633 nm laser light with a power density of 50 μW/cm^2. This power density avoids sample heating. The sample was analyzed at a pressure of 10^{-4} Torr. Figure adapted and reprinted with permission from [80]. © 2019 American Physical Society

References

1. R.K. Wagness, *Electromagnetic Fields*, 2nd edn. (Wiley, New York, 1986), pp. 348, 355 and 370
2. M. Born, E. Wolf, *Principles of Optics*, 7th edn. (Cambridge, Cambridge, 1999), pp. 11–16
3. F. Wooten, *Optical Properties of Solids* (Academic, New York, 1972), pp. 16–29
4. H. Fujiwara,*Spectroscopic Ellipsometry: Principles and Applications*, (Wiley, West Sussex, 2007), pp. 33–48, pp. 179–183, pp. 189–190, pp. 209–246
5. S. Nichols, O. Arteaga, A. Martin, B. Kahr, Measurement of transmission and reflection from a thick anisotropic crystal modeled by a sum of incoherent partial waves. J. Opt. Soc. Amer. A, **32**, 2049 (2015)
6. G.E. Jellison, L.A. Boatner, J.D. Budai, B.-S. Jeong, D.P. Norton, Spectroscopic ellipsometry of thin film and bulk anatase (TiO_2). J. Appl. Phys. **93**, 9537 (2003)
7. G.E. Jellison, J.D. Hunn, H.N. Lee, Measurement of optical functions of highly oriented pyrolytic graphite in the visible. Phys. Rev. B **76**, 085125 (2007)
8. G.E. Jellison, M.A. McGuire, L.A. Boatner, J.D. Budai, E.D. Specht, D.J. Singh, Spectroscopic dielectric tensor of monoclinic crystals: $CdWO_4$. Phys. Rev. B **84**, 195439 (2011)
9. R.M.A. Azzam, N.M. Bashara, *Ellipsometry and Polarized Light*, chaps. 3 and 4 (North Holland, New York, 1989)
10. M. Schubert, Polarization-dependent optical parameters of arbitrarily anisotropic homogeneous layered systems. Phys. Rev. B **53**, 4265–4274 (1996)
11. M. Schubert, Theory and application of generalized ellipsometry, Chap. 9, in *Handbook of Ellipsometry*, ed by H.G. Tompkins, E.A. Irene (William Andrew, New York, 2005), pp. 637–717
12. M. Schubert, *Infrared Ellipsometry on Semiconductor Structures: Phonon, Plasmons, and Polaritons* (Springer, Heidelberg, 2004). (chapter 2)
13. O.S. Heavens, Optical properties of thin films. Rep. Prog. Phys. **23**, (1960). (see equation 53), 1 or E. Hecht, *Optics*, (Reading MA, Addison Wesley, 2002)

14. Y. Li, T.F. Heinz, Two-dimensional models for the optical response of thin films. 2D Mater. **5**, 025021 (2018)
15. L.P. Mosteller Jr., F. Wooten, Optical properties and reflectance of uniaxial absorbing crystals. J. Opt. Soc. Am. **58**, 511–518 (1968)
16. F. Wooten, Reflectivity of uniaxial absorbing crystals. Appl. Opt. **23**, 4226–4227 (1984)
17. D.J. De Smet, Ellipsometry of anisotropic substrates: Reexamination of a special case. J. Appl. Phys. **76**, 2571 (1994)
18. D.E. Aspnes, Approximate solution of ellipsometric equations for optically biaxial crystals. J. Opt. Soc. Am. **70**, 1275–1277 (1980)
19. G.E. Jellison Jr., J.S. Baba, Pseudodielectric functions of uniaxial materials in certain symmetry directions. J. Opt. Soc. Am. A **23**, 468–475 (2006)
20. V.G. Kravets, A.N. Grigorenko, R.R. Nair, P. Blake, S. Anissimova, K.S. Novoselov, A.K. Geim, Spectroscopic ellipsometry of graphene and an exciton-shifted van Hove peak in absorption. Phys. Rev. B **81**, 155413 (2010)
21. M.I. Alonso, M. Garrriga, F. Alsina, S. Pinol, Determination of the dielectric tensor in anisotropic materials. Appl. Phys. Lett. **67**, 596–598 (1995)
22. R. Goldhahn, A.T. Winzer, V. Cimalla, O. Ambacher, C. Cobet, W. Richter, N. Esser, J. Furthmüller, F. Bechstedt, H. Lu, W.J. Schaff, Anisotropy of the dielectric function for wurtzite InN. Superlattices Microstruct. **36**, 591–597 (2004)
23. F.J. Nelson, V.K. Kamineni, T. Zhang, E.S. Comfort, J.-U. Lee, A.C. Diebold, Optical properties of large-area polycrystalline chemical vapor deposited graphene by spectroscopic ellipsometry. Appl. Phys. Lett. **97**(253110), 1–3 (2010)
24. E. McCann, Chapter 8 Electronic properties of monolayer and bilayer graphene, in *NanoScience and Technology*. ed. by G. Nanoelectronics (Berlin, Springer, 2012), pp. 237–275
25. F.J. Nelson, J.-C. Idrobo, J. Fite, Z.L. Mišković, S.J. Pennycook, S.T. Pantelides, J.U. Lee, A.C. Diebold, Electronic excitations in graphene in the 1–50 eV range: the π and $\pi + \sigma$ peaks are not plasmons. Nano Lett. **14**, 3827–3831 (2014)
26. M. Merano, Fresnel coefficients of a two-dimensional atomic crystal. Phys. Rev. A **93**, 013832 (2016)
27. C.H. Lui, L. Liu, K.F. Mak, G.W. Flynn, T.F. Heinz, Ultraflat graphene. Nature **462**, 339–341 (2009)
28. C.R. Dean, A.F. Young, I. Meric, C. Lee, L. Wang, S. Sorgenfrei, K. Watanabe, T. Taniguchi, P. Kim, K.L. Shepard, J. Hone, Boron nitride substrates for high-quality graphene electronics. Nat. Nanotech. **5**, 722–726 (2010)
29. A.C. Diebold, B. Doris, A survey of non-destructive surface characterization methods used to insure reliable gate oxide growth for silicon IC devices. Surf. Inter. Anal. **20**, 127 (1993)
30. H.R. Huff, D.W. McCormack Jr., C. Au, T. Messina, K. Chan, R.K. Goodall, Current status of 200 mm and 300 mm Silicon wafers. Jap. J. Appl. Phys. **37**, 1210–1216 (1998)
31. F. Nelson, A. Sandin, D.B. Dougherty, D.E. Aspnes, J.E. Rowe, A.C. Diebold, Optical and structural characterization of epitaxial graphene on vicinal 6H-SiC(0001)–Si by spectroscopic ellipsometry, auger spectroscopy, and STM. J. Vac. Sci. Technol. B **30**, 04E106 (2012)
32. A. Boosalis, T. Hofmann, V. Darakchieva, R. Yakimova, M. Schubert, Visible to vacuum ultraviolet dielectric functions of epitaxial graphene on 3C and 4H SiC polytypes determined by spectroscopic ellipsometry. Appl. Phys. Lett. **101**, 011912 (2012)
33. W.Y. Liang, Optical anisotropy in layer compounds. J. Phys. C: Solid State Phys. **6**, 551 (1973)
34. H.-L. Liu, C.-C. Shen, S.-H. Su, C.-L. Hsu, M.-Y. Li, L.-J. Li, Optical properties of monolayer transition metal dichalcogenides probed by spectroscopic ellipsometry. Appl. Phys. Lett. 105, 201905 (2014)
35. V.G. Kravets, V.V. Prorok, L.V. Poperenko, I.A. Shaykevich, Ellipsometry and optical spectroscopy of low-dimensional family TMDs. Semi. Phys., Quan. Elec. Optoelec, **20**, 284–296 (2017)
36. Y. Li, A. Chernikov, X. Zhang, A. Rigosi, H.M. Hill, A.M. van der Zande, D.A. Chenet, E.-M. Shih, J. Hone, T.F. Heinz, Measurement of the optical dielectric function of monolayer transition-metal dichalcogenides: MoS_2, $MoSe_2$, WS_2, and WSe_2. Phys. Rev. B **90**, 205422 (2014)

37. A.R. Beal, H.P. Hughes, Kramers-Kronig analysis of the reflectivity spectra of ~H-MoS, 2H-MoSe, and 2H-MoTe. J. Phys. C: Solid State Phys. **12**, 881–890 (1979)
38. G.Y. Jia, Y. Liu, J.Y. Gong, D.Y. Lei, D.L. Wang, Z.X. Huang, Excitonic quantum confinement modified optical conductivity of monolayer and few-layered MoS_2. J. Mater. Chem. C **4**, 8822 (2016)
39. C.H. Hok, Y.S. Huang, K.K. Tiong, P.C. Liao, In-plane anisotropy of the optical and electrical properties of layered ReS_2 crystals. J. Phys.: Condens. Matter **11**, 5367–5375 (1999)
40. Y.V. Morozov, M. Kuno, Optical constants and dynamic conductivities of single layer MoS_2, $MoSe_2$, and WSe_2. Appl. Phys. Lett. **107**, 083103 (2015)
41. D.W. Berreman, Optics in stratified and anisotropic media: 4×4-matrix formulation. J. Opt. Soc. Am. **62**, 502–510 (1972)
42. G.E. Jellison Jr., F.A. Modine, Optical constants for silicon at 300 and 10 K determined from 1.64 to 4.73 eV by ellipsometry. J. Appl. Phys. **53**, 3745–3753 (1982)
43. D.E. Aspnes, A.A. Studna, Dielectric functions and optical parameters of Si, Ge, GaP, GaAs, InP, InAs, and InSb from 1.5 to 6.0 eV. Phys. Rev. B **27**, 985–1009 (1983)
44. G.E. Jellison Jr., Optical functions of silicon determined by two-channel polarization modulation ellipsometry. Opt. Mater. **1**, 41–47 (1992)
45. S.J. O'Mullane, D. Dixit, A.C. Diebold, *Advanced Ellipsometry Measurements for Semiconductor Structures, Characterization and Metrology for Nanoelectronics and Nanostructures*, chap. 3 (Pan Stanford, 2016)
46. G.E. Jellison, Jr., F.A. Modine, Parameterization of the optical functions of amorphous materials in the interband region. Appl. Phys. Lett. **69**, 371 (1996); **69**, 2137 (1996)
47. A.S. Ferlauto, G.M. Ferreira, J.M. Pearce, C.R. Wronski, R.W. Collins, Analytical model for the optical functions of amorphous semiconductors from the near-infrared to ultraviolet: applications in thin film photovoltaics. J. Appl. Phys. **92**, 2424 (2002)
48. H. Ehrenreich, H.R. Phillip, B. Segal, Optical properties of Aluminum. Phys. Rev. **132**, 1918–1928 (1963)
49. V.K. Kamineni, M. Raymond, E.J. Bersch, B.B. Doris, A.C. Diebold, Optical metrology of Ni and NiSi thin films used in the self-aligned silicidation process. J. Appl. Phys. **107**, 093525 1–8 (2010)
50. J.D. Smith, The remarkable Ibn al-Haytham. Math. Gazz. **76**, 189–198 (1992)
51. R. Loudon, The Raman effect in crystals. Adv. Phys. **50**, 813–864 (2001)
52. X. Zhang, Q.-H. Tan, J.-B. Wu, W. Shi, P.-H. Tan, Review on the Raman spectroscopy of different types of layered materials. Nanoscale **8**, 6435–6450 (2016)
53. J. Menéndez, Characterization of bulk semiconductors using raman spectroscopy, chap. 1, in: *Raman Scattering in Materials Science*, ed. by. W.H. Weber, R. Merlin (Springer, New York, Berlin, 2000), pp. 55–103
54. A. Joprio, M. Dresselhaus, R. Saito, G.F. Dresselhaus, *Raman Spectrocopy in Graphene Related Systems* (Wiley-VCH, Weinheim, 2011), pp. 108–111
55. T.R. Hart, R.L. Aggerwal, B. Lax, Temperature dependence of Raman Scattering in silicon. Phys. Rev. B **1**(91970), 638–642
56. R. Merlin, A. Pinzcuk, W.H. Weber, Overview of phonon Raman Scattering in solids, in *Raman Scattering in Materials Science*. ed. by W.H. Weber, R. Merlin (Springer, New York, 2000), pp. 1–29
57. G.P. Srivastava, *The Physics of Phonons* (Taylor and Francis, Abingdon, 1990)
58. M.T. Dove, *Introduction to Lattice Dynamics* (Cambridge, Cambridge, 1993)
59. W.-W. Zhang, H. Yu, S.-Y. Lei, Q.-A. Huang, Modelling of the elastic properties of crystalline silicon using lattice dynamics. J. Phys. D: Appl. Phys. **44**, 335401 (2011)
60. P.Y. Yu, M. Cardona, *Fundamentals of Semiconductors, Physics and Materials Properties* (Springer, New York, 2010), p. 378
61. H. Harima, Properties of GaN and related compounds studies by means of Raman Spectroscopy. J. Phys.: Condens. Matter **14**, R967–R993 (2002)
62. R. Loudon, Theory of the first order Raman effect in crystals. Proc. R. Soc. Lond. A **275**, 218–232 (1963)

63. R. Trommer, M. Cardona, Resonant Raman scattering in GaAs. Phys. Rev. B **17**, 1865–1876 (1978)
64. C. Kranert, C. Sturm, R. Schmidt-Grund, M. Grundmann, Raman tensor formalism for optically anisotropic crystals. Phys. Rev. Lett. **116**, 127401 (2016)
65. T. Sohier, M. Calandra, F. Mauri, Two-dimensional Fröhlich interaction in transition-metal dichalcogenide monolayers: theoretical modeling and first-principles calculations. Phys. Rev. B **94**, 085415 (2016)
66. Y. Zhao, S. Zhang, Y. Shi, Y. Zhang, R. Saito, J. Zhang, L. Tong, Characterization of excitonic nature in Raman spectra using circularly polarized light. ACS Nano. **14**, 10527–10535 (2020)
67. J. Ribeiro-Soares, Private communication
68. J. Ribeiro-Soares, R.M. Almeida, E.B. Barros, P.T. Araujo, M.S. Dresselhaus, L.G. Canc¸ado, A. Jorio, Group theory analysis of phonons in two-dimensional transition metal dichalcogenides. Phys. Rev. B **90**, 115438 (2014)
69. A. Jorio, M. Dresselhaus, R. Saito, G.F. Dresselhaus, Chap. 14 Summary of Raman spectroscopy on nanocarbons, in *Raman Spectroscopy in Graphene Related Systems* (Wiley-VCH, Weinheim, 2011), pp. 327–334
70. L.M. Malard, M.A. Pimenta, G. Dresselhaus, M.S. Dresselhaus, Raman Spectroscopy in graphene. Phys. Rep. **473**, 51–87 (2009)
71. C.H. Lui, Z. Ye, C. Keiser, E.B. Barros, R. He, Stacking-dependent shear modes in trilayer grapheme. Appl. Phys. Lett. **106**, 041904 (2015)
72. C. Cong, T. Yu, K. Sato, J. Shang, R. Saito, G.F. Dresselhaus, M.S. Dresselhaus, Raman characterization of ABA- and ABC-stacked trilayer graphene. ACS Nano **5**, 8760–8768 (2011)
73. D. Yoon, H. Moon, Y.-W. Son, G. Samsonidze, B.H. Park, J.B. Kim, Y.P. Lee, H. Cheong, Strong polarization dependence of double-resonant Raman intensities in graphene. Nano Lett. **8**, 4270–4274 (2008)
74. A. Jorio, E.H. Martins Ferreira, M.V.O. Moutinho, F. Stavale, C.A. Achete, R.B. Capaz, Measuring disorder in graphene with the G and D bands. Phys. Status Solidi B **247**, 2980–2982 (2010)
75. G. Rao, S. Mctaggart, J.L. Lee, R.E. Geer, Study of electron beam irradiation induced defectivity in mono and bi layer graphene and the influence on Raman band position and line-width. MRS Online Proc. Libr. **1184**, 137–142 (2009)
76. M.Y. Han, B. Zyilmaz, Y. Zhang, P. Kim, Energy band-gap engineering of graphene nanoribbons. Phys. Rev. Lett. **98**, 206805 (2007)
77. S. Ryu, J. Maultzsch, M.Y. Han, P. Kim, L.E. Brus, Raman spectroscopy of lithographically patterned graphene Nanoribbons. ACS Nano **5**, 4123–4130 (2011)
78. H. Nakayama, T. Nishino, Y. Hamakawa, An analysis of exciton luminescence of silicon for characterization of the content of impurities. Jap. J. Appl. Phys. **19**, 501–511 (1980)
79. H. Nakayama, K. Onishi, H. Sawada, T. Nishino, Y. Hamakawa, Bound multiexciton luminescence in boron doped silicon: excitation level dependence and recombination kinetics. J. Phys. Soc. Jap. **46**, 553–560 (1979)
80. G. Gupta, K. Majumdar, Fundamental exciton linewidth broadening in monolayer transition metal dichalcogenides. Phys. Rev. B **99**, 085412 (2019)

Chapter 2
Introduction to the Band Structure of Solids

Abstract The basics of electronic band structure are reviewed in this chapter. The chapter emphasizes the tight binding model. Spin orbit interactions are also introduced in this chapter using a semiclassical approach. The $k \cdot p$ method of calculating electronic band structure is also briefly introduced. The $k \cdot p$ method provides a useful method for determining the effective mass of electrons and holes from band curvature. The second quantization formalism is introduced. Spatial symmetry is discussed at the end of the chapter including point group and space group symmetry.

Optical and electrical characterization of materials properties probes the electronic band structure of a material. The absorption of light results in electronic transitions between energy levels in the band structure or excitation of vibrational energy levels. Ellipsometry provides a means of characterizing important aspects of the electronic band structure. Photoluminescence comes from the recombination of electron—hole pairs that result from electrons that were excited to higher energy levels in the electronic band structure by absorption of light or by thermal energy. Although the Raman scattering comes from phonon absorption or emission; it involves electronic energy levels in a virtual electronic transitions.

The chapter emphasizes the tight binding model since this model is used to introduce the electronic band structure of graphene in Chap. 7 and single layer transition metal dichalcogenides in Chap. 8. The second quantization formalism is introduced at the end of this chapter since the tight binding Hamiltonian discussed in Chap. 9 is presented in that formalism. Spin orbit interactions are also introduced in this chapter using a semiclassical approach. Spin orbit interactions strongly influence the electronic band structure of transition metal dichalcogenides and topological insulators. The $k \cdot p$ method of calculating electronic band structure is also briefly introduced. The $k \cdot p$ method provides a useful method for theoretically determining the effective mass of electrons and holes which is used in Chap. 5 in the discussion of excitons. 2D materials have a great variety of point group and space group symmetries, and these topics are introduced at the end of the chapter.

The flow of this chapter is as follows: The topic of the relationship between optical properties and band structure is introduced. An introduction to Bloch's theorem for an electron in the periodic potential of a crystal lattice is presented. This is

A. Diebold and T. Hofmann, *Optical and Electrical Properties of Nanoscale Materials*, Springer Series in Materials Science 318,
https://doi.org/10.1007/978-3-030-80323-0_2

followed by brief introductions to the Brillouin zone and then to the Block wave vector. A derivation of the conduction band for a crystal lattice having the face center cubic structure is presented using an s orbital basis set with nearest neighbor tight binding interactions. This is flowed by a tight binding model for the p orbital valence bands with a nearest neighbor tight binding model. Then, spin orbit interactions are discussed. This is followed by an introduction to $k \cdot p$ theory and then an introduction to the second quantization formalism. The chapter ends with an introduction to point group and space group symmetry.

2.1 Band Structure and Optical Properties

The optical response or dielectric function of semiconductor, metal, and dielectric materials results from the interaction of light with the electronic band structure of the solid. The wavelength dependence of the optical response of the light depends on the band structure of the solid as well as doping, crystal defects, and disorder. In this section, the basics of band structure are introduced using a tight binding approximation for the valence and conduction bands of bulk solids. From a historical perspective, the optical physics community used $k \cdot p$ theory in much of their early publications describing optical transitions in semiconductors. Thus, $k \cdot p$ theory is introduced later in the chapter. As Cardona et al., point out, it is important to remember that the quantum mechanical band structure represents the electronic structure for a perfect crystal at zero Kelvin [1]. It is useful to elaborate on this point since electronic band structure calculations typically use the atomic positions in a crystal lattice at room temperature. In Chap. 5, we will find that the energy and line width of (light) absorption features in the dielectric function called critical points change with temperature [2–5]. The population of phonon states increases as temperature increases. The changes in the optical properties with increase in temperature come from a combination of the increase in the lattice spacing and electron–phonon interactions [4]. Therefore, it is important to understand the key details of the electronic structure calculations used to interpret optical properties. We shall see that the tight binding approach is frequently used as an empirical method in which the matrix elements are determined using experimental data or ab initio results [5].

The concept of the Block wavefunctions for solids is introduced in this chapter. The repeated or periodic structure of a crystal lattice allows us to use wavefunctions that have a periodic part and a localized part. This idea was first presented by Bloch. A number of references and text books were used to motivate this text as noted in [6–9]. This chapter is intended to be an introductory level review.

It is useful to recall the classical Drude model for metals in which the electrons are considered an unbound gas of negative charges (free electron gas) with the well-known relationship between energy and wave vector $\vec{k}$:

$$E = \frac{\hbar^2 \vec{k}^2}{2m_0} \tag{2.1}$$

It is also useful to remember that the momentum is related to the wave vector $\vec{k}$: $\vec{p} = \hbar\vec{k}$.

2.2 Block Theorem

The periodic placement of the atoms of the crystal result in a periodic potential $U(\vec{r})$ so that at a specific location $\vec{r}$, the potential energy function $U(\vec{r})$ will be the same as at a location $\vec{r} + \vec{R}$ that represents a displacement of an exact number of unit crystals. The periodicity of the crystal lattice is described by the Bravais lattice. The Bravais lattice is represented by three primitive lattice vectors: $a_x,\ a_y,\ a_y$. The Bravais lattice repeats throughout the crystal, and any two lattice points P and P′ can be related to each other using the Bravais lattice vector $\vec{R}$ was defined above in terms of the primitive lattice vectors $a_x,\ a_y,\ a_y$ as follows: $P' = P + \left(n_x a_x + n_y a_y + n_z a_z\right)$. $\vec{R}$ is called the Bravais lattice vector and it is defined by $\vec{R} = n_x a_x + n_y a_y + n_z a_z$ where a_x, etc., are the primitive vectors that span the lattice. In other words, $\vec{R}$ has the same periodicity as the crystal lattice.

$$U(\vec{r}) = U\left(\vec{r} + \vec{R}\right) \tag{2.2}$$

The Hamiltonian equation for the static crystal lattice at absolute zero with no nuclear motion is:

$$\left[-\frac{\hbar^2}{2m_0}\nabla^2 + U(\vec{r})\right]\psi(\vec{r}) = E\psi(\vec{r}) \tag{2.3}$$

If $U(\vec{r}) = 0$, then we have the free electron gas where (2.1) provides the energy dispersion with wave vector and the wavefunction is given by:

$$\psi(\vec{r}) = \frac{1}{\sqrt{V}}e^{i\vec{k}\cdot\vec{r}} \tag{2.4}$$

This is a plane wave of wavenumber k with $|k| = 2\pi/\lambda$ where λ is the wavelength of the wave. The $\vec{k}$ vector inside a crystal will be defined below. The time dependent plane wave can be pictured as a traveling series of wavefronts (surfaces of constant phase) which are infinitely parallel planes. The wave function is spread evenly across the entire crystal because probability of finding the electron $\psi^*\psi = 1/V$ is the same at every point in space. In a perfect crystal, every unit cell is identical, and the probability of finding the electron at the same place inside a unit cell is the same for each unit cell. Thus the Block theorem indicates that for a period vector of the crystal lattice $\vec{R}$:

$$|\psi(\vec{r})|^2 = \left|\psi\left(\vec{r} + \vec{R}\right)\right|^2 \tag{2.5}$$

Block's theorem postulated that the wavefunctions that are solutions to a Hamiltonian that has a periodic potential can be written as the product of plane wave and a local wave function, $u_{\vec{k}}(\vec{r})$ that has the periodicity of the lattice as follows:

$$u_{\vec{k}}(\vec{r}) = u_{\vec{k}}\left(\vec{r} + \vec{R}\right) \tag{2.6}$$

and

$$\psi_{\vec{k}}(\vec{r}) = e^{i\vec{k}\cdot\vec{r}} u_{\vec{k}}(\vec{r}) \tag{2.7}$$

Using (2.6) and (2.7), the student can prove (2.8):

$$\psi_{\vec{k}}\left(\vec{r} + \vec{R}\right) = e^{i\vec{k}\cdot\vec{R}} \psi_{\vec{k}}(\vec{r}) \tag{2.8}$$

Equations (2.7) and (2.8) refer to a specific energy level, and are thus often stated as $\psi_{n\vec{k}}\left(\vec{r} + \vec{R}\right) = e^{i\vec{k}\cdot\vec{R}}\, \psi_{n\vec{k}}(\vec{r})$. This is Block's theorem which can be written in a more general form:

$$\Psi\left(\vec{r} + \vec{R}\right) = e^{i\vec{k}\cdot\vec{R}}\, \Psi(\vec{r}) \tag{2.9}$$

A thorough discussion of Block band structure can be found in Solid State Physics texts, see for example Ashcroft and Mermin [9].

2.3 First Brillouin Zone

The reciprocal lattice for a real space periodic lattice is periodic in reciprocal space. The reciprocal space lattice is itself a Bravais lattice. A primitive cell in reciprocal space can be constructed by selecting a high symmetry point and drawing lines to the nearest neighbor atoms. The volume enclosed by the planes that bisect the reciprocal lattice vectors from this central point to the nearest neighbors is the first Brillouin zone. The real space primitive lattice for an fcc crystal is shown in Fig. 2.1 and the First Brillouin Zone for an FCC lattice is shown in Fig. 2.2.

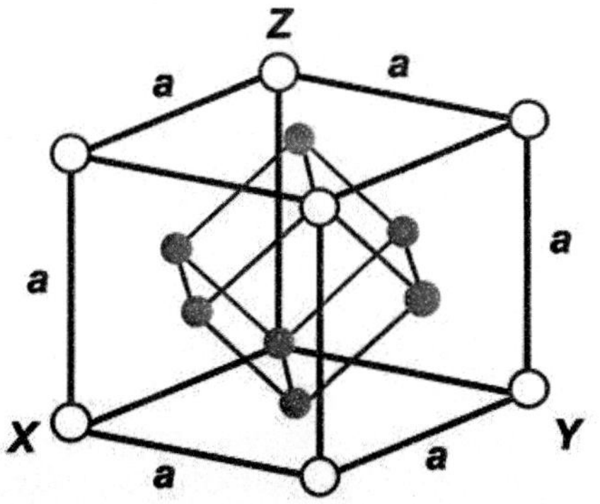

$$\vec{a}_1 = \frac{a}{2}(\vec{y} + \vec{z}) \qquad \vec{a}_2 = \frac{a}{2}(\vec{x} + \vec{z}) \qquad \vec{a}_3 = \frac{a}{2}(\vec{x} + \vec{y})$$

Fig. 2.1 FCC primitive lattice vectors shown with FCC lattice

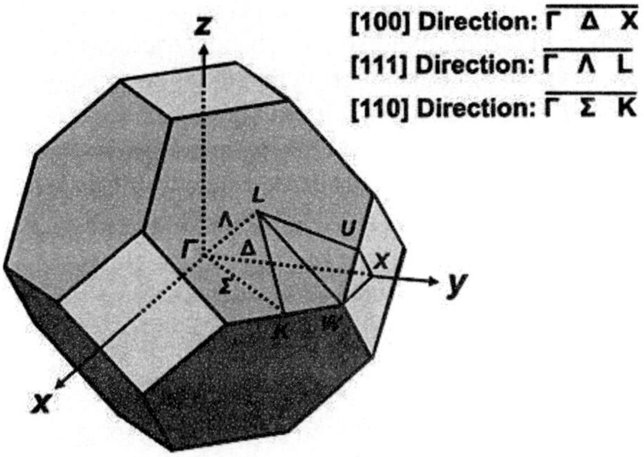

Fig. 2.2 Brillouin Zone of FCC lattice with key crystallographic directions shown. The $\vec{k}_j$ wavevector are labeled $x, \ y, \ z$.

2.4 Block Function Wave Vector $\vec{k}$

The Block wavefunction introduces the $\vec{k}$ vector, which is an important concept that requires careful consideration. The plane wave part of the Bloch wave function, $e^{i\vec{k}\cdot\vec{R}}$, is a function of the $\vec{k}$ vectors and the Bravais lattice vectors $\vec{R}$ translate the plane wave throughout the crystal. For convenience, the wave vectors $\vec{k}$ are vectors that are restricted to be in the first Brillouin zone. The $\vec{k}$ vectors point from the center of the first Brillouin zone toward the edges of the Brillouin zone. The electronic band structure is typically described in terms of high symmetry directions in the Brillouin zone. For example, along the [1 0 0] direction which is from the Γ point at the center of the Brillouin zone to the X point (see Fig. 2.2) the value of k_x varies from 0 to $\frac{2\pi}{a}$. As we discuss below, the $\vec{k}$ vectors that appear in the periodic part of the Bloch wavefunction are quantum numbers that describe the periodic nature of periodic potential of the crystal lattice in reciprocal space.

Below, we will justify why $\vec{p} = \hbar\vec{k}$ refers to the crystal momentum. The crystal momentum is not the so called true electron momentum. It will now become clear why we have stated that the electronic band structure discussion is for a perfect crystal with no defects at absolute zero. We will also find that in the periodic background potential of an idea, defect free crystal, the electron moves without scattering. We have constructed a non-local Bloch wavefunction for the electrons and holes that extends throughout the crystal by using a periodic plane wave, $e^{i\vec{k}\cdot\vec{r}}$. We can further understand the meaning of the $\vec{k}$ vector through a *classical equation of motion.* The electron responds to the forces inside and outside the crystal as follows [6, 9]:

$$\dot{\vec{p}} = \vec{F}(internal) + \vec{F}(external) \;\; classical\ dynamics \tag{2.10}$$

In (2.10), the internal force comes from the periodic crystal lattice. An example of an external force is an electric field such as the electric field of light or a voltage potential across the crystal [9]. Remembering that the time dependent solution to the Hamiltonian is typically calculated using a separation of variables approach so that there is a time independent part of the total wave function $\psi_{\vec{k}}(\vec{r})$ (2.7) multiplied by $e^{-iEt/\hbar}$ with the energy of the electron E being related to the angular frequency ω by:

$$E = \hbar\omega \;\text{ thus }\; e^{-iEt/\hbar} = e^{-i\omega t} \tag{2.11}$$

Localized electron or hole motion is described by a wave packet of wavefunctions centered around the wave vector $\vec{k}$ [6, 9]. The concept of wave packet is critical to understand the Berry phase which is introduced in Chap. 6. The motion of the localized electron or hole is described by the group velocity of the quantum wave packet. The classical group velocity of a wave is the velocity of the overall envelop of amplitudes of the wave. Here we are interested in the group velocity of the wave packet representing the electron or hole. The electron or hole group velocity, $\vec{v}_g \equiv \frac{d\omega}{d\vec{k}}$, for a specific energy level E_n can be related to the angular frequency of the electron using (2.11) as follows [6, 9]:

$$\vec{v}_g \equiv \frac{d\omega}{d\vec{k}} = \frac{1}{\hbar}\frac{dE}{d\vec{k}} = \frac{1}{\hbar}\nabla_{\vec{k}}E_n(\vec{k}) \tag{2.12}$$

Since $\vec{v}_g \neq 0$, this means that in a crystal with a perfect lattice, the electron is continuously moving. In semiclassical descriptions of electron motion, the group velocity is also referred to as the average velocity $\langle v\rangle$.

A semiclassical description of the electron or hole dynamics provides a useful means of distinguishing the crystal momentum from the so called true momentum.

We are interested in the effect of an external force such as an electric field $\vec{\boldsymbol{E}}$, $\vec{F}(external) = e\vec{\boldsymbol{E}}(external)$, *on an electron* in a crystal. For an electron or hole in a crystal, the change in kinetic energy is the work done on the electron or hole. The force acting on an electron or a hole in a crystal is given by the time derivative of the crystal momentum $\vec{p} = \hbar\vec{k}$: $F = \dot{\vec{p}} = \hbar\dot{\vec{k}}$. The time derivative of the group velocity is the acceleration of the electron or hole:

$$\dot{\vec{v}}_g = \frac{1}{\hbar}\frac{d}{dt}\frac{dE}{d\vec{k}} = \frac{1}{\hbar}\frac{d^2E}{d\vec{k}^2}\frac{d\vec{k}}{dt} \tag{2.13}$$

Below, we will define an effective mass for an electron or hole in a crystal using $\frac{\hbar^2}{m^*} = \frac{d^2E_n(\vec{k})}{d\vec{k}^2}$. As discussed below, the effective mass of an electron or hole in a crystal is the curvature of the energy band for the n'th state at $\vec{k}$. Here this relationship allows us to rewrite (2.13) for an external force from an electric field; $-e\vec{\boldsymbol{E}}(external)$:

$$m^*\dot{\vec{v}}_g = \dot{\vec{p}} = -e\vec{\boldsymbol{E}}(external) \tag{2.14}$$

Thus we see that the time derivative of the crystal momentum responds to an external force such as an electric field while an electron or hole in a crystal experiences both external and internal forces.

Another way of obtaining the same result is to consider the infinitesimal change in energy dE [6]. The relationship between infinitesimal changes in work, force and distance can be written as $dE = -e\vec{E}(external)\cdot\vec{v}_g\,dt$. This expression can be related to the infinitesimal change in wave vector $d\vec{k}$ using (2.12) as follows:

$$dE = \left(\frac{dE}{d\vec{k}}\right)d\vec{k} = \hbar\vec{v}_g d\vec{k} \tag{2.15}$$

Comparing $dE = -e\vec{E}(external)\dot{\vec{v}}_g dt$ and (2.15), one can see that [6]:

$$d\vec{k} = -\frac{e\vec{E}(external)}{\hbar}dt \to \hbar\frac{d\vec{k}}{dt} = \dot{\vec{p}} = -e\vec{E}(external) \tag{2.16}$$

The electron or hole in a crystal responds to external forces such as an electric field, and (2.16) is the time dependence of the true momentum. When we compare this to the classical dynamical equation for an electron in a crystal, $\dot{\vec{p}} = \vec{F}(internal) + \vec{F}(external)$, one can see that in a crystal, the momentum $\vec{p} = \hbar\vec{k}$ includes the effect of forces inside the crystal. Thus, $\vec{p} = \hbar\vec{k}$ is not the momentum due to an external force (so called true electron momentum) and is thus called the crystal momentum [6].

Quantum mechanics provides additional insight into the $\vec{k}$ wave vector for a periodic potential and momentum $\vec{p} = \hbar\vec{k}$ for a free electron. The momentum operator $\vec{p} = \frac{\hbar}{i}\vec{\nabla}$ can be used to show that while the *free electron wavefunction has eigenstates of momentum* [9]:

$$\vec{p}\psi(\vec{r}) = \vec{p}\frac{1}{\sqrt{V}}e^{i\vec{k}\cdot\vec{r}} = \frac{\hbar}{i}\vec{\nabla}\psi(\vec{r}) = \frac{\hbar}{i}\vec{\nabla}\frac{1}{\sqrt{V}}e^{i\vec{k}\cdot\vec{r}} = \hbar\vec{k}\psi(\vec{r}) \quad free\ electron \tag{2.17}$$

The Block wavefunction for a periodic potential does not have Eigen states of momentum:

$$\vec{p}\psi_{\vec{k}}(\vec{r}) = \vec{p}\left[e^{i\vec{k}\cdot\vec{r}}u_{\vec{k}}(\vec{r})\right] = \hbar\vec{k}\psi_{\vec{k}}(\vec{r}) + e^{i\vec{k}\cdot\vec{r}}\frac{\hbar}{i}\vec{\nabla}u_{\vec{k}}(\vec{r}) \quad Bloch\ electron \tag{2.18}$$

We recall that $\vec{k}$ can be confined to the first Brillouin zone because any $\vec{k}'$ not in the first Brillouin zone can be written as:

$$\vec{k}' = \vec{k} + \vec{K}$$

where $\vec{K}$ is a reciprocal lattice vector.

2.5 A Simple s Level Conduction Band for a Semiconductor Using a Tight Binding Approximation

Band theory can be presented in countless different ways. Here, we assume that the reader is aware that semiconductors have a band gap between the valence and conduction bands. The atomic structure of silicon is $1s^22s^22p^63s^23p^2$, so the outer shell of $3s^23p^2$ will form the basis set of orbitals for the Block wavefunction that become the valence and conduction bands of silicon. Here, we wish to introduce the tight binding method and avoid the complications of starting with a sp^3 hybrid orbital basis set. We start with some simplifying assumptions, namely that the conduction band retains an s like character while the valence band has a p like character. The tight binding method used here assumes that the atomic orbitals are largely unchanged when they combine to form the bands. The simplest form of tight binding theory assumes that the interaction between atoms is very short ranged and restricts this interaction to nearest neighbors. The nearest neighbor approximation can be corrected by including next nearest neighbor interactions, and next-next nearest neighbor interactions have also been included in some studies. The outer $2s^22p^2$ and the $4s^24p^2$ form the valence and conduction bands of carbon and germanium respectively. We note that Slater and Koster [8] proposed in 1954 that the tight binding method should be used as an empirical method where the matrix elements in the determinant (secular equation) are fit to experimental data. $\left|\langle\Phi_{i,j}|H - E|\psi(k)\rangle\right| = 0$ where the Bra-Ket notation is used to represent the integral over space.

Now, we want to construct a Tight Binding Block wave function for the band from the local atomic wavefunctions $\psi_n(\vec{r})$ that are solutions of the atomic Hamiltonian H_{atomic}:

$$H_{atomic}\psi_n(\vec{r}) = E_n\psi_n(\vec{r}) \tag{2.19}$$

The periodic potential due to the crystal lattice is $U(\vec{R})$. If $U(\vec{R})\cdot\psi_n(\vec{r})$ is small, then the perturbation due to the periodic lattice potential is small. Thus, we can construct wavefunction for one band $\phi(\vec{r})$ using a linear combination of N' atomic wavefunctions (for example for a band with s and p character such as that described by (2.20), N' is the number of types of wavefunctions (s, p, etc.) used to form the band, and not the number of atoms) for the periodic part of the Block wavefunction:

$$\phi(\vec{r}) = \sum_{n=1}^{N'} b_n\psi_n(\vec{r}) \tag{2.20}$$

The N' refers to the different atomic orbitals such as s, p, d, etc. The crystal Hamiltonian is considered to be a small perturbation $U(\vec{R})$ from the atomic state so that:

$$H_{crystal} = H_{atomic} + U(\vec{R}) \tag{2.21}$$

If $U(\vec{R}) = 0$, then for a crystal with N atoms, for each atomic level $\psi_n(\vec{r})$ there would be N levels in the periodic potential with the wavefunctions $\psi_n(\vec{r} - \vec{R})$ for each of the N sites $\vec{R}$ in the crystal.

We want the Bloch wavefunction to be centered at a location with the distance to that origin is given by the vector $\vec{R}$, so we add their together contributions, and the Bloch wavefunction is:

$$\psi_{\vec{k}}(\vec{r}) = \sum_{\vec{R}} e^{i\vec{k}\cdot\vec{R}}\phi\left(\vec{r} - \vec{R}\right) \tag{2.22}$$

One can prove that $\Psi\left(\vec{r} + \vec{R}\right) = e^{i\vec{k}\cdot\vec{R}}\ \Psi(\vec{r})$ for this Block wave function.

$$\begin{aligned}
\psi_{\vec{k}}\left(\vec{r} + \overrightarrow{R'}\right) &= \sum_{\vec{R}} e^{i\vec{k}\cdot\vec{R}}\ \phi\left(\vec{r} - \vec{R} + \overrightarrow{R'}\right) \\
&= e^{i\vec{k}\cdot\overrightarrow{R'}}\left[\sum_{\vec{R}} e^{i\vec{k}\cdot\left(\vec{R}-\overrightarrow{R'}\right)}\ \phi\left(\vec{r} - \vec{R} + \overrightarrow{R'}\right)\right] \\
&= e^{i\vec{k}\cdot\overrightarrow{R'}}\left[\sum_{\vec{R}} e^{i\vec{k}\cdot\left(\vec{R}-\overrightarrow{R'}\right)}\ \phi\left(\vec{r} - \left(\vec{R} - \overrightarrow{R'}\right)\right)\right] = e^{i\vec{k}\cdot\overrightarrow{R'}}\psi_{\vec{k}}(\vec{r})
\end{aligned}$$

Here we use a model s orbital conduction band to demonstrate the tight binding approach. So only s orbitals contribute to $\phi(\vec{r}) = \sum_{n=1}^{N'} b_n \psi_n(\vec{r})$ and thus $\phi(\vec{r}) = \psi_s(\vec{r})$. We use

$$\psi_{\vec{k}}(\vec{r}\,) = \sum\nolimits_{\vec{R}} e^{i\vec{k}\cdot\vec{R}}\; \psi_s(\vec{r} - \vec{R}\,) \tag{2.23}$$

and $\left[H_{atomic} + U(\vec{R}\,)\right]\psi_{\vec{k}}(\vec{r}) = \mathcal{E}(\vec{k}\,)\psi_{\vec{k}}(\vec{r}\,)$ results in:

$$\int \psi_s^*(\vec{r})[H_{atomic} + U(\vec{r})]\sum_{\vec{R}} e^{i\vec{k}\cdot\vec{R}}\psi_s(\vec{r} - \vec{R})\mathrm{d}^3\mathrm{r}$$
$$= \mathcal{E}(\vec{k})\int \psi_s^*(\vec{r})\sum_{\vec{R}} e^{i\vec{k}\cdot\vec{R}}\psi_s(\vec{r} - \vec{R})\mathrm{d}^3\mathrm{r}$$

$$\mathrm{E}_s + \int \psi_s^*(\vec{r})[H_{atomic}]\sum_{\vec{R}\neq 0} e^{i\vec{k}\cdot\vec{R}}\;\psi_s(\vec{r} - \vec{R})\mathrm{d}^3\mathrm{r}$$
$$+ \int \psi_s^*(\vec{r})[U(\vec{r})]\sum_{\vec{R}} e^{i\vec{k}\cdot\vec{R}}\;\psi_s(\vec{r} - \vec{R})\mathrm{d}^3\mathrm{r}$$
$$= \mathcal{E}(\vec{k})\left[1 + \sum_{\vec{R}\neq 0} e^{i\vec{k}\cdot\vec{R}}\int \psi_s^*(\vec{r})\psi_s(\vec{r} - \vec{R})\mathrm{d}^3\mathrm{r}\right] \tag{2.24}$$

The tight binding approximation refers to the assumption that the electrons are tightly bound to the atoms and thus the interaction of an electron bound to an atom with the neighboring nucleus is small and the interaction between the orbitals of neighboring atoms is very small. Thus, we assume that for non-zero lattice vectors $\vec{R}$:

$$\int \psi_s^*(\vec{r}\,)\psi_s(\vec{r} - \vec{R}\,)\,e^{i\vec{k}\cdot\vec{R}}\;\mathrm{d}^3\mathrm{r}\;\approx 0 \tag{2.25}$$

Thus:

$$\mathrm{E}_s + \sum_{\vec{R}} e^{i\vec{k}\cdot\vec{R}}\int \psi_s^*(\vec{r}\,)[U(\vec{r}\,)]\psi_s(\vec{r} - \vec{R}\,)\,\mathrm{d}^3\mathrm{r} = \mathcal{E}(\vec{k}\,)$$

It is useful to separate the interaction at the central atom from the interactions with neighbors.

$$\mathrm{E}_s + \int \psi_s^*(\vec{r}\,)[U(\vec{r}\,)]\psi_s(\vec{r}\,)\,\mathrm{d}^3\mathrm{r} + \sum_{\vec{R}\neq 0} e^{i\vec{k}\cdot\vec{R}}\int \psi_s^*(\vec{r}\,)[U(\vec{r}\,)]\psi_s(\vec{r} - \vec{R}\,)\,\mathrm{d}^3\mathrm{r} = \mathcal{E}(\vec{k}\,)$$

Using the following notation:

$$\alpha(\vec{R}) = \int \psi_s^*(\vec{r})\psi_s(\vec{r} - \vec{R})\, \mathrm{d}^3\mathrm{r} \approx 0 \tag{2.26a}$$

$$\beta_s = -\int \psi_s^*(\vec{r})U(\vec{r})\psi_s(\vec{r})\, \mathrm{d}^3\mathrm{r} \tag{2.26b}$$

$$\gamma(\vec{R}) = -\int \psi_s^*(\vec{r})U(\vec{r})\psi_s(\vec{r} - \vec{R})\, \mathrm{d}^3\mathrm{r} \tag{2.26c}$$

We write:

$$\mathcal{E}(\vec{k}) = E_s - \beta_s - \sum_{\vec{R}\neq 0} \gamma(\vec{R})\, e^{i\vec{k}\cdot\vec{R}}$$

The $\gamma(\vec{R})$ are equivalent for all lattice vectors $\vec{R}$ pointing toward nearest neighbor atoms for fcc and bcc lattices due to symmetry. This allows us to rewrite $\mathcal{E}(\vec{k}) = E_s - \beta_s - \sum_{\vec{R}\neq 0} \gamma(\vec{R})\, e^{i\vec{k}\cdot\vec{R}}$ using this information:

$$\mathcal{E}(\vec{k}) = E_s - \beta_s - \gamma(\vec{R}) \sum\nolimits_{\vec{R}\neq 0} e^{i\vec{k}\cdot\vec{R}} \tag{2.26d}$$

We also note that $\gamma(\vec{R})$ for most potentials will be governed by the overlap between orbitals on different atoms and thus will quickly decrease in value with increasing $\vec{R}$.

Group IV elements such as silicon, germanium, and carbon have the diamond lattice, and III-V materials such as GaAs, etc., have the zinc blende lattice. Both of these structures consists of two offset fcc lattices. For simplicity, we first consider only a fcc lattice. We recall the primitive lattice for an fcc crystal shown in Fig. 2.1.

The nearest neighbors for a FCC lattice are:

$$\left(0, \frac{a}{2}, \frac{a}{2}\right);\ \left(0, \frac{-a}{2}, \frac{a}{2}\right);\ \left(0, \frac{a}{2}, \frac{-a}{2}\right);\ \left(0, \frac{-a}{2}, \frac{-a}{2}\right)$$

$$\left(\frac{a}{2}, 0, \frac{a}{2}\right);\ \left(\frac{-a}{2}, 0, \frac{a}{2}\right);\ \left(\frac{a}{2}, 0, \frac{-a}{2}\right);\ \left(\frac{-a}{2}, 0, \frac{-a}{2}\right)$$

$$\left(\frac{a}{2}, \frac{a}{2}, 0\right);\ \left(\frac{-a}{2}, \frac{a}{2}, 0\right);\ \left(\frac{a}{2}, \frac{-a}{2}, 0\right);\ \left(\frac{-a}{2}, \frac{-a}{2}, 0\right)$$

An important but subtle point is that the integral $\gamma(\vec{R}) = -\int \psi_s^*(\vec{r})U(\vec{r})\psi_s(\vec{r} - \vec{R})\,d^3r$ will have the same value for any of the directions $\vec{R}$ listed above. The crystal potential $U(\vec{r})$ has the full cubic symmetry of the lattice. So interchanging the y value of $\vec{r}$ for x value or changing signs does not change U($\vec{r}$). $\gamma(\vec{R})$ will have the same value for $\vec{R} = \left(\frac{a}{2}, \frac{a}{2}, 0\right)$ or $\left(\frac{-a}{2}, \frac{a}{2}, 0\right)$. That will be true for the integral over all space. For example, we can look at just the x component of the integral for $\vec{R}$ with $\vec{R} = \frac{a}{2}$ or $-\frac{a}{2}$. The s orbitals are a function of only the magnitude of $\vec{r}$, ie, $\Phi(r)$. If we consider a wave function that is just a function of x, $\Phi(x)$ then:

$$\int_{-\infty}^{\infty} \Phi^*(x)U(x)\Phi\left(x - \frac{a}{2}\right)dx = \int_{-\infty}^{\infty} \Phi^*(x)U(x)\Phi\left(x + \frac{a}{2}\right)dx$$

The same will be true for all the values of $\vec{R}$.
Inserting the nearest neighbor positions in (2.26d) gives:

$$\varepsilon(\vec{k}) = E_s - \beta_s - \gamma(\vec{R})\begin{bmatrix} e^{i(k_x+k_y)a/2} + e^{i(-k_x+k_y)a/2} + e^{i(k_x-k_y)a/2} + e^{-i(k_x+k_y)a/2} \\ +e^{i(k_x+k_z)a/2} + e^{i(-k_x+k_z)a/2} + e^{i(k_x-k_z)a/2} + \\ e^{-i(k_x+k_z)a/2} + e^{i(k_y+k_z)a/2} + e^{i(-k_y+k_z)a/2} \\ +e^{i(k_y-k_z)a/2} + e^{-i(k_y+k_z)a/2} \end{bmatrix}$$

Using $\cos a = \frac{e^{ia}+e^{-ia}}{2}$ we get:
$e^{i(k_x+k_y)a/2} + e^{i(-k_x+k_y)a/2} = e^{ik_x a/2}e^{ik_y a/2} + e^{-ik_x a/2}e^{ik_y a/2} = 2\cos\frac{k_x a}{2}\, e^{ik_y a/2}$
which gives

$$\mathcal{E}(\vec{k}) = E_s - \beta_s - 4\gamma(\vec{R})\left[\cos\frac{k_x a}{2}\cos\frac{k_y a}{2} + \cos\frac{k_y a}{2}\cos\frac{k_z a}{2} + \cos\frac{k_x a}{2}\cos\frac{k_z a}{2}\right] \quad (2.27)$$

Remembering that the directions along the Brillouin zone using Fig. 2.2.
The key boundaries of the Brillouin Zone are:

$$\Gamma \text{ point:} (0, 0, 0)$$

$$X \text{ point:} \frac{2\pi}{a}(1, 0, 0)$$

$$L \text{ point:} \frac{\pi}{a}(1, 1, 1)$$

$$K \text{ point:} \frac{2\pi}{a}\left(\frac{3}{4}, \frac{3}{4}, 0\right)$$

The k dispersion of the energy of this s band has a simplified functional dependence along key FCC crystallographic directions. $\gamma(\vec{R})$ becomes a constant for nearest neighbors since R is a constant.

Along the [100] direction, which is from the Γ point to the X point, since the value of k_x varies from 0 to $\frac{2\pi}{a}$:

$$k_y = k_z = 0 \text{ and } k_x = \frac{2\pi\alpha}{a} \text{ with } 0 \le \alpha \le 1$$

$$\begin{aligned}\mathcal{E}(\vec{k}) &= E_s - \beta_s - 4\gamma\left[\cos\frac{2\pi\alpha}{2} + 1 + \cos\frac{2\pi\alpha}{2}\right] \\ &= E_s - \beta_s - 4\gamma[1 + 2\cos\pi\alpha] \text{ along the [100] direction}\end{aligned} \tag{2.28a}$$

Along the [111] direction, which is the Γ point to the L point, since the value of $k_x = k_y = k_z$ vary from 0 to $\frac{\pi}{a}$:

$$k_x = k_y = k_z \text{ and } k_x = \frac{2\pi\alpha}{a} \text{ with } 0 \le \alpha \le 1/2$$

$$\mathcal{E}(\vec{k}) = E_s - \beta_s - 12\gamma\cos^2\pi\alpha \tag{2.28b}$$

Along the [110] direction, which is the Γ point to the K point, since $k_z = 0$ and the value of $k_x = k_y$ vary from 0 to $\frac{3\pi}{2a}$:

$$k_x = k_y \text{ and } k_x = \frac{2\pi\alpha}{a} \text{ with } 0 \le \alpha \le 3/4 \text{ and } k_z = 0$$

$$\mathcal{E}(\vec{k}) = E_s - \beta_s - 4\gamma[\cos^2\pi\alpha + 2\cos\pi\alpha] \tag{2.28c}$$

Now we can draw a Band Structure diagram for the energy vs the position in the Brillouin zone for this Tight Binding s conduction band. It is shown in Fig. 2.3.

Close to Γ, $ka << 1$, and we can use $k_x^2 = k_y^2 = k_z^2 = k^2/3$ and $k_x = \frac{2\pi\alpha}{a}$ gives $\frac{k^2}{3} = \frac{4\pi^2\alpha^2}{a^2}$. Here with a Taylor series expansion of $\cos(x) \sim 1 - x^2/2$:

$$\mathcal{E}(k) = E_s - \beta_s - 4\gamma[1 + 2\cos\pi\alpha] \approx E_s - \beta_s - 12\gamma + \gamma k^2 a^2$$

The effective mass m^* can be determined using;

$$\frac{\hbar^2}{m^*} = \frac{d^2\mathcal{E}(k)}{dk^2} \tag{2.29}$$

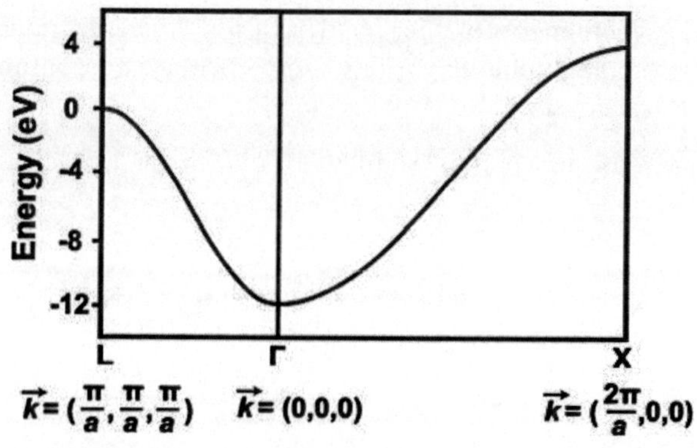

Fig. 2.3 Tight Binding Model s band using $E_s - \beta_s = 0$, and $\gamma = 1$ eV. At X, E(k) = + 4 eV. At Γ, E(k) = − 12 eV, and the band width is 16 eV. The s band forms the conduction band and the minimum is found at the Γ point for this tight binding model

$$m^* = \frac{\hbar^2}{2\gamma a^2}$$

At X, $k_x = \frac{2\pi}{a}$ and $\alpha = 1$, $\mathcal{E}(\vec{k}) = E_s - \beta_s - 4\gamma[1 + 2\cos\pi] = E_s - \beta_s + 4\gamma$.
At Γ, $\alpha = 0$ and $\mathcal{E}(\vec{k}) = E_s - \beta_s - 4\gamma[1 + 2\cos 0] = E_s - \beta_s - 12\gamma$.

Thus the band width, W, is 16γ. This leads to the observation that the overlap integrals $\gamma(\vec{R}) = -\int \psi_s^*(\vec{r})U(\vec{r})\psi_s(\vec{r} - \vec{R})\,d^3r$ determine the band width in the tight binding model. The smaller the overlap integral, the smaller the band width. If we chose $\gamma = 1$ eV and a = 0.4 nm, then

$$m^* \approx 0.2\, m_0$$

The effective mass depends on whether k is set to one at the top or bottom of the band. The derivation of the expression for the effective mass will be discussed later.

It is useful to discuss the meaning of the Block theorem when considering the tight binding method [9]. An electron in a tight binding orbital is delocalized across the entire lattice. Another way of stating this is: an electron will be found with equal probability in each cell of the crystal lattice since it is a Block wave function: $|\psi(\vec{r})|^2 = \left|\psi(\vec{r} + \vec{R})\right|^2$. The reader is referred to [9] for further discussion of tight binding model. They point out that the tight binding levels have a running wave (itinerant) character because the mean velocity of an electron in a Block level with wave vector $\vec{k}$ is given by:

$$v(\vec{k}) = \frac{1}{\hbar}\frac{\partial \mathcal{E}(\vec{k})}{\partial \vec{k}} \tag{2.30}$$

Since energy of the tight binding levels is a function of $\vec{k}$, there is a mean velocity $v(\vec{k})$ as long as there is any overlap, i.e., $\gamma(\vec{R}) = -\int \psi_s^*(\vec{r})U(\vec{r})\psi_s(\vec{r} - \vec{R})\,\mathrm{d}^3\mathrm{r}$. Decreasing the overlap reduces the velocity [9]. The reader may find this useful since the tight binding model is used to describe graphene and transition metal dichalcogenides in subsequent chapters. The carrier dynamics of these 2D materials is further discussed then.

2.6 A Simple p Level Valence Band Using a Tight Binding Approximation

Once again, we make a simplifying assumption to illustrate the application of the tight binding method to the electronic band structure of semiconductors. Here, we assume that the highest energy valence band of a semiconductor is formed from only the *p* orbitals. We can use the tight binding approximation to form three *p* bands for monatomic, fcc semiconductor crystals [9, 10]. It follows that since when only *p* orbitals form the band, the secular equation results in a 3 × 3 determinant, when the *d* orbitals form the band it is a 5 × 5 determinant.

We start with (2.20) based on using *p* orbitals instead of *s* orbitals in (2.24). Here, $i = x$, y, or z and we initially include the overlap integrals for the nine elements of the determinant:

$$\begin{aligned}\mathcal{M}_{ij} &= \mathcal{E}(\vec{k})\delta_{ij} + \sum_{\vec{R}\neq 0}\int \psi_{p_i}^*(\vec{r})\psi_{p_j}(\vec{r} - \vec{R})e^{i\vec{k}\cdot\vec{R}}\mathrm{d}^3\mathrm{r} - E_{pi}\delta_{ij} \\ &+ E_{p_i}\sum_{\vec{R}\neq 0}\int \psi_{p_i}^*(\vec{r})\psi_{p_j}(\vec{r} - \vec{R})e^{i\vec{k}\cdot\vec{R}}\mathrm{d}^3\mathrm{r} \\ &+ \int \psi_{p_i}^*(\vec{r})U(\vec{r})\psi_{p_j}(\vec{r})\mathrm{d}^3\mathrm{r} \\ &+ \sum_{\vec{R}\neq 0}\int \psi_{p_i}^*(\vec{r})U(\vec{r})\psi_{p_j}(\vec{r} - \vec{R})e^{i\vec{k}\cdot\vec{R}}\mathrm{d}^3\mathrm{r}\end{aligned} \tag{2.31}$$

These integrals can be replaced by the following symbols [9] where the integrals ar2.33e over all space:

$$\alpha_{ij}(\vec{R}) = \int \psi_{p_i}^*(\vec{r})\psi_{p_j}(\vec{r} - \vec{R})\,\mathrm{d}^3\mathrm{r} \tag{2.32a}$$

$$\beta_{ij} = -\int \psi_{p_i}^*(\vec{r})U(\vec{r})\psi_{p_j}(\vec{r})\mathrm{d}^3\mathrm{r} \tag{2.32b}$$

$$\gamma_{ij}(\vec{R}) = -\int \psi_{p_i}^*(\vec{r})U(\vec{r})\psi_{p_j}(\vec{r} - \vec{R})\,\mathrm{d}^3\mathrm{r} \tag{2.32c}$$

Above, we mentioned that this initial application of the tight binding method assumes that overlap integral between adjacent atoms is: $\alpha_{ij}(\vec{R}) \approx 0$. Since $\int \psi^*_{p_i}(\vec{r})\psi_{p_j}(\vec{r}-\vec{R})\, e^{i\vec{k}\cdot\vec{R}}\, \mathrm{d}^3\mathrm{r} = e^{i\vec{k}\cdot\vec{R}} \int \psi^*_{p_i}(\vec{r})\psi_{p_j}(\vec{r}-\vec{R})\, \mathrm{d}^3\mathrm{r}$, now (2.31) becomes:

$$\mathcal{M}_{ij} = \mathcal{E}(\vec{k})\delta_{ij} - E_{p_i}\delta_{ij} + \beta_{ij} + \sum_{\vec{R}\neq 0} \gamma_{ij}(\vec{R})e^{i\vec{k}\cdot\vec{R}} \tag{2.33}$$

Again, we consider only the nearest neighbors for a FCC lattice which are provided above.

The p orbitals have an identical radial distribution about each axis (x, y, z): $\psi_{p_{x,y,\text{ or } z}}(\vec{r}) = x\,\Phi(r)$, y $\Phi(r)$, or $z\,\Phi(r)$ and the p band will be a linear combination of these orbitals [9]. The motivation for this approximation is described below. It comes from the functional form of hydrogenic p orbitals described in chemistry texts.

$$\psi_{p_x}(\vec{r}) = x\,\Phi(r);\; \psi_{p_y}(\vec{r}) = y\,\Phi(r);\; \psi_{p_z}(\vec{r}) = z\,\Phi(r) \tag{2.34}$$

Let's consider $\gamma_{xx}(\vec{R}) = -\int_{-\infty}^{\infty} \psi^*_{p_x}(\vec{r})U(\vec{r})\psi_{p_x}(\vec{r}-\vec{R})\,\mathrm{d}^3\mathrm{r}$ for all the permutations of the vector $\vec{R}$ for nearest neighbors of the FCC lattice,

$\psi_{p_x}(\vec{r}) = x\,\Phi(r)$ so $\psi_{p_x}(\vec{r}-\vec{R}) = (x-R_x)\Phi(\|\vec{r}-\vec{R}\|)$ using the notation for the magnitude of vector difference $\left\|\vec{r}-\vec{R}\right\| = \sqrt{(r_x - R_x) + (r_y - R_y) + (r_z - R_z)}$:

$$\sum\nolimits_{\vec{R}\neq 0} \gamma_{xx}(\vec{R})\,e^{i\vec{k}\cdot\vec{R}} = \sum\nolimits_{\vec{R}\neq 0}\left[-\int_{-\infty}^{\infty} x\Phi(r)U(\vec{r})(x-R_x)\Phi(\vec{r}-\vec{R})\,\mathrm{d}^3\mathrm{r}\right]e^{i\vec{k}\cdot\vec{R}} \tag{2.35}$$

For $\left(\frac{a}{2}, \frac{a}{2}, 0\right)$, the integral contribution, $\gamma_{xx}(\vec{R})$, where the integrals are over all space, is:

$$-\int x\,\Phi(r)\left(x-\frac{a}{2}\right)U(\vec{r})\,\Phi\left(\left[\left(x-\frac{a}{2}\right)^2 + \left(y-\frac{a}{2}\right)^2 + (z)^2\right]^{\frac{1}{2}}\right)\mathrm{d}^3\mathrm{r}$$

$$= -\int x\left(x-\frac{a}{2}\right)\Phi(r)\,\Phi\left(\left[\left(x-\frac{a}{2}\right)^2 + \left(y-\frac{a}{2}\right)^2 + (z)^2\right]^{\frac{1}{2}}\right)U(\vec{r})\,\mathrm{d}^3\mathrm{r}$$

Using the locations of the nearest neighbors for an fcc lattice, we can write down all the contributions to $\sum_{\vec{R}\neq 0}\gamma_{xx}(\vec{R})\,e^{i\vec{k}\cdot\vec{R}}$ again the integrals are over all space:

For $\left(\frac{a}{2}, \frac{a}{2}, 0\right)$, $\gamma_{xx}(\vec{R})\,e^{i\vec{k}\cdot\vec{R}}$ is:

$$- e^{i(k_x+k_y)\frac{a}{2}} \int x\left(x - \frac{a}{2}\right)\Phi(r)\,\Phi\left(\left[\left(x - \frac{a}{2}\right)^2 + \left(y - \frac{a}{2}\right)^2 + (z)^2\right]^{\frac{1}{2}}\right)U(\vec{r})\,\mathrm{d}^3\mathrm{r}$$

For : $\left(\frac{a}{2}, -\frac{a}{2}, 0\right) : -e^{i(k_x-k_y)\frac{a}{2}} \int x\left(x - \frac{a}{2}\right)\Phi(r)$

$$\Phi\left(\left[\left(x - \frac{a}{2}\right)^2 + \left(y + \frac{a}{2}\right)^2 + (z)^2\right]^{\frac{1}{2}}\right)U(\vec{r})\mathrm{d}^3\mathrm{r}$$

For : $\left(-\frac{a}{2}, \frac{a}{2}, 0\right) : -e^{i(-k_x+k_y)\frac{a}{2}} \int x\left(x + \frac{a}{2}\right)\Phi(r)$

$$\Phi\left(\left[\left(x + \frac{a}{2}\right)^2 + \left(y - \frac{a}{2}\right)^2 + (z)^2\right]^{\frac{1}{2}}\right)U(\vec{r})\mathrm{d}^3\mathrm{r}$$

For: $\left(-\frac{a}{2}, -\frac{a}{2}, 0\right) : -e^{i(-k_x-k_y)\frac{a}{2}} \int x\left(x + \frac{a}{2}\right)\Phi(r)$

$$\Phi\left(\left[\left(x + \frac{a}{2}\right)^2 + \left(y + \frac{a}{2}\right)^2 + (z)^2\right]^{\frac{1}{2}}\right)U(\vec{r})\mathrm{d}^3\mathrm{r}$$

For : $\left(\frac{a}{2}, 0, \frac{a}{2}\right) : -e^{i(k_x+k_z)\frac{a}{2}} \int x\left(x - \frac{a}{2}\right)\Phi(r)$

$$\Phi\left(\left[\left(x - \frac{a}{2}\right)^2 + (y)^2 + \left(z - \frac{a}{2}\right)^2\right]^{\frac{1}{2}}\right)U(\vec{r})\mathrm{d}^3\mathrm{r}$$

For : $\left(\frac{a}{2}, 0, -\frac{a}{2}\right) : -e^{i(k_x-k_z)\frac{a}{2}} \int x\left(x - \frac{a}{2}\right)\Phi(r)$

$$\Phi\left(\left[\left(x - \frac{a}{2}\right)^2 + (y)^2 + \left(z + \frac{a}{2}\right)^2\right]^{\frac{1}{2}}\right)U(\vec{r})\mathrm{d}^3\mathrm{r}$$

$$\text{For}: \left(-\frac{a}{2}, 0, \frac{a}{2}\right): -e^{i(-k_x+k_z)\frac{a}{2}} \int x\left(x+\frac{a}{2}\right)\Phi(r)$$
$$\Phi\left(\left[\left(x+\frac{a}{2}\right)^2+(y)^2+\left(z-\frac{a}{2}\right)^2\right]^{\frac{1}{2}}\right)U(\vec{r})\mathrm{d}^3\mathrm{r}$$

$$\text{For}: \left(-\frac{a}{2}, 0, -\frac{a}{2}\right): -e^{i(-k_x-k_z)\frac{a}{2}} \int x\left(x+\frac{a}{2}\right)\Phi(r)$$
$$\Phi\left(\left[\left(x+\frac{a}{2}\right)^2+(y)^2+\left(z+\frac{a}{2}\right)^2\right]^{\frac{1}{2}}\right)U(\vec{r})\mathrm{d}^3\mathrm{r}$$

$$\text{For}: \left(0, \frac{a}{2}, \frac{a}{2}\right): -e^{i(k_y+k_z)\frac{a}{2}} \int x^2\Phi(r)$$
$$\Phi\left(\left[(x)^2+\left(y-\frac{a}{2}\right)^2+\left(z-\frac{a}{2}\right)^2\right]^{\frac{1}{2}}\right)U(\vec{r})\mathrm{d}^3\mathrm{r}$$

$$\text{For}: \left(0, \frac{a}{2}, -\frac{a}{2}\right): -e^{i(k_y-k_z)\frac{a}{2}} \int x^2\Phi(r)$$
$$\Phi\left(\left[(x)^2+\left(y-\frac{a}{2}\right)^2+\left(z+\frac{a}{2}\right)^2\right]^{\frac{1}{2}}\right)U(\vec{r})\mathrm{d}^3\mathrm{r}$$

$$\text{For}: \left(0, -\frac{a}{2}, \frac{a}{2}\right): -e^{i(-k_y+k_z)\frac{a}{2}} \int x^2\Phi(r)$$
$$\Phi\left(\left[(x)^2+\left(y+\frac{a}{2}\right)^2+\left(z-\frac{a}{2}\right)^2\right]^{\frac{1}{2}}\right)U(\vec{r})\mathrm{d}^3\mathrm{r}$$

$$\text{For}: \left(0, -\frac{a}{2}, -\frac{a}{2}\right): -e^{i(-k_y-k_z)\frac{a}{2}} \int x^2\Phi(r)$$
$$\Phi\left(\left[(x)^2+\left(y+\frac{a}{2}\right)^2+\left(z+\frac{a}{2}\right)^2\right]^{\frac{1}{2}}\right)U(\vec{r})\mathrm{d}^3\mathrm{r}$$

Remember that $\int_{-\infty}^{\infty}\Phi^*(x)U(x)\Phi\left(x-\frac{a}{2}\right)dx = \int_{-\infty}^{\infty}\Phi^*(x)U(x)\Phi\left(x+\frac{a}{2}\right)dx$, and we define

$$\gamma_2 \equiv -\int x\left(x-\frac{a}{2}\right)\Phi(r)\Phi\left(\left[\left(x-\frac{a}{2}\right)^2+\left(y-\frac{a}{2}\right)^2+(z)^2\right]^{\frac{1}{2}}\right)U(\vec{r})\,\mathrm{d}^3\mathrm{r} \tag{2.36a}$$

Then it follows that all of the γ_2 integral contributions for the nearest neighbor positions: $\left(0,\frac{a}{2},\frac{a}{2}\right)$; $\left(0,\frac{-a}{2},\frac{a}{2}\right)$; $\left(0,\frac{a}{2},\frac{-a}{2}\right)$; $\left(0,\frac{-a}{2},\frac{-a}{2}\right)$; $\left(\frac{a}{2},0,\frac{a}{2}\right)$; $\left(\frac{-a}{2},0,\frac{a}{2}\right)$; $\left(\frac{a}{2},0,\frac{-a}{2}\right)$; $\left(\frac{-a}{2},0,\frac{-a}{2}\right)$; $\left(\frac{a}{2},\frac{a}{2},0\right)$; $\left(\frac{-a}{2},\frac{a}{2},0\right)$; $\left(\frac{a}{2},\frac{-a}{2},0\right)$; $\left(\frac{-a}{2},\frac{-a}{2},0\right)$, are identical:

$$\gamma_2 \equiv -\int x\left(x-\frac{a}{2}\right)\Phi(r)\Phi\left(\left[\left(x-\frac{a}{2}\right)^2+\left(y-\frac{a}{2}\right)^2+(z)^2\right]^{\frac{1}{2}}\right)U(\vec{r})\mathrm{d}^3\mathrm{r}$$

$$=-\int x\left(x+\frac{a}{2}\right)\Phi(r)\Phi\left(\left[\left(x+\frac{a}{2}\right)^2+\left(y-\frac{a}{2}\right)^2+(z)^2\right]^{\frac{1}{2}}\right)U(\vec{r})\mathrm{d}^3\mathrm{r}$$

$$=-\int x\left(x-\frac{a}{2}\right)\Phi(r)\Phi\left(\left[\left(x-\frac{a}{2}\right)^2+\left(y+\frac{a}{2}\right)^2+(z)^2\right]^{\frac{1}{2}}\right)U(\vec{r})\mathrm{d}^3\mathrm{r}$$

$=-\int x\left(x+\frac{a}{2}\right)\Phi(r)\Phi\left(\left[\left(x+\frac{a}{2}\right)^2+\left(y+\frac{a}{2}\right)^2+(z)^2\right]^{\frac{1}{2}}\right)U(\vec{r})\,\mathrm{d}^3\mathrm{r}$, etc.

Also, we define γ_1 for application to γ_{ij}:

$$\gamma_1 \equiv -\int x\left(y-\frac{a}{2}\right)\Phi(r)\Phi\left(\left[\left(x-\frac{a}{2}\right)^2+\left(y-\frac{a}{2}\right)^2+(z)^2\right]^{\frac{1}{2}}\right)U(\vec{r})\,\mathrm{d}^3\mathrm{r} \tag{2.36b}$$

And we note that due to symmetry:

$$\gamma_2 = -\int y\left(y-\frac{a}{2}\right)\Phi(r)\Phi\left(\left[(x)^2+\left(y-\frac{a}{2}\right)^2+\left(z-\frac{a}{2}\right)^2\right]^{\frac{1}{2}}\right)U(\vec{r})\,\mathrm{d}^3\mathrm{r}$$

$$\text{We define } \gamma_0 \equiv -\int\left(x^2-y\left(y-\frac{a}{2}\right)\right)\Phi(r)\Phi \left(\left[(x)^2+\left(y-\frac{a}{2}\right)^2+\left(z-\frac{a}{2}\right)^2\right]^{\frac{1}{2}}\right)U(\vec{r})\mathrm{d}^3\mathrm{r} \tag{2.37}$$

Then $\gamma_0+\gamma_2=-\int x^2\Phi(r)\Phi\left(\left[(x)^2+\left(y-\frac{a}{2}\right)^2+\left(z-\frac{a}{2}\right)^2\right]^{\frac{1}{2}}\right)U(\vec{r})\,\mathrm{d}^3\mathrm{r}$.

$=-\int x^2\Phi(r)\Phi\left(\left[(x)^2+\left(y-\frac{a}{2}\right)^2+\left(z+\frac{a}{2}\right)^2\right]^{\frac{1}{2}}\right)U(\vec{r})\,\mathrm{d}^3\mathrm{r}$, etc.

This gives:

$$\sum_{\vec{R}\neq 0}\gamma_{xx}(\vec{R})e^{i\vec{k}\cdot\vec{R}} = \left[e^{i(k_x+k_y)\frac{a}{2}}+e^{i(-k_x+k_y)\frac{a}{2}}+e^{i(k_x-k_y)\frac{a}{2}}+e^{-i(k_x+k_y)\frac{a}{2}}\right.$$
$$\left.+e^{i(k_x+k_z)\frac{a}{2}}+e^{i(-k_x+k_z)\frac{a}{2}}+e^{i(k_x-k_z)\frac{a}{2}}+e^{-i(k_x+k_z)\frac{a}{2}}\right]\gamma_2$$

$$+\left[e^{i(k_y+k_z)\frac{a}{2}}+e^{i(-k_y+k_z)\frac{a}{2}}+e^{i(k_y-k_z)\frac{a}{2}}+e^{-i(k_y+k_z)\frac{a}{2}}\right](\gamma_0+\gamma_2)$$

$$\sum_{\vec{R}\neq 0}\gamma_{xx}(\vec{R})e^{i\vec{k}\cdot\vec{R}}=\left(4\cos\frac{k_xa}{2}\cos\frac{k_ya}{2}+4\cos\frac{k_xa}{2}\cos\frac{k_za}{2}\right)\gamma_2$$

$$+\left(4\cos\frac{k_ya}{2}\cos\frac{k_za}{2}\right)(\gamma_0+\gamma_2)$$

$$=\left(4\cos\frac{k_xa}{2}\cos\frac{k_ya}{2}+4\cos\frac{k_xa}{2}\cos\frac{k_za}{2}+4\cos\frac{k_ya}{2}\cos\frac{k_za}{2}\right)\gamma_2$$

$$+\left(4\cos\frac{k_ya}{2}\cos\frac{k_za}{2}\right)\gamma_0 \quad (2.38)$$

Now we have enough information to further develop (2.33) into a determinant using:

$$\mathcal{M}_{ij}=\mathcal{E}(\vec{k})\delta_{ij}-E_{p_i}\delta_{ij}+\beta_{ij}+\sum_{\vec{R}\neq 0}\gamma_{ij}(\vec{R})e^{i\vec{k}\cdot\vec{R}}$$

The solution to this equation is a 3 × 3 determinant of the *i,j* components. First let's look at the $ij = xx$ component:

$$\mathcal{M}_{xx}=\mathcal{E}(\vec{k})-E_{p_x}+\beta_{xx}+\sum_{\vec{R}\neq 0}\gamma_{xx}(\vec{R})e^{i\vec{k}\cdot\vec{R}} \quad (2.39)$$

$$\mathcal{M}_{xx}=\mathcal{E}(\vec{k})-E_{p_x}+\beta_{xx}$$
$$+\left(4\cos\frac{k_xa}{2}\cos\frac{k_ya}{2}+4\cos\frac{k_xa}{2}\cos\frac{k_za}{2}+4\cos\frac{k_ya}{2}\cos\frac{k_za}{2}\right)\gamma_2$$
$$+\left(4\cos\frac{k_ya}{2}\cos\frac{k_za}{2}\right)\gamma_0 \quad (2.40a)$$

$$\mathcal{E}_0(\vec{k})=E_{p_x}-\beta_{xx}$$
$$-\left(4\cos\frac{k_xa}{2}\cos\frac{k_ya}{2}+4\cos\frac{k_xa}{2}\cos\frac{k_za}{2}+4\cos\frac{k_ya}{2}\cos\frac{k_za}{2}\right)\gamma_2 \quad (2.40b)$$

$$\mathcal{M}_{xx}=\mathcal{E}(\vec{k})-\mathcal{E}_0(\vec{k})+\left(4\cos\frac{k_ya}{2}\cos\frac{k_za}{2}\right)\gamma_0 \quad \text{for } ij=xx \text{ determinant element} \quad (2.41)$$

The other ij elements of the determinant can be calculated in a similar fashion. We have used the fact that $\beta_{xx}=\beta_{yy}=\beta_{zz}$ *and* $\beta_{xy}=0$. This can be proven as follows:

$$\beta_{ij} = -\int \psi^*_{p_i}(\vec{r})U(\vec{r})\psi_{p_j}(\vec{r})\mathrm{d}^3\mathrm{r}$$

$$\beta_{xx} = -\int x\Phi^*(r)U(\vec{r})x\varphi\Phi(r)\mathrm{d}^3\mathrm{r} = -\int y\Phi^*(r)U(\vec{r})y\varphi\Phi(r)\mathrm{d}^3\mathrm{r} = \beta_{yy} \tag{2.42a}$$

$$0 = \beta_{xx} - \beta_{yy} = -\int \left(x^2 - y^2\right)\Phi^*(r)U(\vec{r})\varphi\Phi(r)\mathrm{d}^3\mathrm{r}$$

$x^2 - y^2 = (x - y)(x + y)$ if we use $x' = x + y$ and **$y' = x - y$, which is a rotation in space), and r is a magnitude so a rotation in space means that** $\Phi(r) = \Phi\left(r'\right)$ then

$$\begin{aligned} 0 = \beta_{xx} - \beta_{yy} &= -\int \left(x^2 - y^2\right)\Phi^*(r)U(\vec{r})\varphi\Phi(r)\mathrm{d}^3\mathrm{r} \\ &= -\int \left(x' \cdot y'\right)\Phi^*\left(r'\right)U\left(\vec{r}\,'\right)\varphi\Phi\left(r'\right)\mathrm{d}^3\mathrm{r} = \beta_{xy} \end{aligned} \tag{2.42b}$$

The determinant for the tight binding p orbital band for fcc is given by [9]:

$$\begin{vmatrix} \mathcal{E}(\vec{k}) - \mathcal{E}_0(\vec{k}) + (4\cos\frac{k_y a}{2}\cos\frac{k_z a}{2})\gamma_0 & -(4\sin\frac{k_x a}{2}\sin\frac{k_y a}{2})\gamma_1 & -(4\sin\frac{k_x a}{2}\sin\frac{k_z a}{2})\gamma_1 \\ -(4\sin\frac{k_y a}{2}\sin\frac{k_x a}{2})\gamma_1 & \mathcal{E}(\vec{k}) - \mathcal{E}_0(\vec{k}) + (4\cos\frac{k_z a}{2}\cos\frac{k_x a}{2})\gamma_0 & -(4\sin\frac{k_y a}{2}\sin\frac{k_z a}{2})\gamma_1 \\ -(4\sin\frac{k_z a}{2}\sin\frac{k_x a}{2})\gamma_1 & -(4\sin\frac{k_z a}{2}\sin\frac{k_y a}{2})\gamma_1 & \mathcal{E}(\vec{k}) - \mathcal{E}_0(\vec{k}) + (4\cos\frac{k_x a}{2}\cos\frac{k_y a}{2})\gamma_0 \end{vmatrix} = 0 \tag{2.43}$$

At the center of the Brillouin Zone $\vec{k} = 0$ so $k_x = k_y = k_z = 0$

$$\begin{vmatrix} \mathcal{E}(\vec{k}=0) - \mathcal{E}_0(\vec{k}=0) + 4\gamma_0 & -4\gamma_1 & -4\gamma_1 \\ -4\gamma_1 & \mathcal{E}(\vec{k}=0) - \mathcal{E}_0(\vec{k}=0) + 4\gamma_0 & -4\gamma_1 \\ -4\gamma_1 & -4\gamma_1 & \mathcal{E}(\vec{k}=0) - \mathcal{E}_0(\vec{k}=0) + 4\gamma_0 \end{vmatrix} = 0 \tag{2.44}$$

The 3 p bands are degenerate at $\vec{k} = 0$. Thus using (2.40):

$$\begin{aligned} 0 &= \mathcal{E}(\vec{k}=0) - E_{p_x} + \beta_{xx} + 12\gamma_2 + 4\gamma_0 \\ &= \mathcal{E}(\vec{k}=0) - E_{p_y} + \beta_{yy} + 12\gamma_2 + 4\gamma_0 \\ &= \mathcal{E}(\vec{k}=0) - E_{p_z} + \beta_{zz} + 12\gamma_2 + 4\gamma_0 \end{aligned}$$

Thus $\mathcal{E}(\vec{k}=0) = E_{p_x} - \beta_{xx} - 12\gamma_2 - 4\gamma_0$.

Along a cube axis (aka, from the Γ point to the X point, since the value of k_x varies from 0 to $\frac{2\pi}{a}$:

$$k_y = k_z = 0 \text{ and } k_x = \frac{2\pi\alpha}{a} \text{ with } 0 \leq \alpha \leq 1$$

$$\begin{vmatrix} \mathcal{E}(\vec{k}) - \mathcal{E}_0(\vec{k}) + 4\gamma_0 & -\left(4\cos\frac{k_x a}{2}\right)\gamma_1 & -\left(4\cos\frac{k_x a}{2}\right)\gamma_1 \\ -\left(4\cos\frac{k_x a}{2}\right)\gamma_1 & \mathcal{E}(\vec{k}) - \mathcal{E}_0(\vec{k}) + \left(4\cos\frac{k_x a}{2}\right)\gamma_0 & -4\gamma_1 \\ -\left(4\cos\frac{k_x a}{2}\right)\gamma_1 & -4\gamma_1 & \mathcal{E}(\vec{k}) - \mathcal{E}_0(\vec{k}) + \left(4\cos\frac{k_x a}{2}\right)\gamma_0 \end{vmatrix} = 0 \quad (2.45)$$

Along a cube axis, there are 2 solutions one of which is degenerate. There will also be degenerate solutions along a cube diagonal axis such as the **[111] direction**, which is the Γ point to the L point, since the value of $k_x = k_y = k_z$ vary from 0 to $\frac{\pi}{a}$.

Tetrahedral bonds in semiconductors are sp^3 hybrids. The s and p character of the bonds can be determined using a linear combination of atomic orbitals and a tight binding calculation. For example, Koster and Slater solved the secular equation $\left|\langle\Phi_{i,j}|H - E|\psi(k)\rangle\right| = 0$ using a linear combination of both s and p orbitals and let the band gap and mixing of the valence and conduction bands come from the solution [8]. Koster and Slater are attributed with being the first to suggest using experimental data and ab initio results to fit the tight binding matrix elements [5, 8]. The p nature of the top of the valence band and the s nature of the bottom of the conduction band are still observed at the Γ point in the Brillouin zone for GaAs and some other semiconductors [5, 8, 10]. The sp^3 hybrid wavefunction for a diamond or zinc blende structure can be written as a linear combination of the four orbitals ($m = 1$ to 4) and the two atoms per unit cell ($l = 1, 2$) the constituent orbitals as follows [5, 8, 10]:

$$\begin{aligned} \psi_{\vec{k}}(\vec{r}) &= \sum_{\vec{R}} e^{i\vec{k}\cdot\vec{R}} \phi(\vec{r} - \vec{R}) \\ &= \sum_{m=1}^{4} \sum_{l=1}^{2} \sum_{\vec{R}} e^{i\vec{k}\cdot\vec{R}} c_{m,j}(\vec{k}) \psi_{m,l}(r - r_l - \vec{R}) \end{aligned} \quad (2.46)$$

When the tight binding method is applied to GaAs and experimental data is used to determine the band structure, significant issues still occur. The tight binding approximation discussed thus far assumes that there is no interaction between electrons or that a small interaction is included in the periodic potential in an average way [9]. The independent electron approximation will fail for partially filled shells such as those found in d and f shells.

As mentioned above, the valence band is spit into one degenerate level and a second non-degenerate level. The degenerate level is further split by spin – orbit coupling which has not been accounted for thus far.

2.7 Hybrid sp^3 Bonding in Semiconductors Versus the Band Picture

As mentioned above, the chemist's picture of sp^3 hybrid orbitals is often used to describe bonding in tetrahedral semiconductor structures. The hybridized sp^3 orbitals used to calculate the band structure of silicon required that both the bonding and antibonding orbital be used in the calculation [7, 8, 10]. The hybridized orbitals from adjacent atoms that point toward each other form a bond using the linear combination of atomic orbitals approach. The tight binding model requires 2nd and 3rd nearest neighbors to reproduce the band structure calculated from more rigorous methods [7, 8, 10]. The p character of the upper most valence band at the Γ point in the Brillouin zone for most semiconductors suggests that the p band approximation for the valence band provides a useful initial guide for our discussion of optical properties [6]. Below, we discuss $k \cdot p$ theory. The $k \cdot p$ solutions for the band structure of both group IV and III–V semiconductors also show the p character of the upper valence bands.

2.8 Spin–Orbit Coupling (A Semiclassical Approach)

Spin–Orbit splitting of the valence band and sometimes of the conduction band has a significant impact on the optical properties of a solid. The spin orbit coupling energy is a small enough to be considered a perturbation of the band energy and can be calculated using a perturbation theory approach. It is the valence band that is found to have spin orbit splitting. The origin of spin orbit coupling has been described using a semiclassical approach [11–13]. Although a rigorous derivation of spin orbit coupling requires relativistic considerations, the semiclassical derivation discussed below provides useful insight. The spin orbit coupling term is typically stated as follows:

$$H_{so} = \lambda \vec{L} \cdot \vec{S} \tag{2.47}$$

Here, λ determines the strength of the interactions, $\vec{L}$ is the orbital angular momentum, and $\vec{S}$ is the electron spin angular momentum. Thomas first proposed that understanding the origin of spin orbit coupling starts by considering the electron in a resting frame of reference with the nucleus rotating around it [11, 12]. The result Thomas obtained for the spin–orbit energy splitting was later found to be in complete agreement with relativistic quantum mechanics [13]. Although Spavieri and Mansuripur [13] discuss the various points of view concerning the validity of the semiclassical approach, it still provides a useful means of introducing spin orbit interactions. Since a moving charge produces a magnetic field, the magnetic field produced by the nucleus couples to the total angular momentum of the p orbital electrons. The total angular momentum comes from the orbital angular momentum and electron spin.

The energy U of a magnetic moment $\vec{\mu}$ in a magnetic field $\vec{\boldsymbol{B}}$ is:

$$U = -\vec{\mu} \cdot \vec{\boldsymbol{B}} \tag{2.48}$$

Thomas used (2.48) to represent the energy of the electron with the effective magnetic field that comes from a relativistic transformation of the electric field of the nucleus $\vec{\boldsymbol{E}}_n$ to the rest frame of the electron [11]. The effective magnetic field $\vec{\boldsymbol{B}}$ in Si units is:

$$\vec{\boldsymbol{B}} = -\frac{\vec{\boldsymbol{v}} \times \vec{\boldsymbol{E}}_n}{\boldsymbol{c}^2}$$

$\vec{v}$ is the velocity of the electron. The first discussions of the spin orbit interaction were for a Bohr atom so the electric field for the electron will be radial for an electron orbiting the nucleus [11]. Thus the electric field $\vec{E}_n$ points in a direction from the nucleus to the electron with unit vector $\hat{r}$, and the force on the electron from the nucleus with charge q_n is $e\vec{E}_n = \frac{eq_n}{r^2}\hat{r}$, the magnitude of the electric field is $E = \frac{eq_n}{r^2}$ with $\hat{r} = \frac{\vec{r}}{r}$, thus $\vec{E} = \frac{E}{r}\vec{r}$, and the classical momentum of the electron $\vec{p} = m_e\vec{v}$. The magnetic field can now be expressed as

$$\vec{\boldsymbol{B}} = -\frac{\vec{\boldsymbol{v}} \times \vec{\boldsymbol{E}}_n}{\boldsymbol{c}^2} = -\frac{\vec{\boldsymbol{r}} \times \vec{\boldsymbol{p}}}{\boldsymbol{m}_e\,\boldsymbol{c}^2}\frac{E}{r}$$

and the electric field is the gradient of the electrostatic potential (here a central potential is assumed) and the potential energy is $U_{potential}(r) = eV(r)$:

$$E = \frac{\partial V(r)}{\partial \mathrm{r}} = \frac{1}{\mathrm{e}}\frac{\partial U_{potential}(\mathrm{r})}{\partial \mathrm{r}}$$

The angular momentum of the electron is $\vec{L} = \vec{r} \times \vec{p}$ magnetic field can be expressed as:

$$\vec{\boldsymbol{B}} = -\frac{\vec{\boldsymbol{r}} \times \vec{\boldsymbol{p}}}{\boldsymbol{m}_e\boldsymbol{c}^2}\frac{E}{r} = \frac{1}{m_e e c^2}\frac{1}{r}\frac{\partial U(r)}{\partial r}\vec{L}$$

The magnetic moment of the electron due to it's spin is:

$$\mu_S = -\frac{g_s\mu_B}{\hbar}\vec{L}$$

where g_s is the electron spin g factor which is classically 1 and a relativistic value of ~ 2. The Bohr magneton is μ_B and $\vec{S}$ is the spin angular momentum vector. Since the constants multiplying $\vec{S}$ are negative, the magnetic moment is anti-parallel to the spin angular momentum vector.

Using $U = -\vec{\mu} \cdot \vec{B}$ we can calculate the Larmor (precession) part of the spin–orbit interaction using μ_S. The Larmor part is due to electron interacting with the magnetic field from the nucleus moving around it in the electron rest frame of reference.

$$H_{Larmor} = -\vec{\mu} \cdot \vec{B} = \frac{2\mu_B}{\hbar m_e ec^2} \frac{1}{r} \frac{\partial V(r)}{\partial r} \vec{L} \cdot \vec{S}$$

The Thomas interaction energy also need to be considered. This energy comes from precession in the rest frame of the electron [13]. The precession rate Ω_{Thomas} of the magnetic moment requires relativistic correction with the Lorentz factor γ [see Appendix A of [13]. $\Omega_{Thomas} = \omega(\gamma - 1)$]. The Hamiltonian for this correction is $H_{Thomas} = \Omega_{Thomas} \cdot \vec{S}$. To first order in $\left(\frac{v}{c}\right)^2$:

$$H_{Thomas} = \Omega_{Thomas} \cdot \vec{S} = -\frac{\mu_B}{\hbar m_e ec^2} \frac{1}{r} \frac{\partial V(r)}{\partial r} \vec{L} \cdot \vec{S}$$

The Thomas precession energy reduces the total interaction energy by ½ which results in a total spin orbit interaction Hamiltonian of:

$$H_{Larmor} + H_{Thomas} = \frac{\mu_B}{\hbar m_e ec^2} \frac{1}{r} \frac{\partial U(r)}{\partial r} \vec{L} \cdot \vec{S} \tag{2.49}$$

Spavieri and Mansuripur [13] describe a semiclassical electric dipole model that corrects spin orbit interaction energy without using Thomas' correction, H_{Thomas}. This semiclassical approach provides same motivation for the spin orbit Hamiltonian of (2.47) $H_{so} = \lambda \vec{L} \cdot \vec{S}$.

Above, we found that the p bands can be formed from ψ_{p_x}, ψ_{p_y}, and ψ_{p_z} orbitals. These bands were found to be degenerate at the center of the Brillouin zone, Γ. It is well known that both heavy and light holes are observed in semiconductors, and the p bands are split by spin orbit interactions as shown in Fig. 2.4. We also know that the bands are formed from hybridized orbitals. The spin–orbit split bands can be formed in two ways. One can use hybridized p orbitals or the bands can be formed from the pure angular momentum states Φ_{j,j_z}. Below we will show that the energy difference between the $j = \frac{3}{2}$ and $j = \frac{1}{2}$ bands is Δ. This can be calculated using $\left\langle \Phi_{\frac{3}{2},\frac{3}{2}} | H_{so} | \Phi_{\frac{3}{2},\frac{3}{2}} \right\rangle - \left\langle \Phi_{\frac{1}{2},\frac{1}{2}} | H_{so} | \Phi_{\frac{1}{2},\frac{1}{2}} \right\rangle = \frac{\Delta}{3} - \left(\frac{-2\Delta}{3}\right) = \Delta$. This motivates our discussion about the relationship between ψ_{p_x}, ψ_{p_y}, and ψ_{p_z} and the pure angular momentum states Φ_{j,j_z} provided below.

The spin orbit interaction for p, d, and f orbitals is non-zero. It is useful to review the basic quantum mechanics associated with angular momentum before relating the p orbitals to Φ_{j,j_z}. We start with the solution for the spherical harmonic operator for the particle on a sphere from the Schrodinger equation for the hydrogen atom:

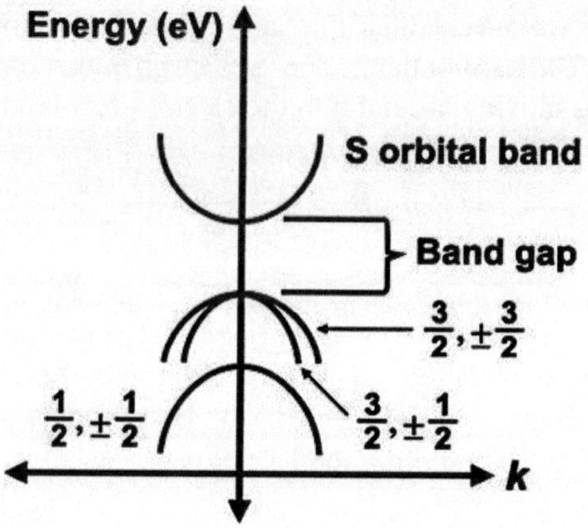

Fig. 2.4 Spin orbit interactions are shown for a direct gap. The spin orbit splitting of the p bands is shown for the angular momentum states (2.52) where the j and m values are listed. The $j, m = \frac{3}{2}, \pm\frac{3}{2}$ are angular momentum values for the heavy hole band $j, m = \frac{3}{2}, \pm\frac{1}{2}$ are the angular momentum values for the light hole band, and the split off band has $j, m = \frac{1}{2}, \pm\frac{1}{2}$. The energy difference between the $j = \frac{3}{2}$ and $j = \frac{1}{2}$ bands is Δ. Away from $k = 0$, the $j, m = \frac{3}{2}, \pm\frac{3}{2}$ are not degenerate

$$\left[\frac{1}{(\sin\theta)^2}\frac{\partial^2}{\partial\varphi^2} + \frac{1}{\sin\theta}\frac{\partial}{\partial\theta}\sin\theta\frac{\partial}{\partial\theta}\right]\psi(\theta,\varphi) + \left(\frac{2I}{\hbar^2}\right)E\psi(\theta,\varphi) = 0$$

$$E_l = \frac{\hbar^2}{2I}l(l+1)$$

Since the classical energy of a rotating body with a moment of inertia I is $\frac{1}{2}I\omega^2$, where ω is the angular frequency of rotation, and $\frac{1}{2}I\omega^2 = \frac{(I\omega)^2}{2I}$ thus $\hbar^2 l(l+1) = (I\omega)^2$ and $I\omega$, the angular momentum is quantized:

$$I\omega = \hbar\sqrt{l(l+1)}$$

It is convenient to use Bra-Ket notation in the discussion of angular momentum. This notation is short hand for an integral over all appropriate values such as over all space. The expectation value for angular momentum states $\psi_{l,s}(r)$ can be stated as follows:

$$\left\langle l,s\middle|\vec{L}\cdot\vec{S}\middle|l,s\right\rangle = \int\limits_{-\infty}^{\infty}\int\limits_{-\infty}^{\infty}\int\limits_{-\infty}^{\infty}\psi_{l,s}^{*}(\vec{L}\cdot\vec{S})\,\psi_{l,s}(r)\mathrm{d}^3\mathrm{r} \tag{2.50}$$

The total angular momentum operator is related to the orbital and spin angular momentum operators by $\vec{J} = \vec{L} + \vec{S}$ and after some algebra $\vec{J}^2 - \vec{L}^2 - \vec{S}^2 = 2\vec{L}\cdot\vec{S}$ [14]. The matrix element for pure angular momentum states with quantum numbers j, l, and s is [14]:

$$\left\langle \vec{L} \cdot \vec{S} \right\rangle = \frac{1}{2}\left\langle \vec{J}^2 - \vec{L}^2 - \vec{S}^2 \right\rangle = \frac{\hbar^2}{2}[j(j+1) - l(l+1) - s(s+1)] \quad (2.51)$$

The $\psi_{p_{x,y,\text{ or } z}}(\vec{r})$ orbitals for an atom are created from the pure angular momentum states which are hydrogen like solutions of the spherical harmonic operator $Y_{l,m}(\theta, \varphi) = \Theta(\theta)\Phi(\varphi)$ [6, 7]. For the p orbitals, $l = 1$ and $m_l = 1, 0, -1$ and the wavefunction $Y_{1,\pm 1}$ refers to $\Phi_{l,\pm m_l}$:

$$Y_{1,\pm 1} = \mp\sqrt{(3/8\pi)}\sin\theta e^{\pm i\varphi} \quad \text{and} \quad Y_{1,0} = \sqrt{(3/4\pi)}\cos\theta$$

$$\psi_{p_x} = \frac{1}{\sqrt{2}}\left(-Y_{1,1} + Y_{1,-1}\right)$$

$$\psi_{p_y} = \frac{1}{\sqrt{2}}\left(Y_{1,1} + Y_{1,-1}\right)$$

$$\psi_{p_{z,}} = Y_{1,0}$$

If one converts the spherical polar coordinates back to Cartesian coordinates, one can see why it was suggested that the p orbitals could be expressed as $\psi_{p_{x,y,\text{ or } z}}(\vec{r}) = x\,\Phi(r)$, y $\Phi(r)$, or $z\,\Phi(r)$ [15]:

$$x = r\sin\theta\cos\varphi$$

$$y = r\sin\theta\sin\varphi$$

$$z = r\cos\theta$$

Now,

$$\psi_{p_z} = Y_{1,0} = \sqrt{(3/4\pi)}\cos\theta = \sqrt{(3/4\pi)}\frac{z}{r} = z\,\Phi(r)$$

$$\begin{aligned}
\psi_{p_x} &= \frac{1}{\sqrt{2}}\left(-Y_{1,1} + Y_{1,-1}\right) \\
&= \frac{1}{\sqrt{2}}\left(\sqrt{(3/8\pi)}\sin\theta e^{i\varphi} + \sqrt{(3/8\pi)}\sin\theta e^{-i\varphi}\right) \\
&= \frac{1}{\sqrt{2}}\sqrt{(3/8\pi)}\sin\theta\cos\phi \\
&= \frac{1}{\sqrt{2}}\sqrt{(3/8\pi)}\frac{x}{r} = x\,\Phi(r)
\end{aligned}$$

Now we must consider that the total angular momentum for the p electrons includes the orbital and spin angular momentum. Following Singh [6], one can state for the p_x orbital for spin up electrons:

$$\psi_{p_x} \uparrow = \frac{1}{\sqrt{2}}(-Y_{1,\,1} + Y_{1,\,-1}) \uparrow$$

The orbital angular momentum must be properly added to the spin angular momentum. The Clebsch-Gordan method is used for the wavefunctions Φ_{j,j_z} [6, 16]. The total angular momentum for the p orbital, $J = l + s$, with the z components J_z for $l = 1$ and $s = \frac{1}{2}$ has values $J = l + s = \frac{3}{2}$ or $J = l - s = \frac{1}{2}$ and the $2J + 1$ eigenvalues of J_z vary from J to $-J$: $J, J-1, \ \ldots, \ -J+1, -J$. The Φ_{j,j_z} can be expressed as linear combinations of orbital and spin angular momentum functions where spin up and spin down is designated by ↑ and ↓ respectively.

$$\Phi_{\frac{3}{2},\frac{3}{2}} = Y_{1,\,1} \uparrow = -\frac{1}{\sqrt{2}}(\psi_{p_{x,}} + i\psi_{p_{y,}}) \uparrow \tag{2.52}$$

$$\Phi_{\frac{3}{2},\frac{1}{2}} = \frac{1}{\sqrt{3}}Y_{1,\,1} \downarrow + \frac{\sqrt{2}}{\sqrt{3}}Y_{1,\,0} \uparrow = -\frac{1}{\sqrt{6}}[(\psi_{p_{x,}} + i\psi_{p_{y,}}) \downarrow -2\psi_{p_{z,}} \uparrow]$$

$$\Phi_{\frac{3}{2},-\frac{1}{2}} = \frac{\sqrt{2}}{\sqrt{3}}Y_{1,\,0} \downarrow + \frac{1}{\sqrt{3}}Y_{1,\,-1} \uparrow = \frac{1}{\sqrt{6}}[(\psi_{p_{x,}} - i\psi_{p_{y,}}) \uparrow +2\psi_{p_{z,}} \downarrow]$$

$$\Phi_{\frac{3}{2},-\frac{3}{2}} = Y_{1,\,-1} \downarrow = \frac{1}{\sqrt{2}}(\psi_{p_{x,}} - i\psi_{p_{y,}}) \downarrow$$

$$\Phi_{\frac{1}{2}'\frac{1}{2}} = -\frac{1}{\sqrt{3}}Y_{1,\,0} \uparrow + \frac{\sqrt{2}}{\sqrt{3}}Y_{1,\,1} \downarrow = -\frac{1}{\sqrt{3}}[(\psi_{p_{x,}} + i\psi_{p_{y,}}) \downarrow +\psi_{p_{z,}} \uparrow]$$

$$\Phi_{\frac{1}{2},-\frac{1}{2}} = -\frac{\sqrt{2}}{\sqrt{3}}Y_{1,\,-1} \uparrow + \frac{1}{\sqrt{3}}Y_{1,\,0} \downarrow = -\frac{1}{\sqrt{3}}[(\psi_{p_{x,}} - i\psi_{p_{y,}}) \uparrow -\psi_{p_{z,}} \downarrow]$$

The spin orbit splitting can be calculated using $H_{so} = \lambda\vec{L}\cdot\vec{S}$; (2.47), and $\left\langle\vec{L}\cdot\vec{S}\right\rangle = \frac{1}{2}\left\langle\vec{J}^2 - \vec{L}^2 - \vec{S}^2\right\rangle = \frac{\hbar^2}{2}[j(j+1) - l(l+1) - s(s+1)]$; (2.51). It is useful to use $H'_{so} = [j(j+1) - l(l+1) - s(s+1)]$ and $H_{so} = \frac{\lambda\hbar^2}{2}H'_{so}$. The pure angular momentum states Φ_{j,j_z} states are orthonormal and $l = 1$ for p orbitals. Using $j = l + s$ one can determine s. The spin orbit splitting between the $j = \frac{3}{2}$ *and* $j = \frac{1}{2}$ bands is $\left\langle\Phi_{\frac{3}{2},\frac{3}{2}}|H_{so}|\Phi_{\frac{3}{2},\frac{3}{2}}\right\rangle - \left\langle\Phi_{\frac{1}{2},\frac{1}{2}}|H_{so}|\Phi_{\frac{1}{2},\frac{1}{2}}\right\rangle = \frac{\Delta}{3} - \left(\frac{-2\Delta}{3}\right) = \Delta$ where $\Delta = \frac{3\lambda\hbar^2}{2}$. This is shown in Fig. 2.4.

One can also invert (2.52) to give $\psi_{p_{x,}} \uparrow$, etc., in terms of wavefunctions that have easily determined values of $J, l,$ and m_l [6]:

$$\psi_{p_x}\uparrow=\frac{1}{\sqrt{2}}\left[-\Phi_{\frac{3}{2},\frac{3}{2}}+\frac{1}{\sqrt{3}}\Phi_{\frac{3}{2},-\frac{1}{2}}-\frac{\sqrt{2}}{\sqrt{3}}\Phi_{\frac{1}{2},-\frac{1}{2}}\right] \tag{2.53}$$

$$\psi_{p_x}\downarrow=\frac{1}{\sqrt{2}}\left[-\frac{1}{\sqrt{3}}\Phi_{\frac{3}{2},\frac{1}{2}}-\frac{\sqrt{2}}{\sqrt{3}}\Phi_{\frac{1}{2},\frac{1}{2}}+\Phi_{\frac{3}{2},-\frac{3}{2}}\right]$$

$$\psi_{p_y}\uparrow=\frac{i}{\sqrt{2}}\left[\Phi_{\frac{3}{2},\frac{3}{2}}+\frac{1}{\sqrt{3}}\Phi_{\frac{3}{2},-\frac{1}{2}}-\frac{\sqrt{2}}{\sqrt{3}}\Phi_{\frac{1}{2},-\frac{1}{2}}\right]$$

$$\psi_{p_y}\downarrow=\frac{i}{\sqrt{2}}\left[\frac{1}{\sqrt{3}}\Phi_{\frac{3}{2},\frac{1}{2}}+\frac{\sqrt{2}}{\sqrt{3}}\Phi_{\frac{1}{2},\frac{1}{2}}+\Phi_{\frac{3}{2},-\frac{3}{2}}\right]$$

$$\psi_{p_z}\uparrow=\left[\frac{\sqrt{2}}{\sqrt{3}}\Phi_{\frac{3}{2},\frac{1}{2}}-\frac{1}{\sqrt{3}}\Phi_{\frac{1}{2},\frac{1}{2}}\right]$$

$$\psi_{p_z}\downarrow=\left[\frac{\sqrt{2}}{\sqrt{3}}\Phi_{\frac{3}{2},-\frac{1}{2}}+\frac{1}{\sqrt{3}}\Phi_{\frac{1}{2},-\frac{1}{2}}\right]$$

The spin orbit interaction between p bands can also be determined using $H_{so}=\frac{\lambda\hbar^2}{2}H'_{so}$, see Singh [6]. For example:

$$\begin{aligned}\langle\psi_{p_x}\uparrow|H_{so}|\psi_{p_y}\uparrow\rangle=\frac{i}{2}\frac{\lambda\hbar^2}{2}\Big[\Big\langle-\Phi_{\frac{3}{2},\frac{3}{2}}\Big|H'_{so}\Big|\Phi_{\frac{3}{2},\frac{3}{2}}\Big\rangle+\frac{1}{3}\Big\langle\Phi_{\frac{3}{2},-\frac{1}{2}}\Big|H'_{so}\Big|\Phi_{\frac{3}{2},-\frac{1}{2}}\Big\rangle\\+\frac{2}{3}\Big\langle\Phi_{\frac{1}{2},-\frac{1}{2}}\Big|H'_{so}\Big|\Phi_{\frac{1}{2},-\frac{1}{2}}\Big\rangle\Big]\end{aligned} \tag{2.54}$$

$$\begin{aligned}&\langle\psi_{p_x}\uparrow|H_{so}|\psi_{p_y}\uparrow\rangle\\&=\frac{i}{2}\frac{\lambda\hbar^2}{2}\left\{-\left[\frac{3}{2}\left(\frac{3}{2}+1\right)-1(1+1)-\frac{1}{2}\left(\frac{1}{2}+1\right)\right]\right.\\&\quad+\frac{1}{3}\left[\frac{3}{2}\left(\frac{3}{2}+1\right)-1(1+1)-\frac{1}{2}\left(\frac{1}{2}+1\right)\right]\\&\quad\left.+\frac{2}{3}\left[\frac{1}{2}\left(\frac{1}{2}+1\right)-1(1+1)-\frac{1}{2}\left(\frac{1}{2}+1\right)\right]\right\}\end{aligned}$$

$$\langle\psi_{p_x}\uparrow|H_{so}|\psi_{p_y}\uparrow\rangle=\frac{i}{2}\frac{\lambda\hbar^2}{2}\left(-\frac{6}{3}\right)=\frac{-i\Delta}{3}\quad\Delta=\frac{3\lambda\hbar^2}{2} \tag{2.55}$$

The optical properties of bulk and 2D materials are impacted by spin orbit interactions. The energy of the valence bands formed from sp^3 hybrid orbitals for group IV and III–V semiconductors are split by spin–orbit interactions [16, 17] as well as for other crystal structures. In Chap. 8, we will see how spin orbit interactions alter the photoluminescence of transition metal dichalcogenides [18]. Spin orbit splitting in

Table 2.1 The spin orbit splitting energy for a variety of materials

Diamond lattice semiconductor	Location in Brillouin Zone	Δ (eV) VB	Direction in Brillouin Zone	Δ_1 (eV) VB
Si [16]	Γ	0.044	Γ to L	0.03
Ge [16]	Γ	0.29	Γ to L	0.19
Zinc Blede lattice semiconductor				
GaAs [17]	Γ	0.351	Γ to L	0.213
InAs [7]	Γ	0.38	Γ to L	0.276
InSb [17]	Γ	0.799	Γ to L	0.497
InP [17]	Γ	0.12	Γ to L	0.126
GaP [6]	Γ	0.094	Γ to L	
Hexagonal trigonal prismatic				
MoS_2 [18]	K, K'	0.148		
$MoSe_2$ [18]	K, K'	0.183		
WS_2 [18]	K, K'	0.426		
WSe_2 [18]	K, K'	0.456		

topological insulators such as Bi_2Se_3 and Weyl semimetals alters the electronic band structure and so called band inversion can occur [19, 20]. Band inversion refers to the valence band taking on a significant s orbital character and the conduction band a p orbital character. In Table 2.1, we show the spin orbit splitting for key locations in the Brillouin zone.

Our discussion about the p orbital bands in a semiconductor has shown us that based on the secular equation, there are three bands two of which are degenerate. However, spin orbit coupling splits one of those bands. The p orbital bands could have been calculated using a linear combination of total angular momentum states (orbital plus spin angular momentum, ie, $\Phi_{\frac{3}{2},\frac{3}{2}}$, etc. The spin orbit splitting of the p band is shown in Fig. 2.4.

The spin orbit splitting of Ge can be observed in the direct gap transition known as the E1 critical point using spectroscopic ellipsometry. The E1 transition is between the valence and conduction bands along the 111 direction in the Brillouin zone. In strained (stressed) SiGe alloys, this splitting is greatly enhanced.

2.9 $k \cdot p$ Theory

In this section, the $k \cdot p$ method of calculating band structure is discussed. This method was widely used by the pioneers in optical physics to calculate the band structure of semiconductors and then the dielectric function. It continues to be used and reading the many of the papers of Cardona [21, 22] and his group requires some familiarity with $k \cdot p$ theory. The reader is referred to Yu and Cardona's [7] discussion of $k \cdot p$

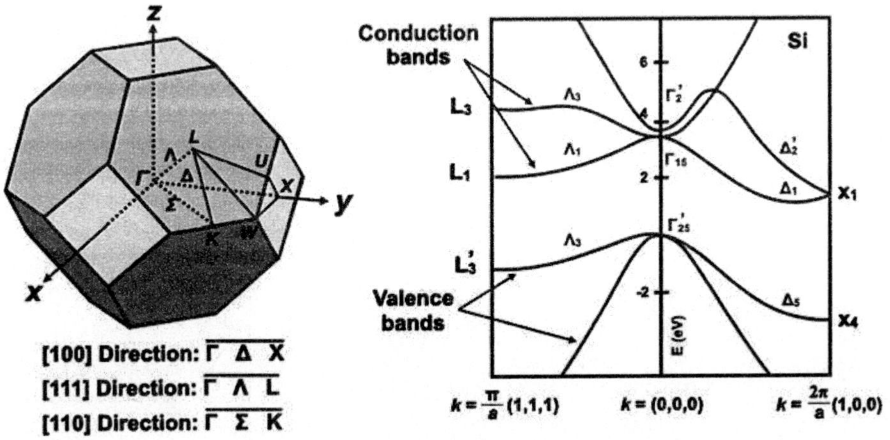

Fig. 2.5 $k \cdot p$ theoretical electronic band structure for silicon and the assocaited Brillouin zone. Figure on right is adapted and reprinted with permission from [21]. © 1966 American Physical Society

theory as well as Dresselhaus [23] who introduced the notation used here and in [7]. It critical to know that the notation for the electronic bands used in $k \cdot p$ calculations refers to the symmetry of the band. For example the lowest lying conduction band in a zinc blende lattice will have the Γ_1 symmetry and the p band will have Γ_4 symmetry [7]. In a diamond lattice, the lowest lying conduction band has the Γ_{15} symmetry and the highest valence band is designated $\Gamma_{25'}$ [7]. The electron band diagrams often use double group notation which includes the symmetry of the spin wave functions [7]. Thus the double group notation for the highest lying valence band is Γ_8 and for the lowest lying conduction band is Γ_6 at the center of Brillouin zone Γ for the band energy diagrams of GaAs. We also note that Voon and Willatzen provides a through exposition of the $k \cdot p$ method [24] this reference covers diamond, zinc blende, and wurtzite lattice structures. In Fig. 2.5, we show the $k \cdot p$ band structure of silicon.

We start with the Schrodinger equation.

$$-\frac{\hbar^2}{2m}\frac{d^2}{dx^2}\psi + V(x)\psi = \mathrm{E}\psi$$

We will develop $k \cdot p$ theory using a one dimensional approach along the x direction.

$$\psi_k(x) = e^{-ikx}u_k(x)$$

$$-\frac{\hbar^2}{2m}\left[\frac{d^2}{dx^2} + 2ik\frac{d}{dx} + k^2\right]u_k(x) + V(x)u_k(x) = Eu_k(x)$$

The momentum operator is $\vec{p} = i\hbar\overrightarrow{\nabla}$

$$\left[-\frac{\hbar^2}{2m}\nabla^2 + \frac{\hbar}{m}k \cdot p + V(x)\right]u_k(x) = \left[E - \frac{\hbar^2}{2m}k^2\right]u_k(x)$$

This is the origin of the famous $k \cdot p$ theory for band structure.
Generalizing to three dimensions:

$$\left[-\frac{\hbar^2}{2m}\nabla^2 + \frac{\hbar}{m}\vec{k} \cdot \vec{p} + V(\vec{r})\right]u_k(\vec{r}) = \left[E - \frac{\hbar^2}{2m}k^2\right]u_k(\vec{r}) \tag{2.56}$$

At $k = 0$

$$\left[-\frac{\hbar^2}{2m}\nabla^2 + V(\vec{r})\right]u_k(\vec{r}) = E\, u_k(\vec{r}) \tag{2.57}$$

Then treat $k \cdot p$ term as a perturbation. Second order terms vanish because of BRILLOUIN ZONE center starting point. Using index n for a non-degenerate energy level (band) n

$$u_{nk} = u_{n0} + \frac{\hbar}{m}\sum_{n' \neq n} \frac{\langle u_{n0}|k \cdot p|u_{n'0}\rangle}{E_{n0} - E_{n'0}} u_{n'0}$$

$$E_{nk} = E_{n0} + \frac{\hbar^2 k^2}{2}\underbrace{\left(\frac{1}{m} + \frac{2}{m^2 k^2}\sum_{n' \neq n} \frac{|\langle u_{n0}|k \cdot p|u_{n'0}\rangle|^2}{E_{n0} - E_{n'0}}\right)}_{1/m*}$$

2.10 Effective Mass

$k \cdot p$ theory allows us to estimate the effective mass of electrons and holes in a specific band. This concept will be important when we consider excitons which have a reduced mass based on the electron and hole masses. In Chap. 5, we will use the parametrized equations for the effective mass of an electron provided below to estimate the properties of excitons. The effective mass is related to the curvature of the band structure as follows:

$$E = \frac{\hbar^2 k^2}{2m} \text{ gives } m_{eff} = \hbar^2\left(\frac{d^2E}{dk^2}\right)^{-1} \tag{2.58}$$

From $k \cdot p$ theory for a direct gap semiconductor, the effective mass for a non-degenerate band is given by [7]:

$$\frac{1}{m_{eff}} = \frac{1}{m} + \frac{2}{m^2k^2} \sum_{n' \neq n} \frac{|\langle u_{n0}|k \cdot p|u_{n'0}\rangle|^2}{E_{n0} - E_{n'0}} \tag{2.59}$$

This is for a non-degenerate band where u_{n0} is the wave function for the nth band. Let's consider the effective mass of the conduction electrons and valence band holes. For direct gap semiconductors having the zinc blende lattice structure such as GaAs, the band gap E_0 can be related to the effective mass for the transition between the bands at the center of the Brillouin Zone between the Γ_{1c} conduction band and the Γ_{4v} valence band:

$$\frac{1}{m_c^*} = \frac{1}{m} + \frac{2}{m^2k^2} \frac{|\langle \Gamma_{1c}|k \cdot p|\Gamma_{4v}\rangle|^2}{E_0} \tag{2.60}$$

Because the symmetry involved, this is further simplified into and expression involving the momentum matrix elements along x, y, and z. Using X, Y, and Z to represent the three $\Gamma_4(p)$ wavefunctions [7]:

$$\langle X|p_x|\Gamma_1\rangle = \langle Y|p_y|\Gamma_1\rangle = \langle Z|p_z|\Gamma_1\rangle = iP \text{ with } \vec{p} = -i\hbar\vec{\nabla} \tag{2.61}$$

The momentum matrix elements for group IV, III-V, and II-VI semiconductors have a constant value. Thus, the effective mass for a conduction electron is:

$$\frac{m}{m_c^*} = 1 + \frac{2P^2}{mE_0} \tag{2.62}$$

$2P^2/m \sim 20$ eV for group IV, III-V, and II-VI semiconductors which is close to that calculated using free electron wavefunctions:

$$P(free\ electron) = 2\pi\hbar/a_0$$

Since $E_0 \sim 2$ eV:

$$\frac{m}{m_c^*} \approx \frac{2P^2}{mE_0} \tag{2.63}$$

The light and heavy hole effective masses and valence band energy levels for both diamond and zinc blende lattice structures can also be determined from $k \cdot p$ theory. The expression for these quantities use terms A, B, and C which are defined using the P matrix elements defined above Γ_1 conduction band and the Γ_4 valence band: E_0 and the Q matrix element defined as a the matrix element between (Γ_4 conduction bands and the valence bands Γ_4: E_0') [7, 21–23]. For zinc blende lattices:

$$\langle X|p_y|\Gamma_{4c}(z)\rangle = \langle Y|p_z|\Gamma_{4c}(x)\rangle = \langle Z|p_x|\Gamma_{4c}(y)\rangle = iQ\ with\ p = -i\hbar\nabla$$

$$\langle X|p_{yz}|\Gamma_{4c}(y)\rangle = \langle Y|p_{zx}|\Gamma_{4c}(z)\rangle = \langle Z|p_y|\Gamma_{4c}(x)\rangle = iQ \ with \ p = -i\hbar\nabla \tag{2.64}$$

The momentum matrix elements Q for diamond lattice semiconductors are determined using the appropriate conduction band wavefunctions [7, 21, 25]. We will use the constants A, B, and C for our discussion of excitons in Chap. 5, and A, B, and C originally come from the Dresselhaus-Kip-Kittel model [23] for the $k \cdot p$ approach for a diamond lattice without spin orbit coupling [24, 25]. The determination of A, B, and C has evolved so that these constants are available for diamond and zinc blende semiconductors [7, 24]. These constants can also be related to the Kohn-Luttinger parameters see for example Yu and Cardona, (2.71a–c) [7]. The energy gaps are defined as follows: E_0 is the energy separation between the Γ_{1c} and the $j = \frac{3}{2}$ valence bands; and E_0' is the difference between the Γ_{4c} and the $j = \frac{3}{2}$ valence bands [7, 24]. When the momentum matrix elements between the Γ_{4v} valence band and the Γ_{3c} conduction band are ignored, the constants A, B, and C are related to the direct band gaps E_0 and E_0' by [7]:

$$\frac{2m}{\hbar^2} A = 1 - \frac{2}{3}\left[\left(\frac{P^2}{mE_0}\right) + \left(\frac{2Q^2}{mE_0'}\right)\right] \tag{2.65}$$

$$\frac{2m}{\hbar^2} B = \frac{2}{3}\left[\left(\frac{-P^2}{mE_0}\right) + \left(\frac{Q^2}{mE_0'}\right)\right] \tag{2.66}$$

$$\left(\frac{2m}{\hbar^2} C\right)^2 = \frac{16P^2Q^2}{3mE_0 mE_0'} \tag{2.67}$$

Using these constants, the heavy hole and light hole band energies are [7, 21]:

$$E_{hh} = -Ak^2 - \left[B^2k^4 + C^2\left(k_x^2k_y^2 + k_y^2k_z^2 + k_z^2k_x^2\right)\right]^{1/2} \tag{2.68}$$

$$E_{lh} = -Ak^2 + \left[B^2k^4 + C^2\left(k_x^2k_y^2 + k_y^2k_z^2 + k_z^2k_x^2\right)\right]^{1/2} \tag{2.69}$$

The hole effective masses are different along [100] and [111] directions are useful in estimating the exciton energies discussed in Chap. 5. They can be determined from tabulated values of A, B, and C:

$$\begin{aligned} k\|(100) \quad \frac{1}{m_{hh}} &= \frac{2}{\hbar^2}(-A + B) \\ \frac{1}{m_{lh}} &= \frac{2}{\hbar^2}(-A - B) \\ k\|(111) \quad \frac{1}{m_{hh}} &= \frac{2}{\hbar^2}\left[-A + B\left(1 + \frac{|C|^2}{3B^2}\right)\right] \end{aligned} \tag{2.70}$$

Table 2.2 The $k \cdot p$ theory parameters A, B, and C. The units for A and B are $\frac{\hbar^2}{2m}$ and for $|C|^2$ are $(\frac{\hbar^2}{2m})^2$ [7, 21, 22]. It should be noted that GaN typically has the Wurtzite crystal structure, but the values in this table are for metastable zinc blende GaN

	A	B	$\|C\|^2$
Si	– 4.28	– 0.68	24
Ge	– 13.38	– 8.5	173
GaAs	– 6.9	– 4.4	43
GaN	– 5.05	– 1.2	34

$$\frac{1}{m_{lh}} = \frac{2}{\hbar^2}\left[-A - B\left(1 + \frac{|C|^2}{3B^2}\right)\right] \tag{2.71}$$

The expressions for the average values are [7]:

$$\frac{1}{m^*_{hh}} = \frac{2}{\hbar^2}\left[-2A + 2B\left(1 + \frac{|C|^2}{15B^2}\right)\right] \tag{2.72a}$$

$$\frac{1}{m^*_{lh}} = \frac{2}{\hbar^2}\left[-2A - 2B\left(1 + \frac{|C|^2}{15B^2}\right)\right] \tag{2.72b}$$

Table 2.2 provides the values of A, B, and C used in Chap. 5 (7). Although the wavefunction symmetry is different for diamond versus zinc blende lattices, the procedure for determining A, B, and C for diamond lattices is similar. The symmetry designated bands for silicon and germanium are listed in Table 2.1 of Cardona and Pollak [21] along with their orbital character.

The symmetry designated bands for silicon and germanium are listed in Table 1 of Cardona and Pollak [21] along with their orbital character. Other reference may be useful. The parameters A, B, and C for Si and Ge based on the Dresselhaus-Kip-Kittel model are related to the expressions for L, M, and N defined in (3.3) in Voon and Willatzen where the values of these parameters are provided in Table 3.3 in [24]. The reader is referred to [7, 24] for further reading.

2.11 Tight Binding Model in the Second Quantization Formalism

The history of second quantization goes back to the early days of quantum mechanics and its founders especially Dirac and Pauli [26, 27]. It was introduced by Dirac [26] for practical reasons because it simplified some calculations [27]. The second quantization formalism is frequently used for theoretical electron band studies. In this book, the second quantization formalism is used to discuss topological physics in Chap. 9. Koch provides an excellent review [28]. Here, we present a brief overview to introduce the topic. The Schrödinger picture of quantum mechanics focuses on the wavefunction of each particle. The second quantization picture focuses on single particle states that are occupied with the other states being empty and can be referred to as an occupation number representation. We start with N lattice sites and an occupation number representation for the wave function:

$$|n_1, \ldots n_j, \ldots.., n_N\rangle \tag{2.73}$$

Next we define operators that create $c_j^\dagger$ or annihilate c_j a fermion from $|n_1, \ldots n_j, \ldots.., n_N\rangle$:

$$\begin{aligned} c_j^\dagger|n_1, \ldots n_j = 0, \ldots.., n_N\rangle &= |n_1, \ldots n_j = 1, \ldots.., n_N\rangle \text{ and} \\ c_j|n_1, \ldots n_j = 1, \ldots.., n_N\rangle &= |n_1, \ldots n_j = 0, \ldots.., n_N\rangle \end{aligned} \tag{2.74}$$

The number of fermions in a state j is given by the operator $\widehat{n_j} = c_j^\dagger c_j$:

$$c_j^\dagger c_j|n_1, \ldots n_j, \ldots.., n_N\rangle = \widehat{n_j}|n_1, \ldots n_j, \ldots.., n_N\rangle \tag{2.75}$$

and the commutation relationships are:

$$\left\{c_i; c_j^\dagger\right\} = \delta_{i,j} \text{ and } \left\{c_i; c_j\right\} = 0 \tag{2.76}$$

The momentum space representation of the second quantization Hamiltonian is very useful. It is useful to describe the creation and annihilation operators in positon space in terms of wave vector space:

$$c_j^\dagger = \frac{1}{\sqrt{N}}\sum_{\vec{k}} e^{-i\vec{k}\cdot\vec{r}_j} c_{\vec{k}}^\dagger \text{ and } c_j = \frac{1}{\sqrt{N}}\sum_{\vec{k}} e^{i\vec{k}\cdot\vec{r}_j} c_{\vec{k}} \tag{2.77a}$$

Or one can write the creation and annihilation operators in wave vector space in terms of position space:

$$c_{\vec{k}}^\dagger = \frac{1}{\sqrt{N}}\sum_{j} e^{i\vec{k}\cdot\vec{r}_j} c_j^\dagger \text{ and } c_{\vec{k}} = \frac{1}{\sqrt{N}}\sum_{j} e^{-i\vec{k}\cdot\vec{r}_j} c_j \tag{2.77b}$$

A simple tight binding Hamiltonian for solid where the electrons with spin $\alpha = +\frac{1}{2}, -\frac{1}{2}$ with wave vector k have energy $E_k = \frac{\hbar^2 k^2}{2m}$ is:

$$H = \sum_{k,\,\alpha} E_k c_{\vec{k}\alpha}^\dagger c_{\vec{k}\alpha} \tag{2.78}$$

Here we added an index for spin state to the creation and annihilation operators. This Hamiltonian can be restated using the relationship between position space and wave vector space for the creation and annihilation operators.

$$H = \frac{1}{N}\sum_{i,j,\ \alpha}\sum_{k} E_{\vec{k}} e^{-i\vec{k}\cdot(r_i - r_j)} c_{i\alpha}^{\dagger} c_{j\alpha}$$

The hopping from site I to site j is given by:

$$\hat{t}_{i,j} = \frac{1}{N}\sum_{k} E_{\vec{k}} e^{-i\vec{k}\cdot(r_i - r_j)} \tag{2.79}$$

And the Hamiltonian is restated as:

$$H = \sum_{i,j,\ \alpha} \hat{t}_{i,j} c_{i\alpha}^{\dagger} c_{j\alpha} \tag{2.80}$$

The $c_{i\alpha}^{\dagger} c_{j\alpha}$ should be interpreted as a fermion at j with spin α jumping from j to i. The nearest neighbor tight binding Hamiltonian can be obtained by assuming that $\hat{t}_{i,j} = -t$ for i, j nearest neighbors and 0 otherwise. First (2.80) is rewritten as:

$$H_{TB} = -t\sum\nolimits_{ij,\ \alpha}\left(c_{i\alpha}^{\dagger} c_{j\alpha} + c_{j\alpha}^{\dagger} c_{i\alpha}\right) \tag{2.81}$$

where ij refers to the different nearest neighbor and next nearest neighbor combinations. Here we only discuss the nearest neighbor contributions. Next, we use the wave vector form of the creation and annihilation operators in (2.77a, b) to incorporate the nearest neighbor interactions directly. The vector from the coordinate system origin to the nearest neighbor is designated as $\vec{n}_i$, and for an atom at *i*, the nearest neighbors j are at $\vec{r}_j = \vec{r}_i + \vec{n}_j$:

$$\left(c_{i\alpha}^{\dagger} c_{j\alpha} + c_{j\alpha}^{\dagger} c_{i\alpha}\right) = \frac{1}{2}\left(\frac{1}{\sqrt{N}}\sum_{\vec{k}} e^{-i\vec{k}\cdot\vec{r}_i} c_{\vec{k}\alpha}^{\dagger} \frac{1}{\sqrt{N}}\sum_{\vec{k}'} e^{i\vec{k}'\cdot(\vec{r}_i + \vec{n}_j)} c_{\vec{k}'\alpha}\right.$$
$$\left. + \frac{1}{\sqrt{N}}\sum_{\vec{k}} e^{-i\vec{k}\cdot(\vec{r}_i + \vec{n}_j)} c_{\vec{k}\alpha}^{\dagger} \frac{1}{\sqrt{N}}\sum_{\vec{k}'} e^{i\vec{k}'\cdot\vec{r}_i} c_{\vec{k}'\alpha}\right)$$

Above, the ½ avoids double counting, and below the $\sum_i e^{-i(\vec{k}-\vec{k}')\cdot\vec{r}_i} e^{i(\vec{k}-\vec{k}')\cdot\vec{r}_i} = N$.

$$= \frac{1}{2N}\left(\sum_{\vec{k},\vec{k}'} e^{-i(\vec{k}-\vec{k}')\cdot\vec{r}_i} e^{i\vec{k}'\cdot\vec{n}_j} c_{\vec{k}\alpha}^{\dagger} c_{\vec{k}\alpha} + \sum_{\vec{k},\vec{k}'} e^{i(\vec{k}-\vec{k}')\cdot\vec{r}_i} e^{-i\vec{k}'\cdot\vec{n}_j} c_{\vec{k}\alpha}^{\dagger} c_{\vec{k}\alpha}\right)$$

$$= \frac{1}{2N}\left(\sum_{\vec{k}} e^{i\vec{k}\cdot\vec{n}_j} c_{\vec{k}\alpha}^{\dagger} c_{\vec{k}\alpha} + \sum_{\vec{k}} e^{-i\vec{k}\cdot\vec{n}_j} c_{\vec{k}\alpha}^{\dagger} c_{\vec{k}\alpha}\right) = \frac{1}{N}\cos\left(\vec{k}\cdot\vec{n}_j\right) c_{\vec{k}\alpha}^{\dagger} c_{\vec{k}\alpha}$$

The wave vector $\vec{k}$ will not change for the hopping interaction, so $\vec{k} = \vec{k}'$. This leads to the following expression for the tight binding Hamiltonian in the second quantization formalism:

$$H_{TB} = -t \sum_{j,\, \vec{k},\alpha} \cos\left(\vec{k} \cdot \vec{n}_j\right) c^{\dagger}_{\vec{k}\alpha} c_{\vec{k}\alpha}$$

The reader is referred to [28] for a more complete discussion of the second quantization formalism.

2.12 Crystal Structure Symmetry - Definitions of Point Groups and Space Groups

Here we provide definitions that will be useful for the discussing Schoenflies point groups and the Space Groups of crystal structures discussed in Chaps. 8 and 9. The reader is referred to more complete discussions of these topics in [29] and [30]. The symmetry of the point group is different from the symmetry of the lattice in that it refers to the symmetry of a specific point in space. For example, there is a point group symmetry for a specific point in the Brillouin zone such as the Γ or M point that depends on the crystal structure. This information is useful for understanding the local electronic band structure. There are 32 point groups which are listed along with the Schoenflies notation in Table 2.3 [31–34]. The symmetry of the crystal structure is known as the Space Group of the crystal structure. The space group is a combination of the point group symmetries with the translation symmetry. The Hermann-Mauguin or International notation is used to designate space groups. There are 230 space groups. The reader is referred to the International Tables for Crystallography for a complete discussion of spatial symmetry along with extensive tables for point groups and space groups which are summarized here. [31–34].

Schoenflies point groups are represented by the Schoenflies notation for the 3D symmetry operations associated with a point in the crystal structure. The 3D symmetry operations of the point group leave the crystal structure unchanged [29, 31]. In the Schoenflies notation, there is a letter symbol with a subscript. For example, a three-fold rotation axis uses the symbol C_3. Here, C_n refers to a cyclic point group with n fold rotation. In addition, C_{nh} refers to C_n with a mirror plane perpendicular to the rotation axis and C_{nv} refers to C_n with n mirror planes through the rotation axis. A summary of the Schoenflies point groups is provided below [29, 30]. The term improper rotation axis refers to a rotation that requires a subsequent inversion operation to return the crystal to the same structure.

Table 2.3 Crystal structure and associated Hermann-Mauguin notation and Schoenflies point group notations are shown for the 32 point groups [29–34]. Figures showing the symmetry operations for each crystal system can be found in [32], and description of space groups and crystal structures in [33, 34]. Often, a superscript is added to the Schoenflies point group symbols to further associate it with a space group

Triclinic		Monoclinic		Tetragonal		Trigonal		Hexagonal		Cubic	
1	C_1	2	C_2	4	C_4	3	C_3	6	C_6	23	T
$\bar{1}$	C_i	$m = \bar{2}$	C_s	$\bar{4}$	S_4	$\bar{3}$	C_{3i}	$\bar{6} = \frac{3}{m}$	C_{3h}		
		$\frac{2}{m}$	C_{2h}	$\frac{4}{m}$	C_{4h}			$\frac{6}{m}$	C_{6h}	$\frac{2}{m}3$	T_h
		Orthorhombic									
		222	D_2	422	D_4	32	D_3	622	D_6	432	O
		$mm2$	C_{2v}	$4mm$	C_{4v}	$3m$	C_{3v}	$6mm$	C_{6v}		
				$\bar{4}2m$	D_{2d}	$\bar{3}\frac{2}{m}$	D_{3d}	$\bar{6}2m$	D_{3h}	$\bar{4}3m$	T_d
		$\frac{2}{m}\frac{2}{m}\frac{2}{m}$	D_{2h}	$\frac{4}{m}\frac{2}{m}\frac{2}{m}$	D_{4h}			$\frac{6}{m}\frac{2}{m}\frac{2}{m}$	D_{6h}	$\frac{4}{m}\bar{3}\frac{2}{m}$	O_h

Table adapted from [32]

Schoenflies Notation for Point Groups

E: identity
i: inversion about the point
σ: reflection about a plane
σ_h: mirror reflection perpendicular to a rotation axis
σ_v: mirror plane containing a rotation axis
C_n: n fold rotation about an axis
C_{nh}: C_n with a mirror plane perpendicular to the rotation axis
C_{nv}: C_n with n mirror planes through the rotation axis
S_n: C_n rotation followed by a reflection about a plane that is perpendicular to the rotation axis
D_n: C_n rotation axis plus n two-fold rotation axis perpendicular to the C_n axis
T: tetrahedral symmetry
T_d: tetrahedral symmetry that includes improper rotations
T_h: tetrahedral symmetry plus an inversion operation
O: is a point group with octahedral symmetry
O_h: octahedral point group with full cubic symmetry.

The Hermann-Mauguin notation for a crystal provides a means of describing the symmetry of the crystal structure. The first part of the Hermann-Mauguin notation is a symbol for the conventional cell of the translation lattice [33, 34]. The subsequent notation describes the relevant symmetry operations for the crystal including rotations, reflections, glide reflections, screw rotations, etc. [33]. The symmetry notation for the necessary symmetry directions are all represented by these symbols [33]. The reader is again referred to the International Crystallographic Tables for a complete discussion [31, 33]. Some examples include:

First Hermann-Mauguin Symbol for 2D Lattice [33]

- p Primitive
- c Centered
- h Hexagonal Centered.

First Hermann-Mauguin Symbol for 3D lattice [33]

- P: Primitive lattice
- F: All Face Centered
- R: Rhombohedral Centered
- H: Hexagonal Centered.

Other Hermann-Mauguin Symbols [33]

- m: Reflection Plane/Mirror Plane
- 2, 3, 4, 6: n-fold counter clockwise rotation axis, the rotation axis is chosen to be the z axis.
- $\bar{1}$: inversion center
- $\bar{2}, \bar{3}, \bar{4}, \bar{6}$: n-fold rotation plus inversion

- $2_1, 3_1, 3_2, 4_1, 4_2, 4_3, \ldots$ n_p fold screw axis with pitch p where the screw vector (p/n) is the shortest translation vector is parallel to the screw rotation axis.

The crystal structure can be related to the Schoenflies point group and the Hermann-Mauguin space group notations which is shown in Table 2.3. We also note the terms symmorphic and non-symmorphic are used to describe the symmetry of some crystal systems in Chap. 9. If the symmetry operations other than lattice translations leave one point fixed, the space group is symmorphic [30, 31]. Symmorphic space groups do not have a Hermann-Mauguin symbols that contain glide or screw operations [31]. Often a superscript is added to the Schoenflies point group symbol to designate the space groups in the same crystal class, e.g., the hexagonal crystal point group D_{6h} can be designated D_{6h}^4 to associate it with a specific space group. The International Crystallographic Tables provide a space group number that is associated with the Hermann-Mauguin symbol [31–35]. An example is provided below. In addition, the point group symbols can be associated with Hermann-Mauguin symbols for the point group as shown in Table 12.1.4.1 of [31]. For example the D_{6h}point group for a hexagonal crystal is associated with the Hermann-Mauguin symbol $\frac{6}{m}\,\frac{2}{m}\,\frac{2}{m}$ where the $\frac{6}{m}$ operation is along the [001] direction and the first $\frac{2}{m}$ operation is along the [100] direction and the second $\frac{2}{m}$ is along the [1 $\bar{1}$ 0] direction [31]. Using the information above, one can determine, for example, that $\frac{2}{m}$ refers to a 180° rotation with a mirror plane. The D_{3h}point group has $\bar{3}\frac{2}{m}$ for the Hermann-Mauguin point group with $\bar{3}$ being along the [001] direction and $\frac{2}{m}$ being along the [100] direction [31]. The Hermann-Mauguin point group symbols are used to determine the full Hermann-Mauguin symbol for the space group of the structure.

An example structure is the 2H phase of MoS_2, $MoSe_2$, WS_2, and WSe_2 shown in Chap. 8 Fig. 2.4. These structures are composed of van der Waals bonded trilayers with the transition metal atom layer in between the chalcogenide atom layers. The symmetry depends on the number of trilayers. Here we add a super script to the Schoenflies symbol since that is the common practice in the literature. The bulk 2H structure has a hexagonal crystal symmetry that s more completely described by the Schoenflies point group D_{6h}^4 symmetry and the associated $P6_3/mmc$ space group (#194) [35, 36]. The Hermann-Mauguin notation begins with a P for hexagonal crystals. A single trilayer of these materials and samples with an odd number of trilayers have the D_{3h}^1point group symmetry and the $P\bar{6}m2$ space group (#187) symmetry [37]. A sample with a small, even number of trilayers has the point group and space group symmetries: D_{3h}^3 and $P\bar{3}m1$ (#164) for a trigonal structure [38]. In Chap. 8, the importance of the symmetry of the point group at the Γ point in the Brillouin zone will become more apparent as the three band tight binding model for the band structure is derived.

There are 12 symmetry operations for the $P\bar{6}m2$ (#187) space group that include: 120° rotation; −120° rotation; 60° rotation plus inversion; −60° rotation plus inversion; and 180° rotation plus inversion all about the [001] axis that is normal to the plane of the trilayers, along with a 180° rotation plus inversion about the [001] axis [37]. Other rotation operations include: 180° rotation about the [$\bar{1}$10], [210], and

Table 2.4 Space group and point group symmetry of Bernal stacked graphene, few layer graphene, and graphite for high symmetry locations in the Brillouin zone

	Space Group	Point Group at Γ	Point Group at K, K'	Point Group at M
Graphene (ML)	$P6/mm$	D_{6h}	D_{3h}	D_{2h}
Even # layers	$P\bar{3}m1$	D_{3d}	D_3	C_{2h}
Odd # layers	$P\bar{6}m2$	D_{3h}	C_{3h}	C_{2v}
Bulk	$P6_3/mmc$	D_{6h}	D_{3h}	D_{2h}

Table adapted from [39]

[120] axes along with 180° rotation plus inversion about the [100], [110], and [010] axes [37]. Mirror planes are also present in the trilayer structure as can be seen in Chap. 8, Fig. 2.4.

There are 12 symmetry operations for the $P\bar{3}m1$ (#164) space group that include 120° rotation; −120° rotation; 120° rotation plus inversion; −120° rotation plus inversion, about the [001] axis that is normal to the plane of the trilayers [38]. There are three 180° rotations and three −180° rotations plus inversion about the [100], [010], and [110] axes [38]. The identity operation and inverse identity operation are also present.

The trilayer stacking configuration for the bulk 2H TMD structure results in a more complex symmetry. There are 24 symmetry operations for $P6_3/mmc$ that include 120° rotation; −120° rotation; 120° rotation plus inversion; −120° rotation plus inversion about the central point in the lattice. 60° rotation plus inversion; −60° rotation plus inversion; and 180° rotation plus inversion all about the [001] axis that is normal to the plane of the trilayers [36]. Other non-unitary symmetry operations involve a shift in axis or a glide operation and the symmetry of $P6_3/mmc$ is described in detail in the International Crystallographic Tables [36].

The space group and point group symmetries of graphene, few layer graphene, and graphite for Bernal stacked graphene are presented in Table 2.4. We note that the space group of graphene has been identified as $P6/mm$ and as $P6/mmm$. Here, we use the designation $P6/mm$ which is provided by [39].

A thorough description of the relationship between crystal structure, material, space groups, and point groups can be found in [40].

References

1. M. Cardona, L.F. Lastras-Martınez, D.E. Aspnes, Comment on "*Ab Initio* calculation of excitonic effects in the optical spectra of semiconductors". Phys. Rev. Lett. **83**, 3970 (1999)
2. P. Lautenschlager, P.B. Allen, M. Cardona, Phys. Rev. B **33**, 5501 (1986)
3. M. Cardona, T.A. Meyer, M.L.W. Thewalt, Temperature dependence of the energy gap of semiconductors in the low-temperature limit. Phys. Rev. Lett. **92**, 196403 (2004)
4. M. Cardona, Electron–phonon interaction in tetrahedral semiconductors. Solid State Commun. **133**, 3 (2005)

5. P. Lautenschlager, M. Garriga, L. Vina, M. Cardona, Temperature dependence of the dielectric function and interband critical points in silicon. Phys. Rev. B **36**, 4821 (1987)
6. J. Singh, *Electronic Structure and Optoelectronic Properties of Semiconductor Structures*, chap. 2. (Cambridge, New York, 2003)
7. P. Yu, M Cardona, *Fundamentals of Semiconductors: Physics and Materials Properties*, 4th edn., chap. 2 (Springer, New York, 2010)
8. J.C. Slater, C.F. Koster, Simplified LCAO method for the periodic potential problem. Phys. Rev. **94**, 1948 (1954)
9. N.W. Ashcroft, N.D. Mermin, *Solid State Physics*, chaps. 8 and 12 (Thompson Learning, U.S., 1976)
10. G.P. Betteridge, Tight binding central interaction model for the band structure of silicon. Aust. J. Phys. **37**, 407–428 (1984)
11. L.H. Thomas, The motion of the spinning electron. Nature **117**, 514 (1926)
12. L.H. Thomas, The kinematics of an electron with an axis. Phil. Mag. **3**, 1–22 (1927)
13. G. Spavieri, M. Mansuripur, Origin of the spin–orbit interaction. Phys. Scrip. **90**, 085501 (2015)
14. L.I. Shiff, *Quantum Mechanics*, 3rd edn, (McGraw-Hill, New York, 1968), p. 437
15. W.J. Moore, *Physical Chemistry*, 4th edn. (Prentice-Hall, Englewood Cliffs, 1972), pp. 639–644
16. P. Lautenschlager, M. Garriga, L. Viña, M. Cardona, Temperature dependence of the dielectric function and interband critical points in silicon. Phys. Rev. B **36**, 4821 (1987)
17. M. Cardona, N.E. Christensen, G. Fasol, Relativistic band structure and spin-orbit splitting of zinc-blende-type semiconductors. Phys. Rev. B **38**, 1806–1827 (1988)
18. Z.Y. Zhu, Y.C. Cheng, U. Schwingenschlogl, Giant spin-orbit-induced spin splitting in two-dimensional transition-metal dichalcogenide semiconductors. Phys. Rev. B **84**, 153402 (2011)
19. Z.-H. Zhu, G. Levy, B. Ludbrook, C.N. Veenstra, J.A. Rosen, R. Comin, D. Wong, Rashba spin-splitting control at the surface of the topological insulator Bi_2Se_3. Phys. Rev. Lett. **107**, 186405 (2011)
20. H. Weng, C. Fang, Z. Fang, B. Andrei Bernevig, X. Dai, Weyl semimetal phase in noncentrosymmetric transition-metal monophosphides. Phys. Rev. X **5**, 011029 (2015)
21. M. Cardona, F.H. Pollak, Energy-band structure of germanium and silicon: the k·P method. Phys. Rev. **142**, 530–543 (1966)
22. F.H. Pollak, M. Cardona, Energy band structure of germanium and gallium arsenide: The ***k.p*** method. J. Phys. Chem. Sol. **27**, 423–425 (1966)
23. G. Dresselhaus, A.F. Kip, C. Kittel, Cyclotron resonance of electrons and holes in silicon and germanium crystals. Phys. Rev. **98**, 368–384 (1955)
24. L.C. Lew Yan Voon, M. Willatzen, *The k * p Method: Electronic Properties of Semiconductors* (Springer, 2009, New York)
25. Ibid, pp. 19–22
26. P.A.M. Dirac, Bakerian lecture—the physical interpretation of quantum mechanics. Proc. Roy. Soc. **A180**, 1 (1942)
27. W. Pauli, On Dirac's new method of field quantization. Rev. Mod. Phys. **15**, 175–207 (1943)
28. E. Koch, Many electron states, chap 2, in *Emergent Phenomena in Correlated Matter*, ed. by E. Pavarini, E. Koch, U. Schollwöck (Deutche National Bibliotek, 2013)
29. M. Grundmann, Crystals, chap. 3, in *The Physics of Semiconductors: An Introduction Including Devices and Nanophysics* (Springer, Berlin, 2006), pp. 29–61
30. N.W. Ashcroft, N.D. Mermin, The crystallographic point groups and space groups, in *Solid State Physics* (Thompson Learning, U.S., 1976), pp. 119–128
31. H. Burzlaff, H. Zimmermann, Point group symbols. Int. Tables Crystal. **A**, Ch. 12.1, 818–820 (2006)
32. Th. Hahn, H. Klapper, Crystallographic and noncrystallographic point groups. Int. Tables Crystal. **A**, Ch. 10.1, 762–803 (2006)
33. B. Souvignier, H. Wondratschek, M.I. Aroyo, G. Chapuis, A.M. Glazer, Space groups and their descriptions. Int. Tables Crystal. **A**, Ch. 1.4, 42–73 (2015)
34. T. Hahn, M.I. Aroyo, Symbols of symmetry elements. Int. Tables Crystal. **A**, CH 2.1.2, 144–149 (2015)

35. G.-B. Liu, D. Xiao, X. Xu, W. Yao, Electronic structures and theoretical modelling of two-dimensional group-VIB transition metal dichalcogenides. Chem Soc. Rev. **44**, 2643 (2015)
36. International Tables for Crystallography (2016). Vol. A, Space group 194, pp. 582–583
37. International Tables for Crystallography (2016). Vol. A, Space group 187, pp. 568–569
38. International Tables for Crystallography (2016). Vol. A, Space group 164, pp. 534
39. A. Jorio, M. Dresselhaus, R. Saito, G.F. Dresselhaus, *Raman Spectroscopy in Graphene Related Systems* (Wiley-VCH, Weinheim, 2011), p. 133
40. U. Műller, *Symmetry Relationships between Crystal Structures: Applications of Crystallographic Group Theory in Crystal Chemistry* (Oxford, Oxford, 2013)

Chapter 3
Instrumentation

In this chapter we provide a brief overview of the instrumentation used for three key optical characterization methods: spectroscopic ellipsometry, Raman spectroscopy, and photoluminescence. We also discuss the relationship between the optical properties determined by each method and the data obtained by the measurement equipment. For example, as discussed in Chap. 1, spectroscopic ellipsometry measures ψ and Δ which are related to the complex refractive index of the measured material or materials through Fresnel reflection coefficients. The energy resolution of the ellipsometric equipment, the uncertainty in ψ and Δ place limits on the complex refractive index determined using a specific ellipsometric system. Similarly, the specific properties of the optical components used in the Raman or photoluminescence equipment place limits on the optical properties such as phonon energy or exciton energy determined by that system.

3.1 Spectroscopic Ellipsometry

The history of spectroscopic ellipsometry goes back to the late 1800s and the work of Drude [1–3]. As Aspnes points out, the term ellipsometry dates back to 1945 [1]. Spectroscopic ellipsometry is known to have excellent sensitivity to very thin films on substrates, and this sensitivity also holds for nanoscale materials. We note that there are several key textbooks [4–6] which provide thorough discussion of ellipsometry and its use, and our purpose here is to provide the reader with enough information for understanding the other chapters in this book. The information present in the scattered light is dependent on the optical elements in the light path. A complete description of the scattered light is provided by Mueller Matrix spectroscopic ellipsometry (MMSE). MMSE is critical for measurement of optically anisotropic materials. We discuss four optical configurations out of the many possible ones as described in [4, 7, 8]. These are: the rotating analyzer ellipsometer; rotating analyzer with compensator; rotating compensator; and dual rotating compensator ellipsometer as

A. Diebold and T. Hofmann, *Optical and Electrical Properties of Nanoscale Materials*, Springer Series in Materials Science 318,
https://doi.org/10.1007/978-3-030-80323-0_3

shown in Fig. 3.1. The addition of just 1 optical element can increase the amount of information as well as decrease the error in the measurement. Other approaches to spectroscopic ellipsometry instruments include piezo-elastic modulators (PEM),

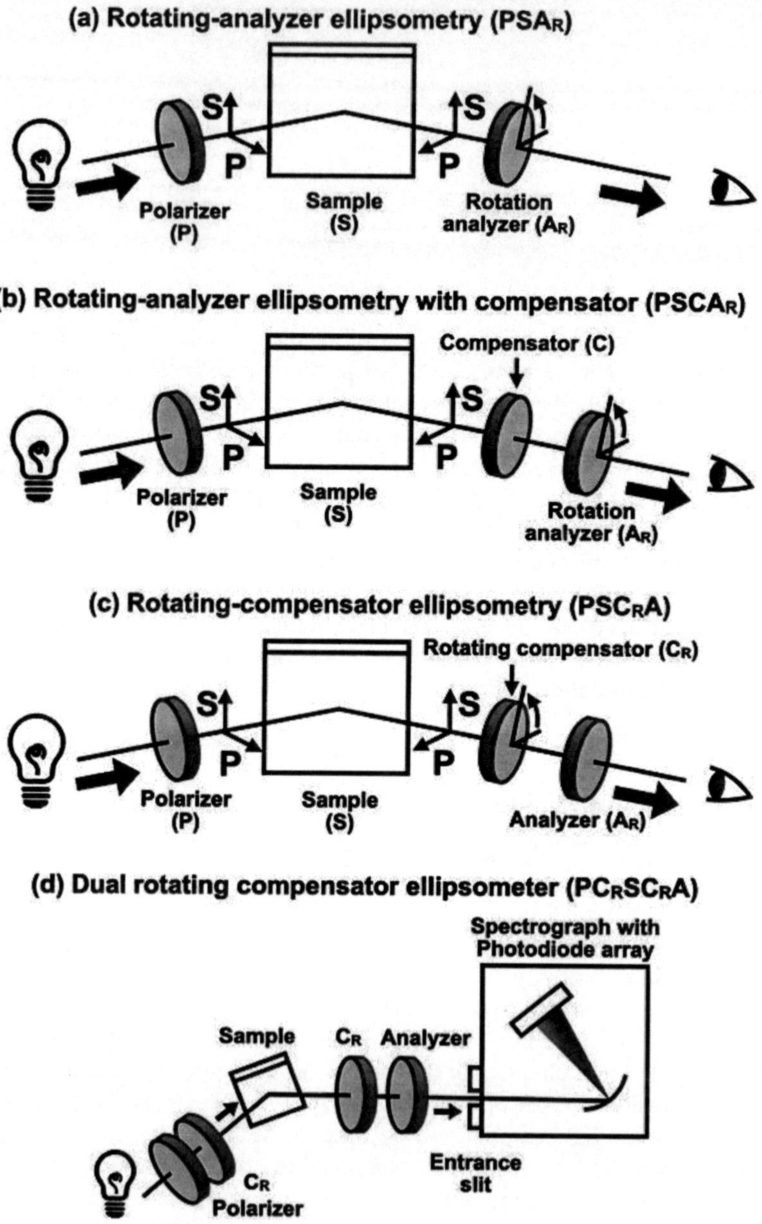

Fig. 3.1 Four example optical paths for spectroscopic ellipsometers. Figure adapted from [4]

which are described in [4]. Jellison describes the advantages of PEM ellipsometers in [9, 10] which include collection of the entire Mueller Matrix spectrum.

Some of the key optical components in the light path are:

- Polarizer—converts unpolarized light to linearly polarized light
- Analyzer—an analyzer is a polarizer that is placed after the sample in the optical path. Only light with the desired polarization is passed to the detector.
- Compensator—optical retarder which ideally introduces a phase shift of 90°, can, depending on its orientation, convert linearly polarized light into any elliptical polarization.
- Variable angle of incidence.

Rotating Analyzer Ellipsometer (PSA_R): The light path for a rotating analyzer ellipsometer consists of a light source, a polarizer (*P*), the sample (*S*), a polarization analyzer continuously rotating at a constant angular frequency (A_R), and a detector as shown in Fig. 3.1a. This optical path is simple and the optical elements are achromatic. The disadvantage of the PSA_R is that the error in the value of Δ increases when Δ Is close to 0° or 180° and the third element of the Stokes vector of the scatter light cannot be measured when $-180° \leq \Delta \leq 0°$.[4] A monochromator (spectrometer) can be placed before (after) the light source as shown in [7, 8].

Rotating Analyzer plus Compensator Ellipsometer ($PSCA_R$): The light path for a rotating analyzer ellipsometer consists of a light source, a fixed polarizer (*P*), the sample (*S*), a fixed compensator (**C**), a polarization analyzer rotating at a constant angular frequency (A_R), and a detector as shown in Fig. 3.1b. The polarization of the incoming light is changed by moving the polarizer to provide different linear polarization settings. The advantages of the RA&CE are that all values of ψ and Δ are accessible, the amount of depolarization can be determined as a function of wavelength, and the sensitivity to changes in ψ and Δ is uniform at a specific wavelength [4]. The additional optical element compared to the RAE results in a longer measurement time and the chromatic (wavelength dependent) nature optical path changes the measurement error with wavelength [4]. A monochromator (spectrometer) can be placed before (after) the light source, and Woollam systems use a monochromator after the light source [7].

Rotating Compensator plus Analyzer Ellipsometer (PSC_RA): The light path for a rotating compensator ellipsometer consists of a light source, a fixed polarizer (*P*), the sample (*S*), a compensator rotating at a constant angular frequency (C_R), a fixed polarization analyzer (*P*), and a detector as shown in Fig. 3.1c. The polarization of the incoming light is changed by moving the polarizer to provide different linear polarization settings. The advantages and disadvantages of the RC&AE are the same as for RA&CE: all values of ψ and Δ are accessible, the amount of depolarization can be determined as a function of wavelength, and the sensitivity to changes in ψ and Δ is uniform at a specific wavelength [4]. When compared to the RAE, the additional optical element results in a longer measurement time and the chromatic (wavelength dependent) nature optical path changes the measurement error with wavelength [4]. A monochromator (spectrometer) can be placed before (after) the light source [7].

Dual Rotating Compensator Ellipsometer (PC_RSC_RA): The light path for a rotating compensator ellipsometer consists of a light source, a fixed polarizer (P), a compensator rotating at a constant angular frequency (C_R), the sample (S), a compensator rotating at a constant angular frequency (C_R), a fixed polarization analyzer and (P), a spectrometer plus detector as shown in Fig. 3.1d. A distinguishing feature of this optical path is that the detector uses a diffraction grating and detects a number of wavelengths simultaneously. This system is often referred to as the RC2 ellipsometer. This optical path provides a complete characterization of all the scattered light providing all sixteen Mueller Matrix elements vs wavelength. The PC_RSC_RA has several advantages over the PSA_R, $PSCA_R$, and PSC_RA configurations including greater accuracy and lower measurement times.

Commercially available implementations all have goniometers that allow collection of data for variable angles of incidence [7, 8]. Woollam variable angle PSA_R ellipsometers (VASE™) use an optical element known as an AutoRetarder™ after the input polarizer and before the sample to improve the accuracy of the rotating analyzer ellipsometer, which is most accurate when the beam entering the detector is nearly circularly polarized [11]. Fitting data obtained from several angles of incidence, typically three, improves the optical model [7–10]. The Woollam Co. RC2™ Dual Rotating Compensator Ellipsometer (PC_RSC_RA) uses a diffractive element before the CCD detector to obtain spectroscopic data. The spectral resolution depends on the optical path. For systems that use a diffraction grating plus CCD (Charge Coupled Detector) such as the PC_RSC_RA **shown in** Fig. 3.1d, the spectral resolution is fixed compared to systems that use a monochromator, for which the bandwidth can be adjusted by changing the entrance and exit slid width at the cost of intensity throughput. The bandwidth on the UV–Vis CCD for the RC2™, for instance, is about 2.7 nm [12].

Reporting Optical Properties in energy versus wavelength. Often, it is useful to determine optical properties over a range of energies with a constant energy spacing. For example, the energy spacing between 200 and 201 nm is ~ 0.031 eV while the energy spacing between 1000 and 1001 nm is ~ 0.001 eV. Spectroscopic ellipsometers that obtain data with constant wavelength spacing require that the data be digitally resampled to obtain a constant energy spacing. The RC2™ gives constant wavelength spacing as it uses a broad band light source and the CCD array is placed inside the detector unit.

Uncertainty in ellipsometric data: Since the complex refractive index (complex dielectric function) is calculated from either ψ and Δ or the Mueller Matrix elements, it is useful to understand the error in these quantities. The error in the Mueller Matrix elements which vary from 1 to $-$ 1 is reported to be $\pm$ 0.002 [7].

One of the most typical applications of ellipsometry is the determination of the thickness of a layer of material or layers of materials on a substrate. Another use of ellipsometry is the determination of the complex refractive index (complex dielectric function) of a material. A few of the many optical models were discussed in Chap. 1, e.g., the Drude and Lorentz models. Some commercial software allows determination of the optical properties using sophisticated algorithms. Kramers–Kronig consistency is an important property of the optical functions used to model a material. When the optical properties of a material are not known, thin film thickness can be determined after the adjustable parameters in the selected optical model are optimized. It is important to check consistency between theoretically calculated and experimentally determined dielectric functions. If there is a large difference between theory and experiment, then the origin of the difference should be determined. For example, in Chap. 4, we will discuss the excitonic effects that shift the energy of key optical transitions for materials such as silicon and graphene. Often the energy (wavelength) resolution of theoretical dielectric function is much higher than experimental data.

3.2 Raman Spectroscopy

In the context of 2D nanoscale materials, Raman spectroscopy provides a means of characterizing the change in vibrational modes with the number of layers of materials. For example, the number of layers of graphene can be characterized using the change in shape of a specific Raman peak (see Chap. 7). The number of trilayers of transition metal change the energy of specific vibrational models (see Chap. 8). In Raman spectroscopy, some of the incident light that scatters from the sample surface gains or losses energy from a phonon producing spectra of the allowed vibrational scattering. As discussed in Chap. 1, phonon modes of solids are either Raman active or Infra-Red active depending on the polarization that occurs during vibration. Both acoustic and optical phonons can be Raman active. Since the acoustic phonons are very low energy and thus the energy separation from the Rayleigh (elastic scattering) peak, observation of acoustic phonons places special requirements on the design of the optical path. The excitation source is typically a laser with a wavelength that is selected to maximize spectra from the desired analysis depth. Since this book emphasizes the characterization of nanoscale materials, the optical path shown in Fig. 3.2 illustrates the use of a microscopic objective. As will be discussed in Chap. 8, the incident light is often circularly polarized to allow selection of a property known as pseudospin which is discussed in Chaps. 7 and 8. It is important to report the wavelength and intensity of the incident light. The sample can be inadvertently heated when too high an intensity is used.

The intensity of the Raman scattered light is approximately 0.1% or less of the incident light. One of the key challenges for Raman instrumentation is removal or reduction of the high intensity reflected beam from the spectra. This is particularly challenging for acoustic phonons due to their low energy and often lower intensity than elastically scattered light of the incident beam. One approach is to use narrow band notch filters in combination with double or triple monochromators. This approach greatly reduces the Raman signal, and other approaches enable measurement of spectra from phonons with low intensity. One can overcome this by using Fourier Transform Raman detection or using holographic gratings.

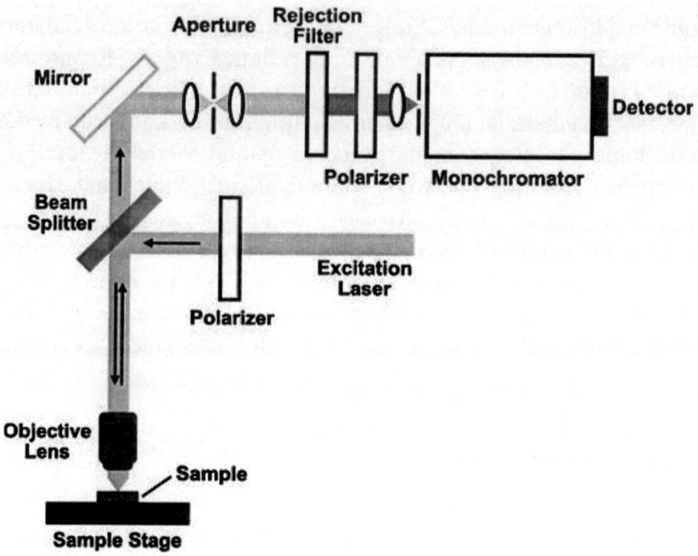

Fig. 3.2 Raman microscope showing the key optical elements

3.3 Photoluminescence Spectroscopy

Photoluminescence (PL) spectroscopy is a nondestructive optical metrology technique that provides information about the detailed electronic structure of nanoscale materials. PL is widely used to characterize thin layers and quantum wells group IV, III–V, and II–VI materials, transition metal dichalcogenides, and topological materials. The incident light has a higher energy than the bandgap of the material resulting in absorption and the production of electron—hole pairs or excitons (see Chap. 5). The electron–hole pairs recombine either radiatively or non-radiatively. The PL system is designed to measure the intensity of the resulting emission of photons after radiative recombination. At high enough concentrations, the excitons interact and emission can arise from the resulting quasiparticle. PL signals are often weak, and they can be increased by cooling the sample to low temperature and using signal amplification methods. Here we describe a photoluminescence system that employs optical beam chopping and a lock-in amplifier to increase signal to noise.

Here, a system designed by Kamineni is used to illustrate a typical PL system [13]. The incident light typically comes from a laser, and in Fig. 3.4 this is illustrated using a 355 nm optically pumped semiconductor laser (OPSL). For nanoscale materials, determining the penetration depth of the light is critical. For *c*-Si quantum well samples (see Chap. 5), one considers the penetration depth of 355 nm UV light in silicon which is ~ 10 nm. The maximum power output of the 355 nm continuous wave (CW) laser shown in Fig. 3.4 is 100 mW and the beam is modulated at 185 Hz using an optical chopper with a 30 slot chopper wheel. The spot size of the laser on the sample needs to be isolated to the measurement are of interest. This is often a challenge for

2D materials that are exfoliated from a bulk sample such a graphene from graphite. The experimental setup shown in Fig. 3.4 has a laser spot size of ~ 1.1 mm when the laser is focused and ~ 50 μm when the laser is not focused. A typical PL optical path for the laser excitation consists of the laser, chopper, mirror, focusing lenses, and a cryostat. The cryostat has windows at appropriate angle that allow reflection and have a window that allows the PL light to exit normal to the sample as shown in Fig. 3.3. Figure 3.3 shows custom sealed quartz windows at a 40° angle to the surface normal for the incident and reflected light. Another sealed quartz window is located along the light path normal to the sample surface. PL data is often obtained at temperatures of a few Kelvin, thus samples are typically cooled down using continuous flow cryostats which do not suffer from mechanical vibrations common in closed loop cryostats. PL spectra can be obtained at any temperature between 4 and 300 K. Often, circularly polarized light is used to excite PL spectra since it can couple directly to pseudospin (see Chap. 8). A circular polarizer can be added to the optical path before incident light interacts with the sample or enters the cryostat window.

The reflected light exits the window after the sample. The PL signal is focused using wide area collection lenses into the monochromator. The system shown in Fig. 3.4 uses a pair of uncoated UV fused silica plano-convex lenses (185–2100 nm) with a diameter of two inches. In addition, a long pass filter with a cut-on wavelength of 600 nm was placed in front of the spectrometer entrance slit to attenuate shorter wavelengths and transmit longer wavelengths. In order to obtain the PL signal as a function of energy, a high-resolution spectrometer is used as shown in Fig. 3.4. One example is a high resolution 750 mm grating equipped with a 1200 groves/mm (500 nm blaze) and 600 groves/mm (1 μm blaze) gratings to cover the visible and near IR wavelengths, respectively. The energy resolution and range of wavelength sensitivity of the detector is critical. One example is a liquid nitrogen-cooled photomultiplier tube (PMT) with an InP/InGaAs photocathode. The near IR PMT, that is described here, has a spectral response that covers a wide range from 300 to 1700 nm

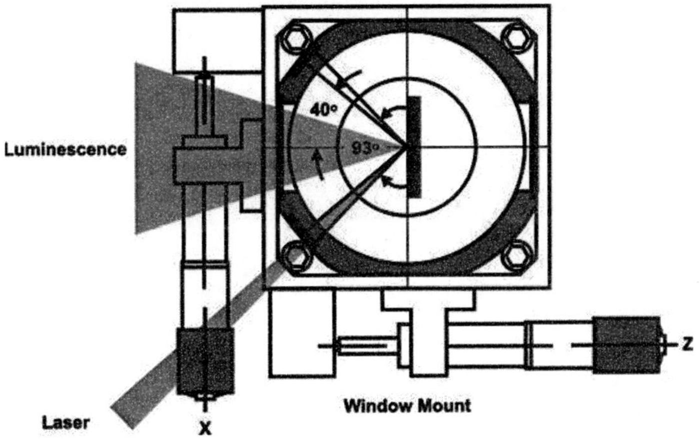

Fig. 3.3 Customized window block layout of the cryostat configured for PL measurements

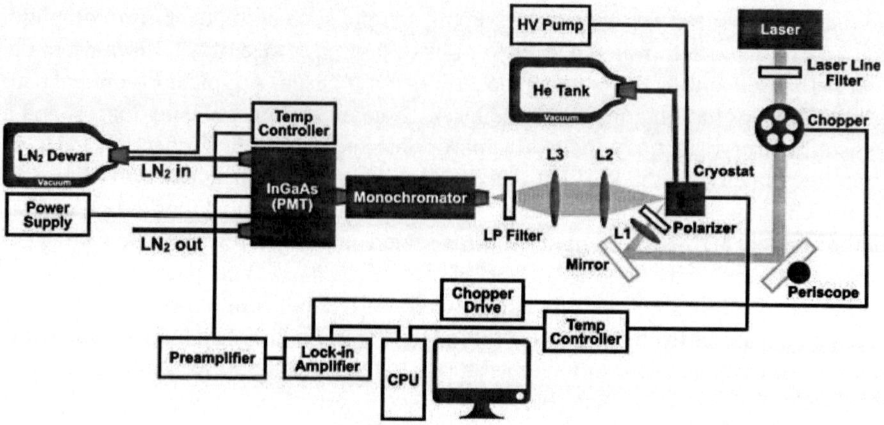

Fig. 3.4 Photoluminescence spectrometer system showing examples of the key components including the laser light source, chopper plus lock-in amplifier, spectrometer, and photodector. Figure adapted from [13]

and a high gain due to 10 stages of Cu-BeO dynodes. Other PMT's that are sensitive to a small range of wavelengths, can have significantly improved signal to noise which results in greater sensitivity to weak PL signals. The current output of the PMT is connected to a preamplifier with a current-to-voltage conversion factor of 4 mV/μA. The output voltage signal of the preamplifier is connected to the input of the lock-in amplifier as shown in Fig. 3.4. The lock-in amplifier's reference frequency synchronizes with the chopper's frequency and is capable of detecting low light levels against a bright background. Thus, the lock-in amplifier behaves likes an extremely narrow band pass filter that separates the luminescence detected by the PMT from its background noise [8]. A computer system controls data acquisition and collection.

The intensity (power per unit area) of the laser both influences the excitation process and potential sample damage. The intensity of the measurement spot on the sample should always be reported, and it is good practice to report the PL spectra as a function of energy density. As will be discussed in Chap. 5, the resulting spectra are strongly influenced by the density of excited electron–hole pairs.

References

1. D.A. Aspnes, Spectroscopic ellipsometry—past, present, and future. Thin Sol. Films **571**, 334–344 (2014)
2. E.A. Irene, A brief history and state of the art of ellipsometry, in *Ellipsometry at the Nanoscale*, ed. by M. Losurdo, K. Hingerl (Springer, Berlin, 2013), pp. 1–30
3. R.M.A. Azzam, The intertwined history of polarimetry and ellipsometry. Thin Sol. Films **519**, 2584–2588 (2011)
4. H. Fujiwara, *Spectroscopic Ellipsometry: Principles and Applications* (Wiley, West Sussex, 2007)

5. R.M.A. Azzam, N.M. Bashara, *Ellipsometry and Polarized Light* (North Holland, New York, 1989)
6. H.G. Tompkis, J.N. Hilfiker, *Spectroscopic Ellipsometry, Practical Application to Thin Film Characterization* (Momentum, New York, 2016)
7. B. Johs, J.A. Woollam, C.M. Herzinger, J.N. Hilfiker, R.A. Synowicki, C.L. Bungay, Overview of variable-angle spectroscopic ellipsometry (VASE): II, in *Advanced applications, Proc. SPIE 10294, Optical Metrology: A Critical Review* (1999), pp. 1029404
8. S.E. Green, C.M. Herzinger, B.D. Johs, J.A. Woollam, System and method for improving data acquisition capability in spectroscopic ellipsometers, US Patent 5757494 (1998)
9. G.E. Jellison, F.A. Modine, Two-modulator generalized ellipsometry: experiment and calibration. Appl. Opt. **36**, 8184–8189 (1997)
10. G.E. Jellison, F.A. Modine, Two modulator generalized ellipsometer for complete Mueller matrix measurement, US patent 5956147 (1999)
11. J.A. Woollam Co., Inc., Guide to Using WVase32, 2010
12. S. Schöche, Private communication (2021)
13. V. Kanineni, Electron-Phonon Interactions and Quantum Confinement Effects on Optical Transitions in Nanoscale Silicon Films, Dissertation submitted to the University at Albany (2011)

Chapter 4
Microscopic Theory of the Dielectric Function

Abstract In this chapter, we derive the dielectric function for direct gap transitions in the electronic band structure using Fermi's Golden Rule. We discuss the relationship between the electronic band structure and locations in the Brillouin zone when there is a high density of transitions with the same energy resulting in a critical point. We show the analysis of the dielectric function at critical points to determine the transition energy and peak broadening. A number of examples of the dielectric functions of semiconductors over the visible wavelength range are presented. We also discuss the effect of doping on the dielectric function.

An understanding of the expected mathematical form of the dielectric function of a material, that is the functional form of the dielectric function versus wavelength, is an essential part of optical characterization. This is true whether one has directly inverted experimental data to arrive at the dielectric function or fit spectroscopic ellipsometric or reflectivity data to a functional form. Furthermore, knowledge of the expected mathematical form is essential for verifying inverted data or fits to parameterized dielectric functions. In this chapter, we discuss the relationship between the electronic band structure and the dielectric function. First, a derivation of the dielectric function (complex refractive index) of crystalline solids is presented to illustrate the relationship between the electronic band structure and the dielectric function. The chapter ends with a discussion of the dielectric function of several key semiconductors in terms of the electronic band structure. The concept of the critical points of a dielectric function is also presented.

Here, the dielectric function for direct gap transitions where the wave vector ($\vec{k}$) is conserved is based on Fermi's Golden Rule. Fermi's Golden Rule provides a concise quantum mechanical formula for determining the transition rate between eigenstates for a weak perturbation such as the electric field of light [1]. Here, the transition rate is the probability of a transition between eigenstates per unit time. Since this book focuses on the optical and electrical properties of crystalline solids, the transitions are between an eigenstate in one of the valence bands to an eigenstate with the same wave vector in one of the conduction bands. The previous discussion of the electronic band structure where the band energy versus wave vector was shown for

A. Diebold and T. Hofmann, *Optical and Electrical Properties of Nanoscale Materials*, Springer Series in Materials Science 318,
https://doi.org/10.1007/978-3-030-80323-0_4

high symmetry parts of the Brillouin zone will be a key aspect of this chapter. The Block wave functions are essential to this derivation.

In order to motivate the derivation, we first describe the key formulas that are used in this chapter in terms of initial and final states. Fermi's Golden Rule can be applied to any system with two states. In the subsequent sections of this chapter, the initial and final states are specified as the valence and conduction bands associated with the transition. Here we use the following relationship between the imaginary part of the dielectric function ε_2 and the transition rate $\mathcal{W}_{ij}$ for a transition from an initial state i to a final state f due to the electric field of light $E(\omega)$ which is derived below [2]:

$$\varepsilon_{2(i\to f)} = 8\pi\hbar\frac{\mathcal{W}_{ij}}{|E(\omega)|^2} = \frac{\boldsymbol{\varepsilon}_1}{\boldsymbol{\omega}}\mathcal{W}_{if} \tag{4.1}$$

Fermi's Golden Rule provides a convenient formula for the transition rate $\mathcal{W}_{ij}$. For optical transitions due to a perturbing light field at angular frequency $\omega = \frac{E}{\hbar}$, the rate for a single transition is related to $H'_{i\to f}$, which is the matrix element of the time dependent perturbation due to the light field between the initial and final states due to the light at angular frequency ω. The total transition rate comes from the number of initial/final state pairs per unit area having that energy difference. The number of these states per unit volume is referred to as the density of states. Fermi's Golden Rule can be stated as [1]:

$$\mathcal{W}_{if} = \left(\frac{2\pi}{\hbar}\right)\left|H'_{i\to f}\right|^2\rho(\omega) \tag{4.2}$$

A semiclassical approach is used for the derivation provided here. Below, the derivation of the formula shown in (4.1) for the dielectric function is presented. Below, we use a state specific transition rate version of Fermi's Golden Rule and integrate over the available states to obtain the destiny of states contribution. As we show in this chapter, a number of valence to conduction band transitions at the same energy must be present for optical absorption to be observable. This requirement is described in the formula for the Joint Density of States.

The flow of this chapter is as follows: The relationship between the dielectric function and optical absorption is presented as a prelude to a semiclassical derivation of the dielectric function using Fermi's Golden Rule. This is followed by a discussion of the dielectric function for parabolic energy bands and then a discussion about the joint density of states, critical points, and van Hove singularities. Then we describe how the critical points are named followed by a section on determination of the energy of the critical points from dielectric function data. The chapter ends with a description of experimental dielectric function data for bulk semiconductors and then a description about the effect of doping on the dielectric function.

4.1 Relationship Between Dielectric Function and Optical Absorption

The amount of absorption of light per unit time from an incoming light beam is the intensity loss per unit time $(-dI/dt)$ can be related to the speed of light in a material c/n where n is the real part of the refractive index and the absorption per unit depth [2]. From Beer's law, $I = I_0 e^{-\alpha x}$ where α is the absorption coefficient, c is the speed of light, I_0 is the initial light intensity, and $dI/dx = -\alpha I_0 e^{-\alpha x} = -\alpha I$ so that [2]:

$$-\frac{dI}{dt} = -\frac{dI}{dx}\frac{dx}{dt} = \alpha I \frac{c}{n} \tag{4.3}$$

Remembering that $\alpha = 4\pi k/\lambda$ where k is the imaginary part of the refractive index and λ is the wavelength, this gives in terms of the angular frequency $\omega = 2\pi c/\lambda$ and imaginary part of the dielectric function $\varepsilon_2 = 2nk$:

$$-\frac{dI}{dt} = \alpha I \frac{c}{n} = -\frac{4\pi kc}{\lambda n} I = \frac{2\omega k}{n} I = \frac{\omega \varepsilon_2}{n^2} I \tag{4.4}$$

The intensity (here we use the time averaged energy flux density) of light in a solid which is $\frac{\varepsilon_1 \varepsilon_0}{2}|E(\omega)|^2$ in CGS units related to the electric field amplitude by [3, 4]:

$$I = \frac{\varepsilon_1 \varepsilon_0}{2}|E(\omega)|^2 \approx \frac{n^2 \varepsilon_0}{2}|E(\omega)|^2 \tag{4.5}$$

For comparison with other texts, the energy density can be converted to Si units by multiplying by $\frac{1}{4\pi\varepsilon_0}$ to give $\frac{n^2}{8\pi}|E(\omega)|^2$ [2]. The energy density is chosen to be the energy of one photon so $\hbar\omega = \frac{\varepsilon_1 \varepsilon_0}{2}|E(\omega)|^2$ where we used $\varepsilon_1 = n^2 - k^2$ and for small k, $\varepsilon_1 \sim n^2$ [4]. Thus,

$$\begin{aligned}
-\frac{dI}{dt} &= \frac{\omega \varepsilon_2}{n^2} I \approx \frac{\omega \varepsilon_2}{n^2}\frac{n^2 \varepsilon_0}{2}|E(\omega)|^2 = \frac{\omega \varepsilon_2 \varepsilon_0}{2}|E(\omega)|^2 \text{ and} \\
-\frac{dI}{dt} &= \hbar\omega \mathcal{W}_{if} = \frac{\varepsilon_1 \varepsilon_0}{2}|E(\omega)|^2 \mathcal{W}_{if} \\
\rightarrow \boldsymbol{\varepsilon_2} &= \frac{\boldsymbol{\varepsilon_1}}{\boldsymbol{\omega}} \mathcal{W}_{if}
\end{aligned} \tag{4.6}$$

Using Fermi's Golden Rule, this formula provides a means of determining the dielectric function from the transition rate.

4.2 Semiclassical Derivation of the Dielectric Function

The interaction of light with a crystalline solid can be determined from (4.6) once we have a formula for the transition rate $\mathcal{W}_{ij}$. In this section, we derive the transition rate using Block wave function for the valence and conduction band wavefunctions using Fermi's Golden Rule which is a time dependent perturbation approach [4, 5]. The use of Fermi's Golden Rule implies that we are using a single electron transition model for the dielectric function. The perturbation matric elements are between single electron states in the crystal. The origin of the term dipole transition will also become apparent.

First, we rewrite the Hamiltonian when Bloch wavefunctions are used.

$$\left[\frac{P^2}{2m} + V(\vec{r})\right]\psi_{Bloch}(\vec{r}) = E\psi_{Bloch}(\vec{r}) \tag{4.7}$$

Here, $P = i\hbar\vec{\nabla}$. Light is described by a time dependent vector potential $\vec{A}(\vec{r}, t)$ and scalar potential $\Phi(\vec{r}, t)$. Gauge invariance is required and the Coulomb gauge $\{\Phi(\vec{r}, t) = 0, \vec{\nabla} \cdot \vec{A}(\vec{r}, t) = 0\}$ simplifies the derivation. In the Coulomb gauge, the electric and magnetic fields are given by:

$$\vec{E}(\vec{r}, t) = -\frac{1}{c}\frac{\partial \vec{A}(\vec{r}, t)}{\partial t} \quad \text{and} \quad \vec{B} = \vec{\nabla} \times \vec{A}(\vec{r}, t) \tag{4.8}$$

Classically, when light is interacting a total charge of Q with the crystal, the momentum P is replaced by $\vec{P} - Q\vec{A}(\vec{r}, t)/c$. For the quantum analog, $\vec{P} - e\vec{A}(\vec{r}, t)/c$ Hamiltonian can be written. The Hamiltonian then becomes:

$$\left[\frac{\left(\vec{P} - e\vec{A}(\vec{r}, t)/c\right)^2}{2m} + V(\vec{r})\right]\psi_{Bloch}(\vec{r}) = E\psi_{Bloch}(\vec{r}) \tag{4.9}$$

$$\begin{aligned} H &= \left[\frac{\left(\vec{P} - e\vec{A}(\vec{r}, t)/c\right)^2}{2m} + V(\vec{r})\right] \\ &= \frac{P^2}{2m} - \frac{\vec{P} \cdot e\vec{A}(\vec{r}, t)/c}{2m} - \frac{(e\vec{A}(\vec{r}, t)/c) \cdot \vec{P}}{2m} + \frac{e\vec{A}(\vec{r}, t) \cdot e\vec{A}(\vec{r}, t)}{2mc^2} + V(\vec{r}) \end{aligned}$$

The second order term is set to zero since we are using low intensity light fields. The $\vec{A}(\vec{r}, t) \cdot \vec{A}(\vec{r}, t)$ is important for light intensity experiments such as second harmonic generation. Since $\vec{P} = i\hbar\vec{\nabla}$, and $\vec{\nabla} \cdot \vec{A}(\vec{r}, t) = 0$, the Hamiltonian becomes:

$$H = \frac{P^2}{2m} - \frac{e\vec{A}(\vec{r}, t) \cdot \vec{P}}{2mc} + V(\vec{r}) \tag{4.10}$$

Thus, the perturbation term

$$H' = -\frac{e\vec{A}(\vec{r},t)\cdot\vec{P}}{2mc} = \frac{-i\hbar e}{2mc}\vec{A}(\vec{r},t)\cdot\vec{\nabla} \tag{4.11}$$

The vector potential $\vec{A}(\vec{r},t)$ can be written using the unit vector $\vec{a}_0$ along the direction of the vector potential and thus the electric field and a scaler magnitude A as follows (4.4)

$$\vec{A}(\vec{r},t) = \frac{A\vec{a}_0}{2}\left[e^{i(\vec{q}\cdot\vec{r}-\omega t)} + e^{-i(\vec{q}\cdot\vec{r}-\omega t)}\right] \tag{4.12}$$

$$H' = \frac{-i\hbar e}{4mc}A\left[e^{i(\vec{q}\cdot\vec{r}-\omega t)} + e^{-i(\vec{q}\cdot\vec{r}-\omega t)}\right]\vec{a}_0\cdot\vec{\nabla} \tag{4.13}$$

The first term in (4.13) is for absorption of light and the second term is for emission. We are considering absorption so here:

$$H' = \frac{-i\hbar e}{4mc}A\left[e^{i(\vec{q}\cdot\vec{r}-\omega t)}\right]\vec{a}_0\cdot\vec{\nabla} = \frac{-i\hbar e}{4mc}A\left[e^{i\vec{q}\cdot\vec{r}}e^{-i\omega t}\right]\vec{a}_0\cdot\vec{\nabla} \tag{4.14}$$

We need to determine $H'_{i\to f}$ so that we can apply Fermi's Golden rule. This matrix element is time dependent due to $e^{-i(\omega t)}$, and the full Hamiltonian is also time dependent. Remembering time-dependent perturbation theory where:

$$\boldsymbol{H\psi(\vec{r},t) = i\hbar\frac{\partial}{\partial t}\psi(\vec{r},t) \to \psi(\vec{r},t)} = \psi_{Bloch}(\vec{r})\,e^{-iEt/\hbar} \tag{4.15}$$

Since we are considering transition between the initial valence band state $\psi_{Bloch\ Valence\ Band}(\vec{r}) = \psi_{BV}(\vec{r})$ and final conduction band $\psi_{Bloch\ Conduction\ Band}(\vec{r}) = \psi_{BC}(\vec{r})$, we can write:

$$H'_{i\to f} = \int_{-\infty}^{\infty}\psi^*_{BCB}(\vec{r})\left[\frac{-i\hbar e}{4mc}A\left[e^{i\vec{q}\cdot\vec{r}}\right]\vec{a}_0\cdot\vec{\nabla}\right]$$

$$\cdot\,\psi_{BVB}(\vec{r})d\vec{r}\int_{-\infty}^{\infty}e^{iE_ct/\hbar}e^{-i\omega t}e^{-iE_vt/\hbar}\,dt \tag{4.16}$$

$$\int_{-\infty}^{\infty}e^{iE_ct/\hbar}\,e^{-i\omega t}\,e^{-iE_vt/\hbar}\,dt \sim \delta\Big(E_c(\vec{k}_c) - E_v(\vec{k}_v) - \hbar\omega\Big) \tag{4.17}$$

$$H'_{i\to f} = \int_{-\infty}^{\infty}\psi^*_{BCB}(\vec{r})\left[\frac{-i\hbar e}{4mc}A\left[e^{i\vec{q}\cdot\vec{r}}\right]\vec{a}_0\cdot\vec{\nabla}\right]\psi_{BVB}(\vec{r})d\vec{r}\ \ \delta(E_c - E_v - \hbar\omega) \tag{4.18}$$

Thus the energy of the photon must equal the energy of the transition between valence and conduction band. The Bloch wavefunctions for a crystal can be written using the volume normalization factor Ω as [4]:

$$\psi_{Bloch\ Valence\ Band}(\vec{r}) = \psi_{BV}(\vec{r}) = \Omega^{-\frac{1}{2}} e^{iK' \cdot \vec{r}} u_{valence}(\vec{r}, \vec{K}') \tag{4.19a}$$

$$\psi_{Bloch\ Conduction\ Band}(\vec{r}) = \psi_{BC}(\vec{r}) = \Omega^{-\frac{1}{2}} e^{iK'' \cdot \vec{r}} u_{conduction}(\vec{r}, \vec{K}'') \tag{4.19b}$$

As discussed in a previous chapter, the periodic Bloch wavefunctions $u_{conduction}$ and $u_{valence}$ can be determined using any one of many quantum theories including tight binding theory, kp theory, and density functional theory. Using $u_{conduction}(\vec{r}, \vec{K}) = u_c(\vec{r}, \vec{K})$ and $u_{valence}(\vec{r}, \vec{K}) = u_v(\vec{r}, \vec{K})$:

$$\int_{-\infty}^{\infty} \psi^*_{BCB}(\vec{r}) \frac{-i\hbar e}{4mc} A\left[e^{i\vec{q}\cdot\vec{r}}\right][\vec{a}_0 \cdot \vec{\nabla}]\psi_{BVB}(\vec{r})d\vec{r}\delta\left(E_c(\vec{k}_c) - E_v(\vec{k}_v) - \hbar\omega\right)$$

$$= \frac{-i\hbar e}{4mc\Omega} A \int_{-\infty}^{\infty} e^{-iK''\cdot\vec{r}} u^*_c(\vec{r}, \vec{K}'') e^{i\vec{q}\cdot\vec{r}} [\vec{a}_0 \cdot \vec{\nabla}] e^{iK'\cdot\vec{r}} u_v(\vec{r}, \vec{K}') d\vec{r} \tag{4.20}$$

$$\times\delta\left(E_c(\vec{k}_c) - E_v(\vec{k}_v) - \hbar\omega\right)$$

Using the formula $\frac{\partial}{\partial x} f(x)q(x) = f'(x)q(x) + f(x)q'(x)$:

$$= \frac{-i\hbar e}{4mc\Omega} A \int_{-\infty}^{\infty} e^{-i(\vec{K}''-\vec{q}-\vec{K}')\cdot\vec{r}} u^*_c(\vec{r}, \vec{K}'')\left[\vec{a}_0 \cdot \vec{\nabla} u_v(\vec{r}, \vec{K}') + i\vec{a}_0 \cdot \vec{K}' u_v(\vec{r}, \vec{K}')\right] d\vec{r}$$

$$\delta\left(E_c(\vec{k}_c) - E_v(\vec{k}_v) - \hbar\omega\right) \tag{4.21}$$

Since the Bloch wavefunctions are orthogonal, $\int_{-\infty}^{\infty} u^*_c(\vec{r}, \vec{K}'')\, u_v(\vec{r}, \vec{K}')\, d\vec{r} = 0$. Any position in the crystal $\vec{r}$ can be found by using a real lattice vector from the origin to the edge of the unit cell, $\vec{R}$, and then a real lattice vector, $\vec{r}'$, from the edge of the unit cell to $\vec{r}$ so that $\vec{r} = \vec{R} + \vec{r}'$. This is convenient because we can then setup the integral in (4.21) as a sum over N unit cells and an integral inside one unit cell: $\int_{-\infty}^{\infty} (\ldots)d\vec{r} \to \sum_N \int inside\ unit\ cell$. Thus:

$$e^{-i(\vec{K}''-\vec{q}-\vec{K}')\cdot\vec{r}} = e^{-i(\vec{K}''-\vec{q}-\vec{K}')\cdot(\vec{R}+\vec{r}')} = e^{i(-\vec{K}''+\vec{q}+\vec{K}')\cdot(\vec{R}+\vec{r}')} \tag{4.22}$$

We chose a reciprocal lattice vector $\vec{\mathcal{K}} = \vec{K}' + \vec{q} - \vec{K}$ so that $e^{i\vec{\mathcal{K}}\cdot\vec{R}} = 1$, and $\sum_N e^{i\vec{\mathcal{K}}\cdot\vec{R}} = \text{N}$. We chose

$$\vec{\mathcal{K}} = \vec{K}' + \vec{q} - \vec{K}'' = 0 \text{ and } \vec{q} \text{ is small } \rightarrow \vec{K}' = \vec{K}''$$

This is a key result since it shows that the momentum is conserved in the direct gap transitions we are considering. Equation 4.21 can now be written using $\sum_N e^{i\vec{\mathcal{K}}\cdot\vec{R}} = N$ as:

$$\frac{-i\hbar e}{4mc\Omega}A\int_{-\infty}^{\infty} e^{-i(\vec{K}''-\vec{q}-\vec{K}')\cdot\vec{r}}u_c^*(\vec{r},\vec{K}'')\vec{a}_0\cdot\vec{\nabla}u_v(\vec{r},\vec{K}')d\vec{r}$$
$$= \frac{-i\hbar e}{4mc\Omega}AN\int_{unit\ cell} u_c^*(\vec{r},\vec{K}'')\vec{a}_0\cdot\vec{\nabla}u_v(\vec{r},\vec{K}')d\vec{r}$$

And the number of unit cells equals the volume of the crystal Ω / the unit cell volume Δ. Using $i\hbar\vec{\nabla}$ as the momentum operator, we restate the matrix element as:

$$\vec{P}_{cv} = -i\hbar\int_{unit\ cell} u_c^*(\vec{r},\vec{K}'')\,\vec{a}_0\cdot\vec{\nabla}u_v(\vec{r},\vec{K}')\,d\vec{r} \tag{4.23}$$

$$H'_{i\rightarrow f} = \frac{-eA}{4mc\Delta}\vec{P}_{cv}\,\delta\Big(E_c(\vec{k}_c) - E_v(\vec{k}_v) - \hbar\omega\Big) \tag{4.24}$$

Restating $\mathcal{W}_{if} = \left(\frac{2\pi}{\hbar}\right)\left|H'_{i\rightarrow f}\right|^2\rho(\omega) \rightarrow \mathcal{W}_{if} = \sum_{all\ k\ states}\mathcal{W}_{if(single\ transition)}$, we have:

$$\mathcal{W}_{if(single\ transition)} = \left(\frac{2\pi}{\hbar}\right)\left|\frac{-eA}{4mc\Delta}\vec{P}_{cv}\right|^2\delta\Big(E_c(\vec{k}_c) - E_v(\vec{k}_v) - \hbar\omega\Big) \tag{4.25}$$

$$\mathcal{W}_{if(single\ transition)} = \left(\frac{\pi}{\hbar}\right)\frac{e^2A^2}{8m^2c^2\Delta^2}\left|\vec{P}_{cv}\right|^2\delta\Big(E_c(\vec{k}_c) - E_v(\vec{k}_v) - \hbar\omega\Big)$$

It can be shown that in cgs units $A^2 = \frac{8\pi\hbar c^2}{\varepsilon_1\omega}$ ([6], Appendix E). Thus:

$$\mathcal{W}_{if(single\ transition)} = \frac{e^2\pi^2}{\varepsilon_1\omega m^2\Delta^2}\left|\vec{P}_{cv}\right|^2$$
$$\delta\Big(E_c(\vec{k}_c) - E_v(\vec{k}_v) - \hbar\omega\Big)$$

A general expression for the imaginary part of the dielectric function is:

$$\varepsilon_2 = \frac{\varepsilon_1}{\omega} W_{if} = \sum\nolimits_{all\ k\ states} \frac{e^2\pi^2}{\omega^2 m^2 \Delta^2} \left|\vec{P}_{cv}\right|^2 \delta\left(E_c(\vec{k}_c) - E_v(\vec{k}_v) - \hbar\omega\right) \quad (4.26)$$

For crystals that are isotropic, the average value of $|P_{cv}|^2 = \langle |P_{cv}|^2 \rangle$ can be restated using $P = i\hbar\nabla$:

$$-i\hbar\Delta^{-1} \int_{unit\ cell} u_c^*\left(\vec{r}, \vec{K}''\right) \vec{a}_0 \cdot \vec{\nabla} u_v\left(\vec{r}, \vec{K}'\right) d\vec{r}$$

$$\Delta^{-1} \int_{unit\ cell} u_c^*\left(\vec{r}, \vec{K}''\right) \vec{a}_0 \cdot \mathrm{P}\vec{\nabla} u_v\left(\vec{r}, \vec{K}'\right) d\vec{r} = \vec{a}_0 K \cdot \vec{\boldsymbol{P}}_{cv}$$

where

$$\vec{\boldsymbol{P}}_{cv} = \Delta^{-1} \int_{unit\ cell} u_c^*(\vec{r}, \vec{K}'')\, \vec{\mathrm{P}}\, u_v(\vec{r}, \vec{K}')\, d\vec{r} \quad (4.27)$$

Next, we assume an isotropic optical response while noting that this will not be the case for 2D materials:

$$\left\langle \left|\vec{a}_0 \cdot \vec{P}_{cv}\right|^2 \right\rangle = \langle |P_{cv}(k_x)|^2 \rangle + \left\langle \left|P_{cv}(k_y)\right|^2 \right\rangle + \langle |P_{cv}(k_z)|^2 \rangle = 3\left\langle \left|\overline{\boldsymbol{P}_{cv}}\right|^2 \right\rangle \quad (4.28)$$

Optical absorption will be observable when there are enough pairs of valence and conduction band states (joint density of states) with the same energy difference. Since $\vec{k}_v = \vec{k}_c$, a useful first approximation is that the values of wavevectors are similar that absorption between energy bands in a localized portion of the Brillouin zone is observable. This approximation allows us to assume that $\overline{P}_{cv}$ is constant when integrating over k states. It is useful to note that this transition can be between any valence band and any conduction band.

We need to determine the density of k states. The volume V *of a single k state* in reciprocal space can be determined using a cube with each side having the length of the reciprocal space vector as follows:

$k_x = \frac{2\pi n_x}{L}$; $k_y = \frac{2\pi n_y}{L}$; $k_z = \frac{2\pi n_z}{L}$ the volume of a cube with (0, 0, 0) as the origin and $n_x = n_y = n_z = 1$ (volume per allowed k) will be $k_x \cdot k_y \cdot k_z = \left(\frac{2\pi}{L}\right)^3$. The number density of k states is then $\left(\frac{L}{2\pi}\right)^3$ and the number per unit volume is $\left(\frac{L}{2\pi}\right)^3 / L^3 = \left(\frac{1}{2\pi}\right)^3$. The $\sum_{all\ k\ states} \rightarrow \int \left(\frac{1}{2\pi}\right)^3 d\vec{k}$ and thus:

$$\sum\nolimits_{all\ k\ states} \frac{e^2\pi^2}{\omega^2 m^2} \left|\vec{a}_0 \cdot \vec{P}_{cv}\right|^2 \delta\left(E_c(\vec{k}_c) - E_v(\vec{k}_v) - \hbar\omega\right)$$

$$\frac{e^2\pi^2}{\omega^2 m^2} \left|\vec{a}_0 \cdot \vec{P}_{cv}\right|^2 \int \left(\frac{1}{2\pi}\right)^3 d\vec{k}\delta\left(E_c(\vec{k}_c) - E_v(\vec{k}_v) - \hbar\omega\right)$$

This provides a useful formula for the imaginary part of the dielectric function using (4.28)**when the k states are equally distributed in 3D**:

$$\varepsilon_2 = \frac{3e^2}{8\pi\omega^2 m^2}\left|\overline{P_{cv}}\right|^2 \cdot \int d\vec{k}\delta(E_c - E_v - \hbar\omega) \tag{4.29}$$

This equation can be converted to SI units by multiplying it by $\left(\frac{1}{4\pi\varepsilon_0}\right)$.

It is important to note that the dielectric function calculated using **ab initio** ***quantum methods, to determine the electronic band structure and thus the wavefunctions, represents the theoretical value at 0 K*** [7]. The room temperature dielectric function will differ slightly for the one calculated from the electronic band structure unless it is corrected for a number of changes including lattice expansion and thermal motion of the atoms in the solid [7]. This is further discussed in Chap. 5.

4.3 The Energy Dependence of the Dielectric Function for Parabolic Bands

It is useful to discuss the energy dependence of the dielectric function for optical transitions that occur between parabolic valence and conduction bands centered at the K = 0 point of the Brillouin zone. The band structure is shown in Fig. 4.1 where $E_{gap} = E_c - E_v$ at $\vec{k} = 0$ and $E_c - E_v = E_{gap} + \frac{\hbar^2 k^2}{2\mu}$ where μ is the reduced mass of the electron–hole pair produced by the transition. This pair is not necessary energetically bound. In this section we are assuming that the reduced mass

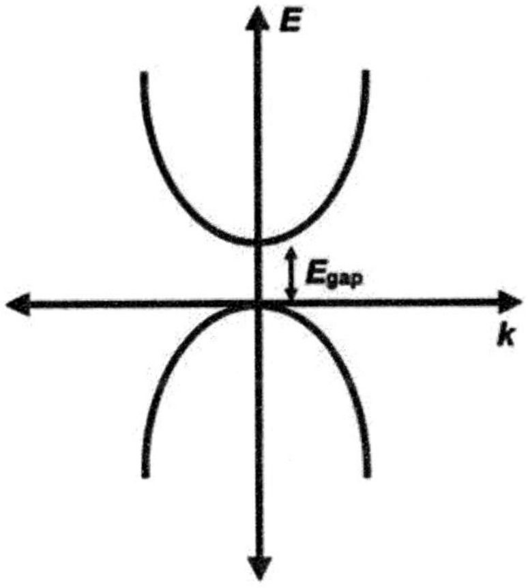

Fig. 4.1 Parabolic electronic band structure with valence band maximum and conduction band minimum at $k = 0$

is isotropic. In the next section we consider a transition where the reduced mass is different along the x, y, and z directions in $\vec{k}$ space.

We need to change the integral over wavevector $\vec{k}$ to an integral over energy. The energy difference between the valence and conduction band is fixed since $\delta(E_c - E_v - \hbar\omega)$. We need to consider a constant energy sphere in $\vec{k}$ space. The volume of constant energy spheres at k and at $k' = k + dk$ where k is the magnitude of the wavevector are:

$$V_k = \frac{4}{3}\pi k^3 \text{ and } V_{k+dk} = \frac{4}{3}\pi(k+dk)^3$$

The number of allowed k states between k and $k + dk$ is:

$$\frac{4}{3}\pi(k+dk)^3 - \frac{4}{3}\pi k^3 = \frac{4}{3}\pi(k^3 + 3k^2dk + 3k(dk)^2 + (dk)^3)$$
$$-\frac{4}{3}\pi k^3 \cong 4\pi k^2 dk$$

From $E_c - E_v = E_{gap} + \frac{\hbar^2 k^2}{2m_r}$ where m_r is the reduced mass, and $E_{cv} = E_c - E_v$ we get:

$$k^2 = \frac{2m_r}{\hbar^2}(E_{cv} - E_{gap}) \text{ and } k = \left(\frac{2m_r}{\hbar^2}\right)^{1/2}(E_{cv} - E_{gap})^{1/2} \text{ with}$$
$$dk = \left(\frac{2m_r}{\hbar^2}\right)^{1/2}\frac{1}{2}(E_{cv} - E_{gap})^{-1/2}\frac{d(E_{cv} - E_{gap})}{dE_{cv}}dE_{cv}$$
$$= \left(\frac{2m_r}{\hbar^2}\right)^{1/2}\frac{1}{2}(E_{cv} - E_{gap})^{-1/2}dE_{cv}.$$

This gives us:

$$4\pi k^2 dk = \frac{4\pi\sqrt{2}m_r^{3/2}}{\hbar^3}(E_{cv} - E_{gap})^{1/2}dE_{cv}$$

Then:

$$\varepsilon_2 = \frac{3e^2}{\pi\omega^2 m^2}\left|\overline{P_{cv}}\right|^2 \int d\vec{k}\delta(E_c - E_v - \hbar\omega)$$
$$= \frac{3e^2}{8\pi\omega^2 m^2}\left|\overline{P_{cv}}\right|^2 \int \frac{4\pi\sqrt{2}m_r^{3/2}}{\hbar^3}(E_{cv} - E_{gap})^{\frac{1}{2}}dE_{cv}\delta(E_c - E_v - \hbar\omega)$$

$$\varepsilon_2 = \frac{3e^2(2m_r)^{3/2}}{2\omega^2 m^2\hbar^3}\left|\overline{P_{cv}}\right|^2(E_{cv} - E_{gap})^{1/2}$$
$$\varepsilon_2(parabolic\ bands) \sim (E_{cv} - E_{gap})^{1/2} \quad (4.30)$$

4.4 Joint Density of States, Critical Points, and Van Hove Singularities

As mentioned in the chapter introduction, an optical transition can be observed when there are enough valence to conduction band transitions having the same energy. In other words, a high Joint Density of States, $\rho(\omega)$. That occurs when the electronic band structure has certain characteristics. Here we provide a more general method for determining that by deriving a general expression for the $\rho(\omega) \rightarrow \rho(E) = \frac{2}{(2\pi)^3} \int d\vec{k}\delta(E_c - E_v - \hbar\omega)$ term in the dielectric function formula [8, 9]. This allows us to characterize interband optical transitions according to the local electronic band structure in the Brillouin zone.

The *Joint Density of States* $\rho(E)$ is the number of wavevectors for the transition energy $E_{cv}(\vec{k}) = E_c(\vec{k}) - E_v(\vec{k})$ in the energy range E_{cv} to $E_{cv} + dE_{cv}$. After accounting for spin, this can be stated as:

$\rho(E)dE = \frac{2}{V}$ [# allowed wavevectors in the nth band between E_{cv} and $E_{cv} + dE_{cv}$].
$\rho(E)dE = \frac{2}{V}$ [Volume of k space in the unit cell/Volume of k space per allowed k].

Remembering that the volume in k space per allowed k is $\left(\frac{2\pi}{L}\right)^3 = \frac{(2\pi)^3}{V}$. For a general energy E, this gives:

$$\rho(E)dE = \frac{2}{V}\frac{\int_{unit\ cell} d\vec{k}}{\frac{(2\pi)^3}{V}} = 2\int_{unit\ cell} \frac{d\vec{k}}{(2\pi)^3} \tag{4.31}$$

$d\vec{k}$ is a volume that is a function of k and can be written as $d\vec{k} = dS\delta k(\vec{k})$. Where $S(E_{cv})$ is a surface of constant energy. $\rho(E)$ is the density of energy levels per unit volume and dE is an infinitesimal energy change so $\rho(E)dE$ is also infinitesimal. Writing (4.31) as a surface integral:

$$\rho(E)dE = \oint \frac{dS}{4\pi^3}\delta k(\vec{k}) \tag{4.32}$$

Now we apply this to a constant energy surface for E_{cv}. Since $S(E_{cv})$ is a surface of constant energy, then $\vec{\nabla}_k E_{cv}$ is a vector normal to the surface. We have:

$$\vec{\nabla}_k E_{cv} = \frac{\partial E_{cv}(\vec{k})}{\partial k_x} + \frac{\partial E_{cv}(\vec{k})}{\partial k_y} + \frac{\partial E_{cv}(\vec{k})}{\partial k_z} \text{ and } |\vec{\nabla}_k E_{cv}|\,\delta k(\vec{k}) = dE_{cv}$$

Thus $E_{cv} + dE_{cv} = E_{cv} + |\vec{\nabla}_k E_{cv}|\,\delta k(\vec{k})$ resulting in $\delta k(\vec{k}) = \frac{dE_{cv}}{|\vec{\nabla}_k E_{cv}|}$ and:

$$\rho(E_{cv})dE = \oint \frac{dS}{4\pi^3}\delta k(\vec{k}) = \oint \frac{dS}{4\pi^3}\frac{dE_{cv}}{|\vec{\nabla}_k E_{cv}|}$$

$$\rho(E_{cv}) = \oint \frac{dS}{4\pi^3}\delta k(\vec{k}) = \oint \frac{dS}{4\pi^3}\frac{1}{|\vec{\nabla}_k E_{cv}|} \tag{4.33}$$

Thus when the gradient of the energy difference $\vec{\nabla}_k E_{cv}$ is 0, the joint density of state diverges. In real crystals, the energy difference is not perfectly constant over the part of the Brillouin zone where the transition is occurring. At these energies E_{cv}, there is a large joint density of states and optical absorption occurs (there is a non-zero value for the imaginary part of the dielectric function. When $\vec{\nabla}_k E_{cv}$ is very small, there is a van Hove singularity. These are also known as critical points.

4.5 The Naming and Energy Dependence of the Critical Points

The designation of the critical points according to the characteristics of the local band structure is used throughout the literature and is reviewed here [4, 8, 9]. The effective masses of the electron m_e^* and hole m_h^* are different along different direction in the Brillouin zone. In order to distinguish the reduced effective mass from the reduced mass μ, here we use the symbol m_r^*. Here we determine the energy dependence of an optical transition where the reduced mass $\frac{1}{m_r^*} = \frac{1}{m_e^*} + \frac{1}{m_h^*}$ is different along the x, y, and z directions of k space. Thus:

$$E_{cv} - E_{gap} = \frac{\hbar^2}{2m_{rx}^*}k_x^2 + \frac{\hbar^2}{2m_{ry}^*}k_y^2 + \frac{\hbar^2}{2m_{rz}^*}k_z^2 = \beta_1 k_x^2 + \beta_2 k_y^2 + \beta_3 k_z^2 \tag{4.34}$$

If we assume that the reduced mass is positive, we have a constant energy surface that is an ellipsoid with:

$$\frac{x^2}{a^2} + \frac{y^2}{b^2} + \frac{z^2}{c^2} = 1 \text{ with volume } V = \frac{4\pi}{3}a \cdot b \cdot c$$

and $a = \left(\frac{E_{cv}-E_{gap}}{\beta_1}\right)^{1/2}, b = \left(\frac{E_{cv}-E_{gap}}{\beta_2}\right)^{1/2}, c = \left(\frac{E_{cv}-E_{gap}}{\beta_3}\right)^{1/2}$ giving $V = \frac{4\pi}{3}a{\cdot}b{\cdot}c = \frac{4\pi}{3}\frac{(E_{cv}-E_{gap})^{3/2}}{(\beta_1\beta_2\beta_3)^{1/2}}$.

Since $\int \rho(E_{cv})dE = total\ number\ of\ energy\ states = volume\ of\ energy\ states \times number\ per\ unit\ volume$

$$\int \rho(E_{cv})dE = \frac{4\pi}{3}\frac{\left(E_{cv}-E_{gap}\right)^{3/2}}{(\beta_1\beta_2\beta_3)^{1/2}}\frac{1}{4\pi^3} = \frac{1}{3\pi^2}\frac{\left(E_{cv}-E_{gap}\right)^{3/2}}{(\beta_1\beta_2\beta_3)^{1/2}}$$

$$\rho(E_{cv}) = \frac{d}{dE}\frac{1}{3\pi^2}\frac{\left(E_{cv}-E_{gap}\right)^{3/2}}{(\beta_1\beta_2\beta_3)^{1/2}} = \frac{1}{2\pi^2}\frac{\left(E_{cv}-E_{gap}\right)^{1/2}}{(\beta_1\beta_2\beta_3)^{1/2}} \tag{4.35}$$

This is the M_0 critical point where $\beta_1, \beta_2,$ and β_3 are all positive. Other critical points can also be derived in a similar fashion for each β being positive or negative. The critical points are designated by M_i, where the value i depends on how many β are negative as shown in Table 4.1. The joint density of states for 2D and 1D materials changes the dielectric function and critical point energy dependence. The Joint Density of States for 2D and 1D materials is listed in Table 4.2 and the imaginary part of the dielectric function shown in Fig. 4.2. The density of states and imaginary part of the dielectric function are further discussed in a subsequent chapter.

It is useful to note that some materials such as gallium arsenide, silicon and alloys of silicon with a large enough silicon concentration will have direct gap transitions which show enhanced absorption beyond that predicted by (4.29). Some III-V and group IV semiconductors exhibit this effect at a critical point designated E_1. This enhanced absorption is a result of "excitonic effects". The topic of bound excitons we introduced in the first chapter. The binding energy of excitons is often less than the thermal energy at room temperature ~ KT. Although this topic is covered in more

Table 4.1 Energy Dependence of 3D Joint Density of States for the Critical Points Centered for direct gap transitions at E_{gap} [4]

Critical point	Type	β_1	β_2	β_3	JDS $E < E_{gap}$	JDS $E > E_{gap}$
M_0	Minimum	+	+	+	0	$\left(E_{cv}-E_{gap}\right)^{1/2}$
M_1 [6]	Saddle point	+	+	−	$C_1 - C_1'\left(E_{cv}-E_{gap}\right)^{1/2}$	C_1
M_2	Saddle point	+	−	−	C_2	$C_2 - C_2'\left(E_{cv}-E_{gap}\right)^{1/2}$
M_3	Maximum	−	−	−	$C_3\left(E_{cv}-E_{gap}\right)^{1/2}$	0

Table 4.2 Joint Density of States for 2D and 1D for the critical points centered for direct gap transitions at E_{gap} [8, 9]

Critical point	Material dimension	JDS $E < E_{gap}$	JDS $E > E_{gap}$
M_0	2D	0	C
M_1 [6]	2D	$-\ln\left(E_{gap}-E_{cv}\right)$	$-\ln\left(E_{cv}-E_{gap}\right)$
M_2	2D	C	0
M_0	1D	0	$(E_{cv}-E_{gap})^{-1/2}$
M_1	1D	$(E_{gap}-E_{cv})^{-1/2}$	0

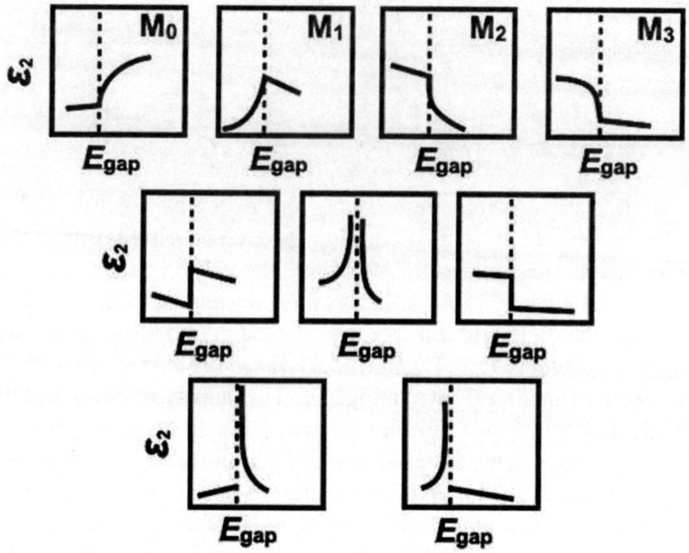

Fig. 4.2 Designation of critical points based on local electronic band structure. The imaginary part of the dielectric function for M_0 through M_3 critical points are shown as function of the transition energy. The transitions are centered at the key transition energy E_{gap} [8, 9]. The top line of plots is for 3D critical points, the middle section is for 2D, and the bottom set of plots is for 1D critical points. Figure adapted from [2]

detail in Chap. 5, it is useful to state that at room temperature, this effect comes from a strong interaction between the excited electron and the hole left in the valence band. These critical points are considered excitonic. In Chap. 7, we will see that an optical transition in graphene between the minimum in the π valence band and the maximum in the π^* conduction band shows a strong excitonic effect which not only increases optical absorption, but shifts the energy of the transition.

4.6 Determining the Critical Point Energy Using Experimental Data

The experimental Critical Point energy and broadening enable comparisons between experimental dielectric function data and help parametrize functional fits to experimental data. Analysis of experimental dielectric function data is also a useful means of understanding the nature of an optical transition. For example, does the experimental data imply that the transition is M_0, M_1, or otherwise. This reflects directly on the electronic band structure along the various directions in the Brillouin zone. The maximum or peaks in the imaginary part of the dielectric function can be described by a general Lorentzian line shape [9–14]. The second derivative of the experimental dielectric function is fit to the second derivative of the Lorentzian line shape function stated in (4.36).

$$\frac{d^2\varepsilon}{dE^2} = \begin{cases} \mu(\mu+1)Ae^{i\beta}(E - E_{gap} + i\Gamma)^{-\mu-2}, & \mu \neq 0 \\ Ae^{i\beta}(E - E_{gap} + i\Gamma)^{-2}, & \mu = 0 \end{cases} \tag{4.36}$$

Here A is the amplitude, β is the phase angle, E_{gap} is the threshold energy, Γ is the broadening and μ is the order of singularity. The phase angle β describes the phase between the real and imaginary parts of dielectric function. The temperature dependent Γ (broadening) expresses the width of the CP absorption. The value of μ is based on the type of CP; it is 1, ½, 0, or −½ for excitonic, one-dimensional (1D), two-dimensional (2D) and three-dimensional (3D) one-electron transitions, respectively [13]. It is important to note that in the case of E_1 CPs for excitonic transitions, its value is 1 due to the discrete excitonic nature of the CP. The temperature dependence of the Critical Point shape has also been investigated experimentally and theoretically [10].

The CP energy and lifetime broadening of a CP can be extracted by fitting (4.36) which is referred to as direct space analysis. The fitting parameters (A, β, E_{gap}, and Γ) in (4.36) are usually determined by using a least-squares regression algorithm to fit a segment of the dielectric function spectra containing the structure of interest. The real and imaginary parts of the dielectric function are shown in Fig. 3a, and the second derivative in Fig. 3b. One can use the Levenberg–Marquardt algorithm

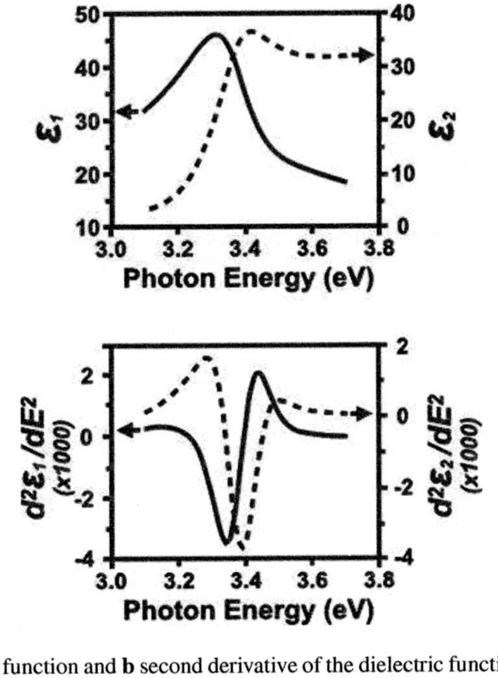

Fig. 4.3 **a** Dielectric function and **b** second derivative of the dielectric function of bulk Si at a local region of the E_1 CP

used for least-squares regression [15, 16]. The second derivative of the experimental data can be obtained by fitting the experimental dielectric function to a low order polynomial in a local region of the curve, and then differentiating the polynomial to obtain the derivative of the curve. As described elsewhere, this method is similar to the Savitzky-Golay smoothing and differentiation of data algorithm [15, 16]. A reciprocal space method for determining the critical point transition energy has been presented by Aspnes [11]. Some excitonic CPs are 2D in nature and a logarithmic form of the dielectric function provides an improved fit to the experimental data. This is described below for CdSe.

4.7 Critical Points in Semiconductors (E_1, E_2, etc.) Review of Si, Ge, GaAs and Other Group IV and III-V Materials

In this section, the relationship between the dielectric function and the electronic band structure of several group IV and III-V semiconductors is explored. In order to motivate that discussion, the Brillouin zone is first reviewed and following that, the origin of the optical transitions is described and the critical point designations presented. As discussed in the introduction presented in Chap. 1, silicon and other group IV semiconductors have an indirect band gap while GaAs and other III-V semiconductors have a direct band gap.

4.7.1 Brillouin Zone of Silicon, Germanium, Tin, and Diamond

Silicon, germanium, and the diamond phase of carbon have the diamond cubic lattice as do many of the alloys of group IV elements. The silicon crystal lattice has a lattice constant of $a_0 = 0.543$ nm with two atoms per primitive cell. The diamond structure has two interlocking face center cubic lattices displaced by ($a_0/4$, $a_0/4$, $a_0/4$). Most group IV elements including silicon and germanium have the FCC Brillouin Zone, and the first Brillouin Zone is a truncated octahedron as shown in Fig. 4.4a. The electronic band structure is typically pictures along the directions of high-symmetry in the Brillouin Zone. These points of high symmetry and vectors connecting these points are labeled in Fig. 4.4a. High-symmetry lines along the [100], [110] and [111] directions from the center of the Brillouin Zone (Γ-point) are labeled as Δ, Σ and Λ respectively.

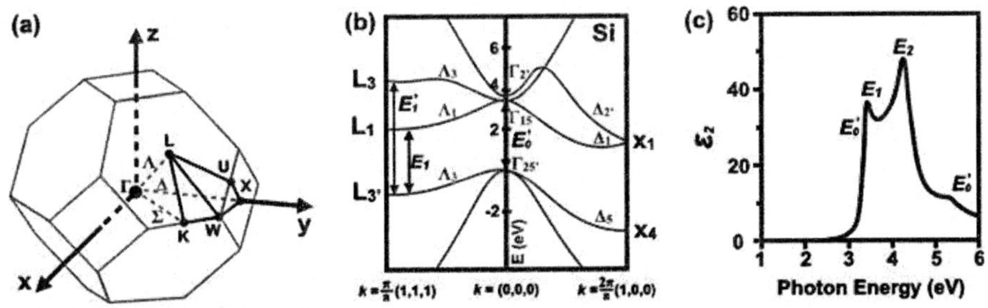

Fig. 4.4 The imaginary part of the dielectric function of silicon measured by spectroscopic ellipsometry is shown. **a** First Brillouin zone of a Si FCC lattice, **b** electronic band structure of Si calculated using $k{\cdot}p$ method [23] and **c** imaginary part of the dielectric function of bulk c-Si at room temperature between 1 and 6 eV [14]. Figure b adapted and reprinted with permission from [23]. © 1966 American Physical Society

4.7.2 Critical Points of Silicon

In the discussion below, it is important to consider that using the results of a band structure calculation to calculate the dielectric function represents most studies represent a theoretical understanding of the optical response at zero Kelvin [7]. The energy dependence of the dielectric function is well known, and is further discussed below and in Chap. 5 [10]. As mentioned above, the dielectric function of semiconductors have several sharp features that can be related to direct band transitions, where there is a constant energy difference between the valence band (VB) and the conduction band (CB). This constant energy difference along the various high-symmetry directions in the electronic band structure of Si is shown in Fig. 4.4b. This high joint density of states results in sharp features in the optical absorption which the critical points (CPs) discussed in previous sections. Because silicon is an indirect gap semiconductor, there is very little light absorption at the ~ 1.1 eV band gap. Indirect optical absorption at the band edge requires momentum (wavevector $\vec{k}$) conservation through phonon scattering during the transition from valence to conduction band. The longitudinal acoustic and transverse acoustic phonons are the primary phonons involved in band edge absorption [17]. The scattering process can involve either phonon emission (Stokes scattering) or phono absorption (antiStokes scattering) [17]. The temperature dependence of each of these contributions and the overall temperature dependence of the absorption coefficient at the band edge has been characterized [10, 18]. Silicon has several CPs between 1 and 6 eV as shown in Fig. 4.4c, that are related to direct transitions in the electronic band structure. The E_1 CP (3.4 eV) arises due to the transition from Λ_3 (VB) to Λ_1 (CB) along $k = (2\pi/a_o)$ (¼,¼,¼) to the edge of the BZ [18]. Rohlfing and Louie [19] showed that the impact of strong electron–hole (exciton) interactions result in an increased oscillator strength and small shift in the energy of the E_1 CP. This information will be useful when we fit experimental dielectric function data to the Lorentzian lineshape as discussed below. The E_1 CP is nearly degenerate with the $E_1 + \Delta$ where Δ is the spin orbit splitting energy. At room temperature, $\Delta \cong 0.03$ eV for silicon and 0.19 eV for germanium. The E_0' CP is due to transitions from the $\Gamma_{25'}$ (VB) to Γ_{15} (CB) at the center of the BZ and the E_1' CP (5.45 eV) is due to transitions from the Λ_3 (VB) to the second CB along the [111] direction. Based on the density of states in electronic band structure of bulk silicon [10] and experimental work [19], the oscillator strength of the E_1 CP has long been considered to be much larger than that of E_0' making it the dominant spectral feature. At low temperature < 75 K, the E_0' and E_1 are separated allowing determination of CP energies (at that temperature), and this is discussed below. The E_2 peak (4.270 and 4.492 eV) in the dielectric function does not correspond to a single, well defined CP and it is due to contributions from a large range of transitions close to the edge of the BZ along the Σ and Δ direction as shown in Fig. 4.5 [8, 20, 21]. The ellipsometric data was corrected for the presence of a surface oxide layer. The CP energies for silicon were re-evaluated by Jellison and Joshi along with the temperature dependence of the absorption edge, and these values are listed in Table 4.3 [17].

The single electron transition approach to determining the dielectric function is an approximation, and a full theoretical treatment for the band structure and for optical transitions requires an ab initio (first principles quantum mechanical) calculation. It

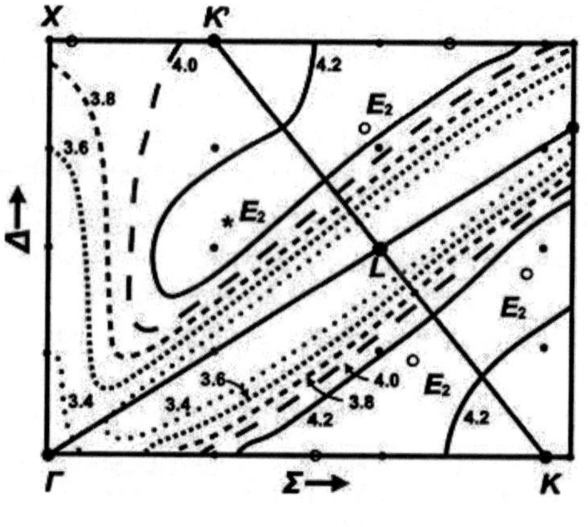

Fig. 4.5 Optical energy contour in the (110) plane for the band structure of silicon. The energy values (eV) are indicated as numbers and stars indicate optical CPs. Figure adapted from [8]

Table 4.3 Silicon critical point energies [17]

Critical point	Jellison and Joshi [17] (eV)
E_{gap}	1.1134
E_0'	3.327 ± 0.002
E_1	3.382 ± 0.005
$E_2(X)$	4.283 ± 0.003
$E_2(\Sigma)$	4.551 ± 0.021
E_1'	5.346 ± 0.023

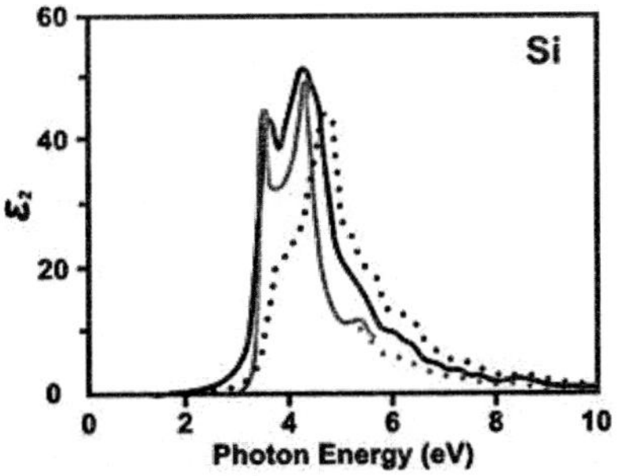

Fig. 4.6 Dielectric function of silicon determined using the three step calculation process by Rolfing and Louis. The theoretical results are shown by (black solid line) and without (black dotted line) excitonic interactions, and the gray line and gray dots are the experimental data [19]. Figure adapted and reprinted with permission from [19]. © 2000 American Physical Society

should be noted that the light perturbs the electronic structure, and for some transitions, the dielectric function calculated from band theory will have significantly reduced oscillator strengths for transitions with an excitonic nature. The excitonic nature of the E_1 CP of silicon was first demonstrated by Pollak's group [23]. Rohlfing and Louie [18] used a three step calculation process that matches well with experimental data for the silicon dielectric function: (a) the ground state band structure is calculated using density functional theory (DFT) with local density approximation (LDA), (b) the self-energy term in a many-body system of electrons is calculated by using Green function Columbic interaction approximation (GWA), and (c) the electron–hole excitation states (excitons) are included using the Bethe–Salpeter equation [18]. Their work is shown in Fig. 4.6. It is important to note that Haug and Koch [24] and Yu and Cardona [25] describe the phenomenological theoretical approach to calculating the dielectric function for direct gap transitions. These theoretical approaches to determining the excitonic nature of the increased oscillator strength are presented in the next chapter. Briefly, light absorption as having an additional component due to the formation of un-bound excitons [24]. This is referred to as a Coulomb enhancement [24].

4.7.3 Critical Points of Germanium and Diamond

Although germanium and carbon (diamond lattice) have many similarities in their band structure, spin–orbit coupling in germanium further splits the nearly degenerate bands along the [111] direction. In addition, the E_1 CP is not excitonic. The difference in spin–orbit spliting between Ge (~0.19 eV) and Si (~0.03 eV) is readily observed by comparing the band structure and their dielectric functions. Recently, Rideau, et al., updated Cardona and Pollak's *k·p* calculations for Si and Ge and calculated the band structure of SiGe alloys using a 30 band approach [27]. In addition, they

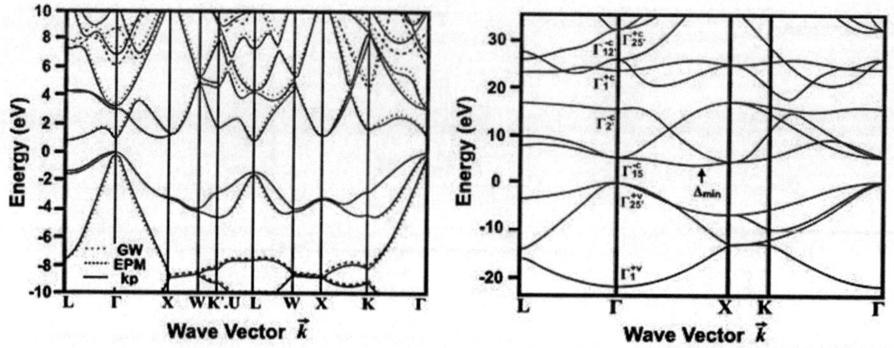

Fig. 4.7 Ge Band Structure from a 30 band $k \cdot p$ calculation shown on the left [27]. MLTO and $k \cdot p$ calculation of band structure of diamond shown on the right. Note that diamond is also an indirect band gap material [28]. Figure on left adapted and reprinted with permission from [27]. © 2006 American Physical Society. Figure on right adapted and reprinted with permission from [28]. © 1994 American Physical Society

determined the effect of strain on Si, Ge, and SiGe alloys [27]. The band structure of Ge [27] and diamond [28] are shown in Fig. 4.7.

The imaginary part of the dielectric function of Ge was first discussed in terms of direct interband transitions was given by Phillips [29]. The E_1 CP is spin orbit split and E_I (2.11 eV) & $E_1 + \Delta$ (2.3 eV) CPs result from transitions in the Λ directions of the Brillouin zone <111>. The E_0' CP also has fine structure due to spin–orbit (SO) splitting. According to Vina, et. al., the E_0' (3.123 eV) CP has been assigned to several different regions in the Brillouin zone: transitions between the $\Gamma_{25'}$ valence band and Γ_{15} conduction band, plus a three-dimensional (3D) minimum critical point (CP) at Γ with a 3D saddle point in the Δ direction of the Brillouin zone, or a large region centered at $(2\pi/a)$ (0.33, 0.24, 0.14) [29]. Many CPs observed in the dielectric function have multiple transitions that contribute at the same energy. Here, these points are close in energy and k space to the Γ_{25} to Γ_{15} gap. Vina et al. [29] point out that the E_2 (4.36 eV) CP has been attributed to a number of different transitions such as an accidental coincidence of an M_1 saddle point at X and an M_2 saddle point in the Σ direction, then to a small region centered at $(2\pi/a)$ (0. 77, 0.29, 0.16), and more recently to a region in the Γ-X-U-L plane near $(2\pi/a)$(3/4, 1/4, 1/4). Transitions between Λ_{45} and $\Lambda_6(E'_1)$ are observed at 5.5 eV [18]. The E_1' CP energy has been reported to have values of 4.35 eV and 4.5 eV [29].

Diamond has received much less attention than Si and Ge. The band gap is indirect and is believed to be a phonon mediated transition of ~ 5.6 eV between the valence-band top at $\Gamma_{25'}$ and the conduction-band minimum along Δ near $(2\pi/a)$(0.76, 0, 0) [30]. The E_1 CP (~ 10 eV) occurs at approximately the same location in the BZ as for Si and Ge.

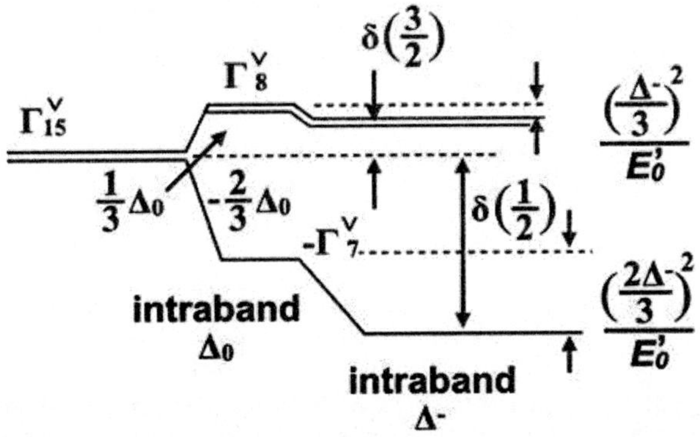

Fig. 4.8 The effect of spin orbit splitting on the Γ_{15}^v band. Figure adapted and reprinted with permission from [32]. © 1990 American Physical Society

4.7.4 Comments on Spin Orbit Splitting and CP Energies for Ge

The value of the spin orbit splitting depends on the bands involved in the optical transition. Spin orbit splitting effect the valence band states which have a p orbital content [31]. (Since Ge has a large spin orbit splitting, this effect is discussed in the literature covering QM calculations [32]. One issue that occurs with comparing theory and experiment is that the observed values of Δ are smaller than those predicted by theory. The spin orbit splitting Δ^- involved in the E_0' CP are the transitions between the Γ_{25} valence band and Γ_{15} conduction band [32]. The spin orbit splitting Δ_0 involved in the E_1 CP is for the transitions between the Λ_3 (VB) to Λ_1 (CB) [32]. The values of the spin orbit splitting are listed in the Table below, and the effect of spin orbit splitting on the valence bands for E_0' is illustrated from [32] in Fig. 4.8.

4.7.5 Critical Points of Sn

α-tin has the diamond crystal structure and the band structure exhibits strong spin–orbit splitting. Based on careful fitting of spectroscopic ellipsometry data, the E_1 CP (1.38 eV at RT) and the $E_1 + \Delta$ (1.853 eV at RT) CPs are due to transitions between the $\Lambda_{4,5}$ upper VB to Λ_6 lowest CB for E_1 and Λ_6 VB and the Λ_6 CB for $E_1 + \Delta$ between $\vec{k} = \frac{\pi}{3a}(111)$ and the L point in the Brillouin zone [33]. This critical point has been fit to a 3D M_0 lineshape and in [32] a 3D M_1 lineshape. The E_0' CP is considered to consist of four transitions and as is seen in Fig. 4.9, the CP is not sharp [32]. The E_2 (3.748 eV at RT) CP is thought to consist of several transitions including a transition between X_4 and X_1 in an extended region near U in the XUK plane. A second contribution comes from what is described as a plateau in the transition energy near $(2\pi/a)(0.75, 0.25, 0.25)$. The E_2 CP lineshape was fit

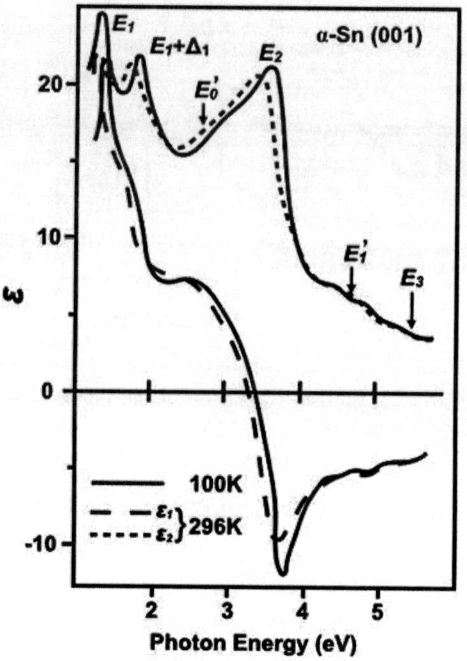

Fig. 4.9 The dielectric function of α–tin determined using spectroscopic ellipsometry. Figure adapted and reprinted with permission from [33]. © 1985 American Physical Society

by a combination of a saddle point (e.g., 3D M_1) and a 2D maximum (e.g., 2D M_1) [33].

4.7.6 Critical Points of GaAs and GaSb

GaAs is a direct band gap material, and the direct gap E_0 CP is observed at the absorption edge. The E_0 (1.517 eV at 300 K) CP transition occurs at the Γ point in the Brillouin zone between the valence band maximum and conduction band minimum as shown in Fig. 4.10. A free exciton is associated with E_0 which strong influences the shape of the dielectric function. At low temperature, the electron and hole of the exciton remain bound and at room temperature the exciton is unbound. The lineshape of excitonic direct gap transitions is further explored in the next chapter. The valence band of GaAs is also split by spin–orbit interactions. This is clearly observed at ~ 3 eV for the E_1 (3.041 eV at 300 K) and $E_1 + \Delta$ ($\Delta = 0.224$ meV at 300 K)CPs. As with silicon, the E_1 CP is observed along the [111] direction and is due to transitions between Λ_3 (VB labeled $L_{4,5}$) to Λ_1 (CB labeled L_6) of the Brillouin zone [34]. The second derivative of the dielectric function of GaAs determined from spectroscopic ellipsometry data obtained at 22 K is shown in Fig. 4.11. The critical point transition

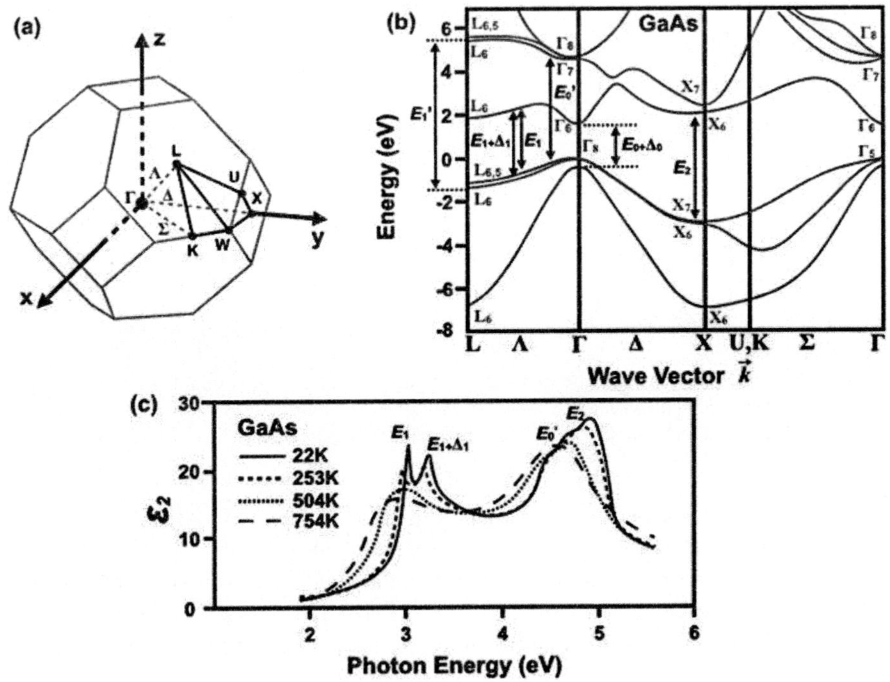

Fig. 4.10 The imaginary part of the dielectric function of GaAs from 22 k to 754 K measured by spectroscopic ellipsometry is shown. The Brillouin Zone is shown in (**a**). The relationship between the electronic band structure and the CP transitions is shown in (**b**). The temperature dependent dielectric function is shown in **c**. Figures (**b**) and (**c**) adapted and reprinted with permission from [34]. © 1987 American Physical Society

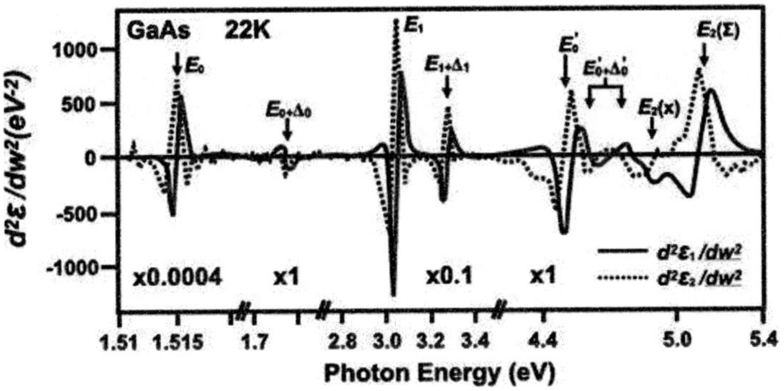

Fig. 4.11 The second derivative of the dielectric function measured by spectroscopic ellipsometry for GaAs is shown. The data is fit to (4.36) to determine the CP energies and broadening. Figure adapted and reprinted with permission from [34]. © 1987 American Physical Society

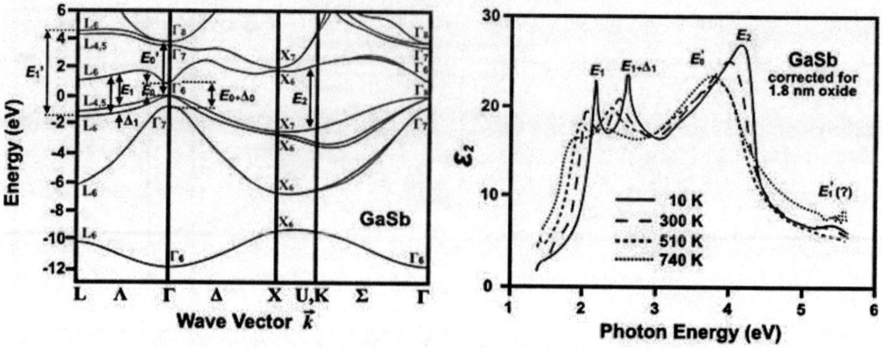

Fig. 4.12 The imaginary part of the dielectric function of GaSb. Figure adapted from [35]. We note that the electronic band structure shown here and in [35] comes from [36]. Figure adapted and reprinted with permission from [35]. © 1991 American Physical Society

energies and the broadening of each transition are shown in [34]. The ellipsometric data was corrected for the presence of a surface oxide layer through optical modeling [34]. The E_0' CP has an energy of 4.509 eV and the E_2 CP has an energy of 5.133 eV [34].

GaSb is also a direct band gap semiconductor. The E_1 (2.184 eV at 10 K/2.052 eV at 300 K) and $E_1 + \Delta_1$(2.618 eV at 10 K /2.494 eV at 300 K) are due to transitions between the $\Lambda_{4,5}$ and Λ_6 VBs and the Λ_6 CB along the [111] direction as shown in Fig. 4.12 [34]. The spin–orbit splitting Δ_1 is 434 meV at 10 K and ~ 0.442 meV at 300 K. These transitions are very prominent at low temperature, and spectra are influenced by the value of the temperature dependence of spin–orbit splitting energy [35]. The E_0' transition is a triplet and is often attributed to transitions between VBs $\left(\Gamma_7^v, \Gamma_8^v\right)$ and the second lowest conduction band $\left(\Gamma_7^v, \Gamma_8^v\right)$, with designations and energies $E_0' = 3.139\text{eV}$, $E_0' + \Delta_0' = 3.391eV$, and $E_0', +\Delta_0 + \Delta_0' = 4.135\text{eV}$ [35]. The second derivative of the dielectric function of GaSb at 10 K is shown in Fig. 4.13. The E_0 and E_1 transitions have a strong excitonic character unless the temperature is above room temperature. A thorough discussion of the CP structure of GaSb can be found in [35].

4.7.7 *Critical Points of GaN*

GaN can be found in both the zinc blende and wurtzite (α-GaN) structure and both forms have direct band gaps of ~ 3.3 eV and 3.5 eV respectively. Schubert provides a thorough treatment of the dielectric function and infrared ellipsometry of anisotropic materials including wurtzite structures [37]. As mentioned in the introductory chapter, the dielectric function is a tensor. The approximation of an isotropic dielectric function used in (4.28) is not valid for the wurtzite structure. The dielectric function of wurtzite GaN and AlN along the c crystal axis versus perpendicular to the

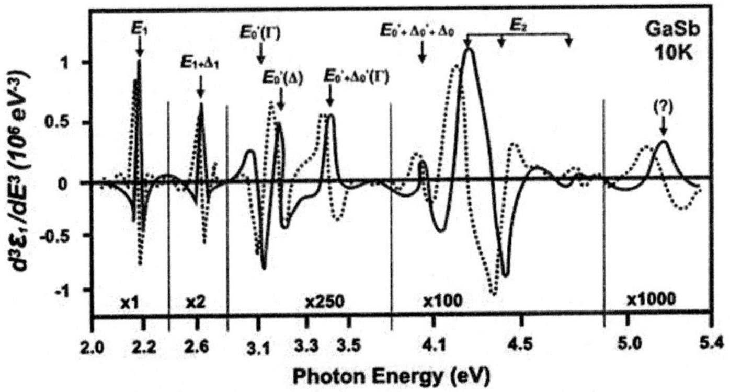

Fig. 4.13 The second derivative of the dielectric function of GaSb determined from spectroscopic ellipsometry. Figure adapted and reprinted with permission from [35]. © 1991 American Physical Society

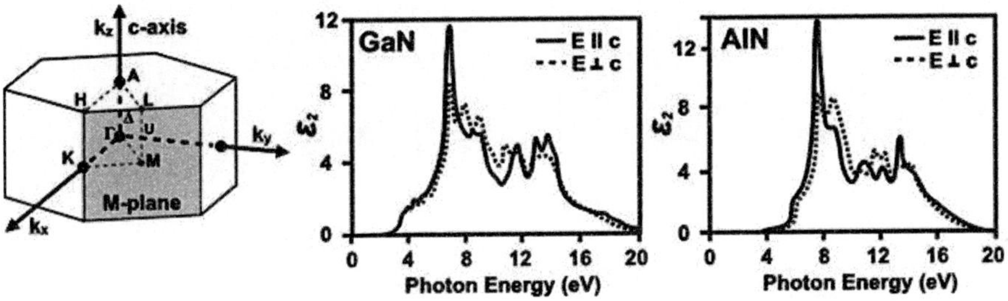

Fig. 4.14 Theoretically calculated dielectric function of wurtzite GaN and AlN along the and perpendicular to the c axis. The Brillouin zone of wurtzite is shown on the right. Dielectric function figures adapted and reprinted from [38]. © 1999 used with the permission of Elsevier

c axis has been theoretically determined as shown in Fig. 4.14 [38]. The experimental data are shown in Fig. 4.15. Another study used an analogy of optical transitions in wurtzite CdSe to indicate that the E_1 CP may be due to transitions from the $\Gamma \to A$ direction in the Brillouin zone [39]. A more recent theoretical and experimental study used the broadband light from a synchrotron to study the dielectric function of zinc blende and wurtzite GaN and wurtzite InN between 14 and 32 eV [40]. The metal d shell is full for both Ga (3d) and In (4d).

4.7.8 *Critical Points of CdSe*

CdSe has a wurtzite structure and has 3 excitonic features in the imaginary part of the dielectric function as shown in Fig. 4.16 [41]. The dielectric function close to the E_0 and E_1 critical points can be fit to a different functional form than (4.36). The "band

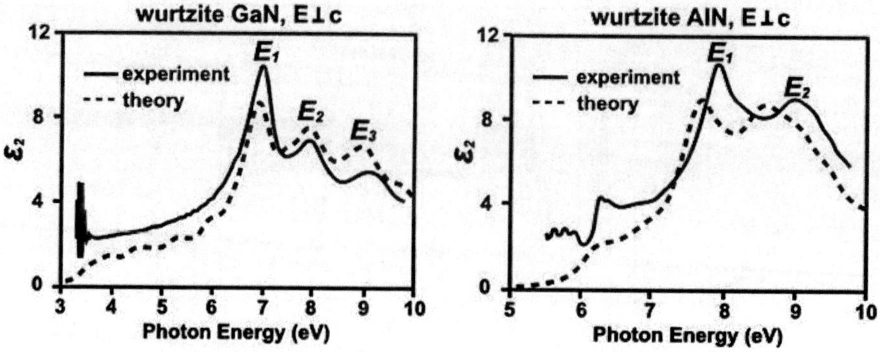

Fig. 4.15 A comparison of the theoretic dielectric function for wurtzite GaN and AlN with experimental data obtained using spectroscopic ellipsometry. Figure adapted and reproduced from [38]. Figure adapted and reprinted from [38]. © 1999 used with the permission of Elsevier

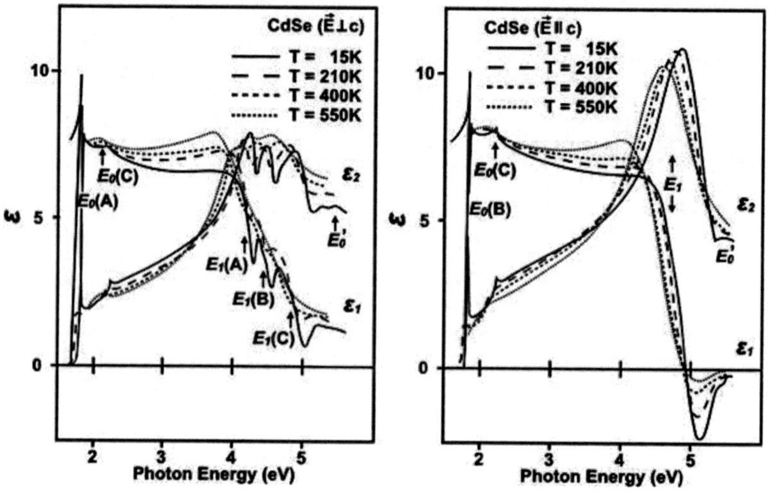

Fig. 4.16 Dielectric function of CdSe along and perpendicular to the c axis of the wurtzite structure as determine by spectroscopic ellipsometry. Figure adapted and reprinted with permission from [41]. © 1986 American Physical Society

edge" E_0 transitions are labeled A, B, and C and the A transition comes from an excitation from the valence band of Γ_9 symmetry and the B and C transitions are due to excitations from the valence band with Γ_7 symmetry. The B and C E_0 transitions are observed for both light polarizations (along and perpendicular to the c axis) while the A E_0 transition is only observed when the light is polarized perpendicular to the c axis: $\vec{E}\perp c$ [41]. The E_1 A, B, and C excitonic transitions are observed when $\vec{E}\perp c$, and the A and B E_1 CP is due to transitions between Γ_5 to Γ_3 along the Γ to A direction in the Brillouin zone [40]. The two different forms for the excitonic lineshape have been used in the literature depending on the quality of the fit [42].

The 2 forms of the dielectric function for excitonic structures frequently used in the literature are:

$$\varepsilon(E) = Ae^{i\beta}(E - E_{gap} + i\Gamma)^{-1} \tag{4.36a}$$

$$\varepsilon(E) = Ae^{i\beta}\ln(E - E_{gap} + i\Gamma) \tag{4.36b}$$

2D Transition

The difference is due to the local band structure. For E_0, the A, B, and C transitions have each been fit to a variation of (4.36b) [41]:

$$\varepsilon(E) = \frac{A_1}{(E - E_{gap} + i\Gamma)} + A_2 e^{i\beta}\ln(E - E_{gap} + i\Gamma) + A_3 \tag{4.37}$$

And the E_1 A, B, and C CPs have been fit to:

$$\varepsilon(E) = C - Ae^{i\beta}\ln(E - E_{gap} + i\Gamma) \tag{4.38}$$

4.7.9 Critical Points of $Si_{1-x}Ge_x$ Alloys

The shape of the dielectric function of $Si_{1-x}Ge_x$ alloys is strongly influenced by the difference in the physics of light absorption between silicon and germanium [43]. As mentioned above, the E_1 CP of silicon is excitonic in nature and the E_1 CP of germanium is not. The change in lattice constant and E_1 and $E_1 + \Delta_0$ and E_2 CP energies as a function of germanium concentration does not follow Vagard's law. Here Δ_0 is the spin orbit splitting of the Λ_3 (VB) as discussed above for germanium. For stress free $Si_{1-x}Ge_x$, the E_1, $E_1 + \Delta_0$, and E_2 CPs energy follow [44]:

$$\begin{aligned} E_1(x) &= 3.395 - 1.287x - 0.153x(1-x)\text{eV} \\ [E_1 + \Delta_0](x) &= 3.428 - 1.132x - 0.062x(1-x)\text{eV} \\ E_2(x) &= 4.372 - 0.00069(1-x)\text{eV} \end{aligned} \tag{4.39}$$

Another influence on the shape of the dielectric function around the E_1 CP for these alloys is stress. Zolner has reviewed the impact of stress (strain) on the electronic band structure of $Si_{1-x}Ge_x$ alloys and the impact of elastic strain on the CP energies [45]. These alloys are often epitaxially deposited on a single crystal substrate such as silicon or germanium, and the lattice mismatch can result in a change in lineshape as shown in Fig. 4.17 for pseudomorphic alloys grown on Si(001) [42]. The biaxial stress results in different optical properties normal versus in the surface plane. Reflective spectroscopic ellipsometry will be sensitive to the in-plane dielectric function. The

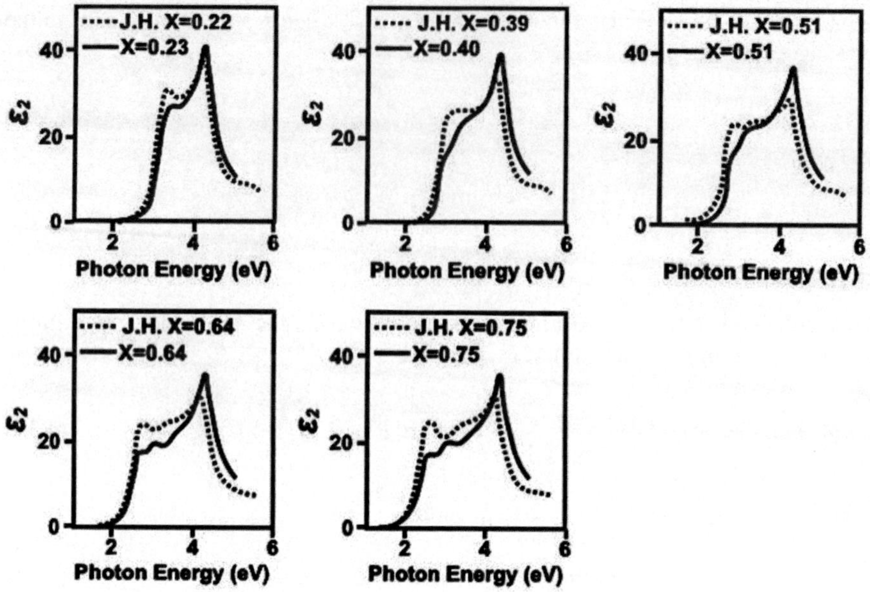

Fig. 4.17 The imaginary part of the dielectric function of $Si_{1-x}Ge_x$ alloys for pseudomorphic and fully relaxed crystals [44] of various concentrations. Figure adapted and reprinted from [43] with the permission of AIP Publishing

change in the energy of E_1 CP versus germanium composition due to the bi-axial compressive stress follows elastic theory. The values of the complex refractive index and the dielectric function for 5% to 75% Ge are tabulated as a function of the energy of the light from 1.24 eV to 5.06 eV in [42]. For example, the energy of E_1 and $E_1 + \Delta_0$ for stress free is 2.70 eV and 2.835 eV and for pseudomorphic $Si_{49}Ge_{51}$ is 2.76 eV and 2.98 eV. $Si_{1-x}Ge_xC_y$ alloys [45] have also been studied.

4.7.10 Critical Points of $Ge_{1-x}Sn_x$ Alloys

The dielectric function of $Ge_{1-x}Sn_x$ ($x \leq 0.11$) pseudomorphic on germanium has also been studied and the imaginary part of the dielectric function is shown in Fig. 4.18 [46]. The biaxial stress induced shift of the E_1 and $E_1 + \Delta$ CPs closely follow the energy shift predicted by elastic theory [47, 48].

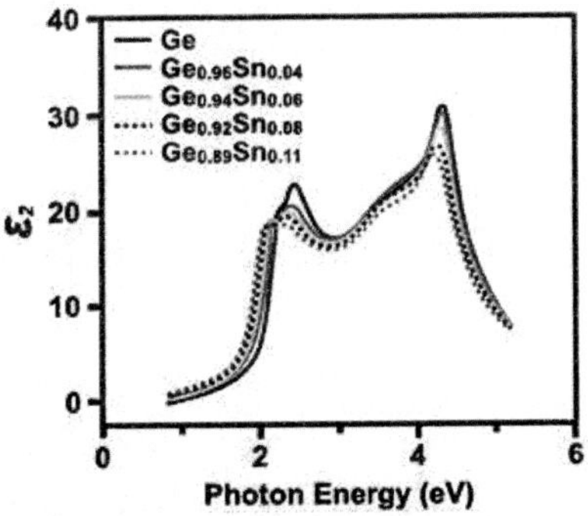

Fig. 4.18 Imaginary part of the dielectric function of pseudomorphic $Ge_{1-x}Sn_x$ for 0% to 11% Sn. Figure adapted and reprinted from [47] with the permission of AIP Publishing

4.8 The Effect of Doping on the Dielectric Function

Doping changes the Fermi level of a semiconductor and thus fills states in the conduction band which blocks optical transitions for the wavevector states that are filled. This effect has been called phase filling since the filled k states in the phase space of allowed conduction band states are occupied [49]. This effect is shown in Fig. 4.19. It is important to restate that valence band is split due to spin orbit coupling Δ_1 and the M1 CP has a contribution from both transitions. The effect of doping on the complex refractive index of heavily doped silicon has also been reported and the complex refractive index and absorption coefficient of As doped Si is shown in Fig. 4.20 [17, 50, 51].

The near IR and IR absorption due to doping is often described by the Drude model discussed in Chap. 1. The carrier concentration can be determined from the absorption coefficient. An approximate expression for the Drude absorption coefficient α when the relaxation time times the angular frequency is less than 1 is:

$$\varepsilon_2 = \frac{\omega\tau\omega_P^2}{\omega^4\tau^2 + \omega^2} \approx \frac{\omega_P^2}{\omega}; \; \omega_P = \frac{Ne^2}{m\varepsilon_0};$$
$$\alpha = \frac{4\pi k}{\lambda} = \frac{4\pi\omega_P^2\tau}{2n\omega\lambda} = \frac{2\pi Ne\mu}{\varepsilon_0 n\lambda\omega} = \frac{Ne\mu}{\epsilon_0\, nc} \tag{4.40}$$

Here, n is the real part of the refractive index, and μ is the carrier mobility.

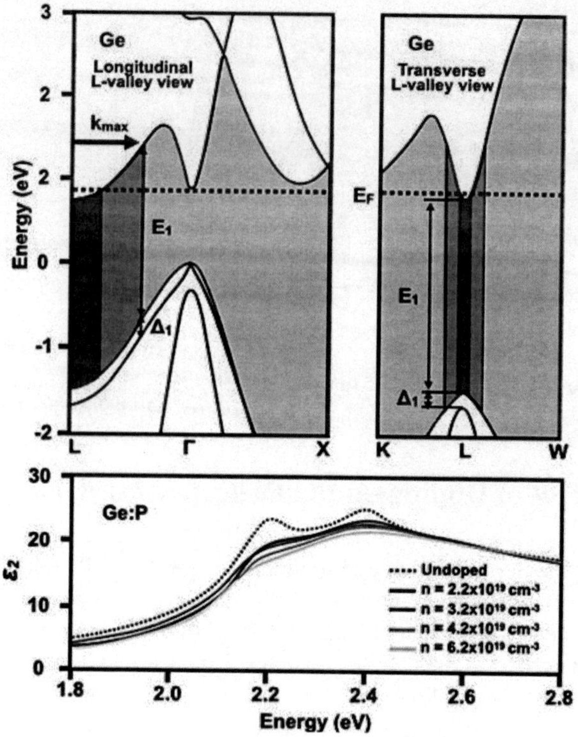

Fig. 4.19 Doping n-type Ge fills the k states in the conduction band close to the L point in the Brillouin zone. The filled states are shown in red for two views of the Brillouin zone. The band structure was determined using $k \cdot p$ theory. The imaginary part of the dielectric function is shown at the bottom for several levels of n type doping. Clearly, doping alters the shape of the M1 CP transition. Figure adapted and reprinted with permission from [49]. © 2017 American Physical Society

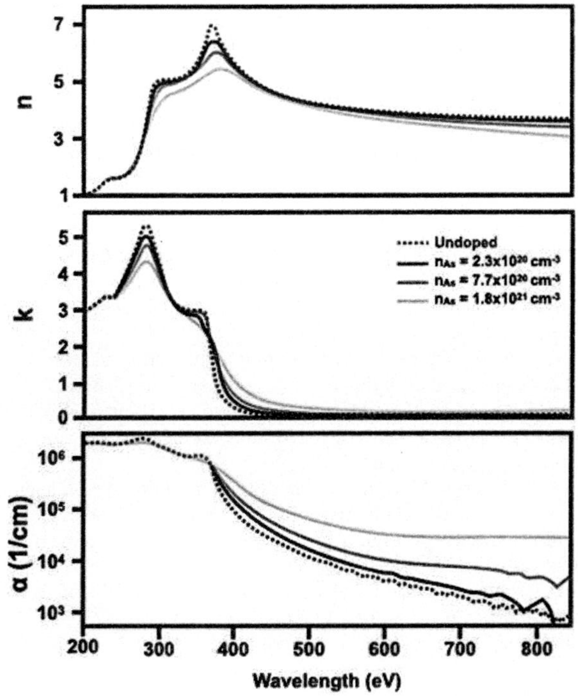

Fig. 4.20 The changes in the complex refractive index of As doped silicon are shown. Figure adapted from [51]

References

1. E. Fermi, *Nuclear Physics* (University of Chicago Press, 1950). ISBN: 978-0226243658, formula VIII.2
2. P.Y. Yu, M. Cardona, *Fundamentals of Semiconductors: Physics and Materials Properties*, 4th edn. (Springer, New York, 2010), p. 260
3. See for example, H.C. Ohanian, J.C. Markert, *Physics for Engineers and Scientists*, 3rd edn. (Norton, New York, 2007), p. 1093
4. F. Wooten, *Optical Properties of Solids* (Academic Press, New York, 1972), pp. 82, 121, 239
5. Note that [2, pp. 243–275, 4, pp. 110–122] provide derivations of the dielectric function from Fermi's Golden Rule. It is useful to remember that the unit systems are different in these derivations
6. F. Wooten, *Optical Properties of Solids, Appendix F* (Academic Press, New York, 1972), p. p240
7. M. Cardona, L.F. Lastras-Martinez, D.E. Aspnes, Comment on "*Ab initio* calculation of excitonic effects in the optical spectra of semiconductors". Phys Rev. Lett. **83**, 3970 (1999)
8. M. Cardona, *Modulation Spectroscopy, Supplement 11 in Solid State Physics* (Academic Press, New York, 1969), pp. 15–23
9. P.Y. Yu, M. Cardona, *Fundamentals of Semiconductors: Physics and Materials Properties*, 4th edn. (Springer, New York, 2010), pp. 261–273
10. P. Lautenschlager, M. Garriga, L. Vina, M. Cardona, Temperature dependence of the dielectric function and interband critical points in silicon. Phys. Rev. B **36**, 4821 (1987)

11. L. Mantese, K.A. Bell, U. Rossow, D.E. Aspnes, Evidence of near-surface localization of excited electronic states in crystalline Si. J. Vac. Sci. Technol. B **15**, 1196 (1997)
12. L. Mantese, K.A. Bell, U. Rossow, D.E. Aspnes, Interpretation of critical point energy shifts in crystalline Si by near-surface localization of excited electronic states. Thin Solid Films **313**, 557 (1998)
13. L.F. Lastras-Martínez, R.E. Balderas-Navarro, J. Ortega-Gallegos, A. Lastras-Martínez, J.M. Flores-Camacho, K. Hingerl, One electron and discrete excitonic contributions to the optical response of semiconductors around E1 transition: analysis in the reciprocal space. J. Opt. Soc. Am. B **26**, 725 (2009)
14. V. Kamineni, *Electron-Phonon Interactions and Quantum Confinement Effects on Optical Transitions in Nanoscale Silicon Films*, PhD Thesis, 2011
15. A. Savitzky, M.J.E. Golay, Smoothing and differentiation of data by simplified least squares procedures. Anal. Chem. **36**, 1627 (1964)
16. H.H. Madden, Comments on the "Savitzky-Golay convolution method for least-squares fit smoothing and differentiation of digital data. Anal. Chem. **50**, 1383 (1978)
17. G.E. Jellison, Jr., P.C. Joshi, Ch. 8: Crystalline silicon solar cells, in *Spectroscopic Ellipsometry for Photovoltaics*. Fundamental Principles and Solar Cell Characterization, vol. 1, ed. By H. Fujiwara, R. Collins (Springer, Cham, 2018), pp. 201–224
18. P. Lautenschlager, P.B. Allen, M. Cardona, Phonon-induced lifetime broadenings of electronic states and critical points in Si and Ge. Phys. Rev. B **33**(8), 5501 (1986)
19. M. Rohlfing, S.G. Louie, Electron-hole excitations and optical spectra from first principles. Phys. Rev. B **62**(8), 4927 (2000)
20. A. Daunois, D.E. Aspnes, Electroreflectance and ellipsometry of silicon from 3 to 6 e V. Phys. Rev. B **18**, 1824 (1978)
21. E.O. Kane, Band structure of silicon from an adjusted Heine-Abarenkov calculation. Phys. Rev. B **146**, 558 (1966)
22. P.Y. Yu, M. Cardona, *Fundamentals of Semiconductors* (Springer, Berlin, 1996), pp. 266–267
23. M. Chandrasekhar, F.H. Pollak, Phys. Rev. B **15**, 2127 (1977)
24. M. Chandrapal, F.H. Pollak, Conclusive evidence for the excitonic nature of the E_1 - $(E_1+\Delta_1)$ optical structure in diamond- and zincblende-type semiconductors at room temperature. Solid State Commun. **18**, 1263 (1976)
25. H. Haug, S.W. Koch, *Quantum Theory of the Optical and Electronic Properties of Semiconductors* (World Scientific, Hackensack, 2005), pp. 186–189
26. P.Y. Yu, M. Cardona, *Fundamentals of Semiconductors* (Springer, Berlin, 1996), pp. 282–292
27. D. Rideau, M. Feraille, L. Ciampolini, M. Minondo, C. Tavernier, H. Jaouen, A. Ghetti, Strained Si, Ge, and $Si_{1-x}Ge_x$ alloys modeled with a first-principles-optimized full-zone $k{\cdot}p$ method. Phys. Rev. B **74**, 195208 (2006)
28. M. Willatzen, M. Cardona, N.E. Christensen, Linear muffin-tin-orbital and k* p calculations of effective masses and band structure of semiconducting diamond. Phys. Rev. B **50**, 18054 (1994)
29. L. Vina, S. Logothetidis, M. Cardona, Temperature dependence of the dielectric function of germanium. Phys. Rev. B **30**, 1979 (1984)
30. S. Zollner, M. Cardona, S. Gopalan, Isotope and temperature shifts of direct and indirect band gaps in diamond-type semiconductors. Phys. Rev. B **45**, 3376 (1992)
31. J. Singh, *Electronic and Optical Properties of Semiconductor Structures* (Cambridge University Press, Cambridge, 2003), p. 62
32. U. Schmid, N.E. Christensen, M. Cardona, Relativistic band structure of Si, Ge, and GeSi: inversion-asymmetry effects. Phys. Rev. B **41**, 5919 (1990)
33. L. Vina, H. Hochst, M. Cardona, Dielectric function of α-tin and it temperature dependence. Phys. Rev. B **31**, 958–967 (1985)
34. P. Lautenschlager, M. Garriga, S. Logothetidis, M. Cardona, Interband critical points of GaAs and their temperature dependence. Phys. Rev. B **35**, 9174–9189 (1987)
35. S. Zollner, M. Garriga, J. Humlicek, S. Gopalan, M. Cardona, Temperature dependence of the dielectric function and the interband critical-point parameters of GaSb. Phys. Rev. B **43**, 4349 (1991)

36. M.L. Cohen, J.R. Chelikowsky, *Electronic Structure and Optical Properties of Semiconductors* (Springer, Berlin, 1981), p. 118ff
37. M. Schubert, *Infrared Ellipsometry on Semiconductor Layer Structures: Phonons, Plasmons, and Polaritons* (Springer, Berlin, 2004), see for example pp. 14–18, 109–143
38. L.X. Benedict, T. Wethkamp, K. Wilmers, C. Cobet, N. Esser, E.L. Shirley, W. Richter, M. Cardona, Dielectric function of wurtzite GaN and AlN thin films. Solid State Commun. **112**, 129–133 (1999)
39. T. Kawashima, H. Yoshikawa, S. Adachi, S. Fuke, K. Ohtsuka, Optical properties of hexagonal GaN. J. Appl. Phys. **82**, 3528–3535 (1997)
40. M. Rakel, C. Cobet, N. Esser, F. Fuchs, F. Bechstedt, R. Goldhahn, W. G.Schmidt, W. Schaff, GaN and InN conduction-band states studied by ellipsometry. Phys. Rev. B **77**, 115120 (2008)
41. S. Logothetidis, M. Cardona, P. Lautenschlager, M. Garriga, Temperature dependence of the dielectric function and the interband critical points of CdSe. Phys. Rev. B **34**, 2458–2469 (1986)
42. L.F. Lastras-Martínez, R.E. Balderas-Navarro, J. Ortega-Gallegos, A. Lastras-Martínez, J.M. Flores-Camacho, K. Hingerl, One electron and discrete excitonic contributions to the optical response of semiconductors around $E1$ transition: analysis in the reciprocal space. J. Opt. Soc. Am. B **26**, 725–733 (2009)
43. G.R. Muthinti, M. Medikonda, T.N. Adam, A. Reznicek, A.C. Diebold, Effects of stress on the dielectric function of strained pseudomorphic $Si_{1-x}Ge_x$ alloys from 0% to 75% Ge grown on Si (001). J. Appl. Phys. **112**, 053519 (2012)
44. J. Humliček, M. Garriga, M.I. Alonso, M. Cardona, Optical spectra of Si_xGe_{1-x} alloys. J. Appl. Phys. **65**, 2827 (1989)
45. S. Zollner, Optical properties and band structure of unstrained and strained $Si_{1-x}Ge_x$ and $Si_{1-x-y}Ge_xC_y$ alloys, Ch 12 in *Silicon-Germanium Alloys Growth Properties and Applications*. Optoelectronic Properties of Semiconductors and Superlattices, vol. 15, ed. by S.T. Pantelides, S. Zollner, M.O. Manasreh (Taylor & Francis, NY, 2002), p. 387
46. R. Lange, K.E. Junge, S. Zollner, S.S. Iyer, A.P. Powell, K. Eberl, Dielectric response of strained and relaxed $Si_{1-x-y}Ge_xC_y$ alloys grown by molecular beam epitaxy on Si(001). J. Appl. Phys. **80**, 4578 (1996)
47. M. Medikonda, G.R. Muthinti, R. Vasic, T. Adam, A. Reznicek, M. Wormington, G. Malladi, A.C. Diebold, The optical properties of pseudomorphic Ge(1-x)Snx ($x = 0$ to 0.11) alloys on Ge(001). J. Vac. Sci. Technol. B **32**, 061805 (2014)
48. V.R. D'Costa, W. Wang, Q. Zhou, E.S. Tok, Y.-C. Yeo, Above-bandgap optical properties of biaxially strained GeSn alloys grown by molecular beam epitaxy. Appl. Phys. Lett. **104**, 022111 (2014)
49. C. Xu, N.S. Fernando, S. Zollner, J. Kouvetakis, J. Menéndez, Observation of phase-filling singularities in the optical dielectric function of highly doped n-type Ge. Phys. Rev. Lett. **118**, 267402 (2017)
50. G.E. Jellison, Jr., F.A. Modine, C.W. White, R.F. Wood, R.T. Young, Optical properties of heavily doped silicon between 1.5 and 4.1 eV. Phys. Rev. Lett. **46**, 1414–1417 (1981)
51. G.E. Jellison, Jr., P.C. Joshi, Ch8 Crystalline silicon solar cells, in *Spectroscopic Ellipsometry for Photovoltaics*, ed. by H. Fujiware, R. Collins (Springer, Switzerland, 2018), pp. 201–225

Chapter 5
Excitons and Excitonic Effects During Optical Transitions

Abstract In this chapter, excitons and excitonic effects during optical transitions are discussed. The transition energy for 1D, 2D, and 3D materials which include the exciton binding energy provide a focal point for the chapter. The well-known Elliott description of optical absorption for 3D, 2D, and 1D materials and the Sommerfeld factor for Coulomb enhancement are discussed. The effect of quantum confinement is described. A quantum mechanical derivation of excitonic effects on direct gap optical transitions which alters the energy dependence of optical absorption and thus the dielectric function is presented. Photoluminescence spectra from semiconductors and quantum wells are also presented.

Understanding excitons and excitonic effects is a critical part of describing optical spectra. Excitons are electron–hole pairs produced by optical absorption. Excitons can be observed in optical data at energies below the band gap and through the luminescence that occurs after the exciton's electron and hole recombine. The strong interaction between the electron and the hole it leaves behind during direct gap optical absorption is referred to as an excitonic effect. In Chap. 4, we discussed the influence of excitonic effects on the E_1 direct gap CP (critical point) of silicon at room temperature. The influence of excitons on the low temperature absorption at the band gap (absorption edge) of gallium arsenide was also discussed. Both of these effects are related and can be observed in the dielectric function using spectroscopic ellipsometry. Excitons are also observed by photoluminescence (PL).

Two types of excitons are frequently discussed: Frenkel excitons and Wannier excitons [1]. A Frenkel exciton is localized to a volume that is typically the size of a crystal lattice unit cell, and the electron and hole interact without screening. For a Wannier exciton, the orbit of the electron with the hole extends over several atoms and the interaction is screened by the local dielectric function of the material. Wannier excitons occur in covalent solids such as silicon, and in CuO the Wannier exciton extends over 14 nm [1]. A number of different terms are often used to describe excitons [1]. Often, an exciton is bound to a defect or impurity. The energy of the exciton is shifted by this state, and the presence and identity of the defect or impurity can be determined through measuring the energy of peaks in PL spectra. Excitons can bind together and terms such as exciton gas [electron hole gas (EHG)], exciton

A. Diebold and T. Hofmann, *Optical and Electrical Properties of Nanoscale Materials*, Springer Series in Materials Science 318,
https://doi.org/10.1007/978-3-030-80323-0_5

liquid [electron hole liquid (EHL)], electron hole plasmas (EHP) when the electron and hole are not bound, and electron–hole droplets are used to describe the low temperature PL spectra of quantum confined semiconductor structures [2–4].

The flow of this chapter is as follows: A hydrogen atom like model for the binding energy between an electron and hole provides and introduction to excitons in 3D materials. The binding energy of excitons in 1D, 2D. and 3D materials is discussed followed by a comparison of the effective masses of electrons and holes determined using $k \cdot p$ theory with experimental results. Next the topic of dielectric quantum wells and the effect of the thickness of silicon nanolayers on Band Gap and Photoluminescence is presented. In order to provide a theoretical background for the effect of bound excitons and excitonic effect on the spectral shape for 1D, 2D, and 3D, a quantum mechanical derivation of excitonic effects on direct gap optical transitions which alters the energy dependence of optical absorption and thus the dielectric function is presented. This is related to the Elliott description of optical absorption and the Sommerfeld factor for Coulomb enhancement. The effect of quantum confinement is presented. Photoluminescence spectra from semiconductors and quantum wells are used to illustrate the effects discussed in this chapter.

5.1 Description of Excitons in 3D, 2D, and 1D

First, we consider the formation of an exciton during optical absorption for a bulk material. Here, we calculate the binding energy between the electron and hole. Here we use a hydrogen atom like model to arrive that the binding energies of the electron and hole in an exciton. The energy of the luminescence of the exciton is approximately the energy of the gap between the conduction and valence band minus the exciton binding energy. Thus, it is very useful to have a quantum mechanical description of the exciton binding energy. In a subsequent section of this chapter, the imaginary part of the dielectric function will be derived so that the resulting expression includes excitons and the Coulomb enhancement of CP absorption due to the electron and hole interaction during absorption. The hydrogen atom like classical picture of the Hamiltonian for the electron and hole starts with the kinetic energy for the electron and hole and the potential energy due to their Coulomb attraction:

$$H = \frac{1}{2} m_e^* \dot{\vec{r}}_e^{\,2} + \frac{1}{2} m_h^* \dot{\vec{r}}_h^{\,2} + V(\vec{r}_e - \vec{r}_h) \tag{5.1}$$

where $V(\vec{r}_e - \vec{r}_h)$ is the potential energy due to the interaction between the electron and hole. m_e^* and m_h^* are the effective masses of the electron and hole in the material. The total mass M, center of mass position $\vec{R}$, and relative position of the electron and hole $\vec{r}$ are:

$$m_e^* \vec{r}_e + m_h^* \vec{r}_h = \left(m_e^* + m_h^*\right) \vec{R} = M \vec{R} \quad \text{and} \quad \vec{r} = \vec{r}_e - \vec{r}_h \tag{5.2}$$

Defining a reduced (effective) mass $\mu = \frac{m_e^* m_h^*}{M}$ one can rewrite (5.1) so that the relative motion of the electron and hole is separated from the relative motion of the electron–hole pair:

$$H = \frac{1}{2}\mu\dot{\vec{r}}^2 + \frac{1}{2}M\dot{\vec{R}}^2 + V(\vec{r}\,) \rightarrow H = \frac{P_r^2}{2\mu} + \frac{P_{CM}^2}{2M} + V(\vec{r}\,) \tag{5.3}$$

Using the correspondence principle and $V(\vec{r}\,) = \frac{-e^2}{\varepsilon|\vec{r}\,|}$ the quantum mechanical Hamiltonian is:

$$H = -\frac{\hbar^2}{2\mu}\vec{\nabla}_r^2 - \frac{\hbar^2}{2M}\vec{\nabla}_R^2 - \frac{e^2}{\varepsilon|\vec{r}|} \tag{5.4}$$

The Hamiltonian for the motion of the center of mass of the particle with wavevector $\vec{K}$ has the same form as that of a free electron:

$$H_{CM}\psi(\vec{R}) = -\frac{\hbar^2}{2M}\vec{\nabla}_R^2\psi(\vec{R}) = E_{CM}\psi(\vec{R}) \quad \text{with} \quad \psi(\vec{R})$$
$$= N^{-1/2}e^{i\vec{K}\cdot\vec{R}} \quad \text{and} \quad E_{CM} = \frac{\hbar^2 K^2}{2M} \tag{5.5}$$

The Hamiltonian for the relative motion of the electron and hole has the form as the hydrogen atom where n is a quantum number and R^* is the Rydberg constant for hydrogen:

$$H_{\vec{r}}\phi(\vec{r}) = \left[-\frac{\hbar^2}{2\mu}\vec{\nabla}_r^2 - \frac{e^2}{\varepsilon|\vec{r}|}\right]\phi(\vec{r}) = E_{\vec{r}}\phi(\vec{r}) \tag{5.6}$$

$$E_{\vec{r}}(n) = E_{\vec{r}}(\infty) - \frac{R^*}{n^2} \quad \text{and} \quad R^* = \frac{\mu e^4}{2\hbar^2\varepsilon_0^2} = 13.6\ \text{eV}\frac{\mu}{m_e\varepsilon_0^2} \tag{5.7}$$

ε_0 is the static dielectric constant of the solid. The energy of an exciton in a 3D material is the gap energy minus the exciton binding energy and its kinetic energy which stated below and the dashed lines in Fig. 5.1 show the exciton energy levels:

$$E_{exciton} = E_{gap} + \frac{\hbar^2 K^2}{2M} - \frac{R^*}{n^2} \tag{5.8}$$

The binding energy of excitons in a semiconductor is typically a few meV. The binding energy of an exciton in GaAs is 4.9 meV. The Bohr exciton radius of GaAs is 13 nm. The thermal energy $K_B T = (8.617 \times 10^{-5}\ \text{eV/K})T$ where K_B is Boltzman's constant. At 300 K, $KT = 25$ meV, and at 21 K, $KT = \sim 2$ meV. Thus excitons are thermally ionized in GaAs at room temperature.

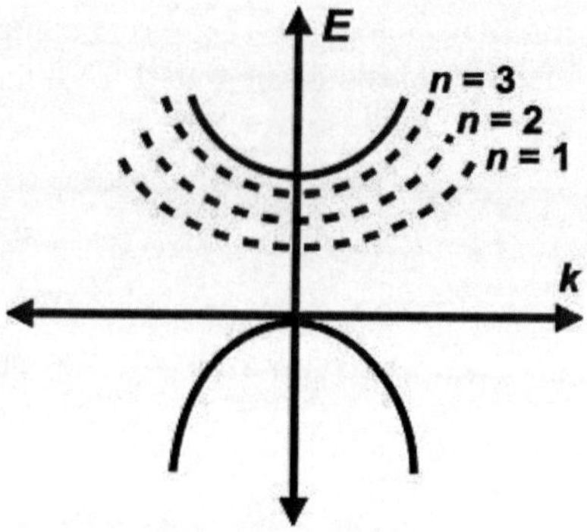

Fig. 5.1 Exciton energy levels for a direct gap transition at a M_0 Critical Point where the energy dependence of the valence and conduction bands on wavevector is parabolic

5.2 Energy of Excitons in 3D, 2D, and 1D

In this section, a phenomenological approach is used to describe the exciton energy for 3D (bulk), 2D (nanofilms), and 1D (nanowires) materials. The allowed values of the wavevector term in (5.8) change from $K^2 = K_x^2 + K_y^2 + K_z^2$ for 3D to $K^2 = K_x^2 + K_y^2$ for 2D nanofilms, and $K^2 = K_x^2$ for 1D nanowires. The energies of excitons in 3D, 2D, and 1D materials can be estimated along with the oscillator strength $f(n)$ for quantum number n and exciton Bohr radius a_{Bohr} and Rydberg constant using [5]:

5.2.1 3D (Bulk Materials)

$$E_{exciton}(3D) = E_{gap} + \frac{\hbar^2(K_x^2 + K_y^2 + K_z^2)}{2M} - \frac{R^*}{n^2}; \tag{5.9}$$

$$f(n) \propto n^{-3};\ a_{ex\ Bohr} \propto a_{hydrogen}\ n$$

$$a_{ex\ Bohr} = \frac{\hbar\varepsilon_0}{\mu e^2}$$

5.2.2 2D (Nanofilms)

The exciton is confined in only one dimension L and the particle in box quantization is assumed: $E_{QC}(nanofilm) = \frac{\hbar^2\pi^2}{2ML}n^2$. The electron–hole binding energy for a 2D film is determined from the equation for radial motion for the hydrogen like electron and hole confined to 2D: $BE = -R^*/\left(n - \frac{1}{2}\right)^2$

$$E_{exciton}(2D) = E_{gap} + E_{QC}(nanofilm) + \frac{\hbar^2 \left(K_x^2 + K_y^2\right)}{2M} - \frac{R^*}{\left(n - \frac{1}{2}\right)^2};$$

$$f(n) \propto \left(n - \frac{1}{2}\right)^{-3}; \quad a_{Bohr} \propto a_{hydrogen} \left(n - \frac{1}{2}\right) \tag{5.10}$$

5.2.3 1D (Nanowire)

Where the exciton is confined in 2 dimensions (L_y and L_z) so that $E_{QC}(nanowire) = \frac{\hbar^2 \pi^2}{2M}\left(\frac{n_y^2}{L_y^2} + \frac{n_z^2}{L_z^2}\right)$ and the energy of the electron hole pair confined to 1D is given by $-\frac{R^*}{\lambda_0^2}$ [6]:

$$E_{exciton}(1D) = E_{gap} + E_{QC}(nanowire) + \frac{\hbar^2 K_x^2}{2M} - \frac{R^*}{\lambda_0^2} \tag{5.11}$$

where λ_0 is determined from boundary conditions so that $\frac{1}{2} + \lambda_0 \ln\left(\frac{2\gamma R}{\lambda_0 a_{hydrogen}}\right) = 0$ and $\gamma \sim 0.3$ [6].

5.2.4 0D (Nanodots)

The exciton is confined in 3 dimensions (L_x, L_y, L_z). $E_{QC}(nanodot) = \frac{\hbar^2 \pi^2}{2M}\left(\frac{n_x^2}{L_x^2} + \frac{n_y^2}{L_y^2} + \frac{n_z^2}{L_z^2}\right)$

$$E_{exciton}(0D) = E_{gap} + E_{QC}(nanodot) - Coulomb\ attraction \tag{5.12}$$

It is useful to explore the excitonic structure observed at the band edge of GaAs. When light is absorbed at the band gap of GaAs, a clear excitonic structure is observed in the dielectric function and absorption coefficient at low temperature as seen in Fig. 5.2 [7]. An important aspect of understanding the line shape of the absorption coefficient is that the CP energy and broadening (lineshape) change with temperature due to thermal expansion of the crystal lattice, single phonon–electron interaction and two phonon–electron interactions [7]. The low temperature (< 5 K) energies of the E_0 and $E_0 + \Delta$ CPs are 1.517 and 1.851 eV. The change in the energy of the E_0 CP at low temperature and the peak broadening Γ with temperature in Kelvin are [7]:

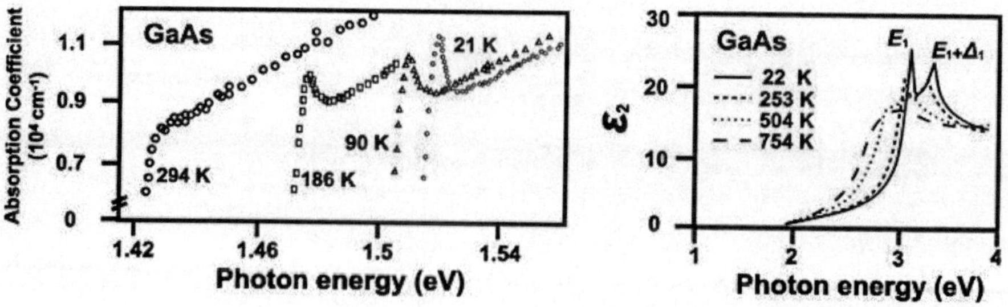

Fig. 5.2 Absorption coefficient and Imaginary part of the dielectric function of GaAs as a function of temperature. The peak due to the bound exciton is visible in the 21, 90, and 186 K absorption coefficient data. Figure on left adapted from [9] which itself was adapted from [49]. Figure on right adapted and reprinted with permission from [7]. © 1987 American Physical Society

$$E_0(T) = 1.517\,\text{eV} - \frac{\left(5.5 \times 10^{-4}\,\text{eV/K}\right)T^2}{(T + 225\,\text{K})} \tag{5.13}$$

$$\Gamma(T) = 23\,\text{meV}\left[1 + \frac{2}{(e^{337\,\text{K}/T} - 1)}\right] - 22\,\text{meV} \tag{5.14}$$

Similar equations are available for $E_0 + \Delta$ [7]. These formulas show that band edge changes from 1.517 eV at 21 K to 1.42 eV at 300 K, and that is evident from the absorption coefficient as shown in Fig. 5.2. It is important to note that the E_1 and other CPs energies and broadening also change with energy and the constants used to describe the energy and broadening changes are available in [7] for GaAs. The binding energy of the exciton can be measured using photoluminescence or from the absorption data. This allows comparison of the Rydberg formula with the measured binding energy and calculation of an experimental effective mass for the exciton. That effective mass can be compared with the effective mass for the specific bands involved in the transition. Formulas for an electron in the conduction band, the heavy hole and the light hole in the valence bands are available along the (001) and (111) directions from $k \cdot p$ theory [8].

Next we use the introductory $k \cdot p$ theory from Chap. 2 to estimate the effective mass of the electron and hole in an exciton to those determined from theoretical data. This highlights the approximations in the presentation of $k \cdot p$ theory discussed in Chap. 2. We note that less approximate $k \cdot p$ theory calculations are available for many materials. From (5.7), one can see that the binding energy (BE) of the exciton can be compared to the calculated BE from the following formula once the effective mass of the electron m_e^* and the hole m_h^* are determined.

$$BE = \frac{13.6\,\text{eV}\frac{\mu}{m_e \varepsilon_0^2}}{n^2} \quad \text{and} \quad \mu = \frac{m_e^* m_h^*}{M} \tag{5.15}$$

As discussed in Chap. 2, the effective mass is different along the (001) and (111) directions of the Brillouin zone, and an average effective mass for the electron and the heavy and light hole must be calculated. We compare the experimental value of the center of mass with that calculated from $k \cdot p$ for an exciton having the hole in the heavy hole band, an exciton having the hole in the light hole band, and an exciton having a hole effective mass that is the average of the light and heavy hole masses. Here we used the effect mass of the conduction band $\frac{m_e}{m_c^*} = 1 + \frac{20\,\text{eV}}{E_{gap}}$ for band gaps less than 2 eV, the average heavy hole and average light hole formulas Chap. 2 (2.68) and (2.69) as well as the following approximate formula for averaging the heavy and light hole [8]:

$$m_h^* = 2\left(\frac{1}{m_{hh}^*} + \frac{1}{m_{lh}^*}\right)^{-1} \tag{5.16}$$

The effective masses for the light hole, heavy hole, averaged hole mass, and conduction band are listed in Table 5.1.

A comparison of the experimental and calculated exciton BE for silicon and germanium show the complexity associated with the above analysis of excitons. This is shown in Table 5.2. This points to the reason that ab initio calculations are done for each materials system.

The previous discussion described an exciton that forms during a direct gap optical transition with hyperbolic bands (M_0) transition. Many transitions occur at CPs that are not perfect hyperbolic bands, for example at M_1 CPs. A more complete theoretical picture of exciton formation would include the photon interacting with the electron in the valence band resulting in the promotion of the electron to the conduction band and its interaction with the resulting hole in the valence band. In addition, many excitons are ionized at room temperature because the thermal energy $K_B T \sim 25$ meV is more than the exciton binding energy. Both discrete and continuum exciton effects

Table 5.1 Effective mass of conduction band electrons, heavy holes and light holes as calculated from $k \cdot p$ theory

	(111)	(111)	AVE	AVE	(001)	(001)	
	m_{hh}^*/m_e	m_{lh}^*/m_e	m_{hh}^*/m_e	m_{lh}^*/m_e	m_{hh}^*/m_e	m_{lh}^*/m_e	m_c^*/m_e
Si	1.3	0.13	−0.9	0.1	0.28	0.2	0.063
GaAs	−0.78	0.07	0.84	0.08	0.40	0.09	0.078
Ge	−0.37	0.03	0.46	0.04	0.20	0.05	0.04

The formula used for determining the conduction band effective mass is most useful for direct gap semiconductors and is used for reference for indirect gaps semiconductors Si and Ge

Table 5.2 Comparison of experimental reduced mass and those calculated from $k \cdot p$ theory

	μ_{lh-c}	μ_{hh-c}	μ_{aveh-c}	exp
GaAs	0.04 m_e	0.0714 m_e	0.051 m_e	0.062 m_e
Si	0.099 m_e	0.169 m_e	0.125 m_e	0.153m_e

can occur in optical transitions and they impact the shape of the dielectric function. This is further explored in the subsequent sections of this chapter.

5.3 Exciton Binding Energy in Semiconductor Dielectric Quantum Wells

It is possible to increase the exciton BE above the thermal energy $K_B T$ by confining the exciton in a quantum well (QW). The increase in BE for dielectric–semiconductor–dielectric quantum wells has been theoretically studied for SiO_2-Si-SiO_2 by Keldysh [10]. This system is interesting since it is experimentally possible to fabricate these QW structures from silicon on insulator (SOI) wafers that are used as substrates from some integrated circuit manufacturing. The theoretical description of the interaction potential between the electron and hole is modified by the presence of the dielectric film above and below the semiconductor slab [10]. For very thin semiconductor slabs, these lines of force travel through the dielectric layer and the relevant dielectric constant for this interaction is that of the dielectric layer. As the slab thickness increases, the lines of force travel through both the semiconductor and the dielectric so the interaction potential needs to involve both dielectric constants. A third thickness range occurs when the semiconductor slab is thick enough so that the lines of force are all contained in the slab. The criteria for separating these three types of interaction has been expressed in terms of the semiconductor slab thickness d, exciton radius in the bulk semiconductor $r_{ex-bulk}$, the ratio of the dielectric constant of the insulator ε_i and the semiconductor ε_s (5.10), and $\mu = \frac{m_e^* m_h^*}{m_e^* + m_h^*}$:

$$d \leftrightarrow \left(\frac{\varepsilon_i}{\varepsilon_s}\right)^2 r_{ex-bulk} \quad \text{where } r_{ex-bulk} = \frac{\varepsilon_s \hbar^2}{e^2 \mu} \tag{5.17}$$

The exciton BE increase has been categorized in three different regimes based on the interaction potential.

The ***first region*** *occurs for* ***very thin films*** *when the film thickness* $d < \left(\frac{\varepsilon_i}{\varepsilon_s}\right)^2 r_{ex-bulk}$.

For Si and SiO_2 for $r_{ex-bulk} = 4\,\text{nm}$, $\varepsilon_i = 3.9$ and $\varepsilon_s = 11.8$ and $d < 0.44\,\text{nm}$. The bulk exciton BE of 15 meV is increased so that the exciton $BE = \left(\frac{2\varepsilon_s}{\varepsilon_i}\right)^2 BE_{bulk} \sim 550\,\text{meV}$. Here, the film thickness is smaller than 0.44 nm which is not practically achievable. Suppose we have HfO_2 as the top dielectric layer, then the dielectric constant is averaged $\varepsilon_i = \frac{3.9+16}{2} = 9.95$ for amorphous HfO_2. Here, $d < 2.8\,\text{nm}$ and BE = ~ 84 meV. Crystalline HfO_2 has a larger dielectric constant and thus the BE increases.

The ***second region*** *is for* ***moderately thin*** *semiconductor slabs* [3, 10]. Here,

$$r_{ex-bulk} > d > \left(\frac{\varepsilon_i}{\varepsilon_s}\right)^2 r_{ex-bulk}$$

$$\text{and } BE(d) = \frac{e^2}{\varepsilon_s d}\left[\ln\left[\left(\frac{\varepsilon_s}{\varepsilon_i}\right)^2 \frac{d}{r_{ex-bulk}}\right] - 0.577\right]\text{eV} \tag{5.18}$$

Thus the $BE \sim 31$ meV for a 1 nm slab and $BE = 58$ meV for a 2 nm slab of Si surrounded by SiO_2 both of which are above KT at room temperature. In the third region, $d > r_{ex-bulk}$ and the slab will increasingly behave like a bulk materials as the thickness increases. It is interesting to note that the BE of excitons for moderately thin slabs of GaAs does not increase above KT at room temperature.

Keldysh also studied semiconductor nanowires surrounded by a dielectric. For ***extremely thin nanowires*** [10]:

$$d < \left(\frac{\varepsilon_i}{\varepsilon_s}\right)^{3/2} r_{ex-bulk} \text{ and } E_{nw} = 4\left(\frac{\varepsilon_s}{\varepsilon_i}\right)^2 BE_{bulk}\left[\ln\left(\frac{\varepsilon_s}{\varepsilon_i}\frac{d}{r_{ex-bulk}}\right)\right]^2$$
$$= \frac{e^2}{d\sqrt{\varepsilon_i \varepsilon_s}}\sqrt{2\ln\frac{\varepsilon_s}{\varepsilon_i}} \tag{5.19}$$

*For **moderately thick nanowires** of diameter d, the **third region**:*

$$r_{ex-bulk} > d > \left(\frac{\varepsilon_i}{\varepsilon_s}\right)^{3/2} r_{ex-bulk} \text{ and } BE_{nw} = \frac{e^2}{d\sqrt{\varepsilon_i \varepsilon_s}}\sqrt{2\ln\left(\frac{\varepsilon_s}{\varepsilon_i}\right)} \tag{5.20}$$

5.4 The Impact of Nanolayer Thickness on Band Gap and Photoluminescence Determination of Exciton Binding Energy

Since excitons are observed as absorption features that alter the absorption spectra at the band edge, it is useful to discuss the change of the band gap that occurs with the decrease in the layer thickness for semiconductor films at this point in the chapter [4]. It is also useful to mention that some indirect band gap materials such as germanium have been predicted to change into direct gap materials [11, 12].

The thickness dependence of photoluminescence (PL) from thin SOI layers has been studied experimentally and theoretically. Here, discuss how to determine the PL peak maximum from the exciton BE for SiO_2-Si-SiO_2 QW and the thickness dependence of the semiconductor slab's band gap. The change in the valence band maximum and conduction band minimum due to many body effects has been parameterized [13, 14] and used to calculate the BE of excitons in thin SiO_2-Si-SiO_2 QWs (layers) [4]. The parameterized VB and CB energies as a function of slab thickness d are [4, 13]:

$$E_{VB}(d) = \frac{K_v}{d^2 + a_v d + b_v} \text{ and } E_C(d) = \frac{K_c}{d^2 + a_c d + b_c} + E_{gap} \tag{5.21}$$

where d is in nm, $K_v = -1326.2$ meV nm^2, $a_v = 1.418$ nm, $b_v = 0.296$ nm^2, $K_c = 394.5$ meV nm^2, $a_c = 0.939$ nm, and $b_c = 0.324$ nm^2. It is useful to note that this parameterization has also been applied to nanowires [13]. Calculation of the optical observed band gap requires inclusion of the electron–hole binding energy for excitons, band gap renormalization, and the phonon required for indirect transitions. Equations (5.9) and (5.10) are approximations. Using (5.21) for the thickness dependent band gap and (5.18) for the exciton BE for moderate silicon slab thickness it has been found that the Pl peak maximum will occur at [4]:

$$E_{PL}(d) = E_C(d) - E_{VB}(d) - BE(d) - \hbar\omega_{TO} \tag{5.22}$$

where $\hbar\omega_{TO}$ is the energy of the transverse optical phonon required for indirect gap transitions. A comparison of the PL peak maximum with experimental PL data is shown in Fig. 5.3.

Photoluminescence characterization of any material requires careful attention to the laser power and wavelength. This is also true for SOI. Photoluminescence measurements of silicon are a function of the excitation energy. Under high enough laser excitation, even relatively thick films (190 nm) show EHL and EHP with the

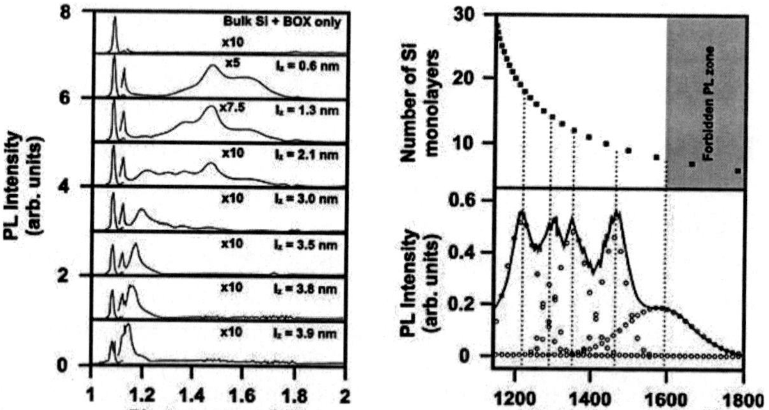

Fig. 5.3 On the left, photoluminescence data from a series of SiO_2-Si-SiO_2 quantum wells are shown and on the right the spectra from nominally 2.1 nm thick SiO_2-Si-SiO_2 quantum well is shown with an expanded energy scale. The number of monolayers contributing to each peak for the 2.1 nm quantum well are shown above the spectra. The monolayer thickness is $\frac{a}{4}$ where a = 0.5431 nm. The QW thickness clearly varies across the area of the QW. The 351 nm Ar^+ ion laser power density was 4 kW/cm^2 and the temperature was ~ 6 K. Figure adapted and reprinted with permission from [4]. © 2005 American Physical Society

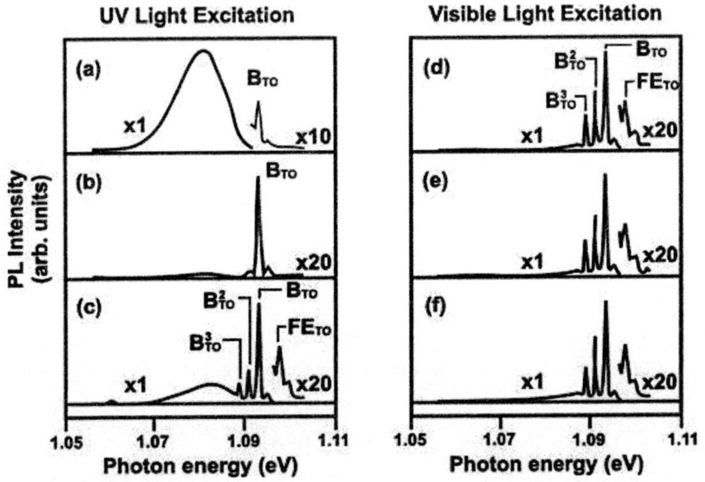

Fig. 5.4 Photoluminescence spectra of thick SOI and bulk silicon are shown as a function of excitation source at 4.2 K. The spectra in (**a**) and (**d**) are from the same SOI (sample A) as are (**b**) and (**e**) (sample B). Figures (**c**) and (**f**) are for bulk silicon. The thickness values of the surface oxide, silicon layer and bottom oxide layer are 630 nm, 75 nm, and 113 nm respectively for sample A, and the surface oxide layer was removed by HF etching for sample B. The UV excitation came from using both the 351 nm and 364 nm lines from an Ar^+ ion laser, and the penetration depth was ~ 0.05 μm. The visible excitation came from the 647 nm line of a Kr^+ ion laser and the penetration depth was ~ 5 μm. The laser power for both excitations was 50 mW focused in 1 mm diameter circle. Figure adapted from [15]. © 1998 The Physical Society of Japan and The Japan Society of Applied Physics

existence and luminescent energy being attributed to lateral confinement. The identification of PL peaks for thicker (non-quantum confined) SOI is useful when thin SOI samples are examined. PL spectra of SOI show excitons bound to boron impurities and several side bands associated with transverse optical (TO), transverse acoustic (TA) and zone center optical (O^{Γ}) phonons as shown in Fig. 5.4 [15]. The spectra in Fig. 5.4 show PL peaks below 1.11 eV, while the data in Fig. 5.3 shows PL peaks from the SiO_2-Si-SiO_2 quantum wells from 1.15 to 1.8 eV. Pauc et al., also observed a strong dependence on the excitation laser wavelength and they observed the EHL and EHP for the SiO_2-Si-SiO_2 quantum wells [3, 4].

Measuring the band gap as a function of thickness is challenging. Not only do the thin silicon films seem to have a variety of thickness values inside a given area, the energy resolution required for observation of a change in band gap is better than 0.1 eV which is challenging for ellipsometry, and the absorption is very small reducing layer contrast. Thus it is a challenge to measure the band gap of thin SOI. One recent study used X-Ray Photoelectron spectroscopy to study the energy of the occupied states in the valence band and synchrotron X-Ray L-edge absorption near edge structure (XANES) to study the conduction band energy [16]. The change in band gap with silicon thickness for SOI is shown in Fig. 5.5. The challenges with the fabrication of the samples and the measurement of VB and CB energies indicate

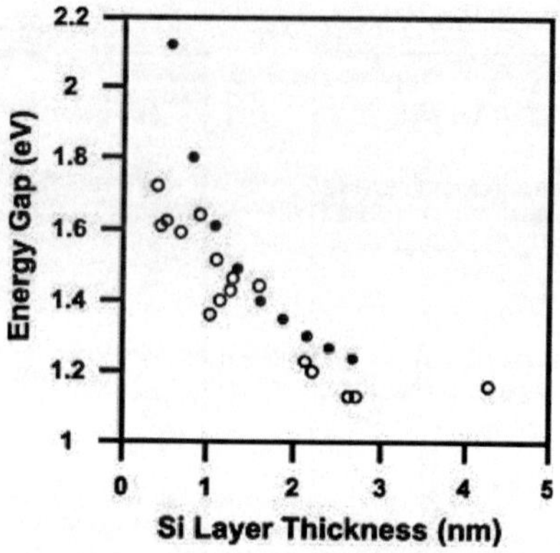

Fig. 5.5 Band gap versus silicon layer thickness of thin silicon on insulator is shown. Figure adapted and reprinted from [16] with the permission of AIP Publishing

that the general trend of increasing band gap as silicon layer thickness decreases in correct.

5.5 Derivation of Dielectric Function Including Excitons and Excitonic Effects

In this section, we will see how the exciton energies from (5.9) to (5.11) are contained in a derivation that includes the electron and hole final state. As discussed in Chap. 4, the E_1 CP transition of germanium and gallium arsenide was identified to excitonic in nature using electroreflectance spectra [17]. This increased absorption around the E_1 CP of diamond and zinc blende semiconductors is also observed in 2D materials at other CP transitions such as the M point transition in graphene [18–20]. Understanding the physics of the lineshape of excitonic CP transitions is useful in interpreting spectra such as the dielectric function. There are two contributions to peak shape of excitonic transitions. At low temperature, both the exciton peaks and the increased absorbance (oscillator strength) due to electron–hole exchange are observed. A good example of this is seen in Fig. 5.2 for GaAs. Below we present two theoretical approaches that describe these effects [21, 22]. The first derivation uses an envelope function to account for the presence of both the electron and the hole and expands on the derivation of the imaginary part of the dielectric function shown in Chap. 4. The second approach due to Elliott [23], results in a Sommerfeld factor that is dimension dependent so that the free excitons and the increased absorption due to electron–hole exchange can be determined for 3D, 2D, and 1D semiconductors. Ab initio quantum mechanical calculations can also be used [24]. These were discussed for silicon in Chap. 4.

5.5.1 *Quantum Mechanical Derivation of Excitonic Effects for a Direct Gap Transition*

In Chap. 4, the imaginary part of the dielectric function was derived using single particle Bloch states for both the initial stated in the valence and final state in the conduction band. First, we expand on that derivation using a final state wavefunction that is a product of the electron wavefunction, the hole wavefunction, and an envelope function describing the relative motion of the electron and hole. At room temperature, optical transitions create excitons that not bound. Thus the BE of the exciton is > 0. For these transitions, the electrons are considered to be excited into continuum states. We expect two contributions to the imaginary part of the dielectric function: a sharp (~ delta function) absorption peak when the electron and hole are bound and a contribution from unbound electron–hole states. Once again, we search for solutions to (5.6), $H_{\vec{r}}\phi(\vec{r}) = \left[-\frac{\hbar^2}{2\mu}\vec{\nabla}_r^2 - \frac{e^2}{\varepsilon|\vec{r}|}\right]\phi(\vec{r}) = E_{\vec{r}}\phi(\vec{r})$. Two types of exciton wavefunctions are possible, Bloch or Wannier. Wannier wavefunctions are localized making them appropriate for excitons when compared to Bloch wavefunctions which are delocalized. Following [22], the Wannier wavefunction $a_n(\vec{r}, \vec{R}_i)$ is the Fourier transform of the Bloch functions in the nth band $\psi_{n\vec{k}}(\vec{r})$ for N unit cells and real space lattice vector $\vec{R}_i$ to the unit cell, $\vec{r} - \vec{R}_i$ is the position inside the unit cell, and wavevector $\vec{k}$ as follows:

$$\psi_{n\vec{k}}(\vec{r}) = \frac{1}{\sqrt{N}} \sum_{\vec{R}_j} e^{i\vec{k}\cdot\vec{R}_i} a_n(\vec{r}, \vec{R}_i) \tag{5.23a}$$

$$a_n(\vec{r}, \vec{R}_i) = \frac{1}{\sqrt{N}} \sum_{\vec{k}} e^{-i\vec{k}\cdot\vec{R}_i} \psi_{n\vec{k}}(\vec{r}) \tag{5.23b}$$

The Wannier wavefunctions are localized in real space as is evident from the real space index $\vec{R}_i$ in (5.23a) and (5.23b). The Wannier wavefunctions are a complete set of orthonormal wavefunctions that are only a function of $\vec{r}_j - \vec{R}_i$ where j is e or h [22]. Thus the exciton wavefunction can be written in two ways using either the Wannier or Bloch functions:

$$\psi(\vec{r}_e, \vec{r}_h) = \frac{1}{\sqrt{N}} \sum_{\vec{k}_e, \vec{k}_h} c(\vec{k}_e, \vec{k}_h)\psi_{\vec{k}e}(\vec{r}_e)\psi_{\vec{k}h}(\vec{r}_h) \tag{5.24a}$$

$$\psi(\vec{r}_e, \vec{r}_h) = \frac{1}{\sqrt{N}} \sum_{\vec{R}_e, \vec{R}_h} \phi(\vec{r}_e, \vec{r}_h)\, a_{ne}(\vec{r}_e, \vec{R}_e)\, a_{nh}(\vec{r}_h, \vec{R}_h) \tag{5.24b}$$

where, $\phi(\vec{r}_e, \vec{r}_h)$ are envelope functions, that is they are the wavefunctions for the relative motion of the bound or unbound electron–hole pairs. The envelope wavefunction will be similar to the hydrogen like wavefunction, and it can have positive (unbound) or negative (bound) energy states. Since the relative motion of the exciton is similar to that of the electron and proton in a hydrogen atom, there will be three quantum numbers: n, l, and m just as for hydrogen. The envelope function for the center of mass position R_{CM} and relative position $\vec{r} = \vec{r}_e - \vec{r}_h$ is:

$$\phi(\vec{r}_e, \vec{r}_h) = \frac{1}{\sqrt{N}} e^{i\vec{k}\cdot\vec{R}_{CM}} R_{n,l}(\vec{r}\,)\, Y_{lm}(\theta, \varphi) \tag{5.25}$$

Since the Bloch and Wannier wavefunctions are Fourier transforms of each other, Wannier wavefunctions for both the electron and hole can be written as a Fourier transform of the Bloch wavefunctions for the electron in the conduction band and hole in the valence band respectively:

$$a_{ne}(\vec{r}_e, \vec{R}_e) = \frac{1}{\sqrt{N}} \sum_{\vec{k}_e} e^{-i\vec{k}_e\cdot\vec{R}_e}\, \psi_{\vec{k}_e}(\vec{r}_e) \tag{5.26a}$$

$$a_{nh}(\vec{r}_h, \vec{R}_h) = \frac{1}{\sqrt{N}} \sum_{\vec{k}_h} e^{-i\vec{k}_h\cdot\vec{R}_h}\, \psi_{\vec{k}_h}(\vec{r}_h) \tag{5.26b}$$

The wavevector for the CM motion with the electron in the conduction band $\vec{K} = \vec{k}_e + \vec{k}_h$, and $\vec{k}_h = -\vec{k}_{VB}$ since the electron and hole have opposite charge. The wavevector for the CM motion will be given by the wavevector of the absorbed light, $k = 2\pi/\lambda$ which for light at 1 eV (1240 nm) is $K = 5.07 \times 10^{-3}\,\mathrm{nm}^{-1}$. The wavevector of light is small which one can see by comparing it to the wavevector at the end of the Brillouin zone along the (100) direction of silicon $\sim 2\pi/(0.5431\ \mathrm{nm})$ or $11.6\,\mathrm{nm}^{-1}$. Thus $\vec{K} \sim 0$ and $\vec{k}_e = \vec{k}_{VB}$. Furthermore, at the moment light is absorbed and the electron and the hole are in the same unit cell $(\vec{R}_e = \vec{R}_h)$. Subsequently, the electron and hole can move to different unit cells.

Putting (5.26a) and (5.26b) and (5.25) into (5.24b) and we get:

$$\psi(\vec{r}_e, \vec{r}_h) = \frac{1}{\sqrt{N}} \sum_{\vec{R}_e, \vec{R}_h} \frac{1}{\sqrt{N}} e^{i\vec{k}\cdot\vec{R}_{CM}} R_{n,l}(\vec{r}) Y_{lm}(\theta, \varphi) \frac{1}{\sqrt{N}} \sum_{\vec{k}_e} e^{-i\vec{k}_e\cdot\vec{R}_e} \psi_{\vec{k}_e}(\vec{r}_e)$$
$$\times \frac{1}{\sqrt{N}} \sum_{\vec{k}_h} e^{-i\vec{k}_h\cdot\vec{R}_h} \psi_{\vec{k}_h}(\vec{r}_h)$$

Choosing the exciton to be in the unit cell at the origin [22], we have: $\vec{R}_e - \vec{R}_h \equiv \vec{r}\,'$ and $\vec{k}_e = \vec{k}_{VB} = \vec{k}'$ with $e^{-i\vec{k}_e\cdot\vec{R}_e} e^{-i\vec{k}_h\cdot\vec{R}_h} = e^{-i\vec{k}_e\cdot\vec{R}_e} e^{i\vec{k}_{VB}\cdot\vec{R}_h} = e^{-i\vec{k}'\cdot(\vec{R}_e - \vec{R}_h)} = e^{-i\vec{k}'\cdot\vec{r}'}$

$$\psi(\vec{r}_e, \vec{r}_h) = \frac{1}{N^{3/2}} \sum_{\vec{R}_e, \vec{R}_h} e^{i\vec{k}\cdot\vec{R}_{CM}} \sum_{\vec{k}', \vec{r}} e^{-i\vec{k}'\cdot\vec{r}'} R_{n,l}(\vec{r}\,)\, Y_{lm}(\theta, \varphi)\, \psi_{\vec{k}_e}(\vec{r}_e)\, \psi_{\vec{k}_h}(\vec{r}_h) \quad (5.27)$$

In Chap. 4 (4.29), $\varepsilon_2 = \frac{3e^2}{\pi\omega^2 m^2} \left|\overline{P}_{cv}\right|^2 \int d\vec{k}\delta(E_c - E_v - \hbar\omega)$ where

$$\vec{P}_{cv} = \Delta^{-1} \int_{unitcell} u_c^*(\vec{r}, \vec{K}'')\, \vec{P} u_v(\vec{r}, \vec{K}')\, d\vec{r}$$

If we again apply the type of derivation for the imaginary part of the dielectric function based on Fermi's Golden rule, we get for a bound or unbound exciton (BE > 0 or BE < 0) with an energy $E_{exciton}(\vec{K}\,) = BE + \frac{\hbar^2 K^2}{2M}$ when the energy $E_{gap} = E_c - E_v$

$$\varepsilon_2 = \frac{3e^2}{\pi\omega^2 m^2} \left|\overline{P}_{cv}\right|^2 \int d\vec{k}\, \delta\Big(E_{exciton}(\vec{K}\,) - E_{gap} - \hbar\omega\Big)$$

where we use (5.27) for the final state wavefunction instead of $u_c^*(\vec{r}, \vec{K}'')$

$$\vec{P}_{cv} = \frac{1}{N^{3/2}} \sum_{\vec{R}_e, \vec{R}_h} e^{i\vec{k}\cdot\vec{R}_{CM}} \sum_{\vec{k}', \vec{r}} e^{-\vec{k}'\cdot\vec{r}'} R_{n,l}(\vec{r}) Y_{lm}(\theta, \varphi) \Delta^{-1}$$
$$\times \int_{unitcell} \psi_{\vec{k}_e}(\vec{r}_e) \psi_{\vec{k}_h}(\vec{r}_h) \vec{P} u_v(\vec{r}, \vec{K}') d\vec{r} \quad (5.28)$$

We can determine $\left|\overline{P}_{cv}\right|^2$ if we use only s final states for $R_{n,l}(\vec{r})Y_{lm}(\theta, \varphi)$. $Y_{lm}(\theta, \varphi) = \frac{1}{2}$ for $l = 0, m = 0$ and the radial wave function for a bound electron–hole $R_{n,l}(\vec{r}) = 2\left(\frac{Z}{a_{ex}}\right)^{3/2} e^{-Zr/a_{ex}}$. At the moment of light absorption, $\vec{R}_e - \vec{R}_h \equiv \vec{r}\,' = 0$, and $\left|R_{n,l}(\vec{r})Y_{lm}(\theta, \varphi)\right|^2$ multiplies the previous result by a constant factor.

This derivation provides a functional form for the excitonic effect on the shape of a direct gap transition in the dielectric function which is shown in Fig. 5.6. For E > 0 the electron–hole states are unbound and the solution for $R_{n,l}(\vec{r})$ is given by [21]:

$$R_{n,l}(\vec{r}) = \frac{(i2kr)^l}{(2l+1)!} e^{\pi|\lambda|/2} \left(\sqrt{\frac{2\pi k^2}{R|\lambda| \sinh\left(\frac{\pi}{2}\right)} \prod_{j=0}^{l} (j^2 + |\lambda|^2)} \right)$$
$$e^{-ikr} F(l + 1 + i|\lambda|; 2l + 2; 2ikr) \quad (5.29)$$

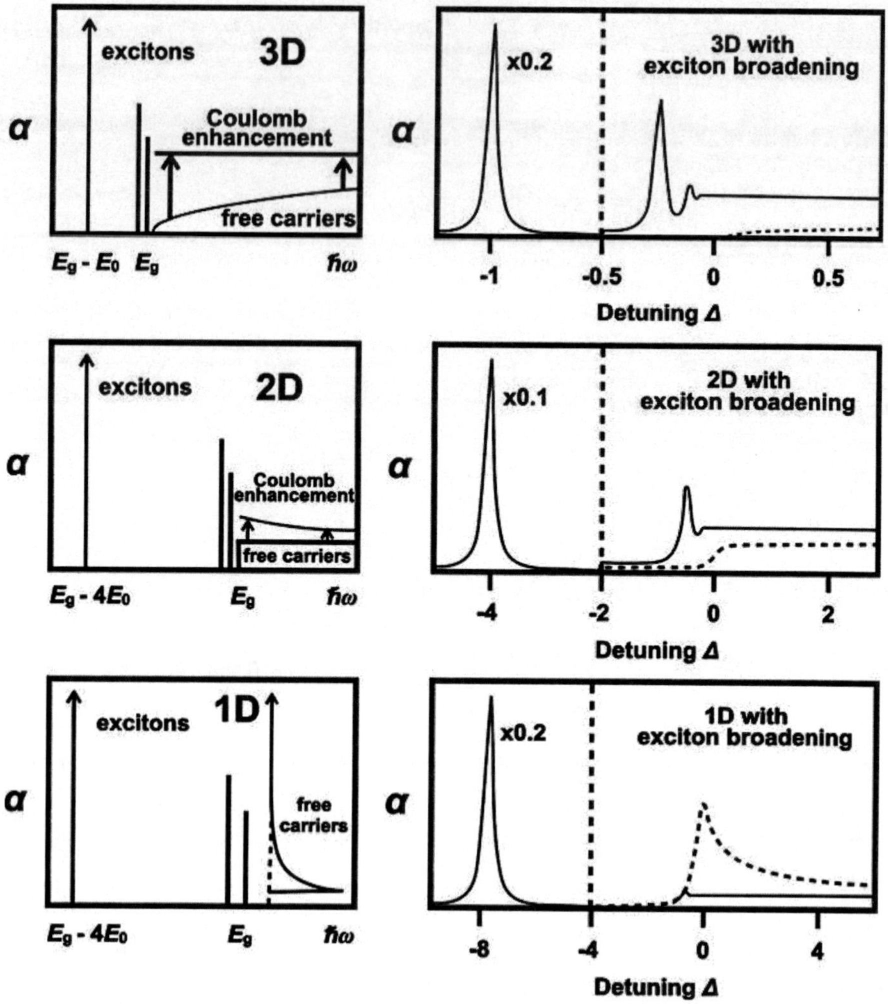

Fig. 5.6 Elliott absorption showing exciton and Coulomb (Sommerfeld) enhancement for 3D (bulk), 2D (Nanofilms), and 1D (nanowires). Broadening of the exciton absorption is shown on the right. Figure adapted from [6]. © 2005 World Scientific

Here $\lambda = \frac{e^2}{\hbar\varepsilon_0}\sqrt{-\mu/2BE}$, k is the wavevector, and BE is the exciton binding energy for the bound state. $\lambda = \sqrt{E_0/BE}$. Also, $F(r = 0) = 1$. This gives a functional form for the imaginary part of the dielectric function when excitonic effects occur:

$$\left|R_{n,l}(\vec{r})Y_{lm}(\theta,\varphi)\right|^2 = \frac{1}{4}e^{\pi/\sqrt{\Delta}}\frac{2\pi k^2}{R|\lambda|\sinh\left(\frac{\pi}{\sqrt{\Delta}}\right)} \quad \text{where } \Delta = \left(\hbar\omega - E_{gap}\right)/E_0 \tag{5.30}$$

5.5.2 Elliott Description of Absorption for 3D, 2D, and 1D and the Sommerfeld Factor for Coulomb Enhancement

The Elliott formula for the absorption coefficient clearly shows the Coulomb enhancement for optical absorption and oscillator strength and how it changes with dimension. This section draws on Haug and Koch [24] who provide an excellent reference for the quantum theory of optical processes. As described above, the Coulomb attraction between the optically excited electron and the hole it leaves in the valence band are observed as both bound exciton peaks and enhanced absorption for direct gap transitions. Implicit in the Elliott expressions is the parabolic nature of the k dependence of the energy of the valence and conduction bands. Also implicit in the formalism shown in this section is the use of the term free carrier absorption for single particle transitions between the valence and conduction band, i.e., it ignores the formation of excitons. The frequency dependent absorption coefficient α is related to the dielectric function through:

$$\varepsilon_2(\omega) = \frac{nc\alpha(\omega)}{\omega}$$

Free carrier absorption: The single particle absorption expression for the imaginary part of the dielectric function derived in Chap. 4 is referred to as free carrier absorption in this section. The free carrier absorption coefficient can be derived from a single particle density matrix for non-interacting electrons and holes [24]. The density matrix is a quantum mechanical description of the statistical state of the particles it describes. This point is important to note since the previous derivation and the Elliott formalism below do not account for the Fermi distribution of the electrons. The 3D, 2D, and 1D frequency dependent absorption coefficient $\alpha(\omega)$ for free carrier absorption in terms of a dimensional dependent constant α_0^D and the band filling factor $A(\omega) = 1 - f_e(\omega) - f_h(\omega)$ and $E_0^D = \frac{\hbar^2}{2\mu}\left(\frac{\pi}{L_c}\right)^2(3 - D)$ is [24]:

$$\alpha(\omega) = \alpha_0^D \frac{\hbar\omega}{E_0}\left(\frac{\hbar\omega - E_{gap} - E_0^D}{E_0}\right)^{\frac{D-2}{2}} \Theta\left(\hbar\omega - E_{gap} - E_0^D\right)A(\omega)$$

$$\alpha_0^D = \frac{4\pi^2|P_{cv}|^2}{\hbar n_b c}\frac{1}{(2\pi a_0)^D}\Omega_D\frac{1}{L_c^{3-D}} \tag{5.31}$$

It is important to note that this approach accounts for the band filling through the Fermi distributions for electrons $f_e(\omega)$ and holes $f_h(\omega)$. D = 3 for 3D (bulk), 2 for 2D (nanofilms), or 1 for 1D (nanowires) in (5.31). Thus for 3D $E_0^D = 0$, 2D $E_0^D = \frac{\hbar^2}{2\mu}(\frac{\pi}{L_c})^2$, and for 1D $E_0^D = 2\frac{\hbar^2}{2\mu}(\frac{\pi}{L_c})^2$. In (5.31) L_c is the confinement dimension, n_b is the background refractive index, $\Omega_D = \omega_{Rabi}\hat{x} - \upsilon_k\hat{z}$ is the vector of the rotation frequency with the Rabi frequency $\omega_{Rabi} = P_{cv}E_0/\hbar$ and the detuning frequency (= 0 at resonance) $\upsilon_k = \varepsilon_{c,\vec{k}} - \varepsilon_{v,\vec{k}} - \omega$ and the unit vectors in Cartesian coordinates $\hat{x}$ and $\hat{z}$. $a_{ex-Bohr} = \hbar\varepsilon_0/\mu e^2$ and $E_0 = \hbar^2/(2\mu a_{ex-Bohr}^2)$. Note that the exciton Bohr radius is not scaled by dimension in (5.31) or (5.32a)–(5.32c). The term α_0^D is given by:

$$\alpha_0^{3D} = \frac{4\pi^2|P_{cv}|^2}{\hbar n_b c}\frac{1}{(2\pi a_{exBohr})^3}\Omega_D \tag{5.32a}$$

$$\alpha_0^{2D} = \frac{4\pi^2|P_{cv}|^2}{\hbar n_b c}\frac{1}{(2\pi a_{exBohr})^2}\Omega_D\frac{1}{L_c} \tag{5.32b}$$

$$\alpha_0^{1D} = \frac{4\pi^2|P_{cv}|^2}{\hbar n_b c}\frac{1}{(2\pi a_{exBohr})}\Omega_D\frac{1}{L_c^2} \tag{5.32c}$$

The energy dependence of the free carrier absorption as a function of dimension is:

$$3\text{D}\ \left(\frac{\hbar\omega - E_{gap}}{E_0}\right)^{\frac{1}{2}};\ 2\text{D}\ \left(\frac{\hbar\omega - E_{gap} - \frac{\hbar^2}{2\mu}\left(\frac{\pi}{L_c}\right)^2}{E_0}\right)^0 = 1;$$

$$\text{and 1D}\ \left(\frac{\hbar\omega - E_{gap} - \frac{\hbar^2}{\mu}\left(\frac{\pi}{L_c}\right)^2}{E_0}\right)^{-\frac{1}{2}}$$

5.5.2.1 Elliott Formula for 3D

Next, the Elliot formula is used to show how the Coulomb enhancement and exciton absorption modify the single particle absorption expression for the imaginary part of the dielectric function discussed in Chap. 4. This formalism describes the quantum mechanical origin of the two parts to the absorption coefficient which were shown phenomenologically in (5.9)–(5.11): absorption due to bound electron hole pairs and the Coulomb enhanced absorption due to valence to conduction band transitions. It is important to note that the Elliott expression predicts a constant energy dependence for the Coulomb enhanced valence to conduction band transitions. This second part of the absorption is also referred to as the continuum part [25]. The Coulomb enhancement factor is determined using (5.31) as follows: $\alpha_{continuum} = \alpha_{freeparticle} C_D(\omega)$ where D = 3D, 2D, or 1D. The Elliott expression for the frequency dependent absorption coefficient [25, 26], for 3D materials including Coulomb enhancement $C_{3D}(\omega)$ can be written as:

$$\alpha_{3D}(\omega) = \alpha_0^{3D}\frac{h\omega}{E_0'}\left[\sum_{n=1}^{\infty}\frac{4\pi}{n^3}\delta\left(\Delta + \frac{1}{n^2}\right) + \Theta(\Delta)\frac{\pi e^{\pi/\sqrt{\Delta}}}{\sinh\left(\frac{\pi}{\sqrt{\Delta}}\right)}\right] \tag{5.33}$$

$$C_{3D}(\omega) = \frac{\pi}{\sqrt{\Delta}}\frac{e^{\pi/\sqrt{\Delta}}}{\sinh\left(\frac{\pi}{\sqrt{\Delta}}\right)}$$

With $E_0' = \frac{e^4\mu}{2\varepsilon_0^2\hbar^2}$; again $\Delta = \left(\hbar\omega - E_{gap}\right)/E_0'$ and $\Theta(\Delta)$ is the Heaviside step function. Again, the first term in the brackets $\sum_{n=1}^{\infty}\frac{4\pi}{n^3}\delta\left(\Delta + \frac{1}{n^2}\right)$ is for the bound exciton peaks and the second term (continuum) is the absorption due to the valence to conduction band absorption increased by Coulomb enhancement. As $\Delta \to 0$, $C_{3D}(\omega) \to \frac{2\pi}{\sqrt{2}}$ thus close to the energy gap the $\left(\hbar\omega - E_{gap}\right)^{1/2}$ energy dependence of single particle absorption picture of Chap. 4 [(4.29) and (4.30)] and this chapter (5.31) is enhanced by $\frac{2\pi}{\sqrt{2}}$. Near the transition energy, the energy dependence of the continuum absorption is constant as shown in Fig. 5.6.

5.5.2.2 Elliott Formula for 2D

The Elliott formula for the absorption coefficient of a 2D material is given by:

$$\alpha_{2D}(\omega) = \alpha_0^{2D}\frac{h\omega}{E_0'}\left[\sum_{n=1}^{\infty}\frac{4}{\left(n+\frac{1}{2}\right)^3}\delta\left(\Delta + \frac{1}{\left(n+\frac{1}{2}\right)^2}\right) + \Theta(\Delta)C_{2D}(\omega)\right] \tag{5.34}$$

$$C_{2D}(\omega) = \frac{e^{\pi/\sqrt{\Delta}}}{\cosh\left(\frac{\pi}{\sqrt{\Delta}}\right)}$$

There is no energy dependence of the free carrier absorption for a nanofilm (2D) $\left[\left(\frac{\hbar\omega - E_{gap} - \frac{\hbar^2}{2\mu}\left(\frac{\pi}{Lc}\right)^2}{E_0}\right)^0 = 1\right]$, while there is an energy dependence to the Coulomb enhancement as can be seen in Fig. 5.6 before broadening. The Coulomb absorption is $C_{2D}(\omega) = 2$ as $\Delta \to 0$ as shown in Fig. 5.6.

5.5.2.3 Elliott Formula for 1D

For 1D nanowires, the Coulomb enhancement is less than 1 and the energy gap of nanowires is considered to not be determinable from linear (low intensity) optical measurements [24, 27, 28]. The hydrogen line equation for the radial motion for electron–hole confided to 1D was described by Loudon [29, 30]. It is also noteworthy that ab initio quantum calculations show that the band structure of small dimension nanowires does not resemble that of a bulk material [31–34]. This is discussed in the next section. The expression for the Coulomb enhancement can be determined by comparison of (5.35) with (5.31).

$$\alpha_{1D}(\omega) = \frac{4\pi\omega|P_{cv}|^2}{n_b c}\frac{2}{E_0}\left[\sum_{\lambda}\left|N_\lambda W^2_{\lambda,\frac{1}{2}}\left({}^{2\gamma R}\!/_{\lambda a_0}\right)\right|^2\pi\delta\left(\Delta-\frac{E_\lambda}{E_0}\right)\right.$$
$$\left.+\frac{1}{\pi a_0}\frac{e^{\pi/\sqrt{\Delta}}}{2\sqrt{\Delta}}\frac{\left[D_0^{(2)}W^{(1)}-D_0^{(1)}W^{(2)}\right]^2}{\left|D_0^{(1)}\right|^2+\left|D_0^{(2)}\right|^2}\right] \tag{5.35}$$
$$C_{1D}(\omega) = \frac{e^{\pi/\sqrt{\Delta}}}{8}\frac{\left[D_0^{(2)}W^{(1)}-D_0^{(1)}W^{(2)}\right]^2}{\left|D_0^{(1)}\right|^2+\left|D_0^{(2)}\right|^2}$$

Here, $W^2_{\lambda,\frac{1}{2}}$ is the Whittaker function [24] and $D_0^{(j)} = \frac{dW^{(j)}(\zeta)}{d\zeta}$ with $\zeta = 2ik\gamma R$. Again, the quantum number λ is determined from boundary conditions so that $\frac{1}{2} + \lambda \ln\left(\frac{2\gamma R}{\lambda a_{hydrogen}}\right) = 0$ and $\gamma \sim 0.3$ [6]. One challenge with the application of (5.35) is that the band structure of small diameter nanowires is very different from that of a bulk material. This is further discussed in the next section.

5.5.2.4 *Comment on Adding Linewidth to the Elliott Formalism*

Urbach broadening and a Lorentzian lineshape can be added to an approximate form of (33) and fit to experimental data for GaAs, InP, and InAs [35]. This approximate method simplifies Kramers–Kronig transformations.

5.6 The Effect of Nanoscale Dimensions on the Band Gap, Band Structure and Exciton Energies of Semiconductors

As discussed above, the band gap of silicon [(001) normal to surface] slabs increases with layer thickness for films less than 25 monolayers thick (< 3.4 nm): see (5.21). The band structure of very thin layers and very small diameter wires has been shown to be different from that of the bulk. For example, the band structure, dielectric function, and exciton energies for a 1.2 nm diameter silicon nanowire has been studied using ab initio quantum calculations as shown in Fig. 5.7 [31]. As one can see from Fig. 5.7, the change in band structure also changes the exciton binding

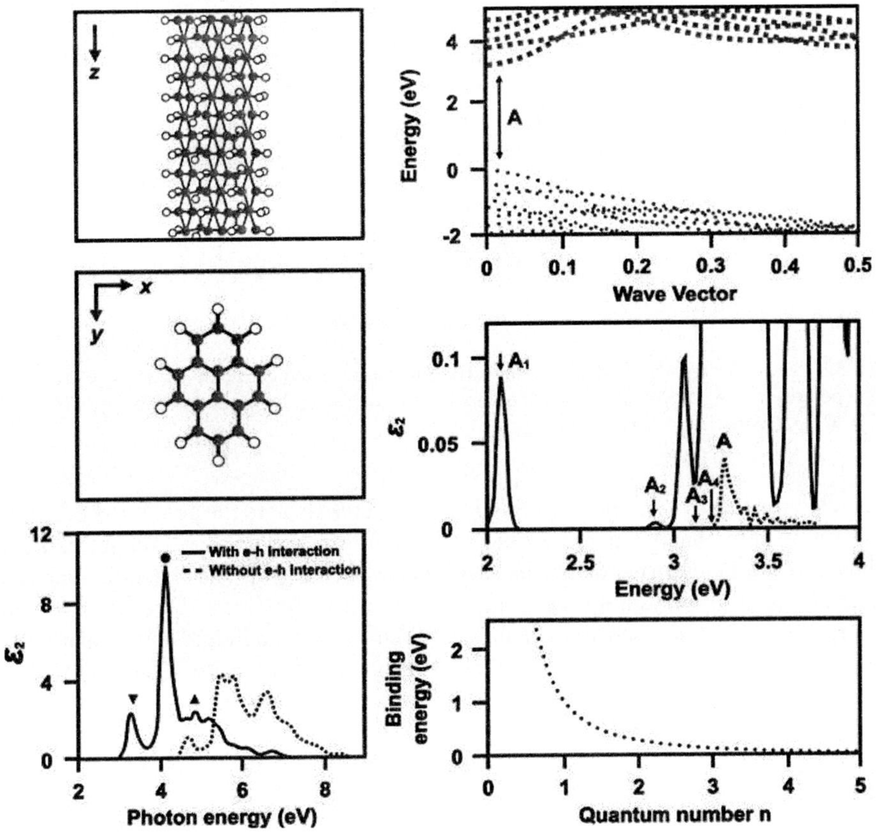

Fig. 5.7 The calculated electronic band structure and exciton energies for a 1.2 nm diameter silicon nanowire. The nanowire crystal structure is shown on the left along with the calculated imaginary part of the dielectric function. On the right, the calculated band structure, the exciton spectra for the transitions at A, and the binding energy of the excitons vs quantum number n are shown. In (**b**) on the right, the dash-dotted line is the one-particle result, and the solid line is the Bethe Salpeter equation (BSE) result. Four excitonic transitions were resulted from the BSE calculation: *A*1, *A*2, *A*3, and *A*4. Figure adapted from [31]. © 2006 L. Yang

energy, and the Coulomb effect described above for 1D Nanowires also changes the CP energies.

The effect of thickness on the band gap of nanoscale germanium films has been explored using first principles quantum calculations [32, 33]. Theoretical studies of the band structure and optical absorption of 0.5–2.4 nm thick germanium films show a clear anisotropy to the fundamental band gap with $E_g(111) < Eg(110) < E_g(100)$ [22]. The explanation for this ordering is taken from the differences in effective mass for the three different crystallographic orientations [32]. The electron effective mass $m*$, determined from the band curvature at the conduction-band minimum follows the same ordering: $m_e(111) = 1.67m_0 > m_e(110) = 0.38\ m_0 > m_e(100) = 0.3m_0$. m_0 is the

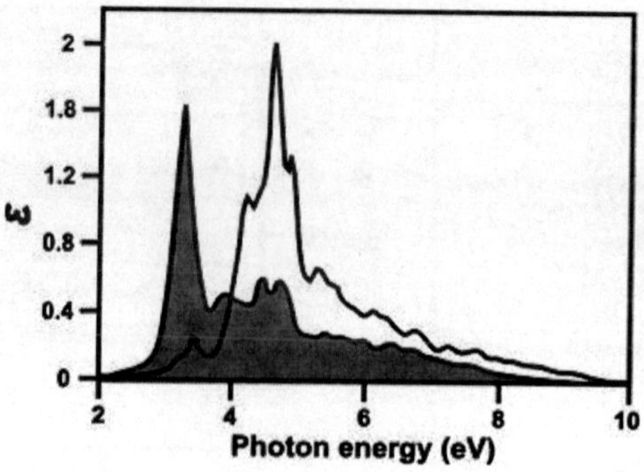

Fig. 5.8 The dielectric function of 0.8 nm (001) silicon (gray) and germanium nanowires with the light polarized along the length of the nanowire. Figure adapted and reprinted with permission from [34]. © 2005 American Physical Society

free-electron mass [32]. The changes in band structure lead to the prediction of direct absorption at the band gap and visible photoluminescence for very thin germanium films [33]. Germanium nanowires have also been studied [34]. The dielectric function is a function of the crystallographic direction of the nanowire, and the combination of oriented nanowires and light polarized along or perpendicular to the nanowire length allow theoretical sampling of the anisotropy of the dielectric function [34]. A comparison of the dielectric function of silicon and germanium nanowires is shown in Fig. 5.8.

5.6.1 The Bandgap of Semiconductor Nanodots

Theoretical calculations based on empirical pseudopotential plane wave methods have been applied to the indirect transition E_0 at the band gap of thin silicon films and nanodots [36, 37]. In these studies, a $1/d^{1.3}[d = nanodotsdiameter]$ dependence was predicted for the change in band gap energy. This study addressed the impact of different dot geometries is discussed, i.e., spherical balls; rectangular boxes with surfaces in the (110), $(1\overline{1}0)$, and (001) directions with edges satisfying $d_x = d_y = d_z/\sqrt{2}$; and cubic boxes with surfaces in the (001), (010), and (100) directions. The quantum mechanical calculations were done for each shape, and the size dependence of the band gap was expressed using a universal curve. The effective size of each dot shape is assumed to be equivalent to a sphere with diameter that encloses the same number of silicon atoms. With ρ = bulk silicon mass density, $d(N_{\mathrm{Si}}) = (3/(4\pi\rho))^{1/3}$ $(N_{\mathrm{Si}})^{1/3}$ and $E_{\mathrm{gap}} = 1.167 + 88.34\ d^{-1.37}$ (eV) with d (Angstroms).

5.6.2 *Thickness Dependence of Exciton Binding Energies in III-V Quantum Wells*

Numerous III-V QW systems have been studied. In this section the goal is not to review that extensive literature but rather for the sake of completeness mention that the energy of excitons in III-V QWs is a function of QW thickness. The exciton radius for the n = 1 exciton in bulk GaAs is ~ 13.5 nm and a BE = 4.2 meV, and since the lattice constant is 0.56 nm the exciton is spread over many lattices and is Wannier in nature [38]. One example is that of $Al_{0.54}Ga_{0.46}As$/GaAs multilayer structure with GaAs QWs of 3, 4, 7, and 10 nm thickness as shown in Fig. 5.9. The exciton BE was determined to be 9.1 meV for the 3 nm QW; 9.2 meV for the 4 nm QW; 6.8 meV for the 7 nm QW; and 6.13 meV for the 10 nm QW [39]. Considering the exciton radius for bulk GaAs, the QWs serve to localize the exciton to 2D. The same theoretical approach used to obtain an Elliott expression for nanofilms and nanowires has been applied to III-V QW structures [40].

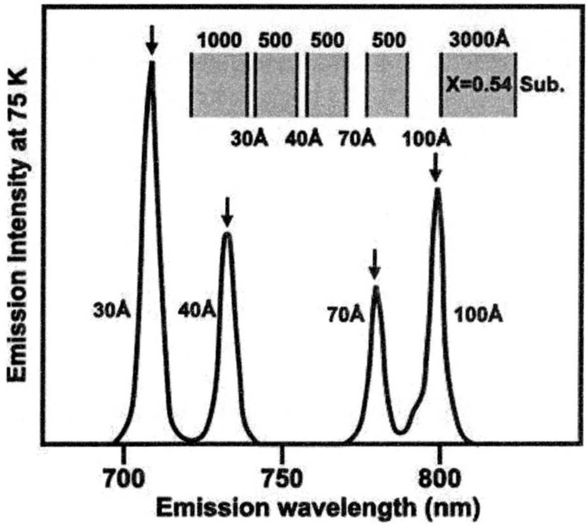

Fig. 5.9 Photoluminescence of $Al_{0.54}Ga_{0.46}As$/GaAs with GaAs quantum wells of 3, 4, 7, and 10 nm thickness at 75 K. The excitation intensity of the 514.5 nm line of the Argon ion laser was 300 mW resulting in 20 W/cm^2. Figure adapted and reprinted from [39] with the permission of AIP Publishing

5.6.3 Electron–Phonon Interactions in Nanoscale SiO_2-Si-SiO_2 Quantum Wells

The energy of the excitonic silicon E_1 CP transition has been shown to shift to higher energies as the thickness of nanoscale SiO_2-Si-SiO_2 QWs decreases from 10 to 2 nm [41] The E_2 CP does not show a shift in energy. Through a study of the temperature dependence of the E_1 CP energy and broadening, this change has been linked to electron phonon interactions and the change in the phonon dispersion with thickness. It is useful to note that as mentioned in Chap. 4, the E_1 CP is a direct gap transition between the Λ_3 VB and Λ_1 CB between the wavevectors $\vec{k} = \frac{2\pi}{a_0}\left(\frac{1}{4}, \frac{1}{4}, \frac{1}{4}\right)$ and $\vec{k} = \frac{2\pi}{a_0}\left(\frac{1}{2}, \frac{1}{2}, \frac{1}{2}\right)$. The change in the E_1 CP of bulk silicon is known to be a function of lattice expansion and electron–phonon interactions [42]. The CP energies and broadening were extracted from the dielectric function by fitting the experimental data to (5.36) Chap. 4 using $\mu = 1$ for excitonic CPs:

$$\frac{d^2\varepsilon}{dE^2} = 2Ae^{i\beta}(E - E_{gap} + i\Gamma)^{-3} \tag{5.36}$$

The temperature dependence of the CP energy shift can be fit to a phenomenological expression that accounts for both thermal expansion and electron–phonon interactions. Here, the average interaction frequency of the phonons is a statistical factor $K_B\theta/\hbar$ and fitting parameters a_B and E_B [43]:

$$\left(\frac{\partial E_{CP}}{\partial T}\right)_P = E_B - a_B\left(1 + \frac{2}{e^{\theta/T} - 1}\right) \tag{5.37}$$

The lower the average phonon frequency θ, the stronger the contribution of acoustic phonons [43]. The temperature dependent data is shown in Fig. 5.10. The thickness dependence of the average phonon frequency θ from (5.37) is listed in Table 5.3. The decrease in θ points to the increased importance of acoustic phonons in the electron–phonon interactions for nanoscale films [41]. Theoretical studies have shown that Free standing, nanoscale silicon films have a number of additional optical phonon branches [44]. The lowest lying optical phonon mode frequency increases with decreasing film thickness d according to $\omega\left(\text{cm}^{-1}\right) = 188.25/\left(\frac{d}{a_0}\right)^{\frac{3}{4}}$ at the zone center. a_0 the lattice constant 0.543 nm. Thus the lowest lying optical phonon mode increases in frequency by a factor of 2 between $\frac{d}{a_0} = 10$ and $\frac{d}{a_0} = 4$. The Bose–Einstein occupation probability for this phonon decreases by more than a factor of two and thus the increasing influence of the acoustic modes on the E_1 CP. The dispersion of the acoustic phonon modes can be altered by changing the stiffness of the dielectric layer on top of the SOI. This was tested by comparing the E_1 CP energies of ~ 5nm thick SOI films with native oxide, SiO_2, and HfO_2 as shown in Fig. 5.11 [41].

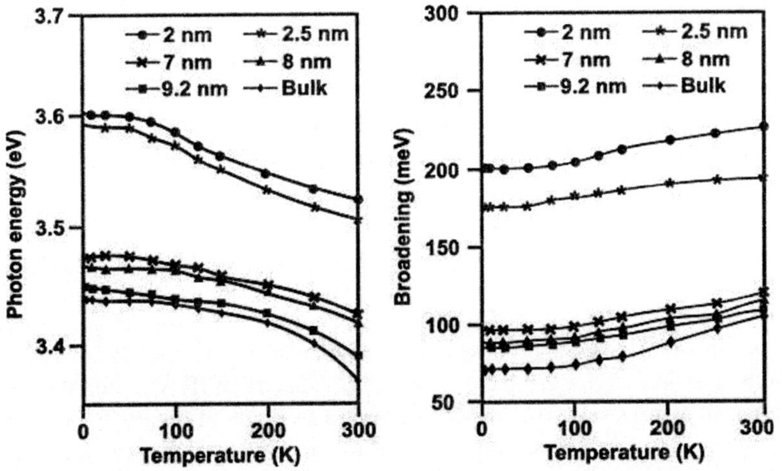

Fig. 5.10 The temperature and thickness dependence of the E_1 critical point energy and broadening for thin silicon films in SiO_2/Si/SiO_2 films is shown. Figure adapted and reprinted from [41] with the permission of AIP Publishing

Table 5.3 Average phonon frequency (90% confidence limits) at various *c*-Si film thicknesses [41]

Thickness (nm)	θ (K)
9.2	406 (167)
8	356 (38)
7	231 (58)
2.5	98.5 (46)
2	90 (65)

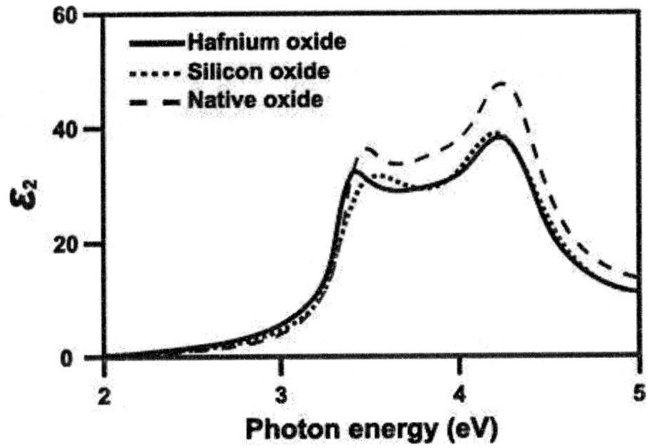

Fig. 5.11 The imaginary part of the dielectric function of ~ 5 nm SiO_2/Si/x QWs with x = native oxide, 20 nm SiO_2, and 10 nm HfO_2. The E_1 CP at ~ 3.4 eV for SiO_2-Si-SiO_2 shifts to lower energies for the HfO_2 covered QW. The Young's modulus of HfO_2 (~ 370 GPa) is greater than that of SiO_2 (~ 75 GPa) [41]. Figure adapted and reprinted from [41] with the permission of AIP Publishing

5.7 Comments on Photoluminescence Lineshape

As first discussed in Chap. 1, the Lorentzian, Gaussian, and Voigt functional forms have all been used to describe the energy dependence of photoluminescence intensity. As discussed in Sect. 1.10, the Lorentzian lineshape is appropriate for a single transition. The lifetime of the transition and electron–phonon scattering both broaden the Lorentzian lineshape. The Gaussian lineshape if appropriate for a peak that is due to multiple transitions, and the Voigt function is a convolution of the Lorentzian and Gaussian functions. In Chap. 1, the lineshape of photoluminescence from a 2D van der Waals materials, WSe_2 was discussed (see Chap. 1, Fig. 1.20). Here, we discuss the Gaussian line shapes and convolutions of Gaussian line shapes are frequently used in the literature to describe the observed PL spectra of III-V materials [45–47]. Lorentzian line shapes are often used for bound excitons in binary III-V semiconductors [48]. These lineshapes can all be fit to experimental spectra using variable amounts of peak broadening. The origin of the peak broadening depends on the material and structure. These sources include thermal broadening, alloy disorder, and defect density.

Photoluminescence lineshape of bulk III-V materials is often modeled as a Gaussian distribution of energies for a central transition i centered at energy E_i [45]:

$$G(\hbar\omega) = \frac{A_i}{\sqrt{2\pi}\Gamma_i} e^{-i\left((\hbar\omega - E_i)^2/2\Gamma_i^2\right)} \tag{5.38}$$

where Γ_i is the broadening due to thermal and other effects. The Gaussian functions for multiple transitions can be added together for closely spaced transitions when modeling multiple transitions.

The exciton linewidths for III-V materials and quantum wells has been modeled by a functional form that includes the effect of exciton scattering from free carriers Γ_0, acoustic phonons aT and optical phonons $b\left(e^{\hbar\omega_{LO}/K_BT} - 1\right)^{-1}$[47]:

$$\Gamma = \Gamma_0 + aT + b\left(e^{\hbar\omega_{LO}/K_BT} - 1\right)^{-1} \tag{5.39}$$

The coefficients a and b are the scattering strengths.

The luminescence broadening due to variability of alloy composition in semiconductor alloys can be separated from thermal broadening at low temperatures below 10 K [48]. The statistical nature of the distribution of Al and Ga on the cation sites results in a statistical fluctuation in the local band gap in an volume the size of exciton. Remembering that the radius of the n = 1 exciton in GaAs is ~ 13.5 nm, the exciton volume contains both Al and Ga cations whose concentration fluctuates with position. A common Pl peak in III-V alloys grown by Molecular Beam Epitaxy comes from the recombination of an electron in the conduction band with a hole bound to an acceptor impurity which is labeled (e, A^0) [48]. Carbon is a common impurity in MBE growth. The luminescent intensity as a function of energy for the (e, A^0) transition has been expressed as a convolution of thermal and compositional disorder based broadening as follows [48]:

$$I(\hbar\omega) \sim \int_{-\infty}^{\infty} \left[\hbar\omega^* - \left(E_g - E_a\right)\right]^{\frac{1}{2}} e^{-i\left(\left[\hbar\omega^* - \left(E_g - E_a\right)\right]/KT\right)} e^{-i\left(\left[\hbar\omega^* - \hbar\omega\right]/\sigma_E\right)} d\left(\hbar\omega^*\right) \tag{5.40}$$

where E_g is the direct gap transition energy, E_a is the acceptor binding energy, and σ_E is the statistical fluctuation in band gap energy. The ratio of the PL intensity of the (e, A^0) transition to the free exciton intensity as a function of growth temperature provides a means of characterizing the amount of trace carbon incorporation [48]. Thermal and compositional fluctuation effects also impact PL from nanoscale structures including quantum wells.

References

1. F. Wooten, *Optical Properties of Solids* (Academic Press, New York, 1972), pp. 149–153
2. N. Pauc, V. Calvo, J. Eymery, F. Fournel, and N. Magnea, Two-dimensional electron-hole liquid in single si quantum wells with large electronic and dielectric confinement. Phys. Rev. Lett. **92**, 236802 (2004)
3. N. Pauc, V. Calvo, J. Eymery, F. Fournel, N. Magnea, Electronic and optical properties of Si/SiO2 nanostructures. I. Electron-hole collective processes in single Si/SiO_2 quantum wells. Phys. Rev. B **72**, 205324 (2005)
4. N. Pauc, V. Calvo, J. Eymery, F. Fournel, N. Magnea, Electronic and optical properties of Si/SiO_2 nanostructures. II. Electron-hole recombination at the Si/SiO_2 quantum-well–quantum-dot transition. Phys. Rev. B **72**, 205325 (2005)
5. C.F. Klingshirn, *Semiconductor Optics* (Springer, New York, 1997), p. 167
6. H. Haug, S.W. Koch, *Quantum Theory of the Optical and Electronic Properties of Semiconductors*, 4th edn. (World Scientific, Singapore, 2005), pp. 124, 179–180
7. P. Lautenschlager, M. Garriga, S. Logothetidis, M. Cardona, Interband critical points of GaAs and their temperature dependence. Phys. Rev. B **35**, 9174–9189 (1987)
8. M. Combescot, P. Nozieres, Condensation of excitons in germanium and silicon. J. Phys. C: Sol. Stat. **5**, 2369 (1972)
9. P.Y. Yu, M. Cardona, *Fundamentals of Semiconductors: Physics and Materials Properties* (Springer, New York, 2010), pp. 286–288
10. L.V. Keldysh, Excitons in semiconductor-dielectric nanostructures. Phys. Stat. Sol. (a) **164**, 3 (1997)
11. A.N. Kholod, Andre´s Saul, J.D. Fuhr, V.E. Borisenko, F. Arnaud d'Avitaya, Electronic properties of germanium quantum films. Phys. Rev. B **62**, 12949 (2000)
12. A.N. Kholod, S. Ossicini, V.E. Borisenko, F. Arnaud d'Avitaya, True direct gap absorption in germanium quantum films. Phys. Rev. B **65**, 115315 (2002)
13. Y.M. Niquet, C. Delerue, G. Allan, M. Lannoo, Method for tight-binding parametrization: application to silicon nanostructures. Phys. Rev. B **62**, 5109 (2000)
14. D.A. Kleinman, Binding energy of the electron-hole liquid in quantum wells. Phys. Rev. B **33**, 2540 (1986)

15. S. Ibuka, M. Tajima, M. Saito, J. Jablonski, M. Warashina, K. Nagasaka, Photoluminescence due to degenerate electron-hole system in silicon-on-insulator wafers under ultraviolet light excitation. Jpn. J. Appl. Phys. **37**, 141–145 (1998)
16. Z.H. Lu, D. Groze, Crystalline Si/SiO_2 quantum wells. Appl. Phys. Lett. **80**, 255 (2002)
17. M. Chandrapal, F.H. Pollak, Conclusive evidence for the excitonic nature of the E_1 - ($E_1+\Delta_1$) optical structure in diamond- and zincblende-type semiconductors at room temperature. Solid State Commun. **18**, 1263 (1976)
18. L. Yang, J. Deslippe, C.H. Park, M.L. Cohen, S.G. Louie, Excitonic effects on the optical response of graphene and bilayer graphene. Phys. Rev. Lett. **103**, 186802 (2009)
19. F.J. Nelson, V.K. Kamineni, T. Zhang, E.S. Comfort, J.-U. Lee, A.C. Diebold, Optical properties of large-area polycrystalline chemical vapor deposited graphene by spectroscopic ellipsometry. Appl. Phys. Lett. **97**(253110), 1–3 (2010)
20. F.J. Nelson, J.-C. Idrobo, J. Fite, Z.L. Mišković, S.J. Pennycook, S.T. Pantelides, J.U. Lee, A.C. Diebold, Electronic excitations in graphene in the 1–50 eV range: the π and $\pi + \sigma$ peaks are not plasmons. Nano Lett. **14**, 3827–3831 (2014)
21. H. Haug, S.W. Koch, *Quantum Theory of the Optical and Electronic Properties of Semiconductors* (World Scientific, Hackensack, 2005), pp. 182, 186–189
22. P.Y. Yu, M. Cardona, *Fundamentals of Semiconductors*, 4th edn. (Springer, Berlin, 2010), pp. 162–166, 282–286
23. M. Rohlfing, S.G. Louie, Electron-hole excitations and optical spectra from first principles. Phys. Rev. B **62**(8), 4927 (2000)
24. H. Haug, S.W. Koch, *Quantum Theory of the Optical and Electronic Properties of Semiconductors*, 4th edn. (World Scientific, Singapore, 2005), pp. 82–83, 163–191
25. R.J. Elliott, Intensity of optical absorption by excitons. Phys. Rev. **108**, 1384–1389 (1957)
26. R.J. Elliott, Introduction to Excitons, in *Polarons and Excitons in Polar Semiconductors and Ionic Crystals*. ed. by J.T. Devreese, F. Peeters (Plenum, New York, 1982), pp. 271–292
27. T. Ogawa, T. Takagahara, Interband absorption spectra and Sommerfeld factors of a one-dimensional electron-hole system. Phys. Rev. B **43**, 14325 (1991)
28. T. Ogawa, T. Takagahara, Optical absorption and Sommerfeld factors of one-dimensional semiconductors: an exact treatment of excitonic effects. Phys. Rev. B **44**, 8138 (1991)
29. R. Loudon, One-dimensional hydrogen atom. Amer. J. Phys. **27**, 649–655 (1959)
30. L. Binyai, I. Galbraith, C. Ell, H. Haug, Excitons and biexcitons in semiconductor quantum wires. Phys. Rev. B **36**, 6099–6104 (1987)
31. L. Yang, First-principles Calculations on the Electronic, Vibrational, and Optical Properties of Semiconductor Nanowires. Dissertation, 2006
32. A.N. Kholod, A. Saul, J.D. Fuhr, V.E. Borisenko, F. Arnaud d'Avitaya, Electronic properties of germanium quantum films. Phys. Rev. **B62**, 12949 (2000)
33. A.N. Kholod, S. Ossicini, V.E. Borisenko, F. Arnaud d'Avitaya, True direct gap absorption in germanium quantum films. Phys. Rev. **B65**, 115315 (2002)
34. M. Bruno, M. Palummo, A. Marini, R. Del Sole, V. Olevano, A.N. Kholod, S. Ossicini, Excitons in germanium nanowires: quantum confinement, orientation, and anisotropy effects within a first-principles approach. Phys. Rev. B **72**, 153310 (2005)
35. E.Y. Lin, T.S. Lay, T.Y. Chang, Accurate model including Coulomb-enhanced and Urbach-broadened absorption spectrum of direct-gap semiconductors. J. Appl. Phys. **102**, 123511 (2007)
36. S. Ogut, J.R. Chelikowsky, S.G. Louie, Quantum confinement and optical gaps in Si nanocrystals. Phys. Rev. Let. **79**, 1770 (1997)
37. A. Zunger, L.-W. Wang, Theory of silicon nanostructures. Appl. Surf. Sci. **102**, 350–359 (1996)
38. R.P. Smith, Quantitative measurements of many-body exciton dynamics in GaAs quantum-well structures. Dissertation, Georgia Tech. 2004, p. 11
39. H. Kawai, K. Kaneko, N. Watanabe, Photoluminescence of AlGaAs/GaAs quantum wells grown by metal organic chemical vapor deposition. J. Appl. Phys. **56**, 463–476 (1984)
40. S. Schmitt-Rink, C. Ell, H. Haug, Many-body effects in the absorption, gain, and luminescence spectra of semiconductor quantum-well structures. Phys. Rev. B **33**, 1183–1189 (1986)

41. V.K. Kamineni, A.C. Diebold, Electron-phonon interaction effects on the direct gap transitions of nanoscale Si films. Appl. Phys. Lett. **99**, 151903 (2011)
42. M. Cardona, Electron-phonon interaction in tetrahedral semiconductors. Solid State Commun. **133**, 3–18 (2005)
43. P. Lautenschlager, M. Garriga, L. Vina, M. Cardona, Temperature dependence of the dielectric function and critical point energies. Phys. Rev. B **36**, 4821 (1987)
44. S.P. Hepplestone, G.P. Srivastava, Lattice dynamics of silicon nanostructures. Nanotechnology **17**, 3288 (2006)
45. M. Grundmann, C.P. Dietrich, Lineshape theory of photoluminescence from semiconductor alloys. J. Appl. Phys. **106**, 123521 (2009)
46. M. Grundmann, *The Physics of Semiconductors* (Springer, New York, 2006), pp. 251–267
47. A. Venu Gopal, Rajesh Kumar, A. S. Vengurlekar, A. Bosacchi, S. Franchi, L.N. Pfeiffer, Photoluminescence study of exciton–optical phonon scattering in bulk GaAs and GaAs quantum wells. J. Appl. Phys. **87**, 1858 (2000)
48. E.F. Schubert, E.O. Gobel, Y. Horikoshi, K. Ploog, H.J. Queisser, Alloy broadening in the photoluminescence spectra of $Al_xGa_{1-x}As$. Phys. Rev. B **30**, 813 (1984)
49. M.D. Sturge, Optical absorption of gallium arsenide between 0.6 and 2.75 eV. Phys. Rev. **127**, 768–773 (1962)

Chapter 6
Hall Effect Characterization of the Electrical Properties of 2D and Topologically Protected Materials

Abstract In this chapter we present the classical, quantum, and topological descriptions of electron transport measurements. Hall measurements are introduced using classical physics. The quantization of the electronic levels due to a magnetic field known as the Landau levels is shown. The observation of the quantization of the conductivity in a 2D electron gas at low temperature and high magnetic field due to the pioneering research of von Klitzing is presented. This leads to the introduction of the Berry phase and topological explanation of the quantized conductance. The relationship between the Kubo formula for conductance and topological quantification due to the research of Thouless, Kohmoto, Nightingale, and den Nijs is presented. The Hall characterization of single layer graphene and the observation of the Berry phase confirming the presence of Dirac carriers is used to demonstrate the topological properties of graphene. The family of Hall effects is also presented.

Understanding the electrical transport properties of 2D and topologically protected materials provides insight into the unexpected phenomena observed in many optical measurements. In addition, the charge transport properties provide a means of categorizing 2D materials. Both carriers of charge, electrons and holes, can be classified as either Schrodinger fermions or Dirac fermions. For example, carriers in the π and π^* band in graphene close to the K and K' point in the Brillouin zone are Dirac fermions because their dynamics is described by the relativistic Dirac equation. We note that both types of carriers can be present in the same material. This leads to the question of which type of carrier is observed during optical characterization. That depends on the states involved in the transition. Another descriptive aspect of carrier dynamics is the presence or absence of the topological correction known as the Berry phase. In Chap. 7, a tight binding model for the electronic band structure of graphene is presented along with the Dirac Fermion description of the high mobility carriers of graphene which have a Berry phase correction to their Hall resistivity. This Berry Phase correction is often considered to be the signature of Dirac Fermions. Another aspect of topological carrier dynamics is *topological* protection from scattering. In this chapter, Hall measurements, the family of Hall effects, and topological carrier dynamics are introduced. The exciting interplay between theory and experiment is

A. Diebold and T. Hofmann, *Optical and Electrical Properties of Nanoscale Materials*,
Springer Series in Materials Science 318,
https://doi.org/10.1007/978-3-030-80323-0_6

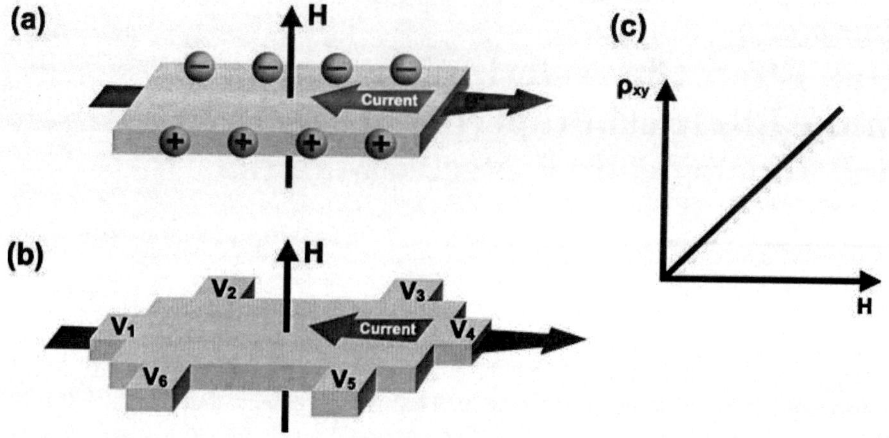

Fig. 6.1 Hall characterization of carriers determines both the carrier density and Hall mobility. Here we show the Hall effect for a metal or a *n* doped semiconductor. **a** Due to the Lorentz force, a Hall voltage results when a magnetic field is applied perpendicular to the current flow. The current flows along the x direction opposite to the electron flow which is shown in the picture, and the magnetic field is applied along the z direction. One commonly used test structure is the Hall bar shown in **b**. In (**b**), the current flows between contacts 1 and 4, and the Hall voltage is measured between contacts 2 and 6 or 3 and 5. The resistivity can be measured between contacts 2 and 3 or 5 and 6 when there is no magnetic field. [Not shown is the van der Pauw test structure with four contacts placed close to the edge [5]. One example is a cloverleaf shaped sample.] The linear relationship between the Hall resistivity and the magnetic field strength is shown in **c**. For a hole (*p*) doped semiconductor with the current flowing in the same direction as the holes, the sign of the Hall voltage changes, and holes build up on the same side as the electrons in a n doped semiconductor [5]

evident in the many publications and awards associated with Hall measurements of bulk, 2D, and topological materials.

Hall Effect measurements have long been used to characterize the mobility and carrier density of semiconductors. Typically, the same sample is used to characterize the resistivity. High electron and hole mobility and the appropriate carrier density are critical for transistor channels and other device characteristics. In a Hall measurement, a material is patterned into a structure that allows current flow and measurement of a voltage transverse to the current flow. Classically, a transverse voltage is measured in the presence of a magnetic field applied perpendicularly to the current flow. The transverse voltage is due to the Lorentz force on the electrons. Figure 6.1 shows the classical Hall Effect measurement. The classical Hall Effect is a good starting point for discussing the wide variety of electrical transport mechanisms that are observed in 2D electron gases (2DEG), 2D materials such as graphene, and materials whose electrical transport is topologically protected from scattering from defects so that there is no energy loss during transport. The Hall Effect is really a family of Hall effects which result from differences in the electronic structure between materials. The family of Hall effects is described in the review article of Chang and Li [1], and this chapter is motivated by their discussion.

It is important to note that the carrier transport characteristics of many nanoscale materials are used to verify that the carriers are Dirac fermions and not Schrodinger electrons or holes. For example, Hall effect measurements of graphene proved the Dirac nature of the carriers justifying the use of the relativistic Dirac equation. In this chapter, we present an introduction to both the physics of Hall measurements and to the topological physics of carrier transport.

The flow of this chapter will be as follows: First it is useful to review classical Hall Effect measurements. Next, the historic observation of the integer quantum Hall effect (IQHE) in 2DEGs by von Klitzing is discussed from the point of view of non-topological quantum mechanics. Then, the topological interpretation of the IQHE will be presented [2–4]. Typically, topological explanations of IQHE and other members of the Hall family rely on edge conduction. Experimental data showing where the conduction for IQHE occurs is also reviewed. Then the fractional QHE will be presented. The anomalous Hall Effect (AHE) and the quantum anomalous Hall effect (QAHE) which do not require a magnetic field are discussed [1]. We note that Hall first observed the AHE when characterizing the ferromagnetic metals Ni and Co. Interestingly, the QAHE was finally observed experimentally in 2013. The spin Hall Effect refers to the separation of spin up from spin down during current transport without a magnetic field. Both the Spin Hall Effect (SHE) and Quantum Spin Hall Effect (QSHE) have been experimentally observed and are discussed. Recently, a thermal version of the Spin Hall effect, the Nernst Spin Hall Effect (NSHE) has been reported. We also discuss the NSHE.

An important point is that these new topologically protected materials are considered to exhibit new electronic states of matter.

6.1 Classical Hall Effect (HE)

In the classical Hall Effect measurement, a magnetic field is applied perpendicularly to the current flow. In Fig. 6.1, we show current flowing along the length of a test structure which is typically called the Hall bar. In the presence of a magnetic field, the Lorentz force results in a Hall voltage.

In the presence of a magnetic field, the total force experienced by an electron flowing along a conducting material (metal or n doped semiconductor) is given by the sum of the force due to the electric field and Lorentz force:

$$\vec{F} = -e(\vec{E} + \vec{v} \times \vec{B}) \tag{6.1}$$

Here, the Lorentz force is the cross product of the velocity of the electron with the magnetic field. e is the charge of the electron, and $\vec{E}$ the electric field. The current is flowing along the x direction. The equation for the steady state condition with no net force along the y direction F_y provides means of simplifying (6.1) for a current flowing along the x direction with electrons having a drift velocity along the x direction v_x (for the configuration shown in Fig. 6.1, $v_x > 0$) and the magnetic field $\vec{B} = (0, 0, B_z)$ along a direction z that is perpendicular to the surface $\vec{v} \times \vec{B} = (v_y B_z, -v_x B_z, 0)$, results in:

$$F_y = -e(E_y - v_x B_z) \tag{6.2}$$

At study state, the force along the y direction is zero:

$$E_y = v_x B_z \tag{6.3}$$

For a current of electrons with electron density n along the x direction, the current density in the absence of the magnetic field is $J_x = -ev_x n$. In the presence of the magnetic field some of the current flows along the y direction, so this is an approximation for J_x. The coupled equations for the current flow are presented below without this approximation and are stated in (6.12). The Hall voltage for a sample of cross-sectional area $A = width(w) \cdot d(thickness)$ can be found by integrating the electric field along the y direction over the width of the sample [5]. Ignoring the small current along y, the total current is $I = J_x A = -ev_x nwd$ and the Hall voltage is given by:

$$V_{Hall} = \int_0^w E_y dy = \int_0^w v_x B_z dy = \int_0^w \frac{-IB_z}{ewdn} dy = \frac{-IB_z}{edn} \tag{6.4a}$$

The transverse *Hall resistance* R_{xy} is determined from $V_{Hall} = V_{xy}$ and the applied current I_x as follows: $R_{xy} = V_{xy}/I_x$, and the Hall resistivity ρ_{xy} is determined from the applied current density j_x as follows: $\rho_{xy} = V_{xy}/j_x$. The Hall coefficient R_H is defined as:

$$V_{Hall} = \frac{-IB_z}{ned} = R_H \frac{IB_z}{d}; \quad \text{for electrons } R_H = \frac{-1}{ne}$$
$$\text{and for holes } R_H = \frac{1}{pq} \tag{6.4b}$$

For a material with uniform carrier concentrations having both electron and hole conduction, it can be shown that the Hall coefficient is a function of the mobility ratio $\frac{\mu_n}{\mu_p}$ [5]:

$$R_H = \frac{\left[\left(p - n\left(\frac{\mu_n}{\mu_p}\right)^2\right) + (\mu_n B_z)^2 (p - n)\right]}{q\left[\left(p + n\left(\frac{\mu_n}{\mu_p}\right)\right)^2 + (\mu_n B_z)^2 (p - n)^2\right]} \tag{6.4c}$$

Equation (6.4c) simplifies for very low and very high magnetic fields. The carrier density n can be determined from a Hall measurement, and using the sheet resistance R_S of the sample where the sheet resistance is the bulk resistivity divided by the sample thickness $R_S = \rho/d = {}^{1}\!/_{\sigma d}$, the Hall mobility μ_{Hall} (mobility determined by a Hall measurement) is defined in (6.5) and is distinguished from the conductivity mobility μ discussed below [5]:

$$\mu_{Hall} \equiv |R_H|\sigma \quad \text{and} \quad \mu_{Hall} = \frac{|V_{Hall}|}{IB_z R_s} \tag{6.5}$$

The Hall angle θ_{Hall} is used to select materials for magnetic device applications and is defined as [5]:

$$\tan(\theta_{Hall}) \equiv \frac{E_y}{E_x} = B_z\mu \tag{6.6}$$

Clearly, the conductivity is different along the current flow than it is transverse to the current flow. At the end of this section, this simple picture of the conductance will be replaced by a discussion of the magneto-conductivity tensor. Classically, the electrons form cyclotron orbits in the presence of a magnetic field. The drift velocity is slower than the classical orbit velocity. The time it takes the electron to complete one orbit is $t = \frac{2\pi r}{v}$. The cyclotron angular frequency $\omega_c = \frac{2\pi}{t}$ can be determined by equating the centripetal force $\frac{mv^2}{r} = m\omega_c^2 r$ and the Lorentz force $ev_x B_z$:

$$m\omega_c^2 r = ev_x B_z = e\omega_c r B_z \quad \omega_c = \frac{eB_z}{m} \tag{6.7a}$$

The cyclotron frequency is a function of the magnetic field strength. The radius of the cyclotron orbit is:

$$r = \frac{mv_x}{eB_z} \tag{6.7b}$$

In the classical picture, when the magnetic field is applied, the electrons or holes drift along the direction of the electric field while undergoing cyclotron orbits. The carriers drift slowly compared to the cyclotron motion.

A useful discussion of resistivity and Hall measurements for semiconductor materials and devices is provided in [5]. Hall measurements of semiconductor samples require further discussion of the mobility and conductivity. Typically the magnetic field strength is between 0.05 and 1 Tesla for semiconductor characterization [5]. The conductivity σ_s of a semiconductor includes the contributions of both the electrons with electron carrier density n and electron (conductivity) mobility μ_e and holes with carrier density p and hole (conductivity) mobility μ_h:

$$\sigma_s = q(n\mu_e + p\mu_h) \tag{6.8a}$$

Typically, transistors are fabricated on doped silicon wafers or doped epilayer surfaces. For heavily p doped silicon $p \gg n$, and $\mu_h = \sigma_s/qp = 1/q\rho p$ where ρ is the resistivity. The carrier density of heavily doped semiconductor is given by [5]:

$$p = \frac{\mathcal{R}}{qR_H}; \quad n = -\frac{\mathcal{R}}{qR_H} \tag{6.8b}$$

Here, $\mathcal{R}$ is the Hall scattering factor which varies from 1 to 2 [5]. $\mathcal{R}$ is dependent on the scattering mechanism, temperature, and magnetic field strength. $\mathcal{R}$ is 1.93 for charged impurity scattering, 1.18 for lattice scattering, and 1 for neutral impurity scattering [5]. A simplified discussion assumes $\mathcal{R}$ to be 1. The Hall mobility for a semiconductor should also account for the scattering factor, and it is given by:

$$\mu_H = \sigma_s|R_H|; \quad \mu_H = \mathcal{R}\mu_e \text{ or } \mu_H = \mathcal{R}\mu_h \tag{6.8c}$$

The cyclotron motion of the electrons along the direction of current flow in a magnetic field gives quantized energy levels known as Landau levels which are discussed below. The quantization occurs in the plane of the current flow when the magnetic field is present, and in 3D systems the electrons move freely normal to this plane. The energy spacing between Landau levels is too small to be observed at room temperature, thus many Hall measurements are done at a few K. In 2DEG, the Hall resistance (conductivity) is also quantized resulting in the quantum Hall effect. The Landau levels in systems confined to two dimensions are critical to understanding the QHE. First, the traditional (non-topological) description of the QHE is presented. This is followed by a topological discussion of the IQHE.

6.1.1 Classical Picture of Edge States

In this sub-section we consider carriers in a thin conducting layer in a sample with edges such as the one shown in Fig. 6.1. In subsequent sections, this conducting layer will be described as a 2D electron gas. In the absence of an electric field, the cyclotron orbits at the physical edge of the sample are not able to complete a cyclotron orbit and keep bouncing (reflecting) off the edge and, when the magnetic field is pointing up, move counter clockwise around the sample. The current on one side of the sample is going in the opposite direction of the other thus cancelling each other. The total current remains zero until an electric field is applied. This is an example of the chirality of the edge states. Here, the reason this is referred to as chiral is that the motion on opposing edges is in the opposite direction. Imagine a mirror in the middle of the sample shown in Fig. 6.1 along the length of the sample and normal to the surface. The moving electrons on one side are not the mirror image of the moving electrons on the other side. In Chaps. 7 and 9, chirality is further discussed.

6.1.2 Classical Picture of Magneto-Conductivity Tensor

The magneto-conductivity tensor describes the conductivity both in the direction of the current flowing along the applied electric field and transverse to the direction of that current flow. The starting point for deriving the magneto-conductivity tensor is the Lorentz equation for a 3D sample in a magnetic field with the electric field in an arbitrary direction in $\vec{r}$ (x, y, and z). We use the coordinate system shown in Fig. 6.1, and we follow the discussion of Yu and Cardona [6]. The derivation that follows is a generalization of the Drude model for conductivity so that the magnetic field is included. The Lorentz equation for the position of the carrier in an electric field $\vec{E}$ with a time τ between collisions is in SI units [6]:

$$m^* \frac{d^2\vec{r}}{dt^2} + \frac{m^*}{\tau}\frac{d\vec{r}}{dt} = -e[\vec{E} + v \times \vec{B}\,] \tag{6.9}$$

The Lorentz Force can be written in CGS units as $= -e[\vec{E} + \frac{v}{c} \times \vec{B}\,]$ when the electron charge, electric field, and magnetic field are given in CGS units. For steady state conditions where the carrier does not accelerate, $\frac{d\vec{r}}{dt}$ is the drift velocity $\vec{v}_d$:

$$\frac{m^*}{\tau}\vec{v}_d = -e[\vec{E} + \vec{v}_d \times \vec{B}\,] \tag{6.10}$$

The x, y, and z components for the Lorentz equation for a magnetic field along the z direction and again using $\vec{v} \times \vec{B} = \left(v_y B_z, -v_x B_z, 0\right)$ are:

$$\begin{aligned} \tfrac{m^*}{\tau} v_{d_x} &= -e\left[E_x + v_{d_y} B_z\right] \\ \tfrac{m^*}{\tau} v_{d_y} &= -e\left[E_y - v_{d_x} B_z\right] \\ \tfrac{m^*}{\tau} v_{d_z} &= -e\left[E_z\right] \end{aligned} \tag{6.11}$$

The current density $\vec{J}$ is minus the electron density n multiplied by the electric charge and the drift velocity, $\vec{J} = -ne\vec{v}_d$. This gives us coupled equations for the x, y, and z components of the current density:

$$\begin{aligned} J_x &= \left(\frac{ne^2\tau}{m^*}\right)E_x - \left(\frac{eB_z\tau}{m^*}\right)J_y \\ J_y &= \left(\frac{ne^2\tau}{m^*}\right)E_y + \left(\frac{eB_z\tau}{m^*}\right)J_x \\ J_z &= \left(\frac{ne^2\tau}{m^*}\right)E_z \end{aligned} \tag{6.12}$$

This can be rewritten using the zero frequency conductivity $\sigma_0 = \left(\frac{ne^2\tau}{m^*}\right)$ and classical cyclotron frequency $\omega_c = \frac{eB_z}{m^*}$:

$$\begin{aligned} J_x &= \sigma_0 E_x - \omega_c \tau J_y \\ J_y &= \sigma_0 E_y + \omega_c \tau J_x \\ J_z &= \sigma_0 E_z \end{aligned} \tag{6.13}$$

These can be solved for the x, y, and z components of the current density:

$$\begin{aligned} J_x &= \frac{1}{1+(\omega_c\tau)^2}\sigma_0[E_x - \omega_c \tau E_y] \\ J_y &= \frac{1}{1+(\omega_c\tau)^2}\sigma_0[E_y + \omega_c \tau E_x] \\ J_z &= \sigma_0 E_z \end{aligned} \tag{6.14}$$

Since $\vec{J} = \boldsymbol{\sigma}\vec{E}$ where $\boldsymbol{\sigma}$ is the magneto-conductivity tensor which is a function of the magnetic field strength $\omega_c(B)$:

$$\boldsymbol{\sigma} = \frac{\sigma_0}{1+(\omega_c\tau)^2}\begin{pmatrix} 1 & -\omega_c\tau & 0 \\ \omega_c\tau & 1 & 0 \\ 0 & 0 & 1+(\omega_c\tau)^2 \end{pmatrix} \tag{6.15}$$

The term magneto-resistance is apparent since there is a decrease in conductivity $1+(\omega_c\tau)^2$ perpendicular to the magnetic field. *It is important to note that the tensor nature of the conductivity must be accounted for when determining the resistivity by inverting the conductivity tensor.* This point will arise when we discuss the anomalous Hall resistivity below. Below, we see that the magneto-conductance is quantized in 2DEG and many 2D materials.

6.2 Integer Quantum Hall Effect (IQHE)

The IQHE was discovered by Klaus von Klitzing on the night of Feb. 4 1980 in the High Magnetic Field laboratory at Grenoble [2]. One of the most amazing and important aspects of the IQHE is that the resistance is quantized to 1 part in 10^9 [2, 3]. The IQHE revolutionized our understanding of transport physics. Because the classical Hall effect can be measured using a constant magnetic field, the first measurements of the IQHE were obtained using a constant magnetic field. Using a MOSFET structure and a high, fixed magnetic field, von Klitzing observed that the Hall resistance (resistance that is transverse to the current flowing from source to drain) showed quantized steps as the carrier density was increased. For the initial observation of the IQHE, von Klitzing used operating conditions that established an inversion layer that formed the 2DEG. He observed that the longitudinal resistance repeatedly rose and fell to zero as the carrier concentration was increased. Today, the Hall effect in 2DEG and most 2D materials is typically characterized by ramping a strong magnetic field. The 2DEG formed in the inversion layer of the transistor in von Klitzing's experiment is also present in other materials systems. For example, epitaxial processes are used to grow hetero-epitaxial film stacks with quantum wells that trap carriers in 2DEGs. As a result of this pioneering research, von Klitzing won the Nobel prize in 1985. This effect was theoretically predicted by Ando, Matsumoto, and Uemura in 1975 [2]. The first observation of the QHE is shown in Fig. 6.2. It is important to note that the Hall resistance (transverse resistance) has plateaus when the longitudinal resistance (along the direction of current flow) is zero.

In discussing the quantum mechanical picture of the IQHE, two key quantizations occur. First the energy of the cyclotron orbits of the 2DEG is quantized, and second, the conductance becomes quantized. As (6.15) shows, in the classical picture conductance is not quantized. Although the quantization of the conductance was theoretically predicted, in his review article, von Klitzing states that even Ando, Matsumoto, and

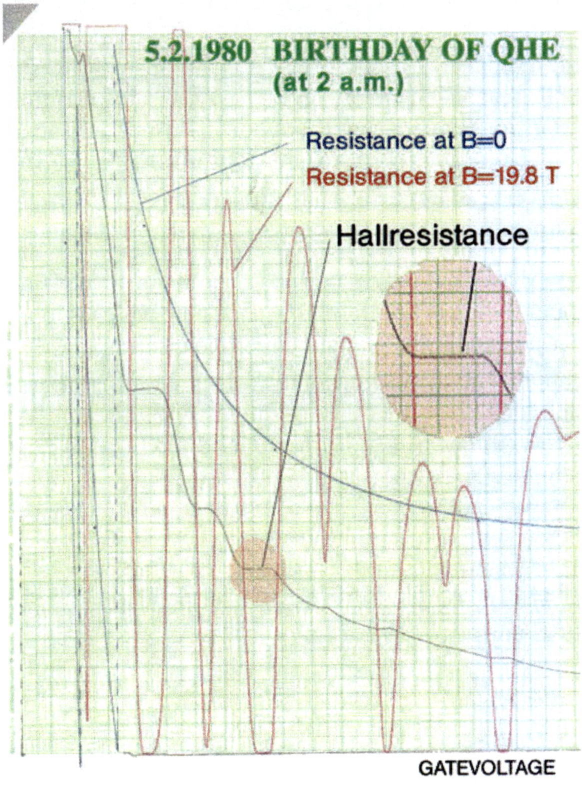

Fig. 6.2 A reproduction of von Klitzing's original data showing the first observation of the IQHE in a silicon MOSFET at low temperature ~ 4.2 K. Figure used with permission from [2]

Uemura were concerned that any unexpected aspect of the physics would mean that their prediction would prove incorrect [2].

The quantization of the Hall resistance $|R_H| = \left(\frac{1}{v^2}\right)\frac{\hbar}{e^2}$ in units of $\frac{\hbar}{e^2}$ has an uncertainty of 10^{-9} for the second and fourth plateau with a value at 90 K of $R_{K\text{-}90} =$ 25,812.807 Ω [3]. This value is not a function of material and carrier density making it a useful calibration standard [3, 4]. Below, we will discuss the IQHE for graphene, and it is important to note that the IQHE is not quantized with the same uncertainty. The theoretical explanation the IQHE in a 2DEG needs to also account for this uncertainty.

6.2.1 Landau Levels—The Quantization of 3D and 2D Carrier Motion in a Magnetic Field

Intuition tells us that the cyclotron orbits can be quantized. However, the idea that the classical orbit becomes quantized in same way as the quantization of a particle moving along a ring of constant diameter is not correct. In this section, we discuss the non-relativistic quantization of the cyclotron orbits. In Chap. 7 on graphene, we discuss the quantization of the relativistic transport Hamiltonian. The motion is quantized along one direction and the quantized energy levels have the same functional relationship between energy level and quantum number as does a harmonic oscillator. The carriers are approximated by the Hamiltonian for free electrons with an effective mass m^*:

$$-\frac{\hbar^2}{2m^*}\nabla^2\psi = E\psi \tag{6.16}$$

The electrons are moving in a uniform magnetic field which is represented by both a vector and scalar potential. The momentum of the electron in CGS units becomes $\vec{P} + \frac{e\vec{A}}{c}$. For a sample with the magnetic field along the z direction and the current running along the x direction, the following holds when the unit vector along z is $\hat{z}$: $\vec{B} = B\hat{z}$ and $\vec{A} = (0, Bx, 0)$ [7]. Here, we used the Landau gauge to select the vector potential, and the scalar potential ϕ of the magnetic field is set to zero [7]. The electron has a magnetic moment due to its spin, and that magnetic moment μ_b interacts with the magnetic field adding a term $g^*\mu_b B_z$ where g^* is the Lande factor to the Hamiltonian proportional to the magnetic moment times the magnetic field along the z direction $\propto \mu_b B_z$[6] which we add later in the derivation. The complete Hamiltonian in CGS units is:

$$\left(\frac{1}{2m^*}\left(\vec{P} + \frac{e\vec{A}}{c}\right)^2 + (g^*\mu_b B_z) + \phi\right)\psi = E\psi$$

ϕ, the scalar potential, is zero in the Coulomb gauge [7]. First we solve the Hamiltonian for the electron momentum in a magnetic field [7, 8]:

$$\frac{1}{2m^*}\left(\vec{P}+\frac{e\vec{A}}{c}\right)^2\psi = E\psi \tag{6.17}$$

where $\vec{A}$ is the vector magnetic potential for the magnetic field, and $\vec{B} = \nabla \times \vec{A}$. Rewriting (6.17) for a 3D sample for the $x, y,$ *and* z components gives:

$$\begin{aligned}&\frac{1}{2m^*}\left(-i\hbar\vec{\nabla}+\frac{e\vec{A}}{c}\right)^2\psi\\ &= -\frac{\hbar^2}{2m^*}\frac{\partial^2}{\partial x^2}\psi + \frac{1}{2m^*}\left(-i\hbar\frac{\partial}{\partial y}+\frac{eBx}{c}\right)^2\psi \quad -\frac{\hbar^2}{2m^*}\frac{\partial^2}{\partial z^2}\psi\\ &= E\psi\end{aligned} \tag{6.18}$$

$$\frac{\partial^2}{\partial x^2}\psi - \left(-i\frac{\partial}{\partial y}+\frac{eBx}{\hbar c}\right)^2\psi + \frac{\partial^2}{\partial z^2}\psi + \frac{2m^*}{\hbar^2}E\psi = 0 \tag{6.19}$$

Separation of variables can be applied to the solution of (6.19). One can write the wavefunction as [7, 8]:

$$\psi(x, y, z) = u(x)e^{ik_y y}e^{ik_z z} \tag{6.20}$$

Solving for the Eigen energies:

$$E_z = \frac{\hbar^2 k_z^2}{2m^*} \quad E' = E - \frac{\hbar^2 k_z^2}{2m^*}$$

$$\begin{aligned}&\left(-i\frac{\partial}{\partial y}+\frac{eBx}{\hbar c}\right)^2 u(x)e^{ik_y y}e^{ik_z z}\\ &= \left(-\frac{\partial^2}{\partial y^2} - i\frac{\partial}{\partial y}\frac{eBx}{\hbar c} - i\frac{eBx}{\hbar c}\frac{\partial}{\partial y} + \left(\frac{eBx}{\hbar c}\right)^2\right)u(x)e^{ik_y y}e^{ik_z z}\\ &= \left(k_y^2 u(x)e^{ik_y y}e^{ik_z z} + 2k_y\left(\frac{eBx}{\hbar c}\right)u(x)e^{ik_y y}e^{ik_z z} + \left(\frac{eBx}{\hbar c}\right)^2 u(x)e^{ik_y y}e^{ik_z z}\right)\\ &= \left(k_y + \left(\frac{eBx}{\hbar c}\right)\right)^2 u(x)e^{ik_y y}e^{ik_z z}\end{aligned} \tag{6.21}$$

Combining (6.19) and (6.21) we get:

$$\frac{\partial^2}{\partial x^2}u(x) - (k_y + \frac{eBx}{\hbar c})^2 u(x) - k_z^2 u(x) + \frac{2m^*}{\hbar^2}Eu(x) = 0$$

$$-\frac{\hbar^2}{2m^*}\frac{\partial^2}{\partial x^2}u(x) + \frac{\hbar^2}{2m^*}(k_y + \frac{eBx}{\hbar c})^2 u(x) = \left(E - \frac{\hbar^2 k_z^2}{2m^*}\right)u(x) = E'u(x)$$

$$-\frac{\hbar^2}{2m^*}\frac{\partial^2}{\partial x^2}u(x) + \frac{m^*}{2}\left(\frac{eB}{m^*c}x + \frac{\hbar k_y}{m^*}\right)^2 u(x) = E'u(x) \tag{6.22}$$

Which can be rewritten using the classical cyclotron angular frequency $\omega_c = \frac{eB}{m^*c}$ and $x_0 = \frac{1}{\omega_c}\frac{\hbar k_y}{m^*}$. As the carriers drift due to the voltage drop between the source and drain, the position x_0 changes.

$$-\frac{\hbar^2}{2m^*}\frac{\partial^2}{\partial x^2}u(x) + \frac{m^*}{2}\omega_c^2(x + x_0)^2 u(x) = E'u(x) \tag{6.23}$$

This is the Schrodinger equation for the harmonic oscillator with a shift of the x coordinate:

$$-\frac{\hbar^2}{2m^*}\frac{\partial^2}{\partial x^2}u(x + x_0) + \frac{m^*}{2}\omega_c^2(x + x_0)^2 u(x + x_0) = E'u(x + x_0)$$

After adding the energies for the electron spin in a magnetic field $g^*\mu_b B_z$ [7, 8], and remembering that the energy E' is the total energy minus the free electron energy along the z direction, the total energy is:

$$E = \left(n + \frac{1}{2}\right)\hbar\omega_c + g^*\mu_b B_z + \frac{\hbar^2 k_z^2}{2m^*} \tag{6.24}$$

The free electron motion along the magnetic field direction z is the same as the classical result for a 3D sample. The energy level solutions in 3D are cylinders with the axis of the cylinder along the magnetic field direction z as shown in Fig. 6.3. The kinetic energy of the electron can still be written as $\hbar^2(k_x^2 + k_y^2)/2m^*$ and thus the radius of the cylinder is proportional to $\sqrt{k_x^2 + k_y^2}$.

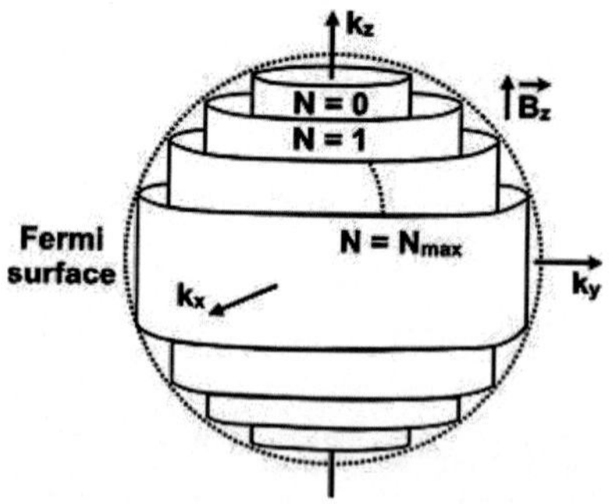

Fig. 6.3 The energy levels for a 3D system using the coordinate system shown in Fig. 6.1 are cylinders in k space with the axis along the direction of the magnetic field. Adapted and reprinted from [56] with the consent of the author

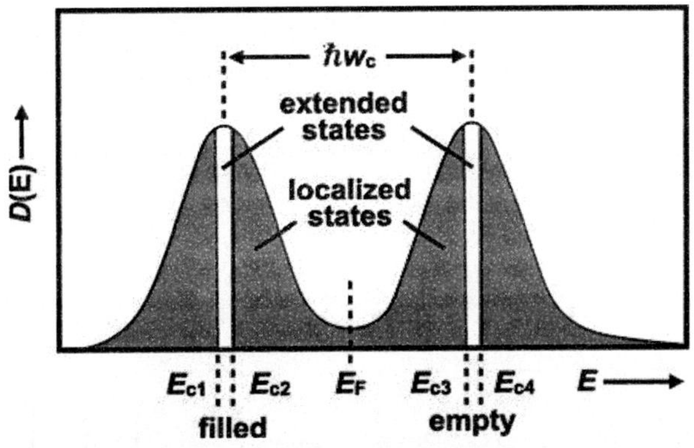

Fig. 6.4 The broadening of the DOS for the Landau levels is shown. The localized Landau states are ascribed to impurities [2]. The prevailing explanation for quantized conductance requires use of the localized states. Figure adapted and reprinted with permission from [4]. © 1986 The Nobel Foundation

For a 2DEG, there is no motion along z. This is the Hamiltonian for a harmonic oscillator, and the energies are independent of k_y and thus are one dimensional in reciprocal space for a 2DEG.

$$E = \left(n + \frac{1}{2}\right)\hbar\omega_c + g^*\mu_b B_z \tag{6.25}$$

The physical picture of carrier motion in a 2DEG is that the electrons drift from source to drain slowly while undergoing rapid in-plane cyclotron motion. The density of states for the Landau levels for a perfect crystal will be a delta function at each energy E because the there is no spread in energy. In a real crystal there are defects and impurities which result in a spread in E'. Even in a high quality, epitaxial grown III–V heterostructures (e.g., GaAs/(AlGa)As/GaAs) there will be a small number of crystalline defects or impurities that result in *localized states*. Observation of the IQHE requires high quality samples. The spread in the DOS due to a small number of defects or impurities is shown in Fig. 6.4. This spread in energies is critical to understanding why each quantized step in conductance occurs over a relatively wide range of magnetic field strengths as seen in Fig. 6.2.

6.2.2 Integer Quantized Transport

The observation of quantized transverse conductance (or conversely quantized Hall resistivity as shown in Fig. 6.2) was the key breakthrough that resulted in von Klitzing's Nobel prize. It is very interesting that the first theoretical explanations for the quantization required that the density of states have a spread in energy due to

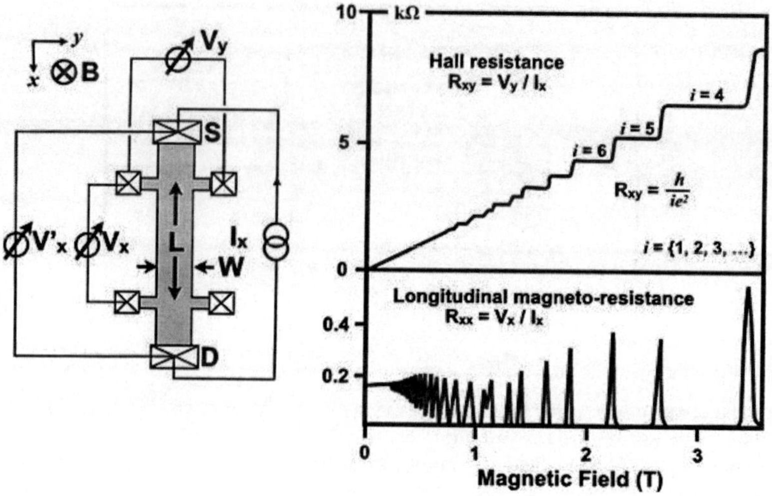

Fig. 6.5 The Integer Quantum Hall effect as observed using the 2DEG in a GaAs/(AlGa)As heterostructures. The longitudinal resistivity R_{xx} goes to zero at magnetic field strengths where the transverse resistivity R_{xy} increases in a quantized step. For example, at 2 T, $R_{xx} = 0$ *and* $R_{xy} = \frac{h}{6e^2}$. Figure adapted and reprinted from [3] with permission. © 2011 The Royal Society (U.K.)

crystal imperfections. Figure 6.5 shows the IQHE in the 2DEG of a Hall bar fabricated using a GaAs/(AlGa)As heterostructures. It is important to note that the longitudinal magneto-resistance goes to zero at the Hall plateaus in resistance (conductance). Thus there is dissipationless current flow along the sample when there is a Hall plateau. However, there is dissipation at the contacts.

Time Reversal Symmetry and the Hall Effect: It is important to note that time reversal symmetry must be broken for any Hall current to flow be it classical or quantized. Time reversal symmetry will be further discussed in Chap. 9. Briefly, time reversal changes several key quantities velocity, current, and magnetic field, but leaves electric field and charge density unaltered as follows:

$$v \rightarrow -v;\ J \rightarrow -J;\ \vec{B} \rightarrow -\vec{B};\ \vec{E} \rightarrow \vec{E};\ \rho \rightarrow \rho$$

Current flow breaks time reversal symmetry while an electric field does not. When current is flowing in the longitudinal direction, energy is dissipated due to the resistivity, and the second Law of Thermodynamics requires that time reversal symmetry is broken. Even when there is no resistance in the longitudinal direction, current is flowing. In the transverse direction, time reversal symmetry must be broken for there to be Hall conductivity. Time reversal symmetry is also broken by the presence of a magnetic field. This is also true when the IQHE is present.

The degrees of freedom of the carriers are not changed by the magnetic field. The degeneracy β multiplied by the number of filled Landau levels must equal the degrees of freedom, thus the Landau levels are highly degenerate [6]. Following von

Klitzing, the degeneracy can be derived from the central point of the cyclotron orbit (center point of the harmonic oscillation) $x_0 = \frac{1}{\omega_0}\frac{\hbar k_y}{m^*}$ [4]. Since width of the sample is L_y, using the same argument as with the particle in a box where the wave function ~ $\sin(k_y L_y)$ is zero at the boundaries giving us allowed solutions $k_y L_y = n\pi$ which have a distance in k space of $\Delta k_y = \frac{\pi}{L_y}$. This result must be multiplied by a factor of 2 due to electron spin giving, $\Delta k_y = \frac{2\pi}{L_y}$. The number of centers of cyclotron orbits in a sample of dimensions $L_x L_y$ is given by

$$\Delta x_0 = \frac{1}{\omega_0}\frac{\hbar \Delta k_y}{m^*} = \frac{m^*}{eB}\frac{\hbar \Delta k_y}{m^*} = \frac{\hbar}{eB}\Delta k_y = \frac{\hbar}{eB}\frac{2\pi}{L_y} = \frac{h}{eB}\frac{1}{L_y} \tag{6.26}$$

The degeneracy factor is $\beta = \frac{L_x}{\Delta x_0} = \frac{L_x L_y eB}{h}$.
and the degeneracy per Landau level per unit area, n_L, is:

$$n_L = \frac{\beta}{L_x L_y} = \frac{eB}{h} \tag{6.27}$$

The magnetic flux quantum is h/e and the area associated with a magnetic flux quantum is: $h/(Be)$ [9]. The flux quantum was introduced as the amount of magnetic flux inside circulating Cooper pairs [9]. Here, the flux quantum is the smallest amount of magnetic flux inside a cyclotron orbit [6]. Thus, the number of magnetic flux quantum in the sample is $L_x L_y\, eB/h$ which is the same as the degeneracy factor. A simple derivation of the IQHE was provided by von Klitzing [2–4]. The classical Hall voltage V_H and Hall resistance R_H for a 2DEG is when the 2D carrier density is $n_{2D} = i n_L$ for an integer number i of levels with the magnetic field pointed along the z axis is:

$$V_{Hall} = \frac{IB_z}{n_{2D}e} \quad \text{and} \quad R_H = \frac{V_H}{I} = \frac{B_z}{n_{2D}e} = \frac{h}{ie^2} \text{ (quantized Hall Resistance)} \tag{6.28}$$

Von Klitzing stated that "More important is the general result, that a discrete energy spectrum with energy gaps exists for an ideal 2DEG in a strong magnetic field and that the degeneracy of each discrete level corresponds to the number of flux quanta $(F \cdot B)/(h/e)$ within the area F of the sample. This corresponds to a carrier density $n_C = e \cdot B/h$ for each fully occupied spin-split energy level E_0, N and therefore to a Hall resistance $R_H = h/(i \cdot e^2)$ for i fully occupied Landau levels as observed in the experiment" [2]. Above, we use n_{2D} instead of n_C.

Another concept that is used in the discussion of the experimental imaging of carrier transport in Hall bar samples is the filling factor υ of the Landau levels [3]. The filling factor υ is defined as the ratio of the electron density n_{2D} to the degeneracy of each Landau level n_L. The degeneracy of a Landau level increase as the magnetic field increases as indicated in (6.27). In other words, more electrons can reside in that level as the magnetic field increases. The chemical potential (Fermi level) and filling factor are functions of magnetic field and temperature. For low temperatures, when the magnetic field is large enough, at specific filling factors υ, the chemical potential will lie between Landau levels. Around integer filling factors, a Hall plateau will occur [3].

Hall resistivity plateau width: The relatively wide range of magnetic fields over which the conductance remains quantized at a plateau is noteworthy. We showed above in Fig. 6.4 that the extended states occurred in the middle of the distribution of energy states and that the energy of *localized states* due to the presence of defects or impurities occurred above and below the energy of the extended states. As the name implies, the localized states are not responsible for charge transport. The extended states can transport charge from one end of the sample to the other (longitudinal conductivity). First we consider the effect of an increasing magnetic field strength. When the magnetic field strength is below that required for completely filling the Landau levels, the localized states are being filled, the occupation of the extended states does not change, and the longitudinal resistivity is zero. The longitudinal current flow is dissipationless. The Hall voltage (resistivity) remains constant until the magnetic field increases to a strength that the Fermi level is inside the allowed range of energies for the extended states. The Hall voltage will again increase as more extended states become populated. This continues until the Fermi energy reaches the next Landau level, and the Fermi level is again inside the energy range for the localized states leading to the next plateau in Hall Voltage.

Another important phenomenon is the Shubnikov de Hass (SdH) oscillations in the longitudinal resistivity R_{xx}. The SdH oscillations can be observed in 3D systems such as metals and in 2D systems such as a 2DEG. The SdH oscillations of a 2DEG can be seen at the bottom right hand plot of R_{xx} in Fig. 6.5. The cyclotron frequency $\omega_c = \frac{eB}{m^*c}$ and harmonic oscillator energy for the Landau levels, $E = \left(n + \frac{1}{2}\right)\hbar\omega_c + const.$, both increase linearly with increasing magnetic field strength. As mentioned above, the number of allowed states also increases linearly with increasing magnetic field. A simple model for the SdH oscillations is to picture the Landau levels increasing in energy and some of the levels that were below the Fermi energy pass through a fixed Fermi level as their energy increases. When the Fermi Energy is between Landau levels, the electrons cannot move away from the Landau levels and there is no scattering and thus no resistivity. When the Fermi energy lies inside the broadened Landau level (see Fig. 6.4), the electrons in that Landau level are free to move to another state, and the scattering that occurs results in a resistivity. The increase in resistivity with increase in magnetic field is due to increase in the number of electrons in a Landau level with increased magnetic field. The SdH oscillations can be used to determine the electron density. The change in the longitudinal resistivity is proportional to the *cos* of the oscillation frequency.

$$\Delta R_{xx} \propto \cos(f) \tag{6.29}$$

6.2.3 *Experimental Microscopic Observation of Carrier Transport and Chemical Potential for the IQHE*

One of the interesting aspects of the QHE and the AQHE is the popular physical picture of the carriers moving at the edges of the top of a Hall Bar sample such as the one fabricated from a GaAs/(AlGa)As heterostructures shown in Fig. 6.6 [1, 3]. This physical picture also accompanies the description of topologically protected carrier transport in other measurements. Experimental observation of carrier transport provides some justification for this picture, but there are exceptions. Von Klitzing characterized the chemical potential using a single-electron transistor (SET) as the local electrometer on top of a 2DEG embedded in a GaAs/(AlGa)As heterostructures [3, 10]. The measurements were done below 2 K in a magnetic field. In addition, a cryogenic scanning force microscope has been used to image transport features by measuring the local potential distribution within the two-dimensional electron system [3, 11]. These methods are described below, and von Klitzing and co-workers interpretation of these measurements is summarized.

Before presenting the experimental imaging of carrier transport, it is useful to describe some of the characteristics of the Hall samples and the electronic structure

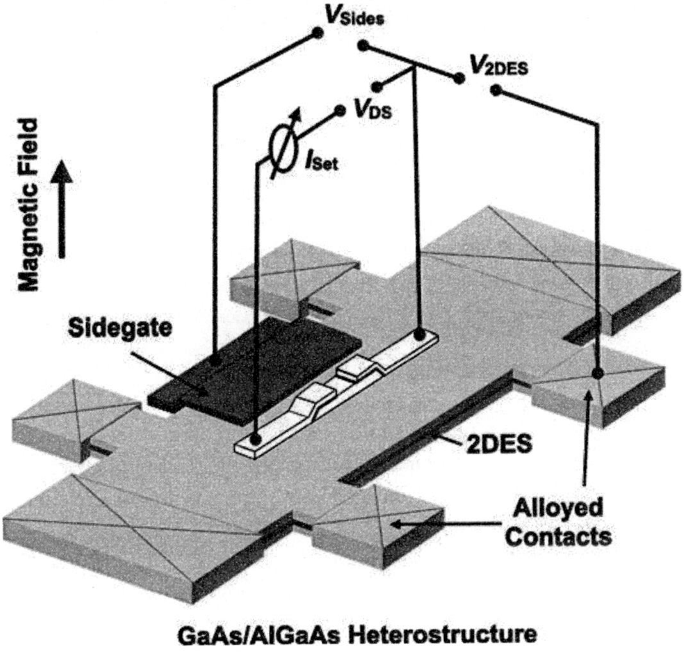

Fig. 6.6 The single electron transistor fabricated on top of the Hall bar made using a GaAs/(AlGa)As heterostructures is used to measure the chemical potential during Hall characterization in a magnetic field at low temperature. Figure adapted and reprinted with permission from [10]. © 1998 American Physical Society

in a magnetic field. Characterization of the Hall conductance requires determination of the carrier density in each sample [3]. The Hall resistivity is proportional to the carrier density in the 2DEG and the Hall resistivity linearly increases with magnetic field at low values of the magnetic field as shown in Fig. 6.5. At these low magnetic fields,

$$|R_H| = \left| {}^{B_z}\!/_{-en_S} \right|. \tag{6.30}$$

The single electron transistor (SET) was fabricated on top of the GaAs/(AlGa)As Hall bar samples, and is shown below in Fig. 6.6 [10]. The purpose of the sidegate is to deplete the 2DEG below the sidegate and push the effective edge of the Hall bar underneath the single electron transistor structure. In this way, the chemical potential can be measured at the edge of the Hall structure as well as in the middle of the Hall structure (when the sidegate voltage is zero) during Hall characterization. This allows carrier flow at the middle and edges of the Hall bar to be characterized using the SET.

Von Klitzing emphasizes [3] that after more than a decade of using cryogenic scanned force microscopy, his group finds that the standard textbook image of current flowing exclusively at the edge is incorrect. The potential variation with magnetic field is divided into three regions based on the variation of Hall potential across the width of the Hall bar. In the first region, the potential drops linearly. In the second region, the Hall potential flattens at the edges of the sample. In the third region, the potential is flat in the middle and drops (or raises) at the edge [3].

The concept of the compressibility of the 2DEG is an essential aspect of describing the integer Hall effect. As previously stated, the filling factor υ is defined as the ratio of the electron density n_s to the degeneracy of each Landau level n_L. For integer filling factors, the chemical potential, μ_{chem}, is between two Landau levels. At low temperatures, Landau levels below the chemical potential are occupied and those above are unoccupied. As stated above, a Hall plateau will occur for filling factors close to an integer value. Under these conditions, the system is incompressible:

$$\frac{\partial \mu_{chem}}{\partial n_S} \to \infty \tag{6.31}$$

The chemical potential increases rapidly with increasing electron density. This occurs as the magnetic field is increased. Regions of the 2DEG are considered to be incompressible when the magnetic field is swept across fields where there is a Hall plateau.

Figure 6.7 shows four snapshots of the 2DEG as the magnetic field at field strengths were the Hall conductance is at a plateau (constant conductance) for a filling factor ($\nu = 2$) where the plateaus are well defined. The magnetic field is sweeping from lower filling factors to higher filling factors. The first Hall bar potential image in Fig. 6.7 shows a potential dropping linearly across the Hall bar. The second Hall bar image shows a flat profile at the edges and drop in the inner regions of the Hall bar which persists from $\nu = 1.96$ to 2.09. The third Hall bar image shows a disordered

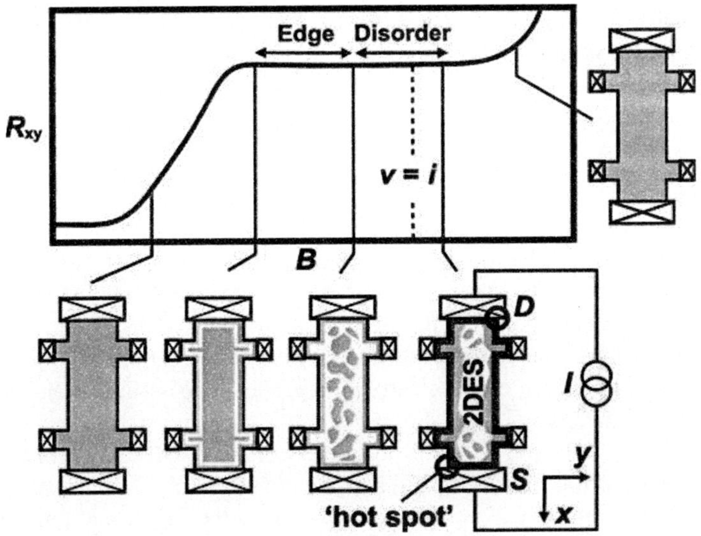

Fig. 6.7 Images of the 2 Degree Electron Gas (2DEG) for different magnetic fields as the magnetic field sweeps across the Hall bar. Incompressible regions form at the edge of the Hall bar. Disorder in the 2DEG appears at higher magnetic fields. Von Klitzing's figure clearly shows that disorder is not required for the Hall plateau for all magnetic fields. Dissipationless current flow does occur along the sample, but close to the current contacts "hot spots" where $R_{xx} \neq 0$ occur as shown above. Here, $R_{xx} \sim R_{Hall}$ resulting in a hot spot [3]. Figure adapted from [3]. Figure adapted and reprinted from [3] with permission. © 2011 The Royal Society (UK)

potential inside the Hall bar. The significance of this data is that the Hall plateau can exist without disorder even though potential disorder appears in the Hall bar as the magnetic field is swept.

6.2.4 Summary for experimental imaging of IQHE

In summary, there is dissipationless transport at magnetic fields which result in a Hall plateau. The often used picture of carrier transport only around the edges is not supported by von Klitzing's experimental imaging of carrier transport. However, transport along the edges of the sample is clearly observed. The concept of incompressible regions in the 2DEG $\frac{\partial \mu_{chem}}{\partial n_S} \to \infty$ is an essential part of the understanding of carrier transport at magnetic fields where there is a Hall plateau. At magnetic fields where the Hall plateau starts, there are incompressible strips in the 2DEG.

6.3 Topological Explanation of the Integer Quantum Hall Effect (IQHE)

The 2016 Nobel Prize in Physics went to David J. Thouless, F. Duncan M. Haldane, and J. Michael Kosterlitz "for theoretical discoveries of topological phase transitions and topological phases of matter." This research paved the way for use of topological concepts to explain superconductivity as well as conductivity in thin conducting layers such as 2DEGs. The topological explanation for the IQHE introduces some key physical concepts and mathematical descriptions that apply to most of the family of quantum Hall effects including the Berry phase of a wavefunction and the Chern number [12]. It is important to note that the Berry Phase correction to the conductance of ½ proves the Dirac nature of the carriers in graphene, $\boldsymbol{\sigma \sim (n + 1/2)e^2/h}$. This is further discussed in Chap. 7.

In 1981, shortly after von Klitzing's discovery of the IQHE, Laughlin proposed a topological explanation for the IQHE [13]. We note that Laughlin was co-recipient of the Nobel Prize in Physics for his contribution to the Fractional Quantum Hall Effect discussed below. This explanation is a key start that has been further developed by subsequent topological theory [12]. Laughlin's concept is that when the 2DEG is at low temperature, the system has quantum coherence and can be described by a single wavefunction with a Hamiltonian that describes its time evolution. In that case, magnetic flux quanta can pump charge from one charge reservoir to a second reservoir as shown in Fig. 6.8. Laughlin's paper indicates that the n electrons transferred result in the quantized conductance (see (11) in [13]). Laughlin uses Gauge invariance to define limits to the vector potential of the magnetic field (see (1) and (2) in [13]). We note that although both classically and quantum mechanically the charge in a reservoir must be an integral number, quantum mechanically, the pump cycles can pump non-integral amounts of charge. The concept of an average charge transfer of integral amounts of charge is introduced through the Chern formula which is discussed below. Laughlin's description of charge pumping is critical to the discussion of the QHE, and charge pumping is part of the lexicon of the QHE.

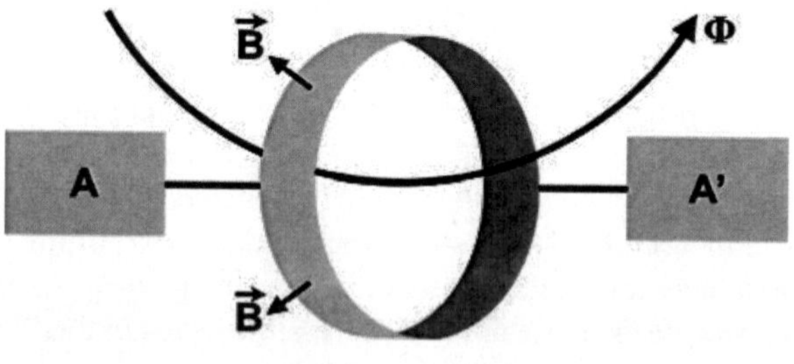

Fig. 6.8 Laughlin's description of the IQHE is that charge pumping from a reservoir A to reservoir B through a loop describing the 2DEG occurs when magnetic flux quanta flow through the loop. Adapted and reproduced from [12], with the permission of the American Institute of Physics

6.3.1 Berry Phase, Berry Curvature, and Berry Potential

The Chern formula is generalization of the Gauss-Bonnet formula relating geometry and topology. Before discussing the Chern formula, the concept of the Berry phase of a wavefunction needs to be introduced. The Berry phase is the phase that the wavefunction picks up after traversing a closed loop. The Berry phase is a key correction to the semiclassical quantum mechanical description of current transport that must be included when a full quantum picture is required. The origin of the Berry phase can be pictured using a continuous path on sphere as shown in Fig. 6.9. It shows that when a wavefunction goes around a closed loop, the vectors that describe that wavefunction point in a different direction, and thus the phase of the wavefunction is different.

In Fig. 6.9, the vectors start at point 1, and the vectors stay tangential to the sphere. After moving around a closed loop that comprises 1/8 of the sphere, the vectors are pointing 90° away from the original direction. Thus there is a relationship between the path length and the direction of rotation of the vectors. This is a geometric effect. If the path covers ¼ of the sphere, the rotation is 180°. If the path is smaller, the rotation is smaller. Rotation angles of a geometric origin are called a Berry phase. The Berry phase is purely geometrical-topological. If we use a flat surface or a cylinder, there is no Berry phase [14, 15]. The rotation angle is related to the integral of the curvature of the surface bounded by the loop.

A useful discussion of the Berry phase for a wavefunction in a magnetic field can be found in Berry's original paper [16], Sakurai's book [15], and Xiao, Chang, and Niu's review [17]. Berry's [16] derivation clearly separates the dynamical and geometric phases of the wavefunction. This results in a time dependent equation for the geometric phase from the Hamiltonian. This geometric phase is a path integral over the closed loop. The derivation of the Berry phase and Berry connection require the adiabatic evolution of the system as the time dependent parameters $R(t)$ representing the system evolve [16, 17]. The path integral is then transformed into a surface integral over the surface enclosed by the path loop on the surface representing

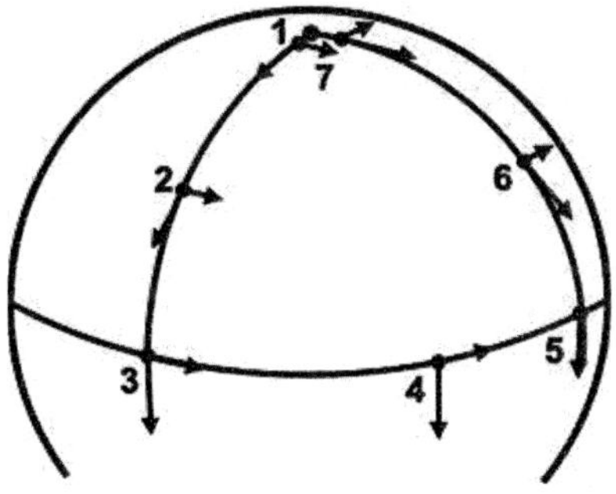

Fig. 6.9 Berry Phase: Wavevectors describing a wavefunction point in a different direction after traversing a closed loop for some surfaces. Figure adapted and used with permission from [14]. © 1999 Nicola Manini

the values of key parameter such as the magnetic field or the allowed wavevector values for wavefunctions in a Block band.

The Hamiltonian $H(R(t))$ of a system with a time dependent parameter $R(t)$ or set of parameters such as the Block wavevector [16, 17] can be written as follows [15]:

$$H(R(t))|n(R(t))\rangle = E_n(R(t))|n(R(t))\rangle \tag{6.32}$$

where $|n(R(t))\rangle$ is normalized and $R(0) = R_0$, and the wavefunctions are instantaneously an ortho-normal basis of eigenstates of $H(R(t))$ [17]. The complete determination of the eigenstates requires a phase factor that depends on $R(t)$. The wavefunction $|\psi_n(t)\rangle$ that includes the phase factor must satisfy the time dependent Hamiltonian [17]:

$$H(R(t))|\psi_n(t)\rangle = i\hbar\frac{\partial}{\partial t}|\psi_n(t)\rangle \tag{6.33}$$

Since the adiabatic theorem requires that a system that is initially in an eigenstate $|n(R(0))\rangle$ will stay as an eigenstate of $H(R(t))$ throughout the adiabatic process [17]. Then we can write the wavefunction so that *both* the dynamical and geometric phase are stated [15–17]:

$$|\psi_n(t)\rangle = e^{i\gamma(t)} e^{-\frac{i}{\hbar}\int_0^t E_n(R(t'))dt'} |n(R(t))\rangle \tag{6.34}$$

where $e^{-\frac{i}{\hbar}\int_0^t E_n(R(t'))dt'}$ is the dynamical phase and $e^{i\gamma(t)}$ is the geometric phase. The time dependent equation for the geometric phase is found by inserting (6.34) into (6.33) with $\frac{\partial}{\partial R(t)} = \nabla_R$ and multiplying by $\langle n(R(t))|$:

$$\frac{d}{dt}\gamma_n(t) = i\langle n(R(t))|\nabla_R|n(R(t))\rangle\frac{\partial R(t)}{\partial t} \tag{6.35}$$

$\gamma_n(t)$ is an additional phase that is acquired as the wavefunction travels along a path and is determined by a path integral in R space in terms of the *Berry vector potential* $\mathcal{A}_n(R(t))$ [16]. The Berry potential is also called the *Berry connection* and is gauge dependent. The vector potential analogy to the vector potential in electrodynamics will be more obvious as the Berry phase is further discussed.

$$\text{Berry vector potential} \quad \mathcal{A}_n(R(t)) = i\langle n(R(t))|\frac{\partial}{\partial R(t)}n(R(t))\rangle \tag{6.36}$$

$$\gamma_n(t) = i\int_{R_0}^{R(t)} \langle n(R(t))|\nabla_R|n(R(t))\rangle\, dR(t') = \int_{R_0}^{R(t)} \mathcal{A}_n(R(t'))dR(t') \tag{6.37}$$

When $R(t)$ changes adiabatically, the geometric phase is a real number. When the path R_0 to $R(t)$ is a closed loop C, the integral can be converted into a surface integral for the surface inside the closed loop C. The geometric phase which has is between 0 and 2π can be determined by an integral over the closed loop:

Berry Phase

$$\gamma_n(t) = i \oint_C \mathcal{A}_n(R(t))dR(t') \tag{6.38}$$

The path integral can be converted into an integral over the surface enclosed by the loop by introducing the Berry curvature [17]:

Berry Curvature of the n'th band

$$\Omega_n(R) = \nabla_R \times \mathcal{A}_n(R) \tag{6.39}$$

Berry Phase (from Berry Curvature)

$$\gamma_n(t) = -\iint_{S(c)} \Omega_n(R) \cdot dS \tag{6.40}$$

The Berry Potential is not observable because it is gauge dependent [15–17]. The Berry curvature is often described as being analogous to a magnetic field in parameter space and the Berry potential as being a vector potential for that magnetic field [17]. As discussed in Chap. 1, the magnetic field is the curl of the vector potential $\vec{A}$: $\vec{B} = \vec{\nabla} \times \vec{A}$. As we show below in (6.44), the Berry curvature is directly related to the IQHE discussed above [18]. It is interesting to note that the Berry Phase can be experimentally observed.

Since the Berry Curvature will be discussed in subsequent chapters in terms of the Bloch wavefunction of 2D materials, it is useful to relate the above discussion to the vector space of Block functions. Below, the Berry Potential and Berry Curvature are presented in terms of the Brillouin zone and momentum space. It is important to note that the Berry Phase and the Berry Curvature are observable because they are gauge invariant. The Berry Potential can be expressed in terms of Block wave functions u_k discussed in Chap. 2 [17, 19].

$$\psi_{n,\vec{k}}(\vec{r}) = e^{i\vec{k}\cdot\vec{r}} u_{n,\vec{k}}(\vec{r}) \quad and \quad u_{n,\vec{k}}(\vec{r}) = e^{-i\vec{k}\cdot\vec{r}} \psi_{n,\vec{k}}(\vec{r}) \tag{6.41}$$

The wavevector dependent Berry curvature for Bloch wavefunctions is [17]:

$$\Omega_{n,\vec{k}}(\vec{k}) = \nabla_{\vec{k}} \times \langle u^*_{n,\vec{k}}(\vec{r})| i\nabla_{\vec{k}} | u_{n,\vec{k}}(\vec{r})\rangle \tag{6.42}$$

The Berry phase for a closed path in momentum space is observable and can be written as:

$$\gamma_n(t) = -\iint\limits_{S(c)} \Omega_{n,\vec{k}}(\vec{k}) \cdot d\vec{k} \tag{6.43}$$

For a system with time reversal symmetry, when $t \to -t$, the energy stays the same but the velocity and wavevector $\vec{k}$ change sign. Thus, for a system with time reversal symmetry, $\Omega_n(-\vec{k}) = -\Omega_n(\vec{k})$ [17]. For a system with spatial inversion symmetry, $\Omega_n\left(-\vec{k}\right) = \Omega_n\left(\vec{k}\right)$. As a result, the Berry phase of a system that has both spatial inversion symmetry and time reversal symmetry is zero [17]. A material with either time reversal or spatial inversion symmetry but not the other can have a Berry phase and curvature. A material with ferromagnetic or antiferromagnetic ordering breaks time reversal symmetry. A magnetic field or electric field can break time reversal symmetry. Some materials have time reversal symmetry but no inversion symmetry. It is important to note that the Berry phase is not always quantized as is the case for ferroelectric materials which display the anomalous Hall effect during transport measurements.

Xiao, Chang, and Niu point out that the Berry curvature is an intrinsic property of the band structure [17]. One way to have a closed path in momentum space is to have a magnetic field generate a cyclotron orbit as is the case in a Hall measurement [17]. This results in a Berry phase. The IQHE can be calculated directly from the Berry curvature of the electronic states in the presence of a magnetic field as described by Thouless, Kohmoto, Nightingale, and den Nijs [20].

6.3.2 *The Kubo Formula for the Conductivity and the TKNN Theory of the IQHE*

The 1 part in 10^9 uncertainty in the quantized resistivity of the 2DEG required a theoretical explanation which was provided by Thouless and then Thouless, Kohmoto, Nightingale, and den Nijs (TKNN) [20]. The nearly free Block electrons in certain semiconductor structures form a 2DEG that is associated with a crystal lattice. Below, we will discuss how the presence of the magnetic field alters the Brillouin zone. TKNN showed that the Kubo formula (linear response theory) could be used derive an equation for the Hall conductivity as a surface integral of the Berry curvature of the occupied bands over the magnetic Brillouin zone as stated in (6.44) [20]. They also showed that the Hall conductivity of the 2DEG is quantized. The Hall conductivity for the jth band is a surface integral over the allowed states of the 2DEG in a magnetic field:

$$\sigma^j_{x,y} = -\frac{e^2}{h}\frac{1}{2\pi i} \iint\limits_{T^2 MBZ} dk_x dk_y [\nabla_k \times \mathcal{A}_n(k)]_z \tag{6.44}$$

The solution to (6.44) is elaborated below. First we introduce the magnetic Brillouin Zone, T^2MBZ. The Hamiltonian for an electron in a 2D crystal with Bravais lattice vectors $\vec{R} = n\vec{a} + m\vec{b}$ with periodic potential $V(\vec{r})$ and a magnetic field normal with Landau gauge $\vec{A}(\vec{r}) = Bx\vec{e}_y$ to the plane of crystal is [21, 22]:

$$\left[\frac{1}{2m^*}\left(\vec{P} + \frac{e\vec{A}(\vec{r})}{c}\right)^2 + V(\vec{r})\right]\psi(\vec{r}) = E\psi(\vec{r}) \tag{6.45}$$

A generalized version of the Block theorem is used to solve this Hamiltonian, and the momentum is referred to as the magnetic momentum [21–23]. First we need a 2D lattice for the crystal with primitive lattice vectors $\vec{a}$ and $\vec{b}$ in the presence of a normal magnetic field that accounts for a rational number α of magnetic flux quanta $\phi_0 = \frac{h}{e}$ in that cell. The area of the primitive lattice is $\vec{a} \times \vec{b}$, and the area of the supercell lattice in the presence of the magnetic field is $q(\vec{a} \times \vec{b})$. The number of flux quantum ϕ_0 per magnetic unit cell is [21–23]:

$$\alpha = \frac{1}{\phi_0}(\vec{a} \times \vec{b})\vec{B} \quad \text{and} \quad \alpha = {}^{p}/_{q} \tag{6.46}$$

The numbers p and q are relative prime numbers. A larger magnetic lattice is chosen so that p flux quanta pass through each unit cell in the 2D lattice for the crystal in the presence of a magnetic field. The lattice vectors are now [21, 22]:

$$\vec{S} = n\vec{a} + m\left(q\vec{b}\right) \tag{6.47}$$

The magnetic crystal momentum for the magnetic lattice is designated $\vec{\Theta}$ to distinguish them from the wavevectors $\vec{k}$ in the absence of the magnetic field. In the absence of a magnetic field, the Block function for a 2D lattice, $u(\vec{r})$, is periodic on the unit cell with boundaries $(a, 0)$ and $(0, b)$, $u(\vec{r} + \vec{\kappa}) = u(\vec{r})$, $\vec{\kappa} = (a, 0)$ or $\vec{\kappa} = (0, b)$. The 2D Block function in the presence of a normal magnetic field with Landau gauge vector potential, $\vec{A} = (0, Bx, 0)$, and the scalar potential ϕ of the magnetic field is set to zero. The Brillouin zone is referred to as the magnetic Brillouin zone [21, 22]. The values of $\vec{\Theta}$ are restricted to the first magnetic Brillouin zone. The Block wavefunction must be periodic in this magnetic unit cell. The wavevectors are said to lie in the Brillouin zone for which $-\frac{\pi}{a} < \Theta_x < \frac{\pi}{a}$; $-\frac{\pi}{qb} < \Theta_y < \frac{\pi}{qb}$. Here, $a = |\vec{a}|$ and $b = \left|\vec{b}\right|$. Below we discuss how the Brillouin zone can be described as a torus which is topologically designated T^2 [23]. Next we show how the surface integral over the Berry curvature $\iint_{T^2MBZ} dk_x dk_y [\nabla_k \times \mathcal{A}_n(k)]_z$ is quantized in units of $2\pi\upsilon$ where $\upsilon$is known as the TKNN or Chern invariant.

6.3.3 Why Topological

As discussed above, the quantization of the carrier transport in the IQHE can also be calculated from the adiabatic curvature of the bands that contribute to the Hall conductance. This topological method is routinely used to calculate transport properties for magnetic materials and topologically protected materials.

The wavefunctions describing the electronic states in a crystal are linear combinations of orthonormal vectors. The space that these vectors span is called Hilbert space. The connection to topology comes from the non-trivial topology for Hilbert space. The non-trivial topology can more readily be seen by discussing the vector space using the Gauss-Bonnet formula. The Gauss-Bonnet formula provides a means of determining the genus of a surface, and the genus is used to distinguish between surfaces with different topologies. The Gauss-Bonnet formula is stated in (6.48):

$$\text{Gauss} - \text{Bonnet Formula} \quad \frac{1}{2\pi} \int_S K(R) dA = 2(1 - g) \tag{6.48}$$

The left side of the integral is an integral of the curvature K of a surface over the surface that represents the allowed values of the variable R. The right side of the equation provides the quantization since the genus, g, is an integer. The curvature of a 3D sphere or torus is defined over the entire surface. The curvature of a sphere of radius R and area $A = 4\pi R^2$ is $K = \frac{1}{R^2}$ which results in the value of 2 for the surface integral over the sphere $\frac{1}{2\pi} \int_S K(R) dA = 2 = 2(1 - g)$. Thus, the genus of a sphere is 0. The genus of a torus (donut) is 1 which can be determined by counting the number of holes in the surface, see Fig. 6.11. In the language of topology, these surfaces can be oriented. If the torus shape is slowly deformed into another shape that still has a single hole such as a coffee cup, that shape also has a genus of 1. Chern generalized the Gauss-Bonnet formula to many dimensions [18, 20]. *It is important to note that the topology of the surfaces of a sphere and torus are an analogy to the topology associated with the electronic states of a crystal.*

The surfaces that we are considering come from the allowed values of key vector variables such as the wavevectors (momentum) of the wavefunctions. It is useful to discuss the surface associated with a 2D Brillouin of a uniformly distributed values of the k_x and k_y wavevector with periodic boundary conditions such as the wavevectors of a 2DEG in a magnetic field. We can use Born von Karmen boundary condition to represent the periodic nature of the wavefunction inside the 2D lattice. Thus, $\psi(\vec{r}) = \psi\left(\vec{r} + \vec{R}\right)$ where $\vec{R}$ is defined in terms of primitive lattice vectors $\vec{a}$ and $\vec{b}$: $\vec{R} = n_1\vec{a} + n_2\vec{b}$. If we consider the allowed values of the wavevectors inside a 2D Brillouin zone with boundary a along k_x and b along k_y, we can shift the allowed k_x wavevectors by $+\frac{\pi}{a}$ and do the same for the k_y wavevectors, then the allowed momentum states for the first Brillouin zone are bounded by $0 \leq k_x \leq \frac{2\pi}{a}; 0 \leq k_y \leq \frac{2\pi}{b}$. The same approach applies to the magnetic Brillouin zone discussed above where

$0 \le k_y \le \frac{2\pi}{qb}$. The wavevectors for any point in the reciprocal lattice can all be linked to the first Brillouin zone using a reciprocal lattice vector as discussed in Chap. 2. The Brillouin zone along k_y wraps back into itself, and this can be visualized as a cylinder. When this is done for k_x, the ends of the cylinder along k_x touch making a torus as shown in Fig. 6.10. One can consider reduced wavevectors k_i so that both the wavevectors vary between $0 \le k_i \le 2\pi$. What this does is map the Brillouin zone or the magnetic Brillouin zone into a square and then uses the periodic boundary conditions to form a torus. The torus is a smooth surface.

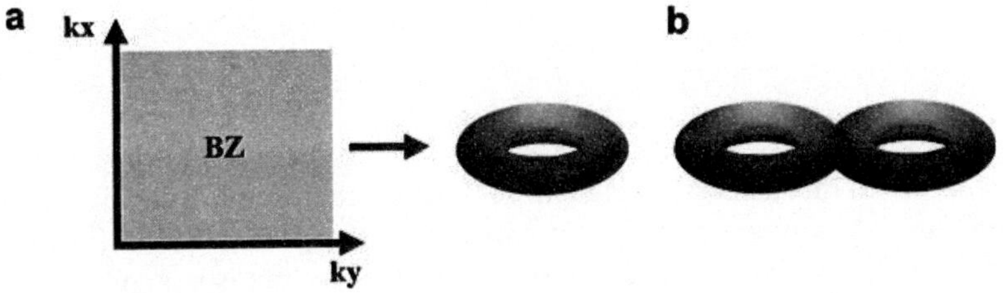

Fig. 6.10 The 2D Brillouin zone for values of k_x, k_y wraps onto itself forming a torus as shown in **a**. The torus has one hole giving it a genus of 1. A surface with two holes and thus a genus of 2 is shown in **b**

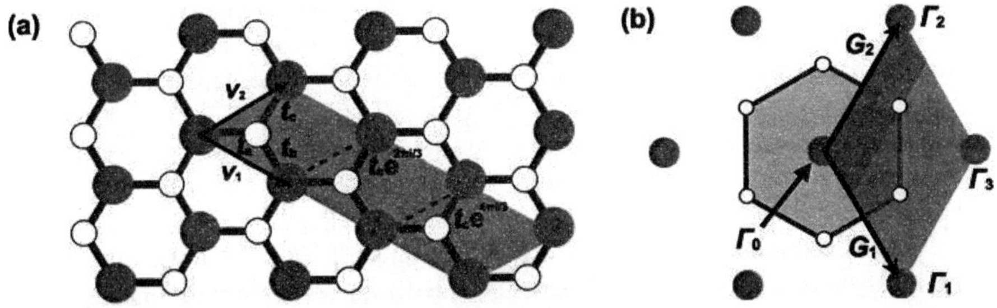

Fig. 6.11 The real space hexagonal lattice showing the two sub-lattices (**a**) and its reciprocal space lattice (**b**) are shown. In this example, $\frac{p}{q} = \frac{1}{3}$. The hexagonal lattice is shown with two inequivalent sub-lattices. Here we show the magnetic supercell and magnetic Brillouin zone for a magnetic flux ϕ of $\frac{\phi_0}{3}$ per unit hexagon and magnetic field of ϕ/S. The area S of the primitive lattice cell formed by the lattice vectors $\vec{v}_1\left(\frac{a\sqrt{3}}{2}, -\frac{a}{2}\right)$ and $\vec{v}_2\left(\frac{a\sqrt{3}}{2}, \frac{a}{2}\right)$ is $S = \frac{a^2\sqrt{3}}{2}$. In (**a**) the shaded area shows the $3\times$ larger hexagonal magnetic supercell in the presence of a magnetic field normal to the plane of the 2D lattice. The dashed red line divides the three primitive cells that comprise the magnetic supercell. In (**b**), the Brillouin zone (light shaded area) and magnetic Brillouin zone (dark shaded area) are shown. The Γ points in the Brillouin zone are also shown. The hopping interactions between nearest and next nearest neighbors are shown by t_a, t_c, $t_c e^{2\pi i/3}$, etc. The two sub-lattices of graphene are discussed in Chap. 7. Figure adapted and reprinted with permission from [29]. © 2006 American Physical Society

6.3.4 *Quantization of the Hall Conductance and the TKNN (Chern) Number*

As the magnetic field is scanned, at some field strengths, the Landau states are filled and there is a energy gap to the next state. Here, we start with the equation for the Hall conductance is an integral of the z component of the vector representing the Berry curvature of a *single filled band* where the wavefunctions have wavevector values over the magnetic Brillouin zone [18, 23]:

$$\sigma_{xy} = \frac{e^2}{h}\frac{1}{2\pi}\int_0^{2\pi/a} dk_x \int_0^{2\pi/qb} dk_y [\nabla_k \times \mathcal{A}(k_x, k_y)]_z = \frac{e^2}{h}\upsilon \tag{6.49}$$

Here, a key to this discussion is the torus surface representing the allowed states is continuous. As mentioned above, since the torus has a genus of 1, the integral over the surface of a torus will be zero unless the Berry curvature is not continuous over the torus. Both bulk and edge states can be used to arrive at the quantization observed in the IQHE. Here bulk refers to states in in the magnetic Brillouin zone. In order to get a non-zero result for the surface integral over the Berry curvature, the magnetic Brillouin zone needs to be separated into at least two sections where the phase of the wavefunctions differs. We note that in addition to proving the relationship between the Kubo formula for the Hall conductivity and the Berry curvature, TKNN also pointed out the topological significance of υ being the first Chern number [20].

In the years since Thouless, Kohmoto, Nightingale, and den Nijs proved that the Hall conductance was quantized; both bulk states and edge states have been show to result in quantized conductance. Here, bulk refers to states in the 2D Magnetic Brillouin zone versus conducting states at the edge of the zone. First, we discuss the bulk state solution using Kohmoto's explanation [23]. The Block wavefunctions $|u(x, y)\rangle$ can be zero somewhere in the magnetic Brillouin zone. In addition, the phase of the wavefunction is uncertain so that phase difference can occur between regions of the magnetic Brillouin zone: $|u'_{k_x,k_y}(x, y)\rangle = e^{i\theta(k_x,k_y)}|u_{k_x,k_y}(x, y)\rangle$. This zero is described as a vortex where around the $u_{k_x,k_y}(x_1, y_1) = 0$, the arrow representing the direction of phase angle rotates by 2π [23]. The vortex is a result of the magnetic field. Kohmoto refers to the area around this vortex as H^1 and the area outside which is the rest of the torus is referred to as H^2. The phase mismatch occurs at the boundary between H^1 and H^2 so that $|u^{H^2}_{k_x,k_y}(x, y) = e^{i\theta(k_x,k_y)}|u^{H^1}_{k_x,k_y}(x, y)\rangle$. The phase $\theta(k_x, k_y)$ is a smooth function along the boundary between H^1 and H^2. The boundary ∂H is a continuous line separating H^1 and H^2. Stokes theorem (really it is Green's theorem in this case) is used to convert the surface integral in each area into a line integral along the boundary [23]. The Berry potential $\mathcal{A}(k_x, k_y)$ in region H^1 is related to $\mathcal{A}'(k_x, k_y)$ in H^2 so that:

$$\mathcal{A}^2(k_x, k_y) = \langle u^{H^2}_{k_x,k_y}(x, y)| i\vec{\nabla} |u^{H^2}_{k_x,k_y}(x, y)\rangle$$

$$= \langle u_{k_x,k_y}^{H^1}(x,y) e^{-i\theta(k_x,k_y)} | i\vec{\nabla} | e^{i\theta(k_x,k_y)} u_{k_x,k_y}^{H^1}(x,y) \rangle$$

$$= -\vec{\nabla}\theta(k_x,k_y) + \mathcal{A}^1(k_x,k_y)$$

Now we separate the integral of the Berry curvature of the Berry potential over the entire torus $\mathcal{A}_n(k)$ of allowed states into an integral for each region of the torus:

$$\iint\limits_{T^2MBZ} dk_x dk_y [\nabla_k \times \mathcal{A}_n(k)]_z$$

$$= \iint\limits_{H^1} dk_x dk_y [\nabla_k \times \mathcal{A}^1(k_x,k_y)]_z + \iint\limits_{H^2} dk_x dk_y [\nabla_k \times \mathcal{A}^2(k_x,k_y)]_z$$

Using Green's theorem to convert the surface integral in each area into a line integral along the boundary ∂H, we get:

$$= \int\limits_{\partial H} d\vec{k}\mathcal{A}^1(k_x,k_y) - \int\limits_{\partial H} d\vec{k}\mathcal{A}^2(k_x,k_y) = \int\limits_{v} d\vec{k}\vec{\nabla}\theta(k_x,k_y) = 2\pi v$$

The integral of the differential of the phase factor over the path ∂H must be 2π since the wavefunction at the start of the path must match the wavefunction after traversing the path ∂H [23]. The phase factor difference is thus $2\pi v$. This results in the quantization observed in (6.49). Here, ***the integer v is known as the TKNN or Chern invariant.*** The concept of a topological invariant is discussed in Sect. 6.3.6 below.

A second solution to (6.49) described by Shen [18] comes an open boundary condition where the Berry connection is defined inside the boundary of the magnetic Brillouin zone and the phase of the wavefunctions must match at the boundary between the magnetic Brillouin zone and the adjacent area. This is an edge state solution. The Berry potential $\mathcal{A}(k_x,k_y)$ is a vector, and the x, y components can be written as $(\mathcal{A}_x(k_x,k_y), \mathcal{A}_y(k_x,k_y))$. We can restate $\vec{\nabla}_k \times \mathcal{A}(k_x,k_y)$ as $\frac{\partial}{\partial k_x}\mathcal{A}_y(k_x,k_y) - \frac{\partial}{\partial k_y}\mathcal{A}_x(k_x,k_y)$:

$$\int\limits_0^{2\pi/a} dk_x \int\limits_0^{2\pi/qb} dk_y [\nabla_k \times \mathcal{A}(k_x,k_y)]$$

$$= \int\limits_0^{2\pi/a} dk_x \int\limits_0^{2\pi/qb} dk_y \left[\frac{\partial}{\partial k_x}\mathcal{A}_y(k_x,k_y) - \frac{\partial}{\partial k_y}\mathcal{A}_x(k_x,k_y) \right]$$

Presuming the Berry curvature exists and is simply connected over the allowed states which are represented by a torus, we can apply Green's theorem to the double integral. We get line integrals around the edges:

$$= \int_0^{2\pi/qb} dk_y[\mathcal{A}_y(2\pi/a, k_y) - \mathcal{A}_y(0, k_y)] - \int_0^{2\pi/a} dk_x[\mathcal{A}_x(k_x, 2\pi/qb) - \mathcal{A}_x(k_x, 0)]$$

A key part of the solution is that the Block wavefunction $|u(k_x, k_y)\rangle$ at $k_x = 0$ and $k_x = 2\pi/a$ can only differ by a phase factor, and that the same is true at $k_y = 0$ and $k_y = 2\pi/qb$. Thus $|u(k_x, 0)\rangle = e^{-i\varphi_x}|u(k_x, 2\pi/qb)\rangle$, and $|u(0, k_y)\rangle = e^{-i\varphi_y}|u(2\pi/a, k_y)\rangle$ [18]. The Berry potential becomes:

$$\mathcal{A}_x(k_x, k_y) = \langle u(k_x, 2\pi/qb)|i\frac{\partial}{\partial k_x}|u(k_x, 2\pi/qb)\rangle = \langle u(k_x, 0)e^{-i\varphi_x}|i\frac{\partial}{\partial k_x}|e^{i\varphi_x}u(k_x, 0)\rangle$$
$$= -\frac{\partial \varphi_x}{\partial k_x} + \langle u(k_x, 0)|\left\{i\frac{\partial}{\partial k_x}\right\}|u(k_x, 0)\rangle = -\frac{\partial \varphi_x}{\partial k_x} + \mathcal{A}_x(k_x, 0)$$

And $\mathcal{A}_y(k_x, k_y) = -\frac{\partial \varphi_y}{\partial k_y} + \mathcal{A}_x(0, k_y)$ then:

$$\int_0^{2\pi/qb} dk_y[\mathcal{A}_y(2\pi/a, k_y) - \mathcal{A}_y(0, k_y)] - \int_0^{2\pi/a} dk_x[\mathcal{A}_x(k_x, 2\pi/qb) - \mathcal{A}_x(k_x, 0)]$$
$$= \int_0^{2\pi/qb} dk_y\left[-\frac{\partial \varphi_y}{\partial k_y}\right] - \int_0^{2\pi/a} dk_x\left[-\frac{\partial \varphi_x}{\partial k_x}\right]$$
$$= \varphi_y(0) - \varphi_y(2\pi/qb) + \varphi_x(2\pi/a) - \varphi_x(0) = 2\pi\upsilon$$

The last equality is because the states at the boundaries are the same states thus [18]:

$$e^{i\varphi_x(0)}|u(0, 2\pi/qb)\rangle. = |u(0, 0)\rangle\ ;\ e^{i\varphi_x(2\pi/a)}|u(2\pi/a, 2\pi/qb)\rangle = |u(2\pi/a, 0)\rangle.$$
$$e^{i\varphi_y(0)}|u(2\pi/a, 0)\rangle. = |u(0, 0);\ e^{i\varphi_y(2\pi/qb)}|u(2\pi/a, 2\pi/qb)\rangle. = |u(0, 2\pi/qb)\rangle.$$

Thus: $|u(0, 0)\rangle = e^{i(-\varphi_y(0)+\varphi_y(2\pi/qb)-\varphi_x(2\pi/a)+\varphi_x(0))}|u(0, 0)\rangle = e^{i(2\pi\upsilon)}|u(0, 0)\rangle$. This approach also results in quantization (6.49): $\boldsymbol{\sigma}_{xy} = \frac{e^2}{h}\frac{1}{2\pi}2\pi\upsilon = \frac{e^2}{h}\upsilon$.

6.3.5 Winding Number and Edge State Quantization in IQHE

The winding number around a point is a mathematical term which refers to the number of times one winds around a point when traversing counter clockwise around a close path that lies on a plane around a point. For example, if one follows a counterclock-wise path around a circle, ellipse, or rectangle with a point in the center, all will have a winding number (or index) of 1. In the above example, the path around the edge of the first MBZ would have a winding number of 1.

The concept of having quantized conduction from edge states is critical to topological materials. The TKNN solution represents the quantization of conduction for a system with a continuous set of periodic wavevectors on the torus surface. Another solution involves a system which has edges along a strip and periodic boundaries at the top and bottom of the strip [24, 25]. The quantized conductance is obtained from the winding number for the edge states when the genus g of the surface associated with the edge states is $g \geq 1$ [24, 25]. Hatsugai has shown that these two approaches are equivalent when the edge states are degenerate with the bulk states [24, 25].

6.3.6 *Brief Introduction to the Topological Description of Electronic Band Structure*

The language of the topology of the electronic band structure is very involved and the reader is referred to [26, 27]. For example, the Gauss–Bonnet theorem expressed in (6.48) is an integral over the manifold such as the sphere of torus discussed above. The integral for the conductance is an integral over the vector bundle associated with the torus. Here we briefly introduce some of the terminology associated with the topologically based discussion of the TKNN quantization and edge state quantization of (6.49) [23].

The magnetic Brillouin zone torus represents the total allowed states and is topologically trivial. Above, we had stated that the Hall conductivity (6.49) would be zero if the wavefunctions (and thus the Berry curvature) were defined over the entire MBZ. We used Komoto's discussion about breaking the MBZ into different sections and matching the wavefunctions at the boundary. This breaks the MBZ into well separated sub-ensemble of states [26]. We matched the sections at the boundary using the phase $e^{i\theta(k_x,k_y)}$. We can now express this process using the terminology of topology. The symmetry group $U(1)$ is a circle group of complex numbers with an absolute value of 1. Here, the phase factor for the wavefunctions is a $U(1)$ symmetry group, $U(1) = e^{i\theta(k_x,k_y)}$, and $U(1)^*U(1) = 1$. A fiber bundle is a space that is locally a product space and has different global structure. A fiber bundle p maps a total space E into a base space B, which is often signified by: $p : E \to B$. We are working with vector bundles which are a type of fiber bundle. The wavefunction $|u_{k_x,k_y}(x, y)\rangle$ forms a $U(1)$ vector bundle on the magnetic Brillouin zone $\left|u'_{k_x,k_y}(x, y)\right\rangle = e^{i\theta(k_x,k_y)}\left|u_{k_x,k_y}(x, y)\right\rangle$ [24, 25]. The integral $\int_0^{2\pi/a} dk_x \int_0^{2\pi/qb} dk_y[\nabla_k \times \mathcal{A}(k_x, k_y)]_z$ is the Chern number which is a topological invariant of the $U(1)$ vector bundle [23, 26, 27] for this 2D system. As mentioned in Sect. 6.3.2, the Berry vector potential (Berry connection) is gauge dependent. Thus different Berry connections can be defined on the same vector bundle. All of these connections result in the same value for the Chern number, and thus it is invariant [27]. *A key aspect of the Chern invariant is that these states do not possess time* *reversal symmetry* which is in contrast to the surface states in Topological Insulators discussed in Chap. 9. This is a very brief introduction to the rich field of topological band theory, and the reader is referred to texts in topological mathematics [28] in addition to topological band theory [26, 27].

6.4 Integer Quantum Hall Effect for Graphene

The electronic band structure, optical, and electrical properties of graphene will be discussed in Chap. 7. Here, we contrast the IQHE of a 2DEG with that of graphene to illustrate how the Berry phase correction provides a signature of the Dirac nature of the carriers in graphene. The previous discussion of the IQHE described the physics of a 2DEG which was formed from electrons or holes in a group IV or III-V semiconductor structure or the 2DEG for perovskite heterostructures. IQHE is observed at low temperature. The nature of the carriers in graphene changes the physics of the IQHE. The carriers in graphene are considered Dirac Fermions due to the unusual band structure of graphene. The relativistic nature of the particles results in a Berry Phase correction that provides a signature for proving the presence of Dirac fermions in 2D nanoscale materials [29–31]. It is noteworthy that the IQHE can be observed at room temperature in graphene [31]. Figure 6.11 shows the magnetic Brillouin zone for graphene, and Fig. 6.12 shows the signature shift in the plateaus associated with the quantized resistivity/conductance [30]. The magnetic Brillouin zone for a hexagonal lattice such as graphene is compared to the real space lattice and magnetic super cell in Fig. 6.11 [29].

For an orthorhombic lattice, a generalized Block wavefunction is used to describe the wavefunctions:

$$\psi(\vec{r}) = \left(e^{i\vec{r}\cdot\vec{\Theta}}\right)u^{\vec{\Theta}}(\vec{r}) \quad \text{where } u^{\vec{\Theta}}\left(\vec{r}+\vec{S}\right) = \left(e^{-i\frac{e}{\hbar}BS_x y}\right)u^{\vec{\Theta}}(\vec{r}) \tag{6.49}$$

The Landau energy levels for a magnetic field B for graphene were theoretically determined using quantum electrodynamics [31]. The Fermi velocity v_F is used in (6.50) instead of the speed of light, and the Landau level index is n:

$$\text{Landau Energy Levels in graphene} \quad E_n = sgn(n)\sqrt{2e\hbar v_F^2|n|B} \tag{6.50}$$

The degeneracy of the Landau levels is $g_s = 4$ due to spin and sublattice (two Dirac cones) degeneracy [31]. The Hall conductance σ_{xy} is shifted from the IQHE for a 2DEG is given by:

$$\sigma_{xy} = \pm g_s\left(n + \frac{1}{2}\right)\frac{e^2}{h} \tag{6.51}$$

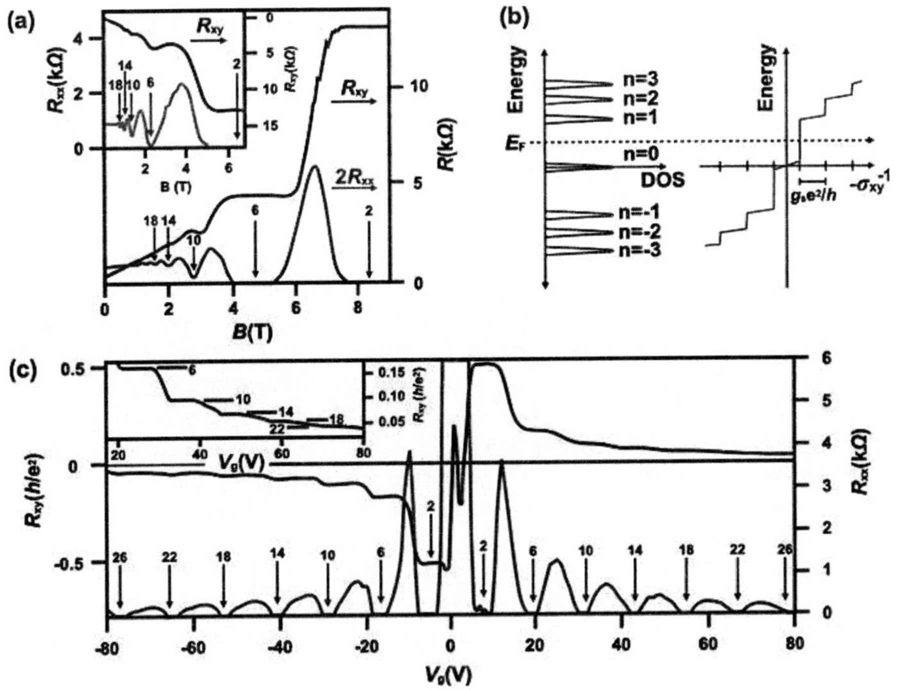

Fig. 6.12 The Integer Quantum Hall Effect for graphene is demonstrated by low temperature, quantized magnetoresistance and Hall resistance shown in this figure. **a** and **c** show the effect of sweeping the magnetic field, and **b** shows the effect of sweeping the gate voltage and thus the carrier density. In (**a**), the Hall resistance R_{xy} and magnetoresistance R_{xx} are shown for a gate voltage of 15 V at 30 mK. In (**b**), the Hall resistance R_{xy} and magnetoresistance R_{xx} are shown for $B = 9$ Tesla at 1.6 K. In (**c**), the Landau levels and corresponding conductance are shown. The Fermi energy is a function of gate voltage. Quantized filling factors $\upsilon = g_s(n + \frac{1}{2})$ with $\upsilon = 2, 6, \mathit{and}\ 10$ ($n = 1, 2, \mathit{and}\ 3$) are resolved. Figure adapted from [30]

The shift in conductance away from zero for n = 0 can be seen in Fig. 6.12 c. The oscillatory change in magnetoresistance ΔR_{xx} is altered from the IQHE result for 2DEG due to the Berry phase correction β for Dirac Fermions. This provides a signature for Dirac Fermions [30]:

$$\Delta R_{xx} = R(B, T)\cos\left(2\pi\left(\frac{B_F}{B} + \frac{1}{2} + \beta\right)\right) \tag{6.52}$$

$R(B, T)$ is the magnetic field and temperature dependent amplitude of the Shubnikov de Haas (SdH) oscillations, and B_F is the frequency of the SdH oscillations. $\beta = \frac{1}{2}$ is the signature of a Dirac particle. Zhang, Tan, Stormer, and Kim emphasizes that the non-zero Berry's phase is photon like behavior of carriers in graphene. The effective mass of a carrier lose to the Dirac point is $m_c \sim 0.007 m_e$ where the carrier density is low $n_s \sim 2 \times 10^{11}/\text{cm}^2$. Here, m_e is the mass of the free electron. We note that resistivity standards made from graphene also show a one part in 10^9 relative uncertainty (1 standard deviation) at $1.5K$ between 10 and 19 Tesla [32].

6.5 Fractional Quantum Hall Effect (FQHE): Many Body Physics in Action

The fractional quantum Hall effect (FQHE) occurs when rational fractional units of conductance are observed instead of integer amounts. It was discovered in 1982 by Tsui, Stormer, and Gossard when the Hall conductivity of modulation doped III-V heterostructures was measured using a magnetic field that is much stronger than that used for the IQHE [33]. Modulation doping reduces carrier scattering due to dopant impurities, and the high mobility $\mu = 90,000cm^2/V - sec$ and low carrier concentration $1.23 \times 10^{11}/\mathrm{cm}^2$ distinguish the first FQHE sample used by Tsui, Stormer, and Gossard from the much lower mobility silicon transistor with variable carrier density used by von Klitzing in the initial observation of IQHE [33]. The goal of this experimental project is interesting and illustrative of the research process. Stormer states in his Noble lecture that the team wanted to search for a Wigner solid (a crystal lattice of electrons) when the sample was in the extreme quantum limit of the lowest Landau level. In 1934, Wigner predicted that at very low temperatures and carrier concentrations, the electrons in a periodic potential will form a crystal lattice [34]. In 3D, a bcc lattice is predicted, and in 2D, a triangular lattice is predicted. Tsui, Stormer, and Gossard observed the IQHE at fields at and below 5 T at ~ 1.5 K. Above 5 T, the 2DEG was in the lowest Landau level [33]. When the field is 3 times higher than that required for a filling factor of 1, the lowest Landau level is 1/3 occupied [33] (as stated above, the filling factor υ is defined as the ratio of the electron density n_s to the degeneracy of each Landau level n_L). Laughlin's approach to the IQHE for i occupied levels resulted in $R_H = \frac{h}{ie^2} = \frac{(h/e)}{ie} = \frac{\phi_0}{ie} = \frac{flux\,quantum}{i \times charge}$. So a hand waving argument using change $q = e = \phi_0/R_H$ gives a 1/3 charge for one plateau in the FQHE data. At the plateau associated with 3 times the magnetic field strength for the lowest Landau level, $R_H = \frac{h}{ie^2} = \frac{h}{\frac{1}{3}e^2}$ the charge q is given by $q = \frac{\varphi_0}{\left(\frac{3h}{e^2}\right)} = \frac{e}{3}$ where φ_0 is the magnetic flux quantum [33]. These devices had some amount of disorder which is key to the FQHE. The physics behind the FQHE is different from the physics described above for the IQHE. The IQHE is understood in terms of the quantized motion of individual electrons in which the Coulomb interaction affects the filling of Landau levels [33]. Understanding the new physics requires a many body approach to the electron system. It is described in terms of strong Coulomb interactions and correlations between electrons. The theoretical explanation of the FQHE introduces the concept of the exchange of fractionally changed quasiparticles. This physics is described below.

The importance of the fractional quantum Hall effect is signified by the 1998 Nobel Prize in Physics which was awarded to Robert B. Laughlin, Horst L. Störmer and Daniel C. Tsui "for their discovery of a new form of quantum fluid with fractionally charged excitations." Stormer eloquently expressed his surprise at the physical phenomena that he and his co-workers observed in 2DEG at his Nobel Prize acceptance speech [33].

"The fractional quantum Hall effect is a very counterintuitive physical phenomenon. It implies that many electrons, acting in concert, can create new particles having a charge *smaller* than the charge of any individual electron. This is not the way things are supposed to be. A collection of objects may assemble to form a *bigger* object, or the parts may remain their size, but they don't create anything *smaller*. If the new particles were doubly charged, it wouldn't be so paradoxical—electrons could "just stick together" and form pairs. But fractional charges are very bizarre indeed. Not only are they smaller than the charge of any constituent electron, but they are exactly 1/3 or 1/5 or 1/7 etc. of an electronic charge, depending on the conditions under which they have been prepared. And yet we know with certainty that none of these electrons has split up into pieces.

Fractional charge is the most puzzling of the observations, but there are others. Quantum numbers—usually integers or half-integers—turn out to be also fractional, such as 2/5, 4/9, and 11/7, or even 5/23. Moreover, bits of magnetic field can get attached to each electron, creating yet other objects. Such composite particles have properties very different from those of the electrons. They sometimes seem to be oblivious to huge magnetic fields and move in straight lines, although any bare electron would orbit on a very tight circle. Their mass is unrelated to the mass of the original electron but arises solely from interactions with their neighbors. More so, the attached magnetic field changes drastically the characteristics of the particles, from fermions to bosons and back to fermions, depending on the field strength. And finally, some of these composites are conjectured to coalesce and form pairs, vaguely similar to the formation of electron pairs in superconductivity. This would provide yet another astounding new state with weird properties."

The IQHE shown in Fig. 6.13 was described in terms of filled Landau levels below the Fermi level and empty levels above the Fermi level. At low temperature the electrons (holes) are not excited into higher levels. The filling and degeneracy of the Landau levels of the 2DEG are a function of the magnetic field. The IQHE has been observed in silicon transistor with a 2DEG inversion layer and III–V samples. The FQHE is observed for 2DEG systems that have a small amount of disorder such as those fabricated using modulation doping where the 2DEG forms away from the doping. In that way, the disorder from the dopant atoms is separated from the 2DEG [33]. Even with remote doping, these systems do have a small amount of disorder. The electrons should spread out evenly throughout the 2DEG to achieve the lowest possible energy state. When these samples are in a high magnetic field that results in a partially filled lowest Landau level, one expects the electrons to be localized due to the disorder and the sample to be insulating [33, 35]. Thus, the observation of transverse current flow is unexpected. Again, transverse $\boldsymbol{\sigma}_{xy}$ current flow occurs when longitudinal current flow along the voltage drop is zero, $\boldsymbol{\sigma}_{xx} = 0$. The theoretical explanation was developed over an extended period of time, and the quasiparticles have been experimentally observed [33, 35, 36].

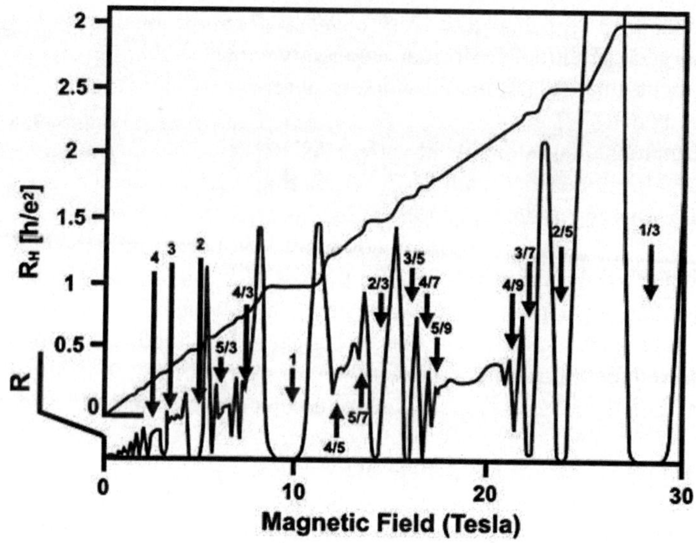

Fig. 6.13 Fractional Quantum Hall Effect. Note the high magnetic field > 10 T used to give fractional filling of the Landau levels. Figure adapted and reprinted with permission from [33]. © 1991 American Physical Society

The physics of the FQHE is complicated and worth further discussion [33]. At low temperatures, the electrons are distributed so that the total energy is minimized. The FQHE is observed below 1 K from 1/10 to 1/100 K using a very high magnetic field [33]. The strong magnetic field creates vortices where the charge density is zero, and there is a fixed amount of magnetic flux, the quantum of magnetic flux, $\phi_0 = \frac{h}{e}$ [33]. So the increase in magnetic field creates more vortices, and an increase in vortices changes the available space for the 2DEG. The ½ and 1/3 FQHE is observed at magnetic fields that produce more vortices than electrons [33]. Thus the fact that the first modulation doped sample had a low carrier density was very serendipitous. There are twice as many vortices as electrons for the ½ FQHE and the electron pairs with two vortices. For the 1/3 FQHE, there are three times as many vortices as electrons and the electron interacts with 3 vortices. The electrons are shielded from interaction with other electrons by the vortices [33]. Figure 6.14 from Stormer's Nobel lecture showing the composite particles for a fractional filling factor of 1/3 is reproduced here in Fig. 6.14. On average, the electron does not reside in close proximity to the nucleus because it is spread out in its orbit. These composite particles are fermions if each electron is paired with an even number of vortices and Bosons if they are paired with an odd number of vortices. Of course, the behaviors of the two types of particles are very different. Fermions do not condense, while Bosons do condense on the lowest energy level available. There is no scattering between the Bosons. Thus there is a prominent drop in the magnetoresistance, σ_{xx}, for odd values of the FQHE [33, 35].

When the magnetic field is slightly away from one that results in an exact match in the number of vortices with electrons (e.g., 2 vortices with one electron), a single

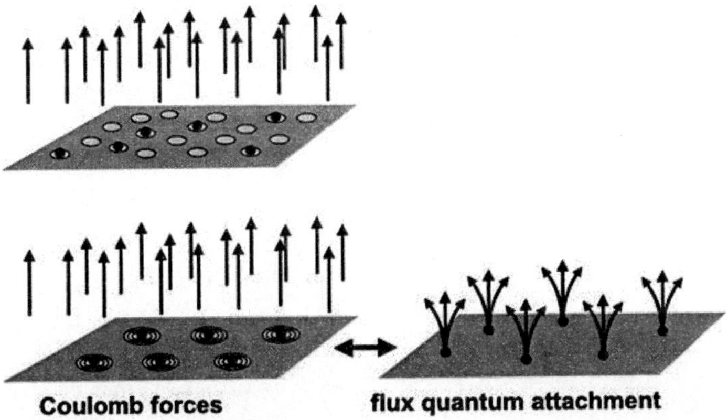

Fig. 6.14 The electrons in the 2DEG are attracted to the vortices enclosing the magnetic flux quantum caused by the high magnetic field for a filling factor = 1/3. The electrons are said to attach to the vortices forming composite particles. Figure adapted and reprinted with permission from [33]. © 1991 American Physical Society

vortex moves through the 2DEG and carries with it a fractional charge. This is the charge observed in the FQHE [33, 35]. Interestingly, Stormer states that the magnetic field does not impact the composite particles because it is already incorporated in the composite particles through the vortex [33].

6.6 Anomalous Hall Effect (AHE)

The anomalous Hall effect was observed by Edwin Hall two years after he discovered the Hall effect. The Hall effect for iron was found to be ten times larger than that observed for a non-magnetic metal [37, 38]. In addition, no magnetic field was required for observing a Hall current. During Hall characterization of ferromagnetic metals like Fe, Co, and Ni, the Hall resistivity $\rho_{xy} = R_0 H + R_s M$ increases quickly with magnetic field strength H and then saturates at approximately the value of the magnetization M as shown in Fig. 6.15a–c [37, 38]. The Hall coefficient R_0 is a function of the carrier density n, $R_0 = -1/ne$, and R_s is a materials dependent resistivity [38]. As the external magnetic field is increased from 0, the magnetization of the sample quickly saturates. Once saturation is reached, the ordinary Hall contribution $R_0 H$ dominates. The AHE is shown in Fig. 6.15, and key mechanisms contributing to the AHE on the left. The current theoretical understanding of the AHE is based on Berry curvature and topological principles. Here, a brief discussion of the AHE is presented. The reader is referred to the experimental and theoretical review of the AHE by Nagoasa, et al. [37]. There are three contributions to the AHE, intrinsic deflection, side jump, and skew scattering [37]. These contributions are shown below

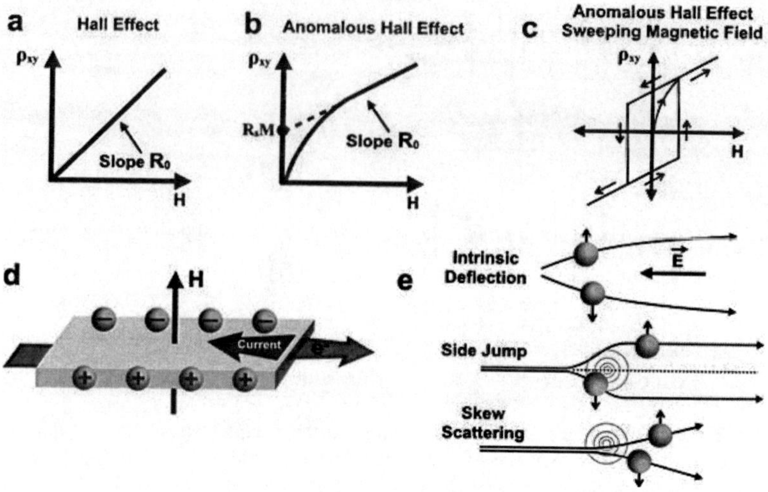

Fig. 6.15 The magnetization of the sample results in a Hall voltage in the anomalous Hall effect. The Hall effect, the increase in Hall voltage with increase magnetic field H for metal or semiconductor is shown in **a**. The anomalous Hall effect $\rho_{xy} = R_0H + R_sM$ for magnetic materials is shown in **b**. The anomalous Hall effect shows hysteresis when the magnetic field is cycled as shown in **c**. Once the magnetization is saturated, $\rho_{xy} = R_sM$ at $H = 0$. The sign of R_s changes between electrons and holes. The separation of charge during electron flow in a magnetic material is shown in **d**. The three contributions to the Anomalous Hall Effect are shown in **e**. (**b**) and (**c**) adapted and reprinted with permission from [39]. © 2015 Taylor & Francis Ltd, www.tandfonline.com. (**e**) adapted and reprinted with permission from [37]. © 2010 American Physical Society

in Fig. 6.15e. The AHE is observed during magneto-transport measurements in many topological materials as discussed in Chapter 9.

The contributions to the AHE can be separated according to powers of the Block state transport lifetime, τ [37]. A perturbation theory approach will have contributions of powers of the scattering rate τ^{-1}. The particles that scatter are quasi-particles. The AHE current, $\boldsymbol{\sigma}_{xy}^{AH}$ can be separated into three parts: skew scattering which is ~ τ or $\boldsymbol{\sigma}_{xx}$, $\boldsymbol{\sigma}_{xy}^{AH-Skew}$; the intrinsic contribution which is ~ τ^0, $\boldsymbol{\sigma}_{xy}^{AH-intrinsic}$; and side jump contribution which is also ~ τ^0, $\boldsymbol{\sigma}_{xy}^{AH-sj}$. The intrinsic contribution is the zero frequency contribution $\omega \to 0$, and $\tau \to \infty$.

$$\boldsymbol{\sigma}_{xy}^{AH} = \boldsymbol{\sigma}_{xy}^{AH-intrinsic} + \boldsymbol{\sigma}_{xy}^{AH-Skew} + \boldsymbol{\sigma}_{xy}^{AH-sj} \tag{6.53}$$

We note that since the conductivity is a tensor as described above and in (6.15). Thus, after inversion of the conductivity tensor, the intrinsic AHE has a contribution of $\boldsymbol{\rho}_{xy} \approx \boldsymbol{\sigma}_{xy}/\boldsymbol{\sigma}_{xx}^2$ and thus proportional to $\sim \rho^2$[37].

Intrinsic Contribution to AHE

Before the AHE was described in terms of Berry phase contributions, Karplus and Luttinger showed that electrons moving in a ferromagnetic conductor with strong spin–orbit coupling can acquire an anomalous velocity perpendicular to the electric field [37]. This intrinsic resistivity $\boldsymbol{\rho}_{xy}$ is proportional to the longitudinal resistivity squared: $\boldsymbol{\rho}_{xy} \propto \boldsymbol{\rho}_{xx}^2$ [37]. This intrinsic contribution is defined as the DC contribution to the interband conductivity [1]. The intrinsic contribution can be calculated from the band structure of a perfect crystal, i.e., defects and impurities do not contribute. Symmetry plays an important role in understanding the intrinsic AHE. The field due to the magnetization of the ferromagnetic metal breaks time reversal symmetry [37]. The Kubo formula for the Hall conductivity of a perfect lattice with Block states $\langle n, \vec{k}|$ for dimension D is: [37]:

$$\boldsymbol{\sigma}_x^{AH-intrinsic} = e^2\hbar \sum_{n \neq n'} \int \frac{d\vec{k}}{(2\pi)^D} \Big[f(\varepsilon_n(\vec{k})) - f(\varepsilon_{n'}(\vec{k})) \Big] \mathrm{Im} \frac{\langle n, \vec{k}|v_i(\vec{k})|n', \vec{k}n', \vec{k}|v_j(\vec{k})|n, \vec{k}\rangle}{[\varepsilon_n(\vec{k}) - \varepsilon_{n'}(\vec{k})]^2} \tag{6.54}$$

where $f(\varepsilon_n(\vec{k}))$ is the Fermi occupation number for the Block state energy $\varepsilon_n(\vec{k})$, and the occupation difference between the n and n' bands must be non-zero. The velocity operator is:

$$v(\vec{k}) = \frac{1}{i\hbar}\Big[r, H(\vec{k})\Big] = \frac{1}{\hbar}\nabla_{\vec{k}} H(\vec{k}) \tag{6.55}$$

Karplus and Luttinger point out that for a metal with d bands, the transitions between the n and n' bands in (6.54) are transitions between d orbital bands [37]. Equation (6.2) can be recast as an integral over the Berry phase. In contrast to a similar use of the Kubo formula for the quantized conductance in the presence of a magnetic field in (6.48), the integral is over the Brillouin zone not the magnetic Brillouin zone.

$$\boldsymbol{\sigma}_{xy}^{AH-intrinsic} = \mathcal{E}_{ijl} \frac{e^2}{\hbar} \sum_n \int \frac{d\vec{k}}{(2\pi)^d} f(\varepsilon_n(\vec{k}))\, b_n^l(\vec{k}) \tag{6.56a}$$

where $\mathcal{E}_{ijl}$ is the antisymmetric tensor, and $b_n^l\left(\vec{k}\right)$ is the Berry Phase curvature which is related to the Berry phase connection for the Block states in the absence of a magnetic field for the energy level n in the crystal as follows:

$$b_n(\vec{k}) = \nabla_{\vec{k}} \times a_n(\vec{k}) \text{ and } a_n(\vec{k}) = i\langle n, \vec{k}|\nabla_{\vec{k}}|n, \vec{k}\rangle \tag{6.56b}$$

The magnitude of the intrinsic contribution to the AHE has been studied in magnetic materials with strong spin–orbit coupling such as oxides and dilute magnetic semiconductors [40].

Here, we are discussion the AHE when it is not quantized. Thus, a band with zero Chern number can have an AHE. A relativistic (Dirac) Hamiltonian with spin–orbit split bands with energy gap 2Δ has been used to provide a theoretical explanation for the AHE in ferromagnetic materials [17]. The Dirac Hamiltonian, $H = \frac{\hbar^2 k^2}{2m} + \lambda(\vec{k} \times \boldsymbol{\sigma}) \cdot e_z - \Delta\sigma^z$, contains a contains a linear energy dispersion for the spit bands which is characterized by the parameter λ and a term for spin–orbit splitting. We further discuss the Dirac Hamiltonian in Chapter 7. The emphasis in this discussion is that the term $\Delta\sigma^z$ breaks time reversal symmetry in spin space. The spin orbit coupling in this Hamiltonian means that time reversal symmetry is also broken in the space of the orbital part of the wavefunctions [17]. The Pauli Matrices $\boldsymbol{\sigma} = (\sigma^x, \sigma^y, \sigma^z)$ are described along with Dirac Hamiltonians in Chaps. 7 and 8. The resulting Berry curvature $\Omega_\pm$ is given by [17]:

$$\Omega_\pm = \mp \frac{\lambda^2 \Delta}{2\left(\lambda^2 k^2 + \Delta^2\right)^{3/2}} \tag{6.57}$$

The integration of this Berry curvature of a fully occupied band is $\pm\pi$. Thus the amount of band filling has a considerable impact on the AHE. If the band filling is such that the Fermi level is in between the energy of the lower band $-\Delta$ and the upper band $+\Delta$, then AHE is resonantly enhanced [17]. If the Fermi energy is $> \Delta$, the contributions from the upper and lower bands cancel.

Skew Contribution (Mott scattering) to AHE

For ferromagnetic materials with strong spin orbit coupling, moving electrons that scatter from a defect will not scatter symmetrically resulting in a contribution to the transverse resistivity [37, 39]. This is known as the skew contribution, and it is proportional to the longitudinal resistivity, $\boldsymbol{\rho}_{xy} \propto \boldsymbol{\rho}_{xx}$. [37]. The skew contribution can be defined as the contribution that is proportional to the Block state transport lifetime ~ τ or $\boldsymbol{\sigma}_{xx}$ [37]. The skew contribution dominates the AHE for nearly perfect crystals. This scattering is due to chiral differences in scattering from disorder for ferromagnets with strong spin–orbit coupling. We note that [37] expresses concern about many textbook derivations of skew scattering from the Boltzmann transport equation. Reference [37] also points to the differences between the transition probability for scattering that is right handed with respect to the magnetization direction vs left handed. This difference first appears in the third order term in perturbation theory derivations. Although the following statement seems to contradict the first, it is useful to state that there are two possible sources of skew scattering. One for a spin- orbit coupled material with linear disorder and one where the disorder itself is also has strong spin–orbit coupling.

Side Jump contribution to AHE

The side jump contribution to the AHE can be considered as the difference between the Hall conductance and the intrinsic plus skew contributions. A semiclassical approach for the side jump conductance describes a carrier as a Gaussian wave packet that scatters from a spherical impurity with strong spin–orbit coupling [37]. The wavepacket of wavevector $\vec{k}$ will scatter transverse to $\vec{k}$ with a displacement of $\frac{1}{6}k\hbar^2/m^2c^2$ [37]. This contribution is independent of scattering lifetime.

Theoretical Interpretation The current understanding of the AHE in transition metals is that the there is a transition from the extrinsic skew scattering mechanism to intrinsic mechanisms as the resistivity increases [37].

6.7 Quantum Anomalous Hall Effect (QAHE)

The Quantum Anomalous Hall Effect (QAHE), quantized conductance due to the magnetization of a material without the presence of an external magnetic field was first predicted by Haldane [41]. The IQHE and QAHE can both be understood in terms of the Berry connection, Berry curvature, and Berry phase [39, 40]. In both cases, the conductivity is associated with edge states. Despite the complicated nature of the edge states responsible for IQHE described above, the IQHE, QAHE, and the quantum spin Hall effect described below are typically pictured in terms of the edge state conduction as shown in Fig. 6.16. Edge state conduction is only in one direction as shown in Fig. 6.16b. Topological insulators (TI) enabled the realization of the QAHE which occurs in magnetically doped TI or when a magnetic ad-layer is added to a TI [1, 39, 40, 42]. The QAHE in 5 quintuple layers of $Cr_{0.15}(Bi_{0.1}Sb_{0.9})_{1.85}Te_3$ is observed below 15 K as shown in Fig. 6.17 [40]. The addition of the magnetic doping results in breaking time reversal symmetry opening a gap in the Dirac cone (surface states) which can result in a number of topologically related phenomena [1, 39, 40, 42]. The existence of the QAHE proves that *quantized* conductance can exist without an external magnetic field [37].

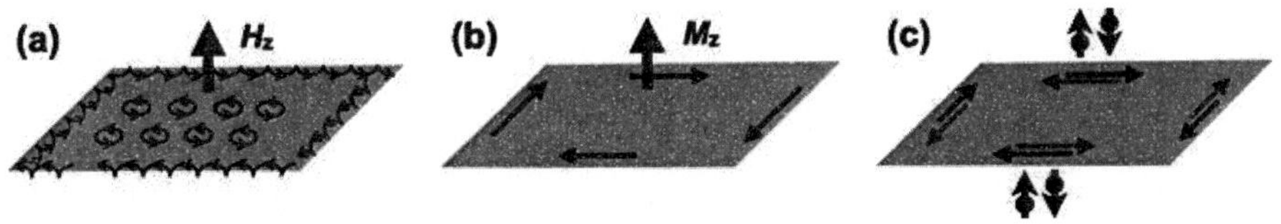

Fig. 6.16 Edge states in the Hall effect. Edge states in quantum Hall effect (QHE) are shown in **a**. The edge state reflections of the charge carriers are responsible for the IQHE. The edge state carrier conduction responsible for the quantum anomalous Hall effect (QAHE) is shown in **b**. The edge state spin conduction responsible for the quantum spin Hall effect QSHE is shown in **c**. Figure adapted and reprinted with permission from [39]. © 2015 Taylor & Francis Ltd., www.tandfonline.com

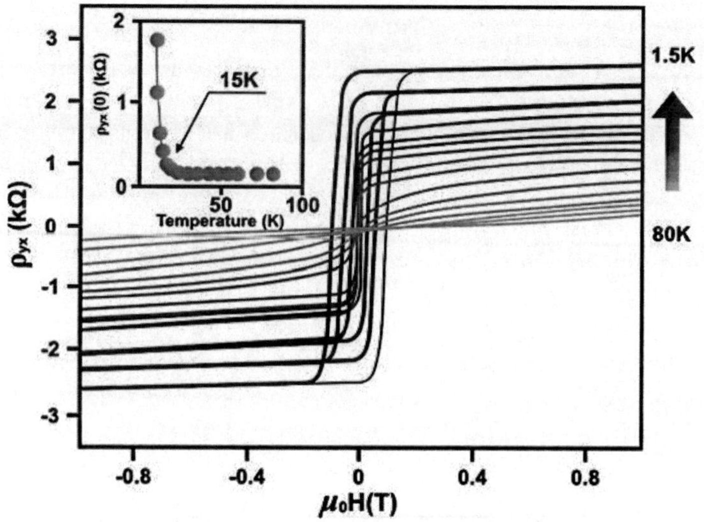

Fig. 6.17 The Quantum Anomalous Hall Effect is observed as the temperature is lowered below 15 K with no magnetic field in $Cr_{0.15}(Bi_{0.1}Sb_{0.9})_{1.85}Te_3$. Figure adapted from [40]. Reprinted with permission from American Association for the Advancement of Science

6.8 Spin Hall Effect (SHE) and Quantum Spin Hall Effect (QSHE)

The SHE effect was first theoretically predicted in 1971 and experimentally observed in 2004 [43]. The SHE refers to the asymmetric deflection of carriers dependent on their spin direction in a non-magnetic material [43]. In the SHE, the spins are polarized perpendicular to the plane in which the current and charge are moving as shown in Fig. 6.16. For a 2D material, the spin polarization (direction of the spin) will be perpendicular to the plane the material lies in. The AHE, SHE and QSHE all share the same three sources of Hall current: intrinsic, skew (Mott scattering), and side jump [43]. One important distinction is that the AHE is observed for magnetic metals, and the SHE and QSHE are observed for non-magnetic materials with strong spin–orbit forces. Again, the intrinsic contribution can be calculated from the Berry curvature which motivated numerous theoretical studies [43–46]. One important difference between the AHE and SHE is that charge is conserved and spin is not. Just as discussed for the AHE, these mechanisms can be distinguished by the dependence of the Hall conductivity on the transport lifetime (momentum relaxation time) τ [43, 46]. Clearly, the SHE in defect free materials can only be due to the spin–orbit force. The SHE has been observed in thin films of GaAs and InGaAs using Kerr rotation an optical method that is sensitive to spin polarization [47]. It is also well known in metals with strong spin–orbit coupling [43–46]. An inverse SHE (ISHE) also exists. In the ISHE, a pure spin current generates a transverse charge current [43].

The three contributions to the SHE and their relationship to τ are [43, 46]:

$$\sigma_{xy}^{SH} = \sigma_{xy}^{SH-intrinsic} + \sigma_{xy}^{SH-Skew} + \sigma_{xy}^{SH-side-jump} \tag{6.58a}$$

$$\sigma_{xy}^{SH-Skew} \propto \tau^{1} \tag{6.58b}$$

$$\sigma_{xy}^{SH-intrinsic} \text{ and } \sigma_{xy}^{SH-Skew} \propto \tau^{0} \text{ also independent of } \sigma_{xx} \tag{6.58c}$$

$$\sigma_{xy}^{SH-side-jump} = \sigma_{xy}^{SH} - \sigma_{xy}^{SH-intrinsic} - \sigma_{xy}^{SH-Skew} \tag{6.58d}$$

The intrinsic contribution to the SHE conductivity of metals has been related to the spin–orbit polarization $\langle l \cdot s \rangle$(coupling) at the Fermi surface [46]:

$$\sigma_{xy}^{SH-intrinsic} \approx \frac{e}{4a} \frac{\langle l \cdot s \rangle_{FermiSurface}}{\hbar^2} \tag{6.59}$$

Here, a is the lattice constant. Remembering (2.51) from Chap. 2, $\langle \vec{L} \cdot \vec{S} \rangle = \frac{1}{2}\langle \vec{J}^2 - \vec{L}^2 - \vec{S}^2 \rangle = \frac{\hbar^2}{2}[j(j+1) - l(l+1) - s(s+1)]$ and Hunds Rule, $\sigma_{xy}^{SH-intrinsic}$ for transition metals is positive when the d band is more than half filled and negative when it is less than half filled [45]. Non-metals also demonstrate the SHE. In the ISHE, both time reversal symmetry and space symmetry are broken [43].

Devices based on spin transport instead of charge transport often use the spin Hall angle to determine how well a material will function in a spin transport device. The spin Hall angle reflects the amount of spin hall conductance relative to change conductance. References [43, 46] provide tables of spin Hall angles which are defined as:

$$\Theta_{SpinHall} = \frac{\sigma_{xy}^{SH}}{\sigma_{xx}^{charge}} \frac{e}{h} \tag{6.60}$$

A large quantum Spin Hall effect has been reported in 2D materials such as graphene [48]. These samples seem to be strongly affected by the copper used to form contacts [48].

6.9 Optical Measurement of Spin and Pseudospin Conductance

In Chap. 7 on graphene, the concept of pseudospin in defined. Pseudospin is due to the symmetry of the crystal lattice and not the electron spin. In Chap. 8 on TMD materials, we discuss the use of circularly polarized light to selectively excite Hall conductance from the two different values of pseudospin.

6.10 Thermal (Nernst) Spin Hall Effect

The Nernst (thermal) Hall Effect and Nernst Spin Hall Effect were predicted in 2008 [49] and observed in 2017 [50]. The effect was predicted for a 2DEG with strong spin–orbit interactions with a thermal gradient along the current direction in a perpendicular magnetic field [49]. Experimental observation was done on a yttrium iron garnet ($Y_3Fe_5O_{12}$, YIG)|Pt bilayer [50].

6.11 Skyrmion Hall Effect

For the sake of completeness, skyrmions are briefly described. A skyrmion is a magnetic quasiparticle that consists of spin texture with a magnetization that is out of plane at the center [51–55]. Magnetic skyrmions are between 1 μm and 100 nm in size and the direction of the electron spin rotates from the edge of the skyrmion to its center as shown in Fig. 6.18. Skyrmions also experience a Hall effect due to current flow. Skyrmions have been observed in thin multilayers containing magnetic films such as Ta/CoFeB/TaOx [53] using polar magneto-optical Kerr effect (MOKE) microscope and Pt(3.2 nm)/CoFeB(0.7 nm)/MgO(1.4 nm) using a time resolved X-Ray microscope [54], and MnSi using positron annihilation spectroscopy [55]. In the skyrmion Hall effect, the direction and density of the electron current play important roles in skyrmion dynamics. At this time, two different explanations of skyrmion dynamics have been proposed [52–54]. The skyrmion is considered to have a topological charge Q of ± 1 which arises from the change in direction of the electron spin from edge to center (see Fig. 6.18) using a two dimensional integral of the local magnetization $\vec{m}$ as follows:

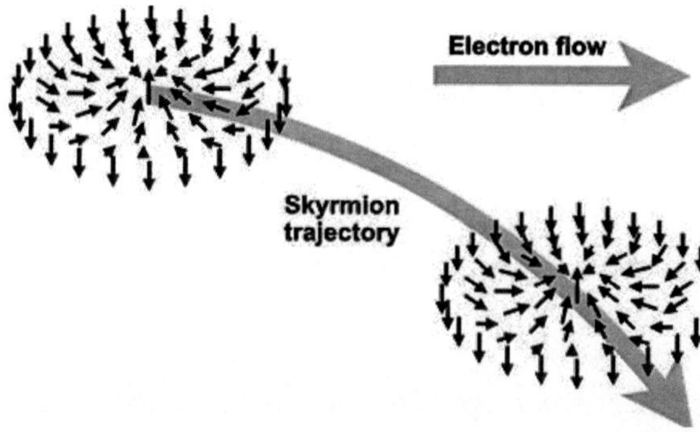

Fig. 6.18 The Hall effect for magnetic skyrmion quasiparticle s is shown. The spin direction changes from the edge of the skyrmion to its center. The magnetization points out of plane at the center. Figure adapted from [52]

$$Q = \frac{1}{4\pi} \int \vec{m} \cdot \left(\frac{\partial \vec{m}}{\partial x} \times \frac{\partial \vec{m}}{\partial y} \right) dxdy$$

The direction of the skyrmion motion depends on the sign of the topological charge. One explanation for the skyrmion Hall effect is a competition between a topological Magnus effect in competition with pinning potentials in the films due to defects or impurities [51, 52]. The skyrmions are considered to be rigid point like Neél particles. Here, the competition between the Magnus like force and pinning results in a temperature and current dependent Hall current. The second explanation requires that the skyrmions are allowed to deform or "breathe" and results in better agreement with spin Hall angles [52, 54]. This approach is based on a field like spin–orbit torque and internal excitation of the skyrmions [52].

6.12 Summary

The family of Hall effects is summarized below in Table 6.1.

Table 6.1 Summary of the various Hall effects, when they were discovered and some key properties of the carriers. The year of experimental discovery is listed

Year discovered	Hall effect	Carrier	Conductivity quantization	Magnetic field or magnetization field	Example materials
1879	Hall effect	Electrons + holes	None	$H \sim 0.05$ to 1 T	Metals, semiconductors
1980	Integer quantum Hall effect	Electrons + holes	$\nu e^2/\hbar : \nu = 1, 2, 3, \ldots$	$H \sim 0.05$ to 10 T	2DEG, graphene, bi-layer graphene Berry phase correction for graphene (π) bi-layer graphene (2π)
1982	Fractional quantum Hall effect	Composite fermions	$\nu e^2/\hbar : \nu = \frac{1}{3}, \frac{2}{5}, \frac{3}{7}, \frac{1}{2}, \frac{2}{3}, \ldots$	$H \sim 0.05$ to 30 T	2DEG, graphene, bi-layer graphene
1881	Anomalous Hall effect	Electrons + holes	None	Intrinsic magnetization	Fe, Ni, Co T_d WTe_2 Predicted VS_2 and VSe_2 Magnetic Weyl Semimetals ($Co_3Sn_2S_2$, Co_2MnGa)
2013	Quantum anomalous Hall effect	Electrons + holes	$\nu e^2/\hbar : \nu = 1, 2, 3, \ldots$	Intrinsic magnetization	Cr and V doped Bi_2Te_3 and Sb_2Te_3
2004	Spin Hall effect	Electron spin		Non-magnetic effect	p doped GaAs and InGaAs
2007	Quantum spin Hall effect	Electron spin		Non-magnetic effect	HgTe/CdTe QWs and InAs/GaSb/AlSb QWs
2017	Nernst spin Hall effect (thermal spin Hall effect)	Electron spin		Thermal gradient	Pt film on Yttrium Iron Garnet

References

1. C.-Z. Chang, M. Lee, Quantum anomalous Hall effect in time-reversal-symmetry breaking topological insulators. J. Phys.: Condens. Matter **28**, 123002 (2016)
2. K. von Klitzing, 25 Years of quantum Hall effect (QHE): a personal view on the discovery, physics and applications of this quantum effect. Seminaire Poincare **2**, 1–16 (2004)
3. J. Weis, K. von Klitzing, Metrology and microscopic picture of the integer quantum Hall effect. Phil. Trans. R. Soc. A **369**, 3954–3974 (2011)
4. K. von Klitzing, The quantized Hall effect, nobel lecture, December 9. Rev. Mod. Phys. **58**(1986), 519–531 (1985)
5. D.K. Schroder, Mobility, in *Semiconductor Materials and Device Characterization*, 3rd edn. (Wiley, New York, 2006), pp. 465–503
6. P.Y. Yu, M. Cardona, Fundamentals of Semiconductors: Physics and Materials Properties, 4th edn. (Springer, New York, 2010), pp. 232–234, 255, 534–538
7. L.D. Landau, E.M. Lifshitz, *Quantum Mechanics (Non-Relativistic Theory)*, 3rd edn. (Butterworth Heinemann, New York, 1977), pp. 456–460
8. J. Singh, *Electronic and Optoelectronic Properties of Semiconductor Structures* (Cambridge, New York, 2003), pp. 460–464
9. F.S. Crawford, Elementary derivation of the magnetic flux quantum. Amer. J. Phys. **50**, 514–516 (1982)
10. Y.Y. Wei, J. Weis, K.V. Klitzing, K. Eberl, Edge strips in the quantum Hall regime imaged by a single-electron transistor. Phys. Rev. Lett. **81**, 1674–1677 (1998)
11. F. Dahlem, E. Ahlswede, J. Weis, K.V. Klitzing, Cryogenic scanning force microscopy of quantum Hall samples: adiabatic transport originating in anisotropic depletion at contact interfaces. Phys. Rev. B **82**, 121305 (2010)
12. J.E. Avron, D. Osadchy, R. Seiler, A topological look at the quantum Hall effect. Phys. Today 38 (2006)
13. R.B. Laughlin, Quantized Hall conductivity in 2 dimensions. Phys. Rev. B **23**, 5632 (1981)
14. http://materia.fisica.unimi.it/manini/berryphase.html
15. J.J. Sakurai, *Modern Quantum Mechanics, Supplement 1* (Addison Wesley Longman, New York, 1994), pp. 464–480
16. M.V. Berry, Quantal phase factors accompanying adiabatic changes. Proc. R. Soc. Lond. A **392**, 45–57 (1984)
17. D. Xiao, M.-C. Chang, Q. Niu, Berry phase effects on electronic properties. Rev. Mod. Phys. **82**, 1959–2007 (2010)
18. S.-Q. Shen, *Topological Insulators: Dirac Equation in Condensed Matters* (Springer, Berlin, 2012), Appendix A1
19. J. Moore, An introduction to topological phases of electrons, Chapter 2, in *Topological aspects of condensed matter physics*. ed. by C. Chamon, M.O. Goerbig, R. Moessner, L.F. Cugliandolo (Oxford, New York, 2007), pp. 3–62
20. D.J. Thouless, Localisation and the two-dimensional Hall effect. J. Phys. C: Solid State Phys. **14**, 3475–3480 (1981). D.J. Thouless, M. Kohmoto, M.P. Nightingale, M. den Nijs, Quantized Hall conductance in a two-dimensional periodic potential. Phys. Rev. Lett.**49**, 405–408 (1982)
21. H.J. Fishbeck, Theory of bloch electrons in a magnetic field. Phys. Stat. Sol. **38**, 11 (1970)
22. S. Janecek, M. Aichinger, E.R. Hernández, Two-dimensional bloch electrons in perpendicular magnetic fields: an exact calculation of the Hofstadter butterfly spectrum. Phys. Rev. B **87**, 235429 (2013)
23. M. Kohmoto, Topological invariant and the quantization of hall conductance. Ann. Phys. **160**, 343–354 (1985)
24. Y. Hatsugai, Chern number and edge states in the integer quantum Hall effect. Phys. Rev. Lett. **71**, 3697–3700 (1993)
25. Y. Hatsugai, Edge states in the integer quantum Hall effect and the Riemann surface of the Bloch function. Phys. Rev. B **48**, 11852 (1993)

26. M. Fruchart, D. Carpentier, An introduction to topological insulators. C. R. Physique **14**, 779–815 (2013)
27. D. Carpentier, Topology of bands in solids: from insulators to dirac matter, in *Dirac Matter. Progress in Mathematical Physics*, vol. 71, ed. by B. Duplantier, V. Rivasseau, J.N. Fuchs (Springer, Basel, 2017), pp. 95–129.
28. G.L. Naber, *Topology, Geometry, and Gauge Fields* (Springer, New York, 2011)
29. Y. Hasegawa, M. Kohmoto, Quantum Hall effect and the topological number in grapheme. Phys. Rev. B **74**, 155415 (2006)
30. Y. Zhang, Y.-W. Tan, H.L. Stormer, P. Kim, Experimental observation of the quantum Hall effect and Berry's phase in graphene. Nature **438**, 201–204 (2005)
31. K.S. Novoselov, Z. Jiang, Y. Zhang, S.V. Morozov, H.L. Stormer, U. Zeitler, J.C. Maan, G.S. Boebinger, P. Kim, A.K. Geim, Room temperature quantum Hall effect in Graphene. Science **315**, 1379 (2007)
32. M. Portail, B. Jouault, F. Schopfer, W. Poirier, Quantum Hall resistance standards from graphene grown by chemical vapour deposition on silicon carbide. Nat. Comm. **6**, 6806 (2015)
33. H.L. Stormer, Nobel lecture: the fractional quantum Hall effect. Rev. Mod. Phys. **71**, 875–889 (1991)
34. E. Wigner, On the interaction of electrons in metals. Phys. Rev. **46**, 1002 (1934)
35. S.M. Girvin, Introduction to the fractional quantum Hall effect. Seminaire Poincare **2**, 53–74 (2004)
36. R.B. Laughlin, Anomalous quantum Hall effect: An incompressible quantum fluid with fractionally charged excitations. Phys. Rev. Lett. **50**, 1395 (1984)
37. N. Nagaosa, J. Sinova, S. Onoda, A.H. MacDonald, N.P, Ong, Anomalous hall effect. Rev. Mod. Phys. **82**, 1539–1592 (2010)
38. A. Kundt, Wied. Ann. 49, On the Hall effect in Ferromagnetic materials, 257 (1893)
39. H. Weng, R. Yu, X. Hu, X. Dai, Z. Fang, Quantum anomalous Hall effect and related topological electronic states. Adv. Phys. **64**, 227–282 (2015)
40. C.-Z. Chang, J. Zhang, X. Feng, J. Shen, Z. Zhang, M. Guo, K. Li, Y. Ou, P. Wei, L.-L. Wang, Z.-Q. Ji, Y. Feng, S. Ji, X. Chen, J. Jia, X. Dai, Z. Fang, S.-C. Zhang, K. He, Y. Wang, L. Lu, X.-C. Ma, Q.-K. Xue, Experimental observation of the quantum anomalous hall effect in a magnetic topological insulator. Science **340**, 167–170 (2013)
41. F.D.M. Haldane, Model for a quantum Hall effect without Landau levels: condensed-matter realization of the "Parity Anomaly" . Phys. Rev. Lett. **61**, 2015 (1988)
42. C.-X. Liu, S.-C. Zhang, X.-L. Qi, The quantum qnomalous Hall effect: theory and experiment. Annu. Rev. Condens. Matter Phys. **7**, 301–321 (2016)
43. J. Sinova, S.O. Valenzuela, J. Wunderlich, C.H. Bach, Spin Hall effects. Rev. Mod. Phys. **87**, 1213–1259 (2015)
44. T. Tanaka, H. Kontani, M. Naito, T. Naito, D.S. Hirashima, K. Yamada, J. Inoue, Intrinsic spin Hall effect and orbital Hall effect in 4d and 5d transition metals. Phys. Rev. B **77**, 165117 (2008)
45. G.Y. Guo, S. Murakami, T.-W. Chen, N. Nagaosa, Intrinsic spin Hall effect in platinum: first-principles calculations. Phys. Rev. Lett. **100**, 096401 (2008)
46. A. Hoffmann, Spin Hall effect in metals. IEEE Trans. Mag. **49**, 5172–5193 (2013)
47. Y.K. Kato, R.C. Myers, A.C. Gossard, D.D. Awschalom, Observation of the spin Hall effect in semiconductors. Science **306**, 5703 (2004)
48. J. Balakrishnan, G.K.W. Koon, A. Avsar, Y. Ho, J.H. Lee, M. Jaiswal, S-J Baeck, J-H Ahn, A. Ferreira, M.A. Cazalilla, A.H. Castro Neto, B. Ozyilmaz, Giant spin Hall effect in graphene grown by chemical vapour deposition. Nat. Comm. **5**, 4748 (2014)
49. S. Cheng, Y. Xing, Q. Sun, X.C. Xie, Spin Nernst effect and Nernst effect in two-dimensional electron systems. Phys. Rev. B **78**, 045302 (2008)
50. S. Meyer, Y. Chen, S. Wimmer, M. Althammer, T. Wimmer, R. Schlitz, S. Geprägs, H. Huebl, D. Ködderitzsch, H. Ebert, G.E.W. Bauer, R. Gross, S.T.B. Goennenwein, Observation of the spin Nernst effect. Nat. Mat. **16**, 997–982 (2017)

51. J. Zang, M. Mostovoy, J. H. Han, N. Nagaosa, Dynamics of skyrmion crystals in metallic thin films. Phys. Rev. Lett. **107**, 136804 (2011)
52. G. Chen, Skyrmion Hall effect. Nat. Phys. **13**, 112–113 (2017)
53. W. Jiang, X. Zhang, G. Yu, W. Zhang, X. Wang, M.B. Jungfleisch, J.E. Pearson, X. Cheng, O. Heinonen, K.L. Wang, Y. Zhou, A. Hoffmann, S.G.E. te Velthuis, Direct observation of the skyrmion Hall effect. Nat. Phys. **13**, 162–169 (2017)
54. K. Litzius, I. Lemesh, B. Krüger, P. Bassirian, L. Caretta, K. Richter, F. Büttner, K. Sato, O.A. Tretiakov, J. Förster, R.M. Reeve, M. Weigand, I. Bykova, H. Stoll, G. Schütz, G.S.D. Beach, M. Kläui, Skyrmion Hall effect revealed by direct time-resolved X-ray microscopy. Nat. Phys. **13**, 170–175 (2017)
55. T. Nakajima, H. Oike, A. Kikkawa, E.P. Gilbert, N. Booth, K. Kakurai, Y. Taguchi, Y. Tokura, F. Kagawa, T-h Arima, Skyrmion lattice structural transition in MnSi. Sci. Adv. **3**, e1602562 (2017)
56. C. Weisbuch, Fundamental properties of III-V semiconductor two dimensional quantized structures, The basis for optical and electronic device applications, Ed. R. Dingle. Semiconduct. Semimetals **24**, 1–133

Chapter 7
Optical and Electrical Properties of Graphene, Few Layer Graphene, and Boron Nitride

Abstract This chapter discusses the unique optical and electrical properties of graphene, bilayer graphene, and boron nitride. The tight binding electronic band structure of the π and π^* bands is presented for both single layer graphene and bilayer graphene. This leads to a discussion of the low energy Dirac carriers present in Dirac cones in graphene. Ellipsometric characterization data for graphene is presented. The Dirac equation is used to show how the optical absorption for the zero mass, low energy carriers in the Dirac cones is directly related to the fine structure constant. The quantization of the Landau levels of the Dirac carriers in graphene is derived. The optical properties and Hall characterization of bilayer graphene and twisted bilayer graphene is discussed. The bands that result from the periodic potential of the moiré lattice due to the twisted alignment of the graphene layers are described. The electronic band structure of boron nitride is presented and heterolayers of graphene and boron nitride are discussed.

In this chapter, we will discuss the unique optical and electrical properties of graphene, bilayer graphene, and boron nitride. As described in this chapter, the electronic band structure of the hexagonal layer of carbon atoms shown in Fig. 7.1 gives graphene its unique optical and electrical properties. Electrons and holes can move through the conduction and valence bands with very high mobility, as high as ~ 1,000,000 cm^2/V s. The optical conductivity and thus absorbance is constant over a wide wavelength range in the IR and near IR. In addition, the magneto-transport properties of graphene prove the Dirac nature of the carriers as was briefly introduced in CH 6. The wealth of new physical phenomena observed for graphene resulted in Novoselov and Geim winning the Nobel Prize in Physics in 2010.

Above and below the planar graphene lattice, p orbitals interact to form the delocalized π valence and π^* conduction bands. The energy dependence of π and π^* bands of graphene can be derived using the tight binding approximation which is presented in this chapter. The π and π^* bands have the same energy (touch) at specific points (K and K') in the Brillouin zone. We will find that the low energy carriers with wavevectors close to K and K' can be described using the relativistic Dirac equation, and it is the massless Dirac fermions that have unique optical and electrical properties. For example, the constant optical absorption of graphene over

A. Diebold and T. Hofmann, *Optical and Electrical Properties of Nanoscale Materials*, Springer Series in Materials Science 318,
https://doi.org/10.1007/978-3-030-80323-0_7

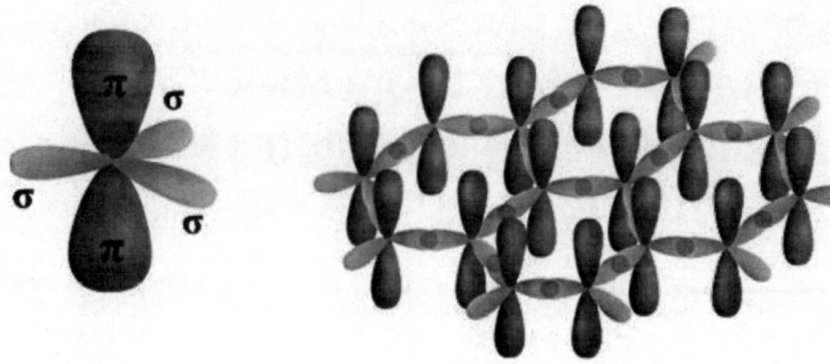

Fig. 7.1 Hexagonal carbon lattice of graphene and the sp^2 hybridized carbon atom. The p_z orbitals (blue) hybridize allowing the electrons to delocalize above and below the plane of carbon atoms. These orbitals form the π and π^* bands

a wide wavelength range in the near IR and IR is seen to be a consequence of the Dirac fermion nature of the carriers. Next the Dirac equation is applied to describing the carrier properties of massless Dirac fermions in a magnetic field. We show that the low temperature Hall characterization of graphene provides proof of the Dirac nature of the low energy carriers in graphene. Some of graphene properties are not due to the Dirac Fermions. For example, the complex refractive index over the visible wavelength range is discussed in terms of optical transitions between π and π^* bands across the entire Brillouin zone. For example the strong absorption at the M point in the Brillouin zone is shifted in energy by the excitonic nature of the transition.

The optical and electrical properties of bilayer graphene are also presented. Graphite is the bulk structure formed by van der Waals bonded many layer graphene. It is the van der Waals binding between graphene layers that allow the layer to layer stacking in bilayer structures to be experimentally altered. Here, we find that new correlated electron phenomena such as superconductivity where the electrons do not act independently. We also discuss boron nitride–graphene heterostructures at the end. The space group and point groups of graphene, few layer graphene, and graphite were presented in Chap. 2.

The flow of this chapter is as follows: Following an introduction of the planar hexagonal graphene lattice and the sp^2 carbon bonding, the nearest neighbor tight binding model for the π and π^* conduction and valence bands is presented. This is followed by an analysis of the unusual linear wavevector dependence of the carrier energy close to the K and K' points in the hexagonal Brillouin zone. Then, the Dirac cone of 2D energy levels is described. The tight binding model including the next nearest neighbors is presented. This is followed by a discussion of pseudospin. Next nearest neighbor tight binding model results are discussed along with the shape of the Dirac cone for nearest neighbor and next nearest neighbor models. Experimental optical spectra are used to show the unusual low energy behavior of the Dirac carriers. This is followed by an introduction to relativistic quantum mechanics and its application to graphene. The Dirac Hamiltonian and Fermi's Golden Rule is used to derive the unusual 2.3% constant optical absorption of the low energy carriers of graphene. The Berry phase of graphene is described which leads into a discussion about the Berry phase correction for the Quantum Hall Effect and Shubnikov de Hass oscillations in graphene. Next, bilayer graphene and its Hall characterization are discussed. Trilayer and tetralayer graphene are briefly mentioned. Then, the optical characterization of graphene and multilayer graphene is presented. This is followed by a discussion of Hall characterization of twisted bilayer graphene. The chapter ends with a description of the tight binding electronic band structure of hexagonal boron nitride and boron nitride–graphene multilayers.

7.1 Hexagonal Graphene

The starting point for this discussion is the hexagonal, one atom thick crystal shape that results from the sp^2 hybridized carbon atoms. Specifically, the carbon $2s$, $2p_x$, and $2p_y$ orbital hybridize into planar sp^2 orbitals with 120° between the central axis of the sp^2 orbitals. Each of the three sp^2 orbitals forms a strong bond known as a σ bond with the neighboring carbon atom resulting in a planar 2D hexagonal structure as shown in Fig. 7.1. The remaining p_z orbital interact above and below the hexagonal sheet with the neighboring carbon atoms resulting in the delocalized π valence and π^* conduction bands which are shown in Fig. 7.2. This is same interaction that is described as a resonance structure for benzene in organic chemistry. There are two shapes that can be used for the primitive lattice, hexagonal and rhombus, and both contain two carbon atoms as shown in Fig. 7.3. The reciprocal lattice is also hexagonal. The valence and conduction bands touch at specific places (K, K' points known as Dirac points) in the Brillouin zone. The energy versus wavevector relationship is linear as it is for photons close to the K and K' points in the Brillouin zone, and to a first approximation the band diagram is a cone close to K and K'.

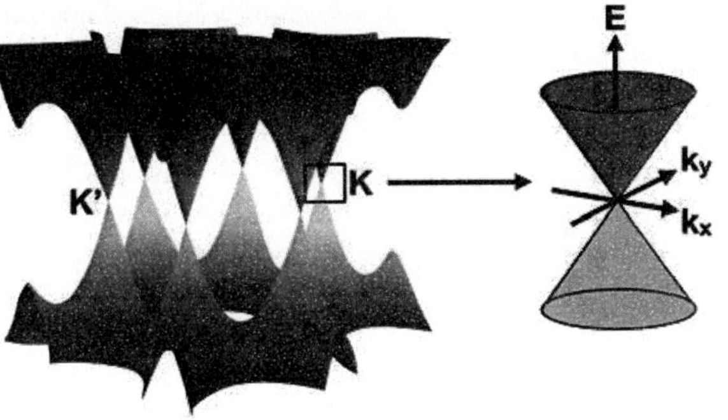

Fig. 7.2 The band diagram (Energy vs. wave vector) is shown for the π and π^* bands of graphene. The valence and conduction bands touch (aka $E = 0$) at the K and K' points in the Brillouin zone. At these points, the energy is a linear function of k forming the Dirac cones. This is in contrast to the $E \sim k^2$ relationship for free electrons in a semiconductor. The graphene band structure is adapted and reproduced from [1] with permission from The Royal Society of Chemistry

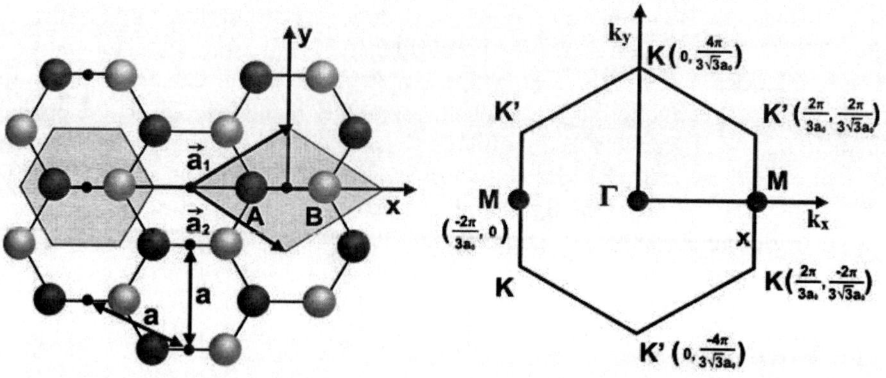

Fig. 7.3 The real space and reciprocal space hexagonal graphene lattice is shown. The light and dark spheres represent carbon atoms. A hexagonal primitive Wigner Seitz cell is shown with two atoms inside each unit cell as shown by the gray hexagon. $\vec{a} = \sqrt{3}\vec{a}_0$ where the distance between carbon atoms, $\vec{a}_0$, is 0.142 nm. A rhombus primitive cell formed from the primitive lattice vectors $\vec{a}_1 = \left(\frac{\sqrt{3}}{2}\hat{x}, \frac{1}{2}\hat{y}\right)a = \left(\frac{3}{2}\hat{x}, \frac{\sqrt{3}}{2}\hat{y}\right)a_0$ and $\vec{a}_2 = \left(\frac{\sqrt{3}}{2}\hat{x}, -\frac{1}{2}\hat{y}\right)a = \left(\frac{3}{2}\hat{x}, -\frac{\sqrt{3}}{2}\hat{y}\right)a_0$ is also shown [3, 4]. The black dots show the center of each rhombus primitive lattice and hexagonal Wigner Seitz lattice. The distance between the central points of the Bravais lattices, $a = 0.248$ nm. The light (dark) spheres form a triangular Bravais sub-lattice. In 2D, the reciprocal lattice vectors $\vec{b}_1$, $\vec{b}_2$ can be determined using $\vec{a}_i \cdot \vec{b}_j = 2\pi\delta_{ij}$. The rhombus used here was first described in [4], and the coordinate system for the real space lattice of graphene is rotated 90° in [2]. Both versions can be found in the literature. The dark and light colored carbon atoms form two inequivalent sub-lattices. It is convenient to choose the K point at $\left(0, \frac{4\pi}{3\sqrt{3}a_0}\right)$ and the K′ point at $\left(0, \frac{-4\pi}{3\sqrt{3}a_0}\right)$ in the reciprocal lattice as a pair of inequivalent K and K′ points in the Brillouin zone when discussing the inequivalent sub-lattices. The other K points in the reciprocal lattice can be connect to this K point by reciprocal lattice vectors and are equivalent. Similarly, the other K′ points can be connected to this K′ point by a reciprocal lattice vector

This region of the band diagram is referred to as a Dirac cone. Below, we present a nearest neighbor, tight binding model for the π and $\pi*$ bands and further discuss the unusual linear energy versus wave vector relationship for carriers close to the Dirac points. One key point is that the band gap is zero for large area graphene because the two atoms in the primitive lattice are chemically identical. This tight binding model introduces a number of key concepts such as *valley splitting* for the conduction band when the two atoms in the hexagonal unit cell have inequivalent positions. The relativistic Dirac Hamiltonian approach for the low energy carriers in graphene is also introduced in this chapter.

7.1.1 *Bravais Lattice of Graphene*

The Bravais lattice of graphene requires careful discussion as will become apparent below. The hexagonal crystal structure of graphene can be described in several ways.

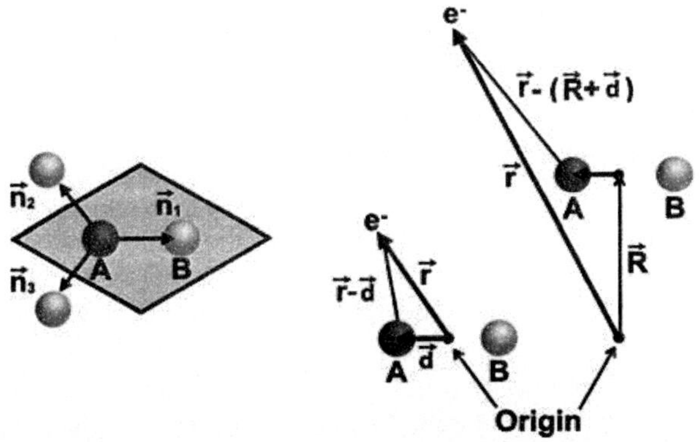

Fig. 7.4 The neighbor atom vectors $\vec{n}_1$, $\vec{n}_2$, $\vec{n}_3$ from the atom A to nearest neighbor B atoms which are used for tight binding approximation for graphene in real space are shown. The vector relationship for the position $\vec{r}$ of the electron with the origin (central point of the primitive lattice is shown for atom A in the primitive lattice and an A atom in a primitive lattice $\vec{R}$ from the origin

First, the hexagonal structure can be described using either a hexagonal or rhombus primitive lattice. The real space and reciprocal space lattice for graphene is shown in Fig. 7.3 [1–4].

Although the carbon atoms are all identical, there are two inequivalent triangular sub-lattices as shown by the dark and light spheres in Fig. 7.3. Both hexagonal and rhombus shaped primitive (Bravais) lattices shown in gray in Fig. 7.3 can be used to describe the crystal structure. In this case, there are two atoms shown as dark and light spheres inside this hexagonal Bravais lattice, each belonging to the dark or light sphere sub-lattice shown in Fig. 7.3. These two atoms are not crystallographically identical. The entire lattice can be reproduced by translations of the central point of the hexagonal Bravais lattice using the vectors $\vec{a}_1$ and $\vec{a}_2$. Here, we note that the orientation of the graphene Bravais lattice in the $x - y$ plane differs between authors. Figure 7.4 shows some of the vectors used in the tight binding calculation of the graphene π and π^* bands. The primitive lattice vectors $\vec{a}_1, \vec{a}_2$, nearest neighbor atom vectors $\vec{n}_1, \vec{n}_2, \vec{n}_3$, and $\vec{d}_1$ and $\vec{d}_2$ the distance from the center of the primitive lattice to the A and B atoms respectively which are used for tight binding approximation for graphene in real space along the unit vectors in *x, y,* and *z* are:

$$\vec{a}_1 = \left(\frac{\sqrt{3}}{2}\hat{x} + \frac{1}{2}\hat{y}\right)a \quad \vec{a}_2 = \left(\frac{\sqrt{3}}{2}\hat{x} - \frac{1}{2}\hat{y}\right)a \quad a = \sqrt{3}a_0 \tag{7.1}$$

$$\vec{d}_1 = -\frac{a}{2\sqrt{3}}\hat{x} \quad \vec{d}_2 = \frac{a}{2\sqrt{3}}\hat{x} \tag{7.2}$$

$$\vec{n}_1 = \frac{a}{\sqrt{3}}\hat{x}\ ;\ \vec{n}_2 = \frac{a}{\sqrt{3}}\left(-\frac{1}{2}\hat{x} + \frac{\sqrt{3}}{2}\hat{y}\right)\ ;\ \vec{n}_3 = \frac{a}{\sqrt{3}}\left(-\frac{1}{2}\hat{x} - \frac{\sqrt{3}}{2}\hat{y}\right)$$

7.2 Tight Binding Approximation for the π Bands of Graphene

The tight binding model provides significant insight into the physical phenomena associated with the π and π^* bands of graphene [2, 4]. In this section, our discussion of the tight binding approximation is based on our previous discussion of the tight binding model for the s and p bands of semiconductors presented in Chap. 2. The initial tight binding description of graphene is further developed below in the discussion about the band structure of bilayer graphene. In the 2 band model for the valence and conduction bands of graphene, each atom contributes one p_z orbital and there are two atoms in each unit cell. As in the discussion of the band structure for 3D semiconductors, we want to construct a tight binding Block wave function for the band from the local atomic wavefunctions $\psi_n(\vec{r})$ that are solutions of the atomic Hamiltonian H_{atomic}:

$$H_{atomic}\,\psi_n(\vec{r}) = E_n\psi_n(\vec{r}) \tag{7.3}$$

If the interaction between p_z orbitals on adjacent atoms is small, i.e., $U(\vec{R})\cdot\psi_n(\vec{r})$ is small, then we are justified in using a linear combination of atomic orbitals to form the wavefunction for the π and π^* bands. In general, as discussed in CH 2, the periodic wave function $\phi(\vec{r})$ is multiplied by the plane wave function and the contributions from each unit cell are summed to give the Block wavefunction. The periodic function can be a linear combination of s, p, d, etc. orbitals as would be the case for a hybridized wavefunction. In that case, $\phi(\vec{r}) = \sum_{n=1}^{N'} b_n\psi_n(\vec{r})$, where we are summing over the N' types of orbitals that contribute. However, for the π and π^* bands, we are using only one p_z orbital from each of the two atoms (A and B) in the reciprocal lattice unit cell as shown in Fig. 7.3. Thus, we replace the $\psi_n(\vec{r})$ in $\phi(\vec{r}) = \sum_{n=1}^{N'} b_n\psi_n(\vec{r})$ with p_z orbitals for each atom in the unit cell in the reciprocal lattice:

$$\phi_A(\vec{r}) = \phi\{p_z\}_A(\vec{r}) \quad \text{and} \quad \phi_B(\vec{r}) = \phi\{p_z\}_B(\vec{r}) \tag{7.4}$$

We state the periodic wavefunction $\phi(\vec{r})$ as a linear combination of the electron being located on the A and B atoms located at $\vec{d}_i;\, i = 1, 2$ with distance of the electron from the atom in that primitive cell in (7.5) is $\vec{r} - \vec{d}_i$. The linear combination coefficients b_n are replaced with wavevector dependent linear combination coefficients $c_{A(B)}(\vec{k})$ [2]:

$$\begin{aligned}\phi(\vec{r}) &= c_A(\vec{k})\,\phi_A(\vec{r} - \vec{d}_1 t) + c_B(\vec{k})\,\phi_B(\vec{r} - \vec{d}_2)\\ &= \sum_{i=1,2} c_i(\vec{k})\,\phi_i(\vec{r} - \vec{d}_i)\end{aligned} \tag{7.5}$$

Note that the origin for the position of the electron $\vec{r}$ is the dot in the center of the rhombus primitive cell shown in Fig. 7.3.

The energy of each orbital $\phi_A(\vec{r})$ or $\phi_B(\vec{r})$ is E_p (i.e., $H_{atomic}\phi_A(\vec{r}) = E_p\psi_n(\vec{r})$) and that energy is the same for each atom in the unit cell. The crystal Hamiltonian is considered to be a small perturbation $U(\vec{R})$ from the atomic state so that:

$$H_{crystal} = H_{atomic} + U(\vec{R}) \tag{7.6}$$

We will use the p_z orbitals of both carbon atoms in the primitive cell from all the N unit cells of graphene in the Block wavefunction $\psi_{\vec{k}}(\vec{r})$. Here, the distance from the center of an atom in a primitive cell with the origin at a distance $\vec{R}$ from the origin of the central primitive cell is $\vec{R}+\vec{d}_i$ and again $\vec{d}_i$; $i = 1, 2$. The distance of an electron from the center of the atom at $\vec{d}_i$ in this primitive unit cell is $\vec{r}-(\vec{R}+\vec{d}_i)$. If $U(\vec{R}) = 0$, then for a crystal with N unit cells, for each atomic level $\psi_n(\vec{r})$ there would be N levels in the periodic potential with the wavefunctions $\psi_n(\vec{r} - \vec{R} - \vec{d}_i)$ for each of the N sites spanned by the Bravais lattice vector $\vec{R}$ in the crystal. Thus, it is convenient to have the Block wavefunction $\psi_{\vec{k}}(\vec{r})$ centered at the middle of the hexagonal unit cell (see Fig. 7.3) at the origin, $\vec{r} = 0$, with the two atoms in the central unit cell located at $\vec{d}_1$ and $\vec{d}_2$. There are contributions to the Block wavefunction from all the unit cells which are located using the vector$\vec{R}$. We add together the contributions, and the Block wavefunction for N unit cells is:

$$\begin{aligned}\psi_{\vec{k}}(\vec{r}) &= \frac{1}{\sqrt{N}}\left(\sum_{\vec{R}\; i=1,2} e^{i\vec{k}\cdot(\vec{R}+\vec{d}_i)}c_i(\vec{k})\,\phi_i(\vec{r}-\vec{R}-\vec{d}_i)\right)\\ &= \frac{\sum_{\vec{R}} e^{i\vec{k}\cdot\vec{R}}}{\sqrt{N}}\left[c_A(\vec{k})e^{i\vec{k}\cdot\vec{d}_1}\,\phi_A(\vec{r}-\vec{R}-\vec{d}_1) + c_B(\vec{k})e^{i\vec{k}\cdot\vec{d}_2}\phi_B(\vec{r}-\vec{R}-\vec{d}_2)\right]\end{aligned} \tag{7.7}$$

This simple tight binding model uses only nearest neighbors which are for atom A at $\vec{n}_1, \vec{n}_2$, and $\vec{n}_3$ from atom A, and for atom B at $-\vec{n}_1, -\vec{n}_2$, and $-\vec{n}_3$ from atom B. Remembering that the wavefunctions ϕ_{p_zA} and ϕ_{p_zB} were selected to be eigenfunctions of the atomic Hamiltonian, H_{atomic}, and that orbitals on different atoms do not overlap to any great extent (see discussion below), and that we are considering only nearest neighbor interactions in our initial consideration of the graphene π bands, the next step is to operate $H_{atomic} + U(\vec{R})$ on $\psi_{\vec{k}}(\vec{r})$ and consider the carbon atom A with 3 nearest neighbor B carbon atoms.

$$\left|H_{atomic} + U(\vec{R})\middle|\psi_{\vec{k}}(\vec{r})\right\rangle = E\psi_{\vec{k}}(\vec{r})\rangle \tag{7.8}$$

$$\psi_{\vec{k}}(\vec{r}) = \frac{\sum_{\vec{R}} e^{i\vec{k}\cdot\vec{R}}}{\sqrt{N}} \left[c_A(\vec{k}) e^{i\vec{k}\cdot d_1} \phi_A(\vec{r} - \vec{R} - \vec{d}_1) + c_B(\vec{k}) e^{i\vec{k}\cdot d_2} \phi_B(\vec{r} - \vec{R} - \vec{d}_2) \right]$$

$$\left| H_{atomic} + U(\vec{R}) \middle| \psi_{\vec{k}}(\vec{r}) \right\rangle = \frac{\sum_{\vec{R}} e^{i\vec{k}\cdot\vec{R}}}{\sqrt{N}} \Big[c_A(\vec{k}) e^{i\vec{k}\cdot\vec{d}_1} E_P \phi_A(\vec{r} - \vec{R} - \vec{d}_1)$$
$$+ c_A(\vec{k}) e^{i\vec{k}\cdot\vec{d}_1} U(\vec{R}) \phi_A(\vec{r} - \vec{R} - \vec{d}_1) + c_B(\vec{k}) e^{i\vec{k}\cdot\vec{d}_2} E_P \phi_B(\vec{r} - \vec{R} - \vec{d}_2)$$
$$+ c_B(\vec{k}) e^{i\vec{k}\cdot\vec{d}_2} U(\vec{R}) \phi_B(\vec{r} - \vec{R} - \vec{d}_2) \Big] \Big\rangle \tag{7.9}$$

Now operate on $|H_{atomic} + U(\vec{R})|\psi_{\vec{k}}(\vec{r})\rangle$ with just the atomic wavefunction for the A atom at $\vec{d}_1$ with respect to the unit cell origin, $\langle \phi_A(\vec{r} - \vec{d}_1)|$. So $\langle \phi_A(\vec{r} - \vec{d}_1)|\phi_A(\vec{r} - \vec{R} - \vec{d}_1)\rangle \approx 0$ unless $\vec{R} = 0$ because *we initially assume that the overlap of the p_z orbitals from two A atoms is negligible since they are not nearest neighbors.* We also initially assume that the overlap integral between nearest neighbor A and B atoms is very small. One can justify this by considering that the distance between carbon atoms is ~ 0.25 nm and that the majority of the electron density of an atomic p_z orbital extends radially about 0.1 nm from the center axis of the carbon atom. Thus electron density from a p_z orbital of an atom interacts little with the electron density from a p_z orbitals from neighboring atoms. It is useful to remember that in this simple tight binding model, the overlap integrals are between the atomic orbitals of (7.3). In a self -consistent theory, the overlap would be between wavefunctions that include the effects of interactions with neighboring atoms. Below when we discuss the band structure of bilayer graphene, we will assume that this overlap integral is small but non-zero. $\langle \phi_A(\vec{r} - \vec{d}_1)|\phi_B(\vec{r} - \vec{d}_2)\rangle = 0.129$ [2]:

$$\left\langle \phi_A(\vec{r} - \vec{d}_1) \middle| H_{atomic} + U(\vec{R}) \middle| \psi_{\vec{k}}(\vec{r}) \right\rangle$$
$$= \Bigg[c_A(\vec{k}) \frac{\sum_{\vec{R}} e^{i\vec{k}\cdot\vec{R}}}{\sqrt{N}} e^{i\vec{k}\cdot\vec{d}_1} E_P \left\langle \phi_A(\vec{r} - \vec{d}_1) \middle| \phi_A(\vec{r} - \vec{R} - \vec{d}_1) \right\rangle$$
$$+ c_A(\vec{k}) \frac{\sum_{\vec{R}} e^{i\vec{k}\cdot\vec{R}}}{\sqrt{N}} e^{i\vec{k}\cdot\vec{d}_1} \left\langle \phi_A(\vec{r} - \vec{d}_1) \middle| U(\vec{R}) \middle| \phi_A(\vec{r} - \vec{R} - \vec{d}_1) \right\rangle$$
$$+ \frac{\sum_{\vec{R}} e^{i\vec{k}\cdot\vec{R}}}{\sqrt{N}} c_B(\vec{k}) e^{i\vec{k}\cdot\vec{d}_2} E_P \left\langle \phi_A\left(\vec{r} - \vec{d}_1\right) | \phi_B(\vec{r} - \vec{R} - \vec{d}_2) \right\rangle$$
$$+ \frac{\sum_{\vec{R}} e^{i\vec{k}\cdot\vec{R}}}{\sqrt{N}} c_B(\vec{k}) e^{i\vec{k}\cdot\vec{d}_2} \left\langle \phi_A(\vec{r} - \vec{d}_1) | U(\vec{R}) | \phi_B(\vec{r} - \vec{R} - \vec{d}_2) \right\rangle \Bigg]$$
$$= \left\langle \phi_A(\vec{r} - \vec{d}_1) | E | \psi_{\vec{k}}(\vec{r}) \right\rangle \tag{7.10}$$

The term $\frac{\sum_{\vec{R}} e^{i\vec{k}\cdot\vec{R}}}{\sqrt{N}} e^{i\vec{k}\cdot\vec{d}_1} E_P \left\langle \phi_A(\vec{r}-\vec{d}_1) | \phi_A(\vec{r}-\vec{R}-\vec{d}_1) \right\rangle$ can be split into a term for the A atom in the central primitive lattice and the terms for the other A atoms which have very little overlap:

$$
\begin{aligned}
&= \frac{1}{\sqrt{N}} e^{i\vec{k}\cdot\vec{d}_1} E_P \left\langle \phi_A(\vec{r}-\vec{d}_1) | \cdot \phi_A(\vec{r}-\vec{d}_1) \right\rangle \\
&\quad + \frac{1}{\sqrt{N}} \sum_{\vec{R}\neq 0} e^{i\vec{k}\cdot(\vec{R}+\vec{d}_1)} E_P \left\langle \phi_A(\vec{r}-\vec{d}_1) | \phi_A(\vec{r}-\vec{R}-\vec{d}_1) \right\rangle \\
&= \frac{1}{\sqrt{N}} e^{i\vec{k}\cdot\vec{d}_1} E_P + 0
\end{aligned}
$$

$$
\begin{aligned}
&\left\langle \phi_A(\vec{r}-\vec{d}_1) \middle| H_{atomic} + U(\vec{R}) \middle| \psi_{\vec{k}}(\vec{r}) \right\rangle \\
&= \left[\frac{1}{\sqrt{N}} e^{i\vec{k}\cdot\vec{d}_1} c_A(\vec{k}) E_P \right. \\
&\quad + 0\ (non\ nearest\ neighbor\ overlap\ term) \\
&\quad + 0\ (non\text{-}nearest\ neighbor\ hopping\ integral) \\
&\quad + 0\ (small\ overlap\ between\ nearest\ neighbor\ atoms\ A\ and\ B) \\
&\quad \left. + \frac{\sum_{\vec{R}_{NN}} e^{i\vec{k}\cdot\vec{R}}}{\sqrt{N}} c_B(\vec{k}) e^{i\vec{k}\cdot\vec{d}_2} \left\langle \phi_A(\vec{r}-\vec{d}_1) | U(\vec{R}) | \phi_B(\vec{r}-\vec{R}-\vec{d}_2) \right\rangle \right] \\
&= \left\langle \phi_A(\vec{r}-\vec{d}_1) | E | \psi_{\vec{k}}(\vec{r}) \right\rangle
\end{aligned}
\tag{7.11}
$$

Using the same approach for $\left\langle \phi_A(\vec{r}-\vec{d}_1) | E | \psi_{\vec{k}}(\vec{r}) \right\rangle$ for we find that $\left\langle \phi_A(\vec{r}-\vec{d}_1) | E | \psi_{\vec{k}}(\vec{r}) \right\rangle \approx E e^{i\vec{k}\vec{d}_1} c_A(\vec{k})$.

Thus:

$$
\begin{aligned}
&\left[\frac{1}{\sqrt{N}} e^{i\vec{k}\cdot\vec{d}_1} c_A(\vec{k}) E_P + \frac{1}{\sqrt{N}} \sum_{\vec{R}} e^{i\vec{k}\cdot(\vec{R}+\vec{d}_2)} c_B(\vec{k}) \right. \\
&\quad \left. \times \left\langle \phi_A(\vec{r}-\vec{d}_1) | U(\vec{R}) | \phi_B(\vec{r}-\vec{R}-\vec{d}_2) \right\rangle \right] = E e^{i\vec{k}\cdot\vec{d}_1} c_A(\vec{k})
\end{aligned}
\tag{7.12}
$$

Multiply both sides by $e^{-i\vec{k}\cdot\vec{d}_1}$ and get

$$
\begin{aligned}
&\left[\frac{1}{\sqrt{N}} c_A(\vec{k}) E_P + \frac{1}{\sqrt{N}} \sum_{\vec{R}} e^{i\vec{k}\cdot(\vec{R}+\vec{d}_2-\vec{d}_1)} c_B(\vec{k}) \right. \\
&\quad \left. \times \left\langle \phi_A(\vec{r}-\vec{d}_1) | U(\vec{R}) | \phi_B(\vec{r}-\vec{R}-\vec{d}_2) \right\rangle \right] = E c_A(\vec{k})
\end{aligned}
$$

The vector $\vec{d}_2 - \vec{d}_1 = \vec{n}_1$ is the vector between the nearest neighbor atoms A and B in the central primitive cell, and using only nearest neighbors for the atom A in the summation (see Fig. 7.4). Selecting the vectors $\vec{R}_2$ and $\vec{R}_3$ to be vectors to the center of the two neighboring unit cells where the other nearest neighbor B atoms are located, we get $\vec{n}_2 = \vec{R}_2 + \vec{d}_2 - \vec{d}_1$ and $\vec{n}_3 = \vec{R}_3 + \vec{d}_2 - \vec{d}_1$ giving us for the nearest neighbor contributions $\frac{\sum_{\vec{R}} e^{i\vec{k}\cdot(\vec{R}+\vec{d}_2-\vec{d}_1)}}{\sqrt{N}} = \frac{e^{i\vec{k}\cdot\vec{n}_1}+e^{i\vec{k}\cdot\vec{n}_2}+e^{i\vec{k}\cdot\vec{n}_3}}{\sqrt{N}}$. Since $\frac{1}{\sqrt{N}}$ appears on both sides of the equation, we get:

$$c_A(\vec{k})E_P + \left[e^{i\vec{k}\cdot\vec{n}_1} + e^{i\vec{k}\cdot\vec{n}_2} + e^{i\vec{k}\cdot\vec{n}_3}\right]c_B(\vec{k})\left\langle\phi_A(\vec{r}-\vec{d}_1)|U(\vec{R})|\phi_B(\vec{r}-\vec{R}-\vec{d}_2)\right\rangle$$
$$= Ec_A(\vec{k}) \tag{7.13}$$

The same operation is now done for atom B at $\vec{d}_2$: $\left\langle\phi_B(\vec{r}-\vec{d}_2)\middle|H_{atomic}+U(\vec{R})\middle|\psi_{\vec{k}}(\vec{r})\right\rangle$. The vectors from the central B atom to the nearest neighbor A atoms to the left, above to the right, and below to the left will be $-\vec{n}_1;\ -\vec{n}_3;\ -\vec{n}_2$:

$$\left[e^{-i\vec{k}\cdot\vec{n}_1} + e^{-i\vec{k}\cdot\vec{n}_2} + e^{-i\vec{k}\cdot\vec{n}_3}\right]c_A(\vec{k})\left\langle\phi_B(\vec{r}-\vec{d}_2)|U(\vec{R})|\phi_A(\vec{r}-\vec{R}-\vec{d}_2)\right\rangle$$
$$+ c_B(\vec{k})E_P = Ec_B(\vec{k}) \tag{7.14}$$

Equations (7.13) and (7.14) form a secular equation and this can be solved using a determinant. Use

$$\left\langle\phi_A(\vec{r}-\vec{d}_1)|U(\vec{R})|\phi_B(\vec{r}-\vec{d}_2)\right\rangle = \left\langle\phi_B(\vec{r}-\vec{d}_2)|U(\vec{R})|\phi_A(\vec{r}-\vec{d}_1)\right\rangle = \gamma_\pi$$

$$\begin{pmatrix} E_P & \gamma_\pi\left[e^{i\vec{k}\cdot\vec{n}_1} + e^{i\vec{k}\cdot\vec{n}_2} + e^{i\vec{k}\cdot\vec{n}_3}\right] \\ \gamma_\pi\left[e^{-i\vec{k}\cdot\vec{n}_1} + e^{-i\vec{k}\cdot\vec{n}_2} + e^{-i\vec{k}\cdot\vec{n}_3}\right] & E_P \end{pmatrix}\begin{pmatrix} c_A(\vec{k}) \\ c_B(\vec{k}) \end{pmatrix}$$
$$= E(\vec{k})\begin{pmatrix} c_A(\vec{k}) \\ c_B(\vec{k}) \end{pmatrix} \tag{7.15}$$

Which we solve using a determinant:

$$\begin{vmatrix} E_P - E(\vec{k}) & \gamma_\pi f(\vec{k}) \\ \gamma_\pi f(-\vec{k}) & E_P - E(\vec{k}) \end{vmatrix} = 0 \tag{7.16}$$

With

$$\left[e^{i\vec{k}\cdot\vec{n}_1} + e^{i\vec{k}\cdot\vec{n}_2} + e^{i\vec{k}\cdot\vec{n}_3}\right] = f(\vec{k}) \text{ and } f(-\vec{k}) = f^*(\vec{k})$$
$$\left(E_P - E(\vec{k})\right)^2 - \gamma_\pi^2 f^*(\vec{k})f(\vec{k}) = 0$$

$$E(\vec{k}) = E_P \pm \gamma_\pi |f(\vec{k})| \text{ with } \gamma_\pi = 3.033\,\text{eV} \tag{7.17}$$

Later we set E_P to zero for convenience. The valence and conduction band separation is maximum at $\vec{k} = 0$ and touch at the K points in reciprocal lattice as we show below. The value of γ_π is discussed by McCann [2]. The impact of $S_{12} = \left\langle \phi_A(\vec{r} - \vec{d}_1) | \phi_B(\vec{r} - \vec{d}_2) \right\rangle \neq 0$ alters (7.10) through (7.13) as follows for nearest neighbor contributions resulting in $E_\pm = \frac{\pm\gamma_\pi \left| f(\vec{k}) \right|}{1 \mp S_{12} \left| f(\vec{k}) \right|}$ with $S_{12} = 0.129$ [2]. This result is derived later in this chapter in the discussion about bilayer graphene.

7.2.1 Another Look at the Reciprocal Lattice of Graphene

The primitive lattice vector of the reciprocal lattice b_1 and b_2 for the rhombus primitive cell shown in Fig. 7.1 are shown in Fig. 7.5 for a different orientation of graphene, and the reciprocal lattice vectors are given by $\vec{b}_1 = \left(\frac{2\pi}{a\sqrt{3}}, \frac{2\pi}{a} \right)$ and $\vec{b}_2 = \left(\frac{2\pi}{a\sqrt{3}}, -\frac{2\pi}{a} \right)$ with each being $\frac{4\pi}{a\sqrt{3}}$ long. Since this orientation of graphene and the one shown in Fig. 7.3 are both found in the literature, they are presented here.

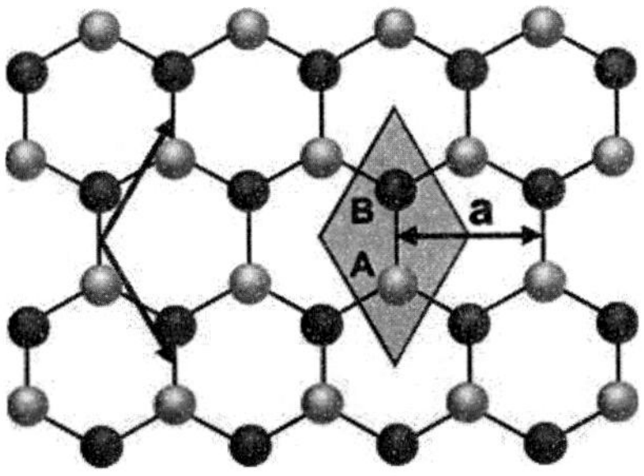

Fig. 7.5 The primitive lattice of graphene are shown using a rhombus primitive cell. In this representation, the primitive lattice vectors are $\vec{a}_1 = \left(\frac{1}{2}\hat{x}, \frac{\sqrt{3}}{2}\hat{y} \right) a$ $\vec{a}_2 = \left(\frac{1}{2}x, -\frac{\sqrt{3}}{2}\hat{y} \right) a$. Alternately, the graphene lattice can be oriented as shown in Fig. 7.3 where the rhombus shaped primitive lattice is rotated 90° from the rhombus in Fig. 7.3a [2]. In this case the primitive lattice vectors are pointed in a different direction than in Fig. 7.3. This change in crystallographic direction simplifies the discussion of pseudospin

7.2.2 Graphene's Π Electronic Band Structure

The energy dependence of the valence and conduction bands of graphene can be calculated using (7.17). A strong absorption feature is seen at the M point in the Brillouin zone. If you use the M point at $\frac{2\pi}{a\sqrt{3}}$ along k_x. as shown in Fig. 7.3.

$e^{i\vec{k}\cdot\vec{n}_1} + e^{i\vec{k}\cdot\vec{n}_2} + e^{i\vec{k}\cdot\vec{n}_3} = f(\vec{k})$ with $\vec{k}_x = \frac{2\pi}{a\sqrt{3}}$ and $\vec{k}_y = 0$ using $e^{i\theta} = \cos(\theta) + i\sin(\theta)$ and the values for $\vec{n}_1$, $\vec{n}_2$, and $\vec{n}_3$ we get $f(\vec{k}) = e^{i\frac{2\pi}{3}} + e^{-i\frac{\pi}{3}} + e^{-i\frac{\pi}{3}} = 0.5 - i0.866$ thus $|f(\vec{k})| = 1$.

At the **M point** using (7.17) we get:

$$E(\vec{k}) = E_P \pm 1\gamma_\pi \quad \text{with} \quad \gamma_\pi = 3.033\ \text{eV} \tag{7.18}$$

It is important to show that the valence and conduction bands have the same energy at the K and K′ points in the Brillouin zone. We can chose the K point at $\vec{k}_x = \frac{2\pi}{a\sqrt{3}}\hat{x}$ and $\vec{k}_y = -\frac{2\pi}{3a}\hat{y}$ and using values for $\vec{n}_1, \vec{n}_2$, and $\vec{n}_3$ we get $f(\vec{k}) = \left[e^{i\vec{k}\cdot\vec{n}_1} + e^{i\vec{k}\cdot\vec{n}_2} + e^{i\vec{k}\cdot\vec{n}_3}\right] = e^{i\frac{i\pi}{3}} + e^{-i\frac{2\pi}{3}} + e^{i0} = 0$. Then $f^*(\vec{k})f(\vec{k}) = 0$ so $|f(\vec{k})| = 0$.

At the **K point** we get:

$$E(\vec{k}) = E_P \tag{7.19}$$

The energy band offset is E_P. Thus, the conduction and valence bands have the same energy (touch) at the K and K' points.

The largest energy difference between the valence and conduction bands is at the **Γ point** where $\vec{k}_x = 0, \vec{k}_y = 0, \vec{k}_Z = 0$. So $f(\vec{k}) = \left[e^{i\vec{k}\cdot\vec{n}_1} + e^{i\vec{k}\cdot\vec{n}_2} + e^{i\vec{k}\cdot\vec{n}_3}\right] = [1 + 1 + 1] = 3$ and using (7.17):

$$E(\vec{k}) = E_P \pm 3\gamma_\pi. \tag{7.20}$$

One can further calculate the bad structure along different directions in the Brillouin zone and the energy dispersion has the shape shown in Fig. 7.6 and is linear close to the K and K' points:

Three is a clear difference between the energy dispersion (Band Energy vs. wavevector) for the carriers near the K and K' points in the Brillouin zone and the $E \sim k^2$ dispersion for free carriers. This linear dispersion is similar to the linear relationship between the wavevector of the massless photon and its energy. That suggests that the carriers around the K and K' points in graphene behave like massless Dirac fermions. Below, we state that the K and K' points are referred to as Dirac points, and further expand on this when relativistic quantum mechanics is introduced below.

We can show that the energy dispersion of the π bands **is linear by starting with** (7.17) $E(\vec{k}) = E_P \pm \gamma_\pi \left|f(\vec{k})\right|$ and $[e^{i\vec{k}\cdot\vec{n}_1} + e^{i\vec{k}\cdot\vec{n}_2} + e^{i\vec{k}\cdot\vec{n}_3}] = f(\vec{k})$. First we expand

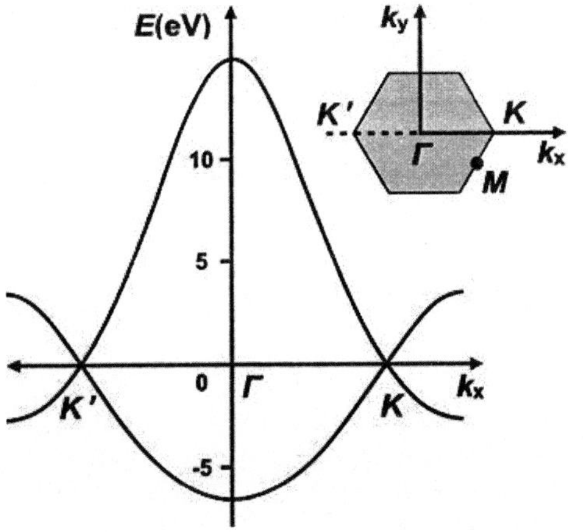

Fig. 7.6 The π *and* π^* electronic band structure of graphene as determined using the NN tight binding approximation. The value of E_p was chosen to be zero, and thus the Fermi level of undoped graphene is 0. Close to the K points, the *change in energy versus wavevector k is linear which is very different from the* k^2 *energy dependence of semiconductors*. This linear energy dispersion means that particles act like Dirac fermions close to the K point. Here, the value of the nearest neighbor overlap integral, $s_0 = 0.129$, was included in the calculation. There are two so called *valleys* in this figure. One at K, and one at K'. Figure adapted from [2]

$f(\vec{k})$ close to the K point at $\vec{\text{K}} = \left(0,\ \frac{4\pi}{3a}\right)$. We will refer wavevectors $\vec{k}'$ close to this K point having wavevector $\vec{\text{K}}$ as $\vec{k}'$ where $\vec{k} = \vec{\text{K}} + \vec{k}'$ As stated in (7.19), and using a series expansion for $e^y \sim 1 + y + \frac{y^2}{2!} + \cdots$ keeping only the linear terms since the wavevector K is very small, thus the k^2 terms are very small. For convenience we assume the Dirac fermion is traveling only along k_y

With

$$k_x = 0,\ f(\vec{k})$$
$$= \left(e^{i0}e^{i\vec{k}'\cdot\vec{n}_1} + e^{i\frac{2\pi}{3}}e^{i\vec{k}'\cdot\vec{n}_2} + e^{-i\frac{2\pi}{3}}e^{i\vec{k}'\cdot\vec{n}_3}\right) \to \Big[1 + (-0.5 + i0.866)\left(1 + ik_y\frac{a}{2}\right) + (-0.5 - i0.866)\left(1 - ik_y\frac{a}{2}\right)\Big]$$
$$= -\frac{a\sqrt{3}}{2}k_y.$$

So near the K point when $k_x = 0$:

$$\sqrt{f(\vec{k})^* \cdot f(\vec{k})} = \pm\frac{a\sqrt{3}}{2}k_y$$

And setting $E_P = 0$, and since the same result would be obtained when $k_y = 0$: $E(\vec{k}) \propto k$

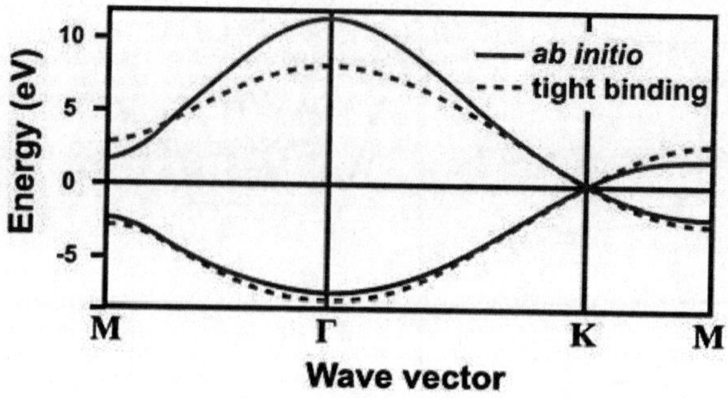

Fig. 7.7 The tight binding π band structure for graphene calculated using nearest neighbors compared to ab-initio results. The tight binding calculation used specific approximations that are described in [5]. Figure adapted and reprinted with permission from [5] © 2002 American Physical Society

$$E(\vec{k}) = \pm\gamma_\pi \frac{a\sqrt{3}}{2}k = \pm\gamma_\pi \frac{3a_0}{2}k$$

7.2.3 *Comparing Nearest Neighbor Graphene Energy Bands to Ab Initio Results*

A number of references have calculated the effect of including next nearest, next-next nearest, and third nearest neighbor carbon atoms on the π band structure. Reference [5] compared a tight binding, nearest neighbor calculation to an ab-initio result for graphene which is shown below in Fig. 7.7 [5]. Close to the K points the calculations provide similar results, but there are significant differences in the band energies close to the Γ point. Thus, the Dirac like $E \propto k$ dispersion holds true and justifies the use of Dirac Hamiltonians for the values of k close to the K and K' points in the Brillouin zone.

7.2.4 *Sub-lattice PseudoSpin (Valley) and the Graphene Band Structure*

As a consequence of the two inequivalent dark (A) atom and light (B) atom sub-lattices in the honeycomb lattice structure, shown in Fig. 7.3, the non-relativistic wavefunctions are linear combinations where the electron density being all on the A atom lattice is combined with the electron density which is all on the B atom lattice. These two degrees of freedom mimic the properties of spin. This property is referred to as *sub-lattice pseudospin* [1] and it has significant consequences for the electrical properties of graphene which we describe after introducing the relativistic Dirac quantum mechanics of graphene below. In the sections that follow the introduction of the Dirac equation, the concept of spinor wavefunctions and valley pseudospin [1] is presented. First we determine the linear combination coefficients for the A and B lattices for *the tight binding two band model* for the π bands. Here we restate (7.9) to show that the wavefunction is a linear combination of electron density on the A atom lattice with that of the B atom lattice [74]:

$$\psi_{\vec{k}}(\vec{r}) = \left\{\frac{\sum_{\vec{R}} e^{i\vec{k}\cdot\vec{R}}}{\sqrt{N}} c_A(\vec{k}) e^{i\vec{k}\cdot d_1} \phi_A(\vec{r} - \vec{R} - \vec{d}_1)\right\}_A + \left\{\frac{\sum_{\vec{R}} e^{i\vec{k}\cdot\vec{R}}}{\sqrt{N}} c_B(\vec{k}) e^{i\vec{k}\cdot d_2} \phi_B(\vec{r} - \vec{R} - \vec{d}_2)\right\}_B \tag{7.21}$$

We can determine $c_A(\vec{k})$ and $c_B(\vec{k})$ using (7.15) and (7.16)

$$\begin{pmatrix} E_P & \gamma_\pi f(\vec{k}) \\ \gamma_\pi f^*(\vec{k}) & E_P \end{pmatrix} \begin{pmatrix} c_A(\vec{k}) \\ c_B(\vec{k}) \end{pmatrix} = E(\vec{k}) \begin{pmatrix} c_A(\vec{k}) \\ c_B(\vec{k}) \end{pmatrix}$$

Here we use $E(\vec{k}) = E_P \pm \gamma_\pi \left|f(\vec{k})\right|$. Now we consider the + solution:

$$E_P c_A(\vec{k}) + \gamma_\pi f(\vec{k}) c_B(\vec{k}) = \left(E_P + \gamma_\pi |f(\vec{k})|\right) c_A(\vec{k})$$

$$\gamma_\pi f^*(\vec{k}) c_A(\vec{k}) + E_P c_B(\vec{k}) = \left(E_P + \gamma_\pi |f(\vec{k})|\right) c_B(\vec{k})$$

The solution is $c_A(\vec{k}) = \frac{1}{\sqrt{2}}$ and $c_B(\vec{k}) = \frac{1}{\sqrt{2}} \frac{f^*(\vec{k})}{|f(\vec{k})|}$ for $E(\vec{k}) = E_P + V_\pi \left|f(\vec{k})\right|$ and $c_A(\vec{k}) = \frac{1}{\sqrt{2}}$ and $c_B(\vec{k}) = \frac{1}{\sqrt{2}} \frac{-f^*(\vec{k})}{|f(\vec{k})|})$ for $E(\vec{k}) = E_P - \gamma_\pi \left|f(\vec{k})\right|$ and as stated above, the eigenvectors for the conduction (+) and valence (−) bands are linear combinations of solutions on the A and B sublattices:

$$\begin{bmatrix} c_A(\vec{k}) \\ c_B(\vec{k}) \end{bmatrix}_+ = \frac{1}{\sqrt{2}} \left[f^*(\vec{k})/|f(\vec{k})| \right] \text{ and } \begin{bmatrix} c_A(\vec{k}) \\ c_B(\vec{k}) \end{bmatrix}_- = \frac{1}{\sqrt{2}} \left[-f^*(\vec{k})/|f(\vec{k})| \right] \tag{7.22}$$

Other hexagonal 2D materials also have sub-lattice pseudospin. Since the basic concept of pseudospin can be understood in terms of non-relativistic quantum mechanics, it is discussed here. Sub-lattice and valley pseudospin will be further discussed in terms of relativistic wavefunctions and Dirac Hamiltonian below.

7.2.5 Dirac Points and Dirac Cones

The two π bands (π^* upper band = conduction and π lower band = valence) have the same number of states. Since each carbon atom contributes one p electron to the π bands and the bands fill with spin up and spin down, the bottom band is completely full. Thus, the Fermi level energy occurs at the place where the two bands touch. The *change in energy versus wavevector k is linear which is very different from the k^2energy dependence of semiconductors. This linear energy dispersion can be described by a Dirac equation and the particles act like Dirac fermionsclose to the K point.* The Dirac point is at the K point in the Brillouin zone, for example $\vec{k}_x = \frac{2\pi}{a\sqrt{3}}$ and $\vec{k}_y = \frac{2\pi}{3a}$. Also, when there is no NNN hopping [1], p 30 the upper and lower bands are symmetric so that:

$$E_C(\vec{k}) = -E_V(\vec{k}) \tag{7.23}$$

The points where the π^* and π touch are called Dirac points and are situated at points in $\vec{k}$ space where the energy dispersion is 0. *We note that the K and K′points are inequivalent and that the states with states with wavevectors close to the K points are sometimes referred to as valleys states (McCann)[2].* The linear energy dispersion of the nearest neighbor model results in a cone shape for the energy versus wavevector for (k_x, k_y) which is referred to as the Dirac cone which is located at the Dirac points in the Brillouin zone as discussed above.

7.2.6 Dirac Cone Shape for Graphene with NNN (Next Nearest Neighbor) Hopping

When next nearest neighbor (nnn) interactions (hopping) is included, graphene's energy dispersion is no longer a perfect cone close to K [1]. This is shown in Fig. 7.8. It is important to note that $E_{C,V}(\vec{k}) \neq E_{C,V}(-\vec{k})$ when NNN interactions are included, and thus time reversal symmetry is broken [1]. The nnn hopping interaction is about a tenth of the nn hoping interaction: $0.1 = t_{nnn}/t_{nn}$. The Dirac points remain at K and K', but the energy at which the valence and conduction band tough is shifted slightly by the nnn hopping interaction [1].

7.2.7 Hexagonal 2D Lattices with Different Atoms at A and B Positions (E.G., Hexagonal Boron Nitride, h-BN)

If the two lattice atoms are different A ≠ B such as B $(1s^2\,2s^2\,2p^1)$ and N $(1s^2\,2s^2\,2p^3)$ the p_z orbitals will form π bands. We note that the total number of p electrons for BN is the same as for graphene. One can extend the nearest neighbor tight binding

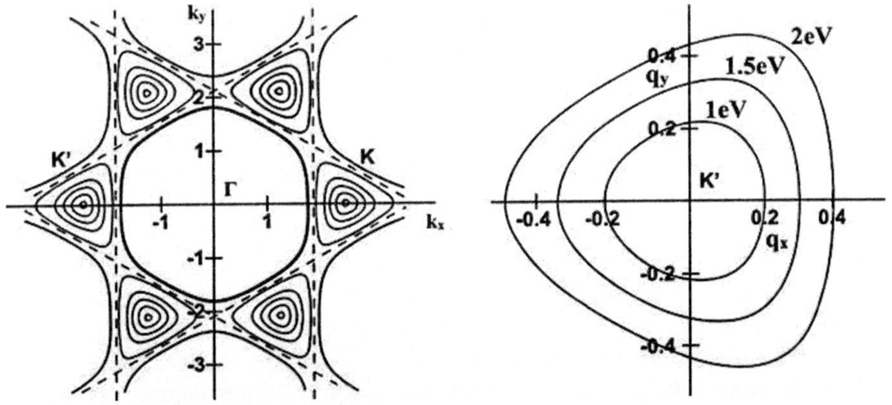

Fig. 7.8 The shape of the Dirac cone after inclusion of NNN. Contours of constant (positive) energy in the wave-vector space. **a** Contours were obtained from the full dispersion relation (see [4], (2.22)). The dashed line corresponds to the energy $t + t_{nnn}$, (t_{nnn} refers to hoping that connects nn sites on the same sublattice) which separates closed orbits around the K *and* K' points (black lines, with energy $\varepsilon < t + t_{nnn}$) from those around the Γ point (gray line, with energy $\varepsilon < t + t_{nnn}$). **b** Comparison of the contours at energy $\varepsilon = 1, 1.5$, and 2 eV around the K′ point. The black lines correspond to the energies calculated from the full dispersion relation ([1], (2.22)) and the gray ones to those calculated to second order within the continuum limit ([1], (2.24)). Figure adapted and reprinted with permission from [1] © 2011 American Physical Society

model of (7.15) [74]:

$$\begin{pmatrix} E_A & \gamma_\pi\left[e^{i\vec{k}\cdot\vec{n}_1} + e^{i\vec{k}\cdot\vec{n}_2} + e^{i\vec{k}\cdot\vec{n}_3}\right] \\ \gamma_\pi\left[e^{-i\vec{k}\cdot\vec{n}_1} + e^{-i\vec{k}\cdot\vec{n}_2} + e^{-i\vec{k}\cdot\vec{n}_3}\right] & E_B \end{pmatrix}\begin{pmatrix} c_A(\vec{k}) \\ c_B(\vec{k}) \end{pmatrix}$$
$$= E(\vec{k})\begin{pmatrix} c_A(\vec{k}) \\ c_B(\vec{k}) \end{pmatrix}$$

Which is easily solved:

$$E(\vec{k}) = \frac{E_A + E_B}{2} \pm \sqrt{\frac{(E_A - E_B)^2}{2} - \gamma_\pi^2 f^*(\vec{k}) f(\vec{k})} \tag{7.24}$$

There is a band gap since $E(p_z)_A E(p_z)_B$. This would be the case for the BN hexagonal lattice where the gap is large ~ 5 to 7 eV. The π and π^* bands for hexagonal BN (hBN) will be similar except for the band gap at the K and K' points in the Brillouin zone. We note that the energy of the first term $\frac{E_A+E_B}{2}$ is half way in between the top of the valence band and bottom of the conduction band. At the end of this chapter, we discuss the tight binding electronic band structure of single layer BN and bilayer BN.

7.3 The Importance of Understanding the Optical and Electrical Properties of Graphene: Proof of Dirac Carriers

The low energy, Dirac fermion, charge carriers in graphene have very unusual optical and electrical transport properties. When we state that the carriers are massless Dirac fermions, we mean that the relativistic Dirac equation provides a more complete physical description of the carriers than the non-relativistic Schrodinger equation. Here we introduce those properties before providing the necessary relativistic quantum mechanical theory required to explain them. In addition to providing an introduction to relativistic quantum mechanics, graphene physics also provides the necessary introduction to topological materials that is necessary for understanding topological insulators, Weyl Metals, and other Dirac materials. As we will discuss in subsequent chapters, separating the optical properties of topological insulators that are topological surface states from interfering signals such as the 2D electron gas that forms due to auto doping and oxidation. The uniaxial optical properties of graphene are first presented in Chap. 1, Sect. 1.5.1 where the low sensitivity of ellipsometry to the out of plane dielectric function is discussed. In this chapter, we show that the in-plane dielectric function over the near IR and visible wavelength range is a result of optical transitions between the π and π^* bands where the wavevectors are confined to the 2D hexagonal structure.

When a particle travels close to the speed of light, it is necessary to use relativistic physics to describe the properties of the particle. The relativistic energy of a particle is $E = \sqrt{m^2c^4 + p^2c^2}$ where the kinetic energy of the particle provides an important correction to the rest energy, $E = mc^2$. The Fermi velocity v_F is the carrier velocity based on having the kinetic energy corresponding to the Fermi level. The group velocity v_g of a carrier in a band is determined from the tight binding energy dispersion using $v_g = \frac{1}{\hbar}\frac{\partial E}{\partial k} = \gamma_\pi \frac{3a_0}{2}$. For the tight binding model of graphene *near the K point*, we can also define $E(\vec{k}) = \hbar v_F = \gamma_\pi \frac{3a_0}{2}k$. Thus $v_g(v_F)$ is not zero at the Fermi level of graphene, and the usefulness of the tight binding model is again seen from the result:$v_F = -\frac{3\gamma_\pi a_0}{2\hbar}$. Clearly, this is due to the Dirac nature of the carriers. Often one reads that the Fermi velocity of an electron in graphene is 1/300 the speed of light. The Fermi velocity of carriers in graphene has been shown to be a function of the carrier concentration [6]. At a carrier concentration of $n < 1 \times 10^{10}$, v_F, v_f has been measured at 3×10^6 m/s which is close to the speed of light $\sim 3 \times 10^8$ m/s. This velocity points to the need to use the Dirac equation for the carriers in the Dirac cone in π and π^*energy bands π^* close to the K and K′ points in the Brillouin zone. The Shubnikov-de-Hass oscillations of suspended graphene were used to determine the Fermi velocity of suspended graphene [6].

Over a large range of wavelengths, a single layer of graphene has a constant ∼ 2.3% absorbance as shown in Fig. 7.9 [7]. The absorbance of each layer of graphene is nearly additive below ~ 1.5 eV as shown in Fig. 7.10 for Bernal stacked several graphene (~10 layer). The first step in characterizing graphene is being able to observe

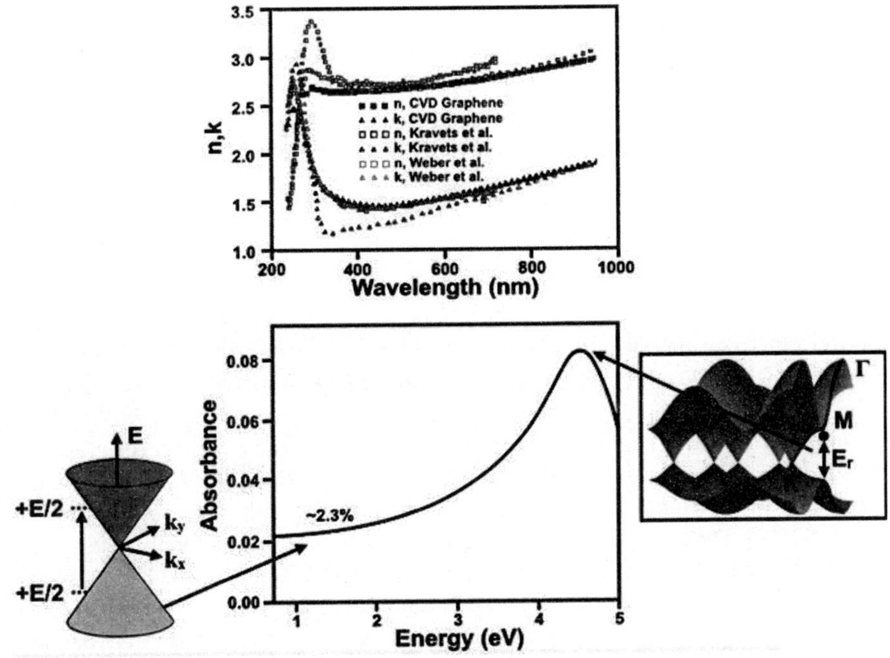

Fig. 7.9 The complex refractive index of CVD and exfoliated graphene and the optical absorbance of graphene as extracted from the complex refractive index of graphene. The constant 2.3% absorbance is observed at near IR wavelengths. The absorbance at near IR wavelengths comes from optical transitions from the π valence band to π** conduction band for small values of k in the Dirac cone. The large absorbance feature at ~ 4.5 eV comes from optical transitions at the M point between π valence band and π* conduction band. You will notice that the experimental value of the transition energy at the M point is 4.5 eV versus the 3 eV obtained from the simple tight binding model. Figure adapted from the dissertation of Florence Nelson and absorbance graph from [7]. Top figure reprinted from [7] with the permission of AIP Publishing

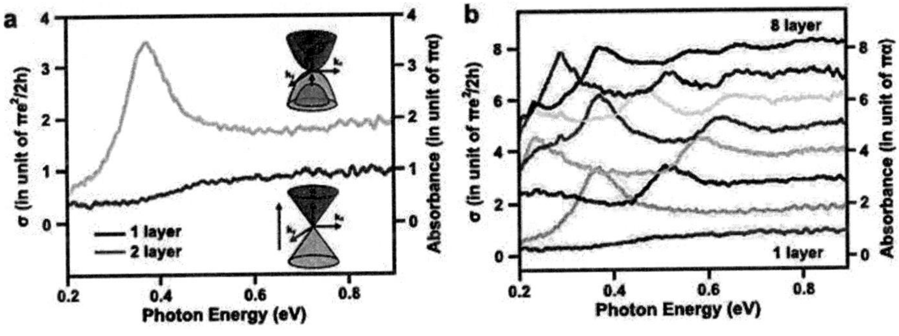

Fig. 7.10 The optical conductivity of multi-layer Bernal stacked graphene is seen to be additive from ~ 0.7 eV to > 1 eV. Note the absorbance for bilayer graphene between 0.3 and 0.4 eV. Figure adapted from [9]

it on a substrate. Bulk graphite (> 20 layer graphene) has a constant refractive index of $\tilde{n} = 2.6 - i\ 1.3$ from 450 nm (~ 2.75 eV) to 700 nm (~ 1.75 eV) [8, 9]. Thus in this wavelength range, graphite, with $k = 1.3$, has an absorption coefficient $\alpha_{abs} = 4\pi k/\lambda$ that clearly varies with wavelength only. This constant refractive index allowed Novoselov, Geim, and coworkers to determine that a 90 nm or 280 nm thickness of SiO_2 on Si facilitated optical observation of graphene using white light [7]. An SiO_2 thickness of 300 nm is typically used since it is readily available. Novoselov, Geim, and co-workers also showed that due to the Dirac nature of the carriers, the absorption of light $A = I/I_0$ is directly related to the fine structure constant: $P = 2.3\% = \pi\alpha_{FS}$, $\alpha_{FS} = {e^2}/{\hbar c} \sim {1}/{137}$. Here, c is the speed of light and e the electrical charge [10]. The transmission of light T is also related to the fine structure constant: $T = \left(1 - \frac{1}{2}\pi\alpha_{FS}\right)^{-2}$, as is the reflectivity $R \ll 1$, $R = \frac{1}{4}\pi^2\alpha_{FS}^2 T$ [10]. The opacity (amount of light that does not get transmitted) is dominated by light absorption [10]. This relationship is derived from the Dirac equation for graphene below. Although the additive layer by layer absorption is found over a technologically useful range of wavelengths, layer to layer interactions do result in absorption features in the far IR [9]. The optical properties of graphene and the observation that the optical spectra of Bernal (AB stacked) and Rhombohedral (ABC stacked) stacked graphene show different interlayer interactions are discussed below.

The optical conductivity is often reported instead of absorbance, refractive index, or dielectric function. Since the imaginary part of the dielectric function is related to the real, n, and imaginary, k, parts of the complex refractive index, $\varepsilon_2 = 2nk$, and the absorption coefficient α is related to the imaginary part of the refractive index, $\alpha = 4\pi k/\lambda$ where λ is the wavelength, it is useful to remember that the optical conductivity is related to all these quantities. The direct relationship between dielectric function and complex optical conductivity $\sigma(\omega)$ (see Chap. 1): $\varepsilon(\omega) = 1 + i\frac{4\pi\sigma(\omega)}{\omega}$. The optical conductivity of the Dirac carriers in graphene is constant: $e^2/4\hbar$. This result is notably different from the classical free electron gas result. The Drude Model for the free electron gas results in a frequency dependent optical conductivity. The zero frequency conductivity is $\sigma_0 = \left(\frac{ne^2\tau}{m}\right) \equiv \omega_p$ the plasmon frequency of the free electron gas. Here, τ is the relaxation time (time between collisions of the electrons. This gives us:

$$\text{Drude } \sigma(\omega) = \frac{\omega_P}{1 - i\omega\tau} = \left(\frac{\omega_P}{1 + \omega^2\tau^2}\right) + i\left(\frac{\omega_P\omega\tau}{1 + \omega^2\tau^2}\right)$$

The optical conductivity of graphene is often reported instead of the complex refractive index or dielectric function. The transmission of normal incidence light through graphene is given by [10]:

$$T \equiv \left(1 + \frac{2\pi\sigma(\omega)}{c}\right)^{-2} = \left(1 + \frac{1}{2}\pi\alpha_{FS}\right)^{-2} \quad \text{and} \quad R \equiv \frac{1}{4}\pi^2\alpha_{FS}^2 T \tag{7.25}$$

Also, $1 - T = 2.3\% = \pi\alpha_{FS}$. A theoretical derivation of the optical properties of single layer graphene is presented below, and for bilayer graphene in [10] and [12].

When one considers the optical properties of graphene from the Far IR to the VUV, the dielectric function (complex refractive index) can be calculated from the band structure over a large wavelength range. As discussed previously, the imaginary part of the dielectric function near a direct interband transition is:

$$\varepsilon_2(\omega) = \frac{e^2}{\pi m^2\omega^2}\int |a_0.p_{CV}|^2\delta(E_C - E_V - \hbar\omega)dk \tag{7.26}$$

where a_0 is the unit polarization vector, p_{CV} is the matrix element of the momentum operator, E_C is energy of the conduction band, and E_V is the energy of the valence band ($E_{CV} = E_C - E_V$). The matrix element of the momentum operator is: $p_{CV} = \frac{ih}{2\pi\Delta}\int u_C^*(r,k)\nabla u_V(r,k)dr$. Here the integral is over the unit cell volume and Δ is the unit cell volume. The real part of the dielectric function can be obtained from (7.26) by performing a Kramers–Kronig transformation. This approximation is valid for low intensity light fields and the dielectric function is often referred to as the linear optical response. However, the transition that occurs at the M point of the Brillouin zone has a strong excitonic character which can be theoretically included in same way as was done for the E1 CP of Silicon in Chap. 4 [13].

Transitions between the valence and conduction σ bands occur at much higher energy than those between the π and $\pi*$ bands. There transitions have been observed by electron energy loss measurements in a scanning transition electron microscope [14]. It is important to note that the character of these transitions changes from single particle like to plasmonic based on the wavevector in the transition. The wavevector changes when the focusing cone of the electron beam changes resulting in more in-plane momentum for the STEM electron beam [15]. In our experiments [14], the results are single particle like due to the use of aberration corrected STEM with a highly focused electron beam [15] (Figs. 7.11 and 7.12).

In Chap. 6, we discussed the observation of the Quantum Hall Effect observed in the low temperature, high magnetic field magnetoresistance and Hall resistance of single layer graphene [11]. The extra phase factor of ½ in the Hall conductivity due to the Berry Phase is considered the signature of a Dirac particle and thus further proof of the linear dispersion of the band energy in the Dirac cone [11]:

$$\text{Hall conductivity} \quad R_{xy}^{-1} = \pm g_s\left(n + \frac{1}{2}\right)\frac{e^2}{h} \tag{7.27}$$

where $g_s = 4$ from the spin up and spin down and the sub-lattice degeneracy (equivalence of K and K'). This is further discussed in Sect. 7.5.

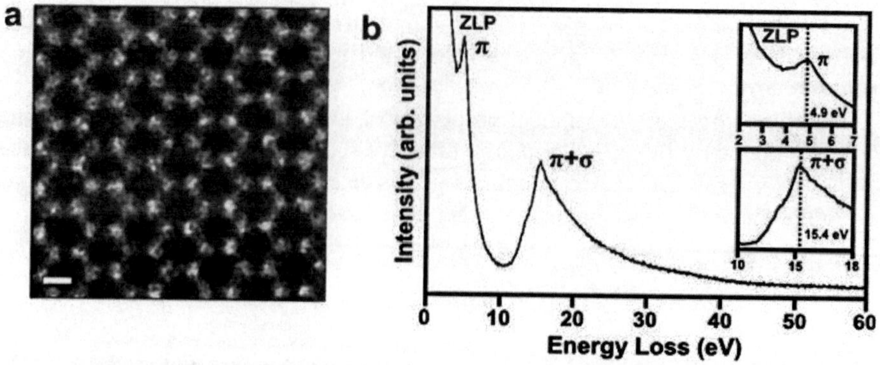

Fig. 7.11 A high resolution, aberration corrected scanning transmission electron microscopy image of graphene and the associated electron energy loss spectra for monolayer graphene showing the π and $\pi + \sigma$ transitions. The spectra can be converteded into the dielectric function of graphene. Figure adapted and reprinted with permission from [14] © 2014 American Chemical Society

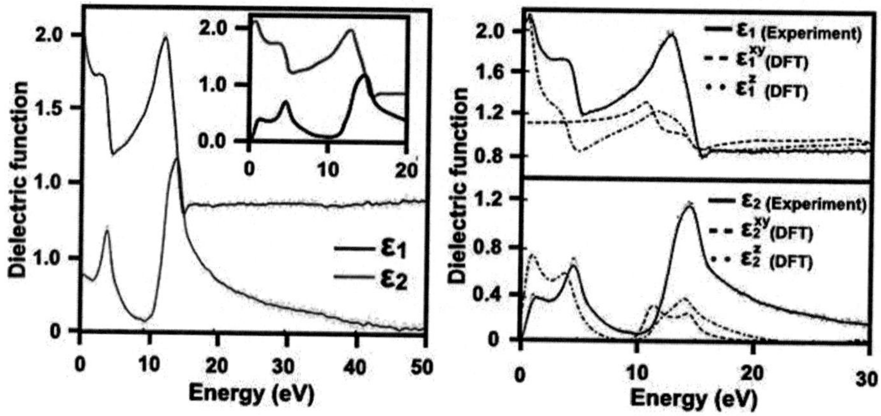

Fig. 7.12 Dielectric function of monolayer graphene. **a** shows ε_1 and ε_2 of monolayer graphene as extracted from the 2D analysis of the inelastic scattering distribution obtained from STEM-EELS. Note that ε_1 stays above zero throughout the whole low-loss region. ε_2 shows two sharp transitions for the π peak, measured experimentally at 1.2 eV and 4.4 eV. The $\pi + \sigma$ peak position was found to be 14.4 eV with a shoulder present at 12.0 eV. **b** Compares experimentally determined ε_1 (upper panel) and ε_2 (lower panel) with DFT calculations. Figure adapted and reprinted with permission from [14] © 2014 American Chemical Society

7.3.1 Electrical Test Structures for Graphene and Graphene Multilayers

The electrical transport properties of single layer graphene, bilayer graphene, and few layer graphene samples are typically tested by encapsulating the graphene structure in hexagonal boron nitride. The dielectric properties and hexagonal crystal structure have proven superior to SiO_2/Si substates. The encapsulated sample is further fabricated into an electrical test structure with bottom and top floating electrodes (gates) above and below the plane of the h-BN—(graphene structure)—h-BN that are used to dope the graphene structure and thus set the carrier concentration. These gates can be set at different voltages so that a displacement field (electrical field) can be established normal to the plane of the graphene structure. Contacts at the side of the graphene structure are used for current flow and measuring Hall voltage. The entire structure can be cooled to a few milli-Kelvin and placed in a magnetic field.

7.4 Introduction to Relativistic Quantum Mechanics for 2D Materials

Relativistic effects need to be included in a complete theoretical description of particles moving at very high velocity. The Dirac equation was developed for that purpose. The starting point for the Dirac equation is a linear relationship between energy and wavevector. Here we assume large area graphene with no band gap. Graphene nanoribbons less than 100 nm in width will have a small, width dependent band gap. Above, we demonstrated this linear relationship near the K and K' points in the Brillouin zone. Here we use the 2D Dirac equation with momentum in only the x and y directions and the Fermi velocity of carriers in graphene,v_F. The Hamiltonian for the 2D Dirac equation is [1, 11]:

$$H^{2D} = \pm v_F \boldsymbol{\sigma} \cdot (-i\hbar\vec{\nabla}) \tag{7.28}$$

where $\boldsymbol{\sigma}$ are the Pauli matrices in 2D:

$$\sigma^x = \begin{pmatrix} 0 & 1 \\ 1 & 0 \end{pmatrix}, \sigma^y = \begin{pmatrix} 0 & -i \\ i & 0 \end{pmatrix} \tag{7.29}$$

The $\pm$ in (7.28) comes from valley pseudospin which is justified below. Near the K points, the $+$ relationship is used, and near the K' points, the – relationship is used. Prior to the Dirac equation, previous attempts to have relativistic Schrodinger equation proved difficult. For example, the Kline-Gordon equation allowed for negative probability densities. Particles with mass (aka, massive particles) are included in the full 2D Dirac equation ([1], (3.16)):

$$H^{2D} = v_F \boldsymbol{\sigma} \cdot (-i\hbar\vec{\nabla}) + M\boldsymbol{\sigma}_z \tag{7.30}$$

where the mass term M will be related to the band gap in Chap. 8 and $\sigma^z = \begin{pmatrix} 1 & 0 \\ 0 & -1 \end{pmatrix}$.

It is important to note that the light like linear relationship between energy and wavevector for carriers close to the K and K′ points in the Brillouin zone motivates the use of $m = 0$ and the substitution of the Fermi velocity for the speed of light c in (7.28).

Remembering that the Block wavefunctions use $e^{i\vec{k}\cdot\vec{r}}$ to build the periodicity into the wavefunctions. Here we use $\vec{\kappa} = \vec{k} - \vec{K}$ or $\vec{\kappa} = \vec{k} - \vec{K}'$. Considering only the plane wave part of the Block wavefunction:

$$\vec{p}\psi(\vec{r}) = \vec{p}\frac{1}{\sqrt{V}}e^{i\vec{k}\cdot\vec{r}} = \frac{\hbar}{i}\vec{\nabla}\psi(\vec{r}) = \frac{\hbar}{i}\vec{\nabla}\frac{1}{\sqrt{V}}e^{i\vec{k}\cdot\vec{r}} = \hbar\vec{k}\psi(\vec{r})$$

Thus in terms of the momentum $\vec{P}$:

$$H^{2D} = \pm\hbar v_F \boldsymbol{\sigma}\cdot\vec{\kappa} = \pm v_F \boldsymbol{\sigma}\cdot\vec{P} \tag{7.31a}$$

$$\begin{aligned} H^{2D} &= \pm\hbar v_F\left(\sigma_x\ \sigma_y\right)\begin{pmatrix} cP_x \\ P_y \end{pmatrix} = \pm\hbar v_F\left(\begin{pmatrix} 0 & P_x \\ P_x & 0 \end{pmatrix} + \begin{pmatrix} 0 & -iP_y \\ iP_y & 0 \end{pmatrix}\right) \\ &= \pm\hbar v_F\begin{pmatrix} 0 & P_x - iP_y \\ P_x + iP_y & 0 \end{pmatrix} \end{aligned} \tag{7.31b}$$

Applying $\vec{P}\psi(\vec{r}) = \hbar\vec{k}\psi(\vec{r})$, the solution for this Hamiltonian is:

$$E = s\hbar v_F|\vec{\kappa}| \tag{7.32}$$

s is either + 1 for the states above the K point (π^* conduction band) or −1 for states below the K point (π valence band). The wavefunctions for the electron and hole states (h) in the K valley for this Hamiltonian with the angle ϑ defined in Fig. 7.13 between the y axis and the vector $\vec{k}$ are:

$$\Psi^K_{e(h)} = \frac{1}{\sqrt{2}}\begin{bmatrix} e^{-i\vartheta/2} \\ +(-)e^{i\vartheta/2} \end{bmatrix} \tag{7.33}$$

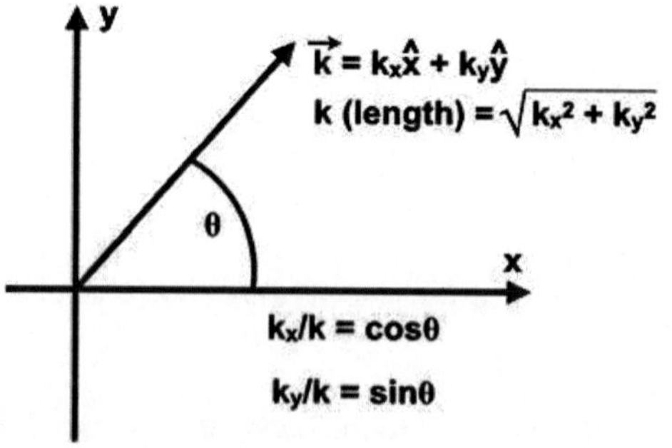

Fig. 7.13 A wavevector can be described by its k_x and k_x components as a length and and an angle between the k_x axis and k_x axis

In (7.33) we leave out the plane wave portion of the wave function, $e^{i\vec{k}\cdot\vec{r}}$ which is not changed by the perturbation caused by light, H' and $\langle e^{-i\vec{k}\cdot\vec{r}}|e^{i\vec{k}\cdot\vec{r}}\rangle = 1$. The optical transition preserves the wavevector $\vec{k}$.

For light interacting with a Dirac fermion, the momentum becomes $\vec{P} \rightarrow \vec{P} - e\vec{A}/c$ where $\vec{A} = \frac{ic\vec{E}(t)}{\omega}$ is the vector potential for light in terms of the electric field of the light $\vec{E} = \frac{1}{2}\left(\vec{E}_0 e^{-i\omega t} + \vec{E}_0^* e^{i\omega t}\right)$ and the ½ is not applicable since we will only use $\vec{E} = \vec{E}_0 e^{-i\omega t}$ for absorption [10]. Thus:

$$H^{2D} = \pm v_F \boldsymbol{\sigma} \cdot \left(\vec{P} - \frac{e\vec{A}}{c}\right) = \pm v_F \boldsymbol{\sigma} \cdot \left(\vec{P} - \frac{e\vec{E}(t)}{i\omega}\right)$$

$$= \pm v_F \boldsymbol{\sigma} \cdot \left(\vec{P} - \frac{e\vec{E}_0}{i\omega} e^{-i\omega t}\right)$$

Which gives a perturbation of $\pm v_F \boldsymbol{\sigma} \cdot \frac{e\vec{E}(t)}{i\omega}$ or $\mp i v_F \boldsymbol{\sigma} \cdot \frac{e\vec{E}(t)}{\omega}$

$$H^{2D} = \pm v_F \boldsymbol{\sigma} \cdot \vec{P} \pm v_F \boldsymbol{\sigma} \cdot \frac{e\vec{E}(t)}{i\omega} = H + H' \tag{7.34}$$

It is important to note that the calculation presented below is for light incident perpendicularly on graphene. The amount of power absorbed by Dirac fermions in graphene can be calculated in the same way that the imaginary part of the dielectric function was calculated from the amount of absorbed power using Block wavefunctions and non-relativistic quantum mechanics. The power that is absorbed is equal to the absorbed energy W_a divided by the incident energy W_i, $P = W_a/W_i$ [10]. The incident energy is $W_i = \frac{c}{4\pi}|E|^2$. The absorbed energy $W_a = \eta\hbar\omega$ where η is the absorption rate (absorption per unit time). Fermi's golden rule allows the calculation of the optical absorption rate. Fermi's golden rule determines η the transition probability/unit time/unit volume from the matrix element of the interaction (perturbation) H' from the initial (valence band) to final state (conduction band). It is usually written:

$$\eta = \frac{2\pi}{\hbar}\left|\langle f|H'|i\rangle\right|^2 \rho(\hbar\omega) \tag{7.35}$$

$\rho(\hbar\omega)$ is the density of states which is a function of the energy of the light.

$$-i v_F \boldsymbol{\sigma} \cdot \frac{e\vec{E}(t)}{\omega} = -i\frac{ev_F}{\omega}\begin{pmatrix}\sigma_x & \sigma_y\end{pmatrix}\begin{pmatrix}E_x(t) \\ E_y(t)\end{pmatrix} = -i\frac{ev_F}{\omega}(\sigma_x E_x(t) + \sigma_y E_y(t))$$

$$= -i\frac{ev_F}{\omega}\begin{pmatrix}0 & E_x - iE_y \\ E_x + iE_y & 0\end{pmatrix} e^{-i\omega t}$$

$$\langle f|H'|i\rangle = i\frac{ev_F}{\omega}\left\langle \psi_h \left| \begin{pmatrix}0 & E_x - iE_y \\ E_x + iE_y & 0\end{pmatrix} e^{-i\omega t} \right| \psi_e \right\rangle$$

$$= i\frac{ev_F}{\omega}\left\langle \frac{1}{\sqrt{2}}\begin{bmatrix} e^{-i\vartheta/2} \\ (-)e^{i\vartheta/2} \end{bmatrix}^* \middle| \begin{pmatrix} 0 & E_x - iE_y \\ E_x + iE_y & 0 \end{pmatrix} e^{-i\omega t} \middle| \frac{1}{\sqrt{2}}\begin{bmatrix} e^{-i\vartheta/2} \\ +e^{i\vartheta/2} \end{bmatrix} \right\rangle$$

$$= i\frac{ev_F}{2\omega}\left\langle \begin{bmatrix} e^{i\vartheta/2} \\ (-)e^{-i\vartheta/2} \end{bmatrix} \middle| \begin{pmatrix} 0 & E_x - iE_y \\ E_x + iE_y & 0 \end{pmatrix} e^{-i\omega t} \middle| \begin{bmatrix} e^{-i\vartheta/2} \\ +e^{i\vartheta/2} \end{bmatrix} \right\rangle$$

$$== i\frac{ev_F}{2\omega}\left\langle [e^{i\vartheta/2} \ (-)e^{-i\vartheta/2}] \middle| \begin{pmatrix} (E_x - iE_y)e^{i\vartheta/2} \\ (E_x + iE_y)e^{-i\vartheta/2} \end{pmatrix} e^{-i\omega t} \right\rangle$$

$$= i\frac{ev_F}{2\omega}\left((E_x - iE_y)e^{i\vartheta} - (E_x + iE_y)e^{-i\vartheta}\right)e^{-i\omega t}$$

$$= \frac{ev_F}{2\omega}(E_y \cos\vartheta - E_x \sin\vartheta)e^{-i\omega t}$$

$$\cos\theta = \frac{e^{i\theta} + e^{-i\theta}}{2}, \ \sin\theta = \frac{e^{i\theta} - e^{-i\theta}}{2i}$$

$$\left|\langle f|H'|i\rangle\right|^2 = \left(\frac{ev_F}{2\omega}\right)^2 \left(E_y^2 \cos^2\vartheta - 2E_xE_y \cos\vartheta \sin\vartheta + E_x^2 \sin^2\vartheta\right)$$

$$\overline{|\langle f|H'|\imath\rangle|^2} = \frac{1}{2\pi}\int_0^{2\pi} \left(\frac{ev_F}{2\omega}\right)^2 \left(E_y^2 \cos^2\vartheta - 2E_xE_y \cos\vartheta \sin\vartheta + E_x^2 \sin^2\vartheta\right)d\vartheta$$

$$= \left(\frac{ev_F}{2\omega}\right)^2 \left(\frac{1}{2}E_y^2 + \frac{1}{2}E_x^2\right) = \frac{1}{2}\left(\frac{ev_F}{2\omega}\right)^2 |\vec{E}|^2 \tag{7.36}$$

Now we need the density of states $\rho(\hbar\omega)$ so that we can complete the determination of η, the absorption rate, from Fermi's golden rule [10]. We need to count states in a 2D annulus of radius k. The area for each point in k in 2D k space is $\left(\frac{2\pi}{L}\right)^2$ (use $L = 1$ since we are not stating the plane wave part of the wavefunction for this discussion) the volume of the annulus is the area of a circle of radius $k + dk$ minus the area of a circle of radius $k = 2\pi k dk$ so the density of states is $2\pi k dk \ /(2\pi)^2$. We note that we should use $2\pi L^2 k dk/(2\pi)^2$ if we include the plane wave part of the wavefunction. We again note that wave part of the wavefunction is not changed by the perturbation caused by the light and furthermore: $\left\langle e^{-i\vec{k}\cdot\vec{r}} | e^{i\vec{k}\cdot\vec{r}} \right\rangle = 1$

$$\rho(\varepsilon)d\varepsilon = \frac{1}{(2\pi)^2}2\pi k dk \quad \varepsilon = \hbar v_F k \quad \rho(\varepsilon)d\varepsilon = \rho(\varepsilon)\hbar v_F dk = \frac{1}{(2\pi)^2}2\pi\frac{\varepsilon}{\hbar v_F}dk$$

$$\text{per spin} \quad \rho(\hbar\omega) = \frac{\omega}{2\pi\hbar v_F^2} \tag{7.37}$$

Thus we need to multiply by 2 to include both spins.

$$\eta = \frac{2\pi}{\hbar}\left|\langle f|H'|i\rangle\right|^2 \rho(\hbar\omega) = \frac{2\pi}{\hbar}\frac{1}{2}\left(\frac{ev_F}{2\omega}\right)^2 |\vec{E}|^2 \frac{\omega}{\pi\hbar v_F^2} = \frac{e^2|\vec{E}|^2}{4\hbar^2\omega} \tag{7.38}$$

The absorption is then [10]:

$$P = \frac{W_a}{W_i} = \frac{\eta\hbar\omega}{\frac{c}{4\pi}|\vec{E}|^2} = \frac{\frac{e^2|\vec{E}|^2}{4\hbar^2\omega}\hbar\omega}{\frac{c}{4\pi}|\vec{E}|^2} = \frac{\pi e^2}{\hbar c} = \pi\alpha \quad \text{i.e. } 2.3\% \tag{7.39}$$

7.4.1 Sub-lattice Pseudospin, Valley Pseudospin, and Chirality for Dirac Fermions in Graphene

This section is intended to be an introduction for our discussion about valley pseudospin and valleys in other 2D materials. Before discussing the Berry phase of carriers in graphene it is useful to discuss pseudospin and chirality. Pseudospin is not a physical property as is the spin of an electron. It is an extra degree of freedom. Above, we described the *sub-lattice pseudospin* for graphene and other hexagonal 2D materials. The wavefunction for both the non-relativistic and relativistic Hamiltonians has a solution for electron density on either the A or B sublattices of graphene as shown in Fig. 7.14 One of these wavefunction could be considered as spin up and the other spin down even though it is not a physical spin. For undoped graphene, the wavefunction can be written as a linear combination of equal parts of the spin up and down solutions. The term *valley pseudospin* comes from the shape of the conduction band at the Dirac points at the K and K' points in the Brillouin zone (see Fig. 7.6) which have the appearance of dual energy valleys. Here, at K $\xi = +$, and at K′ $\xi = -$ for the K, K' pair as shown in Fig. 7.3. The origin of the $\xi = \pm$ valley pseudospin can be seen through a simplified Dirac Hamiltonian argument. (1) Starting with a Dirac interpretation of (7.15) with $E_P = 0$ and the hopping terms γ_π:

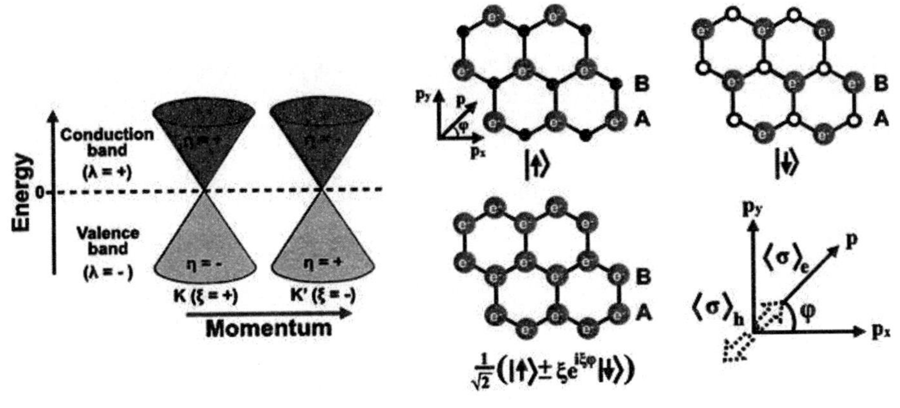

Fig. 7.14 The chiral nature of the lattice pseudospin is shown. On the left, the valley pseudospin and band index are shown in relation to the Dirac cones at K and K'. On the left, the concept of chirality is shown. Adapted from [1] and [2]. Figure on left adapted and reprinted with permission from [1] © 2011 American Physical Society

$$H = \begin{pmatrix} 0 & \gamma_\pi[e^{i\vec{k}\cdot\vec{n}_1} + e^{i\vec{k}\cdot\vec{n}_2} + e^{i\vec{k}\cdot\vec{n}_3}] \\ \gamma_\pi[e^{-i\vec{k}\cdot\vec{n}_1} + e^{-i\vec{k}\cdot\vec{n}_2} + e^{-i\vec{k}\cdot\vec{n}_3}] & 0 \end{pmatrix}$$

Then close to the Dirac points, $\vec{k}$ can written as $\vec{k} = \pm\vec{K} + \vec{q}$ where $\pm\vec{K}$ is the vector to either the K or K′ point (for an inequivalent K, K′ pair, see Fig. 7.3) in the Brillouin zone. Using: $\vec{n}_1 = a_0\hat{x}$ $\quad \vec{n}_2 = \frac{a_0}{2}\left(-\hat{x} + \sqrt{3}\hat{y}\right)$ $\quad \vec{n}_3 = \frac{a_0}{2}\left(-\hat{x} - \sqrt{3}\hat{y}\right)$ and $\vec{K} = \left(0, \frac{4\pi}{3\sqrt{3}a_0}\hat{y}\right)$ and $\vec{K}' = \left(0, \frac{-4\pi}{3\sqrt{3}a_0}\hat{y}\right)$, and keeping first order terms in a series expansion for $e^{i\vec{q}\cdot\vec{n}_i} = \left(1 + i\vec{q}\cdot\vec{n}_i - (\vec{q}\cdot\vec{n}_i)^2 \ldots\right)$, the phase factor $[e^{i\vec{k}\cdot\vec{n}_1} + e^{i\vec{k}\cdot\vec{n}_2} + e^{i\vec{k}\cdot\vec{n}_3}]$ can then be written as (1):

$$\left[e^{i(\pm\vec{K}+\vec{q})\cdot\vec{n}_1} + e^{i(\pm\vec{K}+\vec{q})\cdot\vec{n}_2} + e^{i(\pm\vec{K}+\vec{q})\cdot\vec{n}_3}\right] = (1 + i\vec{q}\cdot\vec{n}_1) + e^{\pm i2\pi/3}(1 + i\vec{q}\cdot\vec{n}_2) + e^{\mp i2\pi/3}(1 + i\vec{q}\cdot\vec{n}_3)$$

Keeping the first order terms, one gets $\mp\frac{3a_0}{2}\left(q_y - iq_x\right)$. This allows one to write a Dirac Hamiltonian with the valley pseudospin term $\xi = \pm$ and the Pauli matrices [1, 2]:

$$H_{\vec{q}}^{\xi} = \xi h v_{Fermi}\left(q_y\sigma^x + \xi q_x\sigma^y\right) \tag{7.40}$$

Defining the Fermi velocity $v_{Fermi} \equiv -\frac{3\gamma_\pi a_0}{2}$. Thus the Hamiltonian has a solution for each valley which is another degree of freedom. Note that when one uses the lattice shown in Fig. 7.5c, $H_{\vec{q}}^{\xi} = \xi\, h v_{Fermi}\left(q_x\,\sigma^x + \xi\,q_y\,\sigma^y\right)$ [1]. The valley degeneracy is said to be indirectly related to the sub-lattice pseudospin [1].

It is useful to describe two terms used to describe Dirac fermions and their wavefunctions: chirality and helicity. Something is chiral when it is not the same as its mirror image, for example, the spin of an electron that is pointing up will be pointing down in its mirror image. A particle with its spin pointed in the direction of its momentum has right handed or positive helicity, and a particle with its spin pointed anti-parallel to its momentum has left handed or negative helicity. For massless particles, chirality and helicity are the same. This is not the case for particles having mass (massive). The chiral nature of the wavefunctions comes from the relationship between the momentum and lattice pseudospin. Following McCann (2), one can write the wavefunction for (7.40) including a planewave for the momentum component with the angle φ between P_x and P_y as:

$$\psi_\pm = \frac{1}{\sqrt{2}}\begin{pmatrix} 1 \\ \pm\xi e^{i\xi\varphi} \end{pmatrix} e^{i\vec{k}\cdot\vec{r}} = \frac{1}{\sqrt{2}}\begin{pmatrix} 1 \\ \pm\xi e^{i\xi\varphi} \end{pmatrix} e^{i\vec{P}\cdot\vec{r}/\hbar} \tag{7.41}$$

Again, ξ is used to distinguish between a K ($\xi = +1$) and K′ ($\xi = -1$) point for an inequivalent pair: $K,\ K'$. The amplitude of the electron density on a specific sub-lattice (A or B) depends on the angle φ which specifies the direction of the momentum. The band index λ for graphene is defines as the valley pseudospin times the chirality η, thus $\lambda = \xi\eta$ [1].

7.4.2 Berry Phase of an Electron in the π Bands of Graphene

The concept of the Berry phase was introduced in Chap. 6. The Berry phase is a key correction to semiclassical quantum mechanical description of current transport that must be included when full quantum picture is required. Figure 6.9 in Chap. 6 shows that when a wavefunction goes around a closed loop over a section of a sphere, the vectors that describe that wavefunction point in a different direction, and thus the phase of the wavefunction is different. In Fig. 6.9, Chap. 6, the vectors start at the top point, and the vectors stay tangential to the sphere. After moving around a closed loop that comprises 1/8 of the sphere, the vectors are pointing 90° away from the original direction. Thus there is a relationship between the path length and the direction of rotation of the vectors. This is a geometric effect. If the path covers ¼ of the sphere, the rotation is 180°. If the path is smaller, the rotation is smaller. The rotation angles of a geometric origin are called a Berry phase. The Berry phase is purely geometrical-topological. If we use a flat surface or a cylinder, there is no Berry phase [16]. The rotation angle is related to the integral of the curvature of the surface bounded by the loop.

The following discussion was motivated by need to show the origin of the Berry phase in the Quantum Hall effect in graphene [11, 17]. The Berry phase can be calculated from two different equations:

$$\theta_n(t) = i \oint c\langle n(R(t))|\nabla_R n(R(t))\rangle dR(t') \tag{7.42a}$$

see (6.37) in Chap. 6 [17, 18] or using A9 in [18]:

$$\theta_n(t) = -i \int_{t_0}^{t} dt' \left\langle \psi(\vartheta(t')) \middle| \frac{d}{dt'} \middle| \psi(\vartheta(t')) \right\rangle \tag{7.42b}$$

$$\theta_n(t) = -i \oint_C dt' \left\langle \psi(\vartheta(t')) \middle| \frac{d}{dt'} \middle| \psi(\vartheta(t')) \right\rangle \tag{7.42c}$$

Equation (7.42a) can be used to calculate the Berry phase using non-relativistic wavefunctions. Equation (7.42b) provides a non-zero Berry phase when $\vartheta(t')$ evolves over a closed path [18]. Equation 7.42 can also be expressed as an integral over a closed loop for $\vartheta(t')$ [17]. First we note that the wavefunction shown in (7.33) is the solution for the Dirac equation for graphene after a gauge transformation [17]. In a close path, the angle of the angle ϑ has a 2π range. Also, we now need the plane wave part of the wavefunction for the following discussion ($\vec{P} = \hbar\vec{k}$):

$$\Psi^K_{e(h)} = \frac{1}{\sqrt{2}}\begin{bmatrix} e^{-i\vartheta/2} \\ +(-)e^{i\vartheta/2} \end{bmatrix} e^{i\vec{P}\cdot\vec{r}/\hbar} \leftarrow \text{gauge transformation} \rightarrow \Psi^K_{e(h)}$$
$$= \frac{1}{\sqrt{2}}\begin{bmatrix} 1 \\ +(-)e^{i\vartheta} \end{bmatrix} e^{i\vec{P}\cdot\vec{r}/\hbar} \tag{7.43}$$

We use $\Psi^K_{e(h)} = \frac{1}{\sqrt{2}}\begin{bmatrix} 1 \\ +(-)e^{i\vartheta} \end{bmatrix} e^{i\vec{P}\cdot\vec{r}/\hbar}$ because they are single valued functions of the angle ϑ. Since $e^{-i\vec{P}\cdot\vec{r}/\hbar} \cdot e^{i\vec{P}\cdot\vec{r}/\hbar} = 1$, $\theta_n(t)$ is:

$$\theta_n(t) = -i\oint_c \frac{1}{2}\left\langle \begin{bmatrix} 1 & e^{-i\vartheta} \end{bmatrix} \middle| \begin{pmatrix} 0 \\ i\left(\frac{\partial\vartheta}{\partial t'}\right)e^{i\vartheta} \end{pmatrix} + \begin{pmatrix} 0 \\ ie^{i\vartheta}\frac{\partial\vec{P}}{\partial t'}\cdot\frac{\vec{r}}{\hbar} \end{pmatrix} + \begin{pmatrix} 0 \\ ie^{i\vartheta}\vec{P}\frac{\partial\vec{r}}{\partial t'}/\hbar \end{pmatrix} \right\rangle dt'$$
$$= -i\oint_c \frac{1}{2}\left(i\frac{\partial\vartheta}{\partial t'} + i\frac{\partial\vec{P}}{\partial t'}\cdot\frac{\vec{r}}{\hbar} + i\vec{P}\frac{\partial\vec{r}}{\partial t'}/\hbar\right)dt'$$

Over a closed loop integrate the angle ϑ over a 2π range, the second two terms are 0, and the first term is:

$$\theta_n(t) = -i\oint_c \frac{1}{2} i\frac{\partial\vartheta}{\partial t'}dt' = \pi \tag{7.44}$$

If you add 2π to ϑ, the phase factor changes by π. It is important to note that the Berry Phase and the Berry Curvature are observable because they are gauge invariant. The spinor rotation of ϑ which will also give a phase factor of π after a 2π rotation (σ_z is the Pauli matrix) [11]:

$$R(\vartheta) = e^{-i\frac{\vartheta}{2}\sigma_z} = \begin{pmatrix} e^{-i\frac{\vartheta}{2}} & 0 \\ 0 & e^{i\frac{\vartheta}{2}} \end{pmatrix} \tag{7.45}$$

An adiabatic rotation of the angle ϑ by 2π will give a phase change of π. Consider $\vartheta + 2\pi$, the 2π rotation gives: $\begin{pmatrix} e^{-i\frac{2}{2}} & 0 \\ 0 & e^{i\frac{2}{2}} \end{pmatrix} = \begin{pmatrix} e^{-i\pi} & 0 \\ 0 & e^{i\pi} \end{pmatrix}$. The Berry phase of an electron in graphene can also be calculated using non-relativistic wavefunctions.

7.5 The Berry Phase Correction for the Quantum Hall Effect and Shubnikov De Hass Oscillations in Graphene

The QHE and SDH oscillations are readily visible in the magneto-transport data for graphene shown in Fig. 6.12 in Chap. 6 [11, 19]. In order to determine the Landau energy levels of graphene, we again start with the Dirac Hamiltonian of (7.31a) and add the effect of a magnetic field as discussed for a non-relativistic Hamiltonian in Chap. 6. In SI units, we have [11]:

$$H^{2D}\psi = \pm\hbar v_F\left(\vec{P} + \frac{e\vec{A}}{c}\right)\cdot\boldsymbol{\sigma}\psi = E\psi \tag{7.46}$$

where $\vec{A}$ is the magnetic field vector potential, and $\vec{B} = \nabla \times \vec{A}$. $\vec{P} = -i\hbar\vec{\nabla}$ and $P_x = -i\hbar\nabla_x$, etc. ψ is a two component vector that we introduced in Sect. 7.4. For CGS units set $c = 1$. The magnetic field is perpendicular to the graphene which is in the $x,\ y$ plane which is also the coordinate system we used above. For convenience, the applied electrical current is directed along the y axis. This will simplify the calculation below. The Landau gauge is again used, so $\vec{A} = -By\hat{x}$. If we use only the + part of (7.46) and remember that the vector $\boldsymbol{\sigma}$ is $(\sigma_x\ \sigma_y)$, $\sigma^x = \begin{pmatrix}0 & 1\\ 1 & 0\end{pmatrix}$, $\sigma^y = \begin{pmatrix}0 & -i\\ i & 0\end{pmatrix}$. Following Kim [11] then:

$$\hbar v_F\left(\vec{P} + \frac{e\vec{A}}{c}\right)\cdot\boldsymbol{\sigma}\psi = \hbar v_F\left((P_x\hat{x},\, P_y\hat{y}) - \frac{eBy\hat{x}}{c}\right)\cdot\left((\sigma_x\ \sigma_y)\cdot\begin{pmatrix}\psi_1\\ \psi_2\end{pmatrix}\right)$$
$$= E\begin{pmatrix}\psi_1\\ \psi_2\end{pmatrix}$$

$$\hbar v_F\left(\left(\left(P_x - \frac{eBy}{c},\, P_y\right)\right)\cdot\begin{pmatrix}\sigma_x\\ \sigma_y\end{pmatrix}\right)\cdot\begin{pmatrix}\psi_1\\ \psi_2\end{pmatrix} = \hbar v_F\left(\left(\left(P_x - \frac{eBy}{c}\right)\sigma_x + P_y\sigma_y\right)\cdot\begin{pmatrix}\psi_1\\ \psi_2\end{pmatrix}\right)$$
$$= \hbar v_F\begin{pmatrix}0 & P_x - \frac{eBy}{c} - iP_y\\ P_x - \frac{eBy}{c} + iP_y & 0\end{pmatrix}\cdot\begin{pmatrix}\psi_1\\ \psi_2\end{pmatrix}$$
$$= \hbar v_F\begin{pmatrix}\left(P_x - \frac{eBy}{c} - iP_y\right)\psi_2\\ \left(P_x - \frac{eBy}{c} + iP_y\right)\psi_1\end{pmatrix}$$
$$= E\begin{pmatrix}\psi_1\\ \psi_2\end{pmatrix}$$

The two resulting equations are:

$$\hbar v_F(P_x - \frac{eBy}{c} - iP_y)\psi_2 = E\psi_1 \tag{7.47}$$

$$\hbar v_F(P_x - \frac{eBy}{c} + iP_y)\psi_1 = E\psi_2$$

Insert $\frac{\hbar v_F(P_x - \frac{eBy}{c} - iP_y)\psi_2}{E} = \psi_1$ into the second equation. $\hbar^2 v_F{}^2(P_x - \frac{eBy}{c} + iP_y)(P_x - \frac{eBy}{c} - iP_y)\psi_2 = E^2\psi_2$

Which results in $\hbar^2 v_F^2\left(P_x^2 - P_x\frac{eBy}{c} - iP_xP_y - \frac{eBy}{c}P_x + \left(\frac{eBy}{c}\right)^2 + i\frac{eBy}{c}P_y + iP_yP_x - iP_y\frac{eBy}{c} + P_y^2\right)\psi_2 = \hbar^2 v_F{}^2(P_x^2 + P_y^2 - 2\frac{eBy}{c}P_x + \left(\frac{eBy}{c}\right)^2 + i\frac{eBy}{c}P_y - i(-i\hbar\nabla_y)\frac{eBy}{c}\Big)\psi_2 = \hbar^2 v_F^2\left(P^2 - 2\frac{eBy}{c}P_x + \left(\frac{eBy}{c}\right)^2 + i\frac{eBy}{c}P_y - \frac{\hbar eB}{c} - i\frac{eBy}{c}P_y\right)\psi_2$
$= \hbar^2 v_F{}^2\left(P^2 - 2\frac{eBy}{c}P_x + \left(\frac{eBy}{c}\right)^2 - \frac{\hbar eB}{c}\right)\psi_2.$

Where $P_x^2 + P_y^2 = P^2$, and $P_y = -i\hbar\nabla_y$ and using the properties of operators:

$$\left(-i(-i\hbar\nabla_y)\frac{eBy}{c}\right)\psi_2 = \left(-\frac{\hbar eB}{c}\psi_2 - i\frac{eBy}{c}(-i\hbar\nabla_y)\psi_2\right) = \left(-\frac{\hbar eB}{c} - i\frac{eBy}{c}P_y\right)\psi_2.$$

Thus

$$\hbar^2 v_F{}^2\left(P^2 - 2\frac{eBy}{c}P_x + \left(\frac{eBy}{c}\right)^2 - \frac{\hbar eB}{c}\right)\psi_2 = E^2\psi_2 \tag{7.48}$$

This equation for E^2 is compared to the non-relativistic equation for the energy E for an electron in a magnetic field. In CH 6: we considered a 2DEG with the magnetic field along the z direction and the current running along the x direction. We can use the analysis of CH 6 when the current runs along the y direction. Again, when the unit vector along z is $\hat{z}$: $\vec{B} = B\hat{z}$ and $\vec{A} = (-By, 0, 0)$. The quantized energy levels for the 2DEG are again $E = \left(n + \frac{1}{2}\right)\hbar\omega_c + g^*\mu_b B_z$ when the energy of the electron spin in a magnetic field is added. Again, $\omega_c = \frac{eB}{m^*}$ The non-relativistic Schrodinger equation for an electron in a 2DEG in a perpendicular magnetic field with the applied electrical current along the y axis is:

$$\frac{P_y{}^2}{2m}\psi + \frac{1}{2m^*}\left(P_x - \frac{eBy}{c}\right)^2\psi = E\psi \tag{7.49}$$

The following restatement of (7.49) is more easily compared to (7.48):

$$\frac{P_y^2}{2m}\psi + \frac{1}{2m}\left(P_x^2 - 2P_x\frac{eBy}{c} + \left(\frac{eBy}{c}\right)^2\right)\psi = E\psi$$

The 2DEG equation (7.49) needs to be restated. Use $P_x{}^2 + P_y{}^2 = P^2$ to get:

$$\frac{1}{2m^*}\left(\vec{P}^2 - 2P_x\frac{eBy}{c} + \left(\frac{eBy}{c}\right)^2\right)\psi = E\psi \tag{7.50a}$$

Equation (7.50a) can be compared to (7.48):

$$\hbar^2 v_F^2\left(P^2 - 2\frac{eBy}{c}P_x + \left(\frac{eBy}{c}\right)^2 - \frac{\hbar eB}{c}\right)\psi_2 = E^2\psi_2 \quad \text{or}$$

$$\hbar^2 v_F^2\left(P^2 - 2\frac{eBy}{c}P_x + \left(\frac{eBy}{c}\right)^2\right)\psi_2 = \left(E^2 + \hbar^2 v_F^2\frac{\hbar eB}{c}\right)\psi_2$$

As discussed in Chap. 6, (6.17) to (6.24), 7.50a results in quantum levels $E = \left(n + \frac{1}{2}\right)\hbar\omega_c$. For a classical massive particle, $\omega_c = \frac{eB}{m^*c}$ see Chap. 6, 6.22) to (6.23. The $\frac{1}{2}\hbar\omega_c$ part of the energy will be missing from the solution to (7.48) due to the extra $\hbar^2 v_F^2\left(-\frac{\hbar eB}{c}\right) \to \frac{1}{2}\hbar\omega_c$ term in (7.48) which is due to the Berry phase of the Dirac carriers. So a solution similar to $E = n\hbar\omega_c$ should be a solution for (7.48) for $\boldsymbol{E}^2$ [11]. The Fermi velocity v_F is used instead of the speed of light in the solution to the Dirac equation for massless particles in a magnetic field, and the Landau level index is n:

Landau Energy Levels in graphene

$$E_n = \text{sgn}(n)\sqrt{2e\hbar v_F^2 |n| B} \tag{7.51}$$

From 7.37, the density of states when there is no magnetic field is $\rho(\hbar\omega) = \frac{\omega}{\pi\hbar v_F^2} = \frac{E}{\pi\hbar^2 v_F^2}$ which is a linear function of energy. The allowed energy states in a magnetic field are given by (7.51). As discussed in Chap. 6 for a 2DEG, we found that the degeneracy factor for the Landau levels is $\beta = \frac{L_x}{\Delta x_0} = \frac{L_x L_y eB}{h}$ and the degeneracy per unit area n_L is $n_L = \frac{\beta}{L_x L_y} = \frac{eB}{h}$.

It follows that the Hall conductance is quantized. Because the Dirac states are topologically protected, the scattering time $\tau \to \infty$. Using Chap. 6, (6.15), the Hall conductance $\sigma_{xy} = \frac{ne}{B_z}$ where the number of electrons is n and again considering a magnetic field B_z that is oriented perpendicular to the plane of the graphene layer. When the value of magnetic field strength fills an integer N Landau levels, then using the degeneracy per unit area n_L, $n = N\frac{\mathbf{e}B_z}{\mathbf{h}}$ and $\sigma_{xy} = N\frac{e^2}{h}$ where h is Plank's constant, and we again have the units of quantized conductance, $\frac{e^2}{h}$. Another way to prove the Quantum Hall Effect for graphene is to calculate the Chern number from the Berry curvature to prove it is not zero. The Hall conductance is the Chern number times $\frac{e^2}{h}$. The Chern number and topological classification of graphene is further explored in Appendix C after the key symmetry properties of topological materials are introduced in Sect. 9.1. At that point, we will discuss the need for breaking time reversal symmetry for there to be a non-zero Chern number.

Magneto-resistance measurements of graphene provide considerable information about the impact of the Berry phase on carrier transport in the π bands. Shubnikov de Hass oscillations are observed in the longitudinal magneto-resistivity along the direction of current flow, R_{xx}, which is altered from the IQHE result for a 2DEG due to the Berry phase correction β for Dirac fermions. This provides a signature for Dirac fermions.

$$\Delta R_{xx} = R(B,T)\cos\left(2\pi\left(\frac{B_F}{B}\right) + \frac{1}{2} + \beta\right) \tag{7.52a}$$

$R(B,T)$ is the magnetic field and temperature dependent amplitude of the Shubnikov de Haas (SdH) oscillations, and B_F is the frequency of the SdH oscillations. $\beta = -\,½$ is the signature of a Dirac particle. The general form of the longitudinal magneto-resistance for 2D systems is discussed in [18] and the references in that paper and stated here in (7.52b). Below we use $\gamma = \frac{1}{2} + \beta$ which we will identify as the phase of the SdH oscillations:

$$\Delta R_{xx} = R(B,T)\cos\left(2\pi\left(\frac{B_F}{B}\right) + \gamma\right) \tag{7.52b}$$

$\gamma = 0$ for graphene. The Berry phase can be understood in terms of the phase of the electron in the cyclotron orbit that results from a magnetic field. We want to contrast the results for graphene with the non-relativistic behavior of a 2DEG. Using Kim's approach [11], we provide a descriptive derivation of 7.49. The quantization of the electron motion in the magnetic field restricts available states and gives rise to Shubnikov de Haas (SdH) oscillations [11]. The semiclassical equation for the quantized dynamics of an electron in a magnetic field is written in terms of the rate of change in the momentum = Lorentz force:

$$\dot{\vec{P}} = \hbar\dot{\vec{k}} = -e(\dot{\vec{R}} \times \vec{B}) = -e(\vec{v} \times \vec{B}) \tag{7.53}$$

Eqaution (7.53) applies to both graphene and 2DEGs. For a 2DEG, the electron's cyclotron orbits are semiclassically pictured as free orbits in 2D. Because $E = \frac{\vec{P}^2}{2m}$ and using the quantum picture for the free electrons in a 2DEG, $\vec{P} = \hbar\vec{k}$. Then $\nabla_k E = \frac{2\hbar^2\vec{k}}{2m} = \frac{\hbar\vec{P}}{m} = \frac{\hbar m\vec{v}}{m} = \hbar\vec{v}$, so $\vec{v} = \frac{\vec{\nabla}_k E}{\hbar}$. No work is done on the electron because the Lorentz force is normal to the momentum, thus E and consequently v are constant. Before applying the 2DEG case, we return to a more general look at electrons that are confined in 2D. One can integrate (7.53) to get the momentum as a function of the real space position $R(t)$:

$$\hbar\left(\vec{k}(t) - \vec{k}(0)\right) = -eB\left(\vec{R}(t) - \vec{R}(0)\right) \times \hat{B} \tag{7.54}$$

Here, $\hat{B}$ is the unit vector along the direction of the magnetic field. The cross-product of $\left(\vec{R}(t) - \vec{R}(0)\right) \times \hat{B}$ rotates $\left(\vec{R}(t) - \vec{R}(0)\right)$ by 90°. So the real space orbit is the k space orbit rotated by 90° and scaled by $\frac{\hbar}{eB}$ [11]. Again using semiclassical quantum mechanics, we use the Bohr-Sommerfeld quantization rule for momentum [11] where the integral over a closed path often refers to the turning points in a classical motion (e.g., the turning points in the vibrational motion of a harmonic oscillator used to model a diatomic molecule's vibrational mode, or the 2π rotational path of a particle on a ring):

$$\oint P \cdot dq = (n + \gamma)2\pi\hbar \tag{7.55}$$

In a magnetic field, the total momentum is $\hbar\vec{k} - \frac{e\vec{A}}{c}$ then:

$$\oint \left(\hbar\vec{k} - \frac{e\vec{A}}{c}\right) \cdot d\vec{R} = (n + \gamma)2\pi\hbar \tag{7.56}$$

This can be separated into two parts and we apply Stokes Theorem to each part:

Stokes Theorem for a vector field F: $\oint F \cdot dR = \oiint_s \nabla \times \vec{F} \cdot d\vec{\sigma}$, which turns a line integral into a surface integral with $d\vec{\sigma}$ being a vector element of surface area directed along the unit normal (to the surface) vector. Using $\vec{B} = \vec{\nabla} \times \vec{A}$,

$$\oint \left(-\frac{e\vec{A}}{c}\right) \cdot d\vec{R} = -\frac{e}{c} \oiint_S (\vec{\nabla} \times \vec{A}) \cdot d\vec{\sigma} = -\frac{e}{c} \oiint_S (\vec{B}) \cdot d\vec{S} \tag{7.57}$$

Using (7.51) and defining $\vec{k}(t) - \vec{k}(0) = \vec{k}'(t)$ and $\vec{R}(t) - \vec{R}(0) = \vec{R}'(t)$

$$\oint \hbar\vec{k} \cdot d\vec{R} = \oint -eB\vec{R}(t) \times \hat{B} \cdot d\vec{R} \tag{7.58}$$

using dot product and cross-product identities for vectors $\vec{a}$, $\vec{b}$, and $\vec{c}$:

$[\vec{a} \cdot \vec{b} = \vec{b} \cdot \vec{a}]$; $[\vec{a} \times \vec{b} = -\vec{b} \times \vec{a}]$; [using $\vec{a} \cdot (\vec{b} \times \vec{c}) = \vec{c} \cdot (\vec{a} \times \vec{b})$ then $\vec{a} \cdot (\vec{b} \times \vec{c}) = (\vec{a} \times \vec{b}) \cdot \vec{c} = -(\vec{b} \times \vec{a}) \cdot \vec{c}$] use $\vec{a} = \vec{B}, \vec{b} = \vec{R}, \vec{c} = d\vec{R}$ thus $-(\vec{b} \times \vec{a}) \cdot \vec{c}$ is $-e\left(\vec{R}(t) \times \vec{B}\right) \cdot d\vec{R}$ and $\vec{a} \cdot (\vec{b} \times \vec{c})$ is $e\vec{B} \cdot \vec{R}(t) \times d\vec{R}$

$$\oint -e\left(\vec{R}(t) \times B\hat{B}\right) \cdot d\vec{R} = \oint -e\left(\vec{R}(t) \times \vec{B}\right) \cdot d\vec{R} = \oint e\vec{B} \cdot \vec{R}(t) \times d\vec{R} \tag{7.59}$$

Using (7.58) and 7.59

$$\oint \left(\hbar\vec{k} - \frac{e\vec{A}}{c}\right) \cdot d\vec{R} = \oint e\vec{B} \cdot \vec{R}(t) \times d\vec{R} - \frac{e}{c} \oiint_S \vec{B} \cdot d\vec{S} = (n + \gamma)2\pi\hbar$$

$$\oint \vec{B} \cdot \vec{R}(t) \times d\vec{R} - \frac{1}{c} \oiint_S \vec{B} \cdot d\vec{S} = (n + \gamma)\frac{2\pi\hbar}{e} = (n + \gamma)\Phi_0 \tag{7.60}$$

where Φ_0 is the magnetic flux quanta $\frac{2\pi\hbar}{e}$. The $\frac{1}{c} \oiint_S \vec{B} \cdot d\vec{S}$ is the magnetic flux through a surface $\Phi = \oint \vec{B} \cdot dS$ where the surface is the electron orbit. $\oint \vec{B} \cdot \vec{R}(t) \times d\vec{R}$ is twice Φ, thus the magnetic flux through an electron orbit is quantized.

$$\Phi = (n + \gamma)\Phi_0 \tag{7.61}$$

It is important to remember that $\Phi = BA$ where A is the area that the magnetic flux passes through. So.

$A = (n+\gamma)\frac{\Phi_0}{B} = (n+\gamma)\frac{2\pi\hbar}{eB}$ in real space. From the discussion of (7.51) stating the real space orbit is the k space orbit rotated by 90° and scaled by $\frac{\hbar}{eB}$, and remembering that the area of a circle is πr^2

$$A_{k\,space} = (n+\gamma)\frac{2\pi\hbar}{eB}\left(\frac{eB}{\hbar}\right)^2 = (n+\gamma)\frac{2\pi eB}{\hbar} \tag{7.62}$$

We can determine γ for a 2DEG and for electrons in graphene. We can do that by determining the area of the circular electron orbit in k space for both systems and then compare to (7.62). For electron in a semiconductor the energy bands are parabolic and $E = \hbar^2k^2/2m^*$ and the energy of the Landau levels are quantized in terms of the cyclotron frequency $\omega_c = eB/m^*$, $E = (n+\frac{1}{2})\hbar\omega_c$. Then for a 2DEG, $k = \sqrt{(n+\frac{1}{2})2\omega_c m^*/\hbar} = \sqrt{2(n+\frac{1}{2})eB/\hbar}$.

For a 2DEG

$$A_{k\,space} = \pi k^2 = 2\pi\left(n+\frac{1}{2}\right)eB/\hbar \tag{7.63}$$

By comparison of (7.62) and (7.63), $\gamma = \frac{1}{2}$ for a 2DEG. For an electron in the Dirac cone of graphene, $E_n = \text{sgn}(n)\sqrt{2e\hbar v_F^2|n|B}$ and $E = \hbar v_F k$ so$k = \sqrt{2e\hbar v_F^2|n|B}/\hbar v_F$, and the area of the circular orbit in k space is.

For graphene,

$$A_{k\,space} = \pi k^2 = |n|2\pi eB/\hbar \tag{7.64}$$

And for graphene, $\gamma = 0$. This is due to Berry phase of π. The magnetic flux through an electron orbit in a 2DEG is quantized $\Phi = \left(n+\frac{1}{2}\right)\Phi_0$ and for graphene in a magnetic field it is $\Phi = n\Phi_0$. As Kim points out [11], the difference of ½ between the Dirac fermions in graphene and the electrons in a 2DEG comes from the π Berry phase that results for a 2π rotation of the pseudospin associated with the electrons in graphene [11, 20]. This is the origin of the difference in the phase of the SDH oscillations. Kim describes a general expression for γ applicable to any band structure in terms of the Berry phase ϕ_B:

$$\gamma - \frac{1}{2} = -\frac{\phi_B}{2\pi} \tag{7.65}$$

In bilayer graphene, the energy dispersion is proportional to k^2 and $\gamma = \frac{1}{2}$. The Shubnikov de Hass oscillations in the low temperature, longitudinal magnetoresistivity shown in Fig. 6.12a, Chap. 6 for graphene prove that $\gamma = 0$ and that the Berry phase is π. At the higher magnetic fields that produce the quantum Hall effect, the longitudinal magneto-resistivity goes to zero and the transverse (Hall) magneto-resistivity R_{xy} (conductivity ${R_{xy}}^{-1}$) for graphene shows plateaus which are integer steps in conductivity ($e^2/\hbar$). The extra ½ is due to the Berry phase.

$$ {R_{xy}}^{-1} = \pm(e^2/\hbar)g_s\left(n + \frac{1}{2}\right) \tag{7.66} $$

The term $g_s\left(n + \frac{1}{2}\right)$ is the filling fraction υ for each graphene LL which was discussed in Chap. 6. The filling factor υ is defined as the ratio of the electron density n_s to the degeneracy of each Landau level n_L. From $n_s = id$, where n_s is the electron density per cm^2 of sample, $d = eB/h$ is the degeneracy of the Landau levels (where h is Plank's constant), and i is an integer, $i = 1, 2, 3, \ldots$. For a sample of given density, n_s, there is a discrete set of magnetic fields, B_i, $i = 1,2,3,\ldots$, for which this condition is satisfied, $B = (nh/e)/i$.

7.6 Electronic Structure of Bilayer Graphene

In this section, we explore how the stacking of two graphene layers on top of each other alters the electronic structure. The alignment between layers has a significant impact on interlayer interactions. In that light, the natural, Bernal stacking order often observed for bilayer samples exfoliated from bulk pyrolytic graphite is first discussed. In Sect. 7.9, we will explore the effect of twisting the alignment of stacking on the electronic structure, transport, and optical properties.

First, we revisit the assumption that the overlap integral between nearest neighbor A and B atoms $\left\langle \phi_A(\vec{r} - \vec{d}_1) \middle| \phi_B(\vec{r} - \vec{R}_{NN} - \vec{d}_2) \right\rangle$ is zero for mono-layer graphene that we used in Sect. 7.2 [1, 2]. We again use only nearest neighbor contributions, e.g., $\left\langle \phi_A(\vec{r} - \vec{d}_1) \middle| \phi_A(\vec{r} - \vec{R} - \vec{d}_1) \right\rangle \approx 0$ unless $\vec{R} = 0$ and $\left\langle \phi_A(\vec{r} - \vec{d}_1) \middle| U(\vec{R}) \middle| \phi_A(\vec{r} - \vec{R} - \vec{d}_1) \right\rangle \approx 0$. Restating (7.10):

$$ \begin{aligned} &\left\langle \phi_A(\vec{r} - \vec{d}_1) \middle| H_{\text{atomic}} + U(\vec{R}) \middle| \psi_{\vec{k}}(\vec{r}) \right\rangle = \left\langle \phi_A(\vec{r} - \vec{d}_1) \middle| E \middle| \psi_{\vec{k}}(\vec{r}) \right\rangle \\ &\frac{1}{\sqrt{N}} e^{i\vec{k}\cdot\vec{d}_1} c_A(\vec{k}) E_P + \frac{\sum_{\vec{R}_{NN}} e^{i\vec{k}\cdot\vec{R}}}{\sqrt{N}} c_B(\vec{k}) e^{i\vec{k}\cdot\vec{d}_2} E_P \left\langle \phi_A(\vec{r} - \vec{d}_1) \middle| \phi_B(\vec{r} - \vec{R} - \vec{d}_2) \right\rangle \\ &+ \frac{\sum_{\vec{R}_{NN}} e^{i\vec{k}\cdot\vec{R}}}{\sqrt{N}} c_B(\vec{k}) e^{i\vec{k}\cdot\vec{d}_2} \left\langle \phi_A(\vec{r} - \vec{d}_1) \middle| U(\vec{R}) \middle| \phi_B(\vec{r} - \vec{R} - \vec{d}_2) \right\rangle \\ &= E(\vec{k}) e^{i\vec{k}\cdot\vec{d}_1} c_A(\vec{k}) + \frac{\sum_{\vec{R}_{NN}} e^{i\vec{k}\cdot\vec{R}}}{\sqrt{N}} c_B(\vec{k}) e^{i\vec{k}\cdot\vec{d}_2} E(\vec{k}) \left\langle \phi_A(\vec{r} - \vec{d}_1) \middle| \phi_B(\vec{r} - \vec{R} - \vec{d}_2) \right\rangle \end{aligned} \tag{7.67} $$

Assume for non NN that $\left\langle\phi_A(\vec{r}-\vec{d}_1)\middle|\phi_B(\vec{r}-\vec{R}-\vec{d}_2)\right\rangle \approx 0$, multiply both sides by $e^{-i\vec{k}\cdot\vec{d}_1}$ and use only NN interactions so that $\left\langle\phi_A(\vec{r}-\vec{d}_1)\middle|U(\vec{R})\middle|\phi_B(\vec{r}-\vec{R}-\vec{d}_2)\right\rangle = \gamma_\pi$ and for NN, $\left\langle\phi_A(\vec{r}-\vec{d}_1)\middle|\phi_1(\vec{r}-\vec{R}-\vec{d}_2)\right\rangle = S_{12}$ and again we use $[e^{i\vec{k}\cdot\vec{d}_1}+e^{i\vec{k}\cdot\vec{n}_2}+e^{i\vec{k}\cdot\vec{n}_3}] = f(\vec{k})$:

$$c_A(\vec{k})E_P + c_B(\vec{k})E_P f(\vec{k})S_{12} + c_B(\vec{k})f(\vec{k})\gamma_\pi = E(\vec{k})c_A(\vec{k}) + E(\vec{k})\,c_B(\vec{k})\,f(\vec{k})\,S_{12} \quad (7.68)$$

$$c_A(\vec{k})\left(E_P - E(\vec{k})\right) + c_B(\vec{k})\left(E_P - E(\vec{k})\right)f(\vec{k})S_{12} + c_B(\vec{k})f(\vec{k})\gamma_\pi = 0 \quad (7.69)$$

And for

$$\left\langle\phi_B(\vec{r}-\vec{d}_1)\middle|H_{\text{atomic}} + U(\vec{R})\middle|\psi_{\vec{k}}(\vec{r})\right\rangle = \left\langle\phi_B(\vec{r}-\vec{d}_1)\middle|E\middle|\psi_{\vec{k}}(\vec{r})\right\rangle:$$

$$c_A(\vec{k})\left(E_P - E(\vec{k})\right)f(-\vec{k})S_{12} + c_A(\vec{k})f(-\vec{k})\gamma_\pi + c_B(\vec{k})\left(E_P - E(\vec{k})\right) = 0 \quad (7.70)$$

$$\begin{vmatrix} E_p - E(\vec{k}) & \left[\left(E_P - E(\vec{k})\right)S_{12} + \gamma_\pi\right]f(\vec{k}) \\ \left[\left(E_P - E(\vec{k})\right)S_{12} + \gamma_\pi\right]f(-\vec{k}) & E_p - E(\vec{k}) \end{vmatrix} = 0$$

$$\left[E_p - E(\vec{k})\right]^2 - \left[\left(E_p - E(\vec{k})\right)S_{12} + \gamma_\pi\right]^2 |f(\vec{k})|^2 = 0 \quad (7.71)$$

Similar to single layer graphene, we set $E_p = 0$ and get $E(\vec{k})\left(1 - S_{12}\left|f(\vec{k})\right|\right) = +\gamma_\pi\left|f(\vec{k})\right|$ and $E(\vec{k})\left(1 + S_{12}\left|f(\vec{k})\right|\right) = -\gamma_\pi\left|f(\vec{k})\right|$

$$E(\vec{k})_\pm = \frac{\pm\gamma_\pi\left|f(\vec{k})\right|}{1 \mp S_{12}\left|f(\vec{k})\right|} \quad (7.72)$$

For bilayer graphene, the results for monolayer graphene were generalized so that there are four p_z orbitals per unit cell [2] and Bernal stacking was used as shown in Fig. 7.15. The Bernal stacking results and the nearest neighbor tight binding approach results in interlayer coupling in the unit cell between only atom A2 and B1 This results in a 4 × 4 determinant for the energy.

The wavefunction for the four basis atoms is:

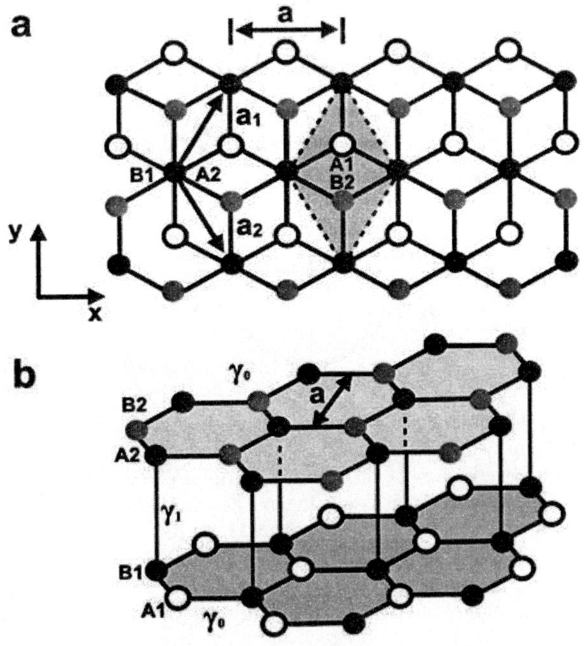

Fig. 7.15 Bernal stacked, bilayer graphene showing a plan (top down) view and a side view. The rhombus primitive cell is shown with the A1 atom on the bottom layer and B2 atom in the top layer. The unit cell has 4 atoms with 2 on each layer, and it should be compared to the unit cells shown above for graphene. An A2 atom will be directly above a B1 atom in the unit cell. The nearest neighbor coupling, γ_0, is shown for each layer. Coupling between layers is shown by γ_1. Bernal stacking results in some of the carbon atoms being directly above (below) an atom in the other layer. These atoms are labeled A1 for the bottom layer and B2 for the top layer, see (**a**). Figure is adapted from [2]

$$\psi_{\vec{k}}(\vec{r}) = \frac{1}{\sqrt{N}}\left(\sum_{\vec{R},\ i=1 \text{ to } 4} e^{i\vec{k}\cdot(\vec{R}+\vec{d}_l)} c_i(\vec{k})\, \phi_i(\vec{r} - \vec{R} - \vec{d}_l)\right)$$

The interlayer coupling term between atoms A1 for the bottom layer and B2 for the top layer has a value $\gamma_1 = 0.39$ eV is:

$$\gamma_1 = -\left\langle \phi_{p_{ZA2}}(\vec{r} - \vec{R}_{p_{zA2}}) \middle| U(\vec{r}) \middle| \phi_{p_{ZB1}}(\vec{r} - \vec{R}_{p_{ZB_1}}) \right\rangle \tag{7.73}$$

One can show by including γ_1 in equations similar to (7.67)–(7.71) that [2]:

$$\begin{vmatrix} E_p - E(\vec{k}) & \left(\left(E_p - E(\vec{k})\right)s_{12} - \gamma_\pi\right) f(\vec{k}) & 0 & 0 \\ \left(\left(E_p - E(\vec{k})\right)s_{12} - \gamma_\pi\right) f(-\vec{k}) & E_p - E(\vec{k}) & \gamma_1 & 0 \\ 0 & \gamma_1 & E_p - E(\vec{k}) & \left(\left(E_p - E(\vec{k})\right)s_{12} - \gamma_\pi\right) f(\vec{k}) \\ 0 & 0 & \left(\left(E_p - E(\vec{k})\right)s_{12} - \gamma_\pi\right) f(-\vec{k}) & E_p - E(\vec{k}) \end{vmatrix} = 0 \tag{7.74}$$

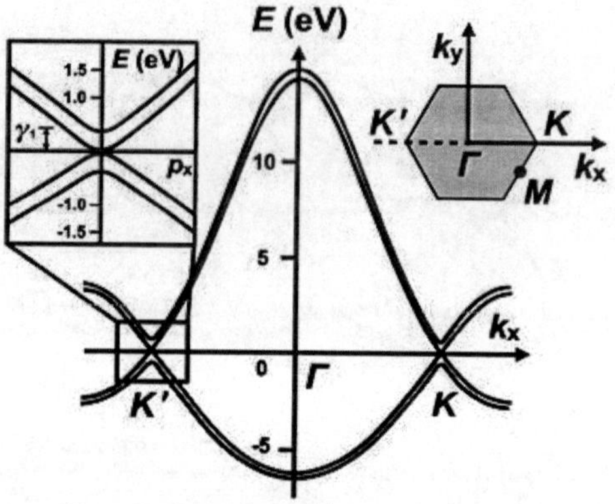

Fig. 7.16 Electronic Band Structure of Bilayer graphene. Interlayer interactions split the valence and conduction bands into two. One set of valence and conduction bands touch at the $K\ and\ K'$ points. The second set of bands is split. The splitting energy is given by the interlayer couplingγ_1. The energy dispersion neat the K point is quadratic. Thus the Dirac equation for the carriers in bilayer graphene is of massive fermions. Figure adapted from [2]

S_{12} is the single layer graphene overlap integral for nearest neighbors. The on-diagonal blocks of both matrices contains the same information as shown in (7.71) and (7.72). The determinant can be solved resulting in the band structure shown in Fig. 7.16. Near the K points, the dispersion is quadratic as we show below and the carriers act as massive Dirac Fermions [2]. If we assume that the overlap integrals $S_{12} \approx 0$, and setting the energy levels to center at zero using $E_p = 0$ and for both layers $\gamma_\pi = 3.033$ eV, (7.74) simplifies:

$$\begin{vmatrix} -E(\vec{k}) & -\gamma_\pi f(\vec{k}) & 0 & 0 \\ -\gamma_\pi f(-\vec{k}) & -E(\vec{k}) & \gamma_1 & 0 \\ 0 & \gamma_1 & -E(\vec{k}) & -\gamma_\pi f(\vec{k}) \\ 0 & 0 & -\gamma_\pi f(-\vec{k}) & -E(\vec{k}) \end{vmatrix} = 0$$

The solution to this 4 × 4 determinant provides the 4 bands for bilayer graphene:

$$E(\vec{k})^2\left(E(\vec{k})^2 - \gamma_\pi^2|f(\vec{k})|^2\right) - \gamma_1^2 E(\vec{k})^2 - \gamma_\pi^2|f(\vec{k})|^2\left(E(\vec{k})^2 - \gamma_\pi^2|f(\vec{k})|^2\right) = 0$$
$$\left(E(\vec{k})^2 - \gamma_\pi^2|f(\vec{k})|^2\right)^2 - \gamma_1^2 E(\vec{k})^2 = 0$$

This results in a quadratic equation for $E(\vec{k})$ which has the solutions [2]:

$$E_{\pm}^{\alpha} = \pm\frac{\gamma_1}{2}\left(\sqrt{1+\frac{4\gamma_{\pi}^2\left|f(\vec{k})\right|^2}{\gamma_1^2}}+\alpha\right), \quad \alpha = \pm 1 \tag{7.75}$$

Remember that for single layer graphene, with $s_0 \sim 0$ and $E_p = 0$ $\gamma_{\pi} = 3.033$ eV which is ~ 3 eV, (7.72) simplifies to $= \pm\gamma_{\pi}\left|f(\vec{k})\right|$. Equation (7.75) provides a 4 band solution which is similar to two single layer graphene 2 band solutions separated by a gap. Using $\gamma_1 = 0.39$ eV then $2\gamma_0\left|f(\vec{k})\right|/\gamma_1 = 6\left|f(\vec{k})\right|/0.39 > 1$ for many values of $\left|f(\vec{k})\right|$ [2]:

$$E_{\pm}^{\alpha} \approx \pm\gamma_{\pi}|f(\vec{k})| \pm \frac{\gamma_1}{2} \quad \text{over much of the Brillouin zone} \tag{7.76}$$

At the K point

$$\left|f(\vec{k})\right| = 0 \text{ so, } E_{\pm}^{\alpha} = 0 \text{ or } \pm\gamma_1 \tag{7.77}$$

So the gap is due to the overlap integral between layers. *The energy of this gap is close to the energy of the absorption feature observed in the optical conductivity of multilayer graphene for 2 layers as shown in* Fig. 7.16 [2].

7.6.1 Massive Dirac Fermions in Bilayer Graphene

Close to the K point, $\left|f(\vec{k})\right| \sim k$ so (7.75) becomes:

$$E_{\pm}^{\alpha} = \pm\frac{\gamma_1}{2}\left(\sqrt{1+\frac{4\cdot 3^2\cdot k^2}{0.39^2}}+\alpha\right)$$

$$\pm\frac{\gamma_1}{2}\left(\sqrt{1+237k^2}+\alpha\right) \sim \pm\frac{\gamma_1}{2}237k^2 \propto k^2 \quad \text{for } \alpha = -1$$

where we used a series expansion and keep only the first two terms when x is small for $\sqrt{1+x} \approx 1+\frac{x}{2}-\frac{x^2}{8}\cdots$ with $x \sim 237k^2$ giving $E_{\pm}^{\alpha}$ a quadratic dependence close to the K and K' points. Since the carriers still move at ~ 1/300 the speed of light, these are massive Dirac particles, and the 2D Dirac equation shown in (7.28) must be modified. It is worth noting that the mobility of bilayer graphene was lower than that of single layer graphene at least in the initial reports of Novoselov, Geim, and coworkers [21]. Proof of the Dirac Fermion nature of the carriers in bilayer graphene comes from low temperature Hall Effect Characterization of the electrical properties.

7.6.2 *The Berry Phase Correction for the Quantum Hall Effect and Shubnikov De Hass Oscillations in Bilayer Graphene*

The coupling between the graphene layers in bilayer graphene results in new and unexpected physics. Although the energy dispersion is quadratic, the carriers are still chiral and have relativistic properties [21]. Bilayer graphene has a Berry phase of 2π instead of the π value observed for single layer graphene [21, 22]. Figure 7.17 shows relationship between energy and wavevector, and the direction of the pseudospin is shown at the bottom. In Fig. 7.18, the experimental data for the Hall characterization of bilayer graphene is shown. The Hall conductivity of bilayer graphene has the expected integer steps in conductance, but the plateau at low carrier concentrations is missing. This missing plateau is unexpected and at these low carrier concentrations and high magnetic fields, the sample is metallic and not insulating. Novoselov, Geim, and coworkers [21] state that: "*The revealed chiral fermions have no known analogues and present an intriguing case for quantum–mechanical studies.*" They found that the Landau level at zero carrier concentration is doubly degenerate resulting in the unusual behavior. More recent theoretical studies have shown that the previously reported 2π for bilayer graphene is really the winding number of the pseudospin [23]. The winding number specifies how many times the vector representing the relative phase on the two different sublattices of carbon atoms

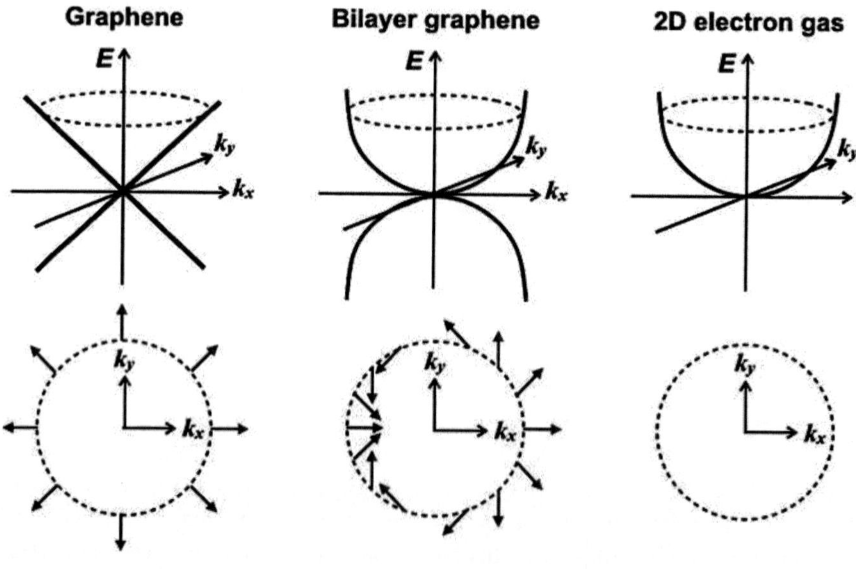

Fig. 7.17 At the top, the electron energy diagram for the 2D structures graphene and bilayer graphene near the *K* point in the Brillouin zone and the energy diagram for a 2 DEG are compared. Below the energy diagrams, the direction of the pseudospin is shown. There is no pseudospin for a 2DEG. Figure adapted and reprinted with permission from [23] © 2011 American Physical Society

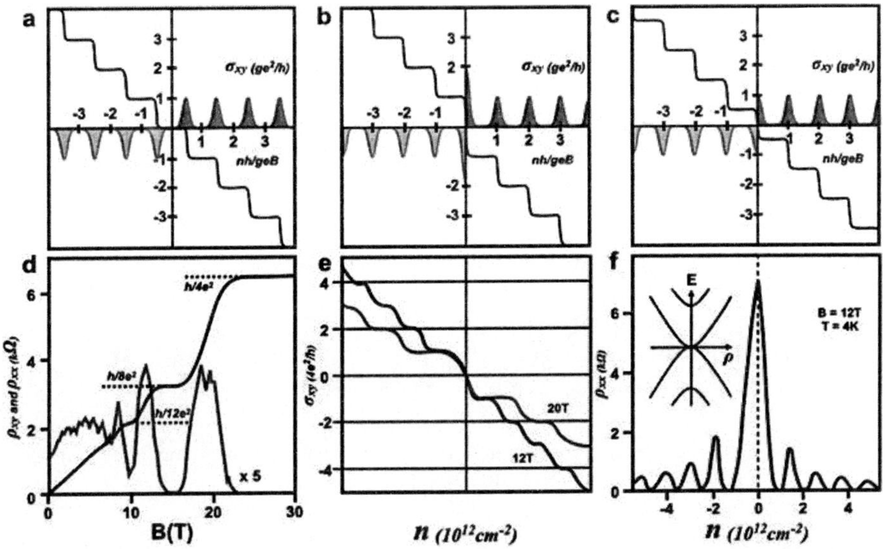

Fig. 7.18 The low temperature, high magnetic field characterization of exfoliated bilayer graphene is compared to that of a 2DEC, and single layer graphene. The factor g in Figures **a**–**c** is the system degeneracy. $\phi_0 = h/e$ is the magnetic quantum flux, and e^2/h is the quantum of conductance. On the left, **a** shows the IQHE of a 2DEG, **b** shows the predicted Hall conductance of bilayer graphene and **c** shows the quantized Hall conductance of single layer graphene. At carrier concentration of zero, the conductance step for bilayer graphene is twice that of single layer graphene. In **d**–**f**, experimental data for bilayer graphene is shown. In (d), the magnetic field dependence of the transverse and longitudinal resistivity is shown. In (**e**), The Hall conductance is shown at 12 T and 20 T versus carrier concentration. In (**f**), the carrier concentration dependence of the longitudinal resistivity is shown as a function of carrier concentration at 4 K and 12 T. The Berry phase is found to be 2π when this data is interpreted. Further interpretation of this data found that the 2π phase comes from the pseudospin winding number [23]. Figure adapted from [21]

(pseudospin vector) rotates when the electronic wave vector undergoes one full rotation around the Dirac point [23]. The 2DEG wavefunctions do not have pseudospin. The interpretation of the Hall measurements for bilayer and multilayer graphene continue to evolve.

7.7 The Electronic Structure of TriLayer and TetraLayer Graphene

The impact of Bernal versus rhombohedral stacking for trilayer (24) and tetralayer graphene (25) has also been studied. The Bernal and rhombohedral structures are shown in Fig. 7.19. The electronic band structure is shown in Fig. 7.20.

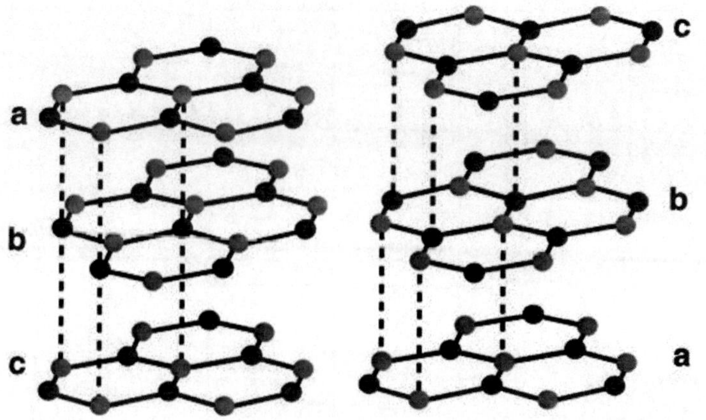

Fig. 7.19 The Bernal structure of trilayer graphene is shown on the left and rhombohedral trilayer graphene is shown on the right. Figure adapted from [2]

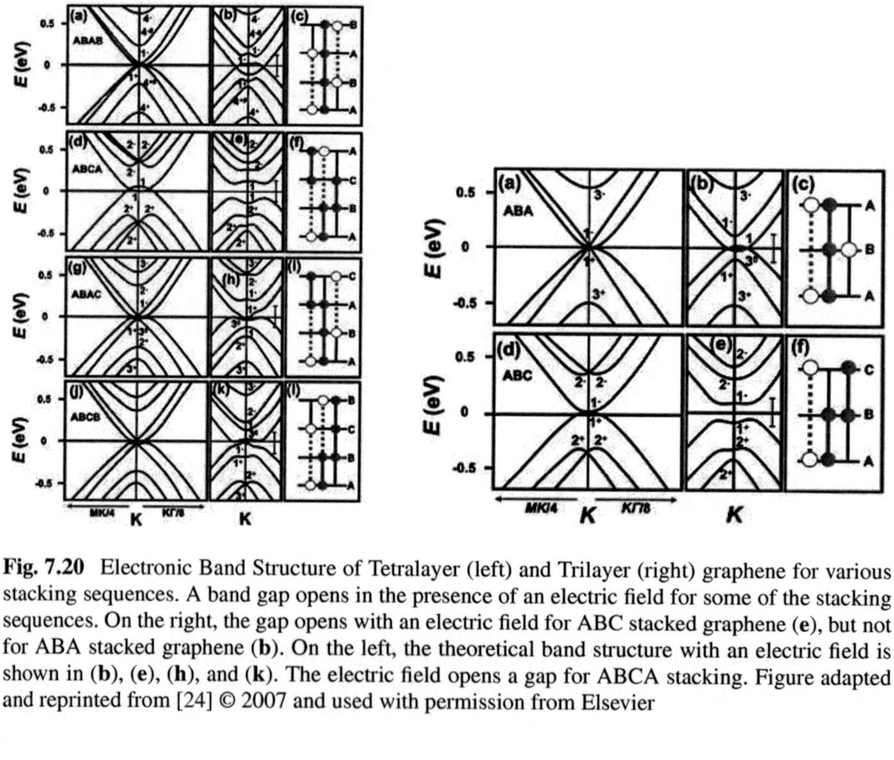

Fig. 7.20 Electronic Band Structure of Tetralayer (left) and Trilayer (right) graphene for various stacking sequences. A band gap opens in the presence of an electric field for some of the stacking sequences. On the right, the gap opens with an electric field for ABC stacked graphene (**e**), but not for ABA stacked graphene (**b**). On the left, the theoretical band structure with an electric field is shown in (**b**), (**e**), (**h**), and (**k**). The electric field opens a gap for ABCA stacking. Figure adapted and reprinted from [24] © 2007 and used with permission from Elsevier

7.7.1 The Berry Phase Correction for the Quantum Hall Effect and Shubnikov De Hass Oscillations in Trilayer Graphene

Here we briefly mention the complicated physics observed for low temperature, high magnetic field measurements of trilayer graphene. As shown in Fig. 7.20, the electronic bands of trilayer graphene are have the combined character of monolayer graphene with a linear energy dispersion at the K points and that of bilayer graphene with a quadratic dispersion and a gap. Thus both massless and massive Dirac fermions are predicted. High quality data was obtained when BN was used as a substrate [26]. The unconventional QHE and Shubnikov de Haas oscillations were both observed [26]. The data is summarized in Fig. 7.21.

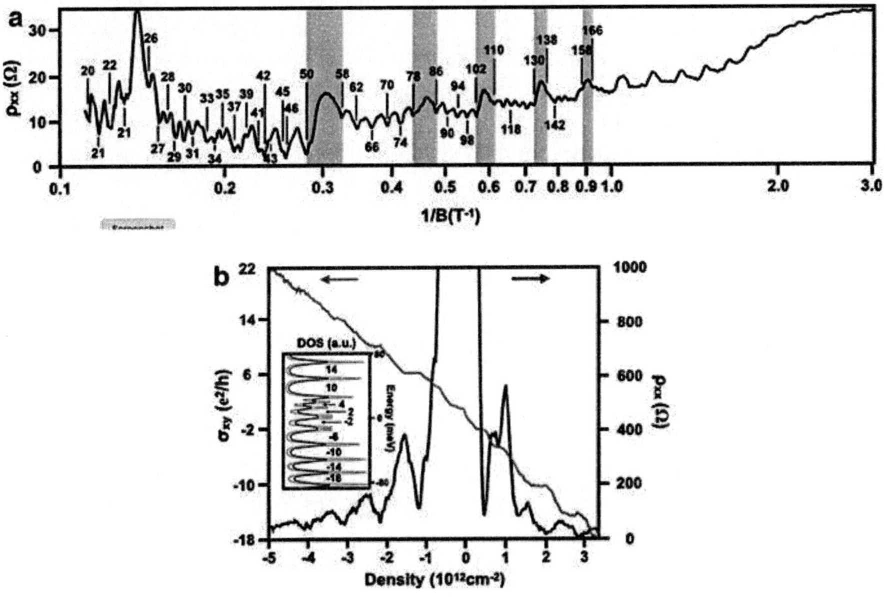

Fig. 7.21 Longitudinal resistivity of Bernal stacked trilayer graphene as a function of inverse magnetic field at 300 mK (**a**), and **b** Hall conductance and longitudinal resistivity as a function of carrier density at 9 T and 300 mK [26]. The filling factors are listed in (**a**) at the minima of the Shubnikov-de-Hass oscillations. Trilayer graphene shows unique Hall measurement data when compared to single and bilayer graphene. For example, the series of plateaus in Hall conductivity is the same at all magnetic fields for single and bilayer graphene. This is due to Landau level crossing. The Hall conductivity and longitudinal resistivity are shown in (**b**) at 9 T and 300 mK. The inset shown the density of states at the energy of the plateaus in the Hall conductivity labeled using the value of the Hall conductivity. Figure adapted from [26]

7.8 Optical Characterization of Graphene and Multilayer Graphene

It is important to note that ellipsometric characterization of 2D materials is sensitive to the in-plane dielectric function of a film that likely has anisotropic optical properties. It is useful to first discuss bulk graphite which is a uniaxial anisotropic material [27]. In uniaxial anisotropic films, the complex refractive indices are the same for the in-plane polarizations of light, x and y, and different for the out-of-plane, or z polarization. In the simple case of an isotropic material, there is no cross-polarized light scattering, and the r_{sp} and r_{ps} elements of the matrix are zero. When uniaxial, anisotropic samples that are positioned so that the optic axis is oriented along an axis of symmetry, there is no cross-polarized light scattering. However, the birefringence of the sample does alter the refracted light as shown in Fig. 7.22. For a typical graphene ellipsometry measurement, the basal plane of carbon atoms is perpendicular to the optic axis, there is no cross-polarized light scattering [27]. However, the birefringence of the sample does split the refracted light beam for graphite. The sensitivity of an ellipsometry measurement to this anisotropy is dependent on the distance the light has to traverse the film, and therefore allow the two refracted beams to separate. In graphene, this path length is severely limited (the nominal thickness of a graphene monolayer is 3.35 Å) [27].

Absorbance and ellipsometric characterization of graphene and multilayer graphene provides further proof of the unique band structure of these materials. The complex refractive index and absorption of graphene from the near IR to the UV is shown in Fig. 7.9, and the optical conductivity for Bernal stacked graphene is shown in the IR in Fig. 7.10. Routine ellipsometric measurement of graphene requires a relatively large area sample even with focusing optics. Most exfoliated samples are often smaller than the focuses spot size. The complex refractive index of exfoliated graphene and CVD graphene is shown in Fig. 7.9. The refractive index data for exfoliated graphene and CVD graphene with relatively large grain size are remarkably similar [7]. The substrate dependence of the noise observed during measurement on silicon wafer substrates with either a native oxide or 300 nm silicon dioxide

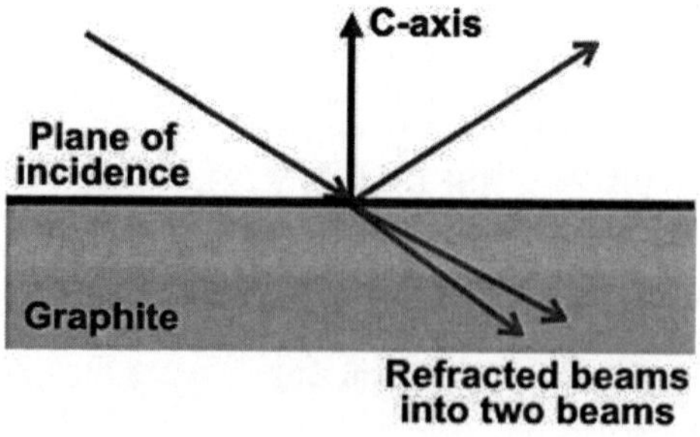

Fig. 7.22 The separation of the ordinary and extraordinary light rays for graphite is shown. Figure adapted from [27]

is frequently not mentioned in the literature. The data shown in Fig. 7.9 for CVD graphene was obtained on a glass slide which was thoroughly cleaned and the back side roughened to prevent backside reflections. During ellipsometric measurement, one can either use the complex refractive index determined from ellipsometry (see Fig. 7.9 and [7]) or a multi-function approach can be used. The absorption feature at ~ 4.5 eV is attributed to absorption between the valence and conduction π bands at the M point in the Brillouin zone, and this transition is known to altered by excitonic effects [7]. These excitonic effects refer to the interaction between the electron that is promoted from the valence band to the conduction band and the hole that is left behind as discussed in CH 5 [7, 28]. Shifts in the energy of this feature have been used to characterize the graphene transfer process [29]. The absorption at 4.5 eV requires the use of either a Fano resonance function or more than one oscillator so that the asymmetric lineshape is reproduced. The origin of the Fano lineshape was described in [27]. The Fano resonance is an interference effect where discrete excitonic states form below the conduction band during optical excitation [28]. The lower density of states results in less screening enabling the formation of the excitonic states. A Fano resonance results in an asymmetric line shape and it red-shifts the position of this peak by ~ 0.5 eV. This is due to a transfer of oscillator strength from interband transitions at 5.1 eV to resonant exciton states at lower energies. This effect can be used in the modeling ellipsometric data through the choice of oscillator line shape incorporated in the optical data modeling [28]. Excitonic effects are also observed for the M point transition of bilayer graphene [13]. In Fig. 7.23 the excitonic effect induced shift in the M point transition is shown for bilayer graphene and for graphite [13].

The impact of layer stacking on the optical spectra of tetralayer graphene was characterized by Heinz and co-workers [25]. In Fig. 7.24, the impact of stacking on the optical conductivity of tetralayer graphene is shown. This group has also verified the impact of an electric field on the band gap of bilayer graphene [30].

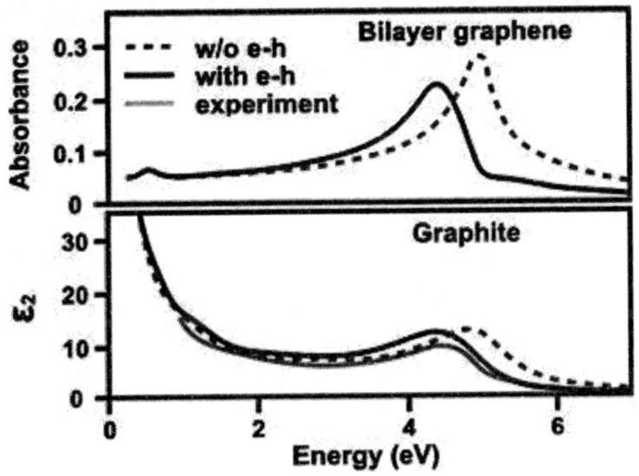

Fig. 7.23 The shift in the M point transition for Bernal stacked bilayer graphene and graphite due to excitonic effects is shown. Figure adapted and reprinted with permission from [13] © 2009 American Physical Society

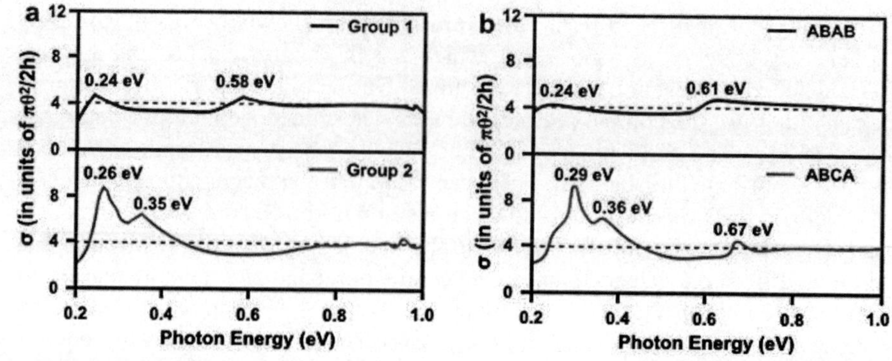

Fig. 7.24 Infrared optical conductivity spectra of tetralayer graphene samples. Two stacking patters are expected, Bernal and Rhombohedral and they were observed in [25] which this figure was adapted from [25]. The two expected responses are shown in (**a**). Theoretically calculated responses for these two stacking patterns are shown in (**b**). The samples were assumed to be undoped and the authors used a phenomenological broadening of 20 meV. Figure adapted and reprinted with permission from [25] © 2010 American Physical Society

Optical spectroscopy can be used to study the bandgap of bilayer graphene. The tune ability of the bad gap of bilayer graphene was experimentally and theoretically studied through photocurrent spectroscopy of hexagonal BN encapsulated bilayer graphene [31, 32]. A typical photocurrent measurement and the associated optical pathway and data is shown in Fig. 7.25 [32]. In Fig. 7.26, the linear relationship between the transition peak energies for P1 and P2 and the electric field displacement (band gap) is shown.

7.9 Effect of Rotational Orietation Between Layers on Bilayer Graphene (Twisted Bilayer Graphene), Monolayer—Bilayer Graphene, and Bilayer-Bilayer Graphene Properties

The effect of rotational orientation on the properties of few layer graphene has been the subject of intensive study. Correlated electron properties, magnetic properties, and reports of superconductivity make transport studies of few layer graphene and exciting topic. Bilayers, monolayer—bilayer, bilayer-bilayer, and even trilayers with rotated middle layers are discussed below. The relative rotational or twist angle between layers of graphene has a strong influence on the transport properties. The moiré pattern forms a moiré lattice which has four allowed states per moiré lattice unit cell which are known as flavors for spin up, spin down, and the two valley pseudospin states (see Sect. 7.2.4). At certain twist angles a moiré pattern forms which results in moiré flat bands and correlated electron interactions are enhanced. The development of hexagonal boron nitride (h-BN) substrates was an essential part

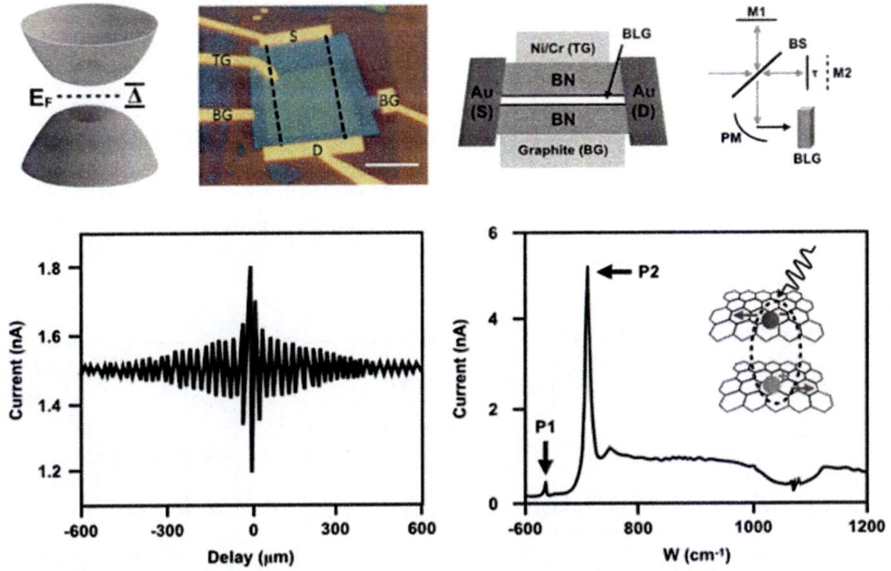

Fig. 7.25 Photocurrent spectroscopy of excitons in Bilayer graphene. **a** Shows the bandgap Δ and Fermi level E_F, **b** is an optical micrograph of the measured device, **c** is a drawing of the cross-section, **d** is the optical setup including the beam splitter (BS) and variable delayτ. **e** is a typical interferogram, and **f** is a photocurrent spectrum showing two sharp peaks. Figure adapted and reprinted from [32] © 2017, American Association for the Advancement of Science

of fabricating electrical structures for characterizing transport properties of these few layer graphene stacks. The electrical test structures are fabricated by encapsulating the multilayer graphene structure in h-BN, and these test structures were described in Sect. 7.3.1. Boron nitride is discussed in Sect. 7.10. In addition, the ability to experimentally set the rotation angle between layers is essential to the studies discuss in both Sects. 7.9 and 7.10.

7.9.1 Twisted Bilayer Graphene

The rotational orientation of a top graphene layer relative to the graphene layer below is fixed for Bernal stacked and other naturally occurring stacking sequences (see Fig. 7.27). Since graphene layers are bonded together using van der Waals forces, the relative rotational orientation, which is often referred to as the twist angle, can be experimentally altered. The electrical and optical properties of graphene—graphite, graphene-graphene, and graphene-hexagonal boron nitride stacks can be altered by changing their relative stacking orientation [33–45]. Certain angular alignments (twist angles) result in a moiré lattice which produces periodic changes in the interlayer interactions as shown in Fig. 7.27. In addition, specific angular alignments

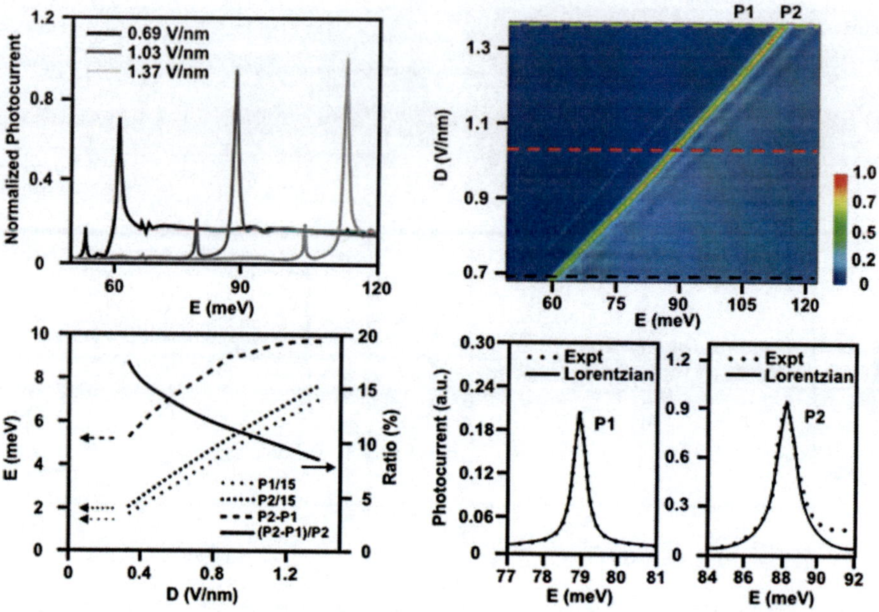

Fig. 7.26 The band gap dependence of the exciton peak energies for bilayer graphene is shown through the peak energy shifts versus electric field displacement. Figure adapted and reprinted from [32] © 2017, American Association for the Advancement of Science

are referred to as "magic" angles because they produce a variety of unexpected correlated electron properties such as superconductivity, correlated insulator, and strange metal phase (described below) [35, 43]. Correlated electron behavior refers to phenomena such as superconductivity that cannot be described by independent electron action. Correlated electron behavior is further discussed in Chap. 9. The transition from a conducting state to a correlated insulator state occurs as the temperature is decreased. The signature of strongly correlated electron behavior is that the property is strongly temperature dependent. The structures displaying unexpected properties are often referred to as *magic angle twisted bilayer graphene* (*MATBG*).

Controlling and reducing the resistivity of electrical contact to 2D layers is critical to fabrication of device structures. The resistivity of graphene-graphite contacts can be as low as 6.6 $\Omega\ \mu m^2$ which is considerably lower than that of metal – graphene contacts [33]. The resistivity has a 60° rotational periodicity, and dips in resistivity were observed at 22° and 39°. These angles are among the commensurate angles predicted for twisted bilayer graphene [33]. Below, we discuss commensurate angles and moiré patterns.

As mentioned above, the electronic band structure and electrical transport properties change as a function of twist angle [34–37]. The initial theoretical description of twisted bilayer graphene considered the interaction between two pristine layers

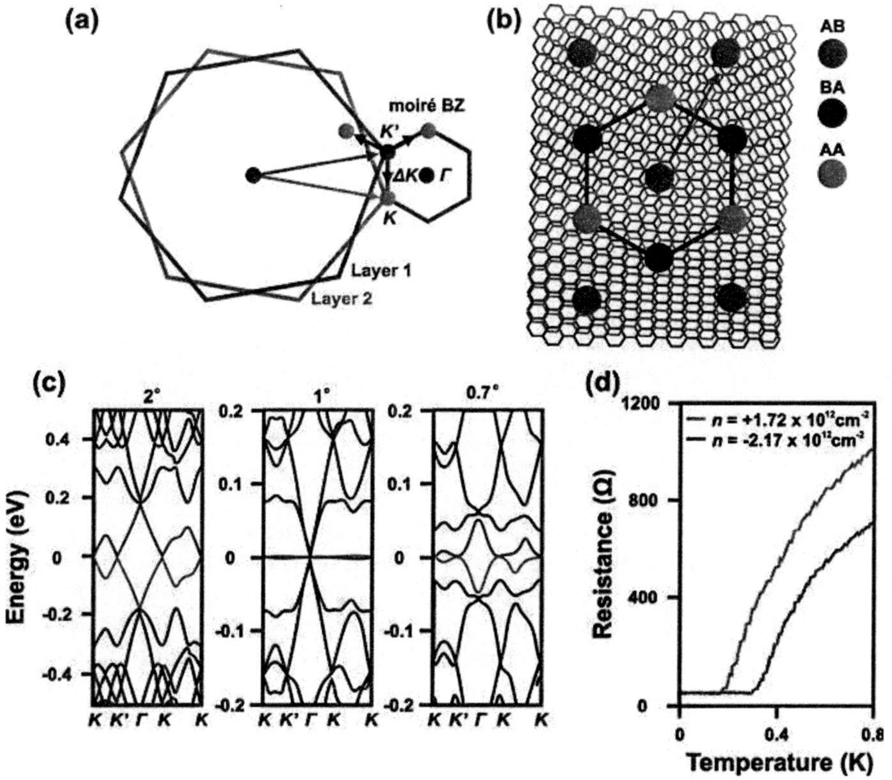

Fig. 7.27 The moiré pattern produced by twisting the angular alignment of bilayer graphene and the band structure and low temperature resistivity are shown. In figure **a**, the two rotated Brillouin zones are shown along with the resulting moiré Brillouin zone. In **b**, the stacking order (AA, AB, BA) at different spatial locations is seen to change resulting in a spatial variation in interlayer coupling. In **c**, the electronic band structure for un-reconstructed bilayer graphene is shown as a function of twist angle [41]. Flat bands (light gray bands close to zero energy) are seen at 1°. In **d**, the superconducting state is observed in the low temperature resistivity data for 1.14° twisted bilayer graphene [36]. Figures 7.27a–c adapted from [41]. Figure 7.27d adapted from [36] and reprinted with permission from American Association for the Advancement of Science

without including the effect of stress and lattice reconstruction [37–40]. The wavelength of the moiré pattern between two pristine graphene layers is a function of twist angle θ_{tw}, $\lambda = a_0/2\sin(\theta_{tw}/2)$ where AA stacking is used to define a twist angle of 0°, see supplemental section of [34] and [47]. The area of the unit cell of the moiré lattice is $A_{ml} = \left(\frac{\sqrt{3}}{2}\right)\lambda^2$ [47]. The smaller the twist angle the longer the wavelength of the moiré pattern. When the twist angle is non-zero, the stacking pattern can have a mixture of AB, BA, and AA stacking so that when viewed from above a moiré pattern appears. It is useful to define the term commensurate as it is applied to the stacking of the atoms in the moiré lattice. Consider AB stacking for bilayer graphene lying horizontally where an atom B_2 from the B sublattice is directly above an atom

A_1 from the A sublattice in the graphene layer below. A commensurate angle occurs when the top layer is rotated about one of the axis made by an A_1B_2 pair, and somewhere else in the lattice there is an $A_1'B_2'$ alignment [37]. A commensurate angle can also be defined using AA stacking and rotation about an A_1A_2 pair. The stacking of two pristine graphene layers will be commensurate when the primitive lattice vectors of the moiré lattice are integer multiples of the single layer graphene primitive lattice. Using the graphene primitive lattice vectors shown in Fig. 7.3, $\vec{a}_1 = \left(\frac{\sqrt{3}}{2}\hat{x}, \frac{1}{2}\hat{y}\right)a$ and $\vec{a}_2 = \left(\frac{\sqrt{3}}{2}\hat{x}, -\frac{1}{2}\hat{y}\right)a$, and an origin having AA stacking, the twist angle can be calculated using the number of integer multiples of the primitive lattice vectors (m, n) that connect the centers of the moiré patterns [38, 39]:

$$\cos\theta_{tw} = \frac{n^2 + 4nm + m^2}{2\left(n^2 + nm + m^2\right)}$$

The local interaction between layers has a clear spatial dependence on the location due to the moiré pattern shown in Fig. 7.27 [40–42], and the magic angles have been theoretically predicted to be a function of the interlayer coupling (hopping) for AB stacking when the interlayer coupling is set to zero for AA stacking [42]. In Fig. 7.27, we show a moiré pattern for un-reconstructed bilayer graphene and the resistivity as a function of temperature and carrier density [36, 41]. Figure 7.27 also shows that the Dirac cone band structure near K and K' flattens and then returns as the twist angle is changed as shown in the band energy diagrams for twist angles of 2°, 1.0° and 0.7° [36, 41, 42].

Theoretical calculations based on the interactions between twisted, unreconstructed graphene layers describe key aspects of the physics. However, the hexagonal graphene lattice is known to distort due to layer to layer interactions for twisted bilayer graphene stacks [34]. Calculation of electronic structure is further complicated by the angular dependence of the lattice reconstruction [34]. There is a competition between commensurate layer alignment and interfacial distortion [34]. A detailed transmission electron microscopy/selected area electron diffraction study [34] showed that as the twist angle is decreased from above to below the characteristic angle of $\sim 1.1^{\circ}$, the incommensurate moiré structure changes to commensurate thus forming domains with soliton boundaries. This study also proves that atomic reconstruction occurs [34]. 1D topological carrier transport is reported to occur at the soliton boundaries between the domains [34].

The filling of the moiré bands for MATBG can result in the appearance of a correlated insulator state as the MATBG is cooled to low temperatures [43]. When flat moiré bands occur (see Fig. 7.27c for a twist angle of $\theta \sim 1°$), electron–electron interactions can occur. At a twist angle of $\theta \sim 1°$, areas of AA, AB, and BA stacking occur. The electron density in the moiré bands was found to be concentrated in the small areas with AA stacking. The h-BN—(bilayer graphene)—h-BN sample is fabricated into an electrical test structure with bottom and top floating gates as described in Sect. 7.3.1. The carrier concentration n is controlled by doping using the dual gate structure. The carrier density for a single electron or hole for each unit cell in the moiré lattice, $n_s = 4/A_{ml} = 4/\left[\left(\frac{\sqrt{3}}{2}\right)\lambda^2\right]$ is function of twist angle θ_{tw} since$\lambda = a_0/2\sin(\theta_{tw}/2)$. Since we defined$n_s = 4/A_{ml}$, the filling factor $\upsilon = (n/n_s)$ for full filling is1. A correlated insulator state believed to be a Mott-like insulator is observed in MATBG when the band is $\frac{1}{4}$, $\frac{1}{2}$, *or* $\frac{3}{4}$ filled [43]. Theoretical study of the explanation for the appearance of the correlated insulator continue [43].

At the magic twist angle of ~ 1.1°, superconductivity has been reported [35]. A magnetic field of 0.4 T greatly reduces the conductivity [35]. Bilayer graphene at a twist angle of 1.16° exhibits superconductivity at 0.3 K while a twist angle of 1.05° exhibits superconductivity at 1.7 K [35]. The superconducting state is observed at filling factors higher but close to half filling for both electrons and holes $|n| > |n_s/2|$ [36]. The correlated electronic properties observed in twisted bilayer graphene [34–36] can also be tuned using both carrier concentration and stress [36]. Superconductivity has been reported for samples using 2.21 GPa pressure at twist angles of 1.27° which differs from samples not under pressure [36]. The superconductivity is dependent on the carrier charge density as shown in Fig. 7.27 [34–36]. Both p and n doped twisted bilayer graphene can be superconducting [36]. The effect of twist angle on the electronic properties of double bilayer graphene has also been studied [44] and is discussed below.

Another correlated electron behavior observed in magic angle twisted bilayer graphene (MA-tBG) (for twist angles between 1° and 1.2°) is the strange metal state. The strange metal state refers to a metal which exhibit a resistivity that is few orders of magnitude higher than a typical metal. The hallmark of this state is a linear dependence of resistivity with temperature from 30 K down to near zero Kelvin [45]. The state appears above a few Kelvin. Once again, the type of correlated electron behavior, superconducting, correlated insulator, or strange metal, is a function of the temperature and filling factor at a specific magic angle. The filling factors where the strange metal behavior occurs is said to be primarily near the correlated insulator phase when $\upsilon = \pm 2 \pm \delta$ where $\delta = 1/2$ [45]. It is also observed at other filling factors, for example near $\upsilon = \pm 1$ and ± 3 in some devices [45]. Although the Drude model for the conductivity of a metal was a purely classical description, the scattering rate τ continues to be used to describe the conductivity of metals and now the strange state of observed for bilayer graphene. For a normal metal, the scattering rate has $\tau \sim T^{-2}$ temperature dependence. In strange metals, $\tau \sim T^{-1}$. The dependence was found to have $\tau = C\hbar/k_B T$ where k_B is Boltzman's constant, and the constant C is dimensionless and close to 1. Due to the presence of Planck's constant, the strange metal behavior is known as Planckian dissipation. The electronic conduction behavior of magic angle bilayer graphene system can described by a filling factor υ discussed in Chap. 6 and temperature depended phase diagram [45].

The electrical test structures used to measure transport properties of twisted bilayer graphene can influence the observed properties. When the relative alignment of the graphene bilayers with one of the encapsulating h-BN layers with an estimated offset angle of 0.6°, the QAHE has been observed [46]. We note that unambiguous determination of the h-BN- graphene offset angle was considered difficult [46]. The

QAHE was observed below 3 K at a twist angle of $1.15° \pm 0.01°$ at a carrier density of $2.37 \times 10^{12}/\text{cm}^2$ with quantization of the Hall resistance to $\rho_{xy} = 1.0010 \pm 0.0002\ ^{h}/_{e^2}$ [46].

The optical response of twisted bilayer graphene has also been studied over a wide range of twist angles that are typically larger than the magic angle where correlated carrier behavior occurs [47–51]. Van Hove singularities form in the electronic structure and significantly alter optical spectra [50]. The understanding of the optical response is based on the interlayer interaction that results from the relative rotation angle of two pristine graphene layers with no reconstruction or stress. Van Hove singularities in the band structure of bilayer graphene arise when the Dirac cones from the top and bottom graphene layers overlap due to the twist angle as shown in Fig. 7.28. The energy gap (transition energy) between the valence and conduction bands at the van Hove singularities is a function of the twist angle θ [50]:

$$E_{vH} = \left(\frac{8\pi\hbar v_F}{3a}\right)\sin\left(\frac{\theta}{2}\right)$$

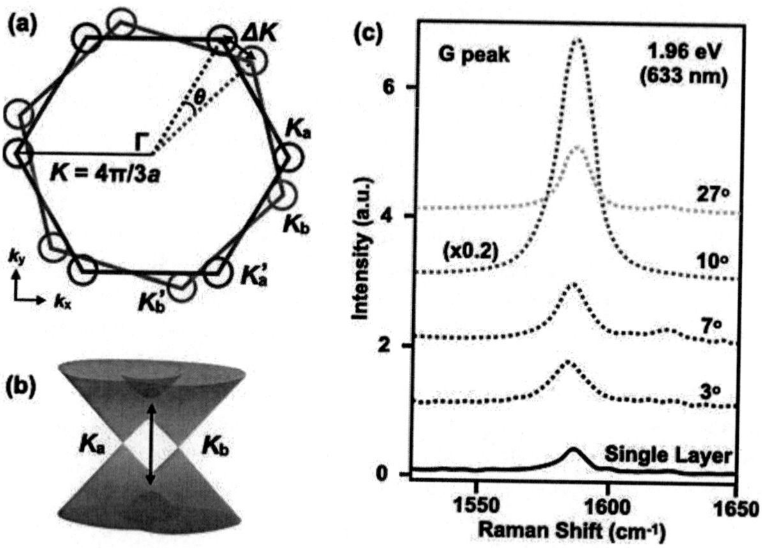

Fig. 7.28 The van Hove singularities form in the electronic band structure when the Dirac cones in the upper and lower layers in twisted bilayer graphene intersect. **a** The upper layer a and lower layer b Brillouin zones are shown. **b** The Dirac cones close to the K point for the upper and lower layers intersect forming the van Hove singularities. The singularities are seen as the quadratic shaped cones between the K_a and K_b Dirac cones. The arrows show transitions between these cones. The transitions are excited by matching the laser probe beam energy to that transition energy. **c** The Raman G peak is shown as a function of twist angle. At a twist angle of 10°, the 1.96 eV laser energy is resonant with the transition at the van Hove singularity. Figure adapted and reprinted with permission from [48] © 2012 American Physical Society

where a is the graphene lattice constant and the Fermi velocity $v_F = 1,000,000\ \text{ms}^{-1}$. The Fermi velocity is reduced by the periodic interlayer interactions. A perturbation theory estimate of the reduced Fermi velocity is $\tilde{v}_F \sim 0.82\ v_F$ (for $\theta = 3.9°$). This is considered an overestimate of the velocity reduction for smaller twist angles [37]. The Fermi velocity for epitaxial, multilayer graphene is $\tilde{v}_F \sim 0.7$ to $0.8 \times 10^6\ \text{ms}^{-1}$[37]. A reduction is also seen for twist bilayer graphene, for example, twist angles between 1.3° and 2° is $v_F \sim 0.6 \times 10^6\ \text{ms}^{-1}$ [47]. A reduced Fermi velocity for bilayer graphene is considered to be an indication of a twisted bilayer. A potential difference between twisted graphene layers is also predicted and observed [47].

The van Hove singularities of twisted bilayer graphene have been studied using Raman spectroscopy [48], optical conductivity [49, 51], and photoluminescence [50]. Photoluminescence is increased by a factor of 2–3 over that of single layer graphene when the laser energy is resonant with E_{vH} [50]. The critical angle θ_c that results in E_{vH} being resonant with the laser energy E_L can be calculated using [48]:

$$\theta_c = 3aE_L/4\pi\hbar v_F$$

Raman characterization of graphene, few layer graphene, and graphite was discussed in Chap. 1 Sect. 1.9.5.3. The G, G', and D peaks are due to resonance scattering involving electronic transitions at the Dirac cones, and G' and D peaks are due to double resonance scattering involving Dirac cones at the K and K'. The Raman spectra for the D and G peaks for bilayer graphene show a significant increase in intensity at critical angles where this resonance occurs as shown for the G peak in Fig. 7.28. The G peak is due to the stretching of the carbon–carbon bond in graphitic materials. The optical conductivity is also a function of twist angle. Optical absorption has been measured at normal incidence with a hyperspectral imaging NIR- visible – DUV microscope [49]. The data from this experiment is plotted as the optical conductivity of twisted bilayer graphene—the optical conductivity of Bernal bilayer graphene versus photon energy in Fig. 7.29. The optical conductivity has been calculated for a few, relatively large twist angles as shown in Fig. 7.29 [51] (Fig. 7.30).

7.9.2 *Monolayer—Bilayer Graphene, Middle Layer—Twist Angle Trilayer Graphene, and Bilayer-Bilayer Graphene*

Twisting the alignment of monolayer graphene placed on Bernal stacked bilayer graphene (tMBG) has produced correlated metallic and insulating states as well as topological magnetic states when an electrical bias potential is applied normal to the graphene plane [52]. In contrast to bilayer graphene and bilayer—bilayer graphene, the lack of symmetry normal to the graphene plane results in properties not observed

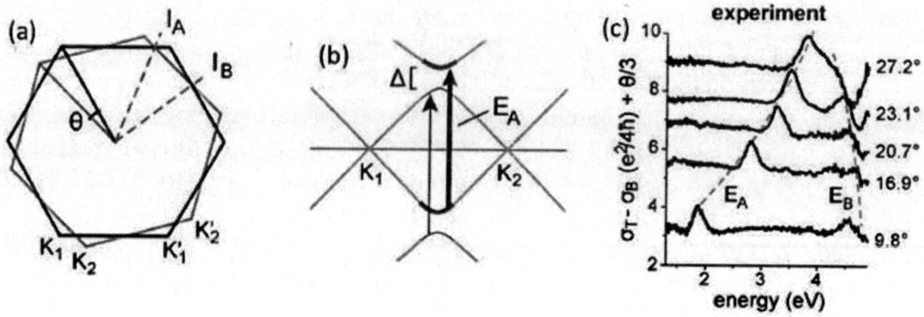

Fig. 7.29 The optical conductivity of twisted bilayer graphene is shown to be a function of twist angle. **a** The Brillouin zone of the top and bottom layers are twisted. **b** The concave down (valence) and concave up (conduction) bands are the van Hove singularity that is seen between K_1 and K_2 when viewing the band structure normal to the line to I_A. The E_A transition is shown. **c** The optical conductivity difference for twisted bilayer graphene – Bernal bilayer grapheme ($\sigma_T - \sigma_B$) with a twist angle dependent offset $\theta/3$ is plotted versus photon energy. Reprinted and adapted with permission from [49] © 2014 American Chemical Society

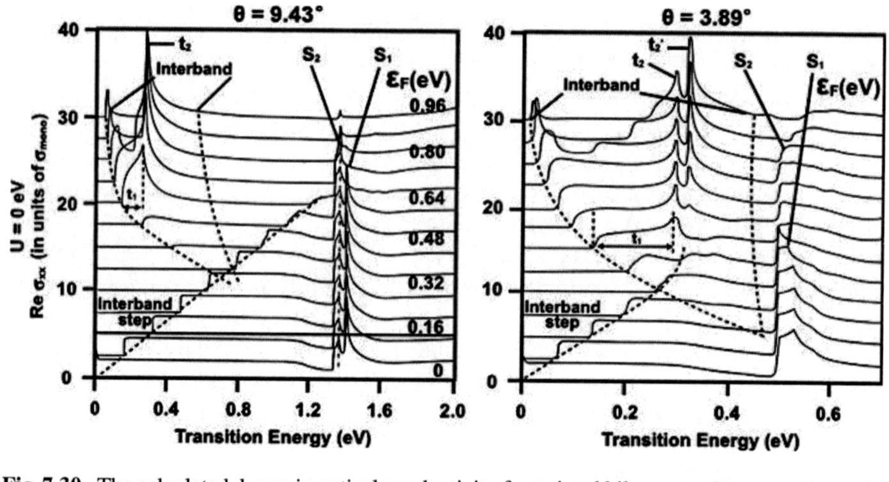

Fig. 7.30 The calculated dynamic optical conductivity for twisted bilayer graphene at twist angles of 9.43° and 3.98° at a variety of Fermi energies. Figure adapted and reprinted with permission from [51] © 2014 American Physical Society

in the higher symmetry multilayer stacks. A dual gate structure with gates above and below the h-BN encapsulated tMBG is as described in Sect. 7.3.1 is used in these studies. As with the MATBG, the dual gate structure was used to dope the trilayer graphene to different carrier densities and provide a displacement field by using a potential difference between the top and bottom gates. The electronic band structure of tMBG has a bias potential dependent band gap of tens of meV where the gap is close to the Γ point n the Brillouin zone. The band structure depends on the direction of the electric field (displacement D) [52]. For the configuration of graphene stacked with a small twist angle on bilayer graphene, when the direction of the electric field is normal to the graphene surface and pointed up from the bilayer graphene toward the monolayer graphene ($D < 0$), the density of states resembles bilayer graphene [52]. When the electric field points down toward the bilayer graphene, the density of states resembles bilayer—bilayer graphene.

tMBG has been studied in the presence of a combined magnetic field and electric field both of which were normal to the graphene plane. Transport measurements were done with current flow along the graphene plane. The metallic, insulating, and magnetic states have twist angle dependence. For a twist angle of $\theta = 1.08°$ at $0.3°\ K$, for small $|D|$, when the magnetic field results in filling factors $v = 0$, the tMBG is semimetallic, and for larger $|D|$, tMBG is a correlated insulator [52]. When the valence or conducting band is fully filled, $v = \pm 4$, correlated insulating states are seen over a wide range of $|D|$ [52]. The superconducting state observed for bilayer graphene is not observed. For another sample with a twist angle of $\theta = 0.89°$ at $0.05°\ K$ over a small range of bias conditions with $D > 0$ the anomalous Hall effect (AHE) was observed at $B = 0$ with a large Hall angle $\rho_{xy}/\rho_{xx} = 10$ [52]. The AHE was not quantized. As mentioned in Chap. 6, topological effects are responsible for the intrinsic contribution to the AHE. Additional information about the transport properties of tMBG can be found in [52].

The electronic band structure of Bernal (AB) stacked bilayer—bilayer graphene (or double bilayer graphene DBG) has been theoretically studied for twist angles up to 22° [44, 53]. Again, a dual gate structure with gates above and below the h-BN encapsulated DBG is used in these studies allowing tuning of carrier density and displacement field. Several studies discuss the correlated insulator state that is observed for small twist angles when an electrical displacement is applied normal the plane of the bilayer-bilayer stack at low temperatures [54–56]. In one study, moiré bands formed at twist angles between 1° and 1.35° ($1.01°,\ 1.10°,\ and\ 1.33°$), and these bands are relatively flat close to the γ point (center) of the moiré band Brillouin zone which promotes electron–electron interactions which can lead to correlated electron behavior [54]. By tuning the electrical displacement normal to the plane of the bilayers, the transport properties can be studied with and without a magnetic field. Twist angle dependent differences are observed which show that correlated insulator states are more likely to appear as the energy separation between the moiré band and the other bands increase [54]. Thus the flatness of the moiré band close to the γ point in the moiré Brillouin zone is not the only criteria for observation of correlated insulator states. The correlated insulator states are observed at $\frac{1}{2}$, $\frac{1}{4}$, and $\frac{3}{4}$ filling of the first conduction band [54]. An in plane magnetic field was used to show that $\frac{1}{4}$, and $\frac{3}{4}$ were spin polarized and the $\frac{1}{2}$ filling was valley polarized. Another study fabricated bilayer-bilayer samples with twist angles of (0.84°, 1.09°, and 1.23°), and the same correlated insulator behavior was observed for the 1.09° and 1.23° samples [55]. For a twist angle of 0.84°, three pairs of flat bands are observed and the correlated insulator states are observed over a wider range of filling factors [55]. Another study finds that the "exceptionally" low resistivity observed at some band fillings [54] which were considered as suggestive of the onset of superconductivity [54], are due to spontaneous symmetry breaking [56]. The change in sign of the Hall coefficient as a function of carrier density and a small residual resistivity provide evidence for spontaneous symmetry breaking in the moiré band due to electron – electron correlation [55]. It is suggested that spontaneous symmetry breaking may be responsible for some of the transport properties observed for transition metal dichalcogenide heterostructures [56].

Twisted middle layer—trilayer graphene displays a wide array of properties including superconductivity [57–59]. The magic angle for twisted middle layer – trilayer graphene is $\theta = 1.60°$, and the magic angle twisted middle layer—trilayer graphene has the acronym MATTG [58]. In one study, the top and bottom graphene layers are aligned and the middle layer is twisted to $\theta = 1.56°$ which theory has pointed to as being a magic angle resulting in flat bands in the moiré band structure [57]. These studies also use devices fabricated with a dual gate structure. By tuning the displacement field, superconductivity was observed at 2.1 K [57]. Both MATBG and MATTG show superconductivity and retain their symmetry at the magic twist angle. They have C_2 and time reversal symmetry [58]. Other twisted few layer graphene systems without this symmetry have not shown superconductivity. In these flat band systems, tuning the displacement field changes the bandwidth which tunes the correlated electron behavior. Superconductivity was also observed in another study where the twist angle for the middle layer is $\theta = 1.57° \pm 0.02°$ around 2 K [58]. The superconducting transition temperature for MATTG is twice that of MATBG which is an indication of the strength of the electron–electron interaction. In this study, at a zero displacement field, quantized conductance was observed at $\sigma_{xy} = 2,\ 4,$ and $10\,\frac{e^2}{h}$ indicating the presence of gapless Dirac states for a filling factor $\leq 4 = 4n/n_s$ where again, n is the carrier density and $n_s = 8\theta^2/\left(a^2\sqrt{3}\right)$ (using a Taylor series expansion for $\sin\theta$) is the superlattice density with the graphene lattice constant $a = 0.246$ nm [58]. Note the definition of the filling factor used in this study, $v = 4n/n_s$ [58]. The estimated band width for the flat bands in MATTG is 100 meV with 40 meV on the hole side and 60 meV on the electron side [59]. This is higher than the estimated 40–60 meV for MATBG, and may enhance correlated interactions. One important finding is that superconductivity in twisted few layer graphene does not seem to be due to the weak electron–electron interactions that occur in Bardeen-Cooper-Schrieffer (BCS) superconductivity [59].

One of the next steps in magic angle few layer graphene is to fabricate quadrilayer graphene that has twisted twisted bilayer-twisted bilayer flat bands [59].

7.10 The Electronic Band Structure and Optical Properties of Hexagonal Boron Nitride (h-BN) and Graphene—h-BN

Single crystal boron nitride can be either hexagonal or cubic. Due to the hexagonal structure, large band gap, and van der Waals interlayer interactions, h-BN plays a critical role in nanoelectronic applications [60]. Single layer, few layer, and bulk hexagonal boron nitride (h-BN) are all highly anisotropic uniaxial materials (see Chap. 1 for a discussion of the Fresnel Reflection coefficients for uniaxial materials) [61, 62]. Typically the c-axis will be normal to the surface for single and few layer samples as well as "flakes" of BN. These BN flakes are used for fabricating BN-graphene-BN multilayers for electronic devices [60]. The dielectric function for monolayer [63] and bulk h-BN [64, 65] is clearly anisotropic. Excitonic effects play a strong role in the electronic band structure and optical properties for single layer and bulk h-BN [66–68]. The tight binding electronic band structure for single layer h-BN is a useful starting point for discussing these properties since it can be compared to the nearest neighbor model for graphene [68]. The nearest neighbor, tight binding model for a single layer hexagonal lattice with two types of atoms was discussed in Sect. 7.2.7. The results of a minimal, tight binding model band structure that includes interlayer hopping between only atoms that are directly below (or on top) of each other is shown in Fig. 7.31.

Band structure calculations show that single layer h-BN is a direct gap material with the gap occurring at the K and K' points [66] while bulk h-BN is an indirect gap material. Close to the gap, the valence electrons are found at N sites and the conduction electrons at B sites [66]. The binding energy of an exciton in bulk h-BN is 0.7 eV which is very high. The binding energy of the first five excitons in single layer h-BN have been calculated and their value depends on the details of the theory used. However, the first exciton has a very large binding energy of ~ 2 eV. The ab initio values for the exciton binding energy are provided in Table 7.1. The

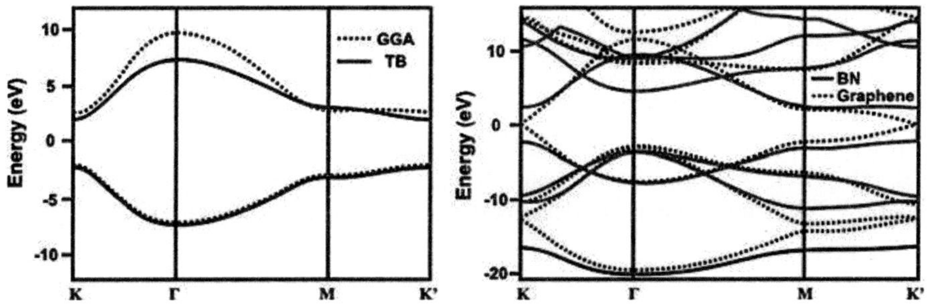

Fig. 7.31 The electronic band structure of single layer h-BN is shown. On the left, the simple tight binding model for the π and π^* bands is shown compared to the Generalized Gradient Approximation for a Density Functional Theory (GAA-DFT) determined band structure. It is obvious that the π and π^* bands do not touch at the K and K' points in the Brillouin zone as they do for graphene. On the left, the band structure of h-BN is compared to that of graphene using GAA-DFT. Figure adapted and reprinted with permission from [68] © 2011 American Physical Society

Table 7.1 Exciton binding energies [64]

	1 (2x)	2 (2x)	3	4	5 (2x)
B.E. (ab initio) (eV)	−1.932	−1.076	−1.045	−0.98	−0.892
Symmetry	E	E	A_1	A_2	E

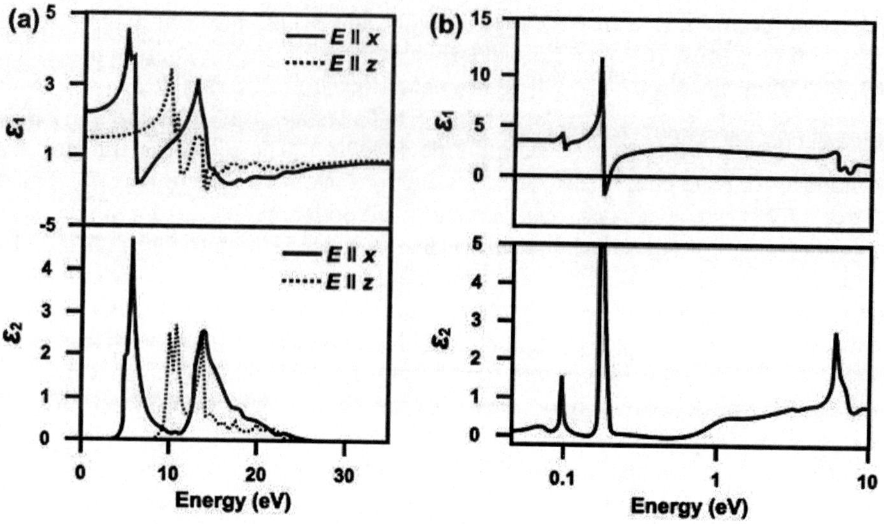

Fig. 7.32 The dielectric function of single layer and bulk BN. Theoretical results are shown in **a** and experimental data in **b**. For (**a**), $E\|x$ is for the electric field of the light along the $x - y$ plane (in-plane dielectric function) and $E\|z$ for the electric field normal to the c face of bulk BN. For (**b**), the near normal incidence reflectance data was inverted Thus, the data in (**b**) is the in-plane dielectric function for single layer BN. Figure adapted from [63] and [65]. Figure 7.32a adapted and reprinted from [63]. [65] © 2015 used with permission from Elsevier. Figure 7.32b adapted and reprinted with permission from. © 1984 American Physical Society

radius of the exciton is localized to nearest neighbors and just beyond. The excitons in single layer h-BN can theoretically described by a tight binding—Wannier method [66]. The excitons in bulk h-BN have more recently been described as a Frenkel type [65]. The dielectric function of single and bulk h-BN is shown in Fig. 7.32, and the relationship between electron band structure and transition energies is shown in Fig. 7.33. The Raman spectra of cubic and hexagonal BN is shown in Fig. 7.34 [69].

7.10.1 Graphene—BN Heterostructures

Graphene—BN heterostructures have a slight mismatch in lattice constants as shown in Fig. 7.35. The infra-red optical absorption spectra of graphene—hexagonal BN multilayers have been theoretically predicted as a function of graphene-BN lattice misalignment, and are shown in Fig. 7.36 [70, 72]. The calculated absorption spectrum is dependent on the assumptions about the coupling between layers [70, 72]. Here we show the absorption coefficient that was determined using a phenomenological Hamiltonian that included the unperturbed graphene Hamiltonian, a term describing the modulation of the graphene—BN potential due to angular alignment, a term for the sub-lattice asymmetry due to the BN substrate, and a term due to hopping

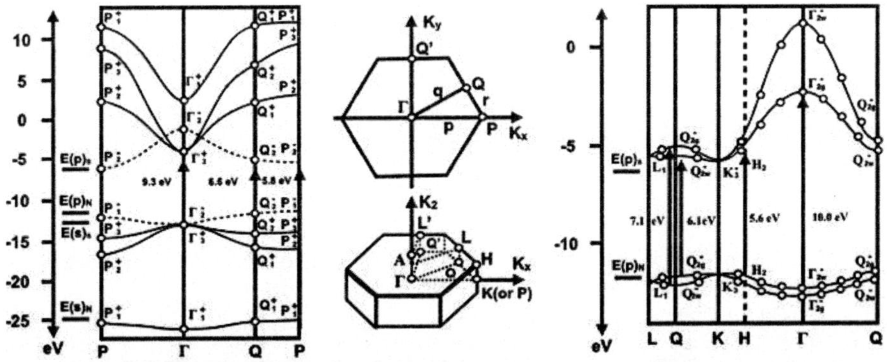

Fig. 7.33 Relationship between optical transitions and the tight binding electronic band structure of h-BN in 2D on the left and for 3D or bulk h-BN on the right. It is important to state that a variety of different calculated and experimental values of the band gap have been reported [59]. A direct gap transition is seen at the P point in the 2D Brillouin zone from P_1^- to P_2^- at 5.8 eV. Several different experimental values have been reported including a reflectance measurement of 5.2 eV [59]. A direct gap transition occurs the calculated band structure at the H point in the Brillouin zone for 3D from H_3 to H_2 at 5.6 eV. Figure adapted and reprinted with permission from [65]. © 1984 American Physical Society

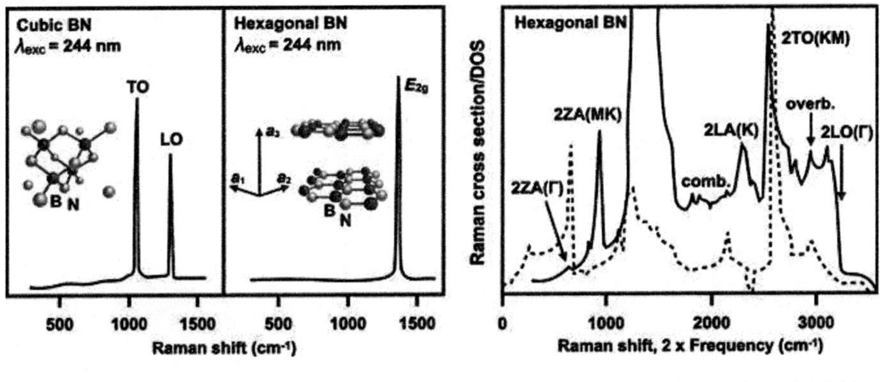

Fig. 7.34 The resonant Raman spectra of c-BN and h-BN are shown on the left excited with 244 nm laser light, and the second order Raman spectra from h-BN on the right. The Density of States (DOS) is also shown on the right using a dashed line with the frequency multiplied by 2 [69]. For h-BN, the laser light was scattered a near normal incidence parallel to the c-axis. The Raman spectra is a function of excitation wavelength and intensity [67]. Figure adapted and reprinted with permission from [69] © 2005 American Physical Society

between graphene and BN due to the pseudo-magnetic field [70, 72]. Reflected electron spectroscopy has shown that there is little interaction between graphene and h-BN and thus their electronic band structures are not altered over a wide energy range [73].

Since h-BN is used to encapsulate graphene and few layer graphene samples in the test structures used for transport measurements, we note that the weak interaction

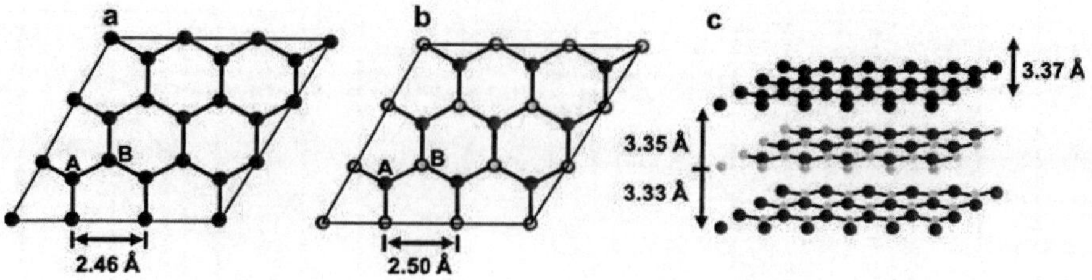

Fig. 7.35 The graphene and BN multilayer stack is not perfectly commensurate due to the small differences in the lattice constants. A graphene—BN—BN multilayer stack is shown on the right with no angular misalignment [70, 71]

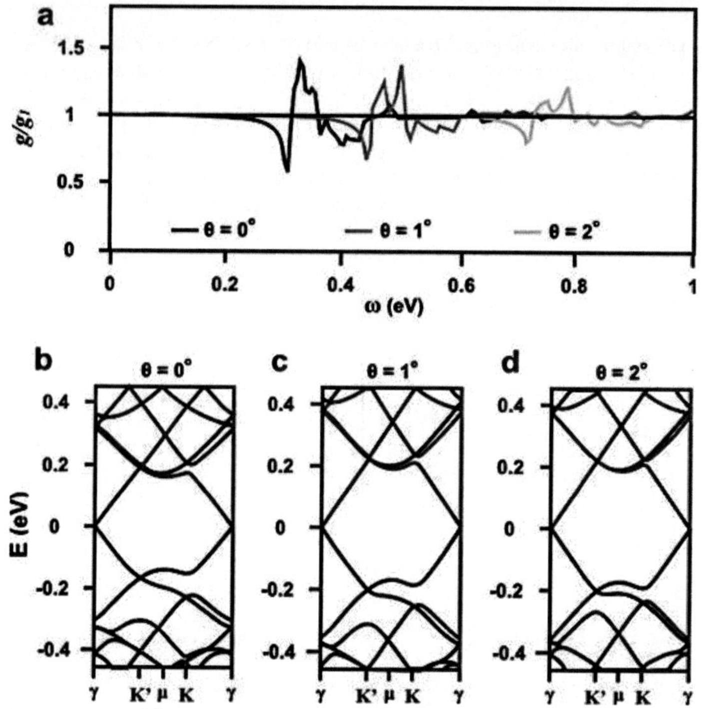

Fig. 7.36 Theoretically predicted Infra-Red optical absorption spectra for graphene—h-BN are shown. Here, g is the optical absorption coefficient, and $g_1 = \pi e^2/\hbar c$ is the optical absorption coefficient of monolayer graphene. The ratio of g/g_1 is shown in **a** for three angular alignments. The electronic band structure for the graphene—BN structure is shown in **b** for these angular alignments. Figure adapted from [72] © IOP Publishing Ltd and Deutsche Physikalische Gesellschaft CC BY 3.0 https://creativecommons.org/licenses/by/3.0/

potential often does not seem to result in an impact on the measured properties. However, as discussed above, alignment of one of the encapsulating h-BN layers with one of the graphene layers in twisted bilayer graphene did enable the observation of the QAHE. Thus, there must be a twist angle dependent interaction between the h-BN and few layer graphene [46].

References

1. See for example: C.N.R. Rao, K. Biswas, K.S. Subrahmanyam, A. Govindaraj, Graphene, the new nanocarbon. J. Mater. Chem. **19**, 2457–2469 (2009); and M.O. Goerbig, Electronic properties of graphene in a strong magnetic field. Rev. Mod. Phys. **83**, 1193–1239 (2011)
2. E. McCann, Electronic properties of monolayer and bilayer graphene, in *NanoScience and Technology*. ed. by G. Nanoelectronics (Berlin, Springer, 2012), pp. 237–275
3. A. Maffucci, G. Miano, Electrical properties of graphene for interconnect applications. Appl. Sci. **4**(2), 305–317 (2014)
4. The first discussion of the tight binding model from graphene can be found in: P.R. Wallace, The band theory of graphite. Phys. Rev. **71**, 622–634 (1947)
5. S. Reich, J. Maultzsch, C. Thomsen, P. Ordejón, Tight-binding description of grapheme. Phys. Rev. B **66**, 035412 (2002)
6. D.C. Elias, R.V. Gorbachev, A.S. Mayorov, S.V. Morozov, A.A. Zhukov, P. Blake, L.A. Ponomarenko, I.V. Grigorieva, K.S. Novoselov, F. Guinea, A.K. Geim, Dirac cones reshaped by interaction effects in suspended graphene. Nat. Phys. **7**, 701–703 (2011)
7. F.J. Nelson, V.K. Kamineni, T. Zhang, E.S. Comfort, J.-U. Lee, A.C. Diebold, Optical properties of large-area polycrystalline chemical vapor deposited graphene by spectroscopic ellipsometry. Appl. Phys. Lett. **97**(253110), 1–3 (2010)
8. P. Blake, E.W. Hill, A.H. Castro Neto, K.S. Novoselov, D. Jiang, R. Yang, T.J. Booth, A.K. Geim, Making graphene visible. Appl. Phys. Lett. **91**, 063124 (2007)
9. K.F. Mak, M.Y. Sfeir, J.A. Misewich, T.F. Heinz, The evolution of electronic structure in few layer graphene revealed by optical spectroscopy. Proc. Natl. Acad. Sci. **107**, 14999–15004 (2010)
10. R.R. Nair, P. Blake, A.N. Grigorenko, K.S. Novoselov, T.J. Booth, T. Stauber, N.M.R. Peres, A.K. Geim, Fine structure constant defines visual transparency of graphene. Science **320**, 1308 (2008)
11. P. Kim, Graphene and relativistic quantum mechanics, in *Dirac Matter* (Springer, New York, 2016), pp. 1–23
12. T. Stauber, N.M.R. Peres, A.K. Geim, Optical conductivity of graphene in the visible region of the spectrum. Phys. Rev. B **78**, 085432 (2008)
13. L. Yang, J. Deslippe, C.H. Park, M.L. Cohen, S.G. Louie, Excitonic effects on the optical response of graphene and bilayer graphene. Phys. Rev. Lett. **103**, 186802 (2009)
14. F.J. Nelson, J.-C. Idrobo, J. Fite, Z.L. Mišković, S.J. Pennycook, S.T. Pantelides, J.U. Lee, A.C. Diebold, Electronic excitations in graphene in the 1–50 eV range: The π and $\pi + \sigma$ peaks are not plasmons. Nano Lett. **14**, 3827–3831 (2014)
15. D. Novko, V. Despoja, M. Sunjic, Changing character of electronic transitions in graphene: from single-particle excitations to plasmons. Phys. Rev. B **91**, 19540 (2015)
16. http://materia.fisica.unimi.it/manini/berryphase.html
17. J. Xue, Berry phase and the unconventional quantum Hall effect in graphene, arXiv:1309.6714v1
18. D. Xiao, M.-C. Chang, Q. Niu, Berry phase effects on electronic properties. Rev. Mod. Phys. **82**, 1959–2007 (2010)

19. Y. Zhang, Y.-W. Tan, H.L. Stormer, P. Kim, Experimental observation of the quantum Hall effect and Berry's phase in graphene. Nature **438**, 201–204 (2005)
20. T. Ando, T. Nakanishi, R. Saito, Berry's phase and the absence of backscattering in carbon nanotubes. J. Phys. Soc. Jpn **67**, 2857 (1998)
21. K.S. Novoselov, E. McCann, S.V. Morozov, V.I. Fal'ko, M.I. Katsnelson, U. Zeitler, D. Jiang, F. Schedin, A.K. Geim, Unconventional quantum Hall effect and Berry's phase of 2π in bilayer grapheme. Nat. Phys. **2**, 177–180 (2006)
22. J.I.A. Li, C. Tan, S. Chen, Y. Zeng, T. Taniguchi, K. Watanabe, J. Hone, C.R. Dean, Even denominator fractional quantum Hall state in bilayer graphene. Science **358**, 648–652 (2017)
23. C.-H. Park, N. Marzari, Berry phase and pseudospin winding number in bilayer grapheme. Phys. Rev. B **84**, 205440 (2011)
24. M. Aoki, H. Amawashi, Dependence of band structures on stacking and field in layered graphene. Sol. State Comm. **142**, 123–127 (2007)
25. K.F. Mak, J. Shan, T.F. Heinz, Electronic structure of few-layer graphene: experimental demonstration of strong dependence on stacking sequence. Phys. Rev. Lett. **104**, 176404 (2010)
26. T. Taychatanapat, K. Watanabe, T. Taniguchi, P. Jarillo-Herrero, Quantum Hall effect and Landau-level crossing of Dirac fermions in trilayer graphene. Nat. Phys. **7**, 621–625 (2011)
27. A.C. Diebold, F.J. Nelson, V.K. Kamineni Spectroscopic ellipsometry of nanoscale materials for semiconductor device applications, in *Ellipsometry at the NanoScale* (Springer, Berlin, 2013), Chap. 16
28. F. Nelson, A. Sandin, D.B. Dougherty, D.E. Aspnes, J.E. Rowe, A.C. Diebold, Optical and structural characterization of epitaxial graphene on vicinal 6H-SiC(0001)–Si by spectroscopic ellipsometry, Auger spectroscopy, and STM. J. Vac. Sci. Technol. B **30**, 04E106 (2012)
29. F.J. Nelson, V. Kamineni, T. Zhang, E. Comfort, J.-U. Lee, A.C. Diebold, Spectroscopic ellipsometry of CVD grapheme. ECS Trans. **35**, 173 (2011)
30. K.F. Mak, C.H. Lui, J. Shan, T.F. Heinz, Observation of an electric-field-induced band gap in bilayer graphene by infrared spectroscopy. Phys. Rev. Lett. **102**, 256405 (2009)
31. C.-H. Park, S.G. Louie, Tunable excitons in biased bilayer graphene. Nano Lett. **10**, 426–431 (2010)
32. L. Ju, L. Wang, T. Cao, T. Taniguchi, K. Watanabe, S.G. Louie, F. Rana, J. Park, J. Hone, F. Wang, P.L. McEuen, Tunable excitons in bilayer graphene. Science **358**, 907–910 (2017)
33. T. Chari, R. Ribeiro-Palau, C.R. Dean, K. Shepard, Resistivity of rotated graphene-graphite contacts. Nano Lett. **16**, 4477–4482 (2016)
34. H. Yoo, R. Engelke, S. Carr, S. Fang, K. Zhang, P. Cazeaux, S.H. Sung, R. Hovden, A.W. Tsen, T. Taniguchi, G-C Yi, M. Kim, M. Luskin, E.B. Tadmor, E. Kaxiras, P. Kim, Atomic and electronic reconstruction at the van der Waals interface in twisted bilayer graphene. Nat. Mat. **18**, 448–453 (2019)
35. Y. Cao, V. Fatemi, S. Fang, K. Watanabe, T. Taniguchi, E. Kaxiras, P. Jarillo-Herrero, Unconventional superconductivity in magic-angle graphene superlattices. Nat. **556**, 43–50 (2018)
36. M. Yankowitz, S. Chen, H. Polshyn, Y. Zhang, K. Watanabe, T. Taniguchi, D. Graf, A.F. Young, C.R. Dean, Tuning superconductivity in twisted bilayer graphene. Science **363**, 1059–1064 (2019)
37. J.M.B. Lopes dos Santos, N.M.R. Peres, A.H. Castro Neto, Graphene bilayer with a twist: electronic structure. Phys. Rev. Lett. **99**, 256802 (2007)
38. E. J. Mele, Commensuration and interlayer coherence in twisted bilayer graphene, Phys. Rev. B **81**, 161405 R (2010)
39. G. Trambly de Laissardíere, D. Mayou, L. Magaud, Numerical studies of confined states in rotated bilayers of grapheme. Phys. Rev. B **86**, 125413 (2012)
40. R. Bistritzer, A.H. MacDonald, Moiré bands in twisted double-layer graphene. Proc. Nat. Acad. Sci. **108**, 12233–12237 (2011)
41. K. Kim, A. DaSilva, S. Huang, B. Fallahazad, S. Larentis, T. Taniguchi, K. Watanabe, B.J. LeRoy, A.H. MacDonald, E.l Tutuc, Tunable moiré bands and strong correlations in small-twist-angle bilayer grapheme. Proc. Nat. Acad. Sci. **114**, 3364–3369 (2017)

42. G. Tarnopolsky, A. Jura Kruchkov, A. Vishwanath, Origin of magic angles in twisted bilayer graphene. Phys. Rev. Lett. **122**, 106405 (2019)
43. Y. Cao, V. Fatemi, A. Demir, S. Fang, S.L. Tomarken, J.Y. Luo, J.D. Sanchez-Yamagishi, K. Watanabe, T. Taniguchi, E. Kaxiras, R.C. Ashoori, P. Jarillo-Herrero, Correlated insulator behaviour at half-filling in magic-angle graphene superlattices, Nature **556**, 80–84 (2018), and U. Zondiner, A. Rozen, D. Rodan-Legrain, Y. Cao, R. Queiroz, T. Taniguchi, K. Watanabe, Y. Oreg, F. von Oppen, A. Stern, E. Berg, P. Jarillo-Herrero, S. Ilani, Cascade of phase transitions and dirac revivals in magic angle grapheme. Nature **582**, 203–208 (2020)
44. N.R. Chebrolu, B. L. Chittari, J. Jung, Flat bands in twisted double bilayer grapheme. Phys. Rev. B **99**, 235417 (2019)
45. Y. Cao, D. Chowdhury, D. Rodan-Legrain, O. Rubies-Bigorda, K. Watanabe, T. Taniguchi, T. Senthil, P. Jarillo-Herrero, Strange metal in magic-angle graphene with near Planckian dissipation. Phys. Rev. Lett. **124**, 076801 (2020)
46. M. Serlin, C.L. Tschirhart, H. Polshyn, Y. Zhang, J. Zhu, K. Watanabe, T. Taniguchi, L. Balents, A.F. Young, Intrinsic quantized anomalous Hall effect in a moiré heterostructures. Science **367**, 900–903 (2020)
47. T-F Chung, Y. Xu, Y.P. Chen, Transport measurements in twisted bilayer graphene: Electron-phonon coupling and Landau level crossing. Phys. Rev. B **98**, 035425 (2018)
48. K. Kim, S. Coh, L.Z. Tan, W. Regan, J.M. Yuk, E. Chatterjee, M.F. Crommie, M.L. Cohen, S.G. Louie, A. Zettl, Raman spectroscopy study of rotated double-layer graphene: misorientation-angle dependence of electronic structure. Phys. Rev. Lett. **108**, 246103 (2012)
49. R.W. Havener, Y. Liang, L. Brown, L. Yang, J. Park, Van Hove singularities and excitonic effects in the optical conductivity of twisted bilayer graphene. Nano Lett. **14**, 3353–3357 (2014)
50. T.V. Alencar, D. von Dreifus, M.G. Cota Moreira, G.S. N. Eliel, C-H Yeh, P-W Chiu, M.A. Pimenta, L.M. Malard, A.M. de Paula, Twisted bilayer graphene photoluminescence emission peaks at van Hove singularities. J. Phys.: Condens. Matter **30**, 175302 (2018)
51. P. Moon, Y-W Son, M. Koshino, Optical absorption of twisted bilayer graphene with interlayer potential asymmetry. Phys. Rev. B **90**, 155427 (2014)
52. S. Chen, M. He, Y.-H. Zhang, V. Hsieh, Z. Fei, K. Watanabe, T. Taniguchi, D.H. Cobden, X. Xu, C.R. Dean, M. Yankowitz, Electrically tunable correlated and topological states in twisted monolayer–bilayer graphene. Nat. Phys. **17**, 374–380 (2021)
53. F. Haddadi, Q.S. Wu, A.J. Kruchkov, O.V. Yazyev, Moiré flat bands in twisted double bilayer graphene. Nano Lett. **20**, 2410–2415 (2020)
54. G.W. Burg , J. Zhu, T. Taniguchi, K. Watanabe, A.H. MacDonald, E. Tutuc, Correlated insulating states in twisted double bilayer grapheme. Phys. Rev. Lett. **123**, 197702 (2019)
55. Y. Cao, D. Rodan-Legrain, O. Rubies-Bigorda, J.M. Park, K.i Watanabe, T. Taniguchi, P. Jarillo-Herrero, Tunable correlated states and spin-polarized phases in twisted bilayer-bilayer grapheme. Nature **583**, 215–220 (2020)
56. M. He, Y. Li, J. Cai, J.Y. Liu, K. Watanabe, T. Taniguchi, X. Xu, M. Yankowitz, Symmetry breaking in twisted double bilayer graphene. Nat. Phys. **17**, 26–30 (2021)
57. Z. Hao, A.M. Zimmerman, P. Ledwith, E. Khalaf, D.H. Najafabadi, K. Watanabe, T. Taniguchi, A. Vishwanath, P. Kim, Electric field–tunable superconductivity in alternating-twist magic-angle trilayer graphene. Science **371**, 1133–1138 (2021)
58. J.M. Park, Y. Cao, K. Watanabe, T. Taniguchi, P. Jarillo-Herrero, Tunable strongly coupled superconductivity in magic-angle twisted trilayer graphene. Nature **590**, 249–255 (2021)
59. A. Yazdani, Magic, symmetry, and twisted matter: alternating magic-angle twist induces unconventional superconductivity. Science **371**, 1098–1099 (2021)
60. C.R. Dean, A.F. Young, I. Meric, C. Lee, L. Wang, S. Sorgenfrei, K. Watanabe, T. Taniguchi, P. Kim, K.L. Shepard, J. Hone, Boron nitride substrates for high-quality graphene electronics. Nat. Nanotech. **5**, 722–726 (2010)
61. M. Schubert, *Infrared Ellipsometry on Semiconductor Structures: Phonon, Plasmons, and Polaritons* (Springer, Heidelberg, 2004)
62. E. Franke, H. Neumann, M. Schubert, T.E. Tiwald, J.A. Woollam, J. Hahn, Infrared ellipsometry on hexagonal and cubic boron nitride thin films. Appl. Phys. Lett. **70**, 1668–1670 (1997)

63. R. Beiranv and S. Valedbagi, Electronic and optical properties of h-BN nanosheet: a first principles calculation. Diamond and Relat. Mat. **58**, 190–195 (2015)
64. R. Mamy, J. Thomas, G. Jezequel, J.C. Lemonnier, Anisotropie des transitions interbandes dans le nitrure de bore hexagonal, J. Phys. (Paris) Lett. **42**, L 473–L 475 (1981)
65. D.M. Hoffman, G.L. Doll, P.C. Eklund, Optical properties of pyrolytic boron nitride in the energy range 0.05–10 eV. Phys. Rev. B **30**, 6051–6056 (1984)
66. T. Galvani, F. Paleari, H. P. C. Miranda, A. Molina-S´anchez, L. Wirtz, S. Latil, H. Amara, F. Ducastelle, Excitons in boron nitride single layer. Phys. Rev. B **94**, 125303 (2016)
67. B. Arnaud, S. Lebe‘gue, P. Rabiller, M. Alouani, Huge excitonic effects in layered hexagonal boron nitride. Phys. Rev. Lett. **96**, 026402 (2006)
68. R.M. Ribeiro, N.M.R. Peres stability of boron nitride bilayers: ground-state energies, interlayer distances, and tight-binding description. Phys. Rev. B **83**, 235312 (2011)
69. S. Reich, A.C. Ferrari, R. Arenal, A. Loiseau, I. Bello, J. Robertson, Resonant Raman scattering in cubic and hexagonal boron nitride. Phys. Rev. B **71**, 205201 (2005)
70. J.R. Wallbank, Electronic Properties of Graphene Heterostructures with Hexagonal Crystals. Springer Doctoral Theses (Springer, New York, 2014)
71. A.J. Green, A.C. Diebold, Thickness and rotational effects in the simulated HRTEM images of graphene on hexagonal boron nitride. Microsc. Microanal. **20**, 1753–1763 (2014)
72. S.L. Abergel, J.R. Wallbank, X. Chen, M. Mucha-Kruczy´nski, V.I Fal'ko, Infrared absorption by graphene–hBN heterostructures. New J. Phys. **15** (2013) 123009
73. J. Jobst, A.J.H. van der Torren, E.E. Krasovskii, J. Balgley, C.R. Dean, R.M. Tromp, S.J. van der Molen, Quantifying electronic band interactions in van der Waals materials using angle-resolved reflected-electron spectroscopy. Nat. Comm. **7**, 13621 (2016)
74. F. Rana, Class notes, Energy bands in graphene: Tight binding and the nearly free electron approach. https://courses.cit.cornell.edu/ece407/Lectures/handout11.pdf

Chapter 8
Optical and Electrical Properties of Transition Metal Dichalcogenides (Monolayer and Bulk)

Abstract In this chapter, we discuss the electronic band structure, electrical, and optical properties of transition metal dichalcogenides. The different crystallographic structures for transition metal dichalcogenides are presented along with a discussion of the chemical bonding. Many of the transition metal dichalcogenides consist of van der Waals bonded monolayers where the monolayers consist of trilayers with a transition metal atom layer between a top and bottom chalcogenide layer. Often these monolayers have a trigonal prismatic arrangement of chalcogenide atoms around the metal atoms. A tight binding model for three of the *d* orbitals of the transition metal atoms provides a useful description of the highest energy valence band and lowest energy conduction bands of trigonal prismatic monolayer transition metal dichalcogenide. The impact of spin orbit coupling on the band structure is shown. We discuss how the electronic band structure due to the honeycomb lattice of many transition metal dichalcogenides monolayers interacts with spin orbit coupling resulting in differences in optical transitions between the K and K' locations in the Brillouin zone. We present photoluminescence spectra demonstrating these differences. We also show theoretical and experimental dielectric function data for a variety of monolayer, multilayer, and bulk transition metal dichalcogenides. We show how Raman spectroscopy is sensitive to the layer structure. We also discuss the observation of superconductivity of TMD materials. A summary of the point group and space group symmetry and Raman Tensors of transition metal dichalcogenides is provided.

The interesting and diverse electronic band structure, electrical, and optical properties of transition metal dichalcogenides (TMD) are reviewed in this chapter. The elements in the seventh column of the periodic table (S, Se, Te, etc.) are referred to as chalcogenides. The transition metals from groups 4 to 10 form MX_2 compounds with chalcogenides that have polar bonds. The metal atoms provide 4 electrons and the chalcogenides accept 2 electrons resulting in oxidation states of M^{+4} and X^{-2} in many TMD materials [1]. TMDs composed of metals from groups 4 to 7 in the Period Table generally form the layered structures while TMDs having metals from groups 8 to 10 typically form non-layered crystals structures [1]. For monolayer TMDs, a single layer consists of three atomic layers with the top and bottom layer being the chalcogenide and the middle layer being the transition metal such as W or Mo.

A. Diebold and T. Hofmann, *Optical and Electrical Properties of Nanoscale Materials*, Springer Series in Materials Science 318,
https://doi.org/10.1007/978-3-030-80323-0_8

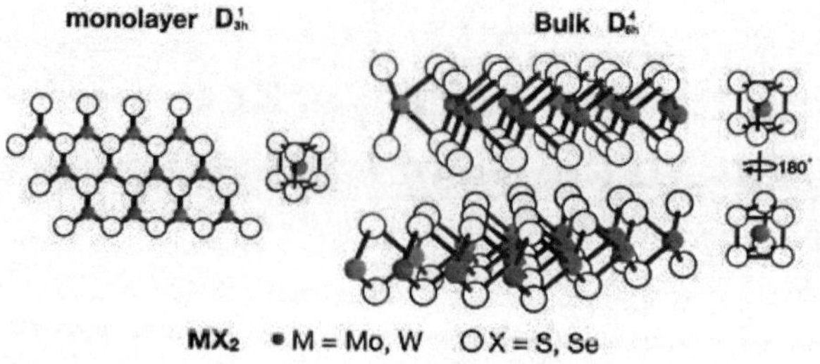

Fig. 8.1 The hexagonal crystal structure with trigonal prismatic metal coordination shown here is one of the stable phases of Transition Metal Dichalcogenides. Because the configuration of chalcogenide atoms around the metal atom is trigonal, this polymorph is $2H_C$. Figure adapted from [2]. © 2015 Royal Society of Chemistry

The chalcogenide atoms will have lone pairs of electrons in the sp^3 orbitals that are pointing away from the metal atoms making the top and bottom layers chemically inert [1]. Since the top and bottom of the layers are inert, van der Waals bonding between layers is required for forming a multilayer solid. Thus, many multilayer TMDs are 2D materials where the single layers can stack in different configurations resulting in structural polymorphs. A number of interesting properties have been observed in TMD materials. The electronic band structure of TMD materials, with a hexagonal crystal lattice with the metal atom in the trigonal prismatic configuration (Fig. 8.1), exhibit interesting optical and electrical properties which are discussed in this chapter [1]. The non-hexagonal crystal structures also exhibit interesting materials properties. One example is orthorhombic WTe_2 with the space group $Pmn2_1$ ($C^7{}_{2v}$), which is a type of topological material known as a type II Weyl semimetal. This is further discussed in Chap. 9. The hexagonal crystal structure of many monolayer TMDs is reminiscent of that of graphene. The 2D hexagonal structure of the layered TMDs have pseudospin states, and spin–orbit coupling results in differences between the optical properties at the K and K' points in the Brillouin zone. In Fig. 8.1, we show the three atomic layer structure for TMD materials having the trigonal prismatic bonding configuration for the transition metal atom. This configuration results in the hexagonal crystal structure which is also shown in Fig. 8.1. The rotational alignment of bilayer TMDs can be altered resulting in twisted bilayers with properties that are different from single monolayers or non-rotated bilayers. This is discussed at the end of the chapter.

Studying the band structure of TMD materials is critical to understanding their optical and electrical properties. For example, the change of band gap from indirect for bilayer and multilayer MoS_2 strongly impacts optical absorption [1–5]. This is shown in Fig. 8.2. The band gap of monolayer MoS_2 is 1.8 eV and is 1.5 eV for $MoSe_2$ [3, 5]. Another interesting aspect of TMD materials is that electrical and optical measurements result in different values for the band gap. This is due to a phenomenon known as band gap renormalization [5]. The magnitude of bandgap renormalization is reported to be an order of magnitude larger than the 10 meV of GaAs [5]. In an optical measurement, the electron and hole interact and an exciton

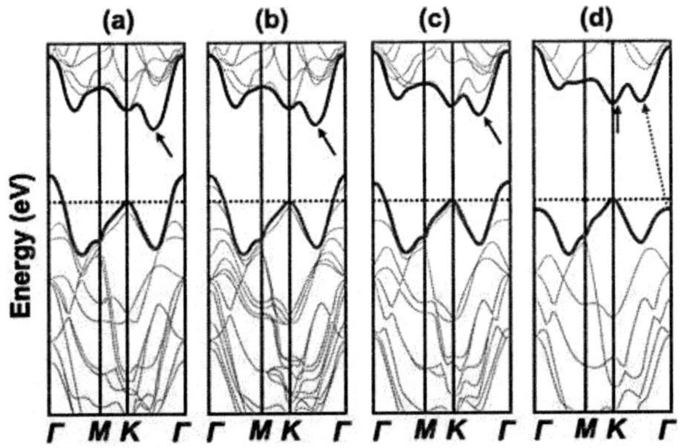

Fig. 8.2 Band structure of the 2H polymorph of bulk MoS_2 to single layer as follows: **a** bulk; **b** four layer; **c** bilayer; and **d** the trigonal monolayer. A direct band gap results from the changes in the energy versus wavevector for both the upper valence band and lower conduction band as the number of layers decreases to one. Figure 8.3 provides an image of the trigonal prismatic monolayer, and Fig. 8.4 provides an image of the bulk 2H polymorph structure. The typical stacking sequence for 2H is AA′ which is shown in Fig. 8.4. Figure adapted and reprinted with permission from [3]. © 2010 American Chemical Society

can be formed. This "excitonic" effect was previous discussed for the silicon E1 transition in Chaps. 4 and 5. The difference in energy between the electrical and optical measurements comes from the binding energy of the exciton. Clearly this effect is important for direct gap transitions, but not for indirect gap transitions. Spin–orbit coupling splits the valence band of some TMD materials resulting in valley dependent optical properties. The pseudospin K and K' valleys of TMD materials can be independently optically excited using circularly polarized light [2, 6, 7]. The results of this difference have been observed by photoluminescence [2, 5], and are further explored after we present a tight binding model for the electronic band structure.

The flow of this chapter is as follows: The chapter starts with a discussion about the structure and bonding for the many different polymorphs of TMD materials. The transition metal atoms in trigonal prismatic monolayers form a honeycomb lattice. Based on this structure, we next use a tight binding approach to arrive at a three band electronic structure for the highest lying valence band and two lowest lying conduction bands for monolayer TMD using the d_{z^2}, d_{xy}, and $d_{x^2-y^2}$ orbitals. Then the splitting of the valence and conduction bands due to spin orbit interactions at the K and K' points in the Brillouin zone is presented. This leads to differences in the electronic structure which can be probed optically. Next, we describe experimental observation of valley pseudospin and valence band spin splitting using circularly polarized light. Then, we turn to the topic of massive Dirac fermions at K and K'. This leads to a discussion about the effect of band gap renormalization. Next, we present experimental data for the complex refractive index (dielectric function) and optical conductivity of monolayer TMD. The topic of twisted bilayer TMD is covered.

Then, we present experimental data for the complex refractive index (dielectric function) of multilayer and bulk TMD. Raman spectroscopy is a means of determining the number of TMD monolayers in a sample, and that is covered next. The theoretical Haeckelite structure is presented to demonstrate the theoretical prediction of new structures. Then, we discuss twisted and hetero-bilayers of TMD with h-BN. This is followed by a discussion of superconducting in TMD materials. Then we discuss ReS_2 and $ReSe_2$ which have the $1T'$ Structure and as a consequence have different properties than the other TMD structures. After a brief discuss about the practical aspects of measuring TMD optical properties, the chapter ends with a summary of space group and point group symmetry and a table of the Raman tensors for different TMD structures.

8.1 Structure and Bonding for TMD Materials

A large number of polymorphs exist for TMD materials. The oxidation state of the metal atom in TMD materials is generally + 4. TMDs with group 6 metal atoms (d^2) typically have the trigonal prismatic structure [1]. The group 5 metal (d^1) TMDs can have either the trigonal prismatic or octahedral structure. Typically group 4 (d^0) and group 10 (d^6) TMDs all have the octahedral structure [1]. Group 7 (d^3) TMDs have a distorted octahedral structure. Since monolayer TMDs exhibit only two polymorphs, trigonal prismatic and octahedral, those are considered first. These two polymorphs are shown in Fig. 8.3.

First we consider the chemical bonding for the group 6 metal (d^2) compound MoS_2. Some of this discussion can be generalized [1, 8]. The S atoms have the electronic configuration of [Ne] $3s^2 3^4$ and the Mo atoms have [Kr] $4d^5 5s^1$. As with many other TMDs, the oxidation state of the metal atom is − 4 and for each of the two chalcogenide atoms it is + 2. Thus, the metal atom is providing 4 electrons for bonding. The outer sulfur electrons hybridize into sp^3 orbitals and bond with the Mo d orbitals. The d orbital splitting is different for each polymorph [1]. The metal–metal bond length for TMDs is 15–25% longer than for elemental transition metal crystals varying from 3.05 to 4.30 Å [1]. Thus the orbitals have less spatial overlap between metal atoms [1]. This is important when considering the band structure of TMD materials since it provides justification for using the tight binding approach. The point group and space group symmetry of single trilayer, multi-trilayer, and bulk TMD materials are discussed in Chap. 2 in Sect. 2.12. Symmetry considerations are further discussed in association with the section on the Tight Binding Model below. Here it is useful to mention that the trigonal prismatic configuration splits the d orbitals into three symmetry related categories which are labeled using the Mulliken notations discussed in Sect. 8.2: A_1' for d_{z^2}; E' for $d_{x^2-y^2}$, d_{xy}, and E'' for d_{xz}, d_{yz}. Symmetry considerations for the transition metal atom layer are used below to motivate d band formation and the mixing of the d orbitals for the 3 band Tight Binding model. Ligand Field Theory provides a useful means of understanding the splitting of the energy of d orbital of the transition metal atoms. If one considers

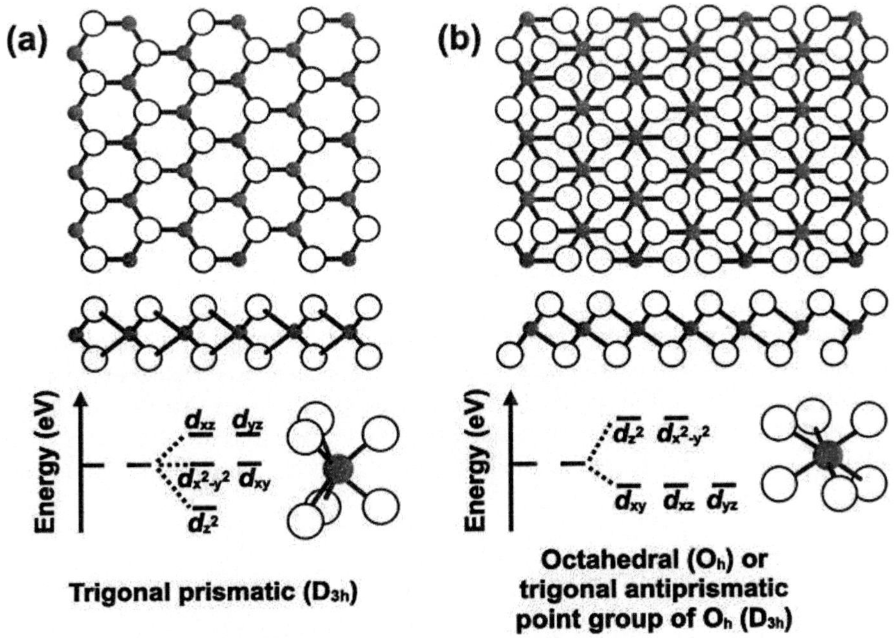

Fig. 8.3 Monolayer TMD materials have two bonding configurations for the transition metal atoms. The trigonal prismatic configuration has the symmetry associated with the D_{3h} point group, and it is shown on the left. The octahedral configuration has the O_h point group, and it is shown in the middle. Ligand (crystal) Field splitting of the trigonal prismatic (**a**) and octahedral (**b**) coordination's for the metal are shown on the right. Top part of figure , adapted from [19]. Bottom part of figure adapted and reprinted from [9] with the permission of AIP Publishing

the electrostatic nature of the metal (positive center)–halcogenide (negative center) interaction, the ligands (chalcogenides) form an electrostatic field around the metal ion that stabilizes the electronic configuration of the transition metal atoms [9]. The crystal field splitting for the d orbital for both the trigonal prismatic and octahedral fields is shown in Fig. 8.3. The d_{xz} and d_{yz} orbitals also participate in bonding the metal atoms to the chalcogenide atoms. The d_{z^2}, $d_{x^2-y^2}$, and d_{xy} are sometimes referred to as "almost" non-bonding [8]. However, it is these three orbitals that to first approximation form the conduction and valence bands that exhibit electronic transport and interesting optical phenomena. The d orbital content of the valence and conduction bands changes across the Brillouin zone. As we show below, at the K point in the Brillouin zone, the d_{z^2} orbitals form one of the conduction bands, and the $d_{x^2-y^2}$ and d_{xy} orbitals hybridize to form a higher energy conduction band and the highest energy valence band. The electronic bands of TMD materials composed of heavy metal atoms will be strongly influenced by spin–orbit coupling. Below, we present a 3 band tight binding model for these bands for a single layer with the trigonal prismatic structure. There we also present a theoretical discussion showing how the degeneracy of these two valence bands is split by spin–orbit coupling. Figure 8.3 also shows that d electrons with octahedral coordination form degenerate sets of orbitals designated e_g for the d_{z^2} and $d_{x^2-y^2}$, and t_{2g} for the d_{xy}, d_{xz}, and d_{yz} orbitals [8].

The structure of bulk TMDs builds on the two monolayer morphs as shown in Fig. 8.4 [10]. The unit cell for the bulk TMDs is shown in Fig. 8.4 with the c axis perpendicular to the layers. The a and b axes are perpendicular to the c axis and the distance a is the minimal chalcogen–chalcogen distance [10]. Most bulk TMD materials have one of three additional structures that are designated using T for tetragonal, H for hexagonal, and R for rhombohedral shown in Fig. 8.4. Their designation depends on the number of trilayers X-M-X that are repeated to form the crystal structure. The 1T structure has tetragonal symmetry with the trilayer repeated once and metal atoms have octahedral symmetry. The 3R refers to rhombohedral symmetry, repeat of three trigonal prismatic trilayers layers to form the unit cell.

Transition metals from Periodic Table groups 4, 5, 6, 7, and 10 all form TMD materials, and more complete discussion be found in Fig. 3.4 of [10]. The d band filling from group 4 to group 10 strongly impacts the electrical properties. TMDs with

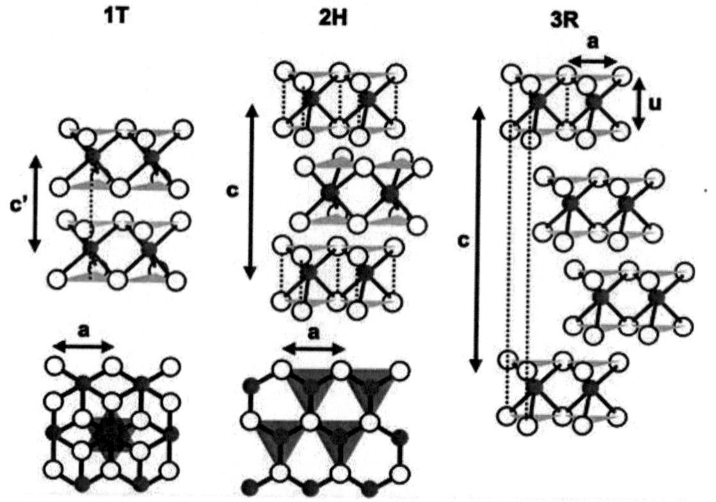

Fig. 8.4 The three most common polymorphs of transition metal dichalcogenide materials are shown. In 1T the metal atoms have octahedral symmetry and the repeated structure consists of one trilayer. Here the $2H_c$ structure is shown. In the $2H_c$ the metal atoms have the trigonal prismatic symmetry and the layers are rotated 180° about the c axis. The $2H_c$ structure has hexagonal symmetry, and the repeat structure has two trilayers. The typical stacking sequence for 2H is referred to as AA'. The Mo and W compounds with S and Se all have the $2H_c$ structure. When looking down the c axis, a transition metal atom is above the two chalcogenide atoms of the trilayer below for the $2H_c$ structure. $NbSe_2$, NbS_2, TaS_2, and $TaSe_2$ have the $2H_a$ structure which is not shown. In the $2H_a$ structure, the layers are all rotated in the same direction. When looking down the c axis, the transition metal atoms of one layer are above the transition metal atoms of the trilayer below. The 3R structure has rhombohedral symmetry and the metal atoms have trigonal prismatic coordination. The repeated 3R structure has three trilayers. The typical stacking sequence for 3R is AB. Also not shown is the $2H_b$ structure for non-stoichiometric $Nb_{1+x}Se_2$ and $Ta_{1+x}Se_2$ with excess metal atoms present between trilayers. Figure adapted from [10]

partially filled d bands such as 2H-$NbSe_2$ and 1T-ReS_2 are metallic [1]. TMDs with filled d bands such as 1T-HfS_2 and 2H-MoS_2 are semiconductors [1]. The structural and electronic properties of TMD materials are surveyed in Table 8.1. A summary of the crystallographic point groups and space groups are described in Sect. 2.12 and summarized for TMD materials is presented in Sect. 8.15 at the end of the chapter.

8.2 Tight Binding Model for Highest Energy Valence Band and Lowest Energy Conduction Bands of Trigonal Prismatic Monolayer TMD

As with graphene, the tight binding approach provides insight into many of the key aspects of the physics of TMD materials. Here we present a 3 band tight binding (TB) model for the d_{z^2}, $d_{x^2-y^2}$, and d_{xy} orbitals where the d orbital interact with the d orbitals of the nearest neighbor metal atoms. The d_{z^2} orbitals form one of the conduction bands and the $d_{x^2-y^2}$ and d_{xy} hybridize to form two of the valence bands [1, 11]. This derivation follows that given in [11]. The 3 band approximation ignores contributions from the chalcogenide atoms. There are 6 nearest neighbor (NN) metal atoms for the trigonal prismatic TMD configuration. The real space and reciprocal space lattice for the metal atoms in trigonal prismatic single tri-layer TMD are shown in Fig. 8.5. The 6 NN metal atoms used to construct the tight binding model are shown in Fig. 8.6 along with the labeling of the metal atoms and the K points in Fig. 8.6. The reciprocal lattice is defined in terms of the metal atom sub-lattice and the real space position of the R_6 metal atom is $(\frac{k_x a}{2}, \frac{\sqrt{3}k_y a}{2})$. Monolayer, trigonal prismatic TMD materials do not have inversion symmetry and thus are not centrosymmetric.

We again use the same type of Block wavefunctions $\Phi_j(\vec{k}, \vec{r})$ used for the TB model of graphene in Chap. 7. Here, the Block wavefunction is a linear combinations of d_{z^2},d_{xy}, and $d_{x^2-y^2}$. *The wavefunction* (8.1)*will be different from that of graphene since each carbon atom in graphene contributes one p_z orbital Block wavefunction* $\Phi_j(\vec{k}, \vec{r}) = \frac{1}{\sqrt{N}} \sum_{i=1}^{N} e^{i\vec{k}\cdot\vec{R}_{j,i}} \phi_j(\vec{r} - \vec{R}_{j,i})$ *and there were two atoms of carbon in each primitive cell that were linearly combined to form the wavefunction in each unit cell:* $\psi(\vec{k}, \vec{r}) = \sum_{l=1}^{2} c_{j,l}(\vec{k}) \Phi_l(\vec{k}, \vec{r})$.

In this 3 Band tight binding model for monolayer, trigonal prismatic TMD each metal atom contributes a linear combination of three d orbitals (8.2). This linear combination is used for the Block wavefunction. For N unit cells:

$$\psi(\vec{k}, \vec{r}) = \frac{1}{\sqrt{N}} \sum_{l=0}^{N} e^{i\vec{k}\cdot\vec{R}_l} \phi_j(\vec{r} - \vec{R}_l) \tag{8.1}$$

Table 8.1 Properties of transition metal dichalcogenides for S, Se, and Te [1]. Here, TP refers to trigonal prismatic and O refers to Octahedral

Column in periodic table	Metal	M^{+4} d^x	Metal coordination	Stacking symmetry	Semiconductor/semimetal	Magnetic state
Group 4	Ti, Hf, Zr	d^0	Typically O	1T TiS_2	Semiconducting 1T-HfS_2 semiconductor	Diamagnetic
Group 5	V, Nb, Ta	d^1	Either TP or O	The S and Se compounds of Nb and Ta are $2H_a^*$ 1T TaS_2 $NbTe_2$:1T′ (monoclinic)	Nd & Ta—all dichalcogenides are metallic NdS_2 & 2H-$NbSe_2$—metallic & low T superconductor—exhibit CDW	Antiferromagnetic, paramagnetic, or diamagnetic
Group 6	Mo, W	d^2	Typically TP	The S and Se compounds of Mo and W are MoS_2: 2H and 3R $2H_c^*$ WTe_2: distorted O	All Mo and W dichalcogenides are semiconducting 2H-MoS_2—semiconductor for few layer WTe_2:1T′	Most are diamagnetic WTe_2 potential Weyl semimetal
Group 7	Tc, Re	d^3	Distorted O	ReS_2 and $ReSe_2$:1T′ (triclinic)	Small gap semiconductors 1T′-ReS_2—metallic	Diamagnetic
Group 10	Pd, Pt	d^6	Typically O		S and Se compounds are semiconductors Te compounds are metallic with $PdTe_2$ a low T superconductor	S and Se compounds are diamagnetic Te compounds are paramagnetic

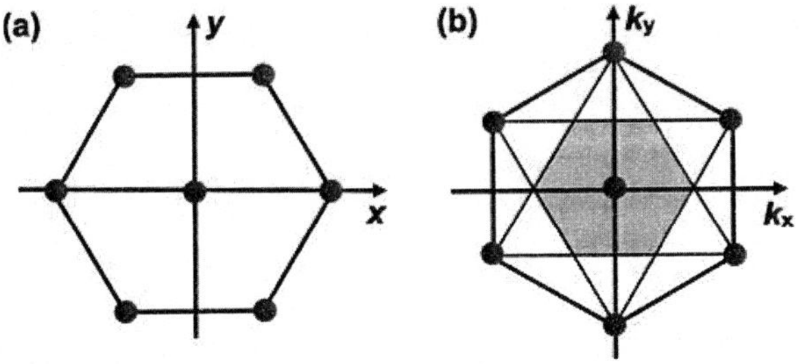

Fig. 8.5 The metal atoms in monolayer, trigonal prismatic, transition metal dichalcogenide lattices form a planar hexagonal lattice that is the middle layer in a chalcogenide–metal–chalcogenide layer structure. In the three band model, the d_{z^2} and $d_{x^2-y^2}$, d_{xy} orbitals interact forming a valence band and two conduction bands. The real space hexagonal lattice for the metal atoms is shown in (**a**) and the reciprocal space is shown in (**b**). The first Brillouin zone is shaded hexagon in (**b**). Note the K point along k_x at $\frac{4\pi}{3a}$, and one of the inequivalent K' $(-K)$ point is at $-\frac{4\pi}{3a}$.

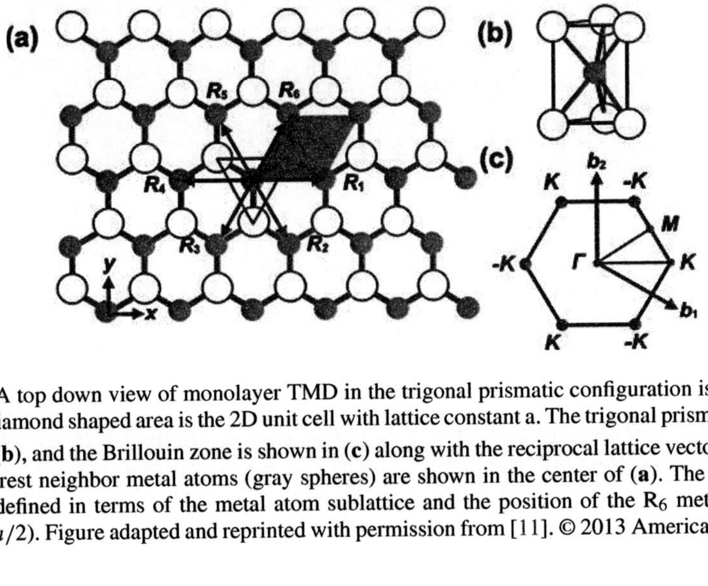

Fig. 8.6 A top down view of monolayer TMD in the trigonal prismatic configuration is shown in (**a**). The diamond shaped area is the 2D unit cell with lattice constant a. The trigonal prismatic cell is shown in (**b**), and the Brillouin zone is shown in (**c**) along with the reciprocal lattice vectors $\overrightarrow{b_1}$, $\overrightarrow{b_2}$. The 6 nearest neighbor metal atoms (gray spheres) are shown in the center of (**a**). The reciprocal lattice is defined in terms of the metal atom sublattice and the position of the R_6 metal atom is $(a/2, \sqrt{3}a/2)$. Figure adapted and reprinted with permission from [11]. © 2013 American Physical Society

where $\vec{R}_0 = 0$ is the central atom. Remember, the central atom contributes to the delocalized wavefunction.

$$\phi_j(\vec{r}) = \sum_{i=1\ to\ 3} c_i \varphi_i(\vec{r}) \tag{8.2}$$

$$H\psi(\vec{k}, \vec{r}\,) = E(\vec{k})\psi(\vec{k}, \vec{r}\,) \tag{8.3}$$

It is also convenient to define another wavefunction Φ_j where:

$$\varphi_1(\vec{r}) = |d_{z^2}\rangle; \quad \varphi_2(\vec{r}) = |d_{xy}\rangle; \quad \varphi_3(\vec{r}) = |d_{x^2-y^2}\rangle;$$

Since we are using the orbitals, we will get 3 energy bands. Here, $\varphi_{1l}(\vec{r} - \vec{R}_i)$ refers to $\varphi_1(\vec{r}) = |d_{z^2}\rangle$ at each unit cell l, etc., then

$$\Phi_j(\vec{k}, \vec{r}) = \frac{1}{\sqrt{N}} \sum_{l=0}^{N} e^{i\vec{k}\cdot\vec{R}_l} \varphi_{jl}(\vec{r} - \vec{R}_l) \text{ and } \psi(\vec{k}, \vec{r}) = \sum_{j=1}^{3} c_j \Phi_j(\vec{r}) \tag{8.4}$$

The l index in (8.4) runs over the all the unit cells in the crystal, and the j index runs from 1 to 3 as indicated in the first part of (8.4). The extra l index is not necessary as the $\varphi_{jl}(\vec{r} - \vec{R}_l)$ are centered at $\vec{R}_l$ so it is dropped.

Here we follow the same approach that we used for the TB model for the π bands of graphene in Chap. 7. There are several ways to arrive at the determinant for the energy. Here we operate on (8.3) with each of the three wavefunctions $\varphi_i(\vec{r})$. It is important to note that we are operating using the wavefunction localized at the central atom in the Brillouin zone $\langle\varphi_1(\vec{r})|$.

$$\langle\varphi_i(\vec{r})|H|\psi(\vec{k}, \vec{r})\rangle = \langle\varphi_i(\vec{r})|E|\psi(\vec{k}, \vec{r})\rangle \tag{8.5}$$

And for each of the three d orbitals *we establish equations that lead to a matrix representation* for the energies:

$$\begin{aligned}
&c_1\langle\varphi_1(\vec{r})|H|\Phi_1(\vec{r}, \vec{k})\rangle + c_2\langle\varphi_1(\vec{r})|H|\Phi_2(\vec{r}, \vec{k})\rangle + c_3\langle\varphi_1(\vec{r})|H|\Phi_3(\vec{r}, \vec{k})\rangle \\
&\quad = \langle\varphi_1(\vec{r})|E|\psi(\vec{k}, \vec{r})\rangle \\
&c_1\langle\varphi_2(\vec{r})|H|\Phi_1(\vec{r}, \vec{k})\rangle + c_2\langle\varphi_2(\vec{r})|H|\Phi_2(\vec{r}, \vec{k})\rangle + c_3\langle\varphi_2(\vec{r})|H|\Phi_3(\vec{r}, \vec{k})\rangle \\
&\quad = \langle\varphi_2(\vec{r})|E|\psi(\vec{k}, \vec{r})\rangle \\
&c_1\langle\varphi_3(\vec{r})|H|\Phi_1(\vec{r}, \vec{k})\rangle + c_2\langle\varphi_3(\vec{r})|H|\Phi_2(\vec{r}, \vec{k})\rangle + c_3\langle\varphi_3(\vec{r})|H|\Phi_3(\vec{r}, \vec{k})\rangle \\
&\quad = \langle\varphi_3(\vec{r})|E|\psi(\vec{k}, \vec{r})\rangle
\end{aligned}$$

This gives us three equations, and the first equation is:

$$\begin{aligned}
&c_1 \frac{1}{\sqrt{N}} \sum_{l=0}^{N} e^{i\vec{k}\cdot\vec{R}_l} \langle\varphi_1(\vec{r})|H|\varphi_1(\vec{r} - \vec{R}_l)\rangle \\
&+ c_2 \frac{1}{\sqrt{N}} \sum_{l=0}^{N} e^{i\vec{k}\cdot\vec{R}_l} \langle\varphi_1(\vec{r})|H|\varphi_2(\vec{r} - \vec{R}_l)\rangle \\
&+ c_3 \frac{1}{\sqrt{N}} \sum_{l=0}^{N} e^{i\vec{k}\cdot\vec{R}_l} \langle\varphi_1(\vec{r})|H|\varphi_3(\vec{r} - \vec{R}_l)\rangle
\end{aligned}$$

$$= E\Big[c_1 \frac{1}{\sqrt{N}} \sum_{l=0}^{N} e^{i\vec{k}\cdot\vec{R}_l} \langle \varphi_1(\vec{r}) | \varphi_1(\vec{r} - \vec{R}_l) \rangle$$

$$+ c_2 \frac{1}{\sqrt{N}} \sum_{l=0}^{N} e^{i\vec{k}\cdot\vec{R}_l} \langle \varphi_1(\vec{r}) | \varphi_2(\vec{r} - \vec{R}_l) \rangle \quad (8.6)$$

$$+ c_3 \frac{1}{\sqrt{N}} \sum_{l=0}^{N} e^{i\vec{k}\cdot\vec{R}_l} \langle \varphi_1(\vec{r}) | \varphi_3(\vec{r} - \vec{R}_l) \rangle\Big]$$

The cylindrical symmetry of d_{z^2} when compared to the $d_{x^2-y^2}$ and d_{xy} orbitals simplifies the solution to (8.5). Where due to orthogonality $= E\big[c_1 \frac{1}{\sqrt{N}} \sum_{l=0}^{N} e^{i\vec{k}\cdot\vec{R}_l} \langle \varphi_1(\vec{r}) | \phi_1(\vec{r} - \vec{R}_l) \rangle + 0 + 0\big]$.

The necessary symmetry considerations for the hopping integrals $\langle \varphi_1(\vec{r}) | H | \varphi_2(\vec{r} - \vec{R}_l) \rangle$ and $\langle \varphi_1(\vec{r}) | H | \varphi_3(\vec{r} - \vec{R}_l) \rangle$ for $d_{x^2-y^2}$ and d_{xy} are discussed below. There are two similar equations for $\langle \varphi_2|$ and $\langle \varphi_3|$. This gives us the matrix equations we are seeking. If we now consider the 6 NN interactions for each term and the symmetry arguments, (8.5) can be further simplified [11]. For now, we assume that the hopping integrals for the d_{z^2} orbitals for each NN at R_l, $\langle \varphi_1(\vec{r}) | H | \varphi_1(\vec{r} - \vec{R}_l) \rangle$ will give the same result as for R_1 (see Fig. 8.4). The utility of this assumption becomes clearer when we consider (8.9). The cylindrical symmetry for the orbital is the same as for the p_z orbital. However this assumption does not work for the $d_{x^2-y^2}$ and d_{xy} orbitals. First we consider the term associated with c_1 noting that $\varphi_1(\vec{r}) = d_{z^2}$:

$$\frac{1}{\sqrt{N}} \sum_{l=0}^{N} e^{i\vec{k}\cdot\vec{R}_l} \langle \varphi_1(\vec{r}) | H | \varphi_1(\vec{r} - \vec{R}_l) \rangle$$

$$= \frac{1}{\sqrt{N}} \Big[\langle \varphi_1(\vec{r}) | H | \varphi_1(\vec{r}) \rangle + e^{i\vec{k}\cdot\vec{R}_1} \langle \varphi_1(\vec{r}) | H | \varphi_1(\vec{r} - \vec{R}_1) \rangle + e^{i\vec{k}\cdot\vec{R}_2} \langle \varphi_1(\vec{r}) | H | \varphi_1(\vec{r} - \vec{R}_2) \rangle$$

$$+ e^{i\vec{k}\cdot\vec{R}_3} \langle \varphi_1(\vec{r}) | H | \varphi_1(\vec{r} - \vec{R}_3) \rangle + e^{i\vec{k}\cdot\vec{R}_4} \langle \varphi_1(\vec{r}) | H | \varphi_1(\vec{r} - \vec{R}_4) \rangle$$

$$+ e^{i\vec{k}\cdot\vec{R}_5} \langle \varphi_1(\vec{r}) | H | \varphi_1(\vec{r} - \vec{R}_5) \rangle$$

$$+ e^{i\vec{k}\cdot\vec{R}_6} \langle \varphi_1(\vec{r}) | H | \varphi_1(\vec{r} - \vec{R}_6) \rangle \Big] \quad (8.7)$$

where the values of $\vec{R}_1$ to $\vec{R}_6$ are:

$$\vec{R}_1 = (a\hat{x}, 0), \vec{R}_2 = \left(\frac{a}{2}\hat{x}, \frac{-\sqrt{3}a}{2}\hat{y}\right), \vec{R}_3 = \left(-\frac{a}{2}\hat{x}, \frac{-\sqrt{3}a}{2}\hat{y}\right)$$

$$\vec{R}_4 = (-a\hat{x}, 0), \vec{R}_5 = \left(-\frac{a}{2}\hat{x}, \frac{\sqrt{3}a}{2}\hat{y}\right), \vec{R}_6 = \left(\frac{a}{2}\hat{x}, \frac{\sqrt{3}a}{2}\hat{y}\right) \tag{8.8}$$

the hopping integrals for the d_{z^2} orbitals for NN at R_l with $l = 2$ to $l = 6$ will give the same result as for R_1 due to the cylindrical symmetry of the d_{z^2} orbitals:

$$\begin{aligned} t_0 &= \langle\varphi_1(\vec{r})|H|\varphi_1(\vec{r} - \vec{R}_1)\rangle = \langle\varphi_1(\vec{r})|H|\varphi_1(\vec{r} - \vec{R}_2)\rangle \\ &= \langle\varphi_1(\vec{r})|H|\varphi_1(\vec{r} - \vec{R}_3)\rangle, \text{ etc.} \end{aligned} \tag{8.9}$$

Below, we will see that there are additional symmetry considerations for the $d_{x^2-y^2}$ and d_{xy}, and we note that d_{z^2} transforms into d_{z^2} after the symmetry rotations and mirror reflections required for $d_{x^2-y^2}$ and d_{xy}. Using $\langle\varphi_1(\vec{r})|H|\varphi_1(\vec{r})\rangle = \epsilon_1$ we obtain:

$$\begin{aligned} \frac{1}{\sqrt{N}}\sum_{l=0}^{N} e^{i\vec{k}\cdot\vec{R}_l}\langle\varphi_1(\vec{r})|H|\varphi_1(\vec{r} - \vec{R}_l)\rangle &= \frac{1}{\sqrt{N}}\Big[\epsilon_1 + e^{iak_x}t_0 \\ + e^{i(\frac{a}{2}k_x - \frac{\sqrt{3}a}{2}k_y)}t_0 + e^{i(-\frac{a}{2}k_x - \frac{\sqrt{3}a}{2}k_y)}t_0 &+ e^{-iak_x}t_0 + e^{i(-\frac{a}{2}k_x + \frac{\sqrt{3}a}{2}k_y)}t_0 \\ + e^{i(\frac{a}{2}k_x + \frac{\sqrt{3}a}{2}k_y)}t_0\Big] \end{aligned} \tag{8.10}$$

Using $e^{i\theta} = \cos(\theta) + i\sin(\theta)$ $e^{i\theta} + e^{-i\theta} = 2\cos(\theta)$ and $\vec{R}_0 = 0$ is the central atom thus this can be rewritten as:

$$\begin{aligned} \frac{1}{\sqrt{N}}\sum_{l=0}^{N} e^{i\vec{k}\cdot\vec{R}_l}\langle\varphi_1(\vec{r})|H|\varphi_1(\vec{r} - \vec{R}_l)\rangle &= \frac{1}{\sqrt{N}}[\epsilon_1 + 2\cos(ak_x)t_0 \\ +2\cos(\frac{a}{2}k_x)e^{i(-\frac{\sqrt{3}a}{2}k_y)}t_0 + 2\cos(\frac{a}{2}k_x)e^{i(+\frac{\sqrt{3}a}{2}k_y)}t_0\Big] \\ = \frac{1}{\sqrt{N}}\left[\epsilon_1 + 2\cos(ak_x)t_0 + 4\cos(\frac{a}{2}k_x)\cos(\frac{\sqrt{3}a}{2}k_y)t_0\right] \end{aligned} \tag{8.11}$$

Defining $\alpha = \frac{a}{2}k_x$ and $\beta = \frac{\sqrt{3}a}{2}k_y$ and removing the common factor $\frac{1}{\sqrt{N}}$, we get:

$$\sum_{l=0}^{N} e^{i\vec{k}\cdot\vec{R}_l}\langle\varphi_1(\vec{r})|H|\varphi_1(\vec{r} - \vec{R}_l)\rangle = [\epsilon_1 + 2\cos(2\alpha)t_0 + 4\cos(\alpha)\cos(\beta)t_0] \tag{8.12}$$

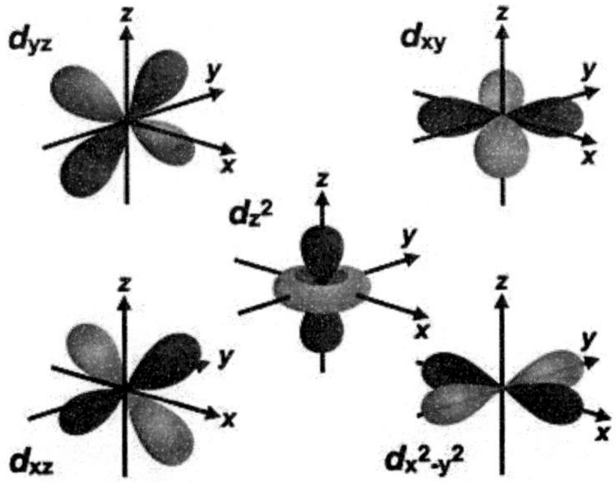

Fig. 8.7 The five d orbitals are shown. The reflection symmetry in the x–y plane allows only the d_{z^2}, $d_{x^2-y^2}$, and d_{xy} to hybridize in the x–y plane. This further justifies using only these orbital in the tight binding model. Also, the sign or phase of an orbital matters when bonds are formed. The different phases for each orbital are represented by dark versus light shading

Since the goal is to simplify the hopping integrals $\langle\varphi_1(\vec{r})|H|\phi_j(\vec{r}-\vec{R}_l)\rangle$ which determine the interaction between neighboring atoms, we use symmetry considerations for the $d_{x^2-y^2}$ and d_{xy} orbitals. First we consider the shape of the d orbitals which is shown in Fig. 8.7.

A more rigorous discussion can be found in [11] which describes the symmetry operations for the trigonal prismatic configurations for the metal atoms. The d_{z^2}, $d_{x^2-y^2}$, and d_{xy} orbitals split into three categories which are labeled using the Mullikan notation for the irreducible representations (IR) for the point group D_{3h} [10]. The IR for d_{z^2} is $A_1^{'}$ and E' for $d_{x^2-y^2}$ and d_{xy}. The other d orbitals that bond with the chalcogenides have the representation of E'' for d_{xz} and d_{yz}. It is worth restating that this 3 band model ignores the contributions of the chalcogenide orbitals to the valence and conduction bands. In that light, the $\varphi_1(\vec{r})$ are more completely specified so that the IR is more apparent. Following the notation used in [11], the orbitals are relabeled as φ_u^j where the j refers to the IR with $A_1^{'}$ being $j=1$ and E' being $j=2$. The u refer to the type of d orbital. We use $j=1\left[A_1^{'}\right]$, $u=1$ for d_{z^2}. We use $j=2$ $[E']$, $u=1$ for $d_{x^2-y^2}$, and $u=2$ for d_{xy} [11]:

$$|\varphi_1^1(\vec{r})\rangle = d_{z^2},\ |\varphi_1^2(\vec{r})\rangle = d_{xy}\ ,\ \text{and}\ |\varphi_2^2(\vec{r})\rangle = d_{x^2-y^2}$$

The symbols shown here for the symmetry operations follow those used to first describe the three band model [11]. Trigonal prismatic TMD monolayers have $\hat{C}_3$ (2π/3) (counter clockwise) rotational symmetry around the z axis which is normal to the $x-y$ plane of the monolayer, and have three mirror planes. The mirror planes are perpendicular to the $x-y$ plane. The $\hat{\sigma}_v$ plane is a perpendicular bisector to the line connecting atoms 1 and 6 as well as atoms 3 and 4. The $\hat{\sigma}_v'$ ($\hat{\sigma}_v''$) are obtained by rotation around the z axis by 2π/3 (4π/3). It is important to note that central point for the rotations and mirror planes used here is the central metal atom shown in Fig. 8.6 [11]. We again need to consider the hopping integrals which represents the interactions between the NN metal atoms at a distance R from the central metal atom for each of the three d orbitals. For example, not only does the d_{z^2} orbital on the central atom interact with d_{z^2} on the atoms at a distance R, but it interacts with the $d_{x^2-y^2}$ and d_{xy} orbitals on the atom at a distance R. There are 6 NN atoms that have this interaction.

$$E_{uu'}^{jj'}(\vec{R}) = \langle \varphi_u^j(\vec{r}) | H | \varphi_{u'}^{i'}(\vec{r} - \vec{R}) \rangle \tag{8.13}$$

The $\vec{R}_i$ will be the positions of the 6 surrounding metal atoms. The equivalence of $E_{u1}^{j1}(\vec{R})$ for each of the vectors $\vec{R}_i$ to each of the 6 NN metal atoms is easily justified for the cylindrical symmetric d_{z^2} orbitals. The $E_{uu'}^{jj'}(\vec{R})$ for each of the six vectors $\vec{R}$ can be generated from $E_{uu'}^{jj'}(\vec{R})$ for one of the $\vec{R}$ vectors using symmetry operations [11]. This justifies using (8.13) for the $d_{x^2-y^2}$ and d_{xy} orbitals. The matrix elements for the Block wavefunctions are:

$$\langle \varphi_u^j(\vec{r}) | H | \Phi_{j'}(\vec{r} - \vec{R}) \rangle = H_{uu'}^{jj'}(\vec{R}) = \sum_{i=1}^{N} e^{i\vec{k} \cdot \vec{R}_i} E_{uu'}^{jj'}(\vec{R}) \tag{8.14}$$

Here we are restating (8.12) using the above terms $E_{u1}^{11}(\vec{R})$ in h_0. The elements of the Hamiltonian matrix (see 8.17) where $\vec{R}_0 = 0$ are listed in (8.15) and (8.16):

$$h_0 = \sum_{l=0}^{N} e^{i\vec{k} \cdot \vec{R}_l} \langle \varphi_1^1(\vec{r}) | H | \varphi_1^1(\vec{r} - \vec{R}_l) \rangle = \epsilon_1 + H_{11}^{11}(\vec{R}) = \epsilon_1 + \sum_{i=1}^{N} e^{i\vec{k} \cdot \vec{R}_l} E_{11}^{11}(\vec{R})$$

$$h_0 = [\epsilon_1 + 2\cos(2\alpha) E_{11}^{11}(\vec{R}) + 4\cos(\alpha)\cos(\beta) E_{11}^{11}(\vec{R})] \tag{8.15}$$

In a similar manner, the other matrix elements for the Hamiltonian can be determined using the orthogonality of the different d orbitals for example:

$$h_1 = \langle \varphi_1(\vec{r}) | H | \Phi_2(\vec{r}, \vec{k}) \rangle = \sum_{l=0}^{N} e^{i\vec{k} \cdot \vec{R}_l} \langle \varphi_1^1(\vec{r}) | H | \varphi_1^2(\vec{r} - \vec{R}_l) \rangle \tag{8.16a}$$

$$h_2 = \langle \varphi_1(\vec{r}) | H | \Phi_3(\vec{r}, \vec{k}) \rangle = \sum_{l=0}^{N} e^{i\vec{k} \cdot \vec{R}_l} \langle \varphi_1^1(\vec{r}) | H | \varphi_2^2(\vec{r} - \vec{R}_l) \rangle \tag{8.16b}$$

Remembering that $|H - E_j S| = 0$ (see Chap. 7 (7.76)), and considering that the overlap integrals S_{ij} will be very small, $|H| \approx 0$ and the energies can be determined by diagonalizing the Hamiltonian. As previously mentioned, the distance between metal atoms in the TMD is 15% to 25% longer than in an elemental single crystal of that metal. The nearest neighbor 2D Hamiltonian is a 3 × 3 matrix as follows [11]:

$$H_{NN}^{2D}\left(k_x, k_y\right) = \begin{pmatrix} h_0 & h_1 & h_2 \\ h_1^* & h_{11} & h_{12} \\ h_2^* & h_{12}^* & h_{22} \end{pmatrix} \tag{8.17}$$

We now turn to the task of determining some of the other matrix elements. The goal is to simplify $\langle\varphi_1^1(\vec{r})|H|\varphi_1^2(\vec{r}-\vec{R}_l)\rangle$ and $\langle\varphi_1^1(\vec{r})|H|\varphi_2^2(\vec{r}-\vec{R}_l)\rangle$ so that these hopping integrals can be related to the hopping integrals between the central metal atom and the metal atom at $\overrightarrow{R_1}$, for example: $\langle\varphi_1^1(\vec{r})|H|\varphi_1^2(\vec{r}-\vec{R}_1)\rangle$. The orbital interactions for the hopping integrals at different NN positions need to be carefully examined. In Fig. 8.8a the orbital configurations for the hopping interaction of $|\varphi_1^1(\vec{r})\rangle = d_{z^2}$ with

$|\varphi_1^2(\vec{r}-\vec{R}_1)\rangle,\ E_{11}^{12}(\vec{R}_1)$

are shown, and in Fig. 8.8b the orbital configurations for the interaction of the central atom $|\varphi_1^1(\vec{r})\rangle = d_{z^2}$ with the metal atom at $\vec{R}_l$ for $|\varphi_2^2(\vec{r}-\vec{R}_l)\rangle,\ E_{12}^{12}$ are shown. In order to meet the symmetry requirements for the orbitals, orbitals without cylindrical symmetry need to be rotated around the axis of the atom on which they reside. The goal is to have the NN interactions between the central atom and the d_{xy} and $d_{x^2-y^2}$ orbitals at NN atoms be the same from $\vec{R}_1$ to $\vec{R}_6$. For example, the central metal atom d_{z^2} overlaps the orbital lobe along $-y$ for $d_{x^2-y^2}$ atom at $\overrightarrow{R_1}$ while the overlap for the central metal atom d_{z^2} with $d_{x^2-y^2}$ before taking symmetry considerations into account at $\vec{R}_5$ is between the $+x$ and $-y$ lobes. This is also true for the overlap of the central metal atom d_{z^2} orbital the d_{xy} at $\vec{R}_1$ versus $\vec{R}_5$.

This issue is overcome by using symmetry. The hopping integrals to atoms 2 through 6 can be expressed in terms of $E_{11}^{12}(\vec{R}_1)$ and $E_{12}^{12}(\vec{R}_1)$ after using the symmetry of the system to rotate the d_{xy} and $d_{x^2-y^2}$ orbitals. The two orbital wavefunctions for d_{xy} and $d_{x^2-y^2}$ can be rotated and mirror plane reflected so that the appropriate overlap is obtained. The rotation operation about the $x-y$ plane on these orbitals will mix them:

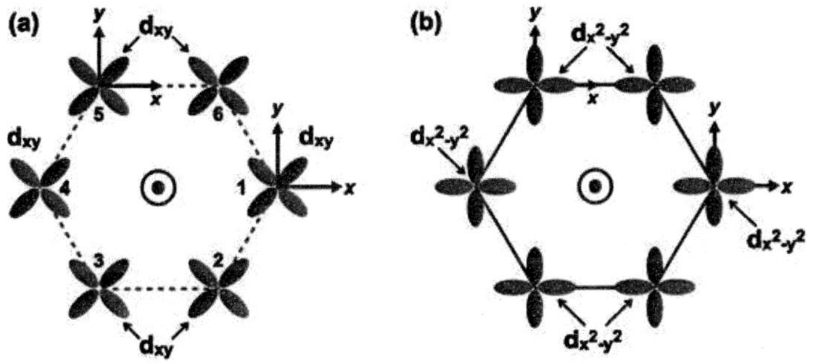

Fig. 8.8 The d_{xy} and $d_{x^2-y^2}$ orbitals are shown for the 6 NN of the central metal atom before rotation. The d_{z^2} orbital of the central metal atom is shown in the top down view as a circle with a dot at its center. In order for orbitals from atoms 2 through 6 to have the same overlap with the orbital at the central atom, the orbitals must be rotated using symmetry operations as discussed in the text

$$\begin{pmatrix} \cos(\theta) & -\sin(\theta) \\ \sin(\theta) & \cos(\theta) \end{pmatrix} \begin{pmatrix} x \\ y \end{pmatrix} \tag{8.18}$$

This represents a counter clockwise rotation of θ. To illustrate the process, the symmetry operation for atom 4 is shown first. The $\widehat{\sigma''}_v$ mirror plane is perpendicular to the $x - y$ plane which is a perpendicular bisector to a line between atoms 5 and 6. The $\widehat{\sigma''}_v$ mirror plane symmetry requires that $d_{xy}(\vec{R}_1) = xy$ and $d_{xy}(\vec{R}_4) = (-x)y$, i.e., $|\varphi_1^2(\vec{R}_4)\rangle = -|\varphi_1^2(\vec{R}_1)\rangle$. The line connecting atom 5 with the central atom is $-120°$ from the line connecting the central *a*tom with atom 1. Thus, the rotation matrix for a $-120°$ rotation gives:

$$\begin{pmatrix} \cos(-2\pi/3) & -\sin(-2\pi/3) \\ \sin(-2\pi/3) & \cos(-2\pi/3) \end{pmatrix} \begin{pmatrix} x \\ y \end{pmatrix} = \begin{pmatrix} -1/2 & \sqrt{3}/2 \\ -\sqrt{3}/2 & -1/2 \end{pmatrix} \begin{pmatrix} x \\ y \end{pmatrix} = \begin{pmatrix} -\frac{1}{2}x + \frac{\sqrt{3}}{2}y \\ -\frac{\sqrt{3}}{2}x - \frac{1}{2}y \end{pmatrix}$$

The $d_{x^2-y^2} = \frac{1}{2}(x^2 - y^2)$ orbital after $-120°$ rotation mixes with $d_{xy} = xy$ as follows:

$$\frac{1}{2}(x'^2 - y'^2) = \frac{1}{2}\left(\left(-\frac{1}{2}x + \frac{\sqrt{3}}{2}y\right)^2 - \left(-\frac{\sqrt{3}}{2}x - \frac{1}{2}y\right)^2\right)$$
$$= -\frac{1}{2}\left[\frac{1}{2}(x^2 - y^2)\right] - \frac{\sqrt{3}}{2}[xy]$$

The $d_{xy} = xy$ orbital after $-120°$ rotation mixes with $d_{x^2-y^2} = \frac{1}{2}(x^2 - y^2)$.

$$x'y' = \left(-\frac{1}{2}x + \frac{\sqrt{3}}{2}y\right)\left(-\frac{\sqrt{3}}{2}x - \frac{1}{2}y\right) = \frac{\sqrt{3}}{4}x^2 + \frac{1}{4}xy - \frac{3}{4}xy - \frac{\sqrt{3}}{4}y^2$$
$$= \frac{\sqrt{3}}{2}\left[\frac{1}{2}(x^2 - y^2)\right] - \frac{1}{2}xy$$

This gives us for d_{xy}: $\psi_1(\vec{R}_5) = -\frac{1}{2}|d_{xy}\rangle + \frac{\sqrt{3}}{2}|d_{x^2-y^2}\rangle$ and for $d_{x^2-y^2}$: $\psi_2(\vec{R}_5) = \frac{\sqrt{3}}{2}|d_{xy}\rangle - \frac{1}{2}|d_{x^2-y^2}\rangle$.

In order to meet the rotation and mirror plan symmetry requirements for the wavefunctions at the NN metal atoms, atoms 3 and 5 are rotated differently than 2 and 6.

First consider the NN d_{xy} interactions with the central atom d_{z^2}. At NN atom 4, the d_{xy} wavefunction is minus the d_{xy} at atom 1 due to $\hat{\sigma}_v''$ mirror symmetry since $\hat{\sigma}_v''$ is a perpendicular bisector to the line between atoms 5 and 6. Thus $E_{11}^{12}(\vec{R}_1) = -E_{11}^{12}(\vec{R}_4)$. Conversely, one can show through rotation or by $\hat{\sigma}_v''$ that $E_{12}^{12}(\vec{R}_1) = E_{12}^{12}(\vec{R}_4)$. In order to duplicate the $d_{xy} : d_{z^2}$ orbital overlap between NN atom 1 and the central atom, we need to transform the d_{xy} at NN atoms 3 and 5 using a $\hat{C}_3$ $(2\pi/3)$ rotation about the z axis. This is accomplished at NN atom 3 by substituting $-x$ for x and rotating the resulting orbital by 60° counter clockwise. The d_{xy} orbital at NN atom 5 is obtained by substituting $-x$ for x and rotating the resulting orbital by 60° clockwise. One can check this by labeling the lobes of d_{xy}, transforming the orbitals and then rotating d_{xy} at NN 1 using $\hat{C}_3$ to positions 5 and then 3. The necessary d_{xy} orbital orientation at NN atom 2 is obtained by rotating by 60° clockwise, and the necessary d_{xy} orbital orientation at NN atom 6 is obtained by rotating by 60° counter clockwise. One can check this by labeling the lobes of d_{xy} and then rotating d_{xy} at NN 4 using $\hat{C}_3$ to positions 2 and then 6.

Now we consider the NN $d_{x^2-y^2}$ orbitals. At NN atom 4, $d_{x^2-y^2}$ remains unchanged by $\hat{\sigma}_v''$, and we remember that $E_{12}^{12}(\vec{R}_1) = E_{12}^{12}(\vec{R}_4)$. To obtain ψ_2, the clockwise rotation matrix $(-\theta)$ is used for atoms 3 (− 240°) and 5 (− 120°), and the counter-clockwise rotation matrix for atoms 2 (300°), 4 (180°) and 6 (60°). One can check this by using $\hat{C}_3$ (see [7, 11]).

The entire set of wavefunctions for d_{xy} and $d_{x^2-y^2}$ for each of the NN atoms becomes (here we use a new symbol for the wave functions after the symmetry operations):

$$\psi_1(\vec{R}_1) \equiv \left|d_{xy}(\vec{R}_1)\right\rangle \text{ and } \psi_2(\vec{R}_1) \equiv \left|d_{x^2-y^2}(\vec{R}_1)\right\rangle$$

$$\psi_1(\vec{R}_2) \equiv -\frac{1}{2}\left|d_{xy}(\vec{R}_2)\right\rangle - \frac{\sqrt{3}}{2}\left|d_{x^2-y^2}(\vec{R}_2)\right\rangle \text{ and}$$
$$\psi_2(\vec{R}_2) \equiv +\frac{\sqrt{3}}{2}\left|d_{xy}(\vec{R}_2)\right\rangle - \frac{1}{2}\left|d_{x^2-y^2}(\vec{R}_2)\right\rangle$$

$$\psi_1(\vec{R}_3) \equiv +\frac{1}{2}\left|d_{xy}(\vec{R}_3)\right\rangle + \frac{\sqrt{3}}{2}\left|d_{x^2-y^2}(\vec{R}_3)\right\rangle \text{ and}$$
$$\psi_2(\vec{R}_3) \equiv +\frac{\sqrt{3}}{2}\left|d_{xy}(\vec{R}_3)\right\rangle - \frac{1}{2}\left|d_{x^2-y^2}(\vec{R}_3)\right\rangle$$

$$\psi_1(\vec{R}_4) \equiv -\left|d_{xy}(\vec{R}_4)\right\rangle \text{ and } \psi_2(\vec{R}_4) \equiv \left|d_{x^2-y^2}(\vec{R}_4)\right\rangle$$

$$\psi_1(\vec{R}_5) \equiv +\frac{1}{2}\left|d_{xy}(\vec{R}_5)\right\rangle - \frac{\sqrt{3}}{2}\left|d_{x^2-y^2}(\vec{R}_5)\right\rangle \text{ and}$$
$$\psi_2(\vec{R}_5) \equiv -\frac{\sqrt{3}}{2}\left|d_{xy}(\vec{R}_5)\right\rangle - \frac{1}{2}\left|d_{x^2-y^2}(\vec{R}_5)\right\rangle$$

$$\psi_1(\vec{R}_6) \equiv -\frac{1}{2}\left|d_{xy}(\vec{R}_6)\right\rangle + \frac{\sqrt{3}}{2}\left|d_{x^2-y^2}(\vec{R}_6)\right\rangle \text{ and}$$
$$\psi_2(\vec{R}_6) \equiv -\frac{\sqrt{3}}{2}\left|d_{xy}(\vec{R}_6)\right\rangle - \frac{1}{2}\left|d_{x^2-y^2}(\vec{R}_6)\right\rangle \tag{8.19}$$

These wavefunctions are used to determine the hopping integrals associated with c_2 and c_3 of (8.6). For c_2 we start with:

$$\frac{1}{\sqrt{N}}\sum_{l=0}^{N} e^{i\vec{k}\cdot\vec{R}_l}\langle\varphi_1(\vec{r})|H|\varphi_2(\vec{r}-\vec{R}_l)\rangle = \frac{1}{\sqrt{N}}[\langle\varphi_1(\vec{r})|H|\varphi_2(\vec{r})\rangle$$
$$+ e^{i2\alpha}\langle\varphi_1(\vec{r})|H|\varphi_2(\vec{r}-\vec{R}_1)\rangle$$
$$+ e^{-I(\alpha-\beta)}\langle\varphi_1(\vec{r})|H|\varphi_2(\vec{r}-\vec{R}_2)\rangle + e^{-i(\alpha+\beta)}\langle\varphi_1(\vec{r})|H|\varphi_2(\vec{r}-\vec{R}_3)\rangle$$
$$+ e^{-i2\alpha}\langle\varphi_1(\vec{r})|H|\varphi_2(\vec{r}-\vec{R}_4)\rangle + e^{-i(\alpha-\beta)}\langle\varphi_1(\vec{r})|H|\varphi_2(\vec{r}-\vec{R}_5)\rangle$$
$$+ e^{i(\alpha+\beta)}\varphi_1(\vec{r})|H|\varphi_2(\vec{r}-\vec{R}_6)\rangle] \tag{8.20}$$

Since we wavefunctions that are involved in bonding to the central metal atom's d_z orbital $\varphi_1(\vec{r})$, we use the symmetry modified wavefunctions $\psi_1(\vec{R}_l)$ for $\varphi_2(\vec{r}-\vec{R}_l)$ at each atom l:

$$= \frac{1}{\sqrt{N}}\Big[\langle\varphi_1(\vec{r})|H|\varphi_2(\vec{r})\rangle + e^{i2\alpha}\langle\varphi_1(\vec{r})|H|\psi_1(\vec{R}_1)\rangle$$
$$+ e^{i(\alpha-\beta)}\langle\varphi_1(\vec{r})|H|\psi_1(\vec{R}_2)\rangle + e^{-i(\alpha+\beta)}\langle\varphi_1(\vec{r})|H|\psi_1(\vec{R}_3)\rangle$$
$$+ e^{-i2\alpha}\langle\varphi_1(\vec{r})|H|\psi_1(\vec{R}_4)\rangle + e^{-i(\alpha-\beta)}\langle\varphi_1(\vec{r})|H|\psi_1(\vec{R}_5)\rangle$$
$$+e^{i(\alpha+\beta)}\langle\varphi_1(\vec{r})|H|\psi_1(\vec{R}_6)\rangle\Big]$$

The wavefunctions of (8.19) allow use of $E^{12}_{11}(\vec{R}_1)$ and $E^{12}_{12}(\vec{R}_1)$:

$$= \frac{1}{\sqrt{N}}\Big[0 + e^{i2\alpha}E^{12}_{11}(\vec{R}_1) + e^{i(\alpha-\beta)}[-\frac{1}{2}E^{12}_{11}(\vec{R}_1) - \frac{\sqrt{3}}{2}E^{12}_{12}(\vec{R}_1)]$$
$$+ e^{-i(\alpha+\beta)}[\frac{1}{2}E^{12}_{11}(\vec{R}_1) + \frac{\sqrt{3}}{2}E^{12}_{12}(\vec{R}_1)] - e^{-i2\alpha}E^{12}_{11}(\vec{R}_1)$$
$$+ e^{-i(\alpha-\beta)}[\frac{1}{2}E^{12}_{11}(\vec{R}_1) - \frac{\sqrt{3}}{2}E^{12}_{12}(\vec{R}_1)] + e^{i(\alpha+\beta)}[-\frac{1}{2}E^{12}_{11}(\vec{R}_1) + \frac{\sqrt{3}}{2}E^{12}_{12}(\vec{R}_1)]\Big]$$

After removing the common $\frac{1}{\sqrt{N}}$ factors for the c's, this can be further rearranged to provide:

$$h_1 = [0 + 2i\sin(2\alpha)E^{12}_{11}(\vec{R}_1) - i\sin(\alpha)e^{-i\beta}E^{12}_{11}(\vec{R}_1) - i\sqrt{3}\sin(\alpha)e^{-i\beta}E^{12}_{12}(\vec{R}_1)$$
$$- i\sin(\alpha)e^{i\beta}E^{12}_{11}(\vec{R}_1) + i\sqrt{3}\sin(\alpha)e^{i\beta}E^{12}_{12}(\vec{R}_1)]$$
$$h_1 = [2i\sin(2\alpha)E^{12}_{11}(\vec{R}_1) - 2i\sin(\alpha)\cos(\beta)E^{12}_{11}\vec{R}_1) - 2\sqrt{3}\sin(\alpha)\sin(\beta)E^{12}_{12}(\vec{R}_1)]$$

Using $-\cos\theta = \cos\theta$ and $t_1 = E^{12}_{11}(\vec{R}_1)$; $t_2 = E^{12}_{12}(\vec{R}_1)$:

$$h_1 = t_1 2i(\sin(2\alpha) + \sin(\alpha)\cos(\beta)) - t_2 2\sqrt{3}\sin(\alpha)\sin(\beta) \tag{8.21}$$

For c_3:

$$\frac{1}{\sqrt{N}}\sum_{l=0}^{N} e^{i\vec{k}\cdot\vec{R}_l}\langle\varphi_1(\vec{r})|H|\varphi_3(\vec{r}-\vec{R}_l)\rangle$$

$$= \frac{1}{\sqrt{N}}\big[\langle\varphi_1(\vec{r})|H|\varphi_3(\vec{r})\rangle + e^{i2\alpha}\langle\varphi_1(\vec{r})|H|\varphi_3(\vec{r}-\vec{R}_1)\rangle + e^{i(\alpha-\beta)}\langle\varphi_1(\vec{r})|H|\varphi_3(\vec{r}-\vec{R}_2)\rangle$$
$$+ e^{-i(\alpha+\beta)}\langle\varphi_1(\vec{r})|H|\varphi_3(\vec{r}-\vec{R}_3)\rangle + e^{-i2\alpha}\langle\varphi_1(\vec{r})|H|\varphi_3(\vec{r}-\vec{R}_4)\rangle$$
$$+ e^{-i(\alpha-\beta)}\langle\varphi_1(\vec{r})|H|\varphi_3(\vec{r}-\vec{R}_5)\rangle + e^{i(\alpha+\beta)}\langle\varphi_1(\vec{r})|H|\varphi_3(\vec{r}-\vec{R}_6)\rangle\big]$$

$$= \frac{1}{\sqrt{N}}\big[\langle\varphi_1(\vec{r})|H|\varphi_3(\vec{r})\rangle + e^{i2\alpha}\langle\varphi_1(\vec{r})|H|\psi_2(\vec{R}_1)\rangle + e^{i(\alpha-\beta)}\langle\varphi_1(\vec{r})|H|\psi_2(\vec{R}_2)\rangle$$
$$+ e^{-i(\alpha+\beta)}\langle\varphi_1(\vec{r})|H|\psi_2(\vec{R}_3)\rangle + e^{-i2\alpha}\langle\varphi_1(\vec{r})|H|\psi_2(\vec{R}_4)\rangle$$
$$+ e^{-i(\alpha-\beta)}\langle\varphi_1(\vec{r})|H|\psi_2(\vec{R}_5)\rangle + e^{i(\alpha+\beta)}\langle\varphi_1(\vec{r})|H|\psi_2(\vec{R}_6)\rangle\big]$$

$$= \frac{1}{\sqrt{N}}\Big[0 + e^{i2\alpha}E_{12}^{12}(\vec{R}_1) + e^{i(\alpha-\beta)}\big[\frac{\sqrt{3}}{2}E_{11}^{12}(\vec{R}_1) - \frac{1}{2}E_{12}^{12}(\vec{R}_1)\big]$$
$$+ e^{-i(\alpha+\beta)}\big[\frac{\sqrt{3}}{2}E_{11}^{12}(\vec{R}_1) - \frac{1}{2}E_{12}^{12}(\vec{R}_1)\big]$$
$$+ e^{-i2\alpha}E_{12}^{12}(\vec{R}_1) + e^{-i(\alpha-\beta)}\big[-\frac{\sqrt{3}}{2}E_{11}^{12}(\vec{R}_1) - \frac{1}{2}E_{12}^{12}(\vec{R}_1)\big]$$
$$+ e^{i(\alpha+\beta)}\big[-\frac{\sqrt{3}}{2}E_{11}^{12}(\vec{R}_1) - \frac{1}{2}E_{12}^{12}(\vec{R}_1)\big]\Big]$$

$$= \frac{1}{\sqrt{N}}\Big[2\cos(2\alpha)E_{12}^{12}(\vec{R}_1) + 2\cos(\alpha)e^{-i\beta}\big[\frac{\sqrt{3}}{2}E_{11}^{12}(\vec{R}_1) - \frac{1}{2}E_{12}^{12}(\vec{R}_1)\big]$$
$$+ 2\cos(\alpha)e^{i\beta}\big[-\frac{\sqrt{3}}{2}E_{11}^{12}(\vec{R}_1) - \frac{1}{2}E_{12}^{12}(\vec{R}_1)\big]\Big]$$

$$= \frac{1}{\sqrt{N}}\big[2\cos(2\alpha)E_{12}^{12}(\vec{R}_1) - i2\sqrt{3}\cos(\alpha)\sin(\beta)E_{11}^{12}(\vec{R}_1)$$
$$- 2\cos(\alpha)\cos(\beta)E_{12}^{12}(\vec{R}_1)\big]$$

$$h_2 = 2t_2\big[\cos(2\alpha) - \cos(\alpha)\cos(\beta)\big] + i2\sqrt{3}t_1\cos(\alpha)\sin(\beta) \tag{8.22}$$

The other matrix elements are determined in a similar manner, and entire set of matrix elements for the Hamiltonian matrix (8.17) is:

$h_0 = [\epsilon_1 + t_0(2\cos(2\alpha) + 4\cos(\alpha)\cos(\beta))]$

$$h_1 = 2it_1(\sin(2\alpha) + \sin(\alpha)\cos(\beta)) - 2\sqrt{3}t_2\sin(\alpha)\sin(\beta)$$

$$h_2 = 2t_2\left[\cos(2\alpha) - \cos(\alpha)\cos(\beta)\right] + i2\sqrt{3}t_1\cos(\alpha)\sin(\beta)$$

$$h_{11} = 2t_{11}\cos(2\alpha) + (t_{11} + 3t_{22})\cos(\alpha)\cos(\beta) + \epsilon_2$$

$$h_{22} = 2t_{22}\cos(2\alpha) + (3t_{11} + t_{22})\cos(\alpha)\cos(\beta) + \epsilon_2$$

$$h_{12} = \sqrt{3}(t_{11} + t_{22})\sin(\alpha)\sin(\beta) + 4it_{12}\sin(\alpha)(\cos(\alpha) - \cos(\beta)) \tag{8.23}$$

$$t_0 = E_{11}^{11}(\vec{R}_1)\ ; \quad t_1 = E_{11}^{12}(\vec{R}_1); \quad t_2 = E_{12}^{12}(\vec{R}_1$$

$$t_{11} = E_{11}^{22}(\vec{R}_1); \quad t_{12} = E_{12}^{22}(\vec{R}_1); \quad t_{22} = E_{22}^{22}(\vec{R}_1) \tag{8.24}$$

The $H_{NN}^{2D}(k_x, k_y) = \begin{pmatrix} h_0 & h_1 & h_2 \\ h_1^* & h_{11} & h_{12} \\ h_2^* & h_{12}^* & h_{22} \end{pmatrix}$ can be solved by diagonalizing the matrix. At the Γ point, $k_x = 0$ and $k_y = 0$ which simplifies the matrix elements providing a diagonal matrix:

$h_0(\Gamma) = [\in_1 + 6t_0]$

$$h_1(\Gamma) = 0$$

$$h_2(\Gamma) = 0$$

$$h_{11}(\Gamma) = 3(t_{11} + t_{22}) + \in_2$$

$$h_{22}(\Gamma) = 3(t_{11} + t_{22}) + \in_2$$

$$h_{12}(\Gamma) = 0 \tag{8.25}$$

At the Γ point there are two degenerate bands with energy $3(t_{11} + t_{22}) + \epsilon_2$ and the third band with energy $[\epsilon_1 + 6t_0]$. The band energies at the high symmetry points in the Brillouin zone are provided in Table 8.2. In addition to the 3 band model, an 11 band [12] Tight Binding model has been developed. These models allow one to determine the orbital character of the bands close to the $K, M, \Gamma,$ and Q points. The Hamiltonian matrix at the K point can also be calculated from (8.23), and it is stated in (8.30).

Table 8.2 The band energies for the 3 Band Tight Binding model for trigonal prismatic monolayer TMDs. The band energies are listed in ascending order. See [11]

	$\Gamma = (0, 0)$	$K = \left(\frac{4\pi}{3a}, 0\right)$	$M = \left(\frac{\pi}{a}, \frac{\pi}{\sqrt{3}a}\right)$
VB	$\epsilon_1 + 6t_0$	$\epsilon_2 - \frac{3}{2}(t_{11} + t_{22}) - 3\sqrt{3}t_{12}$	$f_1 - f_2$
CB	$\epsilon_2 + 3(t_{11} + t_{22})$	$\epsilon_1 - 3t_0$	$\epsilon_2 + t_{11} - 3t_{22}$
CB		$\epsilon_2 - \frac{3}{2}(t_{11} + t_{22}) + 3\sqrt{3}t_{12}$	$f_1 + f_2$

Where $f_1 = \frac{1}{2}(\epsilon_1 + \epsilon_2) - t_0 - \frac{3}{2}t_{11} + \frac{1}{2}t_{22}$ and $f_2 = \frac{1}{2}\sqrt{(\epsilon_1 - \epsilon_2 - 2t_0 + 3t_{11} - t_{22})^2 + 64t_2^2}$

The d_{z^2} nature of the lowest energy conduction band at the K point is apparent from the $\epsilon_1 - 3t_0$ energy of the band since $t_0 = E_{11}^{11}(\vec{R}_1)$. Conversely, the d_{z^2} nature of the highest energy valence band at the Γ point is apparent from the $\epsilon_1 + 6t_0$ energy of the band. The valence bands formed by the interaction between the metal and chalcogenide atoms lie below this. The 3 bands are shown in Fig. 8.9 for MoS_2, $MoSe_2$, $MoTe_2$, WS_2, WSe_2, and WTe_2 as calculated using first principles generated fitting parameters listed in Table 8.3 from [11]. The orbital contributions are shown in Fig. 8.10. The band gap at the K point for monolayer trigonal prismatic TMD calculated using the 3 Band Tight Binding approximations can be compared to experimental data to show that this model provides useful insights.

$$Band\ Gap = \left[\epsilon_1 - 3t_0\right] - \left[\epsilon_2 - \frac{3}{2}(t_{11} + t_{22}) - 3\sqrt{3}t_{12}\right]$$

	Band gap at K using fit to GGA (eV)	FP band gap (eV)	Optical band gap (eV)
MoS_2	1.66	1.8	1.83/1.90
WS_2	1.81		1.95
$MoSe_2$	1.44	1.5	1.66
WSe_2	1.54		1.64
$MoTe_2$	1.07		
WTe_2	1.067		

The 3 Band model describes some of the essential physics that is necessary for understanding optical properties especially at the band edges such as the K points in the Brillouin zone. For example, *the energy dispersion at the K points is quadratic and not linear as it is for single layer graphene.* Below, we show that the addition of spin–orbit coupling splits the NN TB valence band. The addition of third nearest neighbor (TNN) contributions results in a band structure that fits first principles calculations across the Brillouin zone [11]. The TNN and NN 3 band TB models have similar low energy behavior (near the K points) [11]. The addition of spin orbit coupling alters the shape of the bands near the K point which is further discussed below. *The splitting of the valence band is essential to understanding the optical properties of TMD materials.* The addition of even one TMD layer shifts the energy and location in the Brillouin zone of the VB maximum resulting in an indirect band

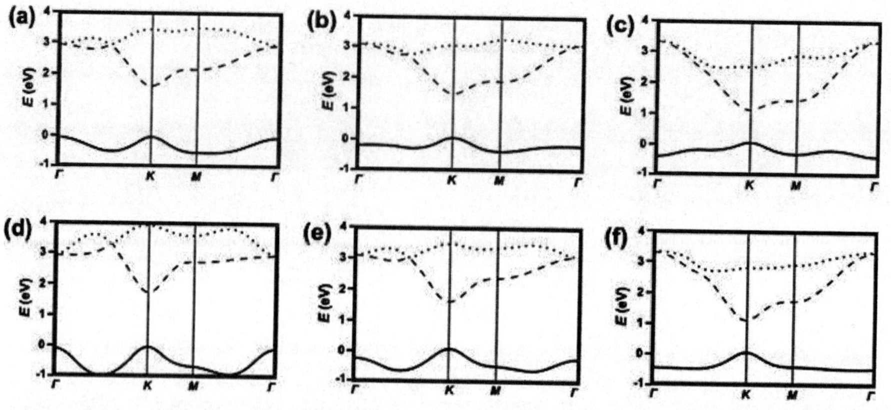

Fig. 8.9 The electronic band structure for monolayer MoS_2, $MoSe_2$, $MoTe_2$, WS_2, WSe_2, and WTe_2 from (a) to (f) calculated using (8.17), (8.23), and (8.24) using fitting parameters listed in Table 8.3 calculated using Tight Binding 3 Band model. The d_{z^2} conduction band is shown in green, and d_{xy} and $d_{x^2-y^2}$ contribute to the valence band shown in blue and to the higher energy conduction band shown in red. Figure adapted from [11]. © L. Szulakowska

gap. This change in VB maximum for bilayer group-6 TMDs is due to the interaction of antibonding p_z orbitals from chalcogenide atoms in the two TMD monolayers, resulting in an increase of the VBM energy at the Γ point [6]. Reference [14] is a computational database for TMD materials [14]. For the sake of completeness, it is noted that the theoretical band structure of $MoS_{2x}Se_{2(1-x)}$ and $Mo_xW_{1-x}S_2$ alloys have been reported [5].

Table 8.3 Tight binding model fitting parameters calculated using first principles generalized-gradient approximation (GGA) and local-density approximation (LDA) cases

	a (Â)	Zx-x (Â)	ε_1	ε_2	t_0	t_1	t_2	t_{11}	t_{12}	t_{22}
GGA										
MoS_2	3.190	3.130	1.046	2.104	−0.184	0.401	0.507	0.218	0.338	0.057
WS_2	3.191	3.144	1.130	2.275	−0.206	0.567	0.536	0.286	0.384	−0.061
$MoSe_2$	3.326	3.345	0.919	2.065	−0.188	0.317	0.456	0.211	0.290	0.130
WSe_2	3.325	3.363	0.943	2.179	−0.207	0.457	0.486	0.263	0.329	0.034
$MoTe_2$	3.557	3.620	0.605	1.972	−0.169	0.228	0.390	0.207	0.239	0.252
WTe_2	3.560	3.632	0.606	2.102	−0.175	0.342	0.410	0.233	0.270	0.190
LDA										
MoS_2	3.129	3.115	1.238	2.366	−0.218	0.444	0.533	0.250	0.360	0.047
ws_2	3.132	3.126	1.355	2.569	−0.238	0.626	0.557	0.324	0.405	−0.076
$MoSe_2$	3.254	3.322	1.001	2.239	−0.222	0.350	0.488	0.244	0.314	0.129
WSe_2	3.253	3.338	1.124	2.447	−0.242	0.506	0.514	0.305	0.353	0.025
$MoTe_2$	3.472	3.598	0.618	2.126	−0.202	0.254	0.423	0.241	0.263	0.269
WTe_2	3.476	3.611	0.623	2.251	−0.209	0.388	0.442	0.272	0.295	0.200

Table reproduced from [11] © American Physical Society.

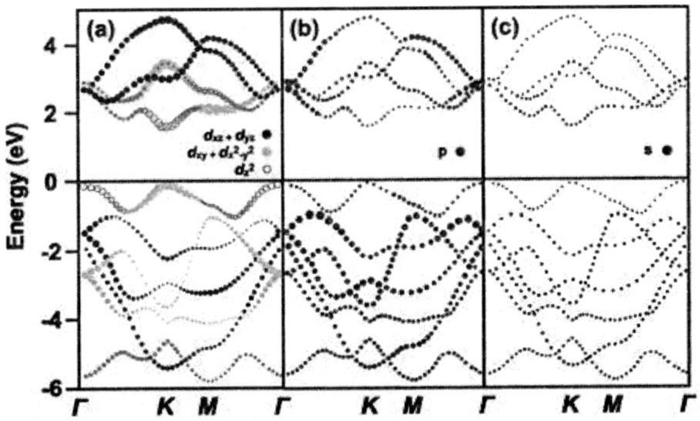

Fig. 8.10 The first principles electronic band structure of MoS_2 is shown. The valence band maximum is set to 0 energy. The orbital character of the highest energy valence band and higher energy conduction band at the K point in the Brillouin zone comes from the mixing of the d_{xy} and $d_{x^2-y^2}$. The lowest energy conduction band comes from the d_{z^2} orbital. Figure adapted and reprinted with permission from [11]. © 2013 American Physical Society

8.2.1 Band Splitting Due to Spin Orbit Coupling

Spin–orbit coupling splits the valence band at the K point changing the optical properties. Spin orbit effects were discussed in Chapter 2 in the introductory section on Band Structure. Spin orbit coupling can be added to Tight Binding calculations using the method developed for bulk semiconductors [13]. The spin orbit effect is added as a perturbation to the Tight Binding Hamiltonian. The spin orbit perturbation is given by:

$$H' = \lambda \vec{L}\ \vec{S} \tag{8.26}$$

where λ determines the strength of the spin orbit interactions, $\vec{L}$ is the orbital angular momentum and $\vec{S}$ the spin angular momentum. Since the metal atoms in monolayer TMD materials are all in the same plane, there is only a z component to the angular momentum. The starting point is to consider the effect at the metal atom on 3 Band Tight Binding model for TMDs using the spin defined orbitals as a bases set [11]:

$$|d_{z^2}\rangle\uparrow;\ |d_{xy}\rangle\uparrow;\ |d_{x^2-y^2}\rangle\uparrow;\ |d_{z^2}\rangle\downarrow;\ |d_{xy}\rangle\downarrow;\ |d_{x^2-y^2}\rangle\downarrow \tag{8.27}$$

$$H' = \lambda\vec{L}\ \vec{S} = \frac{\lambda}{2}\begin{bmatrix} L_z & 0 \\ 0 & -L_z \end{bmatrix} \tag{8.28}$$

$$L_z = \begin{pmatrix} 0 & 0 & 0 \\ 0 & 0 & 2i \\ 0 & -2i & 0 \end{pmatrix}$$

The TB Spin Orbit Hamiltonian in terms of $H_{NN}^{2D}(k_x, k_y)$, (8.17), is [11]:

$$H_{SOC}(\vec{k}) = \begin{bmatrix} H_{NN}^{2D}(k_x, k_y) + \frac{\lambda}{2}L_z & 0 \\ 0 & H_{NN}^{2D}(k_x, k_y) - \frac{\lambda}{2}L_z \end{bmatrix} \tag{8.29}$$

At the K point,

$$H_{NN}^{2D}(k_x, k_y) = \begin{pmatrix} \epsilon_1 - 3t_0 & 0 & 0 \\ 0 & \epsilon_2 - \frac{3}{2}(t_{22} + t_{11}) & -i3\sqrt{3}t_{12} \\ 0 & i3\sqrt{3}t_{12} & \epsilon_2 - \frac{3}{2}(t_{22} + t_{11}) \end{pmatrix} \tag{8.30}$$

After diagonalization there are 3 energy levels: $\epsilon_1 - 3t_0$, $\epsilon_2 - \frac{3}{2}(t_{11} + t_{22}) - 3\sqrt{3}t_{12}$, and $\epsilon_2 - \frac{3}{2}(t_{11} + t_{22}) + 3\sqrt{3}t_{12}$. Leaving:

$$H_{NN}^{2D}(k_x, k_y) = \begin{pmatrix} \epsilon_1 - 3t_0 & 0 & 0 \\ 0 & \epsilon_2 - \frac{3}{2}(t_{22} + t_{11}) - 3\sqrt{3}t_{12} & 0 \\ 0 & 0 & \epsilon_2 - \frac{3}{2}(t_{22} + t_{11}) + 3\sqrt{3}t_{12} \end{pmatrix}$$

Using this NN TB Hamiltonian,

$$
H_{SOC}(\vec{k}) = \begin{bmatrix} \begin{pmatrix} \epsilon_1 - 3t_0 & 0 & 0 \\ 0 & \epsilon_2 - \frac{3}{2}(t_{22} + t_{11}) - 3\sqrt{3}t_{12} & +i\lambda \\ 0 & -i\lambda & \epsilon_2 - \frac{3}{2}(t_{22} + t_{11}) + 3\sqrt{3}t_{12} \end{pmatrix} & 0 \\ 0 & \begin{pmatrix} \epsilon_1 - 3t_0 & 0 & 0 \\ 0 & \epsilon_2 - \frac{3}{2}(t_{22} + t_{11}) - 3\sqrt{3}t_{12} & -i\lambda \\ 0 & +i\lambda & \epsilon_2 - \frac{3}{2}(t_{22} + t_{11}) + 3\sqrt{3}t_{12} \end{pmatrix} \end{bmatrix} \tag{8.31}
$$

Diagonalizing this Hamiltonian results in minimum spin orbit splitting energies for the valence band of $\Delta^{V}_{SOC} = 2\lambda$. Although the NN 3 Band TB model does not indicate that there is conduction band splitting, the conduction bands show a small amount of spitting which is discussed below. References [11, 15] provide a comparison of the spin orbit splitting energies for the valence and conduction bands between this 3 band TB model and density functional theory calculations. Some of this data is shown in Table 8.4. The splitting energies are large especially for WS_2 and WSe_2. The spin orbit splitting at the K point is shown in Fig. 8.11. Time reversal symmetry requires that for each band n, $\Delta_n(\vec{k}) = -\Delta_n(-\vec{k})$ thus the splitting reverses for K points in the Brillouin zone versus the K' points [15] (Fig. 8.12).

It is important to note that the conduction band values are also split [15]. The d_{xz} and d_{yz} orbitals and second order perturbation theory are required for determination of CB spin orbit splitting [11]. First principles calculations showed that the d orbitals mix so that:

$$d_{\pm 2} = \frac{1}{\sqrt{2}}(d_{x^2-y^2} \pm i d_{xy}) \;;\; d_0 = d_{z^2} \;;\; d_{\pm 1} = \frac{1}{\sqrt{2}}(d_{xz} \pm i d_{yz})$$

The orbital energies increase from d_{+2}; d_0; d_{-1}; d_{-2}; d_{+1} and $d_{\pm 1}$. The $d_{\pm 1}$ are included in the spin orbit coupling 2nd order perturbation calculation. The mixed d orbitals are the starting point for many theoretical studies of TMD band structure [7].

Table 8.4 Spin orbit splitting energies for the conduction bands (C) and valence bands (V) of TMD monolayers

	meV	MoS_2	WS_2	$MoSe_2$	WSe_2
DFT + SOC	C	−3	27	−27	38
	V	147	433	186	463
TB + SOC	C	−4	17	−28	−3
	V	147	433	186	463

Data from [15]

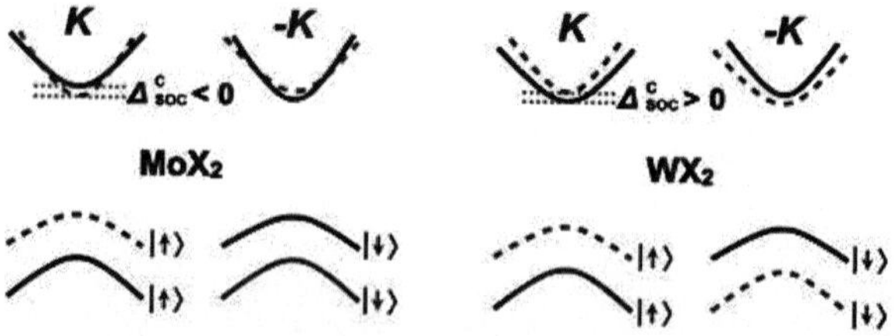

Fig. 8.11 Spin orbit coupling splits the highest energy valence band and the lowest energy conduction band. Time reversal symmetry requires that for each band n, $\Delta_n(\vec{k}) = -\Delta_n(-\vec{k})$ thus the VB splitting reverses for K points in the Brillouin zone versus the K' points (15). The conduction band splitting is larger for W TMD. Figure adapted and reprinted with permission from [11]. © 2013 American Physical Society

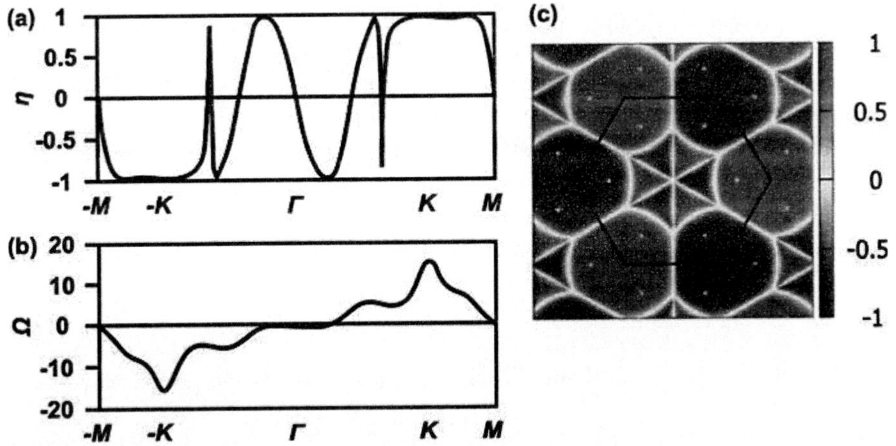

Fig. 8.12 The degree of circular polarization of absorbed light, η, the Berry curvature Ω in units of Angstoms, and a color map of η with the Brillouin zone shown for MoS_2. The value of η versus position is reciprocal space varies from 1 to -1 according the gray scale shown at the right of (**c**). Figures (**a**) and (**c**) clearly show that right and left circularly polarized light will be absorbed selectively at K versus K′. The data was determined using TNN TB theory. Figure adapted and reprinted with permission from [11]. © 2013 American Physical Society

Figure 8.11 shows that the CB splitting reverses at the K point for W TMD while the CB of Mo TMD's shows CB crossing [11].

8.3 Direct Observation of Monolayer TMD Valley Pseudospin and Valence Band Spin Splitting

The direct band gap of monolayer TMD materials greatly increases both light absorption and photoluminescence intensity when compared to the indirect band gap bulk materials. Monolayer MoS_2, $MoSe_2$, and WS_2 (~1 nm thick) absorb 5–10% of incident light [6]. The strong absorption cannot be explained by the single particle transitions, and excitonic effects must be included in first principles calculations [6]. Excitonic peaks are observed in low temperature optical absorption measurements [6]. This results in the electrical band gap being smaller that the optical band gap. Photoluminescence that is excited by circularly polarized light has been used to confirm the spin splitting of the valence band at the K and K' points [16, 17]. The K points and K' points have opposite pseudospin. The photoluminescence lineshape for TMD materials was introduced in Chap. 1.

Here, we introduce the optical selectivity of circularly polarized light to distinguish the K and K' points in the Brillouin zone. In Sect. 8.4 we discuss the origin of the intrinsic Hall conductivity that is excited by circularly polarized light starting with the Dirac Hamiltonian that includes spin orbit interaction. We note that since the electric vector of circularly polarized light is continuously rotating to the right or to the left, circularly polarized light carries angular momentum. The in-plane electric field couples to the Berry phase of the carriers at the K and K' points. The electrons get an "anomalous velocity" that is transverse to the electric field [18] which results in an intrinsic contribution to the Hall conductivity [19]. The intrinsic Hall conductivity due to the Berry curvature for band n in the presence of circularly polarized light is [19]:

$$\sigma^{intrinsic} = \frac{e^2}{\hbar} \int f(\vec{k}) \Omega_n(\vec{k}) \frac{d\vec{k}}{(2\pi)^2}$$

Here, $f(\vec{k})$ is the Fermi–Dirac distribution, and $u_{\vec{k}}(\vec{k})$ is the period part of the Block function.

The main interaction of the circularly polarized light is with the pseudospin. Right circularly polarized light, σ^+, couples to the electronic states close to the K points in the Brillouin zone, and left circularly polarized light, σ^-, couples to the states close to the K' points. The optical field couples only to the orbital part of the wavefunction thus spin is conserved in the optical transition when circularly polarized light excites the transition [19]. The coupling strength between the angular momentum and the right σ_+ and left σ_- circularly polarized light is [19]:

$$\wp_\pm(\vec{k}) = \wp_x(\vec{k}) \pm i\wp_y(\vec{k}) \text{ where } \wp_{i=x,y}(\vec{k}) \equiv m_0 \langle u_c(\vec{k}) | i \vec{\nabla}_{k_i} | u_v(\vec{k}) \rangle$$

Below, we show that the so called coupling strength for circularly polarized light directly selects the valley through the square of the momentum matrix element for circularly polarized light [19]:

$$\left| \wp_\pm(\vec{k}) \right|^2 = \frac{m_0^2 a^2 t^2}{\hbar^2} \left(1 \pm \tau \frac{\Delta'}{\sqrt{\Delta'^2 + 4a^2 t^2 k^2}} \right)^2 \tag{8.32}$$

where a is the lattice constant, t the hopping integral, and Δ' is the band gap. The valley index $\tau = +1$ at K and $\tau = -1$ at K'. $\Delta'^2 \gg 4a^2t^2k^2$. Thus for σ^+ polarization, $\left| P_+(\vec{k}) \right|^2 = 0$ for K', and for σ^-, $\left| P_-(\vec{k}) \right|^2 = 0$ at K.

Use of a resonant frequency allows selection of spin up versus spin down [19]. For example, at the K valley σ^+ polarized light with frequency ω_u having the energy difference between the upper valence band and the conduction band will excite spin up electrons to the conduction band while σ^+ polarized light with frequency ω_d having the energy difference between the lower valence band and the conduction band will excite spin down electrons to the conduction band [19]. At the K' valley due to time reversal symmetry, a ω_u having the energy difference between the upper valence band and the conduction band will excite spin down electrons to the conduction band while σ^- polarized light with frequency ω_d having the energy difference between the lower valence band and the conduction band will excite spin up electrons to the conduction band [19].

In some of the experiments, the photoluminescence of MoS_2 was excited by 1.96 eV (HeNe laser), 2.09 eV and 2.33 eV (532 nm) (solid state laser) light that was circularly polarized by Babinet–Soleil compensator [17]. The excitons are labeled A for the lower energy exciton and B for the higher energy exciton. Circularly polarized light allow for selective excitation of K versus K'. In these experiments, circularly polarized light couples to spin down electrons and result in a spin up electron. Results for MoS_2 are shown in Fig. 8.13. Results for WSe_2 are shown in Fig. 8.14 [18].

In summary, selective excitation of specific valleys (K vs K') and spins is possible using both circularly polarized light and specific frequencies. [19]. The SOC energy difference between spin up an spin down electrons allows selection of a resonance frequency that selects the spin up or spin down electrons from either the K or K' valleys. These selection rules are shown in Fig. 8.15.

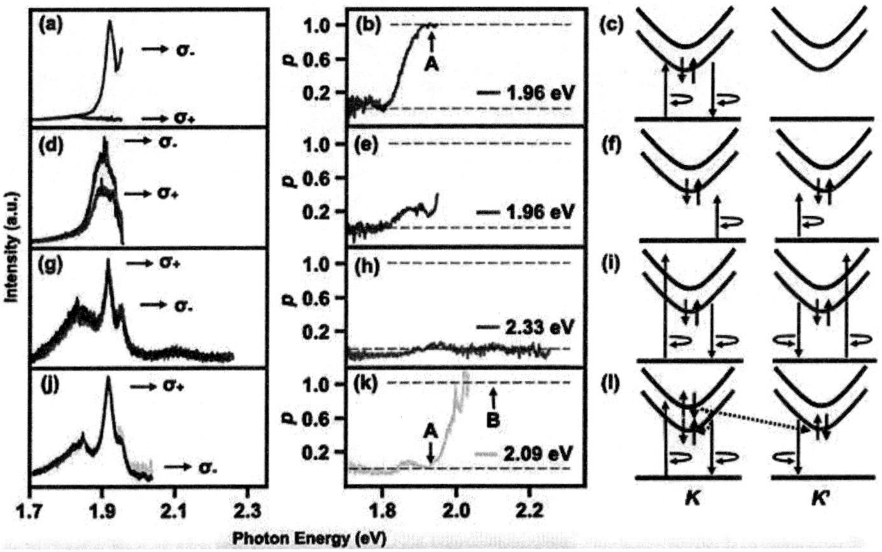

Fig. 8.13 Low temperature circularly polarized photoluminescence from MoS_2. Monolayer data is shown in **(a)**–**(c)** and **(g)**–**(i)**. Bilayer data is shown in **(d)**–**(f)**, and **(j)**–**(l)**. Figures **(a)**–**(f)** are for excitation using 1.96 eV circularly polarized pumping which is on resonance with the A exciton. Figures **(g)**–**(I)** are for 2.33 eV excitation which is off resonance. Figures **(j)**–**(l)** are for 2.09 eV excitation. The photoluminescence spectra are shown in **(a)**, **(d)**, **(g)**, and **(j)**. The helicity of the luminescence is shown in **(b)**, **(e)**, **(h)**, and **(k)**. The helicity parameter is determined using $\rho = (I(\sigma^-) - I(\sigma^+))/[I(\sigma^-) + I(\sigma^+)]$ where $I(\sigma^{\pm})$ refers to the $\sigma^{\pm}$ polarized photoluminescence intensity. The optical processes for circularly polarized absorption and emission are shown in the right column. Figure adapted from [17]

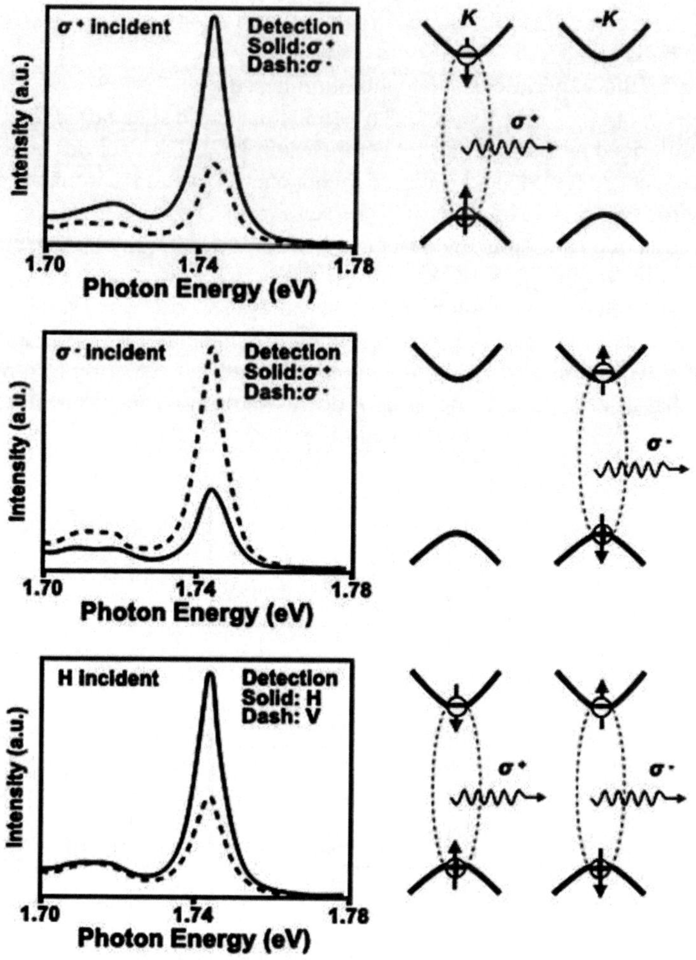

Fig. 8.14 Polarization Resolved Photoluminescence of neutral excitons from monolayer WSe_2. Note that the electron spin is the opposite of the hole spin. The circular polarization of the emission is the same as that of the excitation. Figure adapted from [16]

8.4 Massive Dirac Fermions: Physics and Optical Transitions at the *K* and *K′* Points in the Brillouin Zone

The hexagonal crystal structure of graphene and monolayer trigonal prismatic TMD's both result in interesting physics that can be described in terms of *massive* Dirac fermions [16, 19]. While Dirac fermions in the π bands of monolayer graphene are considered massless due to the linear relationship between energy and wavevector, the VB and CB of TMD materials have a parabolic relationship between energy and wavevector thus the Dirac carriers for TMD materials have a mass. The Fermi

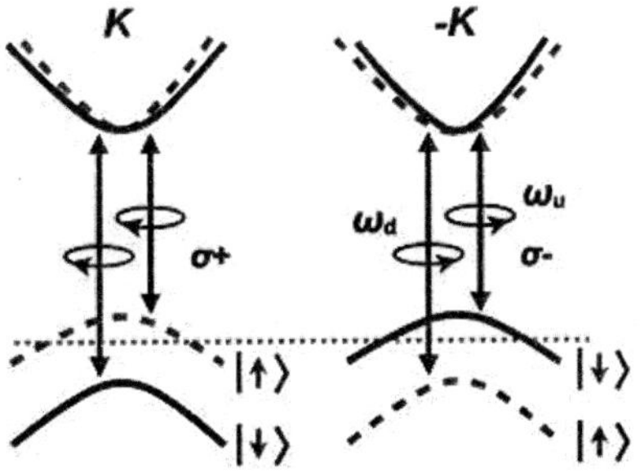

Fig. 8.15 Selection of valley K (K') using circularly polarized light and selection of spin using a resonant frequency are shown. Figure adapted and reprinted with permission from [19]. © 2012 American Physical Society

velocity of these Dirac fermions is not close to the speed of light. A phenomenological approach can be used to construct a Dirac Hamiltonian. A 2 Band model has been used to describe the interaction at the $K(K')$ points using the d orbitals for highest energy valence band labeled v and lowest lying conduction band labeled c and the valley index $\tau = \pm 1$ [19]:

$$|\phi_c\rangle = |d_{z^2}\rangle \text{ and } |\phi_v^\tau\rangle = \frac{1}{\sqrt{2}}\left(|d_{x^2-y^2}\rangle + i\tau|d_{xy}\rangle\right) \tag{8.33}$$

The valence band wavefunction meets the requirement for C_{3h} symmetry. The valence band wavefunctions for the two valleys $|\phi_v^+\rangle$ and $|\phi_v^-\rangle$ are related by time reversal symmetry. The magnetic quantum numbers m_l for the two basis functions are $m_l = 0$ and $m_l = 2\tau$. The Dirac Hamiltonian in this 2 Band basis set must also meet C_{3h} symmetry. The Dirac Hamiltonian for graphene $H^{2D} = \pm\hbar v_F \boldsymbol{\sigma}\cdot\vec{\kappa} = \pm v_F \boldsymbol{\sigma}\cdot\vec{P}$ is a useful starting point. TMD materials have a bandgap which needs to be included along with valley splitting. The Hamiltonian is stated to first order in $\vec{k}$ (see [16, 19]), is:

$$\hat{H} = at\left(\tau k_x\sigma_x + k_y\sigma_y\right) + \frac{\Delta}{2}\sigma_z \tag{8.34}$$

$\boldsymbol{\sigma}$ are the Pauli Matrices, t is the hopping integral, a is the lattice constant, τ is the valley index (+1 or − 1) and Δ is the band gap. The *spin orbit coupling splits the valence band* and needs to be added to (8.34).nnSpin orbit coupling can be added to (8.34) using the Pauli spin matrix $\widehat{S_z}$ resulting in the SOC Hamiltonian [19]:

$$\hat{H} = at\left(\tau k_x\sigma_x + k_y\sigma_y\right) + \left(\frac{\Delta}{2}\sigma_z\right) - \left(\tau\lambda\frac{\sigma_z - 1}{2}\hat{S}_z\right) \tag{8.35}$$

2λ is the spin splitting at the top of the VB at the $K(K')$ point. The spin up and spin down components are decoupled, and s_z remains a good quantum number [19]. In the Dirac picture, spin splitting is a result of breaking inversion symmetry. *Time reversal symmetry requires that the spin splitting be opposite* at the K and K' points in the Brillouin zone.

The Berry curvature in the conduction band, the orbital magnetic moment, and the optical circular dichroism are all related in this massive fermion Dirac model [16]. The Berry curvature is also used to calculate the Valley Hall effect and Spin Hall effect for TMD monolayers [19]. An early discussion of semiclassical electron dynamics in electrical and magnetic fields including the Berry phase can be found in [20]. The Berry phase for the conduction band is calculated using the anomalous velocity that a carrier receives from the interaction between the Berry curvature which acts like a magnetic field and the in-plane electric field from the light $v_{\perp} = -\frac{e}{\hbar}\vec{E} \times \Omega(\vec{k})$[7]:

$$\Omega_n(\vec{k}) = -\sum\nolimits_{n \neq n'} \frac{2\mathrm{Im}\left[\langle u_{n\vec{k}}(\vec{k})|v_x|u_{n'\vec{k}}(\vec{k})\rangle\langle u_{n'\vec{k}}(\vec{k})|v_y|u_{n\vec{k}}(\vec{k})\rangle\right]}{\left[E_{n'}(\vec{k}) - E_n(\vec{k})\right]^2} \tag{8.36}$$

A closed form expression for the Berry phase (curvature) for conduction electrons in a TMD layer with lattice constant a and $\Delta' = \Delta - \tau s_z \lambda$ is [19]:

$$\Omega_c(\vec{k}) = -\tau \frac{2a^2t^2\Delta'}{\left[\Delta'^2 + 4a^2t^2k^2\right]^{3/2}} \tag{8.37}$$

The Berry phase has opposite signs at the K and K' points. The Valley Hall conductivity σ_v^n in terms of the Fermi–Dirac distribution f in units of $e/\hbar$ is [19]:

$$\sigma_v^n = 2\int \frac{d\vec{k}}{2\pi^2}\left[f_{n,\uparrow}\Omega_{n,\uparrow}(\vec{k}) + f_{n,\downarrow}\Omega_{n\downarrow}(\vec{k})\right] \tag{8.38}$$

The integration is done over the states close to one K point. The spin Hall conductivity in units of $e/\hbar$ is [19]:

$$\sigma_s^n = 2\int \frac{d\vec{k}}{2\pi^2}\left[f_{n,\uparrow}\Omega_{n,\uparrow}(\vec{k}) - f_{n,\downarrow}\Omega_{n\downarrow}(\vec{k})\right] \tag{8.39}$$

When there is *moderate hole doping* and $\epsilon_F \ll \Delta - \lambda$ whre ϵ_F is measured from the valence band maximum, the hole conductivities are the same for the spin and valley effects [19]:

$$\sigma_v^h = \sigma_s^h = \frac{1}{\pi}\frac{\epsilon_F}{\Delta - \lambda} \tag{8.40}$$

When there is electron doping, Where $n = e$ for electrons when the Fermi energy is ϵ_F, the Hall and spin Hall conductivities are [19]:

$$\sigma_v^e = \frac{1}{\pi}\frac{\Delta}{\Delta^2 - \lambda^2}\epsilon_F \quad \sigma_s^e = \frac{1}{\pi}\frac{\lambda}{\Delta^2 - \lambda^2}\epsilon_F \tag{8.41}$$

It is important to remember that the spin orbit splitting of the valence band of a W TMD is larger than that of a Mo TMD as indicated in Table 8.4. It is useful to restate that in (8.40) and (8.41), the valence band splitting at the $K(K')$ is 2λ.

The optical circular dichroism $\eta(\vec{k})$ is the degree of circular polarization for a direct gap transition at the wavevector $\vec{k}$ [16]:

$$\eta(\vec{k}) = -\frac{\Omega_n(\vec{k})\ \hat{z}}{\mu_B^*}\frac{e}{2\hbar}\left[\Delta^2 + 4a^2t^2k^2\right]^{1/2} \tag{8.42}$$

$\left[\Delta^2 + 4a^2t^2k^2\right]^{1/2}$ is the direct transition energy at k, m^* is the effective mass at the band edge, and a quantity that resembles the Bohr magneton is: $\mu_B^* = e\hbar/2m^{*2}$ [19]. $\hat{z}$ is the unit vector normal to the TMD plane. The optical circular dichroism or degree of optical polarization can also be calculated from the results of first principles band structure at $\hbar\omega_{cv} = \epsilon_c(k) - \epsilon_v(k)$ using [21]:

$$\eta(\vec{k}, \omega_{cv}) = \frac{\left|P_+^{cv}(k)\right|^2 - \left|P_-^{cv}(k)\right|^2}{\left|P_+^{cv}(k)\right|^2 + \left|P_-^{cv}(k)\right|^2} \tag{8.43}$$

The + sign refers to left handed and the – sign refers to right handed circular polarization. $P_\pm^{cv}(k) = (1/\sqrt{2})\left[P_x^{cv}(k) \pm i P_y^{cv}(k)\right]$ and $P_+^{cv}(k) = \langle\psi_{ck}|\hat{p}|\psi_{vk}\rangle$ [21]. The circular dichroism calculated using the TNN TB model for MoS_2 versus location in the Brillouin zone is shown in Fig. 8.12a. $\eta(\vec{k})$ is -1 at K' and $+1$ at K' [11]. The experimental photoluminescence intensity of MoS_2 drops dramatically away from the resonant transition energy and the circular polarization of the PL matches the predicted circular dichroism from (8.43) as shown in Fig. 8.16 [21]. The PL data shown in Fig. 8.13 also demonstrates the circular dichroism for MoS_2 [17]. The circular dichroism of WSe_2 is apparent in Fig. 8.14 [16].

8.5 Band Gap Renormalization and Photoluminescence Lineshape

The 3 band NN TB model predicted a band gap of MoS_2 of 1.66 eV, however the optical band gap is $\sim$ 1.88 eV [21, 22]. While the three band model provides useful insight into the physics of monolayer trigonal prismatic TMD layers, more rigorous theoretical calculation are necessary for understanding the band gap. The optical transition energies for MoS_2 have been also calculated using GW-Bethe–Salpeter approach [22]. MoS_2 does experience band gap renormalization [22]. The exciton binding energy for MoS_2 is $\sim$ 1 eV, and the GW-Bethe–Salpeter quasiparticle band gap is 2.84 eV [22]. Other TMD monolayers experience band gap renormalization

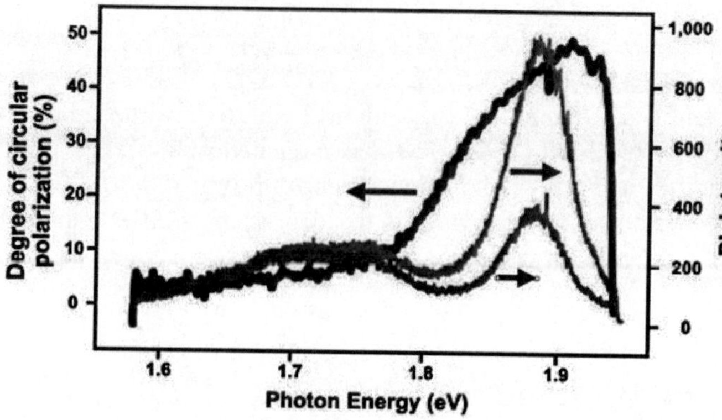

Fig. 8.16 The circular dichroism or preference for optical absorption of right versus left circularly polarized light matches the circular polarization of the photoluminescence for monolayer MoS_2 at 83 K. A wavelength of 633 nm (1.96 eV) was used to excite the PL. Figure adapted from [21]

[23]. For example, scanning tunneling spectroscopy measurements of $MoSe_2$ have been compared to PL measurement to determine that the exciton binding energy of $MoSe_2$ is 0.55 eV [23]. Bilayer graphene and HOPG graphite were used as substrates in these experiments. A comparison of theory and experiments is shown in Fig. 8.17 along with the optical absorption spectra. It is interesting to note the strong absorption peak in Fig. 8.17 due to the excitonic transitions [23]. The exciton binding energy of WS_2 is also large, 0.71 eV, when compared to non TMD materials [24]. The absorbance spectrum of WS_2 also shows strong exciton absorption as shown in Fig. 8.18 [24].

8.6 The Complex Refractive Index (Dielectric Function) and Optical Conductivity of Monolayer TMD

As mentioned in Chap. 1, Sect. 1.5.2 TMD crystals with hexagonal symmetry are uniaxial optical materials. The uniaxial nature of bulk crystals of WS_2, WSe_2, α-$MoTe_2$, NbS_2, and $NbSe_2$ was demonstrated using reflectivity at 300 and 78 K as discussed in Sect. 8.9. However, the complete uniaxial optical properties of monolayer (trilayer) hexagonal TMD films are difficult to measure. Exfoliation of large area samples is difficult, and growth of large area single crystal trilayer films is also difficult. Spectroscopic ellipsometry measurements require large samples relative to those that can be measured with a focused laser beam. In Chap. 1 Figs. 1.7 and 1.8 we showed published data for the in-plane dielectric function, optical conductivity, and absorption of single crystal monolayer MoS_2, $MoSe_2$, WS_2, and WSe_2 between 1.5 and 3 eV. The spin-orbit split exciton transition can be seen in Fig. 1.7a–d. The complex refractive index of monolayer MoS_2, $MoSe_2$, WS_2, and WSe_2 has been characterized over a wide energy range up to ~ 6.5 eV using spectroscopic ellipsometry [25]. These TMD monolayers were deposited on sapphire substrates using chemical

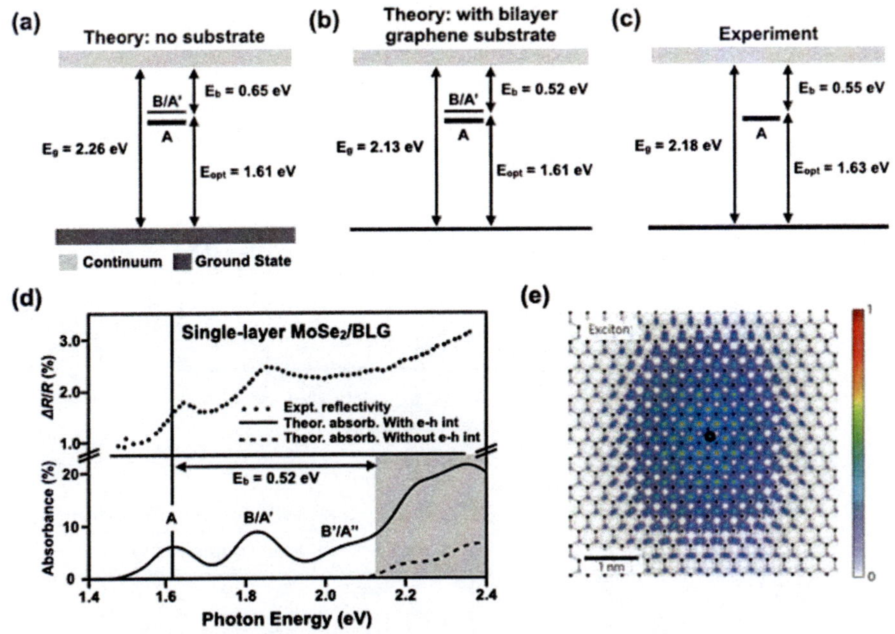

Fig. 8.17 The effect on the absorbance spectra of $MoSe_2$ due to the presence of a bilayer graphene substrate is shown. Theoretical absorbance pathways without a substrate (**a**) and with a bilayer graphene substrate (**b**) are compared to experimental data (**c**). In (**e**), the probability of the location of an exciton is shown to extend over a relatively large area. The probability was determined using the exciton wavefunction. The theoretical and experimental absorbance spectra are shown in (**d**). The exciton binding energies result in a lower optical band gap when compared to the single particle band gap. The bilayer graphene seems to alter the exciton binding energy. Figure adapted from [23]

vapor deposition, and the monolayer nature and surface smoothness verified using atomic force microscopy [25]. Based on the information in [25], the samples do not seem to be single crystal over the measured area. As mentioned in Chap. 1, the complex refractive index seems to have been determined using a isotropic model for the optical properties for the film in [25]. As with graphene, the true extraordinary refractive index for single tri-layer TMD may be close to 1. In that case, there *may* be few differences between the complex refractive index values for single trilayer TMD determined by normal incidence and an isotropic model. These spectra reflect the band gap renormalization observed by PL. The effects of spin orbit splitting are also observed [25]. The complex refractive index for these Mo and W TMD are shown in Fig. 8.19. The refractive index values for all these films are remarkably large. The bandgaps are reported for both the A and B excitons in [25] along with the exciton binding energy. Experimental and theoretical band gaps and exciton energies are compared in Table 8.5. The exciton A band gaps are comparable to theoretical band gaps and the energy difference between the A and B excitons are comparable to calculated spin orbit splitting values. As mentioned above, the NN TB model and TNN TB models are comparable near the low energy $K(K')$ points in the Brillouin

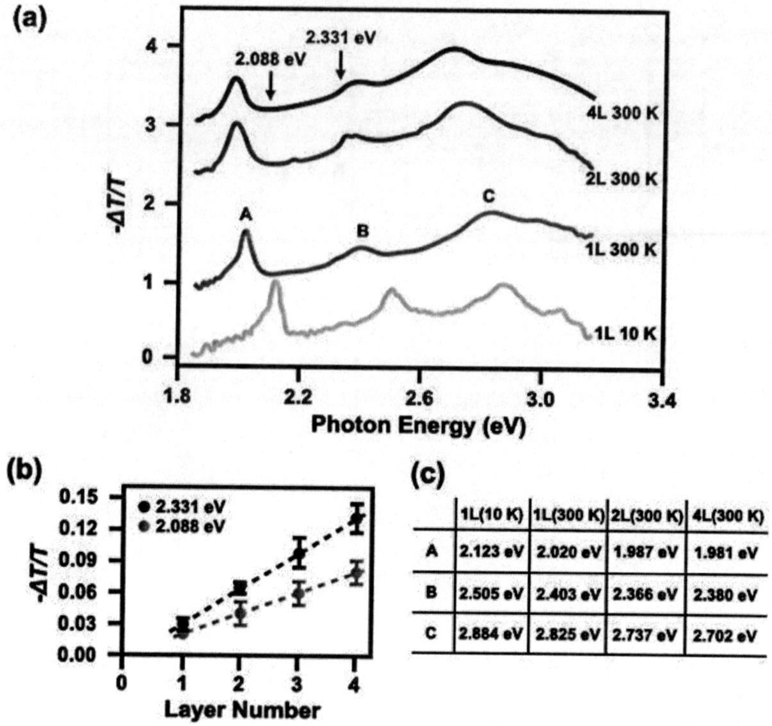

	1L(10 K)	1L(300 K)	2L(300 K)	4L(300 K)
A	2.123 eV	2.020 eV	1.987 eV	1.981 eV
B	2.505 eV	2.403 eV	2.366 eV	2.380 eV
C	2.884 eV	2.825 eV	2.737 eV	2.702 eV

Fig. 8.18 (**a**) Absorbance spectrum of monolayer WS_2 at 10 K and for 1, 2, and 3 layer WS_2 at 300 K. (**b**) The absorbance versus layer for 2.33 and 2.044 eV light. (**c**) The energies of the A, B, and C exciton peaks for different energies and layer numbers. The energy difference between A and B is the same as the spin orbit splitting energy of 0.38 eV. This split does not change with layer number. Figure adapted from [24]

zone, and demonstrate trends in spin orbit splitting that are comparable to the SE and PL values.

8.6.1 Optical Conductivity of Monolayer TMD

Experimental results for the in-plane optical conductivity of monolayer TMD were presented in chapter 1. Here, we discuss comparison between theory and experiment. The optical conductivity of TMD monolayers has been calculated from an 11 band TB model using the Kubo formula [12]. The predicted optical conductivity is shown in Fig. 8.20. The insets in Fig. 8.20 show the low energy onset of absorption. The step structure is due to spin orbit splitting of the valence bands at K and K' [12]. The A and B excitonic peaks in the absorption spectra are attributed to these transitions [12]. The spin orbit coupling for the W will be larger than for Mo and that is reflected in the

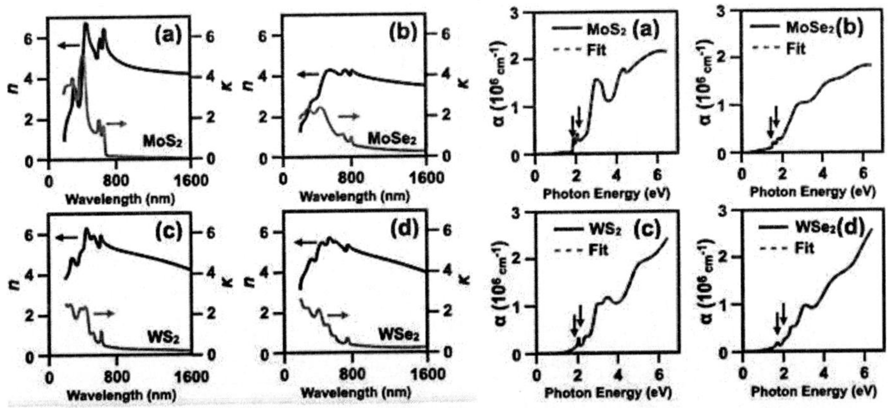

Fig. 8.19 The complex refractive index and optical absorption spectra of monolayer MoS_2, $MoSe_2$, WS_2, and WSe_2. The ellipsometric data was fit assuming an isotropic approximation for the optical properties using multiple angles of incidence. The A and B excitons are visible in the absorption spectra. Figure adapted and reprinted from [25] with the permission of AIP Publishing

Table 8.5 Experimental and theoretical band gaps for monolayer TMD. NNTB refers to the nearest neighbor tight binding model, TNNTB refers to the third nearest neighbor tight binding model, SE is spectroscopic ellipsometry, PL is photoluminescence, and FP refers to first principles quantum calculations

		MoS_2	$MoSe_2$	WS_2	WSe_2		References
Band gap	NNTB	1.66	1.44	1.81	1.54	eV	[11]
Band gap	FP	1.8	1.5				[11]
A exciton band Gap	SE	1.95	1.62	2.11	1.72	eV	[25]
Band gap	PL	1.83	1.55	2.02	−1.68	eV	[21, 24, 26, 27]
Band gap	absorbance 10 K			2.123		eV	[24]
A exciton	absorbance 300 K		1.63			eV	[23]
B exciton band Gap	SE	2.08	1.82	2.45	2.09	eV	[25]
SO VB splitting	NNTB&TNNTB	0.148	0.184	0.43	0.466	eV	[11]
SO splitting	SE A exciton–B excitor	0.13	0.2	0.34	0.37	eV	

relative step size seen in the insets. These transitions are seen in the A and B absorption peaks which were previously discussed. We note that the spectral resolution of the experimental dielectric function data shown in Sect. 1.5.2 smoothes the spectra. The experimental in-plane optical conductivities for monolayer TMD were interpreted using a 2D slab model as discussed in Sect. 1.5.4 and were shown in Chap. 1 Fig. 1.9. We also note that the experimental data shows the excitons discussed above. This illustrates the challenges in comparison between theory and experiment.

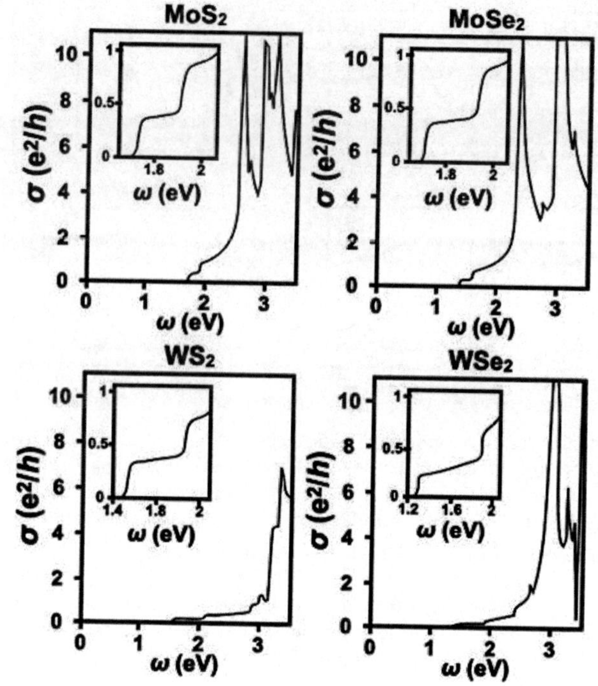

Fig. 8.20 Theoretical real part of the optical conductivity of monolayer TMDs. The conductivity increases from zero after the band gap, and the step structure is due to band splitting from spin orbit coupling. Figure adapted from [12].

8.7 Structure, Electronic Band Structure, and Optical Properties of Bilayer Trigonal Prismatic TMD

As mentioned above, the band gap and optical properties of TMD layers change as the number of layers increases [27]. In Chap. 7, the effect of structural differences in the layer stacking on the optical and electrical properties of graphene was described. The monolayers of naturally occurring TMD crystals stack in a number of configurations see for example Fig. 8.4. Thus, bilayers of trigonal prismatic coordinated monolayers also stack in a number of configurations. The topic of bilayers fabricated with twisted rotational alignment is discussed in Sects. 8.8 and 8.12. The most stable configuration is the 2H coordination where the two layers are rotated 180° from each other, and this sequence is referred to as AA' (see Fig. 8.4) [2, 28]. The 3R configuration is also observed, and the stacking sequence is referred to as AB (see Fig. 8.4) [2]. Bilayers in the 2H configuration have inversion symmetry and thus the valley optical dichroism and valley Hall effect cancel out [2]. The band gap of bilayer TMD materials is indirect [2]. Five different well-ordered stacking configurations are possible as shown in Fig. 8.21. The 2H configuration has AA' stacking, and it has the lowest energy, and

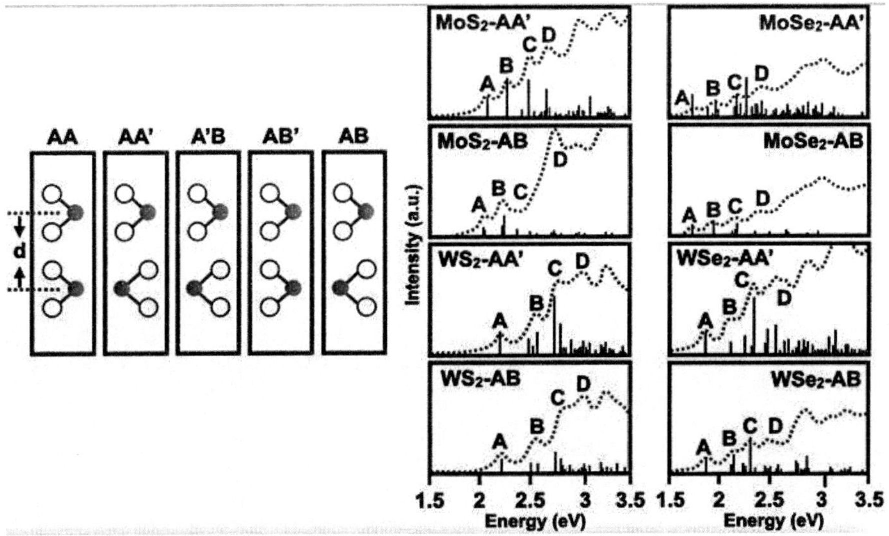

Fig. 8.21 Theoretical imaginary part of the dielectric function (which we presume to be the dielectric function for light parallel to the c-axis) and oscillator strength of AA' and AB stacked MoS_2, $MoSe_2$, WS_2, and WSe_2. The AA' configuration is found in the stable 2H bilayer configuration. Variations of bilayer TMD stacking configurations are shown on the left. Figure adapted and reprinted with permission from [28]. © 2014 American Physical Society

the total ground state energy of the 3R configuration is only slightly different from the 2H configuration [28]. First principles theory has been used to calculate the interlayer distances which are important when using ellipsometry or atomic force microscopy to determine the number of monolayers. These distances are listed in Table 8.6 for the five different stacking orders shown in Fig. 8.21. Theoretical results are known to depend on the details of the calculation procedure. [2, 5, 7, 28]. Calculated optical transition energies for key transitions are also listed in Table 8.6 [28]. The imaginary part of the dielectric function is also shown in Fig. 8.21, and the electronic band structure of Bilayer TMD for the 2H AA' configuration and the 3R AB stacking is shown in Fig. 8.22 [28]. The transition from direct to indirect gap for WS_2 from monolayer to bi, tri, and quad layer is shown in Fig. 8.23 [29]. It is also important to note that strain can change the direct gap of ML TMD into an indirect gap [28].

The interaction between layers changes the band structure for all bilayer and multilayer TMD structures. The effect of interlayer hopping is strong enough to result in the change to an indirect band gap [5, 7]. Band structure calculations of the 2H bilayer WS_2 show that spin orbit splitting of the valence band at the K and K' points remains constant as the number of monolayers increases from 1 to 4 [5]. The interlayer hopping interaction between the p_z orbitals of nearest neighbor chalcogen atoms is reported to be the most significant interaction [5]. Thus, expanding the 3 band, NN and TNN TB models to describe the physics of trigonal prismatic TMD bilayers will not be adequate because interlayer interactions do not include the d_{z^2}, $d_{x^2-y^2}$, and d_{xy}

Table 8.6 BiLayer TMD interlayer distances and relative ground state total energy calculated using the random phase approximation [28]

	Interlayer distance d	Relative ground state total energy	K_c-K_v	K_c-Γ_v	T_c-K_v	T_c-Γ_v
			eV	eV	eV	eV
MoS_2						
AA	6.77	33.3				
AA'	6.27	0	2.41	2.32	2.05	1.96
A'B	6.78	34.3				
AB	6.17	3	2.29	2.12	1.98	1.82
AB'	6.26	10.3				
2H (exp)	5.25/6.15					
3R (exp)	6.14					
WS_2						
AA	6.48	37.9				
AA'	6.24	0	2.48	2.38	2.15	2.05
A'B	6.78	37.1				
AB	6.24	7.3	2.4	2.33	2.05	1.98
AB'	6.27	14.6				
2H (exp)	6.16					
3R (exp)	6.16					
	Interlayer distance d	Relative ground state total energy	K_c-K_v	K_c-Γ_v	T_c-K_v	T_c-Γ_v
		meV/formula unit	eV	eV	eV	eV
$MoSe_2$						
AA	7.18	39.4				
AA'	6.48	0	1.99	1.97	1.68	1.67
A'B	7.12	39.6				
AB	6.47	4.5	1.93	1.93	1.7	1.7
AB'	6.53	13.5				
2H (exp)	6.45					
3R (exp)	6.46					
WSe_2						
AA	7.24	38				
AA'	6.5	0	2.08	2.18	1.71	1.81
A'B	7.24	39.6				
AB	6.54	3.6	2.02	2.15	1.66	1.79
AB'	6.62	14.2				
2H (exp)	6.49					

The gap transition energies were calculated using the G_0W_0 method where G stands for the one-body Green's function *G* and *W* for the dynamically screened Coulomb interaction. In the G_0W_0 method, Density Functional Theory is used to determine the first guess G_0 and W_0 which are then used to calculate the optical transitions [30]. Data from [28]

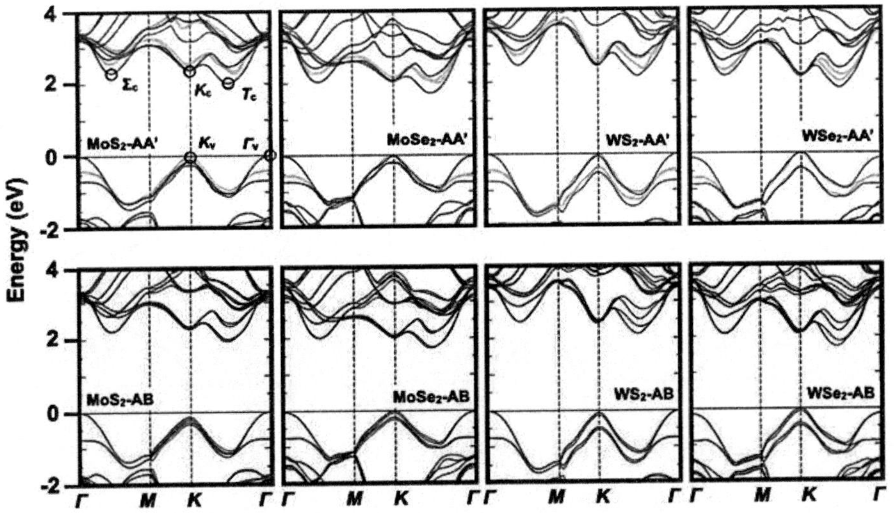

Fig. 8.22 Electronic Band Structure of Bilayer TMD for the 2H AA' configuration and the 3R AB stacking. All of the band gaps are indirect. Figure adapted and reprinted with permission from [28]. © 2014 American Physical Society

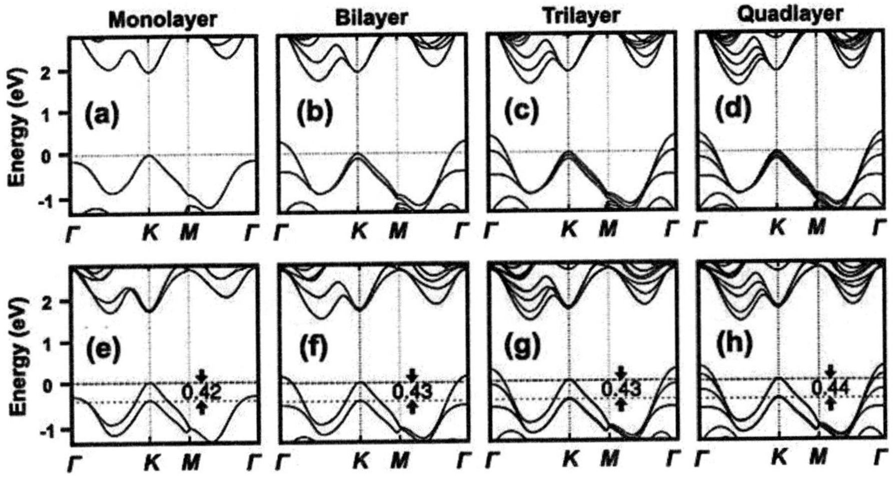

Fig. 8.23 Influence of Spin Orbit Coupling on the Electronic Band Structure of 2H multi-layer WS_2. The transition of the Electronic Band Structure of WS_2 without spin orbit coupling is shown in (a) through (d) and with spin orbit coupling (e) through (h). The bulk lattice constant was used in projector augmented wave method and generalized gradient approximation. Figure adapted from [29]

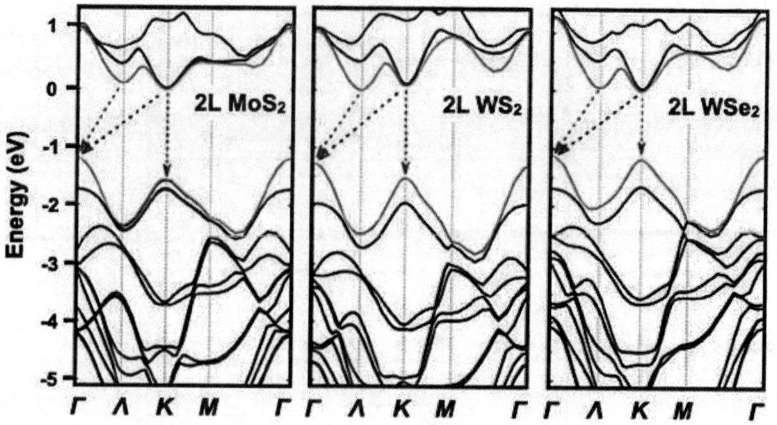

Fig. 8.24 The direct and indirect radiative pathways for photoluminescence are shown for MoS_2, WS_2, and WSe_2 bilayers with AA′ stacking. Figure adapted and reprinted with permission from [31] with permission. © 2013 American Chemical Society

orbitals used in the 3 band model. The van der Waals interactions between layers are difficult to include in band structure calculations [28]. The detailed study of bilayer stacking configurations for MoS_2, $MoSe_2$, WS_2, and WSe_2 [28] required careful consideration of the van der Waals interlayer forces. The reader is referred to He, Hummer, and Franchini [28] for the details of the theoretical procedure. The layer to layer spacing for bilayers is larger than that found for bulk TMD materials with the same stacking configuration [5]. This can complicate ellipsometric determination of film thickness.

The photoluminescence of bilayer MoS_2, WS_2, and WSe_2 and the band structure have been studied [31]. The direct and indirect radiative recombination pathways are shown in Fig. 8.24 [31]. The temperature dependent PL data is shown in Fig. 8.25. The origin of the indirect transition was studies as a function temperature and material [31]. The indirect emission in MoS_2 and WS_2 comes from $\Lambda \rightarrow \Gamma$ transitions, and $K \rightarrow \Gamma$ transitions for WSe_2 [31]. The radiative pathways and PL data for 3 layer, 4 layer and bulk WSe_2 are reported in [31].

8.8 Twisted Bilayer TMD

Correlated electron properties including superconductivity, the band insulator state, and the strange metal state have been observed at specific twist angles known as the magic angle in bilayer graphene as discussed in Chap. 7. Just as with twisted bilayer graphene, the wavelength of the moiré pattern is a function of twist angle. Multilayer structures of other van der Waals materials including TMD are frequently assembled using a variety of experimental procedures [32]. Just as with twisted

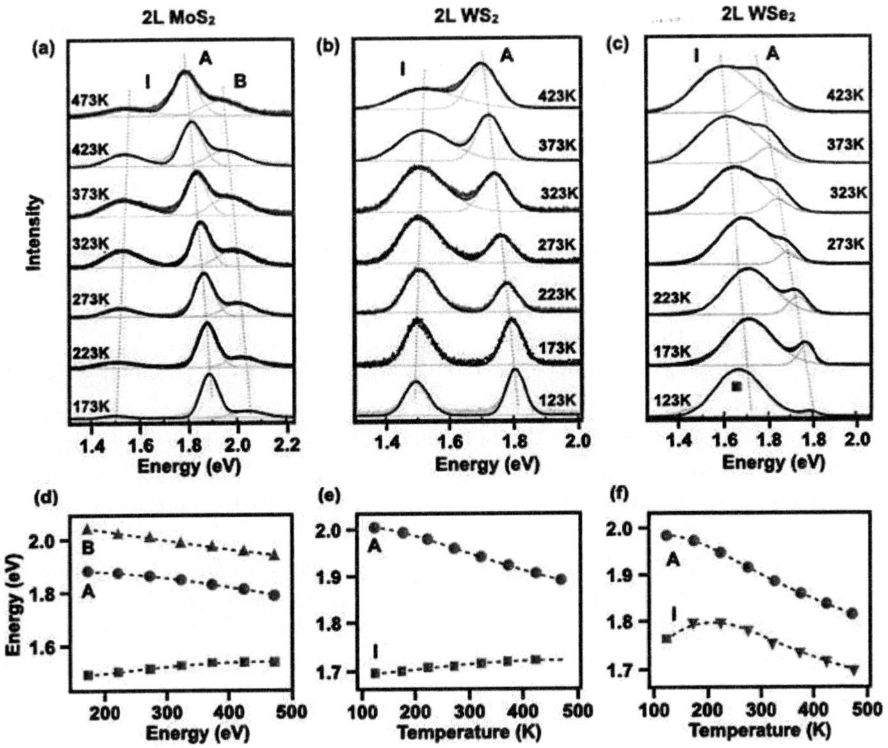

Fig. 8.25 The temperature dependent PL of mechanically exfoliated bilayer MoS_2, WS_2, and WSe_2 is shown. The expected stacking sequence is AA′ for 2H symmetry. Figure adapted and reprinted with permission from [31] with permission. © 2013 American Chemical Society

few layer graphene discussed in Chap. 7, h-BN encapsulation is typically used to fabricate the test structures for transport characterization. The presumption is that the TMD and h-BN have minimal interaction. Flat bands in a moiré band structure can be engineered through the twist angle and the displacement field normal to TMD layer planes. Doping the twisted TMD bilayer is done using the floating gates above and below the TMD plane. The wavelength of the moiré lattice for a twisted TMD bilayer with lattice constant a with a twist angle of θ is $\lambda = a/\sqrt{2(1-\cos\theta)}$, and the area A of the unit cell of the moiré lattice is $A = a^2\sqrt{3}/[4(1-\cos\theta)]$. The lattice constant for WSe_2 is $a = 0.328$ nm. The full filling hole density is $n_s = 2/(\frac{\sqrt{3}}{2}\lambda^2)$, where 2 holes per unit cell were assumed for the samples of WSe_2 used in [32]. The carrier concentration is tuned to observe correlated electron states at low temperature. Tuning the carrier concentration to achieve half filling is done at each twist angle using the dual gate structure. In contrast to bilayer graphene where magic angles occur in a narrow range of twists, bilayer WSe_2 has been found to be a correlated insulator at ½ filling $v = n/n_s = 1/2$ of holes at twist angles between 4° and 5.1° [32]. For example, the device with a twist angle $\theta = 4.2°$, showed correlated insulator behavior below $10K$. Evidence for a superconducting state was reported for an angle of 5.1° when the carrier concentration was tuned to be close to that of the correlated insulator state [32]. A device with a twist angle of 4.8° showed similar behavior. The carrier concentration is tuned to observe the correlated electron state at low temperature.

The band structure of bilayer TMD structures has been theoretically [5, 32] and experimentally [33] studied as a function of twist angle for MoS_2. A twist angle dependent, tight binding, second quantization Hamiltonian for the moiré bands has been used to theoretically study the correlated insulator state [32]. In addition to the twist angle dependent periodic potential of the moiré lattice, both spin–orbit interactions and a Hubbard potential for electron–electron interactions have been included. The Hubbard potential is further discussed as it is applied to other materials systems in Chap. 9. Inversion symmetry is broken in twisted bilayers which means that twisted bilayers have some of the same optical properties as monolayers. The K points do not change much for twisted bilayers while the valence band at the Γ point changes energy and the indirect transition from Γ_v to K_c increases in energy by 0.147 meV [5].

The optical properties of twisted bilayer MoS_2 have been studied over a wide range of twist angles up to 60° [34]. The photoluminescence spectra show that the indirect transition between 1.4 and 1.7 eV is noticeably shifted with twist angle as shown in Fig. 8.26a. The assignment of this transition is based on data from exfoliated bilayers with no twist angle. This is in contrast to the PL peaks due to the direct A transition which is visible between 1.84 and 1.87 eV and is only slightly shifted by twist angle [34]. The monolayer spectra show much stronger PL spectra. In Fig. 8.26b, the differential reflectivity peaks are shown. The peak observed between 1.8 eV and 1.9 eV is also shifted with twist angle. The origin of the changes in energy

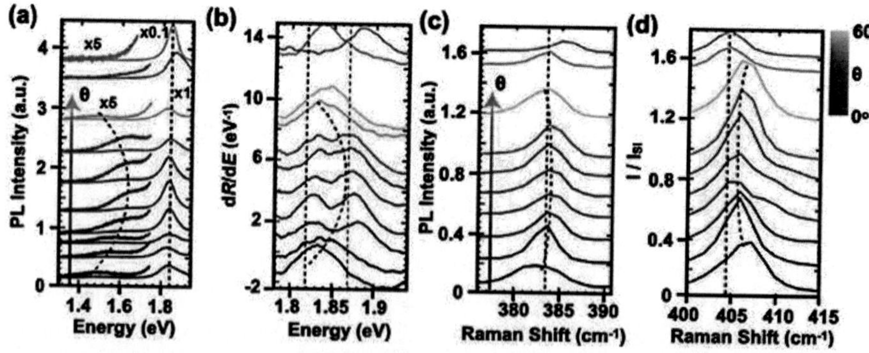

Fig. 8.26 The photoluminescence and Raman spectra of twisted bilayer MoS_2 is shown as a function of twist angle. The photoluminescence spectra are shown in (**a**), differential reflectivity in (**b**), and Raman spectra in (**c**) and (**d**). Differential reflectivity data provides similar information to optical absorption data. The top spectra in (**a**) through (**d**) are for as grown monolayers (top most spectra) and exfoliated monolayers (shown just below the as grown data). Figure adapted and reprinted from [34]. © 2014 American Chemical Society

of the optical transition comes from the change in interlayer separation as function of twist angle [34]. The calculated interlayer separation for AA′ stacking is slightly larger (0.623 nm) than bulk 2H (0.615 nm). The interlayer separation is a maximum at 30° at 0.652 nm which is an increase of 5%. The DFT calculations indicate that states near K do not change in energy while the highest occupied state near Γ is substantially downshifted which increases the $\Gamma - K$ indirect gap by 147 meV [34]. The spacing of the S atoms between layers varies with the place in the moiré pattern which is a function of twist angle. Raman data for the in-plane E_{2g} at 384 nm and out of plane A_{1g} modes are shown in Fig. 8.26(c) and (d) [34]. The out of plane mode is clearly a function of twist angle. Raman spectroscopy of TMD materials is further discussed below.

As discussed in Chap. 7, determination of the number of layers of a 2D material on a substrate such as a SiO_2/silicon wafer is essential for fabrication of electronic devices and test structures. The optical contrast of MoS_2, WSe_2, and $NbSe_2$ on SiO_2/silicon can be used to locate the TMD films. Based on the calculations of [40], SiO_2 thickness values of 90, 140, and 270 nm should provide optical contrasts of greater than 10% using the green color channel of camera while imaging using an optical microscope [40]. Mono, Bi, and Tri layer MoS_2 and WSe_2 can be distinguished using broadband green light illumination [40]. The contrast of $NbSe_2$ was less than 10%.

8.9 The Complex Refractive Index (Dielectric Function) *of Multilayer and Bulk TMD*

Spectroscopic ellipsometry has been used to characterize deposited films of TMD materials. Many of the reported refractive indices that are not for single crystal samples. Some of this data is shown in Figs. 8.27 and 8.28 [35, 36]. Several studies from the mid to late 1970's report reflectivity spectra of 2H $NbSe_2$, 2H $TaSe_2$, 2H TaS_2 and 1T TaS_2 [37]; 2H WSe_2 and 3R WS_2 [38]; and 2H MoS_2, 2H $MoSe_2$, and 2H $MoTe_2$ [39] bulk TMD materials. Normal incidence was used to measure reflectivity from the Basal faces of the materials. The resultant dielectric functions are shown in Figs. 8.29, 8.30, and 8.31.

8.10 The Layer Number Dependence of Raman Scattering from Trigonal Prismatic TMD

As with graphene, Raman spectroscopy has proven to be an effective means of determining the number of TMD layers and their relative orientation [41–45]. In Chap. 1, Raman spectroscopy of TMD materials was introduced along with the Raman tensors that are used to determine the selectivity of polarized light to specific

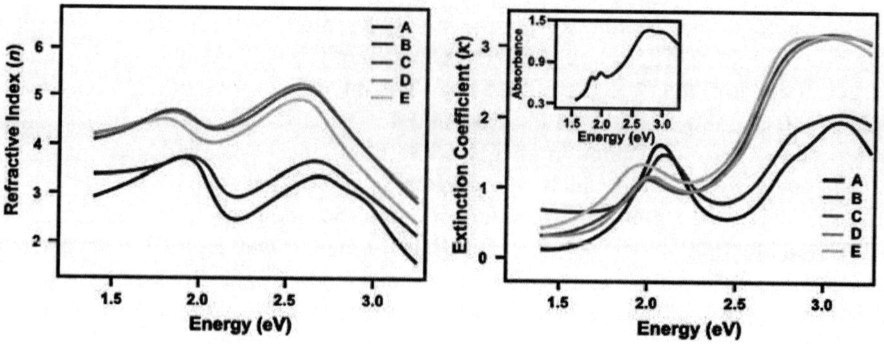

Fig. 8.27 Complex refractive index of multilayer MoS_2 grown by vapor phase sulferization are shown. The reported thickness values of A through E are: 1.99 ± 0.01 nm, 3.016 ± 0.07 nm, 5.536 ± 0.08 nm, 9.836 ± 0.04 nm, and 19.886 ± 0.05 nm as determine by ellipsometry. These values are likely significant only to 0.1 nm. Figure adapted and reprinted from [35] with the permission of AIP Publishing

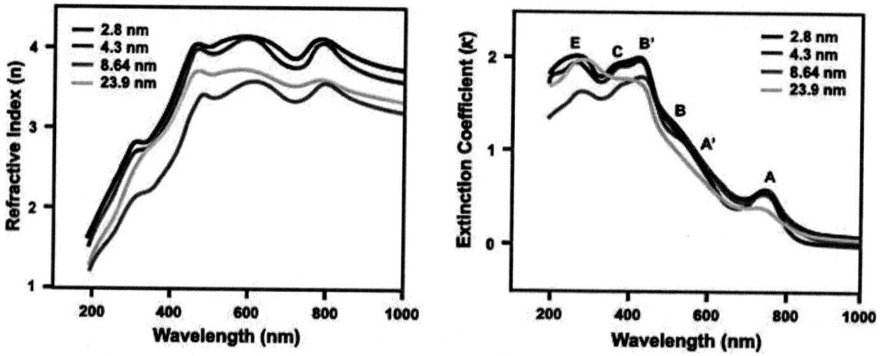

Fig. 8.28 Complex refractive index of Metal Organic Chemical Vapor Deposition grown nanocrystalline WSe_2 on sapphire are shown. Figure adapted from [36].

Raman transitions. Some Raman modes are observed for monolayer, multilayer, and bulk samples, while some low frequency modes are only observed for a few layer samples. Raman is also sensitivity to stacking and rotational twisting of the layers [44]. Here, we provide an overview of Raman scattering especially it application to determination of the number of layers.

First we discuss non-resonant Raman scattering from some representative trigonal prismatic 2H TMD materials: MoS_2, $MoSe_2$, and WSe_2. Some of the symmetry based expected vibrational modes for monolayer and bilayer 2H TMD have not been observed. In Figs. 8.32 and 8.33, the symmetry labeled vibrational modes of mono- and bi- layer MoS_2 are shown [43, 44]. The Raman active, out of plane vibrational mode spectra from monolayer to five layers are shown in Fig. 8.34. The A_{1g}, B^1_{2g}, and E^1_{2g} modes have all been observed with non-resonant scattering for $MoSe_2$ and

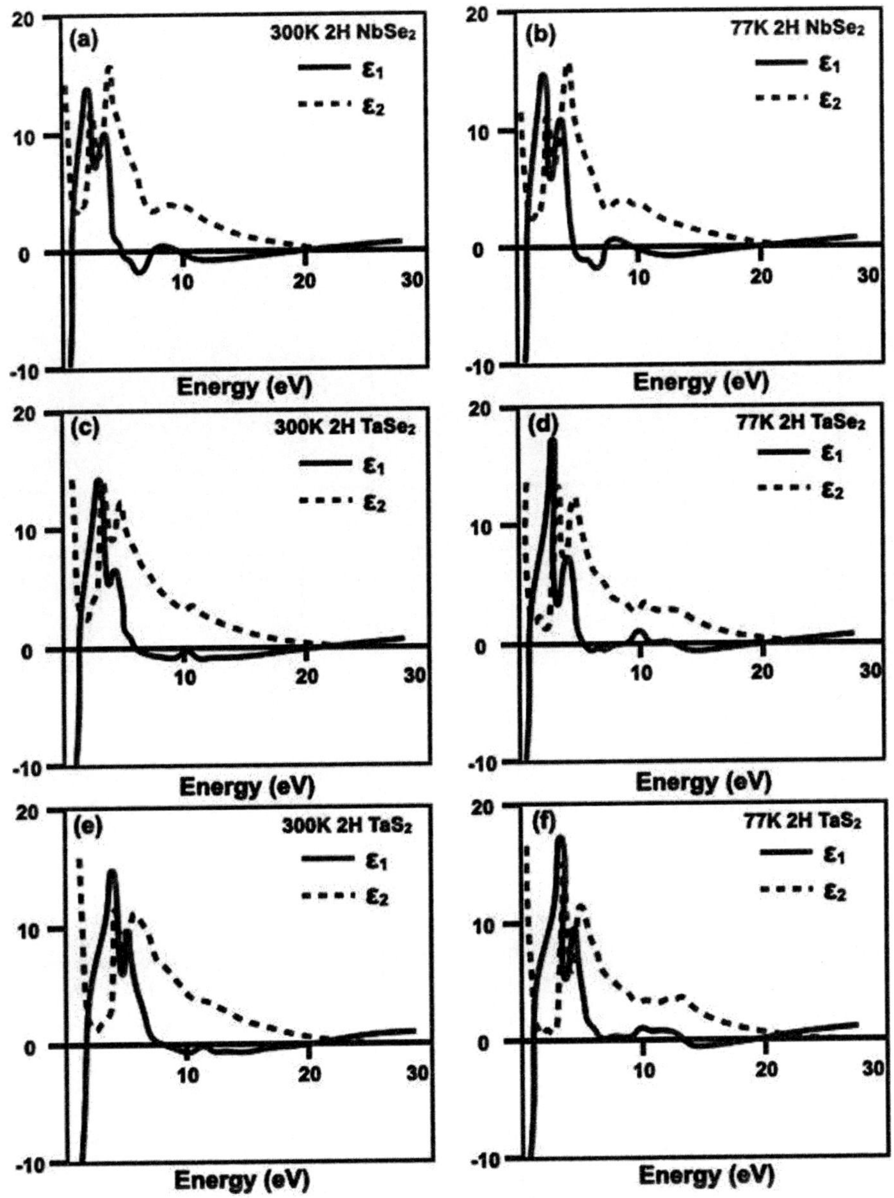

Fig. 8.29 The real and imaginary part of the dielectric function of 2H $NbSe_2$, $TaSe_2$, and TaS_2 from 25 meV to 14 eV. The low energy spectra were measured using a double beam infra-red spectrophotometer. The spectra from 0.5 to 5.9 eV were measured using a double beam spectrophotometer. The high energy spectra from 4. To 14 eV were measured using a vacuum monochrometer. Figure adapted from [37].

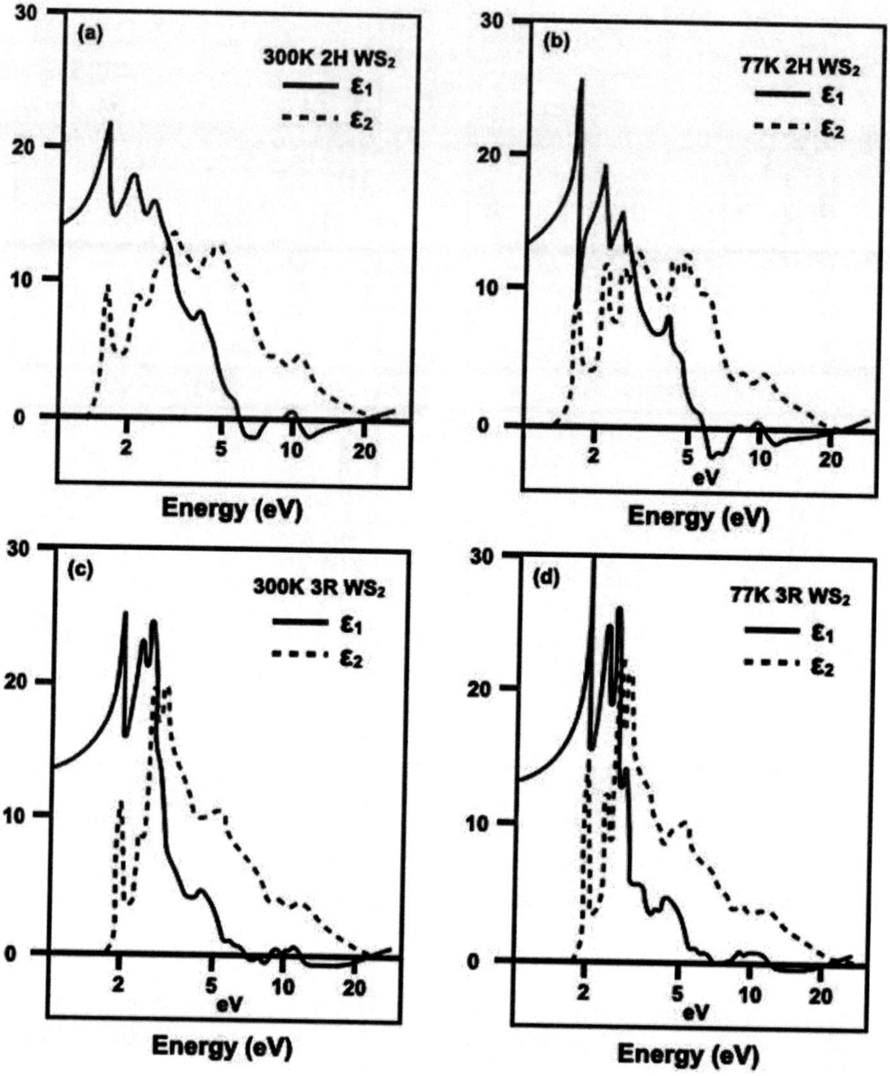

Fig. 8.30 The real and imaginary part of the dielectric function of 2H and 3R WS_2 from 1.5 to 14 eV. The spectra from 1.5 to 5.9 eV were measured using a double beam spectrophotometer. The high energy spectra from 4.5 To 14 eV were measured using a vacuum monochrometer. Figure adapted from [38].

WSe_2 [42]. The selection of a vibrational mode for determination of layer number depends on the material. In Fig. 8.35, the layer dependent Raman data for A_{1g}, B^1_{2g}, and E^1_{2g} for MoS_2, $MoSe_2$, and WSe_2 are shown. The shape and wavelength of the A_{1g} peak of $MoSe_2$ and WSe_2 show easily recognizable changes for 1 through 4 layers, and the peak position of A_{1g} can be used to determine the number of layers of

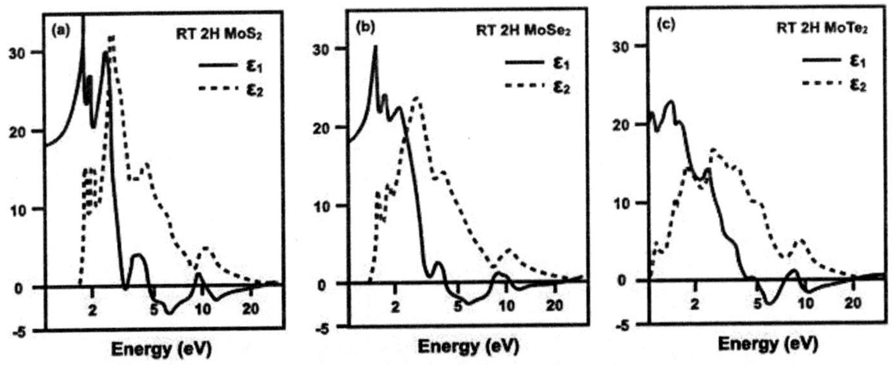

Fig. 8.31 The real and imaginary part of the dielectric function of 2H MoS_2, $MoSe_2$, and $MoTe_2$ from 1 to 14 eV. Figure adapted from [39]. © IOP Publishing. Reproduced with permission. All rights reserved

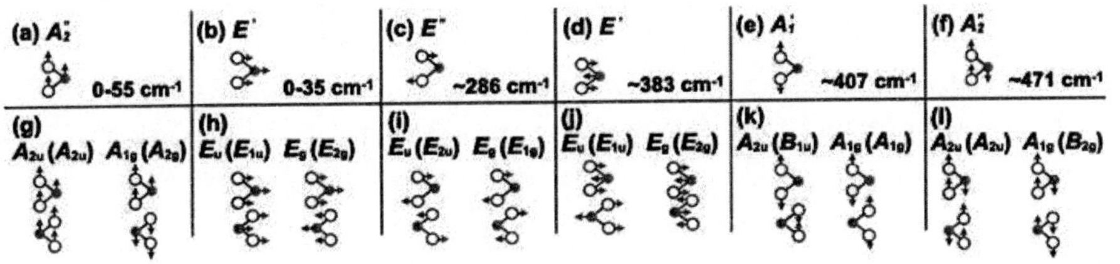

Fig. 8.32 Symmetry allowed vibrational modes for monolayer and bilayer MoS_2. Figure adapted and reprinted with permission from [43]. © 2015 American Physical Society

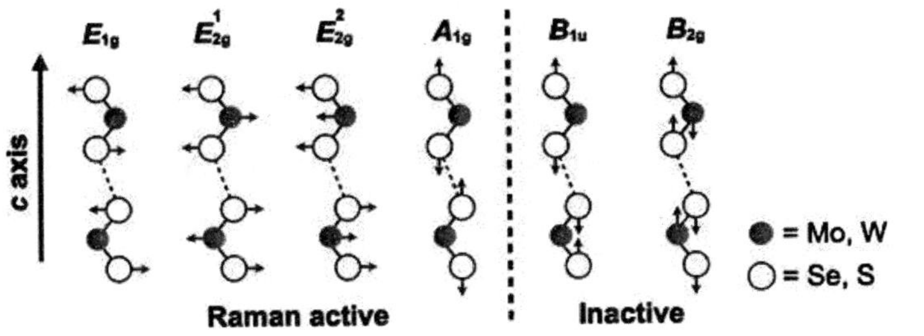

Fig. 8.33 The Raman active modes are shown on the left and the Raman inactive modes on the right for 2H trigonal prismatic TMD monolayers. Figure adapted with permission from [42]. © The Optical Society

MoS_2. The strong E_{2g}^1 inplane vibration peak for monolayer MoS_2 is said to reflect the trigonal prismatic coordination. A summary of the Raman tensor for different TMD structures is provided at the end in Sect. 8.15.

The low frequency shear mode(s) also provide a convenient means of determining the number of monolayers of TMD present. It is useful to note that several key

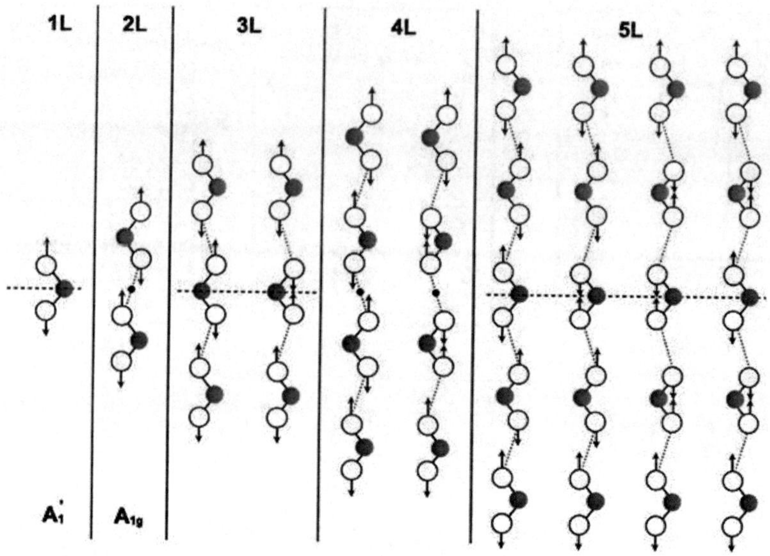

Fig. 8.34 Out of plane Raman active vibrational modes are shown for 1 through 5 TMD layers. Note the interaction between TMD monolayers. Figure adapted from [42]. © The Optical Society

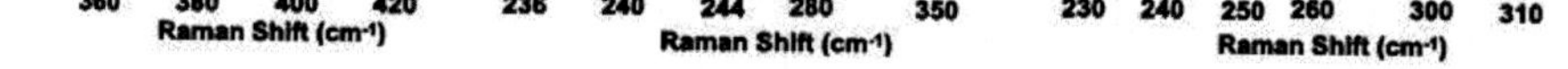

Fig. 8.35 The layer dependent, room temperature Raman data for MoS_2, $MoSe_2$, and WSe_2 are shown. The data for $MoSe_2$ and WSe_2 was taken using an excitation wavelength of 514.5 nm at 68 μW except for bulk WSe_2 where 500 μW was used. The figure on the left is adapted from [41] and the data in the middle and right is adapted from [42]. © The Optical Society

references (e.g., [45]) use the terminology TriLayers (TL) to refer to the three atomic layer thick monolayers of the chalcogenide–metal–chalcogenide structure described in Figs. 8.1 and 8.3. The E^2_{2g} shear mode has a clear blue shift with monolayer number that can easily be used to determine monolayer number in an unknown sample. The E^2_{2g} mode is shown in Fig. 8.36. The change with layer number is shown for MoS_2 and WSe_2 in Fig. 8.37. The layer dependence of the shear mode of $NbSe_2$ is shown in Fig. 8.38.

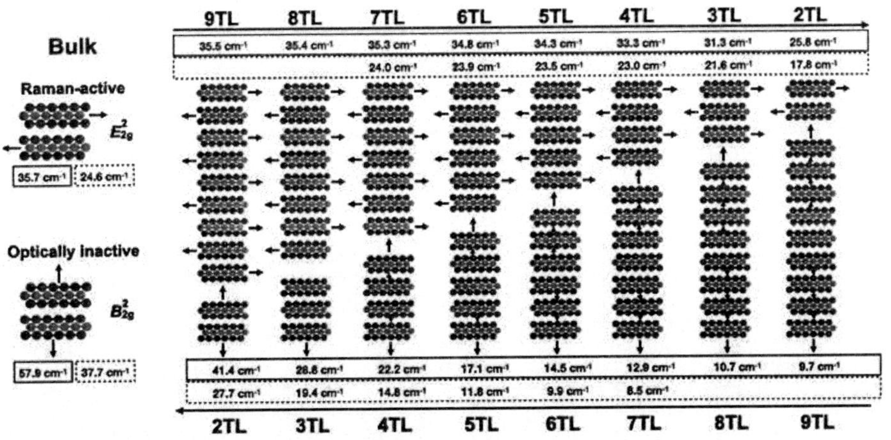

Fig. 8.36 The layer dependence of Raman active E_{2g}^2 shear mode and the optically inactive B_{2g}^2 breathing mode are shown. The mode frequencies for 2H MoS_2 are shown in purple and the frequency of 2H WSe_2 in black. Note that in this figure, the metal atoms are light grey and the chalcogenide atoms are dark grey. Figure adapted with permission from [45]. © 2013 American Chemical Society

8.11 Transition-Metal Dichalcogenide Haeckelites (A Theoretical Material)

The impact of the crystal structure on the properties of 2D materials cannot be understated. Band structure calculations predict that TMD Haeckelite structures made from Mo and W dichalcogenides will exhibit the quantum spin Hall states [47]. This is in sharp contrast to the valley pseudospin for the same monolayer TMD in the trigonal prismatic structure. The Haeckelite TMD structure is obtained from the octahedral configuration when 4–8 defects are incorporated into the lattice [47]. The 2D Haeckelite structure consists of a three atomic layers with the top and bottom layers being the chalcogenide and the middle layer being the transition metal. The WS_2 Haeckelite structure is shown in is shown in Fig. 8.39 along with the Brillouin zone. This structure can be systematically fabricated introducing 4–8 defects using an aberration corrected high resolution transmission electron microscope [47, 48]. The predicted Haeckelite structure distorts into variations of the structure shown in Fig. 8.39 for $MoSe_2$, $MoTe_s$, WSe_2, and WTe_2. There are 4 atomic site in the unit cell, and a simple tight binding model with one d_z orbital per site gives a useful 4 Band model that can incorporate Spin Orbit Coupling. This structure is said to possess both time reversal symmetry and inversion symmetry [47]. Optical investigation of these materials would only be possible if they could be fabricated over a large area.

Traditionally, Bardeen-Cooper-Schrieffer (BCS) superconductivity and ordered magnetic states are considered to be competing properties for bulk materials. For example, a magnetic field does not penetrate into superconducting metal when it is below the superconducting transition temperature unless the magnetic field strength

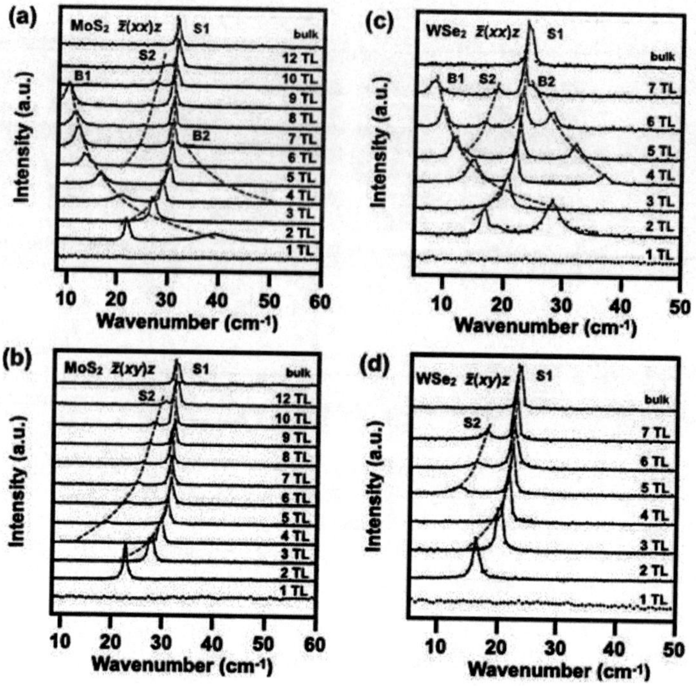

Fig. 8.37 The layer number dependency of the E^2_{2g} shear mode (labeled S1) for 2H MoS_2 and WSe_2. The Damon, Porto, and Tell optical polarization configuration for the Raman measurement is listed in each figure. Thus for incoming light normal $\overline{z}$ to the surface and the scattered light normal to the surface z the (xx) configuration measures the x polarized incoming and x polarized scattered light. The (xy) configuration refers to the measuring the scattered light polarized 90° away from the incoming light. Group theory analysis points to zero scattering intensity for the breathing modes for $\overline{z}(xy)z$, and both modes having non-zero intensity for $\overline{z}(xx)z$ [45]. Figure adapted with permission from [45]. © 2013 American Chemical Society

is very high. A magnetic material has long range spin ordering while superconducting materials have short range spin pairing. One of the theories of ferromagnetism is the Ising model. In order to distinguish the Ising model for ferromagnetism from the Ising model for superconductivity in TMD, we first briefly describe the Ising model for magnetism. The Ising model for ferromagnetism uses a mean field approximation for the energy of a material in a magnetic field pointed along a direction z. The energy of an atom in a magnetic field H is the sum of the electrostatic exchange energy J (~ 1 eV) which is short range that thus mainly between nearest neighbor atoms and the interaction of the electrons with the magnetic field (~ 10^{-4} eV). The total energy would then be the sum over all atoms:

$$\mathcal{E}_i^{atom} = -\frac{J}{2}\sum_{k=1,z} s_k s_i + \mu H s_i \text{ and } \mathcal{E} = \sum_{i=1,N} \mathcal{E}_i^{atom}$$

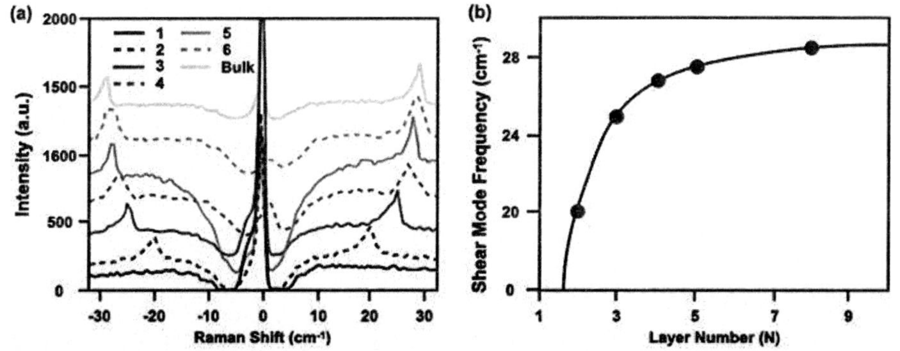

Fig. 8.38 Layer dependence of Raman spectra for 2H-$NbSe_2$. The low frequency interlayer shear mode is first observed for bilayer samples, and the frequency increases with increasing number N of layers as the interlayer interaction increases. The layer number N dependence was fit to $\omega_{S,bulk}\cos\left(\frac{\pi}{2N}\right)$ with $\omega_{S,bulk} = 29\,\mathrm{cm}^{-1}$. The 632.8 nm line from a HeNe laser, and the absorbance at this energy is 0.026. Figure adapted from supplemental section of [46]

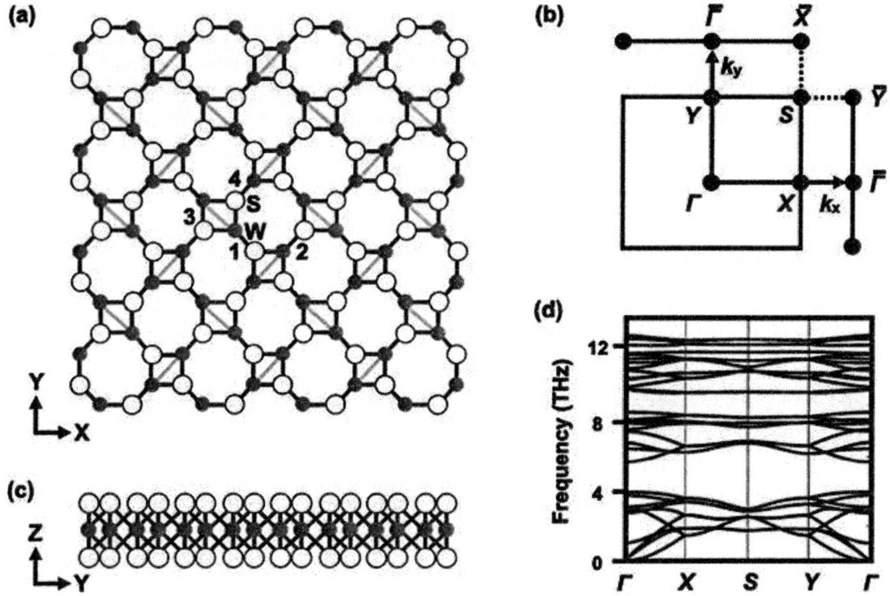

Fig. 8.39 The top view (**a**) and side view (**b**) of the Haeckelite structure of WS_2 is shown. The structure incorporates 4–8 defects in the octahedral antiprismatic configuration (see Fig. 8.3). The Brillouin zone is shown in (**c**), and the phonon dispersion curve is shown in (**d**). Figure adapted and reprinted with permission from [47]. © 2015 American Physical Society

$$s_i = +1(spin\ up) \text{ or } -1(spin\ down) \tag{8.44}$$

The mean spin of the atom is then determined from the Boltzman distribution.

Here we discuss superconducting TMD materials. Bardeen-Cooper-Schrieffer (BCS) superconductivity requires spatial inversion symmetry. As mentioned above, monolayer trigonal prismatic TMD materials such as $NbSe_2$ lack inversion symmetry and are thus not centrosymmetric while bulk TMD trigonal prismatic TMD are centrosymmetric. Many bulk TMD materials are intrinsic superconductors at low temperatures including NbS_2, $NbSe_2$, TaS_2, and $TaSe_2$ [49]. Other bulk TMD materials such as MoS_2, $MoSe_2$, and WSe_2 become superconducting when they are intercalated by alkali metal ions [49]. Curiously, monolayer TMD materials such as 2H-$NbSe_2$, T_d-$MoTe_2$, and electrostatically doped T_d-WTe_2 have also been shown to be superconducting at low temperature [46, 49–55]. This leads to the question of the impact of layer number on superconductivity. The decreasing temperature of the transition to a superconducting state with decrease in 2H–$NbSe_2$ layer number is shown as a function of magnetic field is shown in Fig. 8.40 [46, 52, 53]. The origin of the superconductivity is a direct result of the honeycomb structure and spin–orbit interactions (SOI) resulting in Ising SOI [49]. As mentioned above, another property of BCS superconductivity is that a sufficiently large magnetic field can quench the superconductivity through the orbital and spin Zeeman effects [46]. The observation that monolayer TMD can remain superconducting even with high in-plane magnetic fields is considered strong evidence of that TMD superconductivity is due to Ising SOI [46]. We describe Ising superconductivity in more detail below. Other monolayer TMD materials have exhibited superconductivity. Monolayer T_d-$MoTe_2$ was found to have a 60 fold increase in the superconducting transition temperature from the bulk value of T_c = 120 mK to 7.6 K [54]. The superconducting transition temperature of 2 layer and 3 layer samples were T_c = 2.5 K and 0.5 K respectively

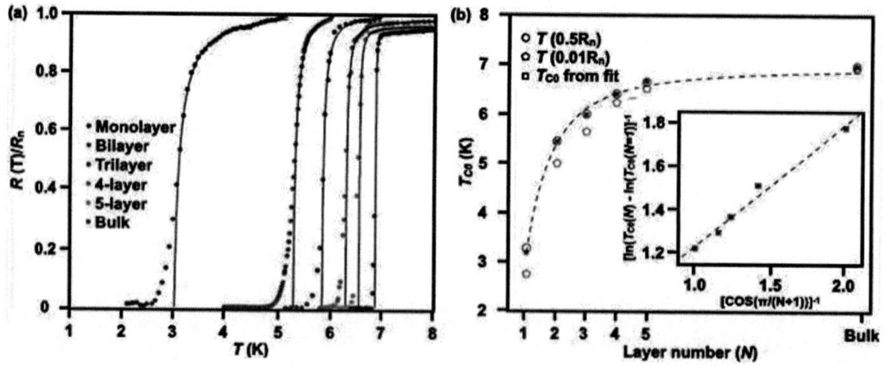

Fig. 8.40 The layer dependence of the superconducting transition temperature for $NbSe_2$ is shown. **a** The layer dependence of the resistance. The layer dependent T_{Co} is the zero field critical temperature for the transition to a superconducting state (bulk ~ 7 K) is shown in (**b**). Figure adapted from [46]

[54]. The monolayer samples were exfoliated from a bulk crystal grown by a self-flux method [54]. The monolayers were encapsulated in h-BN. When T_d-WTe_2 is electrostatically doped to $< 10^{13}\ cm^{-2}$, a superconducting transition temperature of 700 mK was reported [55].

Traditional BCS superconductivity is attributed to s wave, spin singlet electron pairs which have spin angular momentum $S = 0$ and orbital angular momentum $L = 0$ [56]. The electrons have equal and opposite momentum ($\vec{k} = -\vec{k}$) and the spins are opposite. In a metal superconductor, electron–phonon interactions result in the attraction between electrons forming a Cooper pair [56]. The superconductivity of 2D TMD is attributed to Ising spin–orbit coupling (SOC) where the electron spin is locked out of the plane (as shown in Fig. 8.41) and not Rashba SOC [49]. This can be understood as follows. In monolayer TMD the in-plane mirror symmetry is broken, and thus electrons experience and in-plane electric field. An electron with momentum $\vec{k}$ experiences a SOC that is proportional to $\vec{k} \times \vec{E} \cdot \boldsymbol{\sigma}$ where $\boldsymbol{\sigma}$ are the Pauli matrices. For 2D TMD the electric field is in the plane of the monolayer and the SOC $\sim S(\vec{k})\sigma_z$ where $S(\vec{k})$ depends on the crystal lattice structure [49]. *Thus the electron spins are pinned to an out of plane direction which is called Ising SOC to distinguish it from Rashba SOC.* This is in contrast to Rashba SOC where the electron spins are pinned to an in plane direction and the SOC is proportional to $g(\vec{p}) = (k_x\sigma_y - k_y\sigma_x)$ [49, 53]. The Ising SOC can be hundreds of meV, and is ~ 450 meV for WSe_2 [49]. The dependence of the superconductivity on in-plane and out of plane magnetic field strength is shown in Fig. 8.42. In Table 8.4, the SOC

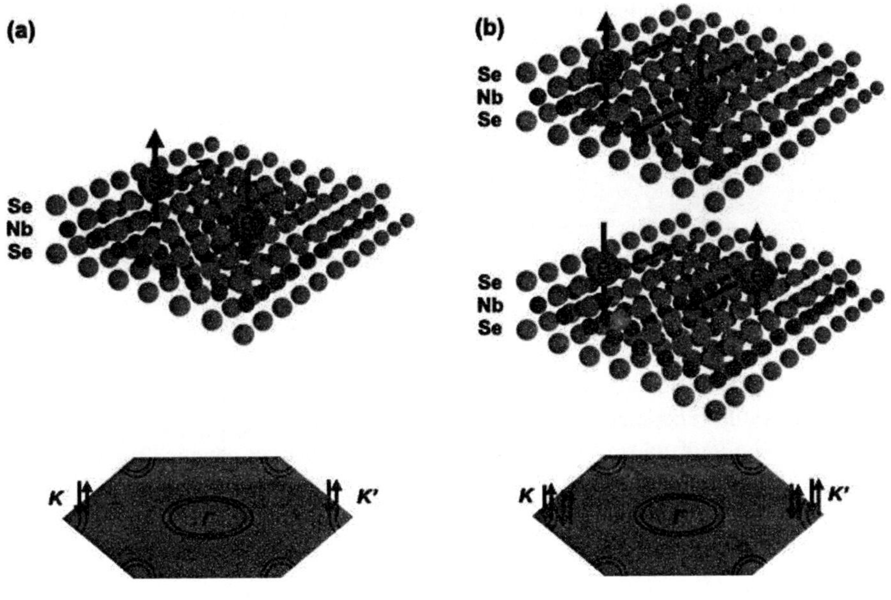

Fig. 8.41 The Ising spin–momentum locking and the Ising pairing are shown for monolayer and bilayer TMD. Figure adapted from [46]

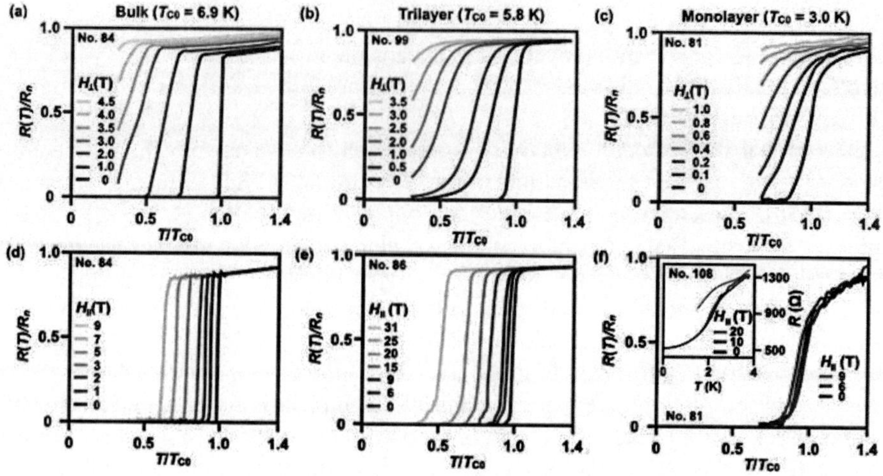

Fig. 8.42 The superconducting transition versus magnetic field for $NbSe_2$ is shown for a variety of in-plane and out of plane magnetic field strengths. The fact that the samples remain superconducting at high magnetic field strengths is considered evidence of the Ising pairing. Figure adapted from [46]

energies for the valence band of WSe_2 and WS_2 were listed without referring to the type of the SOC. As described above, the orbitals that constitute the conduction band minimum are mainly the d_z orbitals of the transition metal. The wavefunction basis set for the Hamiltonian near the K points is assembled from these orbitals. The following Dirac Hamiltonian describes the Ising and Rashba SOC contributions to the energy for carriers around the K and K' points in the Brillouin zone [49]:

$$H(\vec{k} = \vec{p} + \epsilon\vec{K}) = \frac{|\vec{p}|^2}{2m} - \mu + \alpha_R g(\vec{p}) \cdot \boldsymbol{\sigma} \; + \epsilon\beta_{so}\sigma_z \tag{8.45}$$

Here, $\in$ (= + or −) is the valley index, m is the rest mass, μ is the chemical potential at the conduction band bottom in the absence of SOI, α_R is the temperature dependent Rashba coupling strength, and β_{so} is the Ising coupling strength. β_{so} is a constant. For cases where Ising SOI dominates, one can compare (8.34) to (8.45) to see that the band gap at the K points is directly related to β_{so}. For MoS_2, $\beta_{so} =$ 6.2 meV. The Rashba SOC and Ising SOC compete and the Rashba SOC increases as the temperature decreases below T_c [49]. The temperature dependent superconductivity of the spin pairing can be accounted for by invoking the Bogoliubov–de Gennes Hamiltonian [49, 56]. The BdG Hamiltonian is written in terms of the superconducting pairing matrix $\Delta_\Gamma(\vec{k}) = \Delta_0 i\sigma_y$ assuming s wave pairing [49, 56]:

$$H_{BgD}(\vec{k}) = \begin{vmatrix} H_0(\vec{k}) & \Delta_\Gamma(\vec{k}) \\ \Delta_\Gamma^f(\vec{k}) & -H_0^*(-\vec{k}) \end{vmatrix} \tag{8.46}$$

(8.46) can be solved to extract α_R from temperature dependent resistivity data [49, 57]. Values of $\alpha_R p_F$ of 0.9 meV for $T_c = 2.37$ K, 3.0 meV for $T_c = 7.38$ K for T/T_c from ~ 0.2 to 0.7 [49, 57]. The Ising contribution is larger than the Rashba contribution. The interaction between Rashba, Ising (Zeeman), and magnetic field induced spin pairing and thus superconductivity is discussed in [57]. At the time of the authorship of this chapter, the effect of interlayer interaction is under investigation.

Another factor in these studies is that fabricating monolayer samples in usable structures is not always practical. Platinum or gold metal–insulator-superconducting bilayer and trilayer $NbSe_2$ samples were shown to be superconducting at low temperature (< 10 K) [52]. The paramagnetic-limited superconductor–normal metal transition of bilayer samples occurs at 5.8 K, and the few layer results in this study were considered representative of the superconducting monolayers of $NbSe_2$ [52]. As mentioned above, the interlayer interactions alter optical properties causing a transition for direct to indirect band gap. This contrasts with the assertion that the interlayer interaction is weak enough so that observation of the superconducting transition for bi- and tri- layer samples can be attributed to the non-centrosymmetric nature of monolayer $NbSe_2$. In addition, bilayer samples are centrosymmetric while trilayer sample are not. Although the superconducting transition temperature changes with layer number, both centrosymmetric and non-centrosymmetric samples are superconducting as shown in Fig. 8.40. It is important to note that the tight binding model for trilayer, ABA stacked trigonal prismatic TMD can be constructed from the 3 band TB TNN model for monolayer TMD. This model was used to determine the in plane spin susceptibility and superconducting gap (see Sect. 8.6 in supplemental for [51]). The theoretical description of interlayer interactions used to understand the change in optical properties including the change from direct to indirect gap required inclusion of the chalcogenide p_z orbitals and interlayer van der Waals interactions. Further theoretical studies will help our understanding of the role of layer interactions in superconductivity.

This leads to the question of the impact of the superconducting transition on optical properties. The bulk complex refractive index of 2H $NbSe_2$ at 300 and 77 K is shown in Fig. 8.27. A low temperature (from m K to 10 K) study of the complex refractive index/dielectric function of monolayer, bilayer and trilayer $NbSe_2$ would shed light on the impact of super conductivity on its optical properties.

8.12 Twisted and Hetero-Bilayers of Transition Metal Dichalcogenides with graphene and h-BN

Hetero-layers of TMD on graphene have been characterized for small twist angles < 2° [58] and larger twist angles up to 64 [59] using Raman. Hetero-layers of h-BN – (TMD) – h-BN are also being evaluated as potential structures for electrical devices (57). Just as with bilayers of graphene or TMD, moiré patterns are formed as shown in Fig. 8.43 for 30° of relative rotation. The lattice mismatch between graphene and

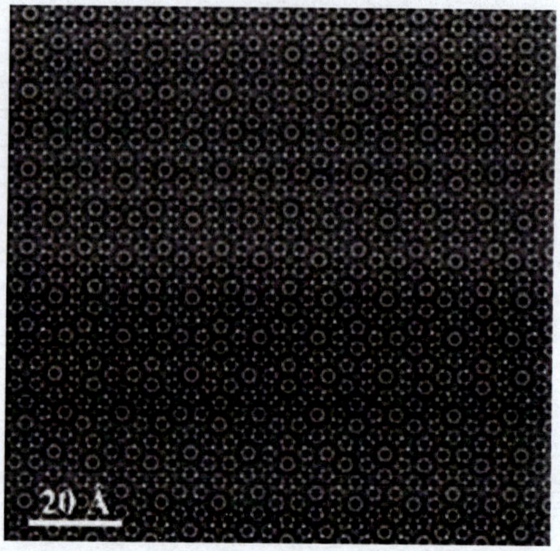

Fig. 8.43 Simulated high resolution—transmission electron microscope (HR-TEM) image of grapheme—h-BN bilayer with 30° of relative rotation. The HR-TEM simulations are described in [60]

hBN is $\delta = 0.017$ (we note that $\delta = 0.018$ has also been used in the literature) [58]. The wavelength λ_M of the moiré superlattice for grapheme—h-BN for relative rotation θ is [59]:

$$\lambda_M = \frac{(1+\delta)a}{\sqrt{2(1+\delta)(1-\cos\theta)-\delta^2}}$$

Here, a is the graphene lattice constant. Raman spectra for different angular rotations are shown in Fig. 8.44. The angular rotation determines both the wavelength of the moiré superlattice and thus the periodicity of the potential that alters the band structure resulting in superlattice Dirac points with energies shifted from zero. The energy of the superlattice Dirac points E_{DSP} is given by [59]:

$$E_{DSP} = \frac{2\pi v_F}{\lambda_M\sqrt{3}}$$

8.13 ReS_2 and $ReSe_2$ with the $1T'$ Structure

The influence of the crystal structure on the electronic band structure and thus the optical and electrical properties of TMD materials becomes evident when 1 T′ $ReSe_2$ is considered [61]. Although there is a 1.6% difference in plane between 3 $ReSe_2$ unit cells and 8 graphene unit cells, when $ReSe_2$ is epitaxially grown on bilayer graphene

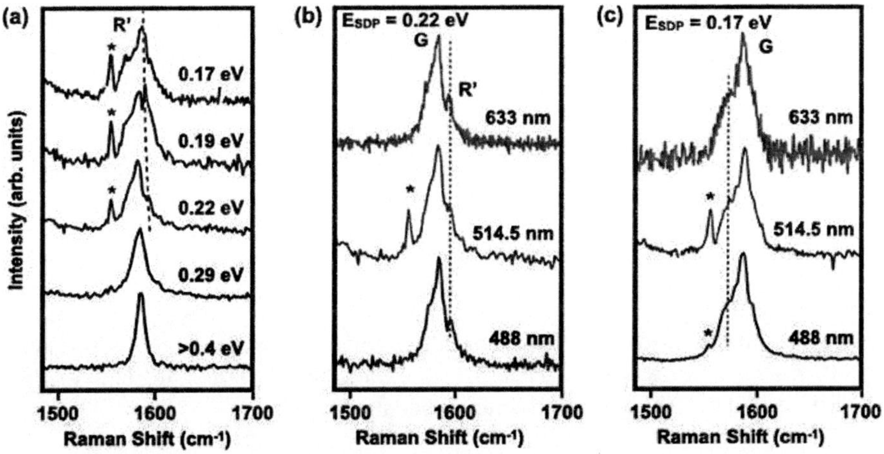

Fig. 8.44 Raman spectra of the G peak of twisted grapheme—h-BN bilayers. **a** The different values of superlattice Dirac points E_{DSP} are determined by the twist angle. **b** and **c** show the laser wavelength dependence of the G peak. Figure adapted from [58]. Adapted with permission from [58]. © 2013 American Chemical Society

on SiC (0001) it retains the $1T'$ structure. The $1T'$ with zig-zag Re arrangement aligned along the $1T'$ $ReSe_2$ has triclinic symmetry with a P1 space group [61]. Re is a group 7 transition metal with one more valence electron than group 6 metals W and Mo. For the 2H TMD materials discussed in Sect. 8.2 (e.g., see Table 8.2), at the high symmetry Γ point in the Brillouin zone, the valence band for the 3 band tight biding model was formed from d_{z^2} orbitals. For the band tight binding model, the valence band also had chalcogenide p orbital character. Using polarized photons in Angle Resolved Photoelectron Spectroscopy, the distorted $1T'$ $ReSe_2$ was found to not have the d_{z^2} and p_z orbital character for the valence band at the Γ point. The interlayer spacing and band gap are increased. The electronic structure is not a strong function of the number of $ReSe_2$ due to the change in orbital character. The band gap of $1T'$ $ReSe_2$ is 1.9 eV and for doped $1T'$ $ReSe_2$ it s 1.4 eV[61]. The band gap for monolayer $1T'$ $ReSe_2$ is indirect [61].

8.14 Practical Aspects of Characterization of TMD Materials Using Spectroscopic Ellipsometry

Although many TMD materials are optically uniaxial, many studies use isotropic models for analysis of data from monolayer materials. One example is the use of Tauc-Lorentz models for bulk samples of WS_2 and MoS_2 in the spectral range of 375–1700 nm. The Tauc-Lorenz model fits for monolayers of WS_2 and MoS_2 from

[62]. Although uniaxial optical models are now available in commercial laboratory ellipsometers, isotropic models can provide a means of estimating bulk sample thickness.

8.15 Symmetry and Space Group Summary for Transition Metal Dichalcogenides

TMD materials often crystalize in more than one phase [10, 62–69]. The various phases are summarized in Table 8.7 and the Raman tensors in Table 8.8 following Ribeiro Soares et al., as discussed in [63, 69]. We note that the structure of monolayers of TMD with a bulk T_d and T' structure is identical.

Table 8.7 Schoenflies point groups and Hermann-Mauguin space groups of the $2H_a$, $2H_c$, $3R$, $1T$, T_d and $1T'$ transition metal dichalcogenide crystal structures

Polytype	2 H_a hexagonal			2 H_c hexagonal			3 R rhombohedral	
	Prevalent structure for MX2 M = Nb, Ta; X = S, Se, Te			Prevalent structure for MX_2 M = Mo, W; X = S, Se; $MoTe_2$			NbS_2	
Number of layers	Bulk	Odd#	Even #	Bulk	odd #	even #	Bulk	
Number of repeated trilayers in a unit cell (see Fig. 8.4)	2	N	N	2	N	N	3	
Schoenflies point group	D_{6h}^4	D_{3h}^1	D_{3h}^3	D_{6h}^4	D_{3h}^1	D_{3h}^3	C_{3v}^s	
Space group	$P6_3/mmc$, # 194	$P\bar{6}m2$, # 187	$P\bar{3}m1$, # 164	$P6_3/mmc$, # 194	$P\bar{6}m2$, # 187	$P\bar{3}m1$, # 164	R3m; # 160	
Polytype	IT tetragonal		Other phases including IT′					Non-layered
	MX_2 M = Ti, Zr, Hf, V; X = S, Se, Te MX_2 M = Nb, Ta; X = S, Se		Distorted monoclinic $NbTe_2$ [64]	monoclinic (IT′) $MoTe_2$ [65, 69]	Triclinic ReS_2 [66] & TcS_2 [67] & $ReSe_2$ [10, 61, 66, 67]	Orthorhombic (T_d) $MoTe_2$ [69] & WTe2 [65, 68] THz light pulses can drive WTe_2 to metastable monoclinic (1T′) [68]	IT′ orthorhombic $ReTe_2$ [67]	Monoclinic $TcTe_2$ [67]
Number of layers	Bulk	Odd and even #	Bulk	Bulk	Bulk, odd and even N	Bulk	Bulk	

(continued)

Table 8.7 (continued)

Polytype	2 H_a hexagonal			2 H_c hexagonal			3 R rhombohedral	
	Prevalent structure for MX2 M = Nb, Ta; X = S, Se, Te			Prevalent structure for MX_2 M = Mo, W; X = S, Se; $MoTe_2$			NbS_2	
Number of repeated trilayers in a unit cell (see Fig. 8.4)	1	N			1, N		1, N	
Schoenflies point group	D_{3h}^{3}	D_{3h}^{3}	C_{2h}^{3}				D_{3h}^{1s}	
Space group	$P\bar{3}m1$, # 164	$P\bar{3}m1$, # 164	12 (C2/m)	$P2_1/m$	$P\bar{1}$	$Pmn2_1$	Pcab	C2/c

Note that several compounds naturally occur in more than one phase, and some compounds can be grown in a non-prevalent phase. The space group numbers are also shown. Crystallographic symmetry is discussed in Chap. 2.12. The ordering mechanisms for TMD are further described in Fig. 3.4 in [10]

Table 8.8 Raman tensor information for $2H_a$, $2H_c$, 1T, T', and Td transition metal dichalcogenide crystal structures [63, 69]

Polytype	$2Ha$ and $2H_b$			1T
Number of layers	Bulk	Odd #	Even #	Bulk odd and even #
Symmetry	D_{6h}^4	D_{3h}^1	D_{3h}^3	D_{3h}^3
Raman tensor	$A_{1g} = \begin{pmatrix} a & 0 & 0 \\ 0 & a & 0 \\ 0 & 0 & b \end{pmatrix}$	$A'_{1g} = \begin{pmatrix} a & 0 & 0 \\ 0 & a & 0 \\ 0 & 0 & b \end{pmatrix}$	$A_{1g} = \begin{pmatrix} a & 0 & 0 \\ 0 & a & 0 \\ 0 & 0 & b \end{pmatrix}$	$A_{1g} = \begin{pmatrix} a & 0 & 0 \\ 0 & a & 0 \\ 0 & 0 & b \end{pmatrix}$
	$E_{1g} = \begin{pmatrix} 0 & 0 & 0 \\ 0 & 0 & c \\ 0 & c & 0 \end{pmatrix}$	$(E')_{(x)} = \begin{pmatrix} 0 & d & 0 \\ d & 0 & 0 \\ 0 & 0 & 0 \end{pmatrix}$	$(E_g)_{(1)} = \begin{pmatrix} c & 0 & 0 \\ 0 & -c & d \\ 0 & d & 0 \end{pmatrix}$	$(E_g)_{(1)} = \begin{pmatrix} c & 0 & 0 \\ 0 & -c & d \\ 0 & d & 0 \end{pmatrix}$
	$E_{1g} = \begin{pmatrix} 0 & 0 & -c \\ 0 & 0 & 0 \\ -c & 0 & 0 \end{pmatrix}$	$(E')_{(y)} = \begin{pmatrix} d & 0 & 0 \\ 0 & -d & 0 \\ 0 & 0 & 0 \end{pmatrix}$	$(E_g)_{(2)} = \begin{pmatrix} 0 & -c & -d \\ -c & 0 & 0 \\ -d & 0 & 0 \end{pmatrix}$	$(E_g)_{(2)} = \begin{pmatrix} 0 & -c & -d \\ -c & 0 & 0 \\ -d & 0 & 0 \end{pmatrix}$
	$E_{2g} = \begin{pmatrix} 0 & d & 0 \\ d & 0 & 0 \\ 0 & 0 & 0 \end{pmatrix}$	$E'' = \begin{pmatrix} 0 & 0 & 0 \\ 0 & 0 & c \\ 0 & c & 0 \end{pmatrix}$		
	$E_{2g} = \begin{pmatrix} d & 0 & 0 \\ 0 & -d & 0 \\ 0 & 0 & 0 \end{pmatrix}$	$E'' = \begin{pmatrix} 0 & 0 & -c \\ 0 & 0 & 0 \\ -c & 0 & 0 \end{pmatrix}$		

Polytype	1T′	T_d
Number of layers	Bulk	Bulk
Symmetry	C_{2k}	C_{2v}^7

(continued)

Table 8.8 (continued)

Polytype	1T′	T_d
Raman tensor	$A_g = \begin{pmatrix} b & 0 & d \\ 0 & c & 0 \\ d & 0 & a \end{pmatrix}$	$A_1 = \begin{pmatrix} a & 0 & 0 \\ 0 & b & 0 \\ 0 & 0 & c \end{pmatrix}$
	$B_g = \begin{pmatrix} 0 & f & 0 \\ f & 0 & e \\ 0 & e & 0 \end{pmatrix}$	$A_2 = \begin{pmatrix} 0 & d & 0 \\ d & 0 & 0 \\ 0 & 0 & 0 \end{pmatrix}$
		$B_1 = \begin{pmatrix} 0 & 0 & e \\ 0 & 0 & 0 \\ e & 0 & 0 \end{pmatrix}$
		$B_2 = \begin{pmatrix} 0 & 0 & 0 \\ 0 & 0 & f \\ 0 & f & 0 \end{pmatrix}$

The transverse optical phonon modes of monoclinic 1T′ $NbTe_2$ have also been reported (in increasing energy: A_g^1,B_g^1, A_g^2,B_g^2, A_g^3, B_g^3, A_g^4, A_g^5, A_g^6, B_g^4, A_g^7, A_g^8) [64]

References

1. M. Chhowalla, H.S. Shin, G. Eda, L.-J. Li, K.P. Loh, H. Zhang, The chemistry of two-dimensional layered transition metal dichalcogenide nanosheets. Nat. Chem. **5**, 263–275 (2013)
2. H. Zeng, X. Cui, An optical spectroscopic study on two dimensional group-VI transition metal dichalcogenides. Chem. Soc. Rev. **44**, 2629–2642 (2015)
3. A. Splendiani, L. Sun, Y. Zhang, T. Li, J. Kim, C.-Y. Chim, G. Galli, F. Wang, Emerging photoluminescence in monolayer MoS_2. Nano Lett. **10**, 1271–1275 (2010)
4. W. Jin, P.-C. Yeh, N. Zaki, D. Zhang, J.T. Sadowski, A. Al-Mahboob, A.M. van der Zande, D.A. Chenet, J.I. Dadap, I.P. Herman, P. Sutter, J. Hone, R.M. Osgood, Direct measurement of the thickness-dependent electronic band structure of MoS_2 using angle-resolved photoemission spectroscopy. J. Phys. Rev. Lett. **111**, 106801 (2013)
5. A.V. Kolobov, J. Tominaga, Electronic band structure of 2D TMDCs, Chap. 6 in *Two-Dimensional Transition-Metal Dichalcogenides*. Springer Series in Materials Science, vol. 239 (Springer, Switzerland, 2016), pp. 165–226. See p. 198
6. M. Bernardi, C. Ataca, M. Palummo, J.C. Grossman, Optical and electrical properties of two dimensional layered materials. Nanophotonics **5**, 111–125 (2016)
7. G.-B. Liu, D. Xiao, X. Xu, W. Yao, Electronic structures and theoretical modelling of two-dimensional group-VIB transition metal dichalcogenides. Chem. Soc. Rev. **44**, 2643 (2015)
8. A.N. Enyashin, G. Seifert, Electronic properties of MoS_2 monolayer and related structures. Nanosyst.: Phys. Chem. Math. **5**, 517–539 (2014)
9. C. Gong, H. Zhang, W. Wang, L. Colombo, R.M. Wallace, K. Cho, Band alignment of two-dimensional transition metal dichalcogenides: application in tunnel field effect transistors. Appl. Phys. Lett. **103**, 053513 (2013)

10. A.V. Kolobov, J. Tominaga, Bulk TMDCs: review of structure and properties, Chap. 3 in *Two-Dimensional Transition-Metal Dichalcogenides*. Springer Series in Materials Science, vol. 239 (Springer, Switzerland, 2016), pp. 29–77
11. G.-B. Liu, W.-Y. Shan, Y. Yao, W. Yao, D. Xiao, Three-band tight-binding model for monolayers of group-VIB transition metal dichalcogenides. Phys. Rev. B **88**, 085433 (2013). For plots of the 3 Band solution from Liu, et al., see L. Szulakowska, Studies of electronic properties of atomically thin nanostructures with tight-binding model and DFT methods. Master's Thesis, Wydział Podstawowych Problemów Techniki
12. J.Á. Silva-Guillén, P. San-Jose, R. Roldán, Electronic band structure of transition metal dichalcogenides from ab initio and Slater-Koster tight-binding model. Appl. Sci. **6**, 284 (2016)
13. M.D. Jaffee, J. Singh, Inclusion of spin orbit coupling into tight binding band structure calculations for bulk and superlattice semiconductors. Solid State Commun. **62**, 399–402 (1987)
14. F.A. Rasmussen, K.S. Thygesen, Computational 2D materials database: electronic structure of transition-metal dichalcogenides and oxides. J. Phys. Chem. C **119**, 13169–13183 (2015)
15. K. Kosmider, J.W. Gonzalez, J. Fernandez-Rossier, Large spin splitting in the conduction band of transition metal dichalcogenide monolayers. Phys. Rev. B **88**, 245436 (2013)
16. X. Xu, W. Yao, D. Xiao, T.F. Heinz, Spin and pseudospins in layered transition metal dichalcogenides. Nat. Phys. **10**, 343–350 (2014)
17. K.F. Mak, K. He, J. Shan, T.F. Heinz, Control of valley polarization in monolayer MoS_2 by optical helicity. Nat. Nanotech. **7**, 494–498 (2012)
18. D. Xiao, M.-C. Chang, Q. Niu, Berry phase effects on electronic properties. Rev. Mod. Phys. **82**, 1959–2007 (2010)
19. D. Xiao, G.-B. Liu, W. Feng, X. Xu, W. Yao, Coupled spin and valley physics in monolayers of MoS_2 and other group-VI dichalcogenides. Phys. Rev. Lett. **108**, 196802 (2012)
20. M.-C. Chang, Q. Niu, Berry phase, hyperorbits, and the Hofstadter spectrum: semiclassical dynamics in magnetic Bloch bands. Phys. Rev. B **53**, 710–7023 (1996)
21. T. Cao, G. Wang, W. Han, H. Ye, C. Zhu, J. Shi, Q. Niu, P. Tan, E. Wang, B. Liu, J. Feng, Valley-selective circular dichroism of monolayer molybdenum disulphide. Nat. Comm. **3**, 887 (2012)
22. D.Y. Qiu, F.H. da Jornada, S.G. Louie, Optical spectrum of MoS2: many-body effects and diversity of exciton states. Phys. Rev. Lett. **111**, 216805 (2013)
23. M.M. Ugeda, A.J. Bradley, S.-F. Shi, F.H. da Jornada, Y. Zhang, D.Y. Qiu, W. Ruan, S.-K. Mo, Z. Hussain, Z.-X. Shen, F. Wang, S.G. Louie, M.F. Crommie, Giant bandgap renormalization and excitonic effects in a monolayer transition metal dichalcogenide semiconductor. Nat. Mater. **13**, 1091–1095 (2014)
24. B. Zhu, X. Chen, X. Cui, Exciton binding energy of monolayer WS_2, Nat. Sci. Rep. **17**, 9218 (2015)
25. H.-L. Liu, C.-C. Shen, S.-H. Su, C.-L. Hsu, M.-Y. Li, L.-J. Li, Optical properties of monolayer transition metal dichalcogenides probed by spectroscopic ellipsometry. Appl. Phys. Lett. **105**, 201905 (2014)
26. S. Tongay, J. Zhou, C. Ataca, K. Lo, T.S. Matthews, J. Li, J.C. Grossman, J. Wu, Thermally driven crossover from indirect toward direct bandgap in 2D semiconductors: $MoSe_2$ versus MoS_2. Nano Lett. **12**, 5576–5580 (2012)
27. W. Zhao, Z. Ghorannevis, L. Chu, M. Toh, C. Kloc, P.-H. Tan, G. Eda, Evolution of electronic structure in atomically thin sheets of WS_2 and WSe_2. ACS Nano **7**, 791–797 (2013)
28. J. He, K. Hummer, C. Franchini, Stacking effects on the electronic and optical properties of bilayer transition metal dichalcogenides MoS_2, $MoSe_2$, WS_2, and WSe_2. Phys. Rev. B **89**, 075409 (2014)
29. H. Zeng, G.-B. Liu, J. Dai, Y. Yan, B. Zhu, R. He, L. Xie, S. Xu, X. Chen, W. Yao, X. Cui, Optical signature of symmetry variations and spin-valley coupling in atomically thin tungsten dichalcogenides. Nat. Sci. Rep. **3**, 1608 (2013)
30. A. Stan, N.E. Dahlen, R. van Leeuwen, Levels of self-consistency in the GW approximation. J. Chem. Phys. **130**, 114105 (2009)

31. W. Zhao, R.M. Ribeiro, M. Toh, A. Carvalho, C. Kloc, A.H. Castro Neto, G. Eda, Origin of indirect optical transitions in few-layer MoS_2, WS_2, and WSe_2. Nano Lett. **13**, 5627–5634 (2013)
32. L. Wang, E.-M. Shih, A. Ghiotto, L. Xian, D.A. Rhodes, C. Tan, M. Claassen, D.M. Kennes, Y. Bai, B. Kim, K. Watanabe, T. Taniguchi, X. Zhu, J. Hone, A. Rubio, A. Pasupathy, C.R. Dean, Correlated electronic phases in twisted bilayer transition metal dichalcogenides. Nat. Mater. **19**, 861–866 (2020)
33. S. Huang, X. Ling, L. Liang, J. Kong, H. Terrones, V. Meunier, M.S. Dresselhaus, Probing the interlayer coupling of twisted bilayer MoS_2 using photoluminescence spectroscopy. Nano Lett. **14**, 5500 (2014)
34. A.M. van der Zande, J. Kunstmann, A. Chernikov, D.A. Chenet, Y.-M. You, X.X. Zhang, P.Y. Huang, T.C. Berkelbach, L. Wang, F. Zhang, M.S. Hybertsen, D.A. Muller, D.R. Reichman, T.F. Heinz, J.C. Hone, Tailoring the electronic structure in bilayer molybdenum disulfide via interlayer twist. Nano Lett. **14**, 3869–3875 (2014)
35. C. Yim, M. O'Brien, N. McEvoy, S. Winters, I. Mirza, J.G. Lunney, G.S. Duesberg, Investigation of the optical properties of MoS_2 thin films using spectroscopic ellipsometry. Appl. Phys. Lett. **104**, 103114 (2014)
36. S.M. Eichfeld, C.M. Eichfeld, Y-C Lin, L. Hossain, J.A. Robinson, Rapid, non-destructive evaluation of ultrathin WSe_2 using spectroscopic ellipsometry. APL Mater. 2, 092508 (2014)
37. A.R. Beal, H.P. Hughes, W.Y. Liang, The reflectivity spectra of some group VA transition metal dichalcogenides. J. Phys. C: Sol. Stat. Phys. **8**, 4236–4248 (1975)
38. A.R. Beal, W.Y Liang, H.P. Hughes, Kramers-Kronig analysis of the reflectivity spectra of 3R-WS_2 and 2H WS_2. J. Phys. C: Sol. Stat. Phys. **9**, 2449–2457 (1976)
39. A.R. Beal, W.Y Liang, H.P. Hughes, Kramers-Kronig analysis of the reflectivity spectra of 3R-WS_2 and 2H WS_2. J. Phys. C: Sol. Stat. Phys. **12**, 881–890 (1979)
40. M.M. Benameur, B. Radisavljevic, J.S. Héron, S. Sahoo, H. Berger, A. Kis, Visibility of dichalcogenide nanolayers. Nanotechnology **22**, 125706 (2011)
41. C. Lee, H. Yan, L.E. Brus, T.F. Heinz, J. Hone, S. Ryu, Anomalous lattice vibrations of single- and few-layer MoS_2. ACS Nano **4**, 2695 (2010)
42. P. Tonndorf, R. Schmidt, P. Böttger, X. Zhang, J. Börner, A. Liebig, M. Albrecht, C. Kloc, O. Gordan, D.R.T. Zahn, S. Michaelis de Vasconcellos, R. Bratschitsch, Photoluminescence emission and Raman response of monolayer MoS_2, $MoSe_2$, and WSe_2. Opt. Express **21**, 4908–4916 (2013)
43. N. Scheuschner, R. Gillen, M. Staiger, J. Maultzsch, Interlayer resonant Raman modes in few-layer MoS_2. Phys. Rev. B **91**, 235409 (2015)
44. A.V. Kolobov, J. Tominaga, Raman scattering of 2D TMDCs, Chap. 7 in *Two-Dimensional Transition-Metal Dichalcogenides*. Springer Series in Materials Science, vol. 239 (Springer, Switzerland, 2016), pp. 227–294
45. Y. Zhao, X. Luo, H. Li, J. Zhang, P.T. Araujo, C.K. Gan, J. Wu, H. Zhang, S.Y. Quek, M.S. Dresselhaus, Q. Xiong, Interlayer breathing and shear modes in few-trilayer MoS_2 and WSe_2. Nano Lett. **13**, 1007–1015 (2013)
46. X. Xi, Z. Wang, W. Zhao, J.-H. Park, K.T. Law, H. Berger, Ising pairing in superconducting $NbSe_2$ atomic layers. L. Nat. Phys. **12**, 138–144 (2015)
47. S.M. Nie, Z. Song, H. Weng, Z. Fang, Quantum spin Hall effect in two-dimensional transition-metal dichalcogenide haeckelites. Phys. Rev. B **91**, 235434 (2015)
48. H.-P. Komsa, J. Kotakoski, S. Kurasch, O. Lehtinen, U. Kaiser, A.V. Krasheninnikov, Two-dimensional transition metal dichalcogenides under electron irradiation: defect production and doping. Phys. Rev. Lett. **109**, 035503 (2012)
49. N.F.Q. Yuan, B.T. Zhou, W.-Y. He, K.T. Law, Ising superconductivity in transition metal dichalcogenides. AAPPS Bul. **26**, 12–19 (2016)
50. Y. Saito, Y. Nakamura, M.S. Bahramy, Y. Kohama, J. Ye, Y. Kasahara, Y. Nakagawa, M. Onga, M. Tokunaga, T. Nojima, Y. Yanase, Y. Iwasa, Superconductivity protected by spin–valley locking in ion-gated MoS_2. Nat. Phys. **12**, 144–150 (2016)

51. W. Shi, J. Ye, Y. Zhang, R. Suzuki, M. Yoshida, J. Miyazaki, N. Inoue, Y. Saito, Y. Iwasa, Superconductivity series in transition metal dichalcogenides by ionic gating. Sci. Rep. **5**, 12534 (2015)
52. E. Sohn, X. Xi, W.-Y. He, S. Jiang, Z. Wang, K. Kang, J.-H. Park, H. Berger, L. Forró, K.T. Law, J. Shan, K.F. Mak, An unusual continuous paramagnetic-limited superconducting phase transition in 2D $NbSe_2$. Nat. Mater. **17**, 504–509 (2015). https://doi.org/10.1038/s41563-018-0061-1 and supplemental
53. R. Winkler, *Spin–Orbit Coupling Effects in Two-Dimensional Electron and Hole Systems* (Springer, Erlangen, 2003), Chapter 6
54. D.A. Rhodes, A. Jindal, N.F.Q. Yuan, Y. Jung, A. Antony, H. Wang, B. Kim, Y.-c. Chiu, T. Taniguchi, K. Watanabe, K. Barmak, L. Balicas, C.R. Dean, X. Qian, L. Fu, A.N. Pasupathy, J. Hone, Enhanced superconductivity in monolayer T_d-$MoTe_2$. Nano Lett. **21**, 2505–2511 (2021)
55. E. Sajadi, T. Palomaki, Z. Fei, W. Zhao, P. Bement, C. Olsen, S. Luescher, X. Xu, J.A. Folk, D.H. Cobden, Gate-induced superconductivity in a monolayer topological insulator. Science **362**, 922–925 (2018)
56. J.-X. Zhu, *Bogoliubov-de Gennes Method and Its Applications* (Springer, Switzerland, 2016), pp. 3–64
57. J.M. Lu, O. Zeliuk, I. Leermakers, N.F.Q. Yuan, U. Zeitler, K.T. Law, J.T. Ye, Two dimensional ising superconductivity in gated MoS_2. Science **350**, 1353–1357 (2015). https://doi.org/10.1126/science.aab2277
58. A. Eckmann, J. Park, H. Yang, D. Elias, A.S. Mayorov, G. Yu, R. Jalil, K.S. Novoselov, R.V. Gorbachev, M. Lazzeri, A.K. Geim, C. Casiraghi, Raman fingerprint of aligned graphene/h-BN superlattices. Nano Lett. **13**, 5242–5246 (2013)
59. R. Ribeiro-Palau, C. Zhang, K. Watanabe, T. Taniguchi, J. Hone, C.R. Dean, Twistable electronics with dynamically rotatable heterostructures. Science **361**, 690–693 (2018)
60. A.J. Green, A.C. Diebold, Thickness and rotational effects in simulated HRTEM images of graphene on hexagonal boron nitride. Microsc. Microanal. **20**, 1753–1763 (2014)
61. B.K. Choi, S. Ulstrup, S.M. Gunasekera, J. Kim, S.Y. Lim, L. Moreschini, J.S. Oh, S.-H. Chun, C. Jozwiak, A. Bostwick, E. Rotenberg, H. Cheong, I.-W. Lyo, M. Mucha-Kruczynski, Y.J. Chang, Visualizing orbital content of electronic bands in anisotropic 2D semiconducting $ReSe_2$. ACS Nano **14**, 7880–7891 (2020)
62. G.A. Ermolaev, D.I. Yakubovsky, Y.V. Stebunov, A.V. Arsenin, V.S. Volkov, Spectral ellipsometry of monolayer transition metal dichalcogenides: analysis of excitonic peaks in dispersion. J. Vac. Sci. Technol. B **38**, 014002 (2020)
63. J. Ribeiro-Soares, R.M. Almeida, E B. Barros, P.T. Araujo, M.S. Dresselhaus, L.G. Cançado, A. Jorio, Group theory analysis of phonons in two-dimensional transition metal dichalcogenides. Phys. Rev. B **90**, 115438 (2014)
64. A.H. Barajas-Aguilar, J.C. Irwin, A.M. Garay-Tapia, T.Schwarz, F. Paraguay Delgado, P.M. Brodersen, R. Prinja, N. Kherani, S.J. Jiménez Sandoval, Crystalline structure, electronic and lattice-dynamics properties of $NbTe_2$. Sci. Rep. **8**, 16984 (2018)
65. B.E. Brown, The crystal structures of WTe_2 and high-temperature $MoTe_2$. Acta Crystallogr. **20**, 268–274 (1966)
66. N.W. Alcock, A. Kjekshus, The crystal structure of $ReSe_2$. Acta Chem. Scand. **19**, 79–94 (1965). https://www.sciencedirect.com/science/article/abs/pii/0022508871901688
67. J.C. Wildervanck, F. Jellinek, The dichalcogenides of technetium and rhenium. J. Less Common Met. **24**, 73–81 (1971)
68. E.J. Sie, C.M. Nyby, C.D. Pemmaraju, S.J. Park, X. Shen, J. Yang, M.C. Hoffmann, B.K. Ofori-Okai, R. Li, A.H. Reid, S. Weathersby, E. Mannebach, N. Finney, D. Rhodes, D. Chenet, A. Antony, L. Balicas, J. Hone, T.P. Devereaux, T.F. Heinz, X. Wang, A.M. Lindenberg, An ultrafast symmetry switch in a Weyl semimetal. Nature **565**, 61–66 (2019)
69. X. Ma, P. Guo, C. Yi, Q. Yu, A. Zhang, J. Ji, Y. Tian, F. Jin, Y. Wang, K. Liu, T. Xia, Y. Shi, Q. Zhang, Raman scattering in the transition-metal dichalcogenides of $1T$-$MoTe_2$, T_d-$MoTe_2$, and T_d-WTe_2. Phys. Rev. B **94**, 214105 (2016)

Chapter 9
Optical and Electrical Properties Topological Materials

Abstract In this chapter, we present an overview of the structure, optical and electrical properties of materials that exhibit, or are predicted to exhibit, topological properties. We note that many of these materials consist of layers that are bonded by van der Waals forces and many have a local hexagonal structure. The materials are divided into those with a band gap and those without a band gap. We present the definitions of many of the key terms used in the topological description of materials as well as a description of the various types of topological materials such as topological insulators, Weyl semimetals, and Dirac semimetals. A useful and important part of this discussion is topological classification. For example, we describe the topological invariant $\mathbb{Z}_2$. We present a brief overview of a tight binding, second quantization Hamiltonian that includes spin orbit and electron–electron interactions. Then we discuss materials with a band gap including the well-known tetradymites such as Bi_2Se_3 starting with a discussion of the crystal and electronic structure and resulting $\mathbb{Z}_2$ classification. Next we present the optical and electrical properties of these materials. Whenever possible, experimental data for the dielectric functions are shown. Photoluminescence and Raman spectra are also shown, and the layer number dependence of the Raman spectra are discussed. This is followed by a similar discussion of gapped materials including Weyl semimetals, Dirac semimetals, and nodal line materials. When possible, experimental data is discussed in terms of whether or not topological properties are observed.

Advances in our understanding of electronic structure have resulted in the discovery of a wealth of new physical phenomena associated with the topological properties of electronic band structure of a new class of materials known as *Topological Materials*. One useful way of starting the discussion about Topological Materials is to consider the behavior of the charge carriers. Often, materials are designated as having either having Schrodinger fermions or Dirac fermions [1]. Carriers that display dynamical properties that are described by the relativistic Dirac Hamiltonian are referred to as Dirac fermions, while Schrodinger fermions are described by the non-relativistic Schrodinger Hamiltonian with an effective mass m^*, $H_S = P^2/2m^*$. Since the Schrodinger fermions, electrons and holes, have different effective masses, their dynamics can be described by separate Schrodinger equations unless the electrons

A. Diebold and T. Hofmann, *Optical and Electrical Properties of Nanoscale Materials*, Springer Series in Materials Science 318,
https://doi.org/10.1007/978-3-030-80323-0_9

are interacting with holes. This is in contrast to Dirac fermions where electrons and holes are interconnected by the same Dirac equations and have the same effective mass [1]. Here we refer to these materials as Dirac Materials. Another distinction is that the spin in the Dirac equation for a Dirac material can be a pseudospin [1]. In this chapter, we build on the discussion about topological physics which started with the Quantum Hall effect in Chap. 6. The topic continued with the properties of Dirac fermions in graphene in Chap. 7. Here, we discuss the optical and electrical properties of specific topological materials including 3D topological insulators along with Weyl semimetals and Dirac semimetals which are also known as Dirac materials. Many of these materials are composed of multi-layers that are bonded by van der Waals forces. Before providing an introductory overview of topological materials, we discuss Dirac fermions in Sect. 9.1.

This chapter builds on two recurrent themes for 2D materials namely: (1) materials with a hexagonal (honeycomb) lattice often display new physical phenomena due to their electronic band structure; and (2) many 2D materials are composed of unit layers that are bonded together by van der Waals forces. The van der Waals bonding allows separation of single layer units from bulk materials so that the unique physics of these 2D materials can be studied. The stacking of these single layer units into bilayer and few layer materials often alters the physical phenomena from that observed in single layer and bulk properties. In Chap. 7 these themes were introduced using single layer, bilayer, twisted bilayer, and multi-layer graphene. In Chap. 8, these themes were extended to transition metal dichalcogenides (TMD) where van der Waals bonding of trigonal prismatic TMD allowed separation of single layer units (trilayers of X–M–X where X = chalcogenide and M = transition metal) each of which has a single atomic layer of transition metal atoms arranged in a honeycomb lattice. Here in this chapter, many of the materials also have a van der Waals bonded layer structure. This allows separation of single layer units and multilayer of the material with a honeycomb lattice that can also be separated and characterized. For example, we will see that tetradymite compounds such as Bi_2Se_3 have a structure composed of Se–Bi–Se–Bi–Se quintuple layers where the bulk material has van der Waals bonding between quintuple layers. The honeycomb lattice structure will become evident in Sect. 9.4.1. Other topological materials with the honeycomb lattice structures are discussed here including materials with sub-units having a honeycomb lattice structure along one crystallographic direction, for example, the pyrochlore iridates $X_2Ir_2O_7$ are discussed in Sect. 9.5.1.1, and Ca_3P_2 discussed in Sect. 9.5.1.3. Another theme from Chaps. 2, 7, and 8 is the usefulness of the tight binding model for understanding the electronic band structure of 2D materials. Tight binding models have also proven useful for describing the properties of topological materials as discussed below. One example is the pyrochlore iridates. Some topological insulators, such as the Heusler and Half Heusler Alloys, do not have a van der Waals bonded layer structure. Other examples of materials that do not have a van der Waals bonded layer structure include the transition metal monopnictides which are Weyl semimetals. These materials are also discussed to emphasize that not all topological materials have the honeycomb lattice structure. It is important to note that the presence or lack of inversion symmetry has a significant influence on topological properties. The symmetry of stacked unit layers of van der Waals materials usually changes with the number of layers as discussed in Chap. 8. It is also noteworthy that many of the non-van der Waals materials also have layered structures which strongly influence their properties. In the case of magnetic materials with the magnetic atom or ion occupying specific layers, the magnetic properties are often strongly influenced by the interaction between layers. A key message of this chapter is the importance of a diverse set of symmetries that impose properties on the electronic band structure which result in the presence of topological phases. Below, we will see that these symmetries such as inversion symmetry which is associated with the space group of the crystal structure, and we introduce new symmetries such as time reversal symmetry and parity which are defined below.

It is important to consider that the measured properties of any material depend on the quality of the sample. The growth of higher quality crystals with increased control over defects and intentional doping continues to improve. In that light, conflicting reports on the materials properties of topological materials are often related to the specific sample being measured. For example, the same material may display different properties depending on the amount of doping. When possible, specific information about the sample discussed in this chapter such as carrier concentration and if the sample is single crystal is stated.

We also note that cleanly separating materials into categories such as Topological Insulators or Weyl and Dirac Semimetals is challenging. Often a material predicted to be in a specific category may display properties in a different category in the presence of combined magnetic and electrical fields. It is also emphasized that sample variability between studies of other topological materials such as $ZrTe_5$ has produced reports of different topological phases. In that light, the method used to grow the crystals and the physical properties such as carrier density are reported along with the electrical data or when appropriate, the optical data.

The structure of this chapter is as follows: The chapter begins with a discussion of topological materials. Here, we provide descriptions of many of the key terms used in the chapter including non-spatial symmetry terms such as time reversal symmetry, particle-hole symmetry, and chirality. This section also describes many of the different types of topological materials such as Topological Insulators, Weyl and Dirac Semimetals, Mott Insulators, and others. Next we discuss a generic tight binding Hamiltonian using the second quantization formalism. The Hamiltonian includes spin orbit and on-site Coulomb (Hubbard) interactions. A 3D Dirac equation is also presented. Then, we provide an overview of the optical and electrical properties of topological materials. The descriptions covering the electronic band structure and properties of topological materials are divided into two sections: first materials with a band gap; and then materials without a band gap. Materials with a band gap are discussed in the section on 3D topological materials. This section is divided into three subsections: first we present the crystal and electronic band structures, next the optical properties, and then the electrical properties. Materials without a band gap are presented in the subsequent section which covers Weyl and Dirac semimetals and related materials. This section is again divided into subsections covering crystal and electronic band structures, then the optical properties, and then the electrical properties. This section covers Type 1 and Type 2 Weyl semimetals and nodal line materials.

9.1 Overview of Topological (Dirac) Materials

As mentioned above, the terminology *Dirac materials* refers to materials where low energy fermionic excitations behave like massless Dirac fermions [1]. Relativistic Dirac fermions and the Hamiltonian for Dirac fermions were first introduced in Chap. 7 in Sects. 7.3 and 7.4 which describe the low energy carriers in large area single layer graphene. In that discussion, we presented the Dirac Hamiltonian H_D for a massless carrier. The concept of a "massive" Dirac fermion was introduced using bilayer graphene in Chap. 7 in Sects. 7.6.1 and 7.6.2. The properties of bilayer graphene depend on the layer stacking and the relative alignment between layers. The complex range of properties associated with twisted bilayer graphene was discussed in Sect. 7.9. It is useful to recall that the Fermi velocity of twisted bilayer graphene is reduced over that of single layer graphene. For example, for twist angles between 1.3° and 2° the Fermi velocity is $v_F \sim 0.6 \times 10^6\,\mathrm{ms}^{-1}$ with an effective mass of $m^* \sim 0.03m_e$ [2]. This about 0.6 of the velocity of the low energy carriers in monolayer graphene at low temperature. Here, we start with Dirac Hamiltonian for a massive fermion. The 3D Dirac Hamiltonian H_D is written in terms of the mass, speed of light c and the Pauli matrices, $\boldsymbol{\sigma} = (\sigma^x, \sigma^y)$ and the spin matrix, σ^z:

$$H_D = c\boldsymbol{\sigma} \cdot \left(-i\hbar\vec{\nabla}\right) + mc^2\sigma^z \tag{9.1}$$

where the Pauli matrices are:

$$\sigma^x = \begin{pmatrix} 0 & 1 \\ 1 & 0 \end{pmatrix}, \quad \sigma^y = \begin{pmatrix} 0 & -i \\ i & 0 \end{pmatrix}, \quad \text{and} \quad \sigma^z = \begin{pmatrix} 1 & 0 \\ 0 & -1 \end{pmatrix} \tag{9.2}$$

The effective mass of a Dirac Fermion is related to the band gap $\Delta = 2mc^2$ [1]. For the solid state systems that we are discussing, the carriers are traveling at the Fermi velocity v_F and not the speed of light. For graphene, $v_F \sim 1{,}000{,}000\,\mathrm{ms}^{-1}$ at low temperature which means that relativistic effects must be accounted for. Thus, the speed of light, $c \sim 300{,}000{,}000\,\mathrm{ms}^{-1}$, is replaced with the Fermi velocity, v_F, and the resulting equation is referred to as the Dirac equation.

$$H_D = v_F\boldsymbol{\sigma} \cdot (-i\hbar\vec{\nabla}) + m{v_F}^2\sigma^z \tag{9.3}$$

It is useful to recall from Chap. 7 that there is no band gap for large area single layer graphene, and the Dirac Fermions in single layer graphene are considered to be massless. We also note that a graphene ribbon is considered to have a band gap. Also recall the discussion in Chap. 7 that there is a linear relationship between wavevector

$k = \sqrt{k_x^2 + k_y^2}$ in 2D $\vec{k}$: (k_x, k_y) space and the energy close to the K and K' points in the Brillouin Zone, and by comparison the relationship between photon energy and wavevector, the mass was considered to be zero. The mass of fermions in some Dirac materials is non-zero, and these fermions are said to be "massive". The mass for the 3D Dirac fermions for both Topological Insulators and Weyl Metals is zero.

It is now useful to review some important terminology that is prevalent in the literature for Dirac materials. The reader is referred to Chaps. 6 and 7 where some of these concepts were introduced.

- *Parity*: The parity operator transforms the coordinates from (x, y, z) to $(-x, -y, -z)$. The parity of a wavefunction is even if the wavefunction is not changed by the parity transformation. If the sign of the wavefunction changes, the parity of the wavefunction is odd. The parity operator gives a value of $+1$ for even functions and -1 for odd functions after operating on the wavefunction.
- *Helicity*: *Helicity* has to do with the connection between the direction of spin and momentum. The helicity is right handed if spin and momentum point in same direction and left handed if opposite. For Topological Insulators, the helicity of the Dirac Fermions is neither right nor left handed since the spin and momentum directions are locked perpendicular to each other.
- *Chirality*: A particle (here an electron or hole) is said to be *chiral* if the mirror image of the particle is not identical. Helicity and chirality are the same for massless particles. The chiral operator determines the chiral symmetry of a wavefunction.
- Topological Electronic States: The designation "topological" refers to the non-trivial topology of the Hilbert vector space that the wave functions describing their electronic states span as discussed in Chap. 6 [3]. The Hilbert Space is a vector space that is said to have the inner product structure. The inner product structure allows the length and angle between vectors to be determined. Wavefunctions are linear combination of orthonormal vectors that form a basis set. In crystalline solids the wavefunction maps from k space into a manifold in Hilbert space. The topology of this Hilbert space impacts the electronic states of the solid [3, 4].
- *Time Reversal Symmetry*: A key symmetry for Topological Insulators and many other Dirac Materials is invariance under time reversal. Time reversal of a quantum system reverses momentum and spin [5]. The impact of time reversal symmetry on the band structure is that for time reversal invariant systems the energy of a state with wave vector $\vec{k}$ equals that for $-\vec{k}$. Time reversal is a complicated subject and the reader is referred to [5].
- *Time Reversal Invariant Momentum points (TRIM)*: TRIM points in the Brillouin Zone are points where the momentum is time reversal symmetry invariant. Below, we use the symbol Λ_i to designate these points.
- *Band Inversion*: Typically the valence band is formed from p orbitals and conduction band from s orbitals for semiconductors. Band inversion occurs when spin orbit coupling changes the band energies so that the valence band takes on the orbital character of the conduction band. The bands involved in band inversion depend on the material. For example, in the tetradymite topological insulators

such as Bi_2Se_3, one of the Bi $6p_z$ orbitals moves below the Fermi level and one of the Se $4p_z$ orbitals moves above the Fermi level. In topological insulators, the inversion occurs at the high symmetry Γ point in the Brillouin zone. In Weyl semimetals the band inversion occurs at the Weyl nodes.

- *Strongly Correlated Electron Behavior*: Strongly correlated electron behavior is present when the electron energy bands in a solid cannot be described by single electron theories. Examples of strongly correlated electron behavior include superconductivity, correlated insulators, and Mott Insulators.
- *Kramers pair*: Kramers pair refers to the degenerate states (pair of states) that Kramers theorem requires when time reversal symmetry is present and the band is ½ filled. The spin ½ systems are degenerate. Kramers theorem applies to the Block functions at TRIM points Λ_i in the Brillouin zone.
- *Particle-Hole (Charge Conjugation) symmetry*: Particle-hole symmetry determines the change in a particle after mixing (changing) the change on a particle. A single particle Hamiltonian will have particle hole symmetry if after operation of the particle hole operator on the Hamiltonian, the particle will have the same transport properties such as mobility and effective mass as well as other properties. For most materials including semiconductors such as silicon, exchanging an electron with a hole gives different properties, so they are antisymmetric. The Particle-Hole symmetry has important implications for topological superconductors.
- *Chern Number*: Chern and Simmons established the mathematical theory behind geometric invariants [6]. Then Thouless, Kohmoto, Nightingale, and den Nijs (TKNN) related this concept to quantized conductance and established the invariant υ which is sometimes referred to as the first Chern number (see Chap. 6, Sect. 6.3.1) [7]. For many systems, υ is equal to the integration of the Berry curvature around the torus shaped Brillouin zone divided by 2π (see Chap. 6, Sect. 6.3.4). Here, the Berry curvature is defined in terms of the Bloch wave functions. A non-zero result depends on the shape of surface that represents the allowed states. The continuous or non-continuous nature of the Berry curvature must be checked over the states covered in the surface integral. The Quantized Hall conductivity is given by $\sigma = \boldsymbol{v}e^2/h$ where e is the charge of the electron and h is Planck's constant. The Chern number of a level or set of levels in the electronic band structure is often calculated to determine if a material has topological properties [3]. A topologically trivial material with inversion symmetry has a Chern number of 0. Non-integer Chern numbers are possible when the fractional quantum Hall effect is present. For the purposes of classifying a material, the Chern number is a member of the integer topological invariants $\mathbb{Z}$.
- *$\mathbb{Z}_2$ topological invariant (materials with a band gap)*: The $\mathbb{Z}_2$ topological invariant Z_2 is often used to describe the *potential* for a time reversal invariant material having a band gap ("gapped") to exhibit topological properties. The $\mathbb{Z}_2$ topological invariant Z_2 was introduced because the Chern number is zero for time reversal invariant systems which includes many of the materials discussed in this chapter. Spin orbit interactions alter the electronic band structure and the $\mathbb{Z}_2$ invariant is determined for band structures that include spin orbit coupling. Fu, Kane, and Mele introduced $\mathbb{Z}_2$ in terms of the matrix elements of Bloch wave functions of

the occupied bands at time-reversal invariant momentum (TRIM) points in the Brillouin Zone [4, 8–10]. It is well suited for describing materials with inversion symmetry. The Chern number is related to the total charge in the quantized transport discussed in Chap. 6. The $\mathbb{Z}_2$ topological invariant Z_2 is linked to the accumulated spin of the Kramer's pair at the edge of a material. Kramer's theorem requires that certain TRIM states be degenerate, and Kramer's theorem is further discussed in Sect. 9.1.2. The $\mathbb{Z}_2$ topological invariant Z_2 has also been used to classify gapless materials.

In mathematics, $\mathbb{Z}$ is the group of integers, and $\mathbb{Z}_2$ is called a quotient group that classifies whether an integer is even or odd [3]. In topological band theory, $\mathbb{Z}_2$ gives a topological index which is generally based on parity. The value of $\mathbb{Z}_2$ is used to determine if a material will be topologically non-trivial. The many different methods for determining $\mathbb{Z}_2$ have been reviewed in [11]. The mathematical theory behind the $\mathbb{Z}_2$ classification is K theory, and there are many other classifications for crystals based on their space group symmetry. These topics are further discussed in Appendix C Topological Periodic Table.

The detailed description of the $\mathbb{Z}_2$ topological invariant depend on the type of material including 2D topological materials, 3D topological insulator, and Weyl and Dirac semimetals [4, 10]. For crystals with time reversal symmetry, there are four TRIM points Λ_i in a 2D Brillouin Zone and eight Λ_i in a 3D Brillouin zone [9, 10]. *A material with time reversal symmetry has one unique $\mathbb{Z}_2$ invariant, υ, in 2D, but four distinct $\mathbb{Z}_2$ invariants* $(\upsilon_0; \nu_1\nu_2\nu_3)$ *in 3D* [9, 10]. *The* $(\upsilon_0; \nu_1\nu_2\nu_3)$ *is also referred to as $\mathbb{Z}_2$ index* [9]. The υ_0 invariant determines if the material is a strong topological insulator. The Chern number and 2D $\mathbb{Z}_2$ *invariant* are related but not identical. The $\mathbb{Z}_2$ invariants are determined from the Block wave functions for the occupied bands using the quantities δ_i which are parity related invariants [9, 10]. For crystals with inversion symmetry, the Block wave functions are Eigen functions of the Parity operator with a value $\xi_{2m}(\Lambda_i)$ at the TRIM points, and δ_i is a product of the $\xi_{2m}(\Lambda_i) = \pm 1$ for the pairs of occupied Kramers pairs (doublets) for $2N$ occupied bands: $\delta_i = \prod_{m=1}^{N} \xi_{2m}(\Lambda_i)$ [9, 10]. The energy band $2m$ is a Kramers pair with the $2m-1$ band. The δ_i are referred to as parity invariants [9]

- $\mathbb{Z}_2$ for 2D crystal: The four TRIM points in the 2D Brillouin zone can be thought of as forming a plane. The value of $\mathbb{Z}_2$ for 2D crystal with inversion symmetry can be determined from the product of the band parity values at the four TRIM points i using $(-1)^{\upsilon} = \prod_{i=1}^{4} \delta_i$. For 2D materials, $\upsilon = 1$ for a strong TI and $\upsilon = 0$ for an insulator.
- $\mathbb{Z}_2$ for 3D crystal: The eight TRIM points can be thought of forming a 3D structure in the Brillouin zone. The value of ν_0 for 3D crystals with inversion symmetry can be determined from the product of the band parity related δ_i at the eight TRIM points, $(-1)^{\upsilon_0} = \prod_{i=1}^{8} \delta_i$. The other invariants ν_1, ν_2, and ν_3 can be determined from the product of the parity of four of the TRIM points [9, 12]. Each weak invariant ν_j is determined from the four TRIM located on the three independent planes in the Brillouin zone using $(-1)^{\upsilon_j} = \prod_{i=1}^{4} \delta_i$ [10]. The parity related

invariants δ_i are related to the band parity $\xi_{2m}(\Lambda_i)$ at the TRIM through: $\delta_i(\Gamma) = \prod_{m=1}^{N} \xi_{2m}(\Lambda_i)$ for $2N$ occupied bands. The location of the 8 TRIM depends on the crystal structure. For example, the TRIM in the Brillouin zone of a crystal with the space group symmetry of $R\bar{3}m$(#166) such as the tetradymites will be different than the TRIM in a crystal with a space group $P2_1/c$ (#14) such as β-Ag_2Te.

- For 3D topological materials, the material is an insulator when all four invariants are zero. When $\nu_0 = 1$, the material is a strong topological insulator [9]. When $\nu_0 = 0$, and one of the other invariants ν_1, ν_2, or ν_3 is 1, the material is a weak topological insulator. In some publications, the value of $\mathbb{Z}_2$ for 3D TI's is reported as a single value of the band parity product index for the eight TRIM points in the Brillouin zone instead of the four distinct $\mathbb{Z}_2$ invariants. For example, for Bi_2Se_3 the $\mathbb{Z}_2$ invariant has been reported as an index (1000) and as a single value -1 [9, 12].

We briefly mention how the invariant, υ_0 can be determined from calculating the Pfaffian using the TRIM points Λ_i: $\delta_i = \frac{Pf[w(\Lambda_i)]}{\sqrt{Det[w(\Lambda_i)]}}$ for crystals with no inversion symmetry [3, 9]. This reduces to the band parity formula when time reversal and inversion symmetry are present. The $w(\Lambda_i)$ is the so called unitary sewing matrix where the matrix elements come from the operation of the time reversal symmetry operator Θ on the wavefunctions of the filled bands. The square of Pfaffian of a $2N \times 2N$ antisymmetric matrix M is equal to the determinant of: $Det(M) = [(Pf(M)]^2$ [3, 9].

- *Mirror Symmetry Invariants and Rotational Invariants*: Other topological invariants for gapped crystals classified with $\mathbb{Z}_2$ have been defined including mirror Chern numbers $n_{\mathcal{M}}$ [13] and invariants such as ones associated with rotational symmetry, glide reflection, etc. [14]. Efforts to provide a complete catalog of the invariants including $\mathbb{Z}$ and $\mathbb{Z}_2$ of the calculated band structure for all the space groups have been reported [15]. We again note that a period table of topological materials is being developed [16], and that topic is further discussed in Appendix C.

The mirror symmetry invariants were initially applied to 3D topological insulators such as Bi and $Bi_{1-x}Sb_x$ with $\mathbb{Z}_2$ indices of (1, 111) having TRIM that are both time reversal and inversion invariant [13]. The presence of additional mirror symmetry further constrains the surface states for the rhombohedral A7 lattice (space group $R\bar{3}m$) [13]. When the mirror Chern numbers $n_{\mathcal{M}} \neq 0$, the surface states will be topologically non trivial. The $n_{\mathcal{M}}$ for non-trivial bands can have two different values ± 1 which are considered to be distinct topological phases [13]. The change in mirror Chern number $\Delta n_{\mathcal{M}}$ at a band inversion is determined by the mirror chirality $\eta = \pm 1$ with free electrons having $\eta = +1$, and anomalous electrons having $\eta = -1$ [13]. The values of the mirror Chern numbers and mirror chirality are used to determine the symmetry based topological properties for each crystal face, (100), (110), (111), and (111)$'$ separately where (111)$'$ are faces that are terminated in the middle of a bilayer [13]. Mirror symmetry plays an important role in locating topologically protected states in all topological materials including Weyl and Dirac semimetals [17] and superconductors [18, 19].

- *$\mathbb{Z}_2$ topological invariant* Z_2 *(materials with no band gap)*: The same method used to determine the strong $\mathbb{Z}_2$ topological invariant Z_2 υ_0 can be applied to materials with a band gap.

The first step in the topological classification of crystals is the presence or absence of a band gap. The evaluation of electronic band structures for all 230 space groups for the presence of topological phases using a more complete set of symmetry properties is discussed in Appendix C. Since the $\mathbb{Z}_2$ index is widely used to evaluate the topological properties of gapped crystalline materials, it is emphasized in the discussion of the material presented in this chapter. The discussion in Appendix C is expanded to include the classification of crystals with electronic band structures with no band gap ("gapless").

- *Majorana fermion*: A Majorana fermion is its own antiparticle. This is in contrast to a Dirac fermion which is not its own antiparticle. It is named for Ettore Majorana. In condensed matter physics, Majorana fermions are thought to appear as a quasiparticle excitation associated with superconductivity [18]. At the time of the writing of this book, the community is working to achieve a consensus concerning reports of experimental observation of Majorana fermions.

9.1.1 Topological Surface States on 3D Topological Insulators

Topological Insulators are materials with a bulk band gap and 2D Dirac fermion surface states [3, 4]. The size of the band gaps is relatively small compared to typical insulators, for example, for Bi_2Se_3 the band gap is 0.3 eV. Surface states connect the valence and conduction band of the topological insulator at the interface between the topological insulator and another insulator such as air. These topological surface states (TSSs) have some similarities to the states in the Dirac cone of graphene. The energy versus wavevector relationship for the TSS is the same linear Dirac cone, but unlike graphene their spins are locked perpendicular to their momentum direction as shown in Fig. 9.1d. One of the distinguishing differences between topological insulators and Weyl semimetals is that the linear band dispersion is 2D in momentum space $(\vec{k}_x, \vec{k}_y)$ versus the 3D k space band dispersion at the Weyl nodes of Weyl semimetals. Spin-momentum locking results in the carriers in the TSS being spin polarized. The surface states have time reversal symmetry which protects them from being scattered. Chirality is preserved for the Dirac fermions in TI materials, and backscattering from $\vec{k}$ to $-\vec{k}$ requires that the spin flip from $\boldsymbol{\sigma}$ to $-\boldsymbol{\sigma}$ which is forbidden [1]. The protection from backscattering is referred to as topological protection. In 2D TI materials the topologically protected states are edge states. The tetradymite Compounds (Bi_2Se_3,

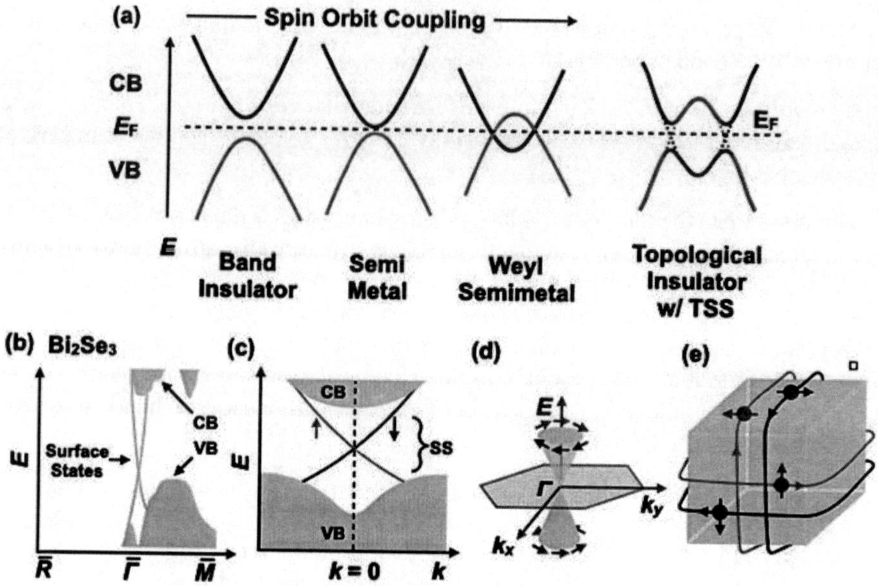

Fig. 9.1 Topologically Protected surface states for a 3D Topological Insulator are pictorially described. **a** The change in bulk band structure from band insulator to Topological Insulator as the spin orbit coupling is increased is illustrated. The orbital character valence and conduction bands switch at the Γ point in the Brillouin zone. The topological surface states (TSS) form at the boundary between the bulk material and insulator (e.g., air) at the surface of the topological insulator. A simplified version of the real band structure of Bi_2Se_3 is shown in (**b**) and the ubiquitous schematic version of (**b**) is shown in (**c**) along with the spin polarization of the surface states is represented by the up and down arrows. (**d**) The 2D Dirac cone structure for the surface states is shown along with the hexagonal Brillouin zone. The spin is seen to be perpendicular to the momentum giving the surface states helical spin polarization. (**e**) The counter propagating surface states with spin and momentum locked perpendicular to each other are shown at the surface of a cube 3D Topological Insulator. (**b**) Adapted from [48]. (**c**)–(**e**) Adapted from [3]. © 2013 the Physical Society of Japan

Bi_2Te_3, $Bi_2Se_xTe_{3-x}$, Sb_2Te_3 and alloys) are prototypical topological insulators and are described in this chapter in Sect. 9.4.

Spin–orbit interaction plays a critical role in TSS. In a semiconductor, the composition of the valence band is dominated by *p* orbitals and the composition of the conduction band by *s* orbitals. In a topological insulator, the orbital character of the valence and conduction bands is altered by spin orbit coupling close to a high symmetry point in the Brillouin zone such as the Γ point. The orbital character of the valence and conduction bands depends on the material itself. For example, in the tetradymite compounds such as Bi_2Se_3 the bands form hybridized Bi 6p and Se 4p orbitals. Band inversion occurs, and one of the hybridized Bi 6p_z orbitals moves below the Fermi level and one of the hybridized Se 4p_z orbitals moves above the Fermi level [12]. The effect of spin orbit coupling on the band inversion is described in Fig. 9.1. In this figure, the Fermi level is located in the band gap of the bulk insulator, and the surface states cross the band gap and the Dirac cones from the upper and lower energy states touch at the Fermi level. The TSS are said to have metallic conduction. In an ideal TI, the Fermi level energy is mid-way in the band gap. However, that is not the case for many TI materials. The TSS, band diagram, and Dirac cone structure is shown in Fig. 9.1. Optical observation of the TSS is very difficult due to their low energy and the presence of other states with the same energy.

9.1.2 Weyl Semimetals and Dirac Semimetals

Weyl semimetals and Dirac semimetals are topological materials that exhibit high carrier mobility and large magnetoresistance [20–22]. Weyl semimetals and Dirac semimetals are closely related materials that are distinguished by symmetry considerations. Either inversion symmetry or time reversal symmetry is broken in a Weyl semimetal. In a Dirac semimetal both inversion and time reversal symmetry are preserved. In Weyl semimetals, the non-degenerate valence and conduction bands touch at an isolated single point or node known as the Weyl node at or close to the Fermi surface at values of the momentum in the first Brillouin zone which are not necessarily located at high symmetry points [22]. The Weyl nodes come in pairs which have opposite chirality. The difference between the Topological Insulator band structure near the 2D surface states and the Weyl semimetal band structure is shown in Fig. 9.2. The bands disperse from the Weyl node linearly in 3D ($\vec{k}_x, \vec{k}_y, \vec{k}_z$) instead of 2D ($\vec{k}_x, \vec{k}_y$) [20, 22]. The 3D Dirac cones in a Weyl semimetal are not degenerate, and the apex of the 3D Dirac cone is called the Weyl node. As with the fermions in graphene and TI materials, these fermions are low energy states. Because Weyl semimetals cannot have both inversion symmetry and time inversion symmetry [22], the role of symmetry in distinguishing Weyl semimetals for Dirac semimetals

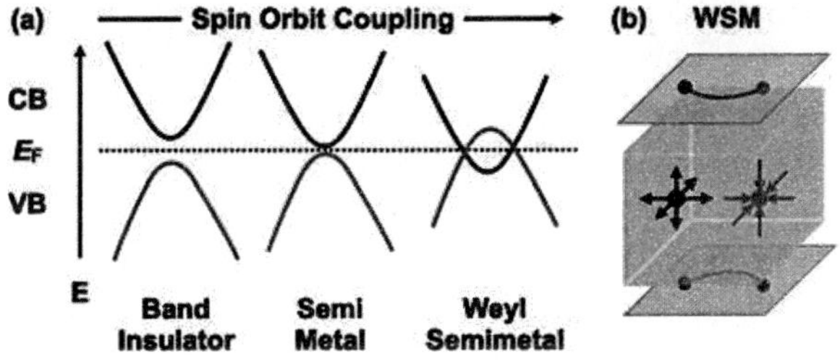

Fig. 9.2 The change in band structure from a band insulator to a Weyl Semimetal (WSM) due to increasing spin orbit coupling is shown in (**a**). Band inversion of the valence band (VB) and conduction band (CB) due to spin–orbit coupling (SOC) changes the orbital character of the valence and conduction bands close to the places in the Brillouin zone where the Weyl nodes occur inside the bulk. In (**b**), a Weyl node pair which has opposite chirality is shown. The Weyl nodes occur at places in the Brillouin zone that are not high symmetry points This image also shows the Fermi arcs the can be observed at the top and bottom surfaces of the crystal. WSMs typically have many Weyl nodes as shown in Fig. 9.4

requires use of the Kramers' theorem about the degeneracy of Eigen states in a band that has time reversal symmetry and has ½ spin. According to Kramers' theorem, these states must be degenerate [23]. *Dirac semimetals* occur when the conduction and valence bands touch in two locations or nodes at the Fermi level for a momentum (pair) in the first Brillouin zone. Again Kramers' theorem applies, and it requires that these points are degenerate. Thus, in a Dirac semimetal, four bands are degenerate (touch) instead of the two bands for Weyl semimetals. In summary, if the crystal has inversion symmetry and time reversal symmetry, then there will be a pair of Weyl nodes and the material is called a Dirac Semimetal [20]. Furthermore, the Dirac cones in a Dirac semimetal are said to be Kramers degenerate. When time reversal symmetry is present and the band is ½ filled, spin ½ systems are degenerate [23]. In contrast to TI materials where the topological properties arise from filled bands, the topological effects for Weyl semimetals occur at the Fermi surface. To date, three types of Weyl semimetals have been reported. The Type 1 Weyl semimetals have 0D nodal points in the bulk with the Dirac cone above and below the Fermi level in momentum space. Type 2 Weyl semimetals have a tilted Dirac cone in momentum space [20]. The third type is a Weyl semimetal with nodal lines [24].

Type 1 and Type 2 Weyl semimetals can be further understood by referring to a Dirac Hamiltonian in momentum space that represents the band structure close to the Weyl points [22] as shown in Fig. 9.3. Burkov [22] starts with a Taylor series expansion about the crystal momentum $\overrightarrow{k_0}$ where the valence and conduction bands touch at the Fermi level ε_o:

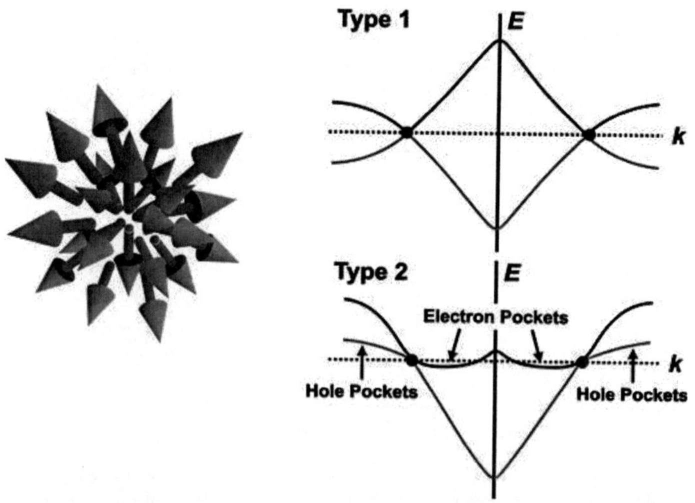

Fig. 9.3 The Berry curvature field for a source Weyl node is shown on the left and Types 1 and 2 Weyl nodes are shown on the right. The wavevector line for the x axis lies at the Fermi level, and the conduction band is shown in black and the valence band in gray for Type 1. For Type 2, the conduction band is shown in gray and the valence band in black. The nodal points are shown as red dots. The Dirac cones that lie above and below the dots for Type 1 and are tilted for Type 2 Weyl semimetals. Figure adapted from [22]

$$H(\vec{k}) = \varepsilon_o\sigma_0 \pm \hbar v_F(\vec{k} - \vec{k}_0) \cdot \boldsymbol{\sigma} \tag{9.4}$$

where v_F is the Fermi velocity, and the unit matrix $\sigma_0 = \begin{pmatrix} 1 & 1 \\ 1 & 1 \end{pmatrix}$ and $\boldsymbol{\sigma} = \left(\sigma_x, \sigma_y, \sigma_z\right)$ are the Pauli matrices. The Hamiltonian is for a massless Fermion with right (+) or left handed (−) chirality. At $\vec{k} = \vec{k}_0$, the bands touch for any value of ε_o and $\vec{k}_0$. The value of $\vec{k}_0$ depends on the band structure of the specific Weyl semimetal, and Weyl semimetals can have several pairs of Weyl nodes as shown in Fig. 9.4 for the transition metal monopnictides. This also requires that the $\vec{k}$ have three dimensions since the Pauli matrices are in all three dimensions. The bands are non-degenerate away from the Weyl node which imposes the requirement that the Weyl metal cannot have both inversion and time reversal symmetry [22]. Equation (9.4) holds when $\vec{k}_0$ is a time reversal invariant momentum (TRIM). When $\vec{k}_0$ is not a TRIM a linear term must be added to preserve the type symmetry that is present in the material as shown in (9.6) [22]. TRIM are located at the Brillouin zone center or at special points at the Brillouin zone boundaries [22].

Often the Weyl nodes are depicted with the Berry curvature vector field pointing toward or away from the Weyl node as shown in Fig. 9.3. The Berry curvature field for the Eigen states for (9.4) can be explicitly stated in terms of the Berry curvature vectors $\vec{\Omega}\left(\vec{k}\right)$ in terms of the crystal momentum toward or away from the Weyl node as [22]:

$$\vec{\Omega}(\vec{k}) = \pm\frac{\vec{k} - \vec{k}_0}{2\left|\vec{k} - \vec{k}_0\right|^3} \tag{9.5}$$

The Weyl nodes can be regarded as sources or sinks of Berry curvature field. The symmetry of the Weyl nodes balances the opposing chiral charge. This balance is broken when an electric and magnetic field are aligned along the same direction along a material. This can result in the quantized conductance discussed in Sect. 9.3. As mentioned before, observation of the Fermi arc is considered proof of the presence of Weyl nodes in a material. The Fermi arc is an edge state in real space present when the edge of the sample is parallel to the line that lies between the two Weyl nodes. The presence of a Fermi arc can be confirmed using X-ray Photoemission spectroscopy. In order for the Weyl properties to be observable and impact the material properties, the Weyl nodes must be close to the Fermi level and there should be few if any other states close to the Fermi level [22].

When $\vec{k}_0$ is not a time reversal invariant momentum (TRIM), a linear term must be added to (9.4) to maintain the type of symmetry present in the material and the Hamiltonian is written as [22]:

$$H\left(\vec{k}\right) = \varepsilon_o\sigma_0 + \hbar\tilde{v}_F\left(\vec{k} - \vec{k}_0\right)\sigma_0 \pm \hbar v_F\left(\vec{k} - \vec{k}_0\right) \cdot \boldsymbol{\sigma}$$

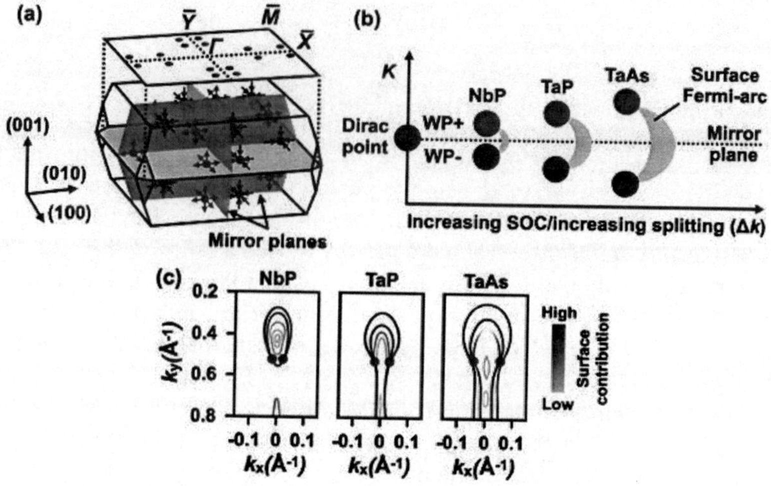

Fig. 9.4 Spin orbit coupling opens the gap in k space between Weyl nodes in Type 1 Weyl semimetals. The Brillouin zone for transition metal monopnictides is shown in (**a**). The surface Fermi arc in the Fermi level is seen to increase as spin orbit coupling increases in (**b**). The surface Fermi arc is shown in k space. Figure adapted from [21]

$$\text{for } \varepsilon_o = 0,\ H\left(\vec{k}\right) = \hbar\tilde{v}_F\left(\vec{k} - \vec{k}_0\right)\sigma_0 \pm \hbar v_F\left(\vec{k} - \vec{k}_0\right)\cdot\boldsymbol{\sigma} \tag{9.6}$$

$\tilde{v}_F$ is the average Fermi velocity, and when $\tilde{v}_F > v_F$, the bands overlap and electron and hole pockets form. This is a Type 2 Weyl semimetal as shown in Fig. 9.3.

Experimentally, the Weyl band structure can be verified using angle resolved photoemission (ARPES) through the Fermi arcs that appear on the surface. A pair of Weyl nodes will have opposite chirality and the nodes are connected by a Fermi arc. The Fermi arc can be thought of as electrons at the Fermi level at the surface of the crystal are found in surface in momentum space that has the shape of an arc. This is shown in Fig. 9.3 [3]. In Fig. 9.4, the impact of spin orbit coupling on the separation of the Weyl nodes and the Fermi arc are shown for transition metal monopnictides [21].

9.1.3 Large Gap Quantum Spin Hall Insulator

Topological edge modes have been identified inside a bulk band gap for transition metal pentatellurides [25]. These modes are distinctly different from the surface modes of topological insulators. These materials will be discussed at the end of this chapter.

9.1.4 Axion and Axion Insulator

Axions are hypothetical subatomic particles that break both charge and parity symmetry. In solid state physics, axion insulators are magnetic topological insulators in which the nontrivial $\mathbb{Z}_2$ index is protected by inversion symmetry instead of time-reversal symmetry. The axion insulator state has been observed in $MnBi_2Te_4$. The $A_2B_2O_7$ pyrochlore and AB_2O_4 spinel crystal structures may be axion insulators as discussed in Sect. 9.5.1. In both structures, the *B*-site atoms form a simple inversion-symmetric pyrochlore lattice, and axions are possible in systems where the metal atom on the *B* is magnetic, while the one on the *A* site is not [1, 26].

9.1.5 Mott Insulator

A Mott insulator is a material with a ½ filled band that is predicted by a simple band theory to be a metal but instead is an insulator. One example is V_2O_3. This phenomenon was discovered by Mott and Peierls when some transition metal oxides were found to be insulating [27]. Mott insulators occur because of electron–electron interaction. In an electronic "band insulator", the energy gap between the filled valence band and empty conduction band is large enough to result in high resistivity. There are many types of Mott insulators such a Mott–Hubbard, Mott-Heisenberg, and Mott-Anderson. These different Mott insulators are distinguished by the presence (Mott-Heisenberg—antiferromagnetic order) or absence (Mott–Hubbard) of long range magnetic order or lattice defect/impurities and electron correlation (Mott-Anderson) [27]. In a strong Mott insulator, the band gap is larger than the energy associated with spin–orbit interactions, and the electrons are considered to be localized to single atoms [27]. Strong Mott insulators are not expected to show topological properties. Some materials transition from a metal to a Mott insulator as temperature is reduced.

9.1.6 Chern Insulator

The Chern insulator is 2D insulator with that lacks momentum states with time reversal symmetry. If the Chern number is not zero, then the material with have edge states which means it can exhibit a quantum spin Hall conductance. The edge states are chiral and thus flow in opposite directions.

9.1.7 Topological Superconductors

Topological superconductors are crystalline superconductors that are protected by mirror symmetry [18, 19, 28]. The electronic band structure may be time reversal symmetric, but Mackenzie and Maeno [28] argue that time reversal symmetry is broken once Cooper pairs are formed below the superconducting transition temperature T_c. T_c varies with the quality of the crystal [28]. Crystal symmetry protects Majorana fermions [18, 19, 28]. In the context of condensed matter physics, Majorana fermions are quasiparticles that are their own antiparticle [3]. The term quasiparticle refers to the collective motion of more than one particle, and a previous example of a quasiparticle is the fractionally charged quasiparticle in FQHE discussed in Chap. 6. In a topological superconductor, Majorana fermions are electron–hole symmetric.

9.2 Tight Binding Hamiltonian with Spin–Orbit and On-Site Coulomb (Hubbard) Interactions and a 3D Dirac Equation

In the emerging field of topological materials, it is useful to have a generic Hamiltonian that describes the important electron–electron interactions that determine whether a material is a topological insulator, Weyl semimetal, Mott insulator, Anion insulator, or normal metal or insulator. Here we first present a tight binding Hamiltonian that provides an intuitive means of understanding the interplay of spin orbit coupling and electron–electron interactions. The literature typically uses the second quantization formalism to express the Hamiltonian to describe materials which display both spin orbit coupling and correlated electron behavior. When correlated electron behavior is present, electron–electron interactions cause a material to be insulating or result in a transition to an insulating state [27]. As discussed in Chap. 7 for twisted bilayer graphene, correlated electron driven transitions are temperature dependent. Next we present a Dirac Hamiltonian that provides an intuitive means of understanding the interplay of mass, intrinsic magnetization, and applied magnetization. In order to apply the Dirac Hamiltonian, the Dirac nature of the material is usually theoretically or experimentally predicted.

Tight Binding Hamiltonian: The tight binding Hamiltonian provides an intuitive means of understanding the physical interactions that produce the phase of a material, and whether or not it will undergo a metal–insulator transition. The second quantization formalism was introduced for a tight binding, nearest neighbor approximation in Chap. 2, Sect. 2.11 to which the reader is referred. Here, we briefly present a tight binding Hamiltonian that includes spin–orbit coupling and on-site Hubbard interactions following Witczak-Krempa et al. [27]. The creation operator $c_{i\alpha}^{\dagger}$, annihilation operator $c_{i\alpha}$, and the occupation number $n_{i\alpha} = c_{i\alpha}^{\dagger}c_{i\alpha}$ are indexed for an orbital α at site i. The model Hamiltonian can be written as the sum of the three key interactions, hopping, spin–orbit, and Hubbard. It uses the site to site hopping amplitude t, the spin orbit interaction strength λ with spin angular momentum $\vec{S}$ and orbit angular momentum $\vec{L}$ and Hubbard repulsion U:

$$H_{TB} = \sum_{i,j;\alpha\beta} [t_{ij,\alpha\beta} c^{\dagger}_{i\alpha} c_{j\beta}] + c.c. + \lambda \sum_{i} \vec{L}_i \cdot \vec{S}_i + U \sum_{i,\alpha} n_{i\alpha}(n_{i\alpha} - 1) \quad (9.7)$$

Here, $c.c.$ refers to the complex conjugate. The first term includes spin independent hopping between sites. This term can be restricted to nearest neighbor site or next nearest neighbor sites can be included. A comparison of the effect of spin–orbit interactions with the on-site Hubbard repulsion is made more apparent by normalizing these terms using the hopping term t, λ/t and U/t. The influence of the spin–orbit interactions relative to the on-site Hubbard repulsion on the state of the system has been schematically presented [27] and is shown here in Fig. 9.5. The spin orbit coupling is small $\lambda/t \ll 1$ for materials having transition metals and a metal–insulator transition can occur across a series of transition metal compounds as U/t varies. Theoretical values for the spin orbit coupling constant λ in units of cm^{-1} (convert to energy units by multiplying by 1.2398×10^{-4} eV/cm^{-1}) for some oxidation states of transition metals can be found in [29]. The spin orbit coupling constant for Ir^{+4} has a value of ~0.4 eV [27, 30]. In contrast, the spin orbit splitting of WS_2 is ~0.4 eV as discussed in Chap. 8, and the atomic SOC for Bi is 1.25 eV [30]. The tight binding Hamiltonian discussed here is frequently modified to include the effects of magnetic order or time reversal symmetry. Time reversal symmetry can be broken using a Zeeman term and magnetic ordering [26]. It is important to note that a series of materials such as the pyrochlore iridates $X_2Ir_2O_7$ can change the spin orbit coupling by changing the metal atom X. This allows the tuning of the electronic structure to enhance the possibility that a material might have topological properties.

3D Dirac Equation: The Dirac equation presented in this section provides an intuitive means of understanding when a material will be a 3D Dirac semimetal, magnetic semiconductor or Weyl semimetal. Burkov et al. [31] and others [32–34] developed a Dirac equation that includes terms for 3D massless Dirac fermions, a mass term m, a term for spin–orbit coupling (an intrinsic Zeeman magnetic field) b along z, and an intrinsic Zeeman magnetic field b' along x [34]. The Dirac Hamiltonian for carriers with $\vec{k} = (k_x, k_y, k_z)$ is:

$$H = \upsilon \tau_x (\boldsymbol{\sigma} \cdot \vec{p}) + m\tau_z + b\sigma_z + b'\sigma_x \quad (9.8a)$$

$$For\ b' = 0: \quad E_{s,u}(k) = s\sqrt{m^2 + b^2 + \upsilon^2 k^2 + u2b\sqrt{\upsilon^2 k_z^2 + m^2}} \quad (9.8b)$$

We note that the effect of an external magnetic field can be added to this Hamiltonian [33]. Here $\boldsymbol{\sigma}$ are the Pauli matrices, τ_i is the $i = x,\ y,$ or z Pauli matrix for

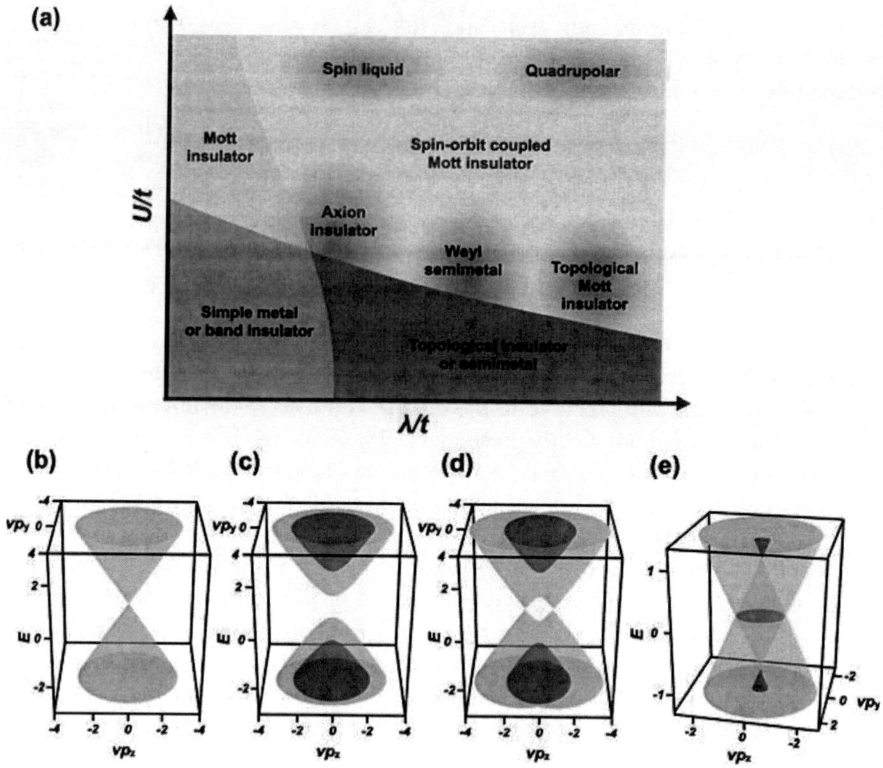

Fig. 9.5 The interplay between physical properties and the resulting phase of a material is shown for the tight binding Hamiltonian in (**a**) and the Dirac Hamiltonian in (**b**) through (**e**). In (**a**), the interplay between spin orbit interaction λ, Hubbard (Coulomb) U, and the temperature determines the state of a material. Both the spin orbit coupling and Hubbard energy are normalized by the tight binding hopping amplitude (energy) t. 'The interplay between mass and intrinsic magnetic field results in a 3D Dirac cone when $m,\ b,\ and\ b'$ are zero as shown in (**b**). The energy is plotted versus (vp_x, vp_y) based on the Hamiltonian of (9.8a). A gapped magnetic semiconductor is shown in (**c**), a Weyl semimetal in (**d**), and a nodal line semimetal in (**e**). (**a**) Adapted and reprinted from [27]. © 2014 by Annual Reviews. (**b**) Adapted and reprinted with permission from [34]. © 2018 American Physical Society

pseudospin, and $s = \pm 1$, $u = \pm 1$. In (9.8b), $k = \left|\vec{k}\right|$. The equation for the energy when $b = 0$ and b' is non-zero is obtained by substituting b' for b in (9.8b). When c are all zero, the Hamiltonian represents a 3D Dirac node with energy having a linear dependence on wavevector. When $b' = 0$ and $|m| > |b|$ and the result is a gapped magnetic semiconductor, when $|m| < |b|$ the result is a Weyl semimetal [34]. The Weyl nodal points in the Brillouin zone are located at $\vec{k} = (0,\ 0,\ \pm\sqrt{b^2 - m^2}/v$. This results can be obtained by setting $E(k) = 0$ and considering $\vec{k} = (0, 0, k_z)$. When the intrinsic magnetic field is along the x axis, and m and b are zero, a ring node semimetal results. The band structures for these limiting results are shown in

Fig. 9.5b, and the reader is referred to [34] for a review of the interplay between Zeman effects along different directions in a material.

9.3 Optical and Electronic Properties of Topological Materials

The electrons and holes in topologically protected states are low energy carriers thus their impact on optical and electrical properties requires careful consideration. Low energy optical characterization data for 2D and topological materials is typically presented in terms of the optical conductivity of a material. Although this section discusses low energy carriers, we do note that the complex dielectric function (complex refractive index) in the visible wavelength range is also reported in this chapter when available. As discussed in Sects. 7.3 and 7.4 for graphene, the real part of the optical conductivity, σ_1, of 2D Dirac Fermions has an unusual constant behavior versus energy (wavelength) in the near IR and IR. To aid in this discussion, the optical conductivity-dielectric function relationship is restated in (9.9) (CGS units).

$$\sigma = \sigma_1 + i\sigma_2; \quad \sigma_2 + i\sigma_1 = \omega[(\varepsilon_1 - 1) + i\varepsilon_2]/4\pi \tag{9.9a}$$

$$\sigma_1 = \frac{\omega\varepsilon_2}{4\pi} \quad and \quad \sigma_2 = \frac{\omega(\varepsilon_1 - 1)}{4\pi} \tag{9.9b}$$

It is important to note that the observed optical conductivity often has contributions from multiple sources including optical transitions between electronic bands such as those discussed in Chap. 4, inter-band transitions at the Dirac cones (2D surface Dirac cones, and 3D bulk Weyl and Dirac nodes with bands linear in k), and excitons. In some cases, the contributions can be separated by fitting function forms for optical transitions and subtracting them as discussed for the pyrochlore iridates in Sect. 9.5.2.

A general rule for the *real part of the optical conductivity of d dimensional Dirac electrons* is [35, 36]:

$$\sigma_1 \sim \omega^{(d-2)/z} \quad \text{and} \quad E(k)\alpha|k|^z \tag{9.10}$$

For a symmetrical Dirac 2D cone or 3D Dirac cone which has a linear dependence on the energy with the magnitude of the wavevector, $|k|$, $z = 1$ [35, 36]. Equation (9.10) gives the expected constant value for σ_1 versus energy for 2D topological surface states, and a linearly increasing σ_1 versus energy ($\hbar\omega$) for 3D topological surface states. The origin of this dependence is the joint density of states. For 3D states there is an extra energy (ω) term [36]. As discussed in Chap. 4, the imaginary part of the dielectric function near a direct interband transition is

$$\varepsilon_2 = \frac{e^2}{\pi m^2\omega^2}\int |a_0 \cdot P_{CV}|^2\delta(E_{CV} - \hbar\omega)d\vec{k} \tag{9.11}$$

where a_0 is the unit polarization vector, P_{CV} is the matrix element of the momentum operator, E_C is energy of the conduction band, and E_V is the energy of the valence band ($E_{CV} \equiv E_C - E_V$). The matrix element of the momentum operator is:

$$p_{CV} = \frac{ih}{2\pi\Delta}\int u_C^*(r,k)\nabla u_V(r,k)dr \tag{9.12}$$

where the integral is over the unit cell volume and Δ is the unit cell volume. For transitions where the interband transition matrix elements P_{CV} do not depend on k, then $\varepsilon_2(\omega) \propto \rho(\hbar\omega)$ the density of states. Thus the optical conductivity can be expressed as follows:

$$\sigma_1 = \frac{\omega\varepsilon_2}{4\pi} \propto \rho(\hbar\omega)/\omega \tag{9.13}$$

The density of states for 2D is $\propto \omega$ and for 3D is $\propto \omega^2$ thus σ_1 is constant for 2D Dirac fermions and linearly increases with ω for 3D Dirac fermions. We recall from Chap. 6, that for the 2D Dirac states of graphene, the optical conductivity σ_1 was constant and we showed that it is related to the fine structure constant. The number of Dirac cones that contribute to this transition and the presence of other transitions in the same energy (wavelength) range all contribute to the intensity of the transitions and thus whether or not the transition can be observed. For graphene, there are 2 sets of Dirac cones (a set has a cone from the valence band touching the cone from the conduction band at a $K(K')$ point) due to the K *and* K' valleys and each one contributes 2 transitions due to up and down spin. For topological insulators, there is one set of Dirac cones at the top surface and one at the bottom [36]. Due to spin-momentum locking, the top set of cones contributes a transition for only one spin. Thus based on the number of Dirac cones, the σ_1 for 3D topological insulators is a factor of 4 less than graphene. That is why the σ_1 of the topological surface states for 3D topological insulators is difficult to observe and as of this time, has not been observed.

For the Type 1 and Type 2 Weyl semimetals, there are many Weyl nodes, and the topological surface state contribution to σ_1 has been reported for some Weyl semimetals. The contribution of the Weyl nodes to the *optical conductivity* at low temperature for samples with no impurities has been theoretically determined for N Weyl nodes to be [34, 37, 38]:

$$\sigma_0^N(\omega) = N\frac{e^2}{12h}\frac{\omega}{v_F}. \tag{9.14}$$

Here, the chemical potential of the material is at the Fermi level, and e is the electrical charge, h is Plank's constant, and v_F is the Fermi velocity [37, 38]. The Weyl semimetals with Nodal rings are predicted to have a constant optical conductivity for low photon energies as discussed in Sect. 9.5.2. The effect of impurities at low temperature on the optical conductivity has been theoretically determined to be:

$$\sigma_0^N(\omega) = N\frac{e^2}{12h}\frac{\omega}{\upsilon_F}\left\{1 - \frac{16N\gamma\omega}{15\pi^2\upsilon_F^3} + \mathcal{O}\left(\frac{N^2\omega^2}{\omega_0^2}\right)\right\}. \tag{9.15}$$

Here, the strength of the impurity potential is $\gamma = \frac{1}{2}n_i v_0^2$ where n_i is the impurity concentration, and v_0 characterizes the strength of an individual impurity [37] and $\omega_0 = 2\pi \upsilon_F^3/\gamma$. In the absence of impurities, (9.15) reverts to (9.14).

The *quantum Hall effect* (*QHE*) was introduced in Chap. 6 using the 2DEG (2D electron gas), and discussed for the Dirac carriers in graphene in Chap. 7. As with graphene and other 2D materials, the carrier transport properties of topologically protected states in topological materials are described by one member of a family of Hall effects. The family of Hall effects are shown in Fig. 9.6 and briefly described in the figure caption. Materials that exhibit magnetization exhibit a Hall effect known as the anomalous Hall effect without the presence of an external magnetic field (see Chap. 6, Sect. 6.6). Reference [39] provides an excellent review of the anomalous Hall effect in topological insulators. Doping a non-magnetic material with magnetic atoms such as Cr can result in a quantum anomalous Hall effect [39]. Spin–Orbit coupling results in both band inversion (discussed above) and changes in spin transport for some materials. These materials can exhibit spin Hall effects. The challenge

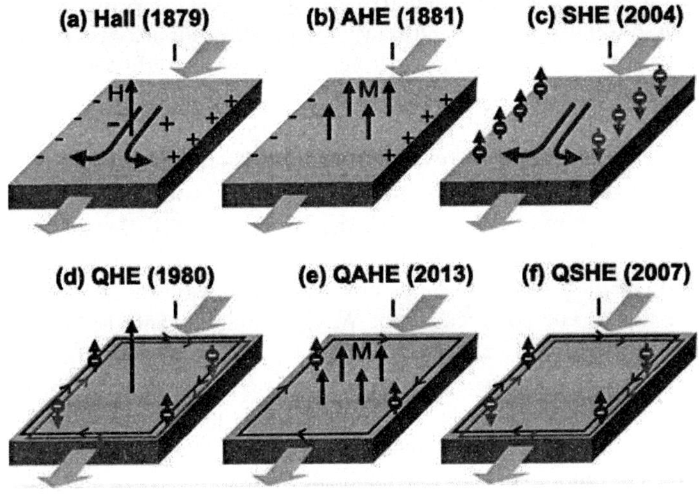

Fig. 9.6 The family of Hall effects is shown along with the date of their discovery. The Hall voltage is proportional to the magnetic field strength for metal as shown in (**a**). For a magnetic metal, the Hall voltage is proportional to the magnetization of the metal and not the magnetic field strength. This is the anomalous Hall effect (AHE) shown in (**b**). Some metals with large spin–orbit coupling exhibit a spin Hall voltage without the presence of a magnetic field as shown in (**c**). The quantum Hall effect (QHE) is shown in (**d**). In the QHE, quantized steps in the resistivity (conductance) are observed in high magnetic fields. The quantum anomalous Hall effect (QAHE) has been observed for 3D topological insulators doped with magnetic atoms such as Fe or Mn. The quantized version of the spin Hall effect (QSHE) has also been observed. Figure adapted from [39].

for 3D topological insulators is separating the transport of bulk carriers from that of the topologically protected states. For example, the 2D topological surface states of 3D TI materials should have quantized transport, but separation of the 2D transport response from that of the bulk transport has proven to be difficult [39]. The Hall transport characterization for each material is discussed below in the section associated with that material.

The literature has example of topological materials exhibiting the AHE, and some materials are expected to exhibit the QAHE. As discussed in Chap. 6, one of the sources that contribute to the AHE comes from the Berry curvature. A *3D QAHE* is predicted for Weyl and Dirac semimetals. Charge transport in Weyl semimetals is attributed to a chiral anomaly. This refers to the non-conservation of chiral charge during carrier transport. The right and left handed Weyl nodes form a nodal pair that are separated in momentum space and connected only through the Fermi arc at the surface of the crystal [22, 40]. The electric and magnetic fields present during a Hall measurement break the symmetry between the nodal pair and result in this chiral anomaly. Changing relative angle between the electric and magnetic field provides a means of characterizing the topological state of a material. Parallel electric and magnetic fields ($\vec{B} \| \vec{E}$) pump charge between the nodes which breaks the chiral charge symmetry. Thus the longitudinal conductivity increases with magnetic field (resistivity decreases). Due to the negative differential resistivity this is referred to as a negative longitudinal resistivity. Ordinary metals show an increase in conductivity with magnetic field, thus the decrease in longitudinal conductivity with magnetic field is unusual [40]. The transverse (Hall) conductivity is observed with parallel electric and magnetic fields. Combined with the negative longitudinal resistivity, the Hall conductance is a signature of a Weyl semimetal [40]. Thus, Weyl semimetal should have a quantized anomalous Hall conductivity that proportional to the separation in wavevector $\Delta\vec{k}$ between Weyl nodes, $\sigma_{xy} = \frac{e^2}{\hbar}\Delta\vec{k}$ [27]. Because the parallel electric and magnetic field along a material breaks time reversal symmetry, Dirac semimetals can display Weyl semimetal properties in the presence of parallel electric and magnetic fields. The conductivity of Weyl semimetals and the chiral nature of the carriers is further discussed in Sect. 9.5.

Magneto-optical measurements have also been used to characterize the low energy optical conductivity or the reflectivity of topological materials [41, 42]. This method is sometimes referred to as magneto–infra red spectroscopy. The magnetic field quantizes the electronic states into Landau levels. The optical conductivity is studied as a function of the magnetic field strength which provides information about the electronic band structure through the changes in the energy of optical transitions at specific points in the Brillouin zone. Thus, it is important to be able to align the magnetic field along a specific crystallographic directions. This method has been used in both the reflection mode which is referred to as the Voigt configuration and the transmission mode which is referred to as the Faraday configuration [42]. The magnetic field has been oriented along the sample in the Voigt configuration and normal to the sample in the Faraday configuration [42]. In the case of biaxial crystals, the crystallographic axis can be oriented thus allowing the magnetic field to break symmetry along a specific axis. The study of isotropic crystals can be done using the Voigt configuration. The Dirac equation (9.8a) has been solved using an external magnetic field [33] and applied to the interpretation of magneto-optical spectroscopy.

9.4 3D Topological Insulators

In this section, the crystal structure, electronic band structure, optical, and electrical transport properties of representative 3D topological insulators and related materials with a band gap including tetradymite compounds (Bi_2Se_3, Bi_2Te_3, $Bi_2Se_xTe_{3-x}$, Sb_2Te_3 and their ternary and quaternary alloys), tetradymite type compounds XA_2B_4 (e.g., $MnBi_2Te_4$), silver chalcogenide β-Ag_2Te, and the transition metal pentatelurides $ZrTe_5$ are presented. The van der Waals bonded tetradymite compounds have been extensively studied since they were the first example of topological insulators described in Sect. 9.1.1. Topological properties have been reported or predicted for the van der Waals bonded multilayer tetradymite type compounds. The axion insulator and Chern insulator states have been observed in $MnBi_2Te_4$. As with other topological materials, the presence of a magnetic field or both an electric field and magnetic field can result in related phenomena such as Weyl semimetal properties.

9.4.1 Crystal and Electronic Band Structure of 3D Topological Insulators and Large Gap Quantum Spin Hall Insulators

Tetradymite Compounds (Bi_2Se_3, Bi_2Te_3, $Bi_2Se_xTe_{3-x}$, Sb_2Te_3 and alloys): The initial search for 3D topological insulators emphasized the bismuth tetradymite compounds. These compounds have many structural similarities to graphene. The tetradymite compounds have a quintuple layer (QL) structure with van der Waals bonding between the QLs as shown for Bi_2Se_3 in Fig. 9.7 [43, 44]. The individual layers have the hexagonal (honeycomb) lattice structure in the (0001) face as shown in Fig. 9.7. The real space lattice vectors for the bulk crystal structure are also shown. The topologically protected 2D surface states have a hexagonal Brillouin zone which is present for all numbers of QLs. The rhombohedral crystal structure has a point group (Hermann-Mauguin space group) of $D_{3D}^5\left(R\bar{3}2/m\right)$ [44]. The short international space group symbol for this space group is $R\bar{3}m$ (#166), and the point group and space group symbols were discussed in Chap. 2, Sect. 2.12. The electronic configuration of Bi is $[Xe]\,4f^{14}5d^{10}6s^26p^3$, and for Se is $[Ar]\,3d^{10}4s^24p^4$. The hybridization of the Bi $6p$ and Se $4p$ orbitals is critical to understanding the electronic band structure. As mentioned in Sects. 9.1 and 9.2, band inversion occurs. One of the hybridized Bi $6p_z$ orbitals moves (assumes an energy) below the Fermi level (for an un-doped material), and one of the hybridized Se $4p_z$ orbitals moves above the Fermi level [12].

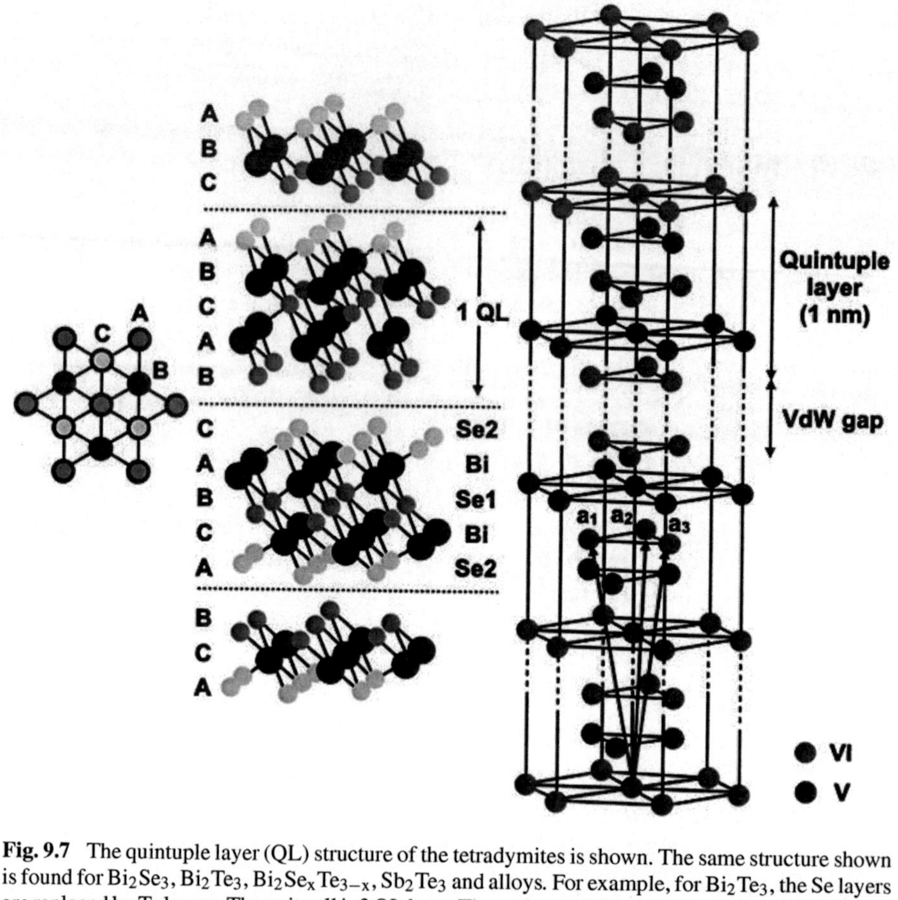

Fig. 9.7 The quintuple layer (QL) structure of the tetradymites is shown. The same structure shown is found for Bi_2Se_3, Bi_2Te_3, $Bi_2Se_xTe_{3-x}$, Sb_2Te_3 and alloys. For example, for Bi_2Te_3, the Se layers are replaced by Te layers. The unit cell is 3 QL long. The reals space lattice vectors for the tetradymite structure is $\vec{a}_1 = \left(-\frac{a}{2}, \frac{\sqrt{3}}{6}a, \frac{c}{3}\right)$, $\vec{a}_2 = \left(\frac{a}{2}, -\frac{\sqrt{3}}{6}a, \frac{c}{3}\right)$, and $\vec{a}_3 = \left(0, \frac{\sqrt{3}}{3}a, \frac{c}{3}\right)$ as shown on the right. The reciprocal lattice vectors are $\vec{b}_1 = \frac{2\pi}{a}\left(-1, -\frac{\sqrt{3}}{3}, \frac{a}{c}\right)$, $\vec{b}_2 = \frac{2\pi}{a}\left(1, -\frac{\sqrt{3}}{3}, \frac{c}{a}\right)$, and $\vec{b}_1 = \frac{2\pi}{a}\left(0, \frac{2\sqrt{3}}{3}, \frac{a}{c}\right)$ are not shown. Figure adapted and reprinted from [43] with the permission of AIP Publishing

One of the challenges in characterizing the topological properties of Bi_2Se_3 and Bi_2Te_3 is accounting for the impact of oxidation, substitutional defects, and the formation of Se or Te vacancies which result in doping thus moving the Fermi level. Certain defects have low formation energies such as doubly charged Se vacancies in the top or bottom QL, and singly charged Bi anti-sites in Bi_2Te_3 [45]. Thus, it is easy to unintentionally dope Bi_2Se_3 and Bi_2Te_3. Bi_2Se_3 is *p* type for both Bi and Se rich growth conditions. Bi_2Te_3 is *n* type for Te rich and a *p* type for Bi rich growth [12, 46, 47]. A theoretical study of the energy of defects in tetradymite compounds is agreement with experimental observations of the relationship between the type of defect and the resulting n or p type doping [45]. Interestingly, $Bi_2Se_1Te_2$ has a specific QL stacking pattern of Te–Bi–Se–Bi–Te and due to the strong Se–Bi bond. This compound does favor Se vacancies and the Fermi level lies in the middle of the band gap [46, 47].

Density Functional Theory (DFT) calculations of the bulk band structure show significant changes when spin orbit interactions are included [12]. The topologically active surface states form at the band gap at the Γ point as shown in Fig. 9.8 for Bi_2Se_3 [12]. A comparison of the theoretical topologically protected surface states between Bi_2Se_3, Bi_2Te_3, Sb_2Se_3 and Sb_2Te_3 shows that if Sb_2Se_3 had the tetradymite structure, it would not form the surface states [12]. In nature, Sb_2Se_3 has the stibnite crystal structure. The location of the Fermi level for these compounds for un-doped material is also shown in Fig. 9.9 [48]. Oxidation and Se (Te) vacancy formation can both result in band bending at the surface which results in the formation of a 2D electron gas (2DEG) whose effects must be accounted for when studying the optical and electrical properties of tetradymite compounds [49]. Angle resolved X-ray photoemission data provides proof of the change in the surface states and existence of the 2DEG as shown in Fig. 9.10.

Some of the tetradymite and tetradymite like topological insulators are 3D topological insulators that have protected 2D surface states. The surface states form at the band gap at the Γ point in the Brillouin zone. The $\mathbb{Z}_2$ index is typically reported to be

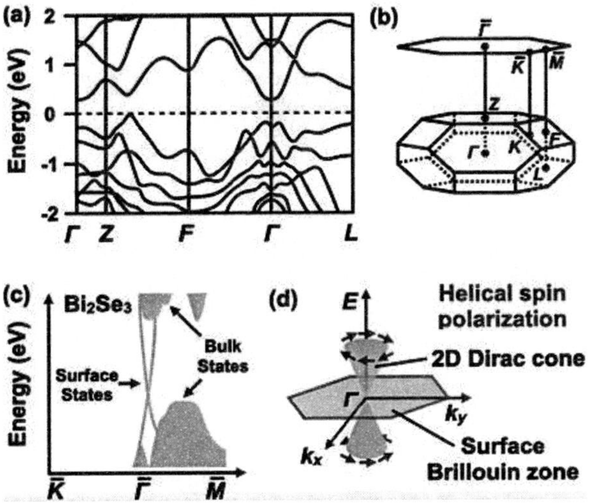

Fig. 9.8 The bulk and surface electronic band structure of Bi_2Se_3 along with the bulk and surface Brillouin zone. The bulk electronic band structure is plotted using the bulk Brillouin zone shown in (**b**). The small band gap at the Γ point is visible. The Fermi level for un-doped Bi_2Se_3 is indicated by the dashed line in the band diagram in (**a**). The surface electronic band structure shown in (**c**) and plotted using the 2D surface Brillouin zone shown in (**b**) has a significantly different appearance from the bulk band structure. The theoretically calculated topological surface states which form at the Γ point are highlighted in (**c**), and the associated Dirac cone is shown in (**d**). The locking of the spin and momentum is also shown. Figure adapted from [12]

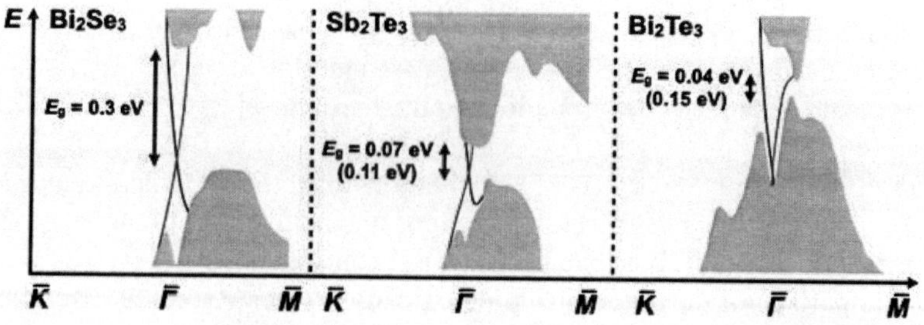

Fig. 9.9 Theoretically calculated surface states for Bi_2Se_3, Bi_2Te_3 and Sb_2Te_3 are shown. The Fermi level is shown by a dotted white line. The spin polarization of the surface states is shown. The value in brackets comes from experimental results. Figure adapted from [48]. © 2017 E. K. de Vries. The data used in de Vries figure was adapted from [12]. The band gap data was obtained from references as described in [48]

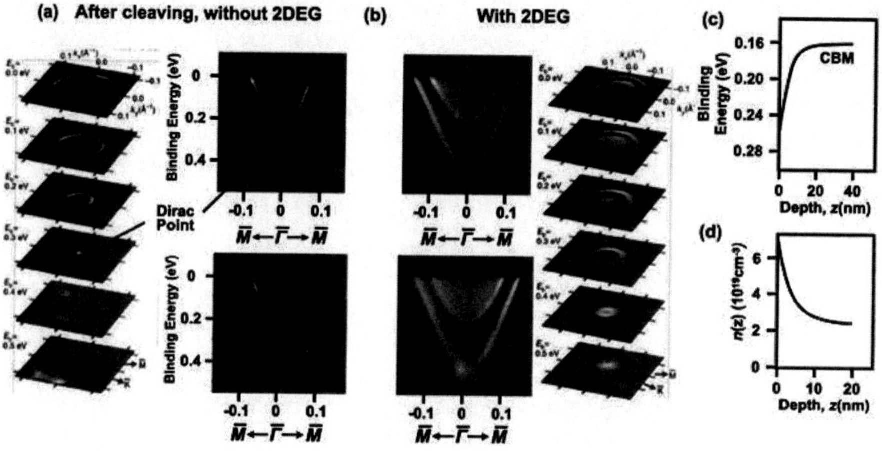

Fig. 9.10 The Dirac cone is clearly visible in the angle resolved X-Ray photoelectron emission (ARPES) data showing the effect of surface oxidation on the Dirac cone and the formation of the 2DEG for Bi_2Se_3. The 2DEG comes from unintentional doping and interferes with the experimental observation of the electronic properties of the topologically protected surface states. Figure adapted from [49]

(1, 000) indicating that these are a strong topological insulators. The first index being the value of ν_0 for 3D crystals with inversion symmetry determined from the product of the band parity related δ_i at the eight TRIM points located at Γ, L, F, and Z for Bi_2Se_3 calculated using $(-1)^{\nu_0} = \prod_{i=1}^{8} \delta_i$. There are 3 TRIM at both L and F and one each at Γ and Z. Since TRS is present, the eight TRIM at Γ, L, F, and Z come in Kramers pairs. Without SOC, Bi_2Se_3 is a trivial insulator. SOC flips the parity at Γ while the parity of the other TRIM remains unchanged resulting in $\nu_0 = 1$ [12]. The other invariants ν_1, ν_2, and ν_3 are determined from the product of the parity of the wavefunctions at the four TRIM points for the three independent planes in the Brillouin zone. Based on the electronic band structure for undoped Bi_2Se_3, Bi_2Te_3,

Table 9.1 The lattice constants of the *Tetradymite Compounds Bi_2Se_3, Bi_2Te_3, and Sb_2Te_3* [50]

	Bi_2Se_3	Bi_2Te_3	Sb_2Te_3
a (nm)	0.4138	0.4383	0.425
c (nm)	2.864	3.049	3.035

and Sb_2Te_3, the $\mathbb{Z}_2$ index is (1, 000) which means that they are expected to be strong topological insulators. The $\mathbb{Z}_2$ index for Sb_2Se_3 is (0, 000). Thus, the topological index predicts that Sb_2Se_3 is an ordinary insulator as is observed experimentally.

The lattice constants of Bi_2Se_3, Bi_2Te_3, and Sb_2Te_3 are provided in Table 9.1 [50]. The reader may ask about other B_2X_3 compounds. As_2X_3 with X=S, Se, or Te all have the orpiment structure. With B=Bi or Sb and X=S, the crystal structure is stibnite [51]. These materials do not have topological surface states.

Tetradymite Type Compounds(XA_2B_4; X=Ge, Sn, Ph, or Mn; A=Sb or Bi; and X=Se or Te): The tetradymite type XA_2B_4 compounds have septuple layers (SL) of B(1)-A(1)-B(2)-X-B(3)-A(2)-B(4) which crystalizes in the rhombohedral structure with point group D^5_{3d} and the space group D^5_{3d} (#166) [52–55]. The symmetry is similar to the tetradymite compounds. The SL are bonded by van der Waals forces which allows for exfoliation. The crystal structure is shown in Fig. 9.11 along with the band structure of 1 through 4 SL of $MnBi_2Te_4$. At the time of the writing of this chapter, information about the characterization of the topological properties of $MnBi_2Te_4$ was available. Classifying the topological phase of $MnBi_2Te_4$ is challenging because it displays different magneto-transport behavior depending on the number of SLs. Due to the close structural relationship with the tetradymite compounds, the tetradymite compounds are described here in Sect. 9.4.

$MnBi_2Te_4$ is believed to be the first reported antiferromagnetic (AFM) topological insulator [53]. Because the Mn^{+2} $3d^5$ electrons are all in a single layer, the spin alignment and thus magnetic moment of the SLs interact [53]. Several theoretical studies have shown that alignment of the spins so that the magnetic moment is normal to the plane of the Mn^{+2} ions has lower energy than alignment along the plane of the Mn^{+2} ions [53, 54]. The spins of one SL interact with the spins of the adjacent SL through long range Anderson exchange [53] resulting in anti-parallel magnetic moments. For example in a two SL material the magnetic moments compensate each other resulting in an anti-ferromagnetic material [53, 54]. Thus, the materials properties and crystal symmetry depend on the presence of an even or odd number of SLs in the sample. The unit cell of ferromagnetic (FM) $MnBi_2Te_4$ is one SL, and the unit cell of antiferromagnetic (AFM) $MnBi_2Te_4$ consists of two SL [54]. An inversion center exists at a Mn site in the crystal. This effect is also shown in Fig. 9.11 as an inset with the electronic band structure for 1 through 4 SL thick examples. The relative alignment of the magnetic moment for even layer samples explains the antiferromagnetic properties of $MnBi_2Te_4$ [53].The experimental observation of axion insulator and Chern insulator states was done using a 6 SL thick sample of $MnBi_2Te_4$ [55]. A single SL would be topologically trivial. Band inversion occurs at

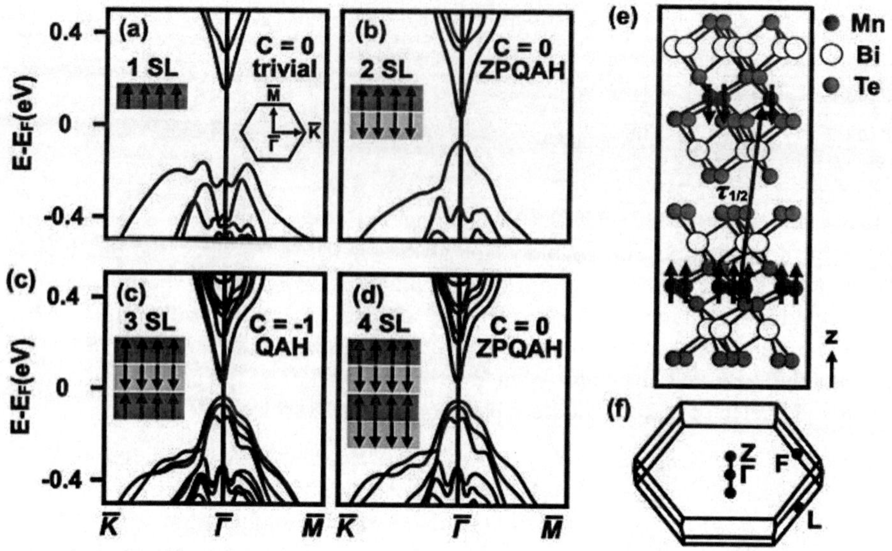

Fig. 9.11 The septuple layer structure of $MnBi_2Te_4$ and the SL dependent electronic band structure are shown. The compensating nature of the magnetic moments between layers from the Mn^{+2} $3d^5$ electron spins is shown for 1–4 SLs in **(a)**–**(d)**. The Chern number C is shown along with the expected quantum anomalous Hall (QAH) state. ZPQAH refers to the "zero plateau QAH state. Hall characterization of $MnBi_2Te_4$ is shown in Fig. 9.27. The ZPQAH is predicted for the 4 and 6 SL thick materials and the QAH state for the 5 and 7 SL thick materials [54]. The SL layer structure is shown in **(e)**. The Brillouin zone is shown in **(f)**. Note that the F points are in the center of the smaller side rectangels and the L points are in the center of the larger side rectangles. Figure adapted and reprinted with permission from [53, 54]. © 2019 American Physical Society

the Γ point in a similar manner as with the Bi_2Se_3 where the Bi one of the $6p_z$ orbital moves below the Fermi level and a Te $5p_z$ orbital moves above the Fermi level [54].

The time reversal invariant momenta (TRIM) can be found at $\Gamma\,(0,\,0,\,0)$, $L\,(0,\,0,\,\pi)$, $F(\pi,\,\pi,\,0)$, $Z(\pi,\,\pi,\,\pi)$ where π represents the end of the Brillouin zone along that direction. The $\mathbb{Z}_2$ invariant was considered in terms of a 2D crystal, and it was calculated using the Γ and 3 F points in the Brillouin zone for bulk crystals [54]. The calculated value of $\mathbb{Z}_2 = 1$ was reported which we interpret as $\upsilon = 1$, indicating that bulk $MnBi_2Te_4$ is topologically non-trivial [54]. In Sect. 9.4.3 data showing that the presence of a magnetic field drives a quantum phase transition from an axion insulator to a Chern insulator is shown.

Silver chalcogenide β-Ag_2Te β: β-Ag_2Te is reported to be a topological insulator with protected surface states [56]. β-Ag_2Te is monoclinic with the space group $P2_1/c$ (#14) [56], and this structure is described as a distorted anti-fluorite structure [56, 57]. At temperatures above ~418 K, Ag_2Te has the α-Ag_2Te anti-fluorite structure [56, 57]. When viewed normal to the plane formed by [010] axis and [100] axis, the tri-layer structure of alternating layers of Ag–Ag–Te then Te–Ag–Ag can be seen in Fig. 9.12 along with the crystal structure and Brillouin zone. The Dirac cone

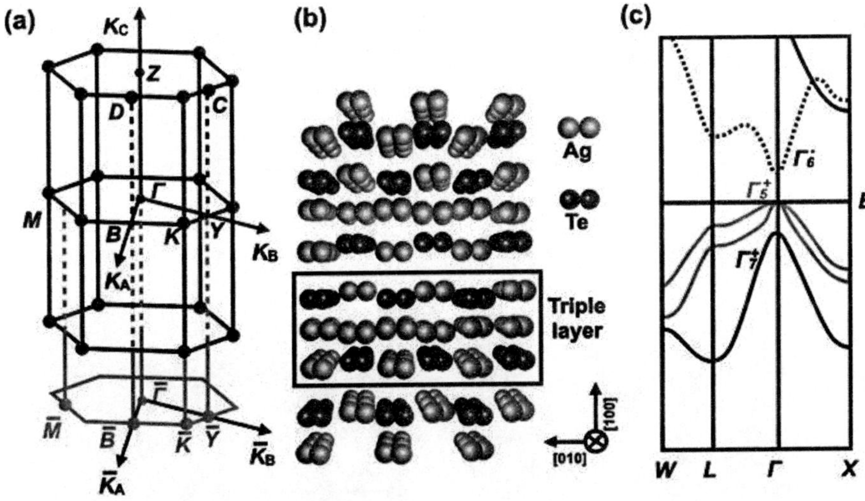

Fig. 9.12 Brillouin zone, triple-layer structure, and band diagram of β-Ag_2Te. The projection of the 3D Brillouin zone onto the 2D surface states shows the distorted structure as the wavevectors along the two axes A and B are not perpendicular. The triple-layer structure Ag–Ag–Te then Te–Ag–Ag is shown in (**b**). The band structure shows the gap where the surface states form. (**a**), (**c**) Adapted and reprinted from [56]. © 2011 American Physical Society. Figure 9.1(**b**) adapted and reprinted with permission from [58]. © 2012 American Chemical Society

is reported to be highly anisotropic. Quantized steps in transverse (Hall) resistivity are not seen in the magnetoresistance data. Instead, a non-saturating longitudinal magnetoresistance is observed [58]. The observation of the Aharanov–Bohm (AB) oscillation is provided as proof of the topological surface states [58].

Transition Metal Pentatelurides; A Large Gap Quantum Spin Hall Insulator: The phase of transition metal pentatelurides has been predicted to be a Weyl semimetal, a 3D Dirac semimetal, and for monolayers a 3D quantum Hall spin insulator [25]. The transition metal pentatelurides of Zr and Hf are orthorhombic layered materials with the Cmcm (#63) (D_{2H}^{17}) space group as shown in Fig. 9.13 [25]. The local bonding configuration of the transition metal atom is trigonal prismatic, and the transition metal atoms do not lie in a plane as do the transition metal atoms in trigonal prismatic TMD. The layer by layer structure of these materials allows exfoliation. Since these materials are metal alloys, the bonding is intermetallic. The bulk crystal has a band gap at or close to the Γ point, and the exact location in the Brillouin zone depends on the lattice constants used in the calculation [25]. Band structure calculations show that $ZrTe_5$ is an exception to the rule in that the band inversion is not due to spin orbit coupling, but to the symmetry of the material. A single layer of $ZrTe_5$ or $HfTe_5$ has space group is $P_{mmn}\left(D_{2h}^{13}\right)$ (#59) with an inversion center that is not located at the origin but at (¼, ¼). The inversion center is shown in Fig. 9.13d. Single layer $ZrTe_5$ has an indirect band gap which is negative in the sense that energy of the lowest occupied conduction band energy lies below that of the highest occupied valence

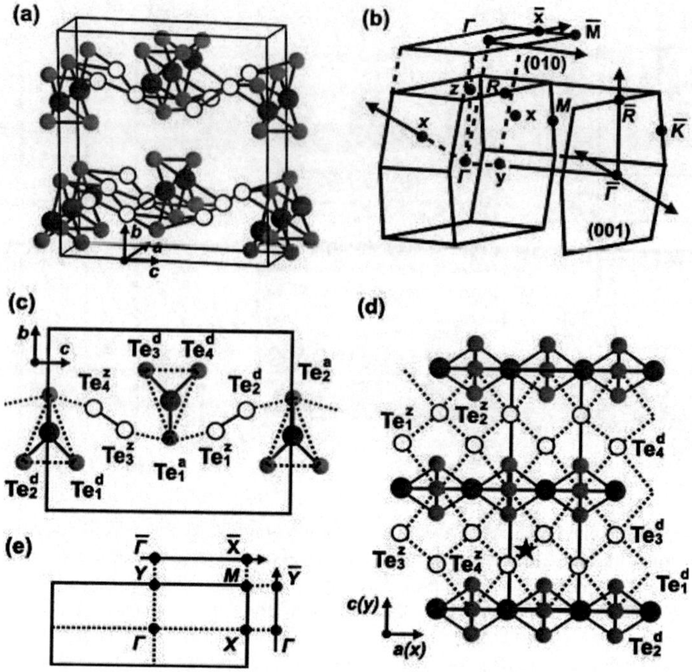

Fig. 9.13 Crystal structure of the transition metal pentatelurides $ZrTe_5$ (a = 0.39876 nm, b = 1.4505 nm, and c = 1.32727 nm) and $HfTe_5$. The layered structure is shown in (**a**). The Brillouin zone is shown in (**b**), and the trigonal prismatic bonding of the Zr atoms is shown in (**c**). The inversion center is shown as a black star in (**d**). The top plane of the Brillouin zone shown in (**b**) is highlighted in (**e**). The trigonal prismatic chains of Zr are shown in (**a**) and (**c**). Figure adapted and reprinted from [25]. © 2014 H. Weng, X. Dai, and Z. Fang. CC BY 3.0 https://creativecommons.org/licenses/by/3.0/

band energy [25]. TRIM for single layer $ZrTe_5$ occurs at (π, 0) and (π, π) in pairs with the opposite parity. In addition, the inclusion of spin orbit interactions results in some theoretical band structure models showing an indirect band gap. The single layer and bulk electronic band structures are said to be similar. The electronic states that exist at the band gap are shown in Fig. 9.14. The band gap of these materials is known to depend on the crystal growth method and conditions.

The TRIM for bulk $ZrTe_5$ at Γ, X, Y, and M were used to calculate the $\mathbb{Z}_2$ index which has been reported as (1, 001) [25] and (1, 110) [59]. Experimental atomic coordinates were used [59] in the DFT calculations of the band structure. When the

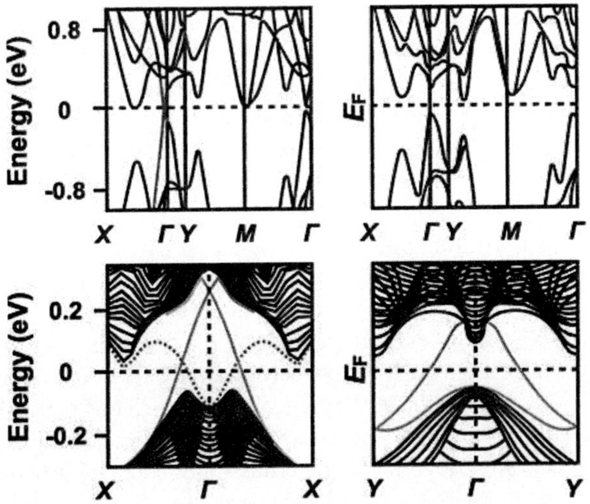

Fig. 9.14 The electronic band structure of $ZrTe_5$. The left top pane shows the bulk band structure without spin orbit coupling and the right pane with spin orbit coupling. The bottom panes show the band structure at the x and y faces of the Brillouin zone shown in Fig. 9.13(**c**) with the topological gap states shown in dashed black and gray. The dashed black lines are for prismatic termination and the gray for zigzag termination. Figure adapted and reprinted from [25].

interlayer spacing was increased in this DTF study, the $\mathbb{Z}_2$ index became (0, 110) [59]. The $\mathbb{Z}_2$ index for single layer $ZrTe_5$ uses the 4 relevant TRIM. Since the TRIM for single layer $ZrTe_5$ at $(\pi, 0)$ and (π, π) occur in pairs with the opposite parity, $\mathbb{Z}_2 = 1$ was reported in Appendix B [25]. Although the stated value of $\mathbb{Z}_2$ is -1, here, we interpret this as meaning that $v = 1$ which results in $(-1)^v = -1$.

9.4.2 Optical Properties of 3D Topological Insulators and Large Gap Quantum Spin Hall Insulators

Tetradymite Compounds (Bi_2Se_3, Bi_2Te_3, $Bi_2Se_xTe_{3-x}$, Sb_2Te_3 and alloys): Tetradymite materials are optically biaxial. The direct band gap of Bi_2Se_3 is temperature dependent, and the 300 meV value occurs at 10 K, however, it lowers as the temperature increases. Two values have been reported at 77 K; 160 meV and 35 meV [51]. The room temperature indirect band gap for Bi_2Te_3 is 130 meV and for Sb_2Te_3 is 210 meV [51]. As discussed in Chap. 1, determination of both the ordinary and extraordinary parts of the dielectric function is possible for bulk samples, but it is difficult for single QL samples. In addition, the optical response of single and few

QL samples will be dominated by the ordinary dielectric function. The optical properties of topological insulators can be separated into two regions. The Dirac fermions are low energy excitations and are observed in Infra-Red range. Visible wavelength transitions are related to the bulk electronic band structure. The low energy optical conductivity of topological insulators is expected to have some contribution from the topologically protected surface states, but to date, it has not been observed. As mentioned above, there is only one Dirac cone contribution to the surface interband optical conductivity for topological insulators. The conductivity needs to be corrected for contributions from the Drude response from the 2DEG when it is present [46, 60, 61]. The position of the Fermi level can be changed by unintentional doping [46, 60–62], intentional doping by Ca [60], and careful growth [62]. In most samples, the Fermi level of Bi_2Se_3 crosses the bulk conduction band and the Fermi level of Bi_2Te_3 crosses the bulk valence band [46]. Thus the bulk conductivity interferes with the observation of surface states in the optical conductivity of Bi_2Se_3 and Bi_2Te_3. Defects often dope these materials unless special growth methods are used [62]. However, Bi_2Se_2Te is noteworthy because the Fermi level between the bulk valence and conduction bands [46]. The reflectivity and optical conductivity of Bi_2Se_3, Bi_2Se_2Te, $Bi_{2-x}Ca_xSe_3$, and Bi_2Te_2Se are shown in Fig. 9.15. The biaxial nature of Bi_2Se_3 is evident from the reflectivity data shown in Fig. 9.16. The low temperature and low energy optical properties of the Bi_2Se_2Te are shown in Figs. 9.17 and 9.18 along with the complex dielectric function. Bi_2Se_2Te. This material is noteworthy because it should not be subject to defect induced doping. The Fermi level for Bi_2Se_2Te lies between the bulk valence and conduction bands [46].

The complex refractive index of Bi_2Se_3 in the visible wavelength range has a peak in ε_2 that is associated with four different joint densities of states as shown in Fig. 9.19. These include $F_{1v} \rightarrow F_{1c}$; $\Gamma_{3v} \rightarrow F_{1c}$; $L_{1v} \rightarrow L_{2c}$; and $Z_{1v} \rightarrow Z_{1c}$ [43,

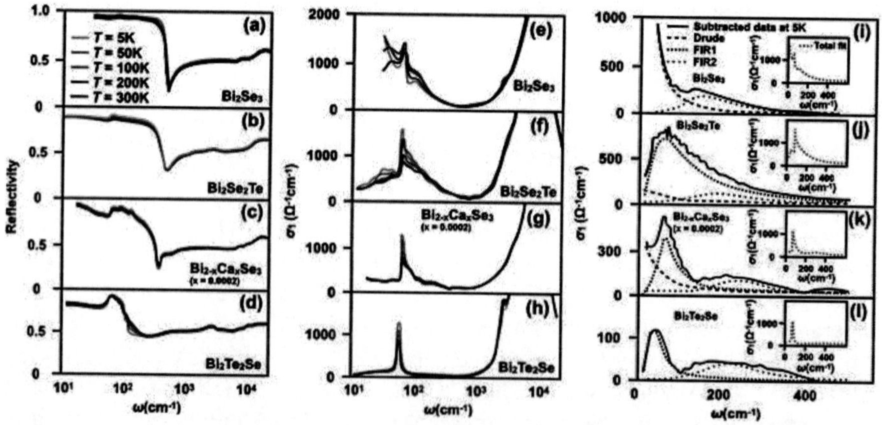

Fig. 9.15 The reflectivity and optical conductivity of four tetradymite topological insulators. The α and β IR active phonons are identified in the middle section. The optical conductivity data at the far right shows the Drude response. Figure adapted and reprinted with permission from [60]. © 2012 American Physical Society

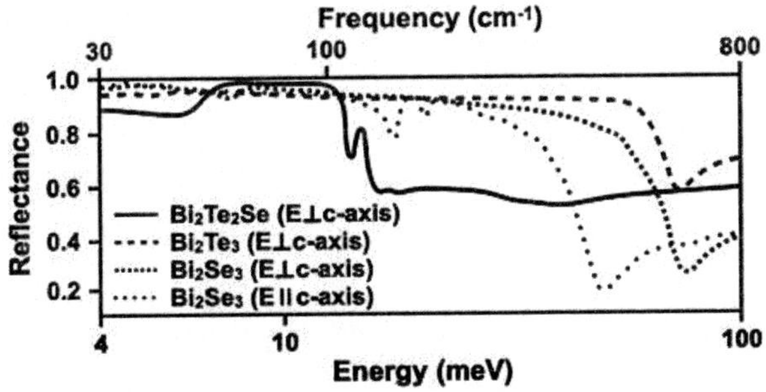

Fig. 9.16 The low energy, polarized reflectance of Bi_2Se_3, Bi_2Te_3, and Bi_2Te_2Se. Figure adapted and reprinted with permission from [46]. © 2014 American Physical Society

61, 63]. Since the surface of the tetradymite compounds oxidizes after a few hours, the complex refractive index of the oxide layer is shown in Fig. 9.19. It is important to note that the value of the ellipsometric angle Δ changes before any oxide is observed at the surface by X-Ray Photoelectron Spectroscopy [43]. The seems to be due to changes in the surface morphology, or possibly Bi segregation to the surface [43].

The complex refractive index of Bi_2Se_3, Bi_2Te_3, and Sb_2Te_3 have been calculated using the full potential linearized augmented plane wave (FP LAPW) method which is based on DFT, and they are shown in Fig. 9.20 [64]. Single crystal samples of Bi_2Se_3, Bi_2Te_3, and Sb_2Te_3 were grown for this study, and FT-IR was used to determine the plasmon frequency. Hall measurements found that the Bi_2Se_3 was n doped with $n = 2.5 \times 10^{19}\,\text{cm}^{-3}$ and FT-IR measured a plasmon frequency of $750\,\text{cm}^{-1} \approx 0.093\,\text{eV}$. Hall measurements and room temperature FT-IR spectroscopy determined that these samples of Bi_2Te_3 and Sb_2Te_3 were p doped with $p = 1.2 \times 10^{19}\,\text{cm}^{-3}$ and $8.1 \times 10^{19}\,\text{cm}^{-3}$ respectively and plasmon frequencies of $306\,\text{cm}^{-1} \approx 0.038\,\text{eV}$ and $1072\,\text{cm}^{-1} \approx 0.133\,\text{eV}$ respectively. In addition, the temperature dependence of the complex refractive index of Sb doped Bi_2Se_3, with the composition $(Bi_{0.8}Sb_{0.2})_2Se_3$, shows a plasmon peak between 0.9 and 1 eV below 250 K [65].

Raman spectroscopy measurements of the tetradymite compounds were discussed in Sect. 1.9.5.2 and shown in Chap. 1, Fig. 1.18. The $2A_{1g}$ and $2E_g$ are Raman active, and the $3A_{2u} + 3E_u$ are Infra-Red active. In few QL samples the forbidden A_{1u} peak is observed in non-resonant Raman scattering [66].

Tetradymite Type Compounds (XA_2B_4; X = Ge, Sn, Ph, or Mn; A = Sb or Bi; and X = Se or Te): Here we report on the optical properties of $MnBi_2Te_4$ which is reported to have topological electrical transport properties as discussed in the section below. The complex dielectric function of the $MnBi_2Te_4$ has been determined using spectroscopic ellipsometry [67] as shown in Fig. 9.21. $MnBi_2Te_4$ is uniaxial, and it has a different dielectric function along and perpendicular to the optical axis. This study found that when the ellipsometric data was measured with the optical (c) axis normal to the surface and the resulting dielectric function was modeled using data from difference angles of incidence, it was not sensitive to the angle

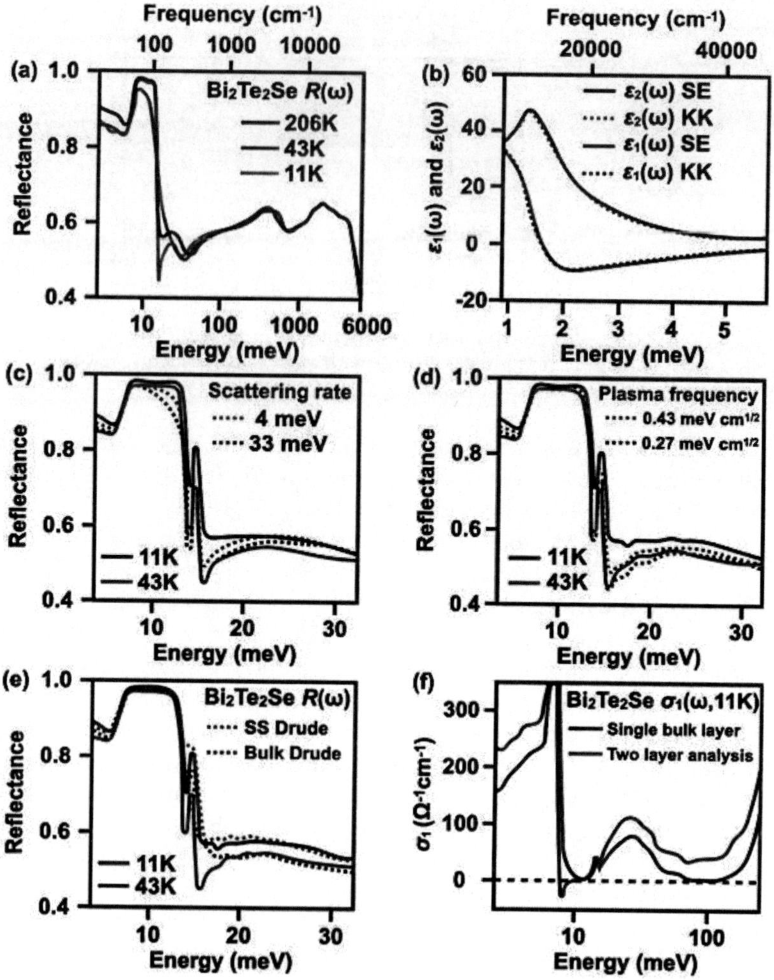

Fig. 9.17 The low energy reflectance, visible wavelength complex refractive index, and optical conductivity of Bi_2Te_2Se. (**a**) Reflectance data over a wide energy range from 1 meV to 6 eV. (**b**) shows the complex dielectric function at room temperature. (**c**), (**d**) show the use of the Drude model to determine scattering rate and plasma frequency from reflectance data. (**e**) shows the relative contribution of the surface and bulk layers when compared to a Drude model. (**f**) shows the optical conductivity at 11 K from a two layer model versus a single bulk layer. The surface layer contribution is modeled by a Drude response. Figure adapted and reprinted with permission from [46]. © 2014 American Physical Society

of incidence [67]. Under these conditions, the experimental determined dielectric function is representative of the dielectric function normal to the optical axis as discussed in Chap. 1.

Transition Metal Pentatelurides: Determining the topological nature of $ZrTe_5$ is challenging. First, we discuss the anisotropic, optical properties of biaxial $ZrTe_5$. The

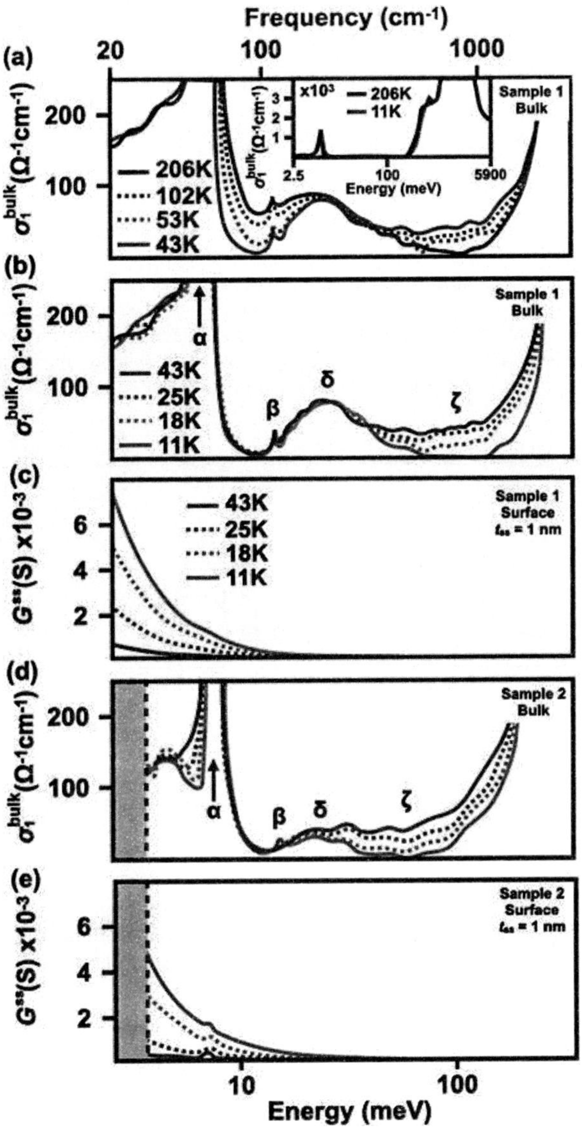

Fig. 9.18 The optical conductivity of the bulk versus surface layer of Bi_2Te_2Se at a variety of temperatures for two different samples. A number of phonon modes are labeled in the spectra. The G^{ss} conductivity comes from a Drude model for the surface layer. We note that the expected temperature dependence is an increase in the Drude response for an electron gas with temperature and not the decrease seen here in (**e**). This implies that the Drude response is from surface states that are suppressed as the temperature increases. Figure adapted and reprinted with permission from [46]. © 2014 American Physical Society

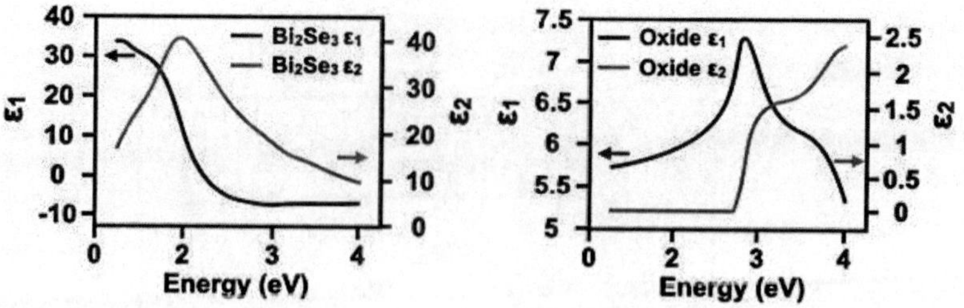

Fig. 9.19 The complex dielectric function of exfoliated Bi_2Se_3 in the visible wavelength range is shown on the left. The dielectric function of the oxide layer after an oxide is allowed to grow on exfoliated Bi_2Se_3 is shown on the right. The pseudo dielectric function of Bi_2Se_3 grown by molecular beam epitaxy can be found in [63]. The sample was oriented with the c axis normal to the surface. Figure adapted and reprinted from [43] with the permission of AIP Publishing

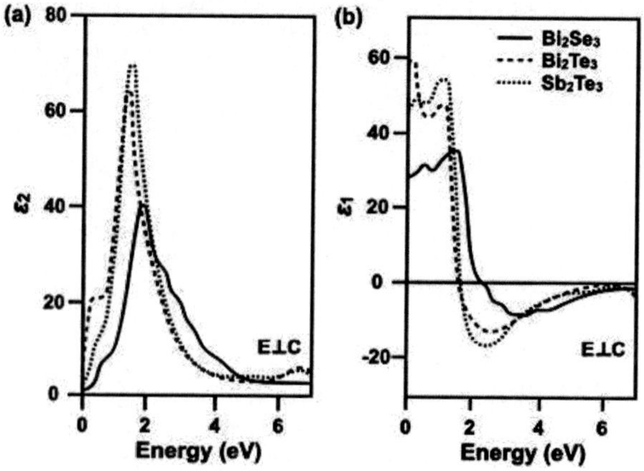

Fig. 9.20 Theoretical complex dielectric functions for Bi_2Se_3, Bi_2Te_3, and Sb_2Te_3. The transverse electric field samples the in-plane optical response of the material (ordinary dielectric function. Figure adapted and reprinted from [64] with the permission of AIP Publishing

dielectric tensor of $ZrTe_5$ in the visible wavelength range and birefringence in the $a\!-\!c$ plane demonstrate optically biaxial nature of the crystal structure shown in Fig. 9.22 [68]. The dielectric tensor was characterized by Mueller Matrix spectroscopic ellipsometry [68]. Here specifying the sample orientation and light beam with respect to the a, b, and c axes are critical to defining the optical spectra of $ZrTe_5$. The band gap of $ZrTe_5$ is small $\sim\leq 10$ meV and sample dependent which means that above 70 K the band gap is effectively zero [69]. The band gap of $ZrTe_5$ cannot be determined from the optical conductivity due to Pauli blocking [69]. Paul blocking occurs when the final states in an optical transition are filled, and the Pauli exclusion principle prevents the transition to those states. The band gap for $ZrTe_5$ can be determined using magneto-optical transmission measurements [69, 70].

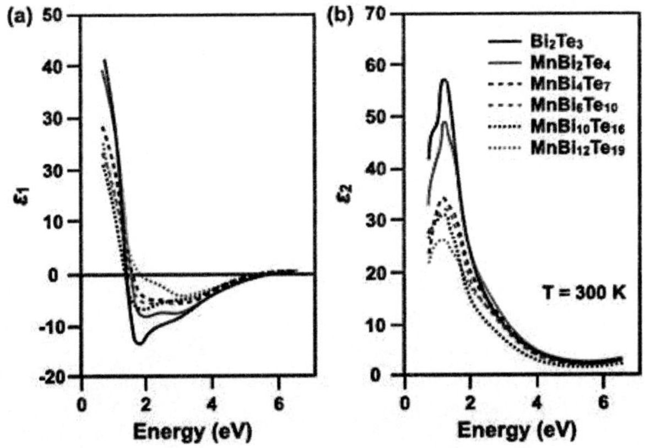

Fig. 9.21 The experimental dielectric function for $MnBi_2Te_4$ and related compounds determined using spectroscopic ellipsometry. Figure adapted and reprinted from [67] with the permission of AIP Publishing

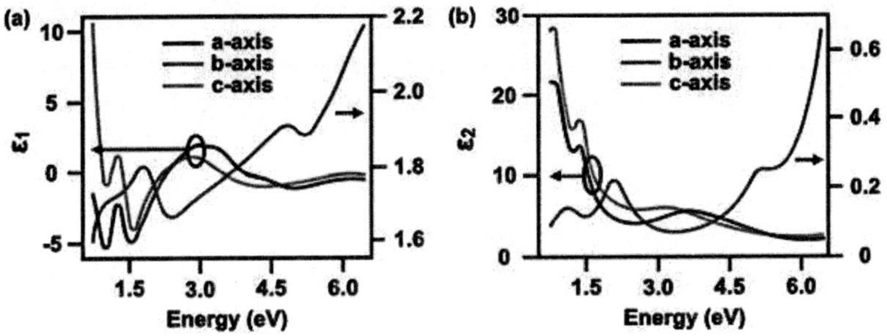

Fig. 9.22 The dielectric tensor along the *a* and *c* axis, and birefringence of anisotropic $ZrTe_5$ is shown. Figure adapted and reprinted from [68] with the permission of AIP Publishing

In Sect. 9.4.3, the temperature-dependent resistivity of $ZrTe_5$ is discussed in more detail. This data often shows a peak in resistivity T_0 along the *a* or *c* crystal axis. The temperature of this peak depends on the crystal growth method and conditions. In order to distinguish samples produced by different growth methods or different growth conditions for a single method, we specify T_0 for each sample when possible.

Next, is useful to consider that the reflectance spectra from $ZrTe_5$ exhibit significant sample dependence [68, 70–73]. The source of these differences has been attributed to differences in carrier concentration due to impurities and in some cases defects. The infra-red optical conductivity obtained from reflectivity data has been studied for evidence of a Weyl or Dirac semimetal.

First we discuss the reflectivity and optical conductivity of single crystal samples grown using the Te flux method and having a $T_0 \sim \pi$ 60 K along the *a* axis. Temperature dependent reflectivity data are shown in Fig. 9.23. The data in Fig. 9.23(a), (c) represent the reflectivity when the electric field of the light is polarized along the *a*

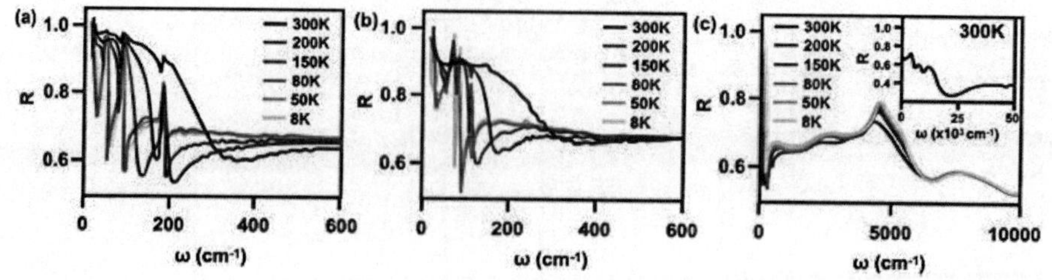

Fig. 9.23 The temperature dependent reflectivity for $ZrTe_5$. Figure (**a**) shows the reflectivity for light polarized along the a axis and (**b**) for light polarized along the c axis. Both (**a**) and (**b**) show the low energy reflectivity observed for wavenumbers below 600 cm^{-1}. (**c**) Shows the reflectivity to 10,000 cm^{-1} for light polarized along the a axis. Figure adapted and reprinted with permission from [70]. © 2015 American Physical Society

axis and the data in (b) is the reflectivity for light polarized along the c axis. This study found a linear increase in the optical conductivity in the $a{-}c$ plane with frequency, $\sigma_1(\omega) \sim \omega$ and concluded that $ZrTe_5$ is a Dirac Semimetal. We recall that for a 3D conical band (Dirac cone), the expected real part of the optical conductivity is given as a function of the number of Dirac cones N, $\sigma_1(\omega) = N\omega e^2/(12hv_F)$ where h is Plank's constant and v_F is the Fermi velocity. However, the optical conductivity along the b axis needs to be considered. Next, we discuss research that demonstrates the anisotropic nature of the low energy optical conductivity.

The anisotropic nature of the low energy optical conductivity in the $a{-}c$ plane versus along the b axis has been measured using the same experimental conditions for two different types of samples [71, 72]. Two growth methods were used in [71], a self-flux crystal growth [$T_0 = 80$ K, *band gap* ~ 6 meV, $n = 3 \times 10^{16}$, $\mu_A = 0.45 \times 10^6\ cm^2/(Vs)$] and chemical vapor transport ($T_0 = 145$ K, *band gap* $\sim$ 10 meV) [71]. The real part of the optical conductivity of metallic samples of $ZrTe_5$ with low carrier concentration have a $\sqrt{\omega}$ dependence along both the a and c crystallographic axis directions and thus show a 2D conical response as shown in Fig. 9.24. The real part of the optical conductivity along the b axis has a $\omega^{3/2}$ dependence [71–74]. This data was interpreted as evidence that the optical properties are not due to a 3D Dirac cone [71]. The same samples used in [71] were found to have anisotropic resistivity, and it was found that from ~0 to 300 K, the resistivity along the b axis is ~80 times larger than along the a axis. In addition, the resistivity along the c axis is ~2 times that along the a axis [72]. This is consistent with the differences in the optical conductivity between the a and b axes. Two theoretical studies have concluded that the real part of the optical conductivity of $ZrTe_5$ in both the band insulator phase and Weyl semimetal phase will have a $\sigma_1(\omega) \sim \sqrt{\omega}$ dependence in the $a{-}c$ plane and a $\sigma_1(\omega) \sim \omega^{3/2}$ dependence along the b axis. Distinguishing between these two phases using optical methods is very difficult [73–75]. One theoretical study found that the magneto-optical conductivity can be used to determine whether the phase of $ZrTe_5$ is a gapped insulator or a Weyl semimetal [73]. Reference [74] notes that at very low doping, dc transport can be used to distinguish between the Weyl semimetal and band insulator phases. gives different ratios of the a axis to b axis resistivity for the

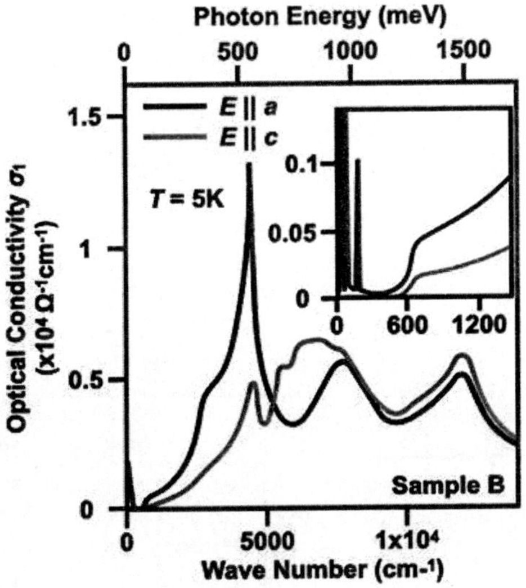

Fig. 9.24 The real part of the optical conductivity of $ZrTe_5$ below 15,000 cm^{-1}. The electric field of the light lies in the $a-c$ plane either parallel to the a or c axis. The low temperature, low energy conductivity in the $a-c$ plane is proportional to $\sqrt{\omega}$. The exact functional dependence depends on the direction of the electric field as shown in the inset. The optical conductivity along the b axis has a linear dependence on frequency, $\sim\omega$. The interpretation of this data is discussed in the text. The data for sample A shown in this figure has an optical band gap of 28 meV, a band gap 6 meV as determined using magneto-optical transmission, a low carrier concentration, $n = 3 \times 10^{16}\,cm^{-3}$, and a high mobility $\mu_{Hall} = 0.6 \times 10^6\,cm^2/Vs$. Sample B has an optical band gap of 74 meV. Figure adapted and reprinted with permission from [71]. © 2019 American Physical Society

Weyl $\left[R_{a-b} \sim v_{F(a\,axis)}/v_{F(b\,axis)}\right]$ and gapped $\left[R_{a-b} \sim n^{-2/5}\right]$ cases. Achieving the necessary low carrier concentration required to enhance $R_{a-b} \sim n^{-2/5}$ signal for the gapped case is difficult. Below, we discuss low temperature (0.6 K) Hall characterization of a sample with $n < 0.5 \times 10^{17}\,cm^{-3}$, that has been interpreted as showing an anisotropic 3D Dirac quantum Hall effect. The measured Fermi surface is highly anisotropic with a real space period along the b axis much larger than the lattice constant for b. This discussion points to the challenges in interpreting optical data in terms of a topological response. We also note that $ZrTe_5$ was predicted to be a quantum spin Hall insulator at low temperature [70] which emphasizes the difficulty in designating the topological phase of $ZrTe_5$.

The highly anisotropic crystal structure of $ZrTe_5$ results in Raman spectra that are dependent on crystallographic direction. The 12 atoms in the unit cell results in 12 $\times$ 3 = 36 phonon modes at the Γ point in the Brillouin zone [76, 77]. 18 of these modes are Raman active. Of the 18 Raman active modes, the A_g and B_{2g} modes are detected [76, 77] using the backscattering geometry discussed in Chap. 1. The room temperature Raman spectra and the azimuthal angular dependence of some of the A_g and B_{2g} modes are shown in Fig. 9.25. The low temperature carrier concentration

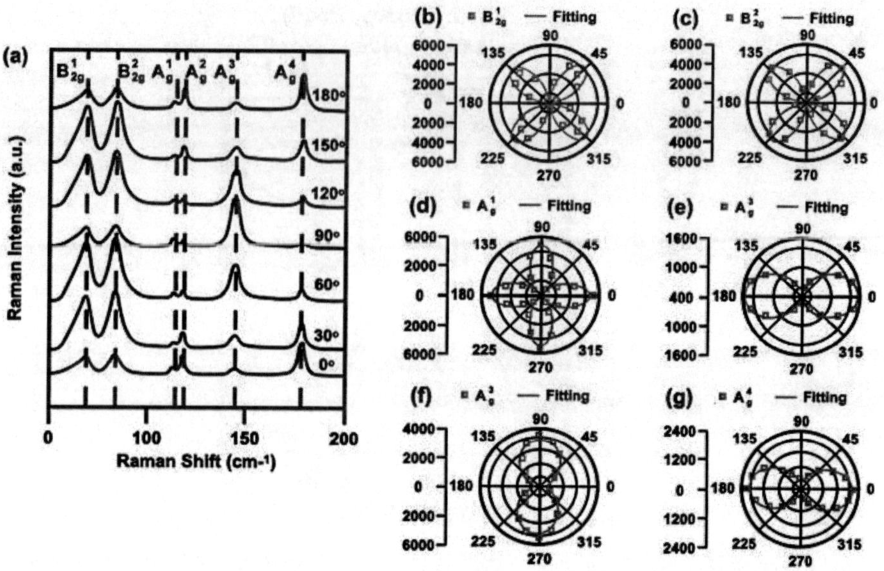

Fig. 9.25 The room temperature Raman spectra of several layer $ZrTe_5$ obtained 633 nm He–Ne laser incident along the b axis and polarized parallel to the a axis resulting in the $ac(aa)ac$ configuration for 0°. The sample was rotated away from this configuration while keeping the beam normal to the (010) surface. One of the A_g modes is beyond the spectral range of the system used to obtain the spectra. Figure adapted and reprinted from [76]. © 2016 American Chemical Society

and the peak in resistivity of the samples used in this study were $n \sim 10^{13}$ cm^{-3} and ~130 K respectively [76].

9.4.3 Electrical Properties of 3D Topological Insulators and Large Gap Quantum Spin Hall Insulators

Tetradymite Compounds (Bi_2Se_3, Bi_2Te_3, $Bi_2Se_xTe_{3-x}$, Sb_2Te_3 and alloys): 3D topological insulators are candidate materials for spintronic applications [39, 78]. Despite this promise, experimental observation of the topologically protected electronic transport is difficult due to the signal from bulk carriers [51]. The observation of the quantum anomalous Hall effect (QAHE) in 3D topological insulators was enabled by doping tetradymite TI materials with Fe or Mn [39, 51]. As shown in Fig. 9.26, quantized conduction (or equivalently, quantized Hall resistance) is not observed for Bi_2Se_3. The presence of the topological surface states was confirmed by ARPES (angle resolved photoemission spectroscopy) as shown in Fig. 9.10.

Tetradymite Type Compounds (XA_2B_4; X = Ge, Sn, Ph, or Mn; A = Sb or Bi; and X = Se or Te): Here we discuss the magneto-transport data for $MnBi_2Te_4$ which has

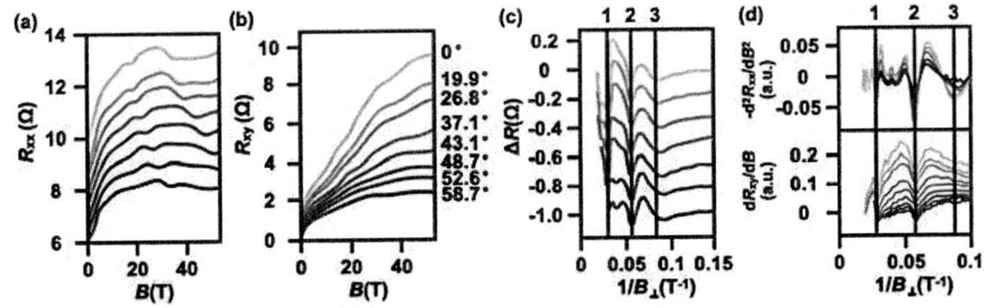

Fig. 9.26 Hall characterization of Bi_2Se_3 samples with a carrier concentration of 4×10^{16} cm^{-3} at 1.5 K is shown. The low carrier concentrations were achieved by substituting small amounts of Sb for Bi. *The presence of bulk carriers makes observation of quantized conductance from the topological surface states difficult.* The longitudinal resistance R_{xx} is shown in (**a**). The data is obtained with the orientation of the magnetic field with respect to the sample at several θ angles ($B_\perp = B\cos\theta$) as shown in (**b**). The linear response of the resistivity below 4 T is indicative of the 3D quantum limit. In (**c**), the R_{xx} data from (**a**) is fit to a polynomial providing a background signal that is subtracted from each curve. This emphasizes the oscillatory behavior of resistivity. In (**d**) the second derivative of the longitudinal resistivity with respect to the magnetic field strength and in (**e**) the second derivative of the transverse resistivity versus the magnetic field strength are shown. The vertical lines in (**c**), (**d**), and (**e**) show the first three Landau levels. Figure adapted from [78]

been reported to have topological electrical properties. Typical flux growth methods are used to obtain single crystal samples [55, 79, 80]. As discussed in Sect. 9.4.1, even numbers of SLs allow for the magnetic moment of Mn^{+2} layers in adjacent SLs to compensate an thus be antiferromagnetic while odd numbers of SLs do not.

First we discuss the characterization of 6 SL and bulk single crystal $MnBi_2Te_4$. Here we recall the earlier statement that sample variability between studies of other topological materials produced reports of different topological phases. The samples in this study were grown by reaction of a 1:1 mixture of high-purity Bi_2Te_3 and MnTe in a sealed silica ampoule under a continuously pumped vacuum [55]. The mixture was first heated to 973 K then slowly cooled down to 864 K where it was annealed for a prolonged period. Scotch tape exfoliation was used to obtain flakes with a few SLs. The 6 SL sample in this study displays the axion insulator state at zero magnetic field and the Chern insulator phase in the presence of a magnetic field [55]. The properties were tuned using a gate to apply a voltage bias to the transport channel. Samples can be distinguished by the Neil temperature which indicates the transition temperature for antiferromagnetic ordering. The 6 SL sample used in this study had the same Neil temperature as bulk $MnBi_2Te_4$ $T_N = 25$ K, and an electron density and mobility of $n = 1.2 \times 10^{20}$ cm^{-3} and $\mu = 74$ cm^2/(Vs) *at* 1.6 K [55]. The low temperature 1.6 K magneto-transport data shows a large longitudinal resistance at the zero transverse (Hall) resistivity plateau at gate voltages between 22 V *and* 30 V in Fig. 9.27(a). Here the resistivity is plotted as a function of magnetic field normal to the sample surface. The large longitudinal resistivity in conjunction with the zero Hall resistivity plateau are signatures of the axion insulator state [55]. The data from (a) is replotted to show the resistivity versus gate voltage at zero magnetic field in

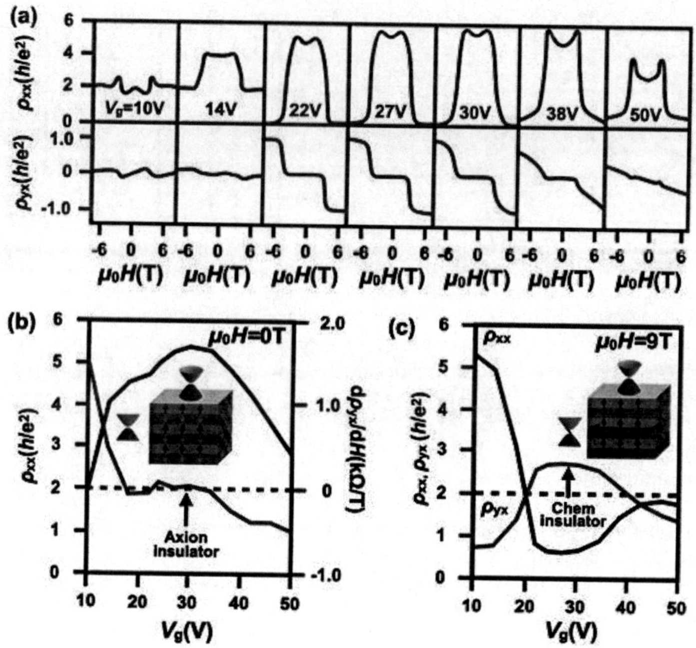

Fig. 9.27 The Hall transport measurements for 6 septuple layer $MnBi_2Te_4$ show quantized steps in the transverse (Hall) resistivity at 1.6 K. The gate voltage determines the carrier density, and the Fermi level lies in the bulk band gap between 33 and 30 V. Figure adapted from [55]

Fig. 9.27(b). The resistivity versus gate voltage is shown at 9 T in (c). At a magnetic field of 9 T for the same range of gate voltages, steps in the quantized transverse Hall resistance of h/e^2 are observed while the longitudinal resistivity is zero, and the sample is in the Chern insulator phase [55]. This shows the conditions under which an axion or Chern insulator is observed. The sample used in this study was thought to be electron doped [55].

Samples with an odd number of SL have also been studied, and the quantum anomalous Hall effect has been observed at 1.4 K for 5 SL samples [79]. The samples used in this study were also grown from a melt the few SL flakes were exfoliated from the resulting single crystal [79]. The SL number was determined optically using the Beer-Lambert law. The Neil temperature was function of SL number with $T_N = 23\,\text{K}$, $21\,\text{K}$, and $18\,\text{K}$ for the 5, 4, and 3 SL samples. The presence of a magnetic field aligns the magnetic moment as it sweeps from positivity to negative [79]. The Hall resistivity data for 4 SL sample [79] shows the same behavior as the 6 SL sample in [55]. A finite, SL layer number dependent, transverse (Hall) resistivity $<1\,h/e^2$ is observed for the 3 and 5 SL samples. The value changes from positive to negative depending on whether the magnetic field is swept from negative to positive or positive to negative values as the magnetic moment of the layers aligns during the sweep. There are other reports that also describe the topological behavior of 5 SL

$MnBi_2Te_4$ [80]. One study discusses the positive AHE as being due to a competition between skew scattering and the intrinsic Berry curvature (see Chap. 6) [80].

Transition Metal Pentatelurides: Here we discuss the electrical transport properties of $ZrTe_5$. As discussed above, $ZrTe_5$ is considered to be a candidate for being either a Dirac or Weyl semimetal. The temperature dependence of the resistivity of $ZrTe_5$ has been studied along all three crystallographic directions [72]. There is peak in the resistivity for all three crystallographic directions that has been attributed to charge density waves. First we discuss the work reported in [72]. Here, the transition temperature T_0 was shown to be 78 K for the sample obtained using self-flux growth method and 145 K for chemical vapor transport method. This demonstrates that the temperature at which this peak appears is dependent on the crystal growth method and growth temperature which suggests carrier density dependence [72]. A pressure dependence study of the resistivity points to a temperature dependent shift in the chemical potential which is accompanied by a switch from electron dominated conduction at low temperature to hole dominated conduction at a higher temperature [72].

The anomalous Hall effect has been observed along with the negative longitudinal magnetoresistivity [81]. One study of bulk single crystals of $ZrTe_5$ determined the angular dependence of the alignment of the magnetic field for transport properties when the current flowed along the so called chain of trigonal prismatic Zr or a axis. The temperature dependent resistivity along the Zr chain axis a of the samples used in this study [81] shows a maximum in resistivity at $T_0 \leq 5\,\text{K}$ which is a much lower T_0 than that observed for the samples used in the study discussed in [72]. The combination of the anomalous Hall effect and the negative longitudinal magnetoresistivity observed at 2.5 K is considered confirmation of Weyl nodes. This study points to differences in the ARPES spectra for samples having at $T_0 = 135\,\text{K}$ versus $T_0 \leq 5\,\text{K}$. The samples with $T_0 = 135\,\text{K}$ have two electron pockets while the samples with $T_0 \leq 5\,\text{K}$ have a single electron pocket close to the Γ point in the Brillouin zone [81]. The lack of a magnetic ordering in these samples was confirmed thus ruling out a ferromagnetic origin for the AHE. Another ARPES study of polycrystalline $ZrTe_5$ finds that the details of band structure depend on the incident photon energy 5.9 eV versus 5.64 eV which points to the influence of final state effects on the spectra [82]. This study classified $ZrTe_5$ as a weak topological insulator.

In another study shown in Fig. 9.28, the signature of the 3D quantized Hall conductivity was observed in the bulk of $ZrTe_5$ by studying the magneto-transport properties at low temperature (0.6 K) and low magnetic fields so that only the lowest Landau level is occupied [83]. The temperature dependent resistivity along the Zr chain axis a of the samples used in this study [83] shows a peak in resistivity and change in carrier type at $T_0 = 95\,\text{K}$. This is the same sort of behavior in resistivity as seen the data of [72]. When the magnetic field is normal to the surface, $\sigma_{xy} = \frac{e^2}{2\pi h} G_z$ is dependent on the reciprocal lattice vector G_z {which is the z component of a reciprocal lattice vector associated with the period of the 3D system along the magnetic field [83]} of the system, not the crystal. In this publication the x axis lies along a crystallographic axis, and z axis lies along the b crystallographic axis. The plateau of

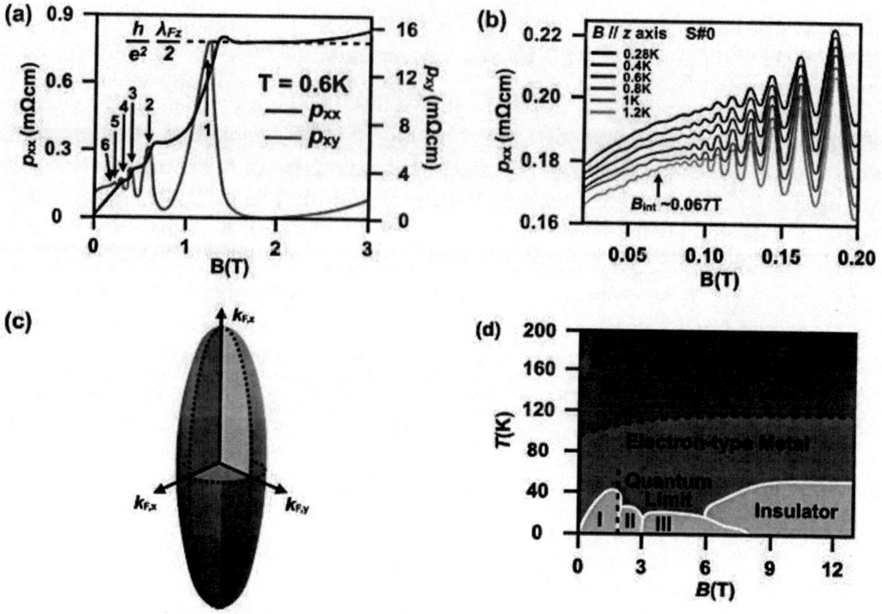

Fig. 9.28 The longitudinal magnetoresitivity and Hall resistivity of the Large Gap Quantum Spin Hall Insulators $ZrTe_5$ are shown at 0.6 K. The quantized resistivity shown in (**a**) is for a sample with and electron density of $<0.5 \times 10^{17}\,cm^{-3}$ (see Fig. 9.1(d) in [83]). The The Shubnikov–de–Haas oscillations in the longitudinal resistivity is shown in (**b**). The Fermi surface for the 3D electron gas is shown in (**c**). The surface has different vector magnitudes along a, b, and c crystallographic axes. The shape of the Fermi surface was determined using Shubnikov–de–Haas oscillations measured along the three different crystallographic directions. A phase diagram for $ZrTe_5$ is shown in (**d**). Region is thought to harbor charge density wave (CDW) or Fermi instabilities. Region II is conformed to have 3D QHE and CDW. Region III may have fractional $\upsilon = 1/3$ quantum Hall states. Figure adapted from [83]

the magneto-resistivity data provides the real space period λ_Q of system $\rho_{xy} = \frac{h}{e^2}\lambda_Q$. The plateau shown in Fig. 9.28(a) allows determination of the vector G_z for the Fermi surface of the 3D electron gas for one sample of $ZrTe_5$. The $\lambda_Q = 5.8$ nm for one of the samples studied in [83] can be compared to the lattice constant along the z direction of 1.45 nm [83]. By rotating the direction of the magnetic field at 1.5 K, the dimensions of the Fermi surface were determined along the a, b, and c axis directions. The frequency of the Shubikov-de Hass oscillations $B_{F,i}$ is related to the area in wavevector space $S_{F,i}$ by $B_{F,i} = S_{F,i}\frac{h}{2\pi e}$ where i refers to the x, y, and z direction. The S-d-H oscillations occur at very low magnetic field strengths (<0.2 T) which is considered to be due to a very high carrier mobility [83].

Each study of $ZrTe_5$ uses a different sample set with different carrier concentrations and defect densities. This might explain the study dependent differences in the Hall data for $ZrTe_5$. That suggests that each sample be distinguished by T_0, mobility, carrier concentration, and High-Resolution X-Ray rocking curves to confirm crystal quality [69]. We note that scanning tunneling microscopy studies of a polycrystalline chemical vapor transport have designated $ZrTe_5$ as a weak topological insulator on the basis of the observation of edge states [84].

9.5 Weyl, Dirac Semimetals, and Related Materials

The unusually high carrier mobility and magnetoresistance of Weyl semimetals has sparked a strong interest in these materials. The topological properties of the electronic band structure of these materials results in Weyl nodes or nodal lines in the bulk and a Fermi arc of topological surface states as described above in the introduction to Topological Materials. Unlike the hexagonal crystal structure with trigonal prismatic metal coordination that is naturally occurring for many Transition Metal Dichalcogenide monolayers, a number of different crystal structures have Weyl semimetal properties. Weyl metals can be identified through measurement of the Fermi surface using ARPES and can be inferred through their electrical and optical properties. As described in Sect. 9.3, the observation of a chiral anomaly in the presence of a parallel electric and magnetic fields ($\vec{B} \| \vec{E}$) pump is used to confirm the presence of Weyl nodes. In this configuration, the longitudinal conductivity increases with magnetic field (resistivity decreases). Combined with the negative longitudinal resistivity, the Hall conductance is a signature of a Weyl semimetal [40].

The chiral nature of the Weyl fermions can be directly probed using circularly polarized, mid-IR light by measuring the resulting photocurrent [85]. This effect is also known as the circular photogalvanic effect. The use of circularly polarized light at low energy ($\lambda = 10.6\,\mu$m, $\sim$120 meV) along the $\vec{a}$ axis of the crystal and perpendicular to the $\vec{c}$ axis excites current flow along the $\vec{b}$ axis. Weyl fermions can be probed with little or no interference from other transitions [85]. The direction of the current flow along the $\vec{b}$ axis is determined by the chirality of the Weyl node as shown in Fig. 9.29.

As discussed in Sect. 9.3, a unique property of Weyl and Dirac semimetals is a linear increase in optical conductivity with optical frequency below energies that excite interband transitions [37, 70]. We recall that in Chap. 7 the low energy excitations of Dirac fermions in graphene have a constant optical conductivity versus frequency that is related to the fine structure constant was described. The conductivity is dependent on the absence or presence of impurities resulting in free carriers and when Coulomb interactions are present [37]. When free carriers are present, a temperature dependent Drude response is observed [37, 70]. We expect the intensity of the Drude response increases with temperature. The optical conductivity $\sigma_0^N(\omega)$ of a Weyl or Dirac Semimetal with N isotropic Weyl nodes with N/2 nodes of each chirality with no impurities and no Coulomb interaction is [37, 70]:

$$\sigma_0^N(\omega) = N\frac{e^2}{12h}\frac{\omega}{v_F} \tag{9.16}$$

where h is Plank's constant, v_F is the Fermi velocity, and e is the electrical charge. When impurities are present, a Drude response is observed at low frequency. The Drude response decreases with temperature as the number of free carriers decreases. Equation (9.16) can be compared to the optical conductivity of d dimensional Dirac

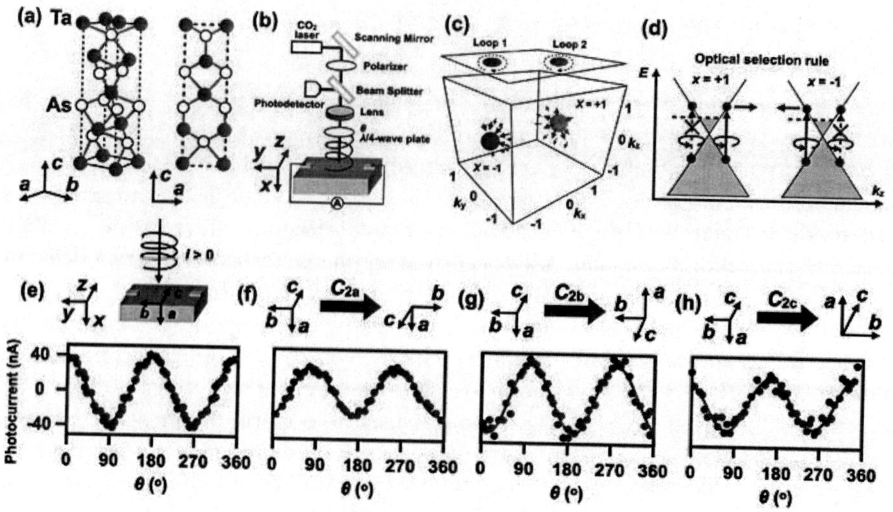

Fig. 9.29 Determination of the chirality of a specific Weyl node of TaAs through the circular photogalvanic effect. The crystal axis of TaAs are shown in (**a**). The apparatus used for reflection of circularly polarized light from and oriented single crystal is shown in (**b**). The $\chi = \pm 1$ chirality of the Weyl nodes is shown in relation to the x, y, and z axis of the wavevectors in (**c**). The optical selection rules for right handed circularly polarized (RHCP) light traveling along the $+\hat{z}$ axis are illustrated in (**d**). The RHCP excites electrons at the Weyl node with chirality $\chi = +1$ with $+k_z$ and $\chi = -1$ with $-k_z$. Figures (**e**)–(**h**) show the changes photocurrent signal due to rotating the crystal about the a, b, or c crystal axis. Figure adapted from [85]

electrons $\sigma_1 \sim \omega^{(d-2)/z}$ and $E(k)\alpha|k|^z$ as discussed in Sect. 9.3, here $z = 1$ and $d = 3$.

Quantum Spin Hall Insulators such as the transition metal pentatellurides demonstrate an anomalous Hall effect (AHE) which was introduced in Chap. 6. AHE materials have a Hall current without the presence of a magnetic field. The origin of the AHE in Weyl semimetals is strong spin orbit coupling and not the magnetic character of the material. The anomalous Hall conductivity is related to the Berry curvature $\Omega^z_{i,\vec{k}}$ [81]:

$$\sigma_{xy}^{AHE} = \sum_{i,\vec{k}} n_{i,\vec{k}} \Omega^z_{i,\vec{k}} \tag{9.17}$$

where the index i is over all Weyl nodes, and $n_{i,\vec{k}}$ is the occupation number. We also recall that band structure determines the location of the Weyl nodes in the Brillouin zone. The Weyl nodes are said to act as magnetic monopoles in $\vec{k}$ space [81]. Thus, the properties of Weyl semimetals are clearly related to topological properties of the electronic band structure. Because spin orbit coupling is involved, the AHE is spin dependent and is sometimes referred to as quantum spin Hall insulators [43].

9.5.1 Structure, Bonding, and Electronic Band Structure of Weyl, Dirac Semimetals, and Related Materials

In this section, we survey the crystal structures of the different materials that have been experimentally confirmed to have Weyl semimetal properties. The section is further divided into three subsections: Type 1 Weyl Semimetals [20, 22], Type 2 Weyl Semimetals [20, 86], and Weyl Semimetals with nodal lines [24].

9.5.1.1 Type 1 Weyl Semimetals

Transition Metal Monopnictides: The transition metal monopnictides TaAs, TaP, NbAs, and NbP in the body center tetragonal structure are considered to be potential Weyl semimetals. These materials lack inversion symmetry but have time reversal symmetry. It is noted that some publications refer to the crystal structure as face center tetragonal, perhaps incorrectly. The crystal structure is body center tetragonal. The Hall characterization data presented in Sect. 9.5.2.1 supports conflicting interpretations concerning confirmation of the presence of Weyl node. The crystal structure (shown in Fig. 9.30) lacks inversion symmetry and the space group is $I4_1md$ (#109, point group C_{4v}^1) [87, 88]. There are 2 cations and 2 anions per unit cell as the 4 implies in the space group notation. There is a high degree of ionicity effectively resulting in filled p^6 shells for the anions and a d^2 shell for the transition metal cations [88]. The upper valence bands are formed from the p^6 and d^2 orbitals, and the unoccupied d orbitals for the lower conduction bands [88]. The spatial extent of the Ta d orbitals is expected to lead to hybridization with the As p orbitals resulting in semimetal behavior [89]. X-Ray diffraction characterization data for TaAs can be found in [90], and the lattice constants for all the monopnictides in Table 9.2. Band

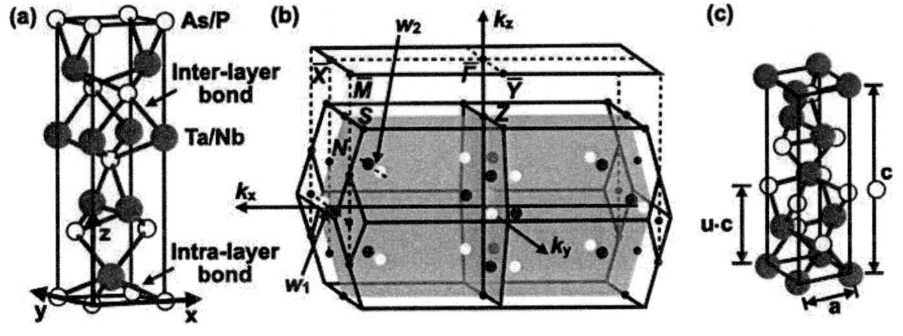

Fig. 9.30 Crystal structure and Brillouin Zone of transition metal monopnictides TaAs, TaP, NbAs, NbP. The crystal structure is body center tetragonal and the space group is I4md (#109, $4C_v$). The crystal structure lack inversion symmetry, but the electronic states retain time reversal symmetry. The tetragonal axis z is shown in (**a**). (**c**) Adapted from [88]. (**a**), (**b**) Adapted and reprinted with permission from [87]. © 2015 American Physical Society

Table 9.2 The lattice constants for the body center tetragonal lattice for the transition metal monopnictides

	TaAs	TaP	NbAs	NbP
a (nm)	0.3466 (0.3437)	0.3339 (0.3318)	0.3477 (0.3452)	0.3350 (0.3334)
c (nm)	1.175 (1.1656)	1.1486 (1.1363)	1.1752 (1.1679)	1.1424 (1.1376)
u	0.417 (0.417)	0.414 (0.416)	0.422 (0.417)	0.416 (0.416)

The values are from ab initio calculations of the minimum energy for the lattice. The values shown in parenthesis are experimental. Data from [90]

structure calculations predict 12 nodal pairs (24 Weyl nodes) in the Brillouin zone for the transition metal monopnictides. The electronic band structures of TaAs, TaP, NdAs, and NdP are shown in Fig. 9.31.

Rare Earth Monopnictides: Rare earth monopnictides are considered topological semimetals and/or topological insulators that break time reversal symmetry. Some of the rare earth monopnictides such as CeSb are predicted to be both. We note that CeX (X=P, As, and Sb) exhibit a wide variety of magnetic phases [91]. The trend as the monopnictides changes from N to Bi is for the materials to be more

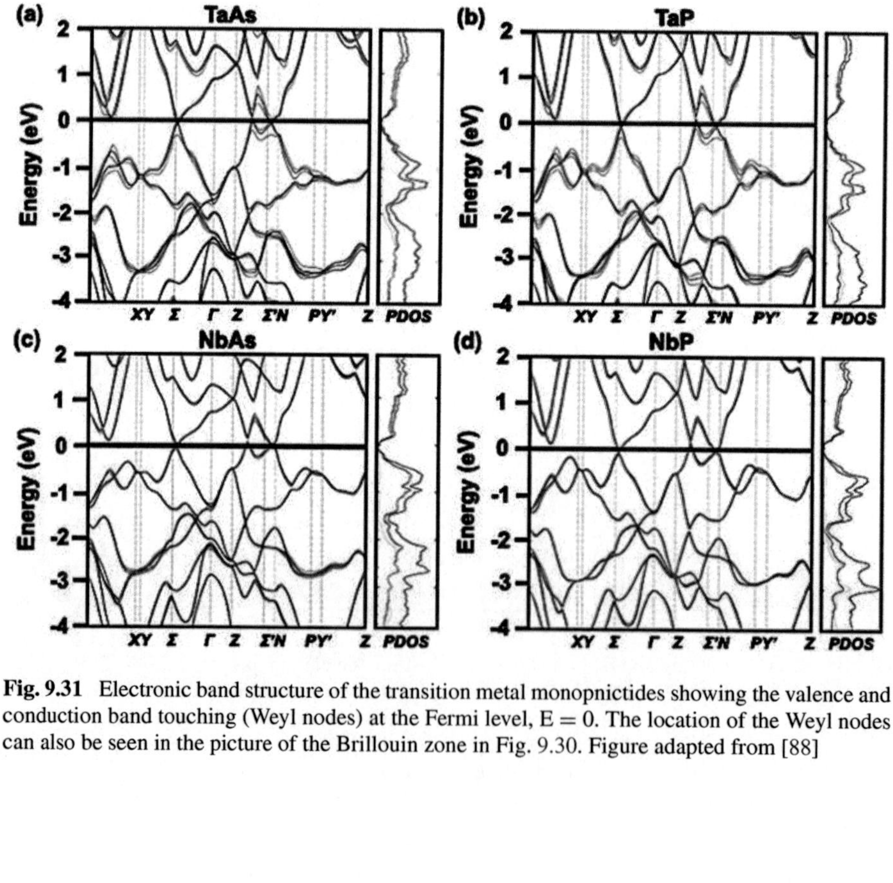

Fig. 9.31 Electronic band structure of the transition metal monopnictides showing the valence and conduction band touching (Weyl nodes) at the Fermi level, E = 0. The location of the Weyl nodes can also be seen in the picture of the Brillouin zone in Fig. 9.30. Figure adapted from [88]

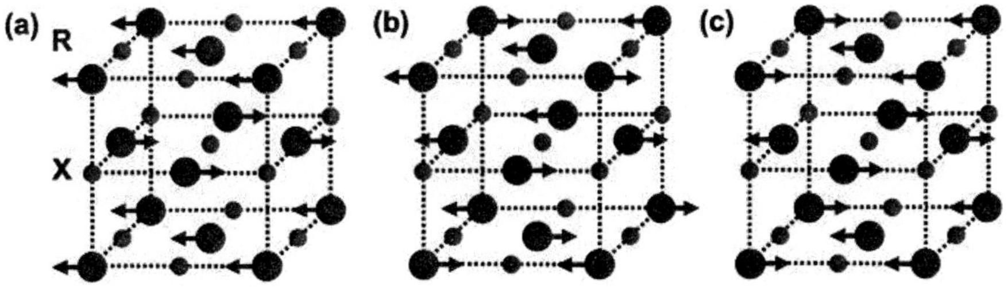

Fig. 9.32 The crystal structure is the NaCl structure, and here the rare earth is labeled R and pnictide X. Three common types of antiferromagnetic ordering found in rare earth monopnictides [92, 95, 96]. AFM-I where the magnetic moments are all aligned in the (001) planes but every other layer is anti-aligned (AFM) as shown in (**a**). AFM-II has aligned $(0\bar{1}1)$ chains but anti-aligned (AFM) between chains as shown in (**b**). All nearest neighbors are anti-aligned in AFM-III as shown in (**c**). Figure adapted from [92].

metallic [92]. It is interesting that most rare earth monopnictides crystallize in the NaCl crystal structure (space group $Fm\bar{3}m$) instead of the body center tetragonal structure found for the transition metal monopnictides as shown in Fig. 9.32. Thus rare earth monopnictides have inversion symmetry. The exception to this includes EuAs, EuSb, and EuBi which do not necessarily crystalize in either structure. The NaCl crystal structure has inversion symmetry, and the f electrons of the rare earth play a key role in the electronic structure and thus materials properties of the rare earth monopnictides. The conduction band is formed from the rare earth t_{2g} f electrons close to the X point in the Brillouin zone and the valence bands are formed from the pnictide $^{3/2}p$ state close to the Γ point [91]. Ab initio band structure calculations show that LaN is a nodal ring semimetal and that LaX (X=P, As, Sb, and Bi) are 3D topological insulators [93]. ARPES measurements show experimental evidence of a 3D topological insulator state for LaBi [94]. The localization of the rare earth f electrons and the resulting strong repulsion between f electrons requires the use of a localization term U discussed in Sect. 9.3 in band structure calculations. The use of the local spin density approximation in the band structure calculation is considered inadequate to describe the band structure [92].

Magnetic ordering of the rare earth f electrons plays a role in the band structure and materials properties. However, the magnetic ordering is considered to be complicated and material dependent [92]. Many of the rare earth monopnictides have antiferromagetic (AFM) ordering and three type types are shown in Fig. 9.32. Other rare earth monopnictides, especially CeX (X=P, As, Sb) exhibit a large number of different magnetic states [91, 92, 95, 96]. LaSb, LaBi, CeSb and CeBi, NdSb, GdSb, and GdBi all exhibit extreme magnetoresistance that depends on the sample [91]. LaSb, LaBi, and YSb are non-magnetic [91]. CeSb exhibits magnetic ordering which splits the single Ce t_{2g} f orbital energies [91, 95]. There are at least fourteen different magnetic phases for CeSb [90, 95]. LaSb has a resistivity plateau versus temperature at a fixed magnetic field where carrier mobility is very high and other exotic behavior that is reminiscent of a topological insulator [97]. The temperature and magnetic field dependence of the magnetoresistivity are discussed in Sect. 9.5.3.1.

The $\mathbb{Z}_2$ topological invariant of LaAs has been calculated to be $\upsilon_0 = 0$ at atmospheric pressure and transition to $\upsilon_0 = 1$ at 6 GPa [98]. The $\mathbb{Z}_2$ topological index of CeBi has been characterized as having $\upsilon_0 = 1$ [99].

Pyrochlore Iridates $X_2Ir_2O_7$ and related materials: Materials containing Ir^{+4} $5d^5$ oxides are candidates for a variety of material phases that arise from cooperative electron interactions including Weyl semimetals and Mott insulators. Here, we emphasize pyrochlore iridates and briefly mention Ir^{+4} $5d^5$ materials having the Ruddlesden-Popper sequence of pseudocubic and planar perovskites structures with $Sr_{n+1}Ir_nO_{3n+1}$ ($n = 1, 2, \infty$) or the hexagonal insulators $(Na/Li)_2IrO_3$. [27]. Because some members of the pyrochlore iridates family have been predicted to be Weyl semimetals and some may also be magnetic axion insulators, they are discussed along with other Weyl semimetals [26, 100–102]. We note that the observed electrical and optical behavior is used to motivate the designation of Weyl semimetal. For example, $Eu_2Ir_2O_7$ is reported to be a Weyl semimetal based on its low energy optical conductivity as discussed in Sect. 9.5.2 [101].

The X in $X_2Ir_2O_7$ is either Yttrium or a lanthanide element. The electronic configuration of Ir is $Xe\,4f^{14}\,5d^7\,6s^2$ which becomes $Xe\,4f^{14}\,5d^5$ for the + 4 oxidation state and $Xe\,4f^{14}\,5d^4$ for the + 5 oxidation state. The oxidation states for pyrochlore compounds can be either $(X^{+2})_2(Ir^{+5})_2(O^{-2})_7$ or $(X^{+3})_2\,(Ir^{+4})_2(O^{-2})_7$. For $Eu_2Ir_2O_7$ and $Y_2Ir_2O_7$, the $(X^{+3})_2\,(Ir^{+4})_2(O^{-2})_7$ oxidation states are observed [100, 101]. In general, the chemical formula for pyrochlores $A_2B_2O_7$ can also be expressed as $(A_2O')(B_2O_6)$ which emphasizes the octahedron structure around the B atoms which plays a key role in the electronic band structure when B is Ir [103]. The cubic crystal structure of $X_2Ir_2O_7$ has the $Fd\bar{3}m$ (O_h^7) (#227) space group with 8 formula units in a unit cell and 4 inequivalent O atom sites [26, 103]. The local structure of the Ir is approximately octahedral, and the distortions to the octahedral crystal field of the oxygen are discussed below. The lattice parameters of the lanthanide pyrochlore iridates can be found in [102]. The rare earth pyrochlore iridates $R_2Ir_2O_7$ (R=Y, Eu, Sm, Nd, Pr, and B) are said to have a geometrically frustrated lattice and theoretical calculations show that increasing the size of the cation drives a metal–insulator transition [104]. This study used a combined DFT and embedded dynamical mean-field theory (DMFT) methods to predict that materials from the $R_2Ir_2O_7$ series, that are both magnetic and insulating, would be topologically trivial [104]. This is in contrast to other theoretical studies discussed below [100]. The transition between metal and insulator occurs at a R cation radius between that of Nd and Pr which is shown in Fig. 9.33(a) using the free energy difference between the paramagnetic state and AIAO magnetic ordering. This transition occurs when an ordered magnetic state become present when the spins on the Ir atoms in the all-in all-out arrangement is present as shown using the magnetic moment versus cation radius in Fig. 9.33(b).

The pyrochlore iridates crystal structure results in crystal field splitting of the Ir $5d$ orbital energies. Spin orbit coupling and Coulomb (Hubbard) localization further interact with the $5d$ orbitals further splitting the orbitals. We also note that the filling

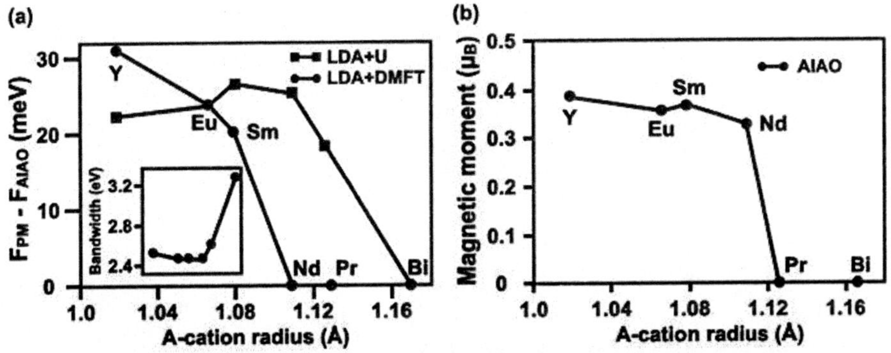

Fig. 9.33 The Metal–Insulator transition for the rare earth pyrochlore iridates $R_2Ir_2O_7$ (R=Y, Eu, Sm, Nd, Pr, and B) as a function of R cation size is shown. In (**a**), the free energy difference F between the paramagnetic state and the all-in-all out ordered magnetic state is shown as a function of the R cation size. Two theoretical methods are compared the local density approximation with the Hubbard interaction (onsite Coulomb interaction) U (LDA-U) (described below) versus a combined DFT and embedded dynamical mean-field theory (DMFT) indicated as (LDA-DMFT). In (**b**), the magnetic moment versus cation size shows the transition from metal to insulator between Nd and Pr. Figure adapted and reprinted with permission from [104]. © 2017 American Physical Society

of the Ir 5*d* levels depends on the Ir oxidation state. Through these interactions, the different R cations drive interesting materials behavior. First, we discuss present the combined impact of crystal field splitting and spin orbit splitting using the diagram shown in Fig. 9.34. The crystal structure for the iridates is shown in Fig. 9.35.

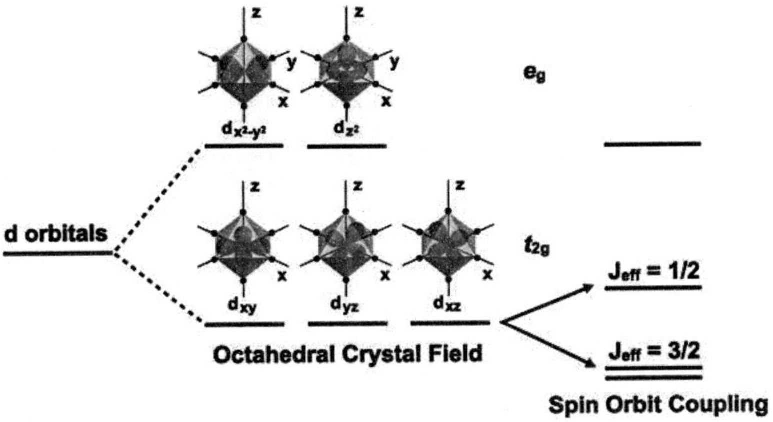

Fig. 9.34 The Ir $5d^5$ orbitals are split by both the crystal field and spin orbit interaction. A perfect octahedral arrangement of the ligands is shown. Orbitals pointing directly at ligands are repelled more than those pointing in between them. The crystal field splitting for the $5d^5$ Ir is ~2 eV. In the pyrochlore iridates, the octahedral are trigonally distorted (rotational symmetry is preserved) by a compression along the (111) direction. The distorted structures are sometimes referred to as cubic. Figure adapted and reprinted with permission from [105]. © 2018 American Physical Society

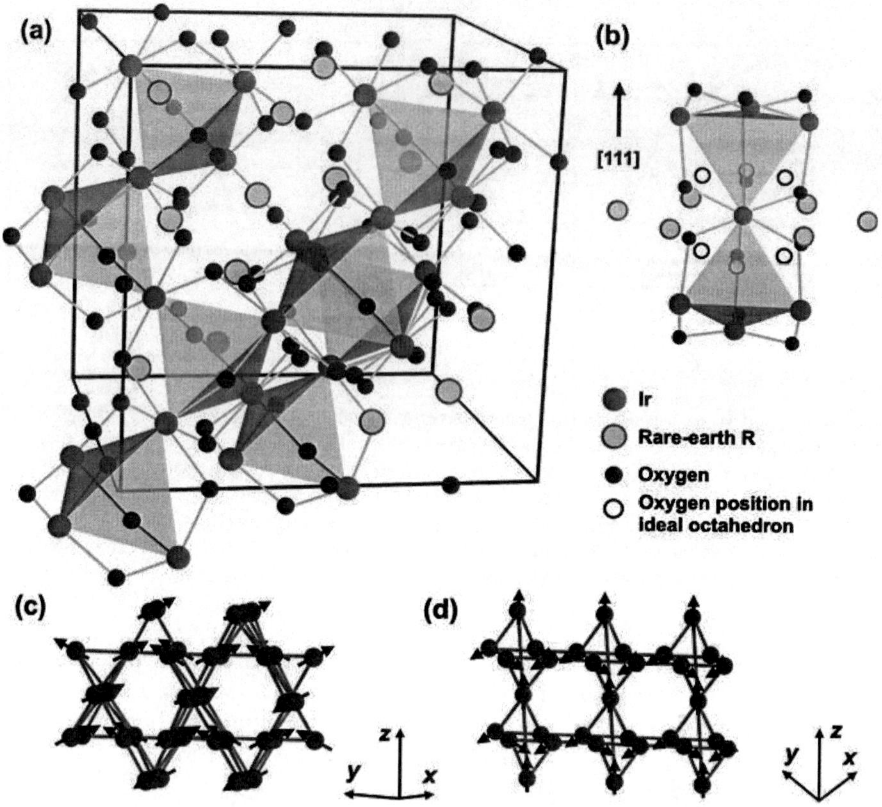

Fig. 9.35 The crystal structure of pyrochlore iridates with the Fd$\bar{3}$m space group. The crystal lattice for the $A_2B_2O_7$ structure is shown in (**a**). The A atoms are the large lighter gray spheres in the corner, and the B atoms are shown using the large darker gray. The oxygen atoms are shown using the smaller gray spheres. In (**b**), in order to show how the lattice positions are shifted, the ideal positions for the oxygen are shown as open circles. The Ir^{+4} ions with 5 electrons in their 5d orbitals are in a corner sharing network in (**a**). The all in all out (AIAO) arrangement for antiferromagnetic ordering is shown in (**c**) and (**d**). The in/out designation refers to whether or not the magnetic moments of the Ir atoms point to the center of tetrahedron or away from the center. The atomic arrange of slabs of Ir ions is shown along the (001) axis in (**c**) and along the (111) axis is shown in (**d**). The layers of the slabs are stacked so that they system has inversion symmetry. The 2D Brillouin zone along the (001) is tetragonal and along the (111) slabs is hexagonal. (**a**), (**b**) adapted and republished with permission of Annual Reviews © 2014 from [27]. Permission conveyed through Copyright Clearance Center, Inc. (**c**), (**d**) adapted and reprinted with permission from [26]. © 2018 American Physical Society

Both the X and Ir cations form a network of corner sharing tetrahedral, and this structure has inversion symmetry [26]. For the structures described here, the Ir^{+4} ions are surrounded by 6 O^{-2} ions with an octahedral symmetry which results in cubic crystal field splitting of the Ir 5d electrons into a $e_g(d_{x^2-y^2}, d_{z^2})$ doublet and t_{2g} (d_{xy}, d_{yz}, d_{xz}) triplet states [26, 27, 105, 106]. The octahedral share the corners

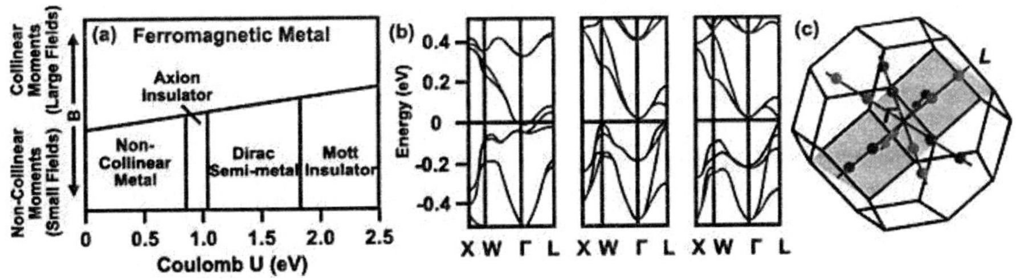

Fig. 9.36 Phase diagram and Electronic band structure of pyrochlore iridates is shown in (**a**). The band structure calculations were done using Local Density Approximation (LDA) plus the Hubbard interaction U (LDA-U) which corrects for the strong Coulomb interaction between localized electrons [108]. A value of U = 0 eV results in a metallic state and U = 1.5 eV or 2 eV results in an insulator with a small band gap. The band diagrams in (**b**) show the electronic structure for U = 0 eV for (**a**), 1.5 eV for (**b**) and 2 eV for (**c**). A value of U = 1.5 eV is reported to provide the lowest energy ground state for $Y_2Ir_2O_7$. The Weyl nodes located within the $\Gamma - X - L$ plane of the Brillouin zone as shown in (**c**). The *predicted* Weyl points for chirality +1 are shown in dark gray and for chirality −1 are shown in light gray. Figure adapted and reprinted with permission from [100]. © 2011 American Physical Society

of the unit cell as shown in Fig. 9.35a. The crystal field splitting is large, and it results in the e_g levels being ~2 eV above the Fermi level. X-Ray absorption spectroscopy measurements point to the presence of a strong spin orbit interaction [105, 107]. The addition of spin–orbit coupling further splits the t_{2g} into an effective pseudospin $J_{eff} = \frac{1}{2}$ doublet and a $J_{eff} = \frac{3}{2}$ quadruplet. The Ir t_{2g} $5d_{xy}$ orbitals hybridize with the basal $2p_x$ and $2p_y$ O orbitals and the Ir t_{2g} $5d_{xz}$ and $5d_{yz}$ hybridize with the apical (i.e. apex) $2p_x$ and $2p_y$ O orbitals. The Ir e_g $5d_{x^2-y^2}$ orbitals hybridize with the basal $2p_x$ and $2p_y$ O orbitals, and the Ir e_g $5d_{z^2}$ orbitals hybridize with the $2p_z$ O orbitals [106]. Thus, the bands near the Fermi level are a combination of the Ir $5d$ t_{2g} orbitals and the O $2p$ orbitals [26, 27]. The calculated electronic band structure for one iridate, the insulator $Y_2Ir_2O_7$, is shown in Fig. 9.36. Below we discuss a simple nearest neighbor tight binding model can be formed from t_{2g} $J_{eff} = \frac{1}{2}$ level because the $J_{eff} = \frac{3}{2}$ quadruple is far from the Fermi level [26, 27].

Here, we discuss two theoretical approaches to predicting whether an Ir pyrochlore will be an axion insulator, Dirac semimetal, Mott insulator, or ferromagnetic metal. One is a tight binding model that includes nearest neighbor interactions which was presented in the second quantization formalism [26, 27]. The Hamiltonian for the tight binding model of the pyrochlore iridates has three contributions. The spin independent hopping between nearest neighbors H_t, the Hamiltonian for spin–orbit interactions H_λ, and the Hamiltonian for magnetic ordering, H_Δ were solved using a spinor degree of freedom per site [26]. The spin–orbit interaction strength is λ, and the magnetic ordering is included by a Zeeman term whose strength is a function of the spin–orbit splitting energy Δ. The inclusion of the magnetic ordering breaks time reversal symmetry (14). The tight binding electronic band structure close to the Fermi level was studied by emphasizing the Ir $J_{eff} = \frac{3}{2}d$ orbitals and symmetry of the Ir octahedral. Many theoretical studies of the Ir pyrochlore materials assume that there is no mixing of the e_g ($d_{x^2-y^2}$, d_{z^2}) doublet and t_{2g} (d_{xy}, d_{yz}, d_{xz}), but there is evidence that mixing does occur [106]. Since the spin–orbit splitting of the t_{2g} states is relatively large and the $J_{eff} = 1/2$ level is ½ full, electronic structure calculations start by assuming that a d orbital forms the band closest to the Fermi level that accounts for key materials properties [26, 27, 105, 106]. The crystal structure and the sub-units of Ir_2O_6 are shown in Fig. 9.35, and the calculated electronic band structure of $Y_2Ir_2O_7$ is shown in Fig. 9.23. The atomic arrange of slabs of Ir ions is shown along the (001) axis in Fig. 9.35c and along the (111) axis as shown in Fig. 9.35d. It is important to state that both the X and Ir cations form a network of corner sharing tetrahedral, and this structure has inversion symmetry [26]. Thus, the Weyl properties of the pyrochlore iridates require that time reversal symmetry be violated [26]. The Bravais lattice for the slab is fcc with four sites per unit cell that form a tetrahedron [26, 109]. A second approach was based on band structure calculations done using Local Density Approximation (LDA) and the LDA-U approach where U (Hubbard) corrects for the strong Coulomb repulsion between localized electrons on the same atom. The results of this approach are shown in Fig. 9.36b. Specifically, the repulsion U was used to describe the repulsion of the Ir 5d electrons and thus their localization. A value of U = 1.5 eV results in the lowest energy state and 24 Weyl nodes shown in the right pane of Fig. 9.36. This theoretical study concludes with a prediction that at low temperature, $Y_2Ir_2O_7$ is a Weyl semimetal and that Eu, Sm, and Nd iridates may also be Weyl semimetals [100]. Research aimed at searching for topological properties in pyrochlore iridates continues at the time of the publication of this book.

Variation of the composition of the pyrochlore Iridates as exemplified by formula such as $(Y_{2-x}Bi_x)Ir_2O_7$ [110] and $(Y_{2-x}Pr_x)Ir_2O_7$ [106] enables tuning the band structure and magnetic ordering through changing structural details or though magnetic doping. For $(Y_{2-x}Bi_x)Ir_2O_7$, the addition of Bi^{+3} stabilizes the antiferromagnetic correlation and enhances electronic conductivity.

Magnetic doping of pyrochlore Iridates is done through addition or substitution of X^{+3} with a Lanthanide in $X_2Ir_2O_7$. The +3 oxidation state of the Lanthanides Pr, Pb, and Ho have an even number of f electrons while Nd, Sm, Gd, Dy, Yb have an odd number of f electrons. The large spin–orbit interaction leads to integer total angular momentum for Pr, Pb, and Ho, and half odd-integer total angular momentum for Nd, Sm, Gd, Dy, Yb. The X ions also experience a crystal field which splits these angular momentum states. The angular momentum of the Lanthanide impacts the $f - d$ exchange interaction [111]. The $(Y_{2-x}Pr_x)Ir_2O_7$ series between x = 0 and x = 2 is a very interesting series since 4 f^2 Pr^{+3} magnetically dopes the insulator $Y_2Ir_2O_7$. The Fd3m (O_h^7) (#227) space group is observed for all Pr concentrations, and there is a transition from insulator to the paramagnetic metallic $Pr_2Ir_2O_7$ at x = 0.8 [106]. The cooperation of the $f - d$ exchange interaction with Hubbard (Coulomb) is theoretically predicted to stabilize the Weyl Semimetal state in some of the Lanthanide iridates [111]. $Pr_2Ir_2O_7$ exhibits the metallic state and an anomalous Hall effect without observable magnetic ordering between 0.3 and 1.5 K [111]. This is in contrast to $Nd_2Ir_2O_7$ which exhibits magnetic ordering below the metal–insulator transition temperature of 36 K. Evidence that thin pyrochlore iridate films grown along the (111) direction show a film layer dependent higher temperature

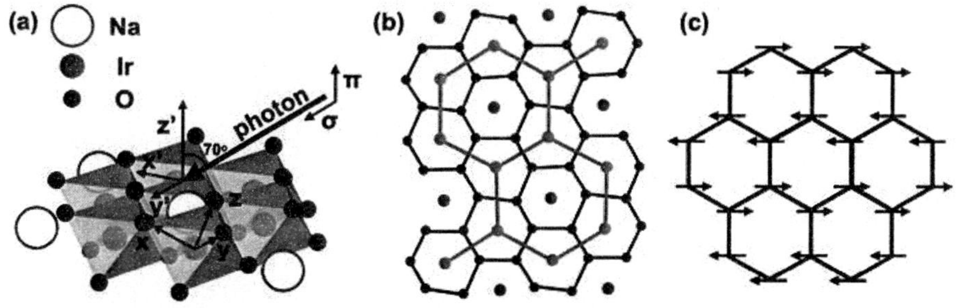

Fig. 9.37 The hexagonal arrangement of the Ir atoms in Na_2IrO_3 is shown. A side view of the octahedral oxygen arrangement around the hexagonally arranged Ir atoms is shown in (**a**). A top down view of the Ir atoms is shown in (**b**). The zigzag arrangement of the magnetic moments for the hexagonal Ir atom lattice is shown in (**c**) Figure adapted and reprinted with permission from [114–116]. Copyrights © 2010, 2011, and 2013 respectively American Physical Society

onset of the anomalous Hall effect provides a potential pathway to fabricating other materials that display topological properties [112]. We also note that femtosecond optical pumping of pyrochlore iridates is predicted to drive the appearance of a stable magnetic Weyl semimetal state [113].

Hexagonal Insulator Iridium oxides (Na/Li)$_2$*IrO*$_3$ *and Perovskites* $Sr_{n+1}Ir_nO_{3n+1}$ ($n = 1, 2, \infty$): The topological properties of Ir^{+4} $5d^5$ materials such as the planar honeycomb structures found in Na_2IrO_3 and Li_2IrO_3 and the Ruddlesden-Popper sequence of pseudocubic and planar perovskite structures $Sr_{n+1}Ir_nO_{3n+1}$ ($n = 1, 2, \infty$) have also been studied [27]. As discussed in Sect. 9.2, the interplay between spin orbit interactions and localization due to Coulomb (Hubbard) repulsion can result in a material being a Mott insulator. Both Na_2IrO_3 and Li_2IrO_3 are Mott insulators [114–116]. The planar hexagonal arrangement of the Ir atoms with the planes separated by Na layers is shown in Fig. 9.37. The Neel temperature for magnetic ordering for Na_2IrO_3 is $T_N = 13.3$ K [116]. Magnetic X-ray scattering at the Ir L_3 absorption edge indicated that the zigzag arrangement is present below the Neel temperature, and this arrangement is shown in Fig. 9.37c and the stripe arrangement of the magnetic moments are both consistent with the scattering data [115]. A band structure calculation using LDA + SO + U combined with the resonant X-Ray scattering has shown that the zigzag arrangement of the magnetic moments is the most likely configuration [16]. The band structure is compared to the optical conductivity data in Fig. 9.38.

The properties of other Ir^{+4} containing crystals also result from electron correlation and spin orbit coupling [117–123]. Relatively small changes in the crystal structure that surrounds the Ir results in significant changes in materials properties as exemplified by the semimetal (nearly band insulator) $SrIrO_3$ (see [7] in [120]), Mott insulator Sr_2IrO_4having 5 electrons in the t_{2g} orbitals [118, 119], and the weakly insulating magnetic $Sr_3Ir_2O_7$ [117, 120, 121]. Except for $SrIrO_3$, these materials are from the Ruddlesden-Popper sequence of pseudocubic and planar perovskite structures $Sr_{n+1}Ir_nO_{3n+1}$ ($n = 1, 2, \infty$) which are designated by the number (n) of Ir–O layers in the unit cell. The crystal structure of $Sr_3Ir_2O_7$ was recently determined to be

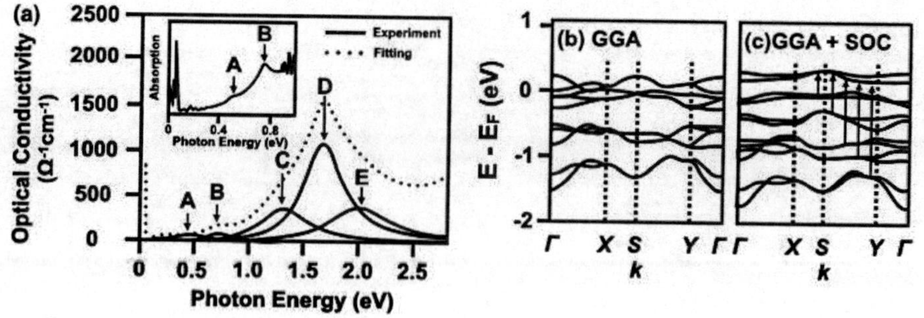

Fig. 9.38 The optical conductivity (**a**) and associated interband transitions (**c**) are shown for Na_2IrO_3. Figure (**b**) shows the band structure calculated without spin orbit coupling. The optical conductivity is fit to five Lorentz oscillators (**a–e**) and the relative contributions are shown in (**a**). Figure adapted and reprinted with permission from [116]. © 2013 American Physical Society

the double layer perovskite tetragonal *I4/mmm* structure by synchrotron based resonant X-Ray diffraction of a single crystal sample [117]. The low temperature crystal structure of Sr_2IrO_4 was reported to transform from tetragonal *I4/mmm* to $I4_1$/acd [118, 119]. Both of these crystal structures are shown in Fig. 9.39, and the associated electronic band structure of Sr_2IrO_4 in Fig. 9.40 where the interplay between spin

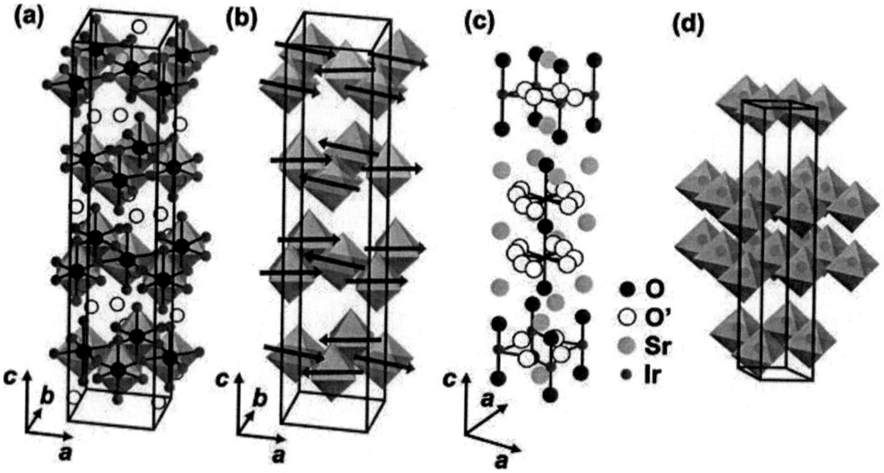

Fig. 9.39 The crystal structures of Sr_2IrO_4 and $Sr_3Ir_2O_7$ are shown. The $I4_1$/acd (#142) structure of Sr_2IrO_4 at 4 K determined by neutron scattering is shown in (**a**). The alignment of the magnetic moments of the IrO_6 octahedra is shown in (**b**) was also determined using neutron scattering [119]. The double layer perovskite tetragonal *I4/mmm* of $Sr_3Ir_2O_7$ structure determined by X-Ray diffraction is shown in (**c**) along with the IrO_6 octahedra [117]. The O′ sites have a 50% occupancy and are shifted by 0.04 nm to fit the diffraction data. Figure adapted and reprinted with permission from [117] 119. © 2012 and 2013 respectively American Physical Society

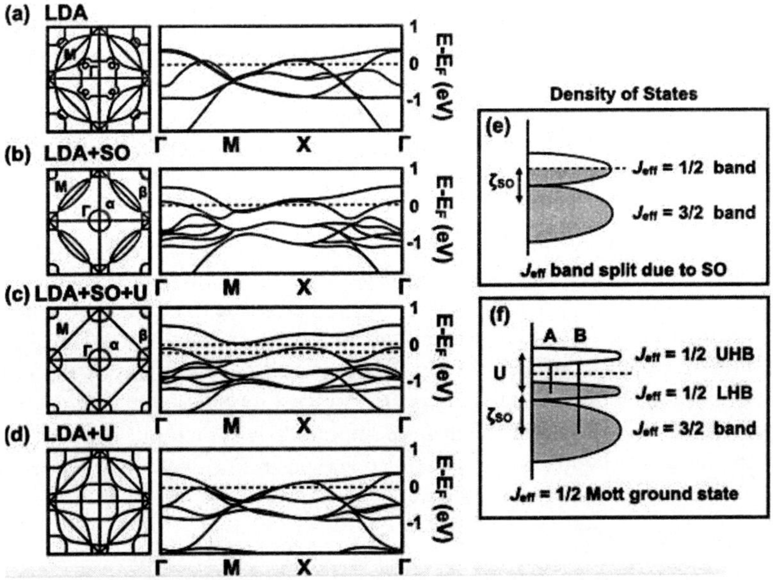

Fig. 9.40 The electronic band structure and density of states of Sr_2IrO_4 are shown. In band structure is shown with **a** local density approximation (LDA), (**b**) LDA plus spin orbit coupling, (**c**) LDA + SOC + U where U is the Coulomb (Hubbard) energy, and (**d**), LDA + U. The LDA + SOC + U results in the band gap that is experimentally observed. The density of states is shown with SOC in (**e**) and with SOC + U in (**f**). The optical transitions A and B observed in the optical conductivity data are shown in (**f**). (**e**), (**f**) Show that the $J_{eff} = \frac{1}{2}$ state is split by the Coulomb interaction. Figure adapted and reprinted with permission from [118]. © 2008 American Physical Society

orbit coupling and Coulomb (Hubbard) localization energy is clearly seen. The Neel temperature of $Sr_3Ir_2O_7$ is $T_N = 224$ K.

Sr_2RuO_4Perovskite: Sr_2RuO_4 is a topological 2D superconductor having the perovskite tetragonal *I4/mmm* (#139) space group as shown in Fig. 9.32. The structure does not change between room temperature and 100 mK [19]. Sr_2RuO_4 has the point group symmetry D_{4h} and the crystal is invariant with regard to an x–y plane M_{xy}, and in the presence of a magnetic field along the z direction, the crystal symmetry becomes C_{4h} [19]. The Ru d^4 has an even number of electrons. The mirror plane symmetry protects the Majorana fermions involved in the superconductivity.

Heusler and Half Heusler Alloys: The Heusler and half Heusler alloys are the subject of an intensive theoretical and experimental search for topological properties [124–127]. One of the challenges in classifying these materials is that some alloy compositions such as RPtBi (R=Gd and Nd) exhibit Weyl semimetal properties only in the presence of a magnetic field [128]. The Heusler alloys are zero gap semiconductors that can have a composition of X_2YZ. X and Y can be transition or rare earth metals [124–127]. The half Heusler alloys are non-magnetic, ternary compounds with a stoichiometry of MM′X. The compositional combinations include M=Lu, La, Sc, or

Y; M′=Pt, Pd, Au, or Ni; and X=Sb, BI, or Sn [124]. Both Heusler and half Heusler compounds have a strong covalent character to their bonding, and theoretical band structure calculations must include spin orbit coupling. Both have a cubic crystal structure. The cubic structure of many of the Heusler alloys is $L2_1$ with a space group of $Fm\bar{3}m$ (#225), while the cubic crystal structure is $C1_b$ for the half Heusler alloys with a space group of $F\bar{4}3m$ (#216) [126, 127]. There is no inversion symmetry for the half Heusler crystal structure. The results of theoretical studies of the half Heusler compounds are described in terms of the band inversion strength for inversion of the Γ_8 and Γ_6 bands at the Γ point in the Brillouin zone, $\Delta = \left[E_{\Gamma_8} - E_{\Gamma_6}\right]$ and a quantity $t = (Z_{M'} + Z_x)V$ which is said to describe the effects of band inversion on the M′X backbone of the ternary half Heusler compounds [124]. $Z_{M'}$ *and* Z_x are the atomic numbers of the M′ and X atoms and V is volume of the unit cell [124]. Compounds with $\Delta > 0$ are expected to have topologically protected states. The crystal structure of the half Heusler compounds is shown in Fig. 9.41, and the change in character (band inversion) is also shown in Fig. 9.41. The $F\bar{4}3m$ (#216)

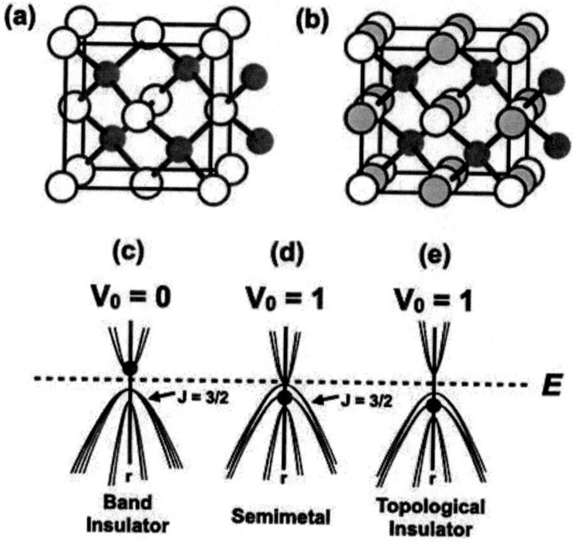

Fig. 9.41 The crystal structure of the MM′X half Heusler compounds is shown along with the change in band structure due to band inversion. The zinc blende structure is shown for comparison in (**a**). The half Heusler crystal structure is shown in (**b**). The M atoms are light gray spheres, the M′ atoms are dark gray spheres, and the clear spheres are the X atoms. There is no inversion symmetry for (**a**) or (**b**). The possible band structures for half Heusler alloys at high symmetry points in the Brillouin zone are shown in (**c**), (**d**) and (**e**). The half Heusler alloys with band structures suggesting topological protection are expected to be gapless semimetals and not gapped topological insulators. When band inversion occurs, the *s* states move to the valence band. The band structures are distinguished by the $\mathbb{Z}_2$ topological invariant Z_2 . Here, we interpret the reported $\mathbb{Z}_2$ as referring to the value of $(-1)^{\upsilon_0}$ where υ_0 is the first topological invariant for 3D crystals. When $\upsilon_0 = 1$, the band structure calculations predict material is a candidate for being a strong topological insulator. Figure adapted from [125, 126]

crystal structure can be thought of as three interpenetrating fcc lattices [128]. The atomic layers are well ordered. For example, along the 111 direction, the crystal has successive atomic layers of R, Pt, Bi for rare-earth Pt Bi half Heusler alloys. The electronic band structure of topologically non-trivial half Heusler compounds is shown in Figs. 9.42 and 9.43.

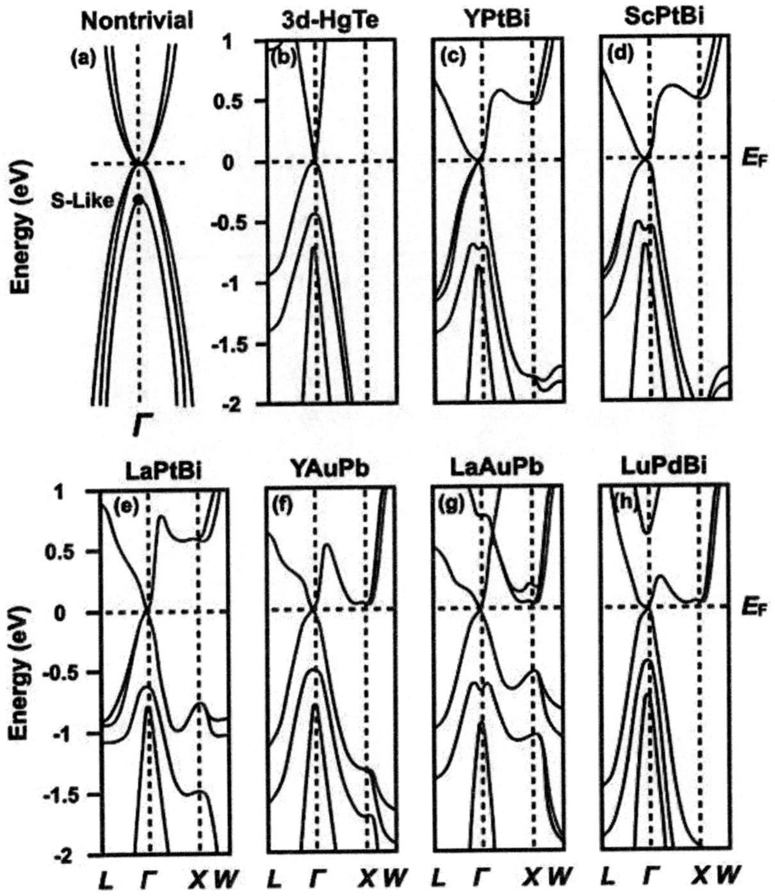

Fig. 9.42 The theoretical electronic band structure of several Bi containing Half Heusler alloys is shown. The band structure predicts these alloys to be topologically non-trivial. The band structure is compared to that of HgTe. The first band structure at the top left shows the band inversion with the s states switching with the p orbital band. Figure adapted and reprinted with permission from [124]. © 2010 American Physical Society

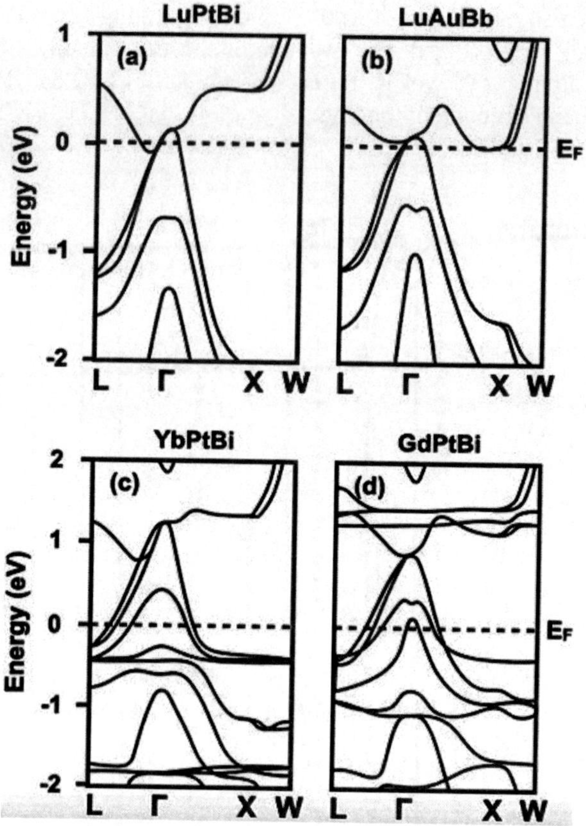

Fig. 9.43 Topologically non-trivial half Heusler alloys with metallic electronic band structures are shown. Figure adapted and reprinted with permission from [124]. © 2010 American Physical Society

9.5.1.2 Type 2 Weyl Semimetals

WTe_2: The TMD's (transition metal dichalcogenide) WSe_2 and WTe_2 are candidates for Type 2 Weyl semimetals [86]. WTe_2 is a layered, non-magnetic semimetal with a distorted octahedral structure referred to as the 1 T′ that is different from the 1 T structure described in Chap. 8 [129–131]. The space group for WTe_2 is the $Pmn2_1$ (#31). This space group is nonsymmorphic (see Chap. 2, Sect. 2.12) which provides topological protection at the $Z = \left(0, 0, \frac{\pi}{c}\right)$, point in the Brillouin zone [129]. A zigzag chain of W atoms provides a 1D conduction path. The interest in WTe_2 comes from the large non-saturating, anisotropic, low temperature magnetoresistance which is unusual for a non-magnetic material [131]. The zigzag chain of W atoms is along the *a* direction, and the *c* direction is normal to the TMD layers just as in 2D TMD materials [131]. The *p* and *n* carriers have the same concentration. The electronic structure of bulk WTe_2 has been studied after cleaving in ultra-high

vacuum using synchrotron based soft X-ray ARPES [129]. Both ARPES and band structure calculations show the presence of electron and hole pockets close to the Γ point in the Brillouin zone, the Weyl points with 3D electronic states [131]. Below, we contrast this paper with an optical study which was interpreted as not proving the presence of Weyl nodes.

9.5.1.3 Weyl Semimetals with Nodal Lines

Another category of Weyl semimetals sometimes known as Dirac semimetals has Weyl (nodal) rings instead of points [20, 132, 133]. Reflection symmetry (mirror plane) or a combination of time reversal and inversion symmetry can result in a topologically protected Weyl (Dirac) ring [133]. The second material has mirror symmetry for one of planes of atoms which allows nodal ring formation. The nodal ring is shown in Fig. 9.44 for Ca_3P_2.

Ca_3P_2: Ca_3P_2 crystallizes in a structure having the hexagonal $P6_3/mcm$ (#193) space group. This structure has inversion and mirror symmetry and the electronic

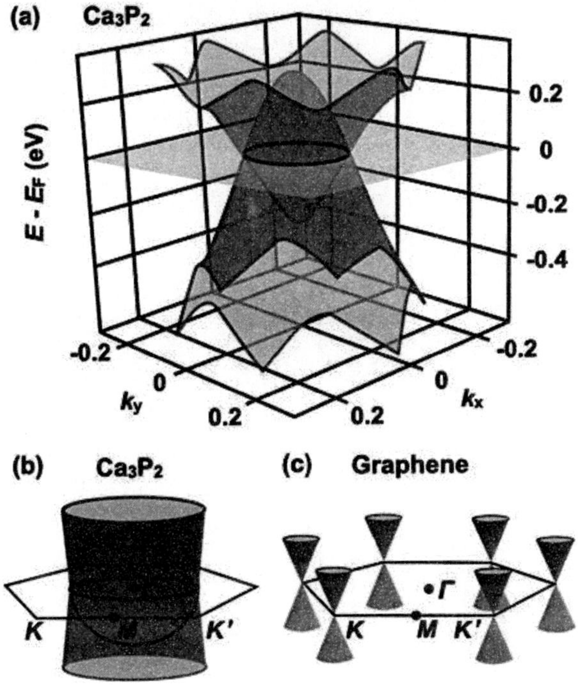

Fig. 9.44 The Weyl (nodal) ring in the band structure of Ca_3P_2. The ring forms at the crossing of the valence and conduction bands. Band inversion is also seen in (**a**). The 2D projection of the Brillouin zone at the Fermi level is shown in (**b**). The Dirac cones around the Brillouin zone of graphene are shown in (**c**). Figure adapted and reprinted from [132].

band structure has time reversal symmetry [133]. The crystal structure and Brillouin zone are shown in Fig. 9.45 and the electronic band structure in Fig. 9.46.

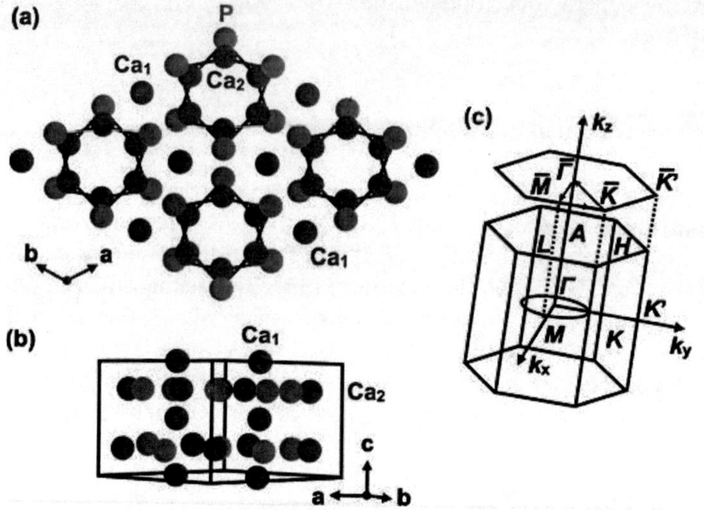

Fig. 9.45 The crystal structure and Brillouin zone of Ca_3P_2 in the hexagonal $P6_3/mcm$ (#193) space group. The Weyl nodal ring is shown in the center of the Brillouin zone in (**c**). Figure (**c**) adapted from [132, 133]. (a), (b) Adapted and reprinted from [132] 2015 L. S. Xie, et al. CC BY 3.0 https://creativecommons.org/licenses/by/3.0/. (c) Adapted and reprinted with permission from [133]. © 2016 American Physical Society

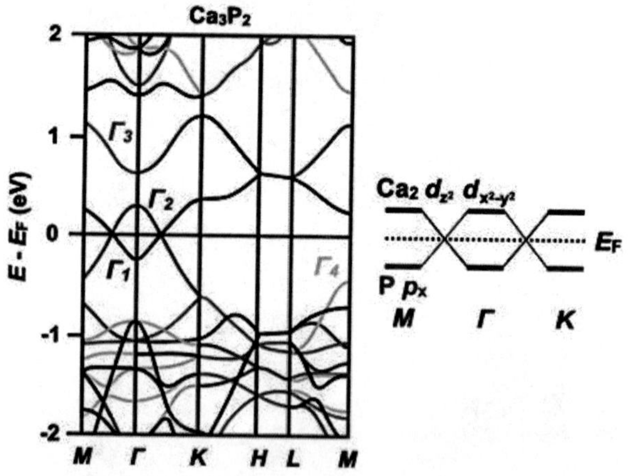

Fig. 9.46 Electronic band structure of Ca_3P_2 showing the tilted Dirac cones and the Weyl nodes $E - E_F = 0$ between the Γ and M points and Γ and K points in the Brillouin zone. The band inversion is shown on the right. Figure adapted and reprinted from [132]. © 2015 L. S. Xie, et al. CC BY 3.0 https://creativecommons.org/licenses/by/3.0/

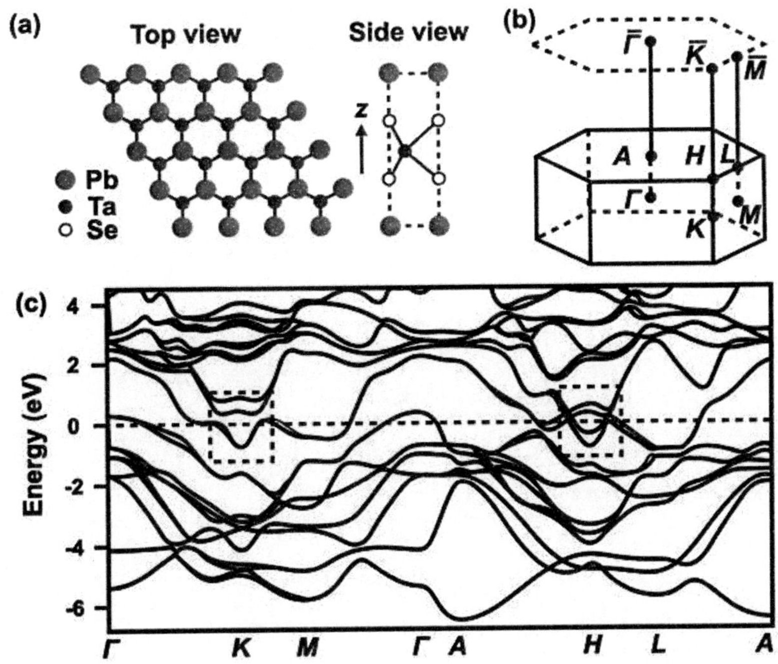

Fig. 9.47 The $PbTaSe_2$ P6m2 (187) crystal structure is shown along with the Brillouin zone and electronic band structure. The band structure shown in (**c**) has the two regions of band crossing at the Fermi level outlined. The areas enclosed by dashed lines are expanded in (**a**) through (**d**) in Fig. 9.48. The figure adapted from [24]

$PbTaSe_2$: $PbTaSe_2$ crystallizes with the P6m2 (#187) structure which does not have inversion symmetry as shown in Fig. 9.48 along with the Brillouin zone [24]. The atoms Pb, Ta, and Se are each in their own hexagonal layer. The Ta atoms have mirror plane symmetry so that a Ta atom at z is over another Ta atom at $-z$ [24]. This mirror plane is said to preserve the nodal line in the Fermi surface. The electronic band structure of $PbTaSe_2$ has two locations in the Brillouin zone that contribute to the Weyl nodal ring structure. One is close to the Γ point and the other at the H point as shown in Fig. 9.48. At the H point the valence and conduction bands cross each other with the Pb $6p_x$ and $6p_y$ levels forming two conduction bands and the Ta $5d_{xy}$ and $5d_{x^2-y^2}$ forming valence bands. $PbTaSe_2$ is also thought to be a superconductor.

9.5.1.4 Dirac Semimetals

Cd_3As_2: Cd_3As_2 is a small band gap semimetal from a family of II_3V_2 materials including Cd_3P_2, Zn_3As_2, and Zn_3P_2 [134]. Of these materials, only Cd_3As_2 has band inversion [134]. The details of the band structure of have been the subject of considerable discussion [34], and here we provide an overview. The band structure

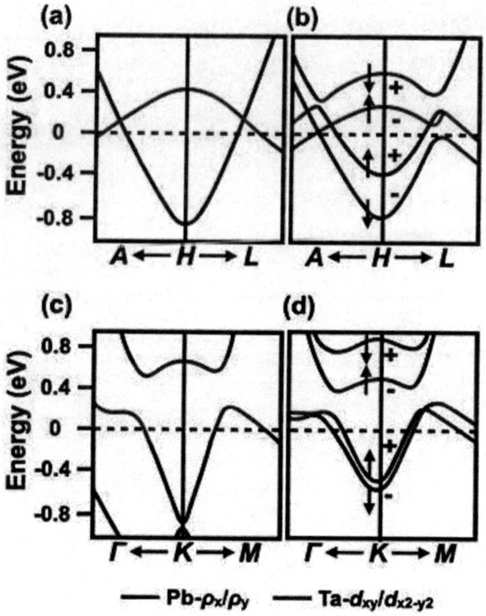

Fig. 9.48 Expanded view of the electronic band structure of $PbTaSe_2$ shown in the dashed enclosed areas in Fig. 9.47c. **a, c** have not spin orbit interactions while **b, d** show the influence of SOC. The spin up and spin down bands are also labeled using arrows. The figure adapted from [24]

of Cd_3As_2 is believed to be inverted at the Γ point in the Brillouin zone (p type As states above s type Cd states) while that is not the case for Cd_3P_2, Zn_3As_2, and Zn_3P_2. The band gap associated with this inversion is often referred to as a negative gap. Cd_3As_2 forms a number of different crystal structures depending on the temperature [35, 134, 135]. Two of these structures were reported to be stable at room temperature, $P4_2/nmc$ and the noncentrosymmetric $I4_1/acd$ [134]. The new structure not described in previous work, $I4_1/acd$, was observed for crystals grown from a melt [35]. $I4_1/acd$ **(#142)** is centrosymmetric and has an inversion center. This structure has important implications for the band structure. Spin splitting is not allowed for this structure, and the bands are degenerate [35, 135]. Knowledge of the correct crystal structure is critical for theoretical calculation of the electronic band structure [35, 134, 135]. The transition between crystal structures from $Fm\bar{3}m$ to $P4_2/nmc$ (#137) to $I4_1/acd$ is with temperature is shown in Fig. 9.49 [135]. The ratio of ionic radii is close to that expected for tetrahedral coordination of the metal ion in an ideal calcium fluoride like structure CaX_2 where r(M)/R(X) = 0.15 to 0.414. $r(Cd^{+2})/r(As^{-3}) = 0.092\,nm/0.222\,nm = 0.41$ [35]. The Cd_3As_2 formula is missing 1/4th of the Cd atoms necessary for the Cd_4As_2 antifluorite formula [35, 135]. Thus, the formally Cd^{+2} have a nearly fourfold coordination with As, and the formally As^{-3} have a nearly fourfold coordination with the Cd [135]. The Cd $5s$ electrons form the conduction band and the As $4p$ electrons form the upper valence band [134]. The electronic band structure for $I4_1/acd$ is shown in Fig. 9.50.

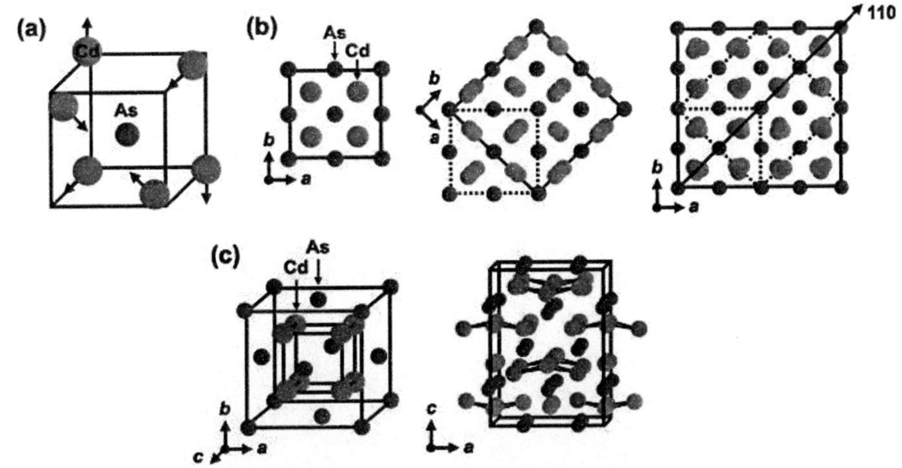

Fig. 9.49 The transition of the Cd_3As_2 crystal structure from the high temperature **Fm-3 m** phase to the intermediate temperature $P4_2/nmc$ phase to the lower temperature $I4_1/acd$ phase is shown. The $I4_1/acd$ phase was identified as the phase of crystal grown from a melt is a variation of the $I4_1$ cd phase reported by other workers [35, 134]. Figure adapted and reprinted with permission from [135]. © 2014 American Chemical Society

This band structure was theoretically determined using a DFT approach which is described in [135]. The band structure close to the Γ point in the Brillouin zone (see Fig. 9.50) is predicted to host two Dirac points. We note that as of the writing of this book, there does not seem to be a consensus concerning the crystal structure and thus the presence or lack of inversion symmetry is not resolved [136]. The high low temperature mobility $10^6\,\mathrm{cm}^2/(\mathrm{V\,s})$ and room temperature mobility of $>10^4\,\mathrm{cm}^2/(\mathrm{V\,s})$ indicate the presence of Dirac carriers in either Dirac nodes or Weyl nodes [136]. If inversion symmetry is not present, then the Dirac nodes will be Weyl nodes.

The presence of massless electrons is generally accepted, but the type of electron is a topic of discussion [136]. Magneto-optical characterization of Cd_3As_2 points to the existence of both 3D Dirac and Kane electrons [41]. The term Kane electron is described below. Here, we discuss the band structure for the low energy carrier close to the Γ point. The Kane Hamiltonian [137, 138] is a $k \cdot p$ approach that includes the spin orbit interaction which is very useful for determining the band structure close to the Γ point. The Kane Hamiltonian for the periodic function $u_{n,k}(\vec{r})$ for the Block wavefunction can be written as [137, 138]:

$$\left\{\left(\frac{P^2}{2m_0} + V(\vec{r})\right) + \left(\frac{\hbar}{m_0}\vec{k}\cdot\vec{p}\right) + \left(\frac{\hbar}{4m_0^2c^2}\nabla V(\vec{r}) \times \vec{p}\cdot\sigma\right)\right\} u_{n,k}(\vec{r})$$
$$= \left(E_n(\vec{r}) - \frac{\hbar^2k^2}{2m_0}\right)u_{n,k}(\vec{r}) \tag{9.18}$$

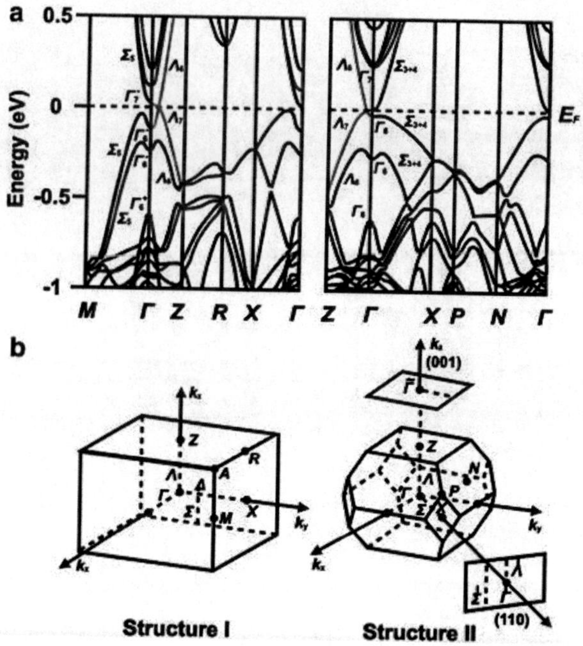

Fig. 9.50 The theoretical electronic band structure of (**a**) tetragonal D_{4h}^{15} $P4_2/nmc$ Cd_3As_2 and (**b**) the experimentally measured body center tetragonal C_{4v}^{12} $I4_1/acd$ phase [135]. This phase is centrosymmetric and band crossing at the Fermi level is allowed and observed in this graph. Band crossing is observed close to the Γ point. Figure adapted and reprinted with permission from [134]. © 2013 American Physical Society

Here, the spin orbit interaction is included using Pauli matrices σ. A four band Kane Hamiltonian for relativistic electrons has been developed [136, 139]. A key contribution to the Kane model that is necessary for its application to Cd_3As_2 is the addition of the effect of the tetragonal crystal field splitting δ by Bodnar [134, 139]. The relativistic Kane Hamiltonian for the low energy carriers close to the Γ point with the c crystallographic axis in Fig. 9.36 being along the z axis using a basis set of s, p_x, p_y, p_z orbitals with wavefunctions:

$$
\begin{array}{llll}
i|s\downarrow\rangle, & \dfrac{1}{\sqrt{2}}|(X-iY)\downarrow\rangle, & -\dfrac{1}{\sqrt{2}}|(X+iY)\downarrow\rangle, & |Z\uparrow\rangle \\
i|s\uparrow\rangle, & \dfrac{1}{\sqrt{2}}|(X-iY)\uparrow\rangle, & -\dfrac{1}{\sqrt{2}}|(X+iY)\uparrow\rangle, & |Z\downarrow\rangle
\end{array}
$$

can be written as (see supplemental [136]):

$$H = \begin{pmatrix} & P_\perp k_- & -P_\perp k_+ & 0 & 0 & 0 & 0 & P_\parallel k_z \\ P_\perp k_+ & 0 & 0 & 0 & 0 & 0 & 0 & 0 \\ -P_\perp k_- & 0 & -\frac{2\Delta}{3} & \frac{\sqrt{2}\Delta}{3} & 0 & 0 & 0 & 0 \\ 0 & 0 & \frac{\sqrt{2}\Delta}{3} & -\left(\delta + \frac{\Delta}{3}\right) & P_\parallel k_z & 0 & 0 & 0 \\ 0 & 0 & 0 & P_\parallel k_z & E_g & P_\perp k_+ & P_\perp k_- & 0 \\ 0 & 0 & 0 & 0 & P_\perp k_- & 0 & 0 & 0 \\ 0 & 0 & 0 & 0 & P_\perp k_+ & 0 & -\frac{2\Delta}{3} & \frac{\sqrt{2}\Delta}{3} \\ P_\parallel k_z & 0 & 0 & 0 & 0 & 0 & \frac{\sqrt{2}\Delta}{3} & -\left(\delta + \frac{\Delta}{3}\right) \end{pmatrix} \tag{9.19}$$

This Kane Hamiltonian includes Bodnar' s tetragonal crystal field splitting term δ [134, 136]. The band gap is E_g, Δ is the spin orbit splitting energy, the wavevectors $k_\pm = \frac{1}{\sqrt{2}}(k_x \pm ik_y)$, and $P_\perp$ and $P_\parallel$ are terms from the second order perturbation theory model of Dresselhaus, Kip, and Kittel and the reader is referred to [140].

The distinctive energy dependence of the band energy motivates the use the term massless Kane electrons for carriers with properties that fit this model. The origin of both types of carriers can be understood as being due to Bodnar's modification of the Kane model [41, 136, 139] to include the tetragonal crystal field splitting δ [134, 139]. In addition to the Kane type band structure, two tilted Dirac cones occur close to the Γ point at $k_z^{Dirac} = \sqrt{\delta |E_g| / P^2}$ where P is the momentum. Several different ab initio calculations also find the two Dirac nodes in the band structure at very low energy [136].

Using experimental data to determine the band structure is challenging. The energy dependence of the carriers in the presence of a magnetic field has been derived thus enabling analysis of the magneto-optical characterization data [136, 139]. Akrap, et al., found the $\sqrt{\vec{B}}$ dependence of massless Kane electrons for the Landau quantized cyclotron resonances of Cd_3As_2 [41, 136]. The infra-red magneto optical measurements characterize carriers in the energy range associated with the Kane electrons, thus other methods must be used to observe the Dirac nodes [41]. Numerous angle resolved photoemission (ARPES), scanning tunneling spectroscopy/scanning tunneling microscopy (STS/STM), and Hall characterization results have all been interpreted as showing massless carriers, and some of these studies report the presence of 3D Dirac nodes [136]. As with other topological materials, the quality of the sample and the presence of unintentional doping explain some of the observed differences. It is noteworthy that magneto-transport data for 20 nm thick Cd_3As_2 epitaxial films grown by molecular beam epitaxy displays the QHE [141]. The magneto-transport data at 1 K has distinctive quantized steps in the transverse (Hall) resistivity with associated near zero longitudinal resistivity at the magnetic fields where the steps occur. The transverse (Hall) resistivity is zero and zero magnetic field, thus there is no anomalous Hall effect. This material has low carrier density compared to samples obtained from bulk crystals. As discussed below, magneto-transport data for thicker epitaxial samples of Cd_3As_2 does not show the QHE.

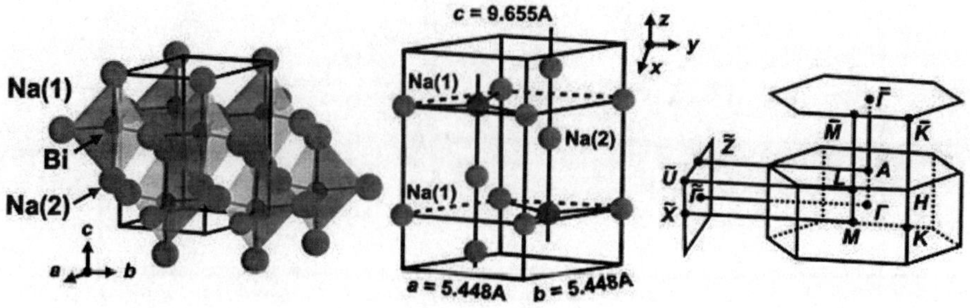

Fig. 9.51 The $P6_3/mmc$ crystal structure of Na, K, and Rb A_3Bi is shown along with the hexagonal Brillouin zone. The lattice parameters for Na_3Bi are shown in the central primitive lattice image. The projected (001) and (010) surface Brillouin zones are shown on the right. Figure on left adapted and reprinted from [143] with the permission of AIP Publishing. Figures on the right adapted and reprinted with permission from [142]. © 2012 American Physical Society

A_3Bi (A = Na, K, Rb): The Na, K, and Rb A_3Bi compounds crystallize in the hexagonal space group $P6_3/mmc$ (#194) crystal structure shown in Fig. 9.51, and the Cs and Li A_3Bi compounds crystallize in the $Fm\bar{3}m$ structure [142, 143]. The Na, K, and Rb A_3Bi compounds have both time reversal and inversion symmetry [142]. The Weyl nature of the Na, K, and Rb compounds have been either predicted or verified [142, 143]. The hexagonal layers are stacked along the c axis. Band inversion results in the valence band having Bi $6p$ character and the conduction band having A (Na $3s$, K $4s$, Rb $5s$) s character [142] The oxidation state of the A is + 1 and Bi is − 3 resulting in filled $6p$ orbitals. Time reversal and inversion symmetry are present for $P6_3/mmc$ (#194) [142]. The electronic band structure is shown in Fig. 9.52. Magnetoresistivity data for Na_3Bi provided evidence for the Weyl nodes and a chiral anomaly as shown in Sect. 9.5.3.4.

Since time reversal and inversion symmetry are present the Fermi points are four fold degenerate [142]. The $\mathbb{Z}_2$ index was reported to be $\mathbb{Z}_2 = 1$ which since Na_3Bi is stated to be topologically non-trivial, here we interpret to mean that the index $\upsilon_0 = 1$ and thus $(-1)^{\upsilon_0} = -1$ [142]. The $\mathbb{Z}_2$ was determined using the wavefunction information at the TRIM points observed in the band structure calculation. The band structure was calculated using experimental atomic positions [142].

9.5.2 *Optical Properties of Weyl, Dirac Semimetals, and Related Materials*

9.5.2.1 Type I Weyl Semimetals

Transition metal monopnictides: The complex dielectric function and optical conductivity of transition metal monopnictides TaAs, TaP, NbAs, and NbP has been calculated using DFT for an energy range from 0 to 15 eV [88, 144]. Since the transition

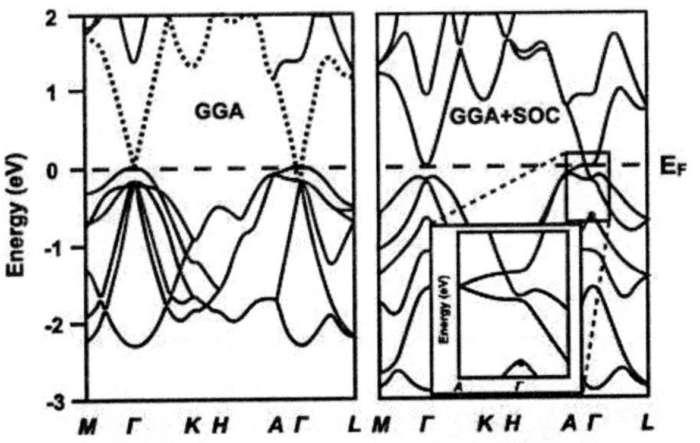

Fig. 9.52 Electronic band structure of Na_3Bi with and without spin orbit coupling. The calculated band structure predicts that Weyl nodes occur at (0, 0, k_z^c = ∼±0.26 × $\frac{\pi}{c}$). Figure adapted and reprinted with permission from [142]. © 2012 American Physical Society

metal monopnictides have 24 Weyl nodes, the effect of the Weyl nodes on optical data may be observable as discussed below. The results for the imaginary part of the dielectric function, ε_2, are shown in Fig. 9.53, and the real part of the dielectric

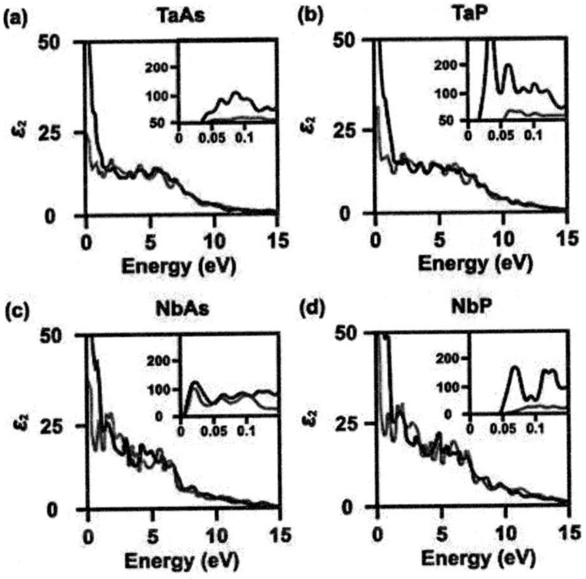

Fig. 9.53 The calculated imaginary part of the dielectric of transition metal monopnictides TaAs, TaP, NbAs, and NbP is shown. The black lines are for light polarized perpendicular to the tetragonal axis and the gray line is for light polarized parallel to the tetragonal axis. A Lorentz broadening of 0.05 eV was used for the main figures and a higher k point density and a Lorentz broadening of 0.01 eV for the inset figures. Note the different behavior at low energies for the higher k point density. Figure adapted from [88]

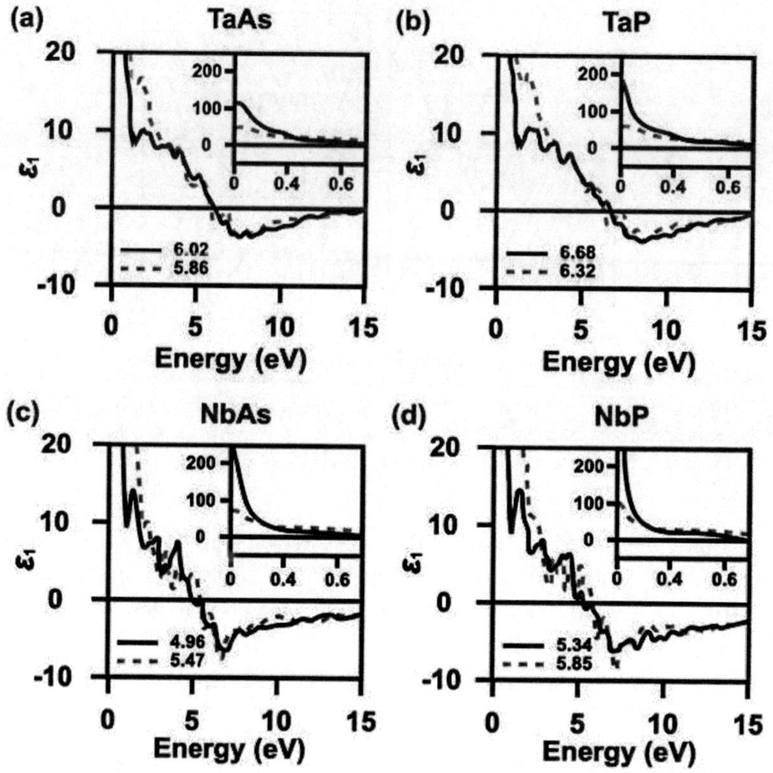

Fig. 9.54 The calculated real part of the dielectric of transition metal monopnictides TaAs, TaP, NbAs, and NbP is shown. The black lines are for light polarized perpendicular to the tetragonal axis and the dashed line is for light polarized parallel to the tetragonal axis. A Lorentz broadening of 0.05 eV was used for the main figures and a higher *k* point density and a Lorentz broadening of 0.01 eV for the inset figures. Note the different behavior at low energies for the higher k point density. The values in each figure indicate where the real part of the dielectric function is zero, e.g. 5.86 versus 6.02 eV. Figure adapted from [88]

function ε_1 in Fig. 9.54. The data show the highly anisotropic nature of the dielectric function and the large value of ε_2. The energy resolution of experimental obtained dielectric functions will be much smoother than the calculated ones.

The unusual electronic structure of Weyl metals contributes to the absorption behavior as seen in the low energy dielectric function between 0 and 0.2 eV. The specific contribution of the Weyl points described above is shown for the Weyl nodes labeled W1 and W2 in Fig. 9.55. The contributions of the Weyl nodes can be separated using this information. The importance of including a Drude term in the low energy optical conductivity has also been discussed [88, 144]. The optical conductivity of TaAs has been attributed to the Weyl nodes as described in (9.11). As shown in Fig. 9.56, there are two Weyl node response regions in the spectra arising from the number of Weyl nodes contributing to the conductivity [145]. The

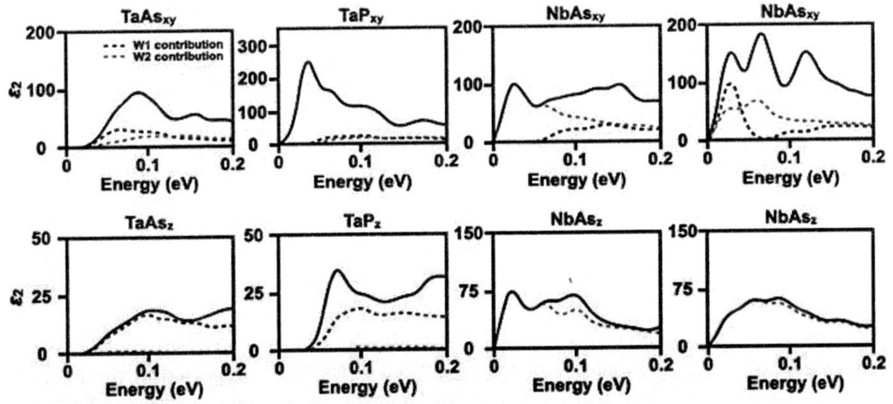

Fig. 9.55 The contribution of the W1 and W2 Weyl nodes to the anisotropic imaginary part of the dielectric function is shown for the transition metal monopnictides TaAs, TaP, NbAs, and NbP. The tetragonal axis is the z axis in these figures. The results for light incident normal along the tetragonal axis is shown in the xy data and light perpendicular to the tetragonal axis is shown in z. Figure adapted from [88]

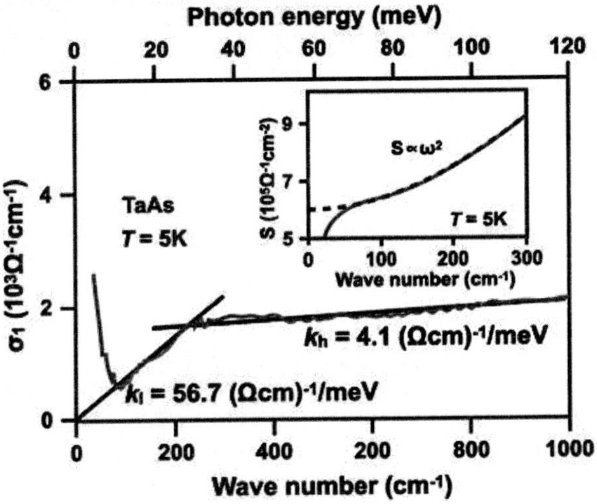

Fig. 9.56 The optical conductivity of TaAs at 5 K is shown. The surface of this crystal is long the (107) plane [145] meaning that both optical axis contribute the optical response. The linear region below 230 cm^{-1} is attributed to interband transitions from 4 Weyl nodes close to the Fermi energy. The linear region above 230 cm^{-1} is attributed to interband transitions that involve all 24 Weyl nodes and interband transitions from non-topological bands. Figure adapted and reprinted with permission from [145]. © 2016 American Physical Society

Weyl node contributions to the measured optical conductivity of NbP and TaP are masked by other transitions. For example, the Weyl node optical conductivity of NbP was reported to be masked by the Drude response due to free carriers and transitions between parallel bands below 100 meV ($\sim$806 cm^{-1}) [146]. The optical conductivity of NbP between 200 meV ($\sim$1613 cm^{-1}) and 300 meV ($\sim$2419 cm^{-1}) has been attributed to transitions between parallel bands. A similar analysis was used for the optical conductivity of TaP [147]. The separation of the W1 and W2 Weyl nodes has been studied by both theory and experiment, and the value of the separation energy varies but is considered to be 60–80 meV [147]. Ab initio DFT theory with the density gradient approximation was used to calculate the low energy optical conductivity of TaP which is shown in Fig. 9.31(**b**). Magneto-optical characterization found a complex set of transitions associated with the W1 nodes including a set of transition whose energy decreased with increasing magnetic field strength [147] (Fig. 9.57).

Rare Earth Monopnictides: The reflectivity of Sm, Pr, Ce, and La antimony monopnictides was measured from 0.001 to 20 eV using synchrotron generated IR radiation [148]. All these monopnictides had peaks in the reflectivity and imaginary part of the dielectric function at 6 eV and 12 eV due to transitions from p to d states [148]. An additional feature at $\sim$1 eV was found to be proportional to the number of f electrons.

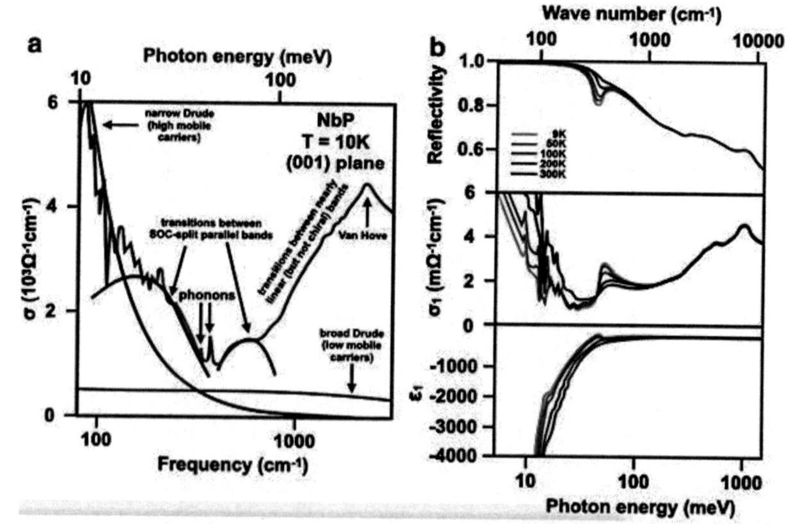

Fig. 9.57 The optical conductivity of NbP at 10 K in (**a**) and TaP in (**b**). The origins of the optical transitions are shown (**a**). Figure (**b**) also shows the reflectivity and real part of the dielectric function for TaP. The same analysis shown in (**a**) for NbP was used for TaP [147]. Figure adapted and reprinted with permission from [146, 147]. © 2018 and 2020 respectively American Physical Society

Pyrochlore iridates and related Ir compounds: The magneto-transport properties of pyrochlore iridates point to the evidence of Weyl nodes and a metal–insulator transition. An experimental study of the optical conductivity of the rare earth plus yttrium pyrochlore iridates $R_2Ir_2O_7$ (R=Pr, Nd, Sm, Eu, Dy, and Y) demonstrated the effect of the ionic radius on the metal–insulator phase transition [149]. This study used polycrystalline samples prepared using a solid state reaction at high pressure and temperature (e.g., 3 GPa at 1200 °C). A phase diagram for the paramagnetic insulator, paramagnetic metal, and antiferromagnetic insulator phases was developed and is shown here in Fig. 9.58 along with optical conductivity data for R=Y, Dy, Eu, and Sm. The optical conductivity data can be separated into the response above and below the metal–insulator transition temperature. The metal–insulator transition temperature of $R_2Ir_2O_7$ for R=Nd, Sm, and Eu has been reported to be 36 K, 117 K, and 120 K respectively [150]. The metal–insulator transition temperature of $Nd_2Ir_2O_7$ has also been reported to be 27 K [149]. Below metal–insulator transition temperature, the real part of the optical conductivity σ_1 is characterized by a linear increase in σ_1 at low energy $<\sim$0.2 eV (see Fig. 9.58f–h). Above 0.2 eV, σ_1 show transitions on top of a broad background $\sigma_1(E)\alpha\sqrt{E - E_g}$ [149]. The linear increase in σ_1 versus energy below 0.2 eV is attributed to the gapless 3D bands from the Weyl points that are linear in k [149]. The broad background is due to typical interband transitions. The exciton–magnon side bands can be seen in the optical conductivity data Fig. 9.58e, f at 10 K after the broad background is subtracted. This data can be fit to two Lorentz oscillators (9.20) with peak A being attributed to the exciton and peak B attributed to the first magnon sideband of the exciton [149]:

$$\sigma(E) = \frac{1}{4\pi} \sum_{i=A,B} \frac{S_i E^2 \Gamma_i}{\left(E_i^2 - E^2\right) + \Gamma_i^2 E^2} \tag{9.20}$$

where E_i, Γ_i, and S_i are the peak energy, dampening (broadening), and strength of the A, or B peaks. The peak energies of A and B along with the energy difference between the A and B peaks are plotted as a function of the ionic radius of the R in Fig. 9.58e. The energy difference is plotted versus $1/E_A$ in Fig. 9.58i. The energy of the exciton peak A is governed by the strength of the Hubbard interaction U discussed in Sect. 9.5.1.1 and in Fig. 9.24. The Hubbard interaction energy was introduced in Sect. 9.2. The energy difference between the A and B peaks is the energy of the magnons which is proportional to the angular momentum, $\frac{5}{3}J$, and the Heisenberg model for these materials relates the angular momentum, the Hubbard energy, and the hopping integral: $J \sim t^2/U$ [149]. The Hubbard energy changes with the ionic radius of the rare earth atom R.

The real part of the optical conductivity of the pyrochlore iridate $Eu_2Ir_2O_7$ shows the linear dependence predicted by (9.14) for the low energy data below 100 cm^{-1}. The Drude and phonon contributions to the real part of the optical conductivity are temperature dependent and are minimized at low temperature. A Fermi velocity of 3.4×10^7 cm/s results from a fit of (9.14) to the data for wavenumbers less than

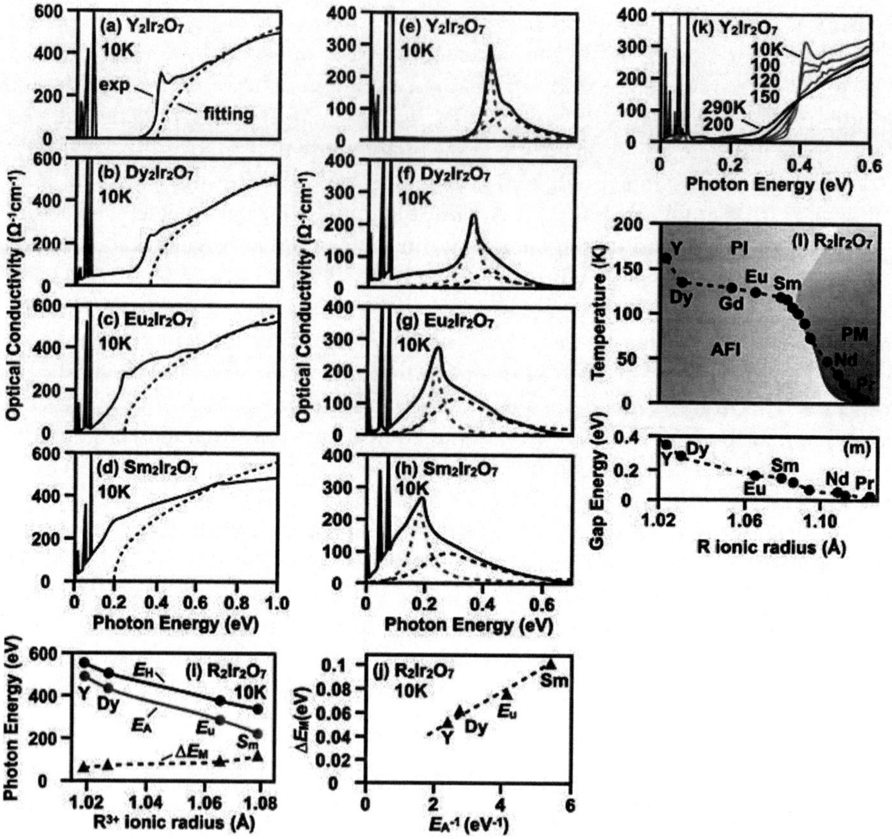

Fig. 9.58 Real part of the low energy optical conductivity at 10 K and phase diagram for pyrochlore iridates $R_2Ir_2O_7$ (R=Sm, Eu, Dy, and Y). The data is for polycrystalline samples. The background in the real part of the optical conductivity is shown as a dashed line in **(a)**–**(d)**. The optical conductivity shown in **(a)**–**(d)** is replotted in **(f)**–**(i)** with the broad background removed to highlight the spectral response attributed to the exciton–magnon side bands. The A peak (light gray dashed line) and B peak (dark gray dashed line) in **(f)**–**(i)** are fit to Lorentzian functions and the peak energies and the difference between peak energies ΔE_M plotted in **(e)** versus ionic radius and ΔE_M versus $1/E_A$ in **(i)**. E_A is used to approximate the true energy gap. The temperature dependent differences above and below the metal insulator transition temperature in the optical conductivity of $Y_2Ir_2O_7$ are shown in **(j)**. A phase diagram and the gap energy are plotted as a function of ionic radius in **(k)**. Figure adapted and reprinted with permission from [149]. © 2016 American Physical Society

100 cm^{-1} [101]. The data also point to a value of $z < 1$ for $E(k)\ \alpha\ |k|^z$. This deviation for a perfect Dirac dispersion is further explored for Cd_3As_2 below.

Hexagonal Insulator Iridium oxides (Na/Li)$_2$IrO$_3$ and Perovskites Sr$_{n+1}$Ir$_n$O$_{3n+1}$ (n = 1, 2, ∞): The optical conductivity determined using spectroscopic ellipsometry and the electronic band structure calculations for Na_2IrO_3 were shown in Fig. 9.38 [118]. The different transitions were fit to Lorentz oscillators to determine

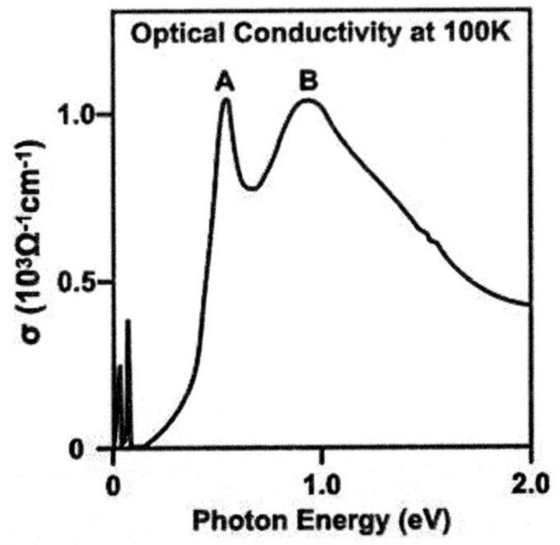

Fig. 9.59 The real part of the optical conductivity of Sr_2IrO_4 is shown. The A and B peaks in conductivity arise from transitions between the Lower Hubbard $J_{eff} = 1/2$ upper Hubbard $J_{eff} = 1/2$ states, and the B peak is due to transitions from the $J_{eff} = 3/2$ and upper Hubbard $J_{eff} = 1/2$ states as shown in Fig. 9.40. Figure adapted and reprinted with permission from [118]. © 2008 American Physical Society

the transition energies. Na_2IrO_3 is an example of the mixing of the $5d^5$ $J_{eff} = 1/2$ and $J_{eff} = 3/2$ discussed in Sect. 9.5.1.1. The optical conductivity of Sr_2IrO_4 is shown in Fig. 9.59. The optical transitions associated with the Lower Hubbard $J_{eff} = 1/2$ upper Hubbard $J_{eff} = 1/2$ states, and the B peak is due to transitions from the $J_{eff} = 3/2$ and upper Hubbard $J_{eff} = 1/2$ states are clearly visible in the spectra.

9.5.2.2 Type II Weyl Semimetal

Transition Metal Dichalcogenides: Although TMD materials were discussed in Chap. 8, TMD materials that are potentially Weyl semimetals are discussed here in this chapter. As discussed in Sect. 9.5.1.2, the crystal structure of WTe_2 is distorted and there are three inequivalent crystallographic directions. The zigzag chain of W atoms is along the a direction, and the c direction is normal to the TMD layers. We note that the Type 2 Weyl response is not easily observed in the low energy optical spectra of WTe_2 [130]. The optical spectra were studied by near normal incidence FT-IR [130]. WTe_2 shows strong anisotropy along the a versus b crystallographic axis. The ratio of plasma frequencies $\omega_{p,a}/\omega_{p,b} = \sim 1.5$ can be interpreted as being due to the anisotropy of the effective masses m^* at low temperature since $\omega_p^2 \sim 1/m^*$[130]. The reflectivity of light polarized along the a and b axes as well the low temperature dielectric function at 10 K along the a and b axes are shown in Fig. 9.60. In Fig. 9.61, the band to band transitions that contribute to the optical conductivity are shown. This theoretical interpretation indicates that the origin of the optical conductivity needs to be carefully understood, and is not always due to transitions near the Weyl point [130]. The band structure calculations discussed in [130] show that Weyl nodes lay

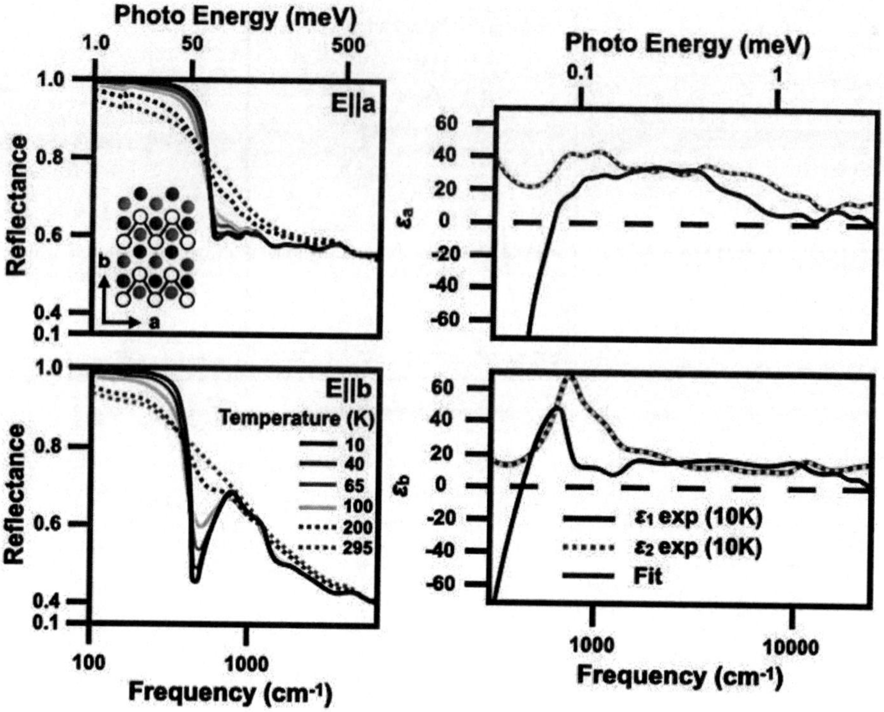

Fig. 9.60 The reflectivity and dielectric function of the Type 2 Weyl semimetal WTe_2. The zigzag chain of W atoms is along the *a* direction, and the *c* direction is normal to the TMD layers. The *b* direction is perpendicular to *a*. The low temperature reflectivity shows a plasma edge which depends on which crystallographic axis the light is polarized along. The dielectric function is also shown along the a axis at the top right and the b axis along the bottom left at 10 K. The real part of the dielectric function crosses 0 at 600 cm^{-1} for the *a* axis and 440 cm^{-1} for the *b* axis. The anisotropy ratio of $\omega_{p\,a}/\omega_{p,b} = \sim 1.5$ can be interpreted in terms of the difference in the effective mass difference between the *a* and *b* crystallographic axis. Figure adapted and reprinted with permission from [130]. © 2017 American Physical Society

above the Fermi level and since optical transitions are from occupied levels to unoccupied levels. Thus, this study finds that the Weyl nodes of WTe_2 do not contribute to the optical conductivity [130]. Also as mentioned in Sect. 9.5.1.2, other band structure calculations, that include an onsite Hubbard energy of $U = 2$ eV, in conjunction with ARPES spectra [129] indicate that the electronic states close to the Γ point are 3D in nature and are Weyl nodes. This data was taken between 400 eV and 800 eV to increase bulk sensitivity. The study also calculated the conductivity using this band structure and found close agreement with experimental data. These different conclusions point out the challenges in verifying the presence of Weyl nodes.

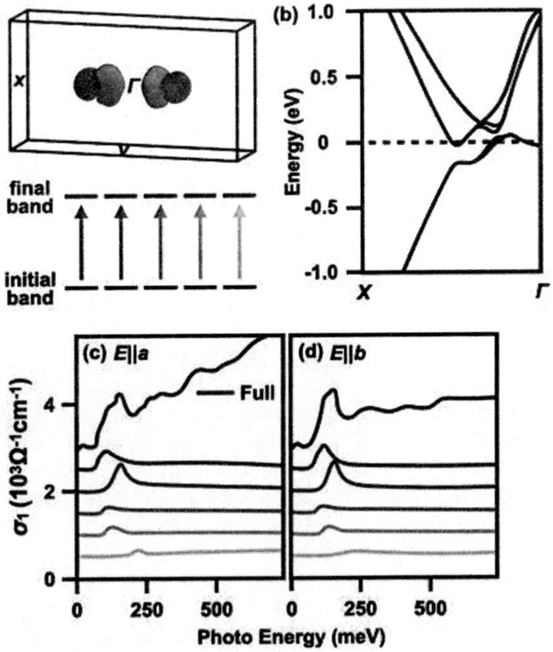

Fig. 9.61 The band to band optical transitions that contribute the real part of the optical conductivity of Type 2 Weyl semimetal WTe_2 are shown. (**a**) The Fermi surface in the First Brillouin Zone is shown the electron pockets in dark gray and hole pockets in light gray. The gray scale coding for the transitions for the bands shown in (**c**) and (**d**) is defined in (**a**). In (**c**) and (**d**) the individual contributions to the optical conductivity along the a and b crystallographic axes due to the five different optical transitions are shown. This study finds that the optical conductivity of WTe_2 is due to band to band transition and not the Weyl nodes. That is in contrast to the combined ARPES and band structure calculations of [129] which reports the presence of Weyl nodes. Figure adapted and reprinted with permission from [130]. © 2017 American Physical Society

9.5.2.3 Nodal Line (Ring) Weyl Semimetals

At the time of the writing of this book, experimental optical conductivity data for nodal line Weyl semimetals does not seem to be available. However, the interband optical conductivity of nodal line Weyl semimetals has been theoretically predicted to depend on the diameter of the Weyl nodal ring in momentum space. The conductivity is predicted to be constant at low energies as is true for 2D Dirac materials such as graphene as shown in Fig. 9.62 [151].

9.5.2.4 Dirac Semimetals

Cd_3As_2: As discussed in Sect. 9.5.1.4, the low energy electron band structure exhibits massless carriers, and some theoretical band studies predict the presence of both Kane

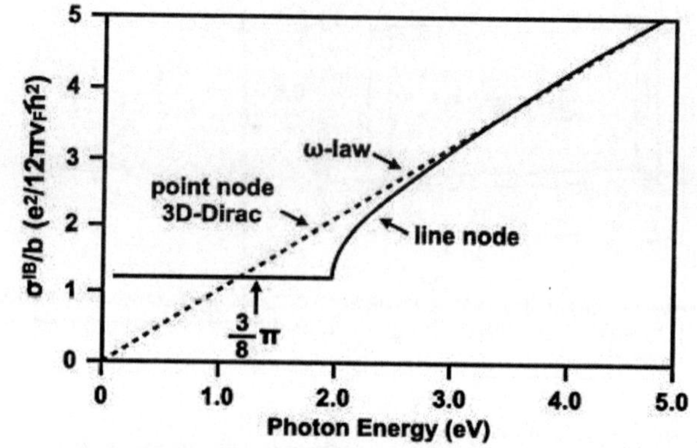

Fig. 9.62 The theoretical interband optical conductivity of nodal ring Weyl semimetals versus normalized photon energy. The optical conductivity of a Weyl semimetal with nodal points is shown in the dashed line. The normalization factor b is the radius of the Weyl nodal circle in momentum space. A discussion is provided in Sect. 9.3 along with (9.8). Figure adapted from [151].

and Dirac electrons. Here we discuss reflectivity [35] and magneto-optical characterization [41] results which builds on the introduction to optical characterization in Sect. 9.3. First we discuss the reflectivity measurements. The optical conductivity was determined from polarized Fourier transform, infra-red measurements of a polished (001) surface of a vapor transport grown crystal. The interband optical conductivity of the Dirac semimetal Cd_3As_2 can be expressed in terms of the Fermi velocity and N the number of Dirac cones as stated for Weyl semimetals in (9.14), $\sigma_0^N(\omega) = N\frac{e^2}{12h}\frac{\omega}{v_F}$ [35]. This equation can be rewritten when the sample is doped, so that the Fermi level E_F is not at zero energy, using the Heaviside step function θ follows:

$$\sigma_0^N(\omega) = N\frac{e^2}{12h}\frac{|\omega|}{v_F}\theta[\hbar\omega - 2E_F] \tag{9.21}$$

In order to better fit the low temperature conductivity data, this equation has been modified to account for self-energy term which is approximated by $\frac{\hbar\omega_g}{2}$. The self-energy is a correction to the mean field approximations used in band structure calculation methods such as DFT. Self-energy is due to many body effects from the electron's interactions with the surrounding medium [152].

$$\sigma_0^N(\omega) = N\frac{e^2}{12hv_F}\frac{(\omega-\omega_g)^2}{\omega}\theta\left[\hbar\omega - 2\max\left\{E_F, \frac{\hbar\omega_g}{2}\right\}\right] \tag{9.22}$$

The optical conductivity of a sample of n doped Cd_3As_2 with a roughly temperature independent electron density of $n = 6 \times 10^{17}\,\mathrm{cm}^{-3}$ and a mobility of $8 \times 10^4\,\mathrm{cm}^2/(\mathrm{V\,s})$ at 12 K is shown in Fig. 9.63 [35]. The optical conductivity at

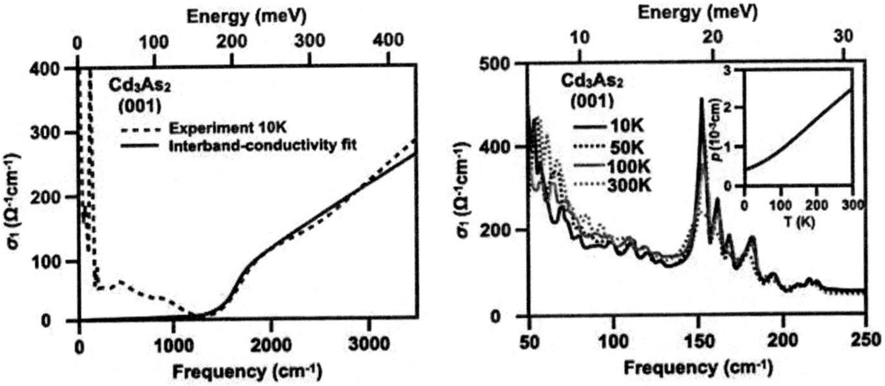

Fig. 9.63 The low energy optical conductivity of Cd_3As_2 obtained from Fourier transform infrared reflectivity data is shown. The arrows in the bottom right pane indicate the location of IR active phonon modes. Although not shown here, the real part of the dielectric function crosses 0 at the screened plasma frequency. Figure adapted and reprinted with permission from [35]. © 2016 American Physical Society

10 K was fit to (9.14) as shown in Fig. 9.63a. The frequency dependent Fermi velocity can be estimated by from the slope of (9.21)–(9.22). The optical conductivity between 2000 cm^{-1} and 8000 cm^{-1} for 10−300 K has a power law dependence on the frequency of $\sigma_0^N(\omega) \sim \pi k^{1.6}$. Recalling that the optical conductivity of d dimensional Dirac electrons has been related to the real part of the optical conductivity σ_1 and to the band dispersion of the Dirac bands by $\sigma_1 \sim \omega^{(d-2)/z}$ and $E(k)\ \alpha\ |k|^z$, this suggests that $z \sim 0.6$ [35]. Thus the energy dispersion factor z is less than the expected value of 1. This result is valid for energies more than 250 meV $(\sim 2000\,\text{cm}^{-1})$ [35]. The intraband optical conductivity can be expressed as a Drude contribution. This discrepancy was attributed to the fact that there is asymmetry between the valence and conduction bands for Cd_3As_2 which is not accounted for in (9.22). The dip in the optical conductivity was traditionally attributed to an optical gap, however, here it is has been reinterpreted as being the Fermi level which is above the Dirac points due to doping in the measured samples [35]. In order to better fit the low temperature conductivity data, this equation has been modified to account for self-energy term which is approximated by $\frac{\hbar\omega_g}{2}$.

Magneto-optical characterization of n doped (001) and (112) oriented Cd_3As_2 points to the presence of massless Kane electrons [41]. 3D Dirac electrons and massless Kane electrons have different magnetic field dependences for the cyclotron resonances [41]. The magnetic field dependence of the energy of Dirac electrons in the n Landau level is $E_n^D = \pm v_F\sqrt{2e\hbar Bn + \hbar^2k^2}$ for Fermi velocity v_F, and for massless Kane electrons is $E_{\xi n\sigma}^{Kane} = \pm\xi v_{Kane}\sqrt{e\hbar B(2n-1+\sigma) + \hbar^2k^2}$ where $\xi = -1$, 0, or 1 and $\sigma = \pm\frac{1}{2}$. The velocity Kane velocity $v_{Kane} = \sqrt{\frac{2}{3}}\frac{P}{\hbar}$ is determined from the anisotropic Kane interband matrix element. Measurements along different crystallographic directions will have different velocity and thus Landau

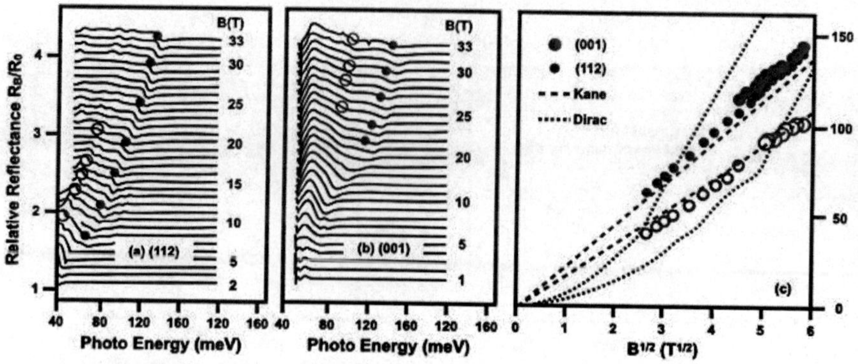

Fig. 9.64 Magneto-reflectivity characterization of (001) and (112) faces of Cd_3As_2 is shown. The dominant (open circle) and weak (filled in circle) cyclotron resonances for the (112) face is shown in (**a**) and the (001) face in (**b**). The resonance features for both crystal faces show $\sqrt{B}$ dependence of the dominant (open circle) and weak (filled in circle) cyclotron resonances expected from Kane massless electrons as shown in (**c**). Figure adapted and reprinted with permission from [41]. © 2016 American Physical Society

level energies. All cyclotron resonances of Kane electrons have a $\sqrt{B}$ magnetic field dependence including $n = 0$ Landau level. For the $n = 0$ Landau level of Dirac electrons, there is no dependence on the magnetic field. The transition energy between the 0 and 1 Landau levels for Dirac electrons has a low energy absorption edge given by $L_{0\to1} = \sqrt{v^2 2e\hbar B + E_F^2} - E_F$ [41]. This absorption edge does not have a $\sqrt{B}$ magnetic field dependence.

Low temperature Fourier transform infra-red spectroscopy in the presence of a magnetic field was used to characterize both a (001) oriented sample that was cut and polished from a bulk crystal and a (112) oriented sample which had a lower estimated electron density of $\sim 10^{17}$ cm^{-3}, and it was grown by a thermal evaporation method. The reflectance data for both samples shows two absorption features with $\sqrt{B}$ magnetic field dependence as shown in Fig. 9.64.

In summary, both optical conductivity and magneto-optical conductivity data point to the presence of massless carriers. The nature of the carrier depends on the energy range of the measurement. Magneto-optical characterization has been interpreted in terms of massless Kane electrons, and investigation of the potential presence of lower energy Dirac electrons was not possible because the energy of these electrons was below the energy range of this measurement. One of the challenges associated with all characterization of Cd_3As_2 is the lack of spatial uniformity in carrier density up to 30% at 5 K have been observed by spatially resolved infra-red reflectivity for samples prepared by multiple methods [75, 136]. The length scale of the charge puddles is $\sim$100 μm, and it becomes more pronounced at low temperature. This phenomenon has the potential to result in spatial dependence for other measurements such as ARPES and STM characterization of Cd_3As_2. This study also found that polishing the sample increased the Drude scattering rate and the high frequency reflectivity.

9.5.3 *Electrical Properties of Weyl, Dirac Semimetals, and Related Materials*

Weyl and Dirac semimetals are expected to have an anomalous Hall conductivity which means that the Hall voltage is proportional to the magnetization of the material and not the magnetic field strength. The magnetization of the Weyl and Dirac semimetals can be increased using a magnetic field, but as the data presented below shows, the magnetization quickly reaches a maximum amount as the magnetic field is increased. This behavior is distinctly different from most metals. This conductivity is due to presence of Berry curvature $\Omega_{i,\vec{k}}$ when an electric field is present, and for i Weyl nodes is given by [81]:

$$\sigma_{xy}^{AHE} = \sum_i n_{i,\vec{k}} \Omega_{i,\vec{k}}^z \tag{9.23}$$

where $n_{i,\vec{k}}$ is the occupation number of the ith Weyl node. The anomalous Hall resistivity is $1/\sigma_{xy}^{AHE}$. The longitudinal magnetoconductance of a normal metal or semimetal is predicted to decrease with B^2: $\sigma_{xx} \sim 1/B^2$ and the transverse Hall conductivity: $\sigma_{xy} \sim 1/B$ [39, 81].

As discussed in Sect. 9.3, charge transport in Weyl semimetals is attributed to a chiral anomaly [40]. Parallel electric and magnetic fields ($\vec{B} \| \vec{E}$) pump charge between the nodes breaking the chiral charge symmetry resulting in a Hall conductance. Thus the longitudinal conductivity increases with magnetic field (resistivity decreases). The presence of the transverse Hall conductivity and negative longitudinal resistivity (decrease in longitudinal resistivity with increasing magnetic field) is a signature of a Weyl semimetal [40]. The charge transport in doped Weyl semimetals (occasionally referred to as Weyl metals) has been theoretically predicted to be quantized and a function of the distance in momentum space between Weyl nodes in a Weyl node pair [153]. In addition, a large spin Hall conductivity has been predicted in the TaAs Weyl semimetal family by ab initio calculations [154].

9.5.3.1 Type 1 Weyl Semimetals

Transition Metal Monopnictides: Here, we discuss the transport data for several transition metal monopnictides—Type 1 Weyl Semimetals. Some of the publications assert that the Hall data is due to the topological properties of the material while others do not [155, 156]. Analysis of the Shubnikov-de Hass oscillations in the longitudinal magnetoresistivity data of TaAs result in a π Berry Phase for the cyclotron orbits [155, 156]. Recall that the $e^{-i\vartheta/2}$ phase factor for a wavefunction changes by π when the angle changes by 2π [see (7.41)–(7.43) in Chap. 7]. Observation of the Berry Phase in transport data is used as evidence of the Weyl nodes. The electron mobility at 10 K is 1.8×10^5 $cm^2/(V\ s)$. The electrical characterization data for TaAs is shown in Fig. 9.65. When the magnetic field is perpendicular to the current flow, the

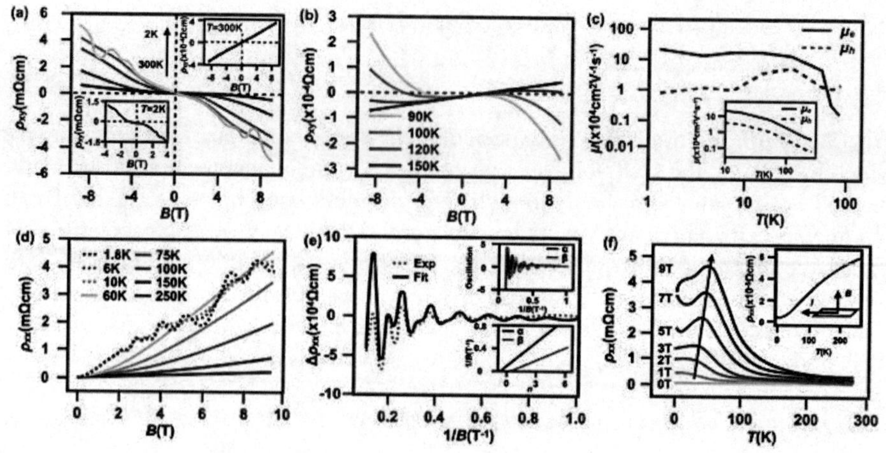

Fig. 9.65 Hall characterization data for TaAs. The Hall (transverse) magnetoresistivity versus temperature is shown in (**a**). The Hall data for four specific temperatures is shown in (**b**). The carrier mobility is shown in (**c**), and the longitudinal magnetoresistance versus temperature in (**d**). The Shubnikovb–de Hass oscillations shown in (**e**) are a sign of the high quality of the sample. Analysis of the SdH oscillations results in a π Berry phase which is considered to be a sign of the presence of Weyl nodes. The temperature dependence of the longitudinal resistivity as a function of the perpendicular magnetic field strength is shown in (**f**). The inset in (**f**) shows the zero field longitudinal resistivity (note the resistivity of the inset is in units of 10^{-5} Ω cm). Figure adapted and reprinted from [155]. © 2015 X. Huang, et al. CC BY 3.0 https://creativecommons.org/licenses/by/3.0/

magnetoresistance is very high at 9 T, see Fig. 9.65f. When it is parallel to the current flow, it becomes negative between 1 T and 9 T [156]. The transport data for NbAs is described as being attributable to either the material being a semimetal or being due to topological origins. The carrier density and mobility of the single crystal sample associated with the data in Fig. 9.66 is 1.8×10^{19} cm^{-3} and 3.5×10^5 cm^2/(V s) at 2 K [156]. The transverse and longitudinal magnetoresistance remain positive at 9 T for NbAs in contrast to TaAs. The anomalous Hall effect is not observed nor is quantized conductance. The data in [110] is considered to not offer proof of the presence of topological band effects. The difference in the properties of these two materials is due to the larger spin orbit coupling for Ta versus Nb. The unusually large magnetoresistance for $\vec{B} \perp \vec{E}$ of NbAs does not saturate up to 9 T [156]. The data for NbAs is shown in Fig. 9.66. At low temperature, the magnetoresistance is linear with magnetic field.

Rare Earth Monopnictides: Although certain rare earth monopnictides display unusual magnetotransport behavior, we are not aware of reports that confirm the presence of 3D Dirac or Weyl nodes. Here we show data for systems which display unusual magnetic layer ordering which results in extreme magnetoresistance. At low temperature the longitudinal magnetoresistance increases with increasing magnetic field strength and does not saturate. There is variability in the amount of extreme magnetoresistivity reported in various publications [91]. One way to distinguish

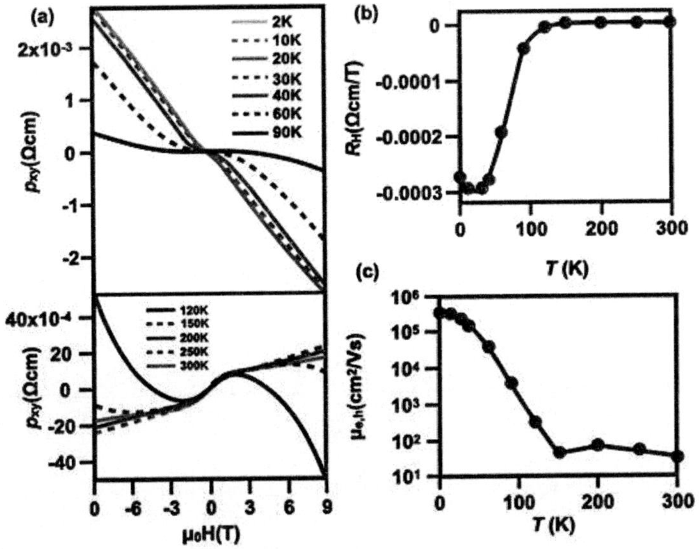

Fig. 9.66 Hall characterization of NbAs. The Hall resistivity is shown in (**a**), the Hall coefficient in (**b**), and the carrier mobility versus temperature in (**c**). Based on this data, the transport properties of NbAs do not show topological band effects. Figure adapted from [156].

between samples is to use the magnetoresistivity at 2 K and 9 T. Despite the variability of the magnetoresistivity with different samples, materials differences can still be traced to the number of f electrons and the amount of orbital hybridization between the pnictide p and rare earth f orbitals. For example, the Ce $4f^1$–Sb $5p^3$ hybridization is stronger than the Gd $4f^7$–Bi $6p^3$ hybridization [91].

The data for the single crystal sample of CeSb shown in Fig. 9.67 was obtained using a Sn flux method [91]. The magnetoresistivity the sample shown in Fig. 9.67 is over 1,672,000% at 2 K and 9 T, and the other three CeSb samples in this study [91] had lower values of the magnetoresistivity at 2 K and 9 T: 480,000%, >230,000%, and >128,000%. One can see the unusual temperature and magnetic field dependence of the longitudinal and transverse (Hall) magnetoresistance for a sample of CeSb [91] exhibiting the greatest increase in longitudinal resistivity for magnetic field sweeps from zero to 9 T as shown in Fig. 9.67. Other rare earth monopnictides with large magnetoresistances include LaBi, LaSb, YSb, GdSb, and GaBi [91]. As mentioned previously, CeSb has at least fourteen different magnetic phases including antiferromagnetic (AF), antiferroferromagnetic (AFF_n), ferromagnetic (FM), antiferroparamagnetic (AFP_n), ferroparamagnetic (FP_n), and paramagnetic (PM) phases [91]. In many of these structures, planes of Ce atoms have the magnetic moments of the f^1 electrons aligned and this interaction results in the overall magnetic structure of the CdSb. For example, there are two additional antiferroferromagnetic configurations: AFF1; repeating stacks with the stacking sequence: ↑↑↓↓↑↓; and AFF2 with repeating stacks having layers stacked with alignments ↑↑↓↓↑↓↓↑↑↓ [91, 96].

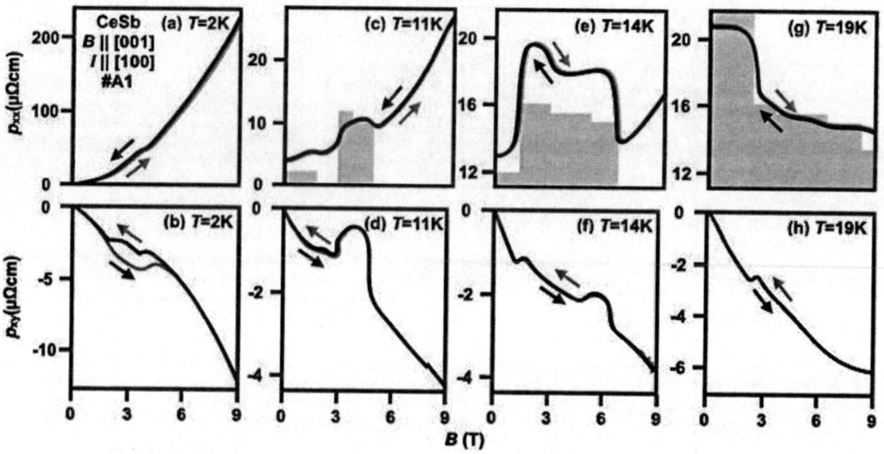

Fig. 9.67 The temperature dependence of the longitudinal and transverse magnetoresistance of CeSb. At different temperatures, the transitions between magnetic states cause changes (kinks) in the longitudinal and transverse magnetoresistance. For example, at 11 K, the changes in magnetoresistance between 3 and 6 T are due to transition from AF to AFF_n to FM (ferromagnetic) phases. The vertical dashed lines in the transverse magnetoresistance represent transitions between different magnetic states as the magnetic field is decreased. This data is from one of several samples fabricated for the study described in [91]. The presence of easily observable hysteresis in the transverse magnetoresistivity is noteworthy. Figure adapted and reprinted with permission from [92]. © 2018 American Physical Society

Other stacking sequences involve paramagnetic layers. Transitions between these phases can be observed at certain temperatures as shown in Fig. 9.67. CeP and CeAs also have many magnetic phases [91]. The role of Weyl nodes in the magnetoresistance behavior of some of the magnetic phases of CeSb is under further study [91].

Pyrochlore Iridates $X_2Ir_2O_7$ and related materials: The Hall transport properties of the pyrochlore iridates have been examined for signs of the presence of Weyl nodes. The magneto-electrical properties of some of the pyrochlore iridates are predicted to exhibit anomalous Hall behavior and charge density waves depending on the material [157]. As discussed in Sect. 9.5.1.1, the nature of the metal A in $A_2Ir_2O_7$ can change the material from metallic (A=Pr) to insulating (correlated insulator) (A=Y, and Yb to Eu) at room temperature [104, 158]. The change from metallic to insulating is attributed to the size of the A metal radii with the large radii resulting in a metallic material. $Pr_2Ir_2O_7$ is metallic down to 30 mK. $Pr_2Ir_2O_7$ exhibits a spontaneous Hall effect at zero magnetic field [159]. We also note that other authors have stated that A=Pr, Nd, Sm, and Eu are metallic while A=Gd, Tb, Dy, Ho, Yb, and Y are semiconducting [150]. This difference in perspective depends on the temperature range that is being considered in the discussion since a metal–insulator transitions have been observed at low temperatures for the lanthanide pyrochlore iridates $Ln_2Ir_2O_7$ (Ln=Nd, Sm, Eu) at 36 K, 17 K, and 120 K respectively [150]. The lanthanide 4f electrons are considered localized while the conducting electrons come from the Ir 5*d* orbitals [150]. When the A metal ion is magnetic, the magneto-electrical response of the material becomes more complicated. The materials properties result from interplay between spin orbit coupling, the topology of the electronic bands, and electronic correlations [158]. $Nd_2Ir_2O_7$ lies between the metallic and insulating states, and the exact stoichiometry of a specific sample can alter the materials response. The low temperature Hall response of polycrystalline $Nd_2Ir_2O_7$ transitions at 100 K into a magnetically disordered phase and then another transition into a magnetically ordered phase below 6 K [158]. This behavior is attributed to the greater magnetic interaction of Nd^{+3} versus Pr^{+3} [158]. Hysteresis is observed in the Hall data of this study. The temperature and magnetic field dependent resistivity and transverse Hall resistivity of $Nd_2Ir_2O_7$ is shown in Fig. 9.68. The reversible Hall resistivity was obtained by calculating ½ the sum of the forward and backward magnetic field sweep,

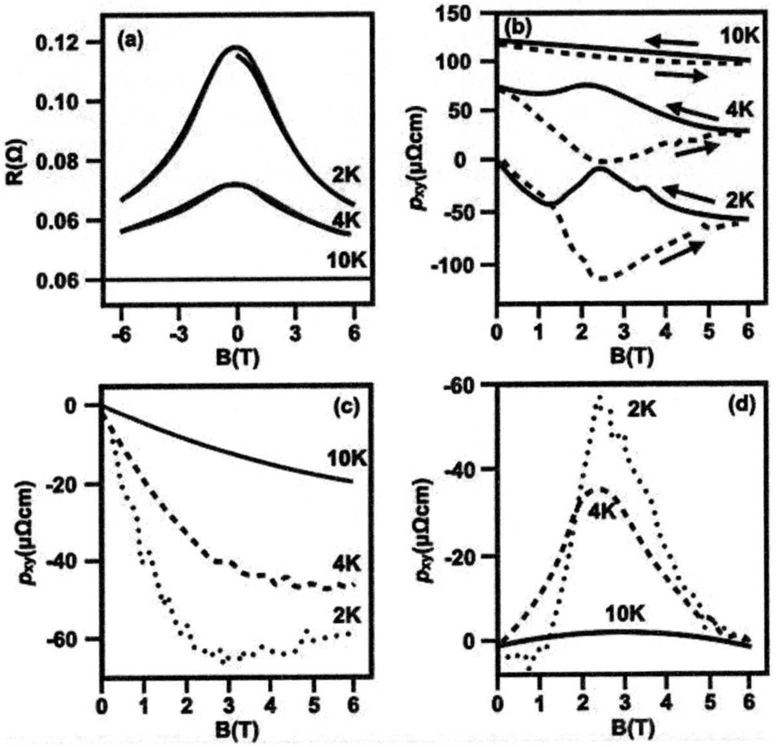

Fig. 9.68 The temperature and magnetic field dependent resistivity and transverse Hall resistivity of $Nd_2Ir_2O_7$ is shown. The resistance versus temperature of the sample is shown in (**a**). The data is seen to decrease with magnetic field at 4 K and below. The transverse magnetoresistivity (Hall) exhibits hysteresis as shown in (**b**). The reversible Hall resistivity (**c**) and irreversible Hall resistivity (**d**) are shown. The reversible Hall resistivity shown in (**b**) has been interpreted as having the magnetic field dependence expected of the anomalous Hall effect expected for Weyl semimetals [158]. Figure adapted and reprinted with permission from [158]. © 2013 American Physical Society

and for the irreversible Hall resistivity, the difference between forward and backward magnetic field sweeps was taken. The reversible Hall resistivity shows the magnetic field dependence expected of the anomalous Hall effect expected for Weyl semimetals [158]. This is attributed to a magnetic ordering in the Ir 5d electrons at the transition temperature [150]. At low temperature, $Pr_2Ir_2O_7$ exhibits a zero magnetic field, spontaneous transverse (Hall) conductance due to broken time-reversal symmetry in the absence of overall magnetic dipole order as shown in Fig. 9.69 [158]. At low temperature, $Pr_2Ir_2O_7$ is said to be a spin ice. The presence of long range magnetic order in of $Eu_2Ir_2O_7$ is inferred from the resistivity and magnetic susceptibility data shown in Fig. 9.70. The metal–insulator transition of $Eu_2Ir_2O_7$ is also seen the data resistivity of Fig. 9.70 [27].

Heusler and Half Heusler Alloys: As mentioned in Sect. 9.5.1.1, some Heusler alloys become Weyl semimetals in the presence of a magnetic field. This is in contrast to the indication in some publications that some members of this family might be topological insulators. GdPtBi and NdPtBi are antiferromagnetic below the Neil temperature of 9 K and 2.1 K respectively [128]. The negative magnetoresistance and observation of the anomalous Hall effect is evidence for presence of Weyl nodes when a magnetic field is applied as shown in Fig. 9.71c. The peak in anomalous Hall conductivity, $\Delta\sigma_{xy}$ and the dip in the longitudinal magnetoresistance when the magnetic and electric fields are parallel at a field strength of 2 T is due to the anomalous Hall effect and is sometimes observed in ferromagnetic materials [128]. As evidenced by the data for GdPtBi, non collinear antiferromagnets with electronic structures having non-zero Berry curvatures can also display the anomalous Hall effect [128]. The negative magnetoresistance and anomalous Hall effect are shown for NdPtBi in Fig. 9.71.

9.5.3.2 Type 2 Weyl Semimetals

WTe_2: A quantum spin Hall state was observed on a monolayer of 1 T′ WTe_2 through a combination of scanning tunneling microscopy (STM)/scanning tunneling spectroscopy (STS) and synchrotron based ARPES [160]. The sample used in this study was grown by molecular beam epitaxy on bi-layer graphene on 6H-SiC [160]. The sample was found to have a small bulk band gap of 55 ± 20 meV by ARPES and 56 ± 14 meV by STS. STS observed localized edge states which are believed to be topologically non-trivial [160].

9.5.3.3 Weyl Semimetals with Nodal Lines

$PbTaSe_2$: Chemically, $PbTaSe_2$ is a non-centrosymmetric transition metal dichalcogenide that is a superconductor [161, 162]. As discussed above, the crystal structure is different from other TMD materials in that the $TaSe_2$ monolayers are sandwiched

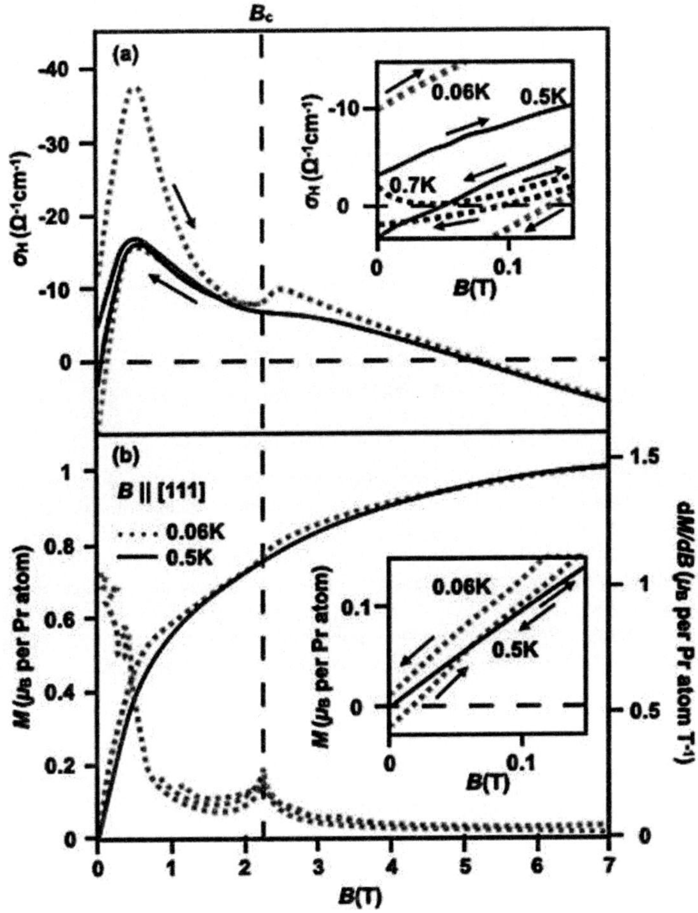

Fig. 9.69 The Hall conductance and magnetization of $Pr_2Ir_2O_7$ versus magnetic field strength is shown. As you sweep from 0 to 7 T, the starting Hall conductance is non-zero for 0.7, 0.5, and 0.06 K. The non-zero Hall conductance means that $Pr_2Ir_2O_7$ displays an anomalous Hall effect. The macroscopic magnetization is zero at zero magnetic field. It is important to note the hysteresis seen in the conductance data, and that the magnetization curves along the [001] and [110] directions do not show the hysteresis seen in the inset nor the anomaly seen in the dM/dB curve just above 2 T at 0.06 K. Recall the different symmetries along the [001] and [111] directions shown in Fig. 9.35. The anomalous Hall effect has been attributed to a chiral antiferromagnetic spin configuration [159]. Figure adapted from [159]

between hexagonal Pb layers [24, 161]. There is a competition between superconductivity and charge density waves (CDW). The Pb sheets slide over the $TaSe_2$ monolayers in a CDW [161, 162]. Experimental confirmation of the topological nature of the surface states is difficult because the Dirac point is estimated to be several hundred meV above the Fermi level making it inaccessible to angle resolved photoemission (ARPES) [162]. These states were imaged using a variation of STS

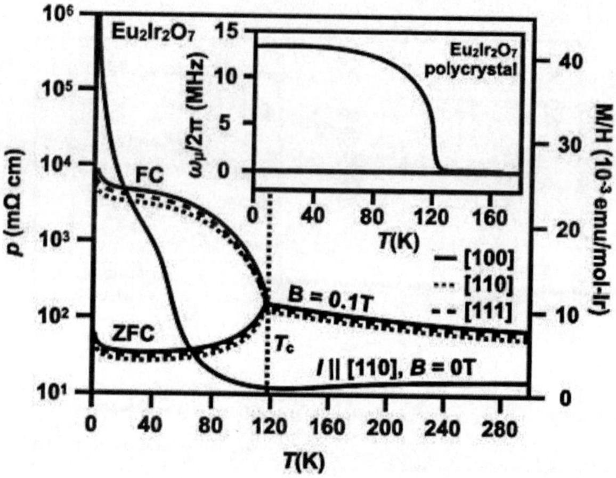

Fig. 9.70 The resistivity and magnetic susceptibility of $Eu_2Ir_2O_7$ is shown as a function of temperature. The critical temperature T_c for the transition from metal to insulator is 120 K. Since the magnetic susceptibility does not show hysteresis, $Eu_2Ir_2O_7$ is not considered to be ferromagnetic. It is inferred that the transition is to an antiferromagnetic insulator. Here, ZFC refers to the temperature dependence with no magnetic field, and FC refers to cooling in the presence of the magnetic field. The difference between the ZFC and FC magnetic susceptibilities is an indication of long range magnetic ordering. Figure adapted and republished with permission of Annual Reviews © 2014 from [27]. Permission conveyed through Copyright Clearance Center, Inc

known as quasi particle scattering interference imaging. This is done using Fourier transform STS [162]. As of the writing of this chapter, Hall data proof of topological surface states has not been found in the literature.

9.5.3.4 Dirac Semimetals

Cd_3As_2: There have been several magneto-transport studies of Cd_3As_2. As mentioned in Sect. 9.5.1.4, specifying the carrier density and sample quality is an important means of understanding why different studies provide different results [136]. The QHE has been observed at 1 K in high quality 20 nm thick Cd_3As_2 oriented along the (112) direction grown by MBE on insulating GaSb on a GaAs substrate having a low carrier concentration 2.8×10^{11} cm^{-2} [141]. When converted to a volume carrier concentration for comparison of the carrier concentration with that of bulk crystal samples discussed in the literature, this sample has a factor of 10 lower carrier density [141]. This data is shown in Fig. 9.72 below. The steps in transverse resistivity are quantized by filling factor $R_{xy} = h/(\nu e^2)$. This study was interpreted as demonstrating the presence of 2D surface states that dominate transport measurements at low temperature [141]. Thicker films do not show the QHE [141]. Magneto-transport data for bulk Cd_3As_2 with a higher carrier concentration is dominated by the bulk 3D Dirac carriers [141, 163]. Thicker films show a non-saturating longitudinal magnetoresitivity with increasing magnetic field strength and a negative transverse magnetoresitivity [163]. Determination of the Berry phase from oscillations in Cd_3As_2

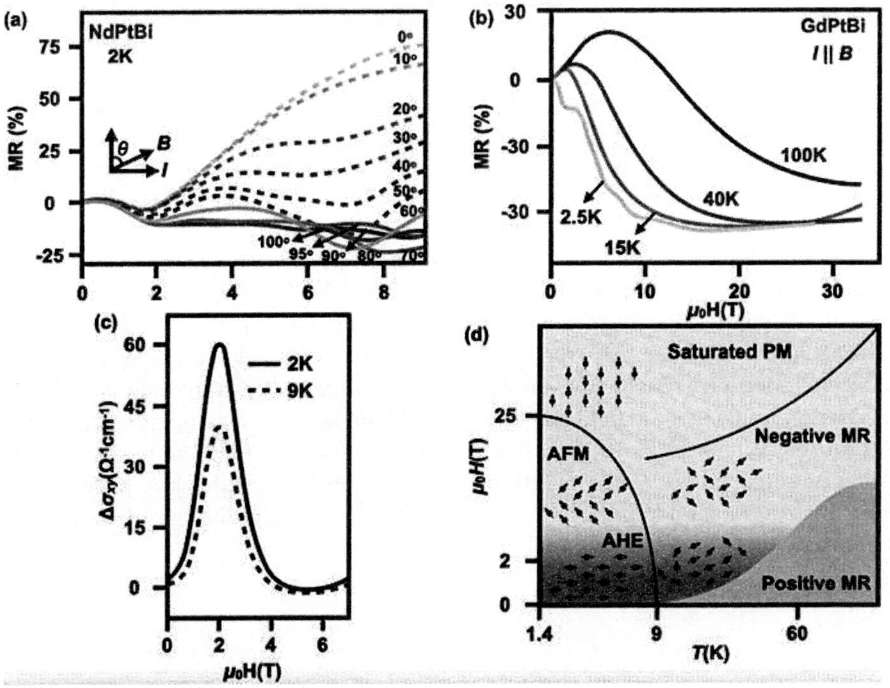

Fig. 9.71 The negative magnetoresitivity and anomalous Hall effect for GdPtBi and NbPtBi are shown along with a H–T (magnetic field–temperature) diagram for GdPtBi. In (**a**), the magnetoresitivity of NdPtBi and in (**b**) the magnetoresistivity of GdPtBi are shown. The extracted anomalous Hall conductivity, $\Delta\sigma_{xy}$ for GdPtBi is shown in (**c**). The H-T diagram showing the change in magnetic order versus magnetic field and temperature for GdPtBi is shown in (**d**). Figure adapted from [128]

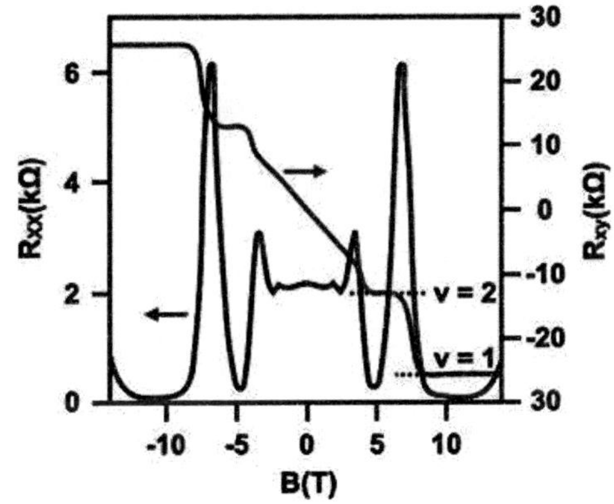

Fig. 9.72 Magneto-transport data for a 20 nm thick sample of Cd_3As_2 showing the quantum Hall effect. At higher temperature, the Shubnikov–de Hass oscillations are observed in the longitudinal resistivity. Figure adapted and reprinted with permission from [141]. © 2018 American Physical Society

transport data is difficult due to the superimposed the linear resistivity response [141].

A_3Bi (A = Na, K, Rb): Here we discuss the electrical transport properties of Na_3Bi. The transport properties of the Dirac semimetal Na_3Bi depend on the quality of the crystal. Crystals grown using a Na flux have been compared to crystals grown from a stoichiometric melt [143, 164, 165]. The results of this comparison are shown in Fig. 9.73. The quality of the flux grown sample is evident from the Shubnikov–de Hass oscillations in the longitudinal magnetoresistivity data shown in Fig. 9.73c. The scanning tunneling microscopy dI/dV data was used to locate the Dirac point [143]. The $\vec{B} \perp \vec{E}$ longitudinal magnetoresistance linearly increases with magnetic field. The same behavior has been observed for Cd_3As_2, and we note that this behavior is rarely observed for typical conductors [164]. The longitudinal magnetoresistance for $\vec{B} \parallel \vec{E}$ decreases with magnetic field and thus is described as a negative magnetoresistance [165]. As mentioned above, this is considered a signature of a Weyl semimetal and Dirac semimetal and is due to current flow between Weyl nodes [165]. The Dirac Fermions are separated into right or left handed groups (chirality) until the electric and magnetic field mix them breaking chirality [113].

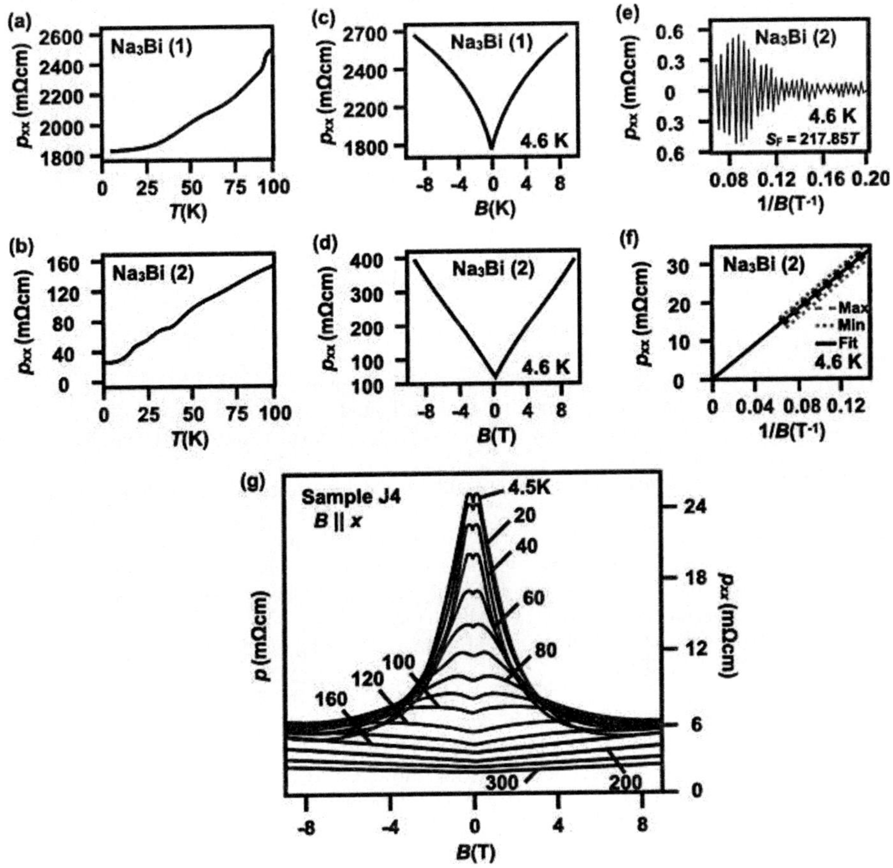

Fig. 9.73 The resistivity and magnetoresistivity of the Dirac semimetal Na_3Bi is shown for two samples. The longitudinal resistivity versus temperature with no magnetic field is shown in (**a**) and (**b**). The longitudinal resistivity versus magnetic field at 4.6 K is shown in (**c**) and (**d**). The data on the left is for $\vec{B} \perp \vec{E}$. The Shubnikov de Hass oscillations shown in (**e**) were obtained by subtracting the smooth background resistivity. The occupation number of the Landau levels versus 1/magnetic field are shown in (**f**). On the right, the negative (decreasing resistance with increasing magnetic field) longitudinal magnetoresistivity for $\vec{B} \| \vec{E}$ is shown. This is considered to be the signature of a Weyl semimetal. **a**–**f** adapted and reprinted from [143]. **g** adapted and reprinted from [165].

References

1. T.O. Wehling, A.M. Black-Schaffer, A.V. Balatsky, Dirac materials. Adv. Phys. **63**, 1–76 (2014)
2. T.-F. Chung, Y. Xu, Y.P. Chen, Transport measurements in twisted bilayer graphene: electron-phonon coupling and Landau level crossing. Phys. Rev. B **98**, 035425 (2018)
3. Y. Ando, Topological insulator materials. J. Phys. Soc. Jpn. **82**, 102001 (2013)
4. M.Z. Hasan, C.L. Kane, *Colloquium*: topological insulators. Rev. Mod. Phys. **82**, 3045–3067 (2010)
5. M.S. Dresselhaus, G. Dresselhaus, A. Jorio, Time reversal symmetry, in *Group Theory* (Springer, Berlin, 2008), pp. 403–430 (Chapter 16)
6. S.S. Chern, J. Simons, Characteristic forms and geometric invariants. Ann. Math. **99**, 48–69 (1974)
7. D.J. Thouless, M. Kohmoto, M.P. Nightingale, M. den Nijs, Quantized Hall conductance in a two-dimensional periodic potential. Phys. Rev. Lett. **49**, 405–408 (1982)
8. C.L. Kane, E.J. Mele, Z_2 topological order and the quantum spin hall effect. Phys. Rev. Lett. **95**, 146802 (2005)
9. L. Fu, C.L. Kane, Topological insulators with inversion symmetry. Phys. Rev. B **76**, 045302 (2007)
10. A. Bansil, H. Lin, T. Das, Colloquium: topological band theory. Rev. Mod. Phys. **88**, 021004 (2016); C.L. Kane, Topological band theory and the Z_2 invariant, in *Topological Insulators*, ed. by M. Franz, L. Molenkamp. Contemporary Concepts of Condensed Matter Science, vol. 6 (Elsevier, Amsterdam, 2013), pp. 1–324 (Chapter 1)
11. R.M. Kaufmann, D. Li, B. Wehefritz-Kaufmann, Notes on topological insulators. Rev. Math. Phys. **28**, 1630003 (2016)
12. H. Zhang, C.-X. Liu, X.-L. Qi, X. Dai, Z. Fang, S.-C. Zhang, Topological insulators in Bi_2Se_3, Bi_2Te_3 and Sb_2Te_3 with a single Dirac cone on the surface. Nat. Phys. **5**, 438–442 (2009)
13. J.C.Y. Teo, L. Fu, C.L. Kane, Surface states and topological invariants in three-dimensional topological insulators: application to $Bi_{1-x}Sb_x$. Phys. Rev. B **78**, 045426 (2008)
14. Z. Song, S.-J. Huang, Y. Qi, C. Fang, M. Hermele, Topological states from topological crystals. Sci. Adv. **5**, eaax2007 (2019)
15. M.G. Vergniory, L. Elcoro, C. Felser, N. Regnault, B. Andrei Bernevig, Z. Wang, A complete catalogue of high-quality topological materials. Nature **566**, 48–485 (2019)
16. A. Kitaev, *Periodic Table for Topological Insulators and Superconductors* (2009). arXiv:0901.2686v2 [cond-mat.mes-hall]
17. B.-J. Yang, N. Nagaosa, Classification of stable three-dimensional Dirac semimetals with nontrivial topology. Nat. Comm. **5**, 4898 (2014)
18. F. Zhang, C.L. Kane, E.J. Mele, Topological mirror superconductivity. Phys. Rev. Lett. **111**, 056403 (2013)
19. Y. Ueno, A. Yamakage, Y. Tanaka, M. Sato, Symmetry-protected majorana fermions in topological crystalline superconductors: theory and application to Sr_2RuO_4. Phys. Rev. Lett. **111**, 087002 (2013)
20. B. Yan, C. Felser, Topological materials: Weyl semimetals. Ann. Rev. Cond. Mat. Phys. **8**, 337–354 (2017)
21. Z.K. Liu, L.X. Yang, Y. Sun, T. Zhang, H. Peng, H.F. Yang, C. Chen, Y. Zhang, Y.F. Guo, D. Prabhakaran, M. Schmidt, Z. Hussain, S.-K. Mo, C. Felser, B. Yan, Y.L. Chen, Evolution of the Fermi surface of Weyl semimetals in the transition metal pnictide family. Nat. Mat. **15**, 27–31 (2016)
22. A.A. Burkov, Topological semimetals. Nat. Mat. **15**, 1145–1148 (2016)
23. M.J. Klein, On a degeneracy theorem of Kramers. Am. J. Phys. **20**(65), 65–71 (1952)
24. G. Bian, T.-R. Chang, R. Sankar, S.-Y. Xu, H. Zheng, T. Neupert, C.-K. Chiu, S.-M. Huang, G. Chang, I. Belopolski, D.S. Sanchez, M. Neupane, N. Alidoust, C. Liu, B.K. Wang, C.-C. Lee, H.-T. Jeng, C. Zhang, Z. Yuan, S. Jia, A. Bansil, F. Chou, H. Lin, M.Z. Hasan, Topological nodal-line fermions in spin-orbit metal $PbTaSe_2$. Nat. Comm. **7**, 10556 (2016)

25. H. Weng, X. Dai, Z. Fang, Transition-metal pentatellurides $ZrTe_5$ and $HfTe_5$: a paradigm for large-gap quantum spin hall insulators. Phys. Rev. X **4**, 011002 (2014)
26. N. Varnava, D. Vanderbilt, Surfaces of axion insulators, Phys. Rev. B **98**, 245117 (2018)
27. W. Witczak-Krempa, G. Chen, Y.B. Kim, L. Balents, Correlated quantum phenomena in the strong spin-orbit regime. Annu. Rev. Condens. Matter Phys. **5**, 57–82 (2014)
28. A.P. Mackenzie, Y. Maeno, Superconductivity of Sr_2RuO_4 and the physics of spin-triplet pairing. Rev. Mod. Phys. **75**, 657–712 (2003)
29. S. Koseki, N. Matsunaga, T. Asada, M.W. Schmidt, M.S. Gordon, Spin−orbit coupling constants in atoms and ions of transition elements: comparison of effective core potentials, model core potentials, and all-electron methods. J. Phys. Chem. A **123**, 2325–2339 (2019)
30. C. Mertens, M. Aichhorn, S. Biermann, Coulomb correlations in 4d and 5d oxides from first principles—or how spin–orbit materials choose their effective orbital degeneracies. Phys.: Condens. Matter. **29**, 263001 (2017)
31. A.A. Burkov, M.D. Hook, L. Balents, Topological nodal semimetals. Phys. Rev. B **84**, 235126 (2011)
32. M. Koshino, I.F. Hizbullah, Magnetic susceptibility in three-dimensional nodal semimetals. Phys. Rev. B **93**, 045201 (2016)
33. R.Y. Chen, Z.G. Chen, X-Y Song, J.A. Schneeloch, G.D. Gu, F. Wang, N. Wang, Magneto-infrared spectroscopy of Landau levels and Zeeman splitting of three-dimensional massless Dirac Fermions in $ZrTe_5$. Phys. Rev. Lett. **115**, 176404 (2015)
34. N.P. Armitage, E.J. Mele, A. Vishwanath, Weyl and Dirac semimetals in three dimensional solids. Rev. Mod. Phys. **90**, 015001 (2018)
35. D. Neubauer, J.P. Carbotte, A.A. Nateprov, A. Lohle, M. Dressel, A.V. Pronin, Inter-band optical conductivity of the [001]-oriented Dirac semimetal Cd_3As_2. Phys. Rev. B **93**, 121202(R) (2016)
36. K. Burch, Private communication
37. P. Hosur, S.A. Parameswaran, A. Vishwanath, Charge transport in Weyl semimetals. Phys. Rev. Lett. **108**, 046602 (2012)
38. P.E.C. Ashby, J.P. Carbotte, Chiral anomaly and optical absorption in Weyl semimetals. Phys. Rev. B **89**, 245121 (2014)
39. C.-Z. Chang, M. Li, Quantum anomalous Hall effect in time-reversal-symmetry breaking topological insulators. J. Phys.: Condens. Matter **28**, 123002 (2016)
40. S. Jia, S.-Y. Xu, M. Zahid Hasan, Weyl semimetals, Fermi arcs and chiral anomalies. Nat. Mat. **15**, 1140–1144 (2016)
41. A. Akrap, M. Hakl, S. Tchoumakov, I. Crassee, J. Kuba, M.O. Goerbig, C.C. Homes, O. Caha, J. Novák, F. Teppe, W. Desrat, S. Koohpayeh, L. Wu, N.P. Armitage, A. Nateprov, E. Arushanov, Q. D. Gibson, R.J. Cava, D. van der Marel, B.A. Piot, C. Faugeras, G. Martinez, M. Potemski, M. Orlita, Magneto-optical signature of Massless Kane electrons in Cd_3As_2. Phys. Rev. Lett. **117**, 136401 (2016)
42. Y. Jiang, J. Wang, T. Zhao, Z.L. Dun, Q. Huang, X.S. Wu, M. Mourigal, H.D. Zhou, W. Pan, M. Ozerov, D. Smirnov, Z. Jiang, Unraveling the topological phase of $ZrTe_5$ via magneto-infrared spectroscopy. Phys. Rev. Lett. **125**, 046403 (2020)
43. A.J. Green, S. Dey, Y.Q. An, B. O'Brien, S. J. O'Mullane, B. Thiel, A.C. Diebold, Surface oxidation of the topological insulator Bi_2Se_3. J. Vac. Sci. Technol. A **34**, 061403 (2016)
44. S. Roy, H.L. Meyerheim, K. Mohseni, A. Ernst, M.M. Otrokov, M.G. Vergniory, G. Mussler, J. Kampmeier, D. Grützmacher, C. Tusche, J. Schneider, E.V. Chulkov, J. Kirschner, Atomic relaxations at the (0001) surface of Bi_2Se_3 single crystals and ultrathin films. Phys. Rev. B **90**, 155456 (2014)
45. L.-L. Wang, M. Huang, S. Thimmaiah, A. Alam, S.L. Bud'ko, A. Kaminski, T.A. Lograsso, P. Canfield, D.D. Johnson, Native defects in tetradymite $Bi_2(Te_xSe_{3-x})$ topological insulators. Phys. Rev. B **87**, 125303 (2013)
46. A.A. Reijnders, Y. Tian, L.J. Sandilands, G. Pohl, I.D. Kivlichan, S.Y. Frank Zhao, S. Jia, M.E. Charles, R.J. Cava, N. Alidoust, S. Xu, M. Neupane, M. Zahid Hasan, X. Wang, S.W. Cheong, K.S. Burch, Optical evidence of surface state suppression in Bi-based topological insulators. Phys. Rev. B **89**, 075138 (2014)

47. Z. Ren, A.A. Taskin, S. Sasaki, K. Segawa, Y. Ando, Large bulk resistivity and surface quantum oscillations in the topological insulator Bi_2Te_2Se. Phys. Rev. B **82**, 241306(R) (2010)
48. E.K. de Vries, *Thesis: Taking Topological Insulators for a Spin: Towards Understanding of Spin and Charge Transport in Bi_2Se_3* (Rijksuniversiteit Groningen, Groningen, 2017)
49. M. Bianchi, D. Guan, S. Bao, J. Mi, B. Brummerstedt Iversen, P.D.C. King, P. Hofmann, Coexistence of the topological state and a two-dimensional electron gas on the surface of Bi_2Se_3. Nat. Comm. **1**, 128 (2010)
50. R.W.G. Wyckoff, *Crystal Structures*, vol. 2 (Wiley, New York, 1964)
51. J.P. Heremans, R.J. Cava, N. Samarth, Tetradymites as thermoelectrics and topological insulators. Nat. Rev. Mat. **2**, 1–21 (2017)
52. D.S. Lee et al., Crystal structure, properties and nanostructuring of a new layered chalcogenide semiconductor, Bi_2MnTe_4. CrystEngComm **15**, 5532–5538 (2013)
53. M.M. Otrokov, I.P. Rusinov, M. Blanco-Rey, M. Hoffmann, A.Y. Vyazovskaya, S.V. Eremeev, A. Ernst, P.M. Echenique, A. Arnau, E.V. Chulkov, Unique thickness-dependent properties of the van der Waals interlayer antiferromagnet $MnBi_2Te_4$ films. Phys. Rev. Lett. **122**, 107202 (2019)
54. D. Zhang, M. Shi, T. Zhu, D. Xing, H. Zhang, J. Wang, Topological axion states in the magnetic insulator $MnBi_2Te_4$ with the quantized magnetoelectric effect. Phys. Rev. Lett. **122**, 206401 (2019)
55. C. Liu, Y. Wang, H. Li, Y. Wu, Y. Li, J. Li, K. He, Y. Xu, J. Zhang, Y. Wang, Robust axion insulator and Chern insulator phases in a two-dimensional antiferromagnetic topological insulator. Nat. Mat. **19**, 522–527 (2020)
56. W. Zhang, R. Yu, W. Feng, Y. Yao, H. Weng, X. Dai, Z. Fang, Topological aspect and quantum magnetoresistance of -Ag_2Te. Phys. Rev. Lett. **106**, 156808 (2011)
57. A. Sulaev, P. Ren, B. Xia, Q.H. Lin, T. Yu, C. Qiu, S.-Y. Zhang, M.-Y. Han, Z.P. Li, W.G. Zhu, Q. Wu, Y.P. Feng, L. Shen, S.-Q. Shen, L. Wang, Experimental evidences of topological surface states of β-Ag_2Te. AIP Adv. **3**, 032123 (2013)
58. S. Lee, J. In, Y. Yoo, Y. Jo, Y.C. Park, H.-J. Kim, H.C. Koo, J. Kim, B. Kim, K.L. Wang, Single crystalline β-Ag_2Te nanowire as a new topological insulator. Nano Lett. **12**, 4194–4199 (2012)
59. G. Manzoni, L. Gragnaniello, G. Autès, T. Kuhn, A. Sterzi, F. Cilento, M. Zacchigna, V. Enenkel, I. Vobornik, L. Barba, F. Bisti, Ph. Bugnon, A. Magrez, V. N. Strocov, H. Berger, O.V. Yazyev, M. Fonin, F. Parmigiani, A. Crepaldi, Evidence for a strong topological insulator phase in $ZrTe_5$. Phys. Rev. Lett. **117**, 237601 (2016)
60. P. Di Pietro, F. M. Vitucci, D. Nicoletti, L. Baldassarre, P. Calvani, R. Cava, Y. S. Hor, U. Schade, S. Lupi, Optical conductivity of bismuth-based topological insulators. Phys. Rev. B **86**, 045439 (2012)
61. Y. Sharma, P. Srivastava, A. Dashora, L. Vadkhiya, M.K. Bhayani, R. Jain, A.R. Jani, B.L. Ahuja, Electronic structure, optical properties and Compton profiles of Bi_2S_3 and Bi_2Se_3. Solid State Sci. **14**, 241 (2012)
62. L.A. Walsh, A.J. Green, R. Addou, W. Nolting, C.R. Cormier, A.T. Barton, T.R. Mowll, R. Yue, N. Lu, N. Kim, M.J. Kim, V.P. LaBella, C. Ventrice, S. McDonnell, W.G. Vandenberghe, R.M. Wallace, A.C. Diebold, C.L. Hinkle, Fermi level manipulation through native doping in the topological insulator Bi_2Se_3. ACS Nano **12**, 6310–6318 (2018)
63. J. Humlíček, D. Hemzal, A. Dubroka, O. Caha, H. Steiner, G. Bauer, G. Springholz, Phys. Scr. **T162**, 14007 (2014)
64. N.T. Mamedov, E.H. Alizade, Z.A. Jahangirli, Z.S. Aliev, N.A. Abdullayev, S.N. Mammadov, I.R. Amiraslanov, Y.-G. Shim, K. Wakita, S.S. Ragimov, A.I. Bayramov, M.B. Babanly, A.M. Shikin, E.V. Chulkov, Infrared spectroscopic ellipsometry and optical spectroscopy of plasmons in classic 3D topological insulators. J. Vac. Sci. Technol. B **37**, 062602 (2020)
65. P.K. Das, T.J. Whitcher, M. Yang, X. Chi, Y.P. Feng, W. Lin, J.S. Chen, I. Vobornik, J. Fujii, K.A. Kokh, O.E. Tereshchenko, C.Z. Diao, J. Moon, S. Oh, A.H. Castro-Neto, M.B.H. Breese, A.T.S. Wee, A. Rusydi, Electronic correlation determining correlated plasmons in Sb-doped Bi_2Se_3. Phys. Rev. B **100**, 115109 (2019)

66. K.M.F. Shahil, M.Z. Hossain, V. Goyal, A.A. Balandin, Micro-Raman spectroscopy of mechanically exfoliated few-quintuple layers of Bi_2Te_3, Bi_2Se_3 and Sb_2Te_3 materials. J. Appl. Phys. 111, 054305 (2012)
67. Z.A. Jahangirli, E.H. Alizade, Z.S. Aliev, M.M. Otrokov, N.A. Ismayilova, S.N. Mammadov, I.R. Amiraslanov, N.T. Mamedov, G.S. Orudjev, M.B. Babanly, A.M. Shikin, E.V. Chulkov, Electronic structure and dielectric function of Mn–Bi–Te layered compounds. J. Vac. Sci. Technol. B 37, 062910 (2019)
68. Z. Guo, H. Gu, M. Fang, B. Song, W. Wang, X. Chen, C. Zhang, H. Jiang, L. Wang, S. Liu, Complete dielectric tensor and giant optical anisotropy in quasi-one-dimensional $ZrTe_5$. ACS Mater. Lett. **3**, 525–534 (2021)
69. A. Akrap, Private communication
70. R.Y. Chen, S.J. Zhang, J.A. Schneeloch, C. Zhang, Q. Li, G.D. Gu, N.L. Wang, Optical spectroscopy study of the three-dimensional Dirac semimetal $ZrTe_5$. Phys. Rev. B 92, 075107 (2015)
71. E. Martino, I. Crassee, G. Eguchi, D. Santos-Cottin, R.D. Zhong, G.D. Gu, H. Berger, Z. Rukelj, M. Orlita, C.C. Homes, A. Akrap, Two-dimensional conical dispersion in $ZrTe_5$ evidenced by optical spectroscopy. Phys. Rev. Lett. 122, 217402 (2019)
72. D. Santos-Cottin, M. Padlewski, E. Martino, S. Ben David, F. Le Mardelé, F. Capitani, F. Borondics, M. D. Bachmann, C. Putzke, P. J. W. Moll, R. D. Zhong, G. D. Gu, H. Berger, M. Orlita, C.C. Homes, Z. Rukelj, A. Akrap, Probing intraband excitations in $ZrTe_5$: a high-pressure infrared and transport study. Phys. Rev. B **101**, 125205 (2020)
73. Y.-X. Wang, F. Li, Examining the validity of the two-dimensional conical model to describe the three-dimensional $ZrTe_5$. Phys. Rev. B **101**, 195201 (2020)
74. Z. Rukelj, C.C. Homes, M. Orlita, A. Akrap, Distinguishing the gapped and Weyl semimetal scenario in $ZrTe_5$: insights from an effective two-band model. Phys. Rev. B **102**, 125201 (2020)
75. I. Crassee, E. Martino, C.C. Homes, O. Caha, J. Novák, P. Tückmantel, M. Hakl, A. Nateprov, E. Arushanov, Q.D. Gibson, R.J. Cava, S.M. Koohpayeh, K.E. Arpino, T.M. McQueen, M. Orlita, A. Akrap, Nonuniform carrier density in Cd_3As_2 evidenced by optical spectroscopy. Phys. Rev. B **97**, 125204 (2018)
76. G. Qiu, Y. Du, A. Charnas, H. Zhou, S. Jin, Z. Luo, D.Y. Zemlyanov, X. Xu, G.J. Cheng, P.D. Ye, Observation of optical and electrical in-plane anisotropy in high-mobility few-layer $ZrTe_5$. Nano Lett. **16**, 7364–7369 (2016)
77. J. Wu, S. Zhang, L. Tong, J. Zhang, Raman spectroscopy of anisotropic two-dimensional materials, in *Raman Spectroscopy of Two-Dimensional Materials*, ed. by P.H. Tan (Springer Nature, Singapore, 2019), pp. 53–80 (Chapter 3)
78. J.G. Analytis, R.D. McDonald, S.C. Riggs, J.-H. Chu, G.S. Boebinger, I.R. Fisher, Two-dimensional surface state in the quantum limit of a topological insulator. Nat. Phys. **6**, 960–964 (2010)
79. Y. Deng, Y. Yu, M.Z. Shi, Z. Guo, Z. Xu, J. Wang, X.H. Chen, Y. Zhang, Quantum anomalous Hall effect in intrinsic magnetic topological insulator $MnBi_2Te_4$. Science **367**, 895–900 (2020)
80. S. Zhang, R. Wang, X. Wang, B. Wei, B. Chen, H. Wang, G. Shi, F. Wang, B. Jia, Y. Ouyang, F. Xie, F. Fei, M. Zhang, X. Wang, D. Wu, X. Wan, F. Song, H. Zhang, B. Wang, Experimental observation of the gate-controlled reversal of the anomalous hall effect in the intrinsic magnetic topological insulator $MnBi_2Te_4$ device. Nano Lett. **20**, 709–714 (2019)
81. T. Liang, J. Lin, Q. Gibson, S. Kushwaha, M. Liu, W. Wang, H. Xiong, J.A. Sobota, M. Hashimoto, P.S. Kirchmann, Z.-X. Shen, R.J. Cava, N.P. Ong, Anomalous Hall effect in $ZrTe_5$. Nat. Phys. **14**, 451–455 (2018)
82. H. Xiong, J.A. Sobota, S.-L. Yang, H. Soifer, A. Gauthier, M.-H. Lu, Y.-Y. Lv, S.-H. Yao, D. Lu, M. Hashimoto, P.S. Kirchmann, Y.-F. Chen, Z.-X. Shen, Three-dimensional nature of the band structure of $ZrTe_5$ measured by high-momentum-resolution photoemission spectroscopy. Phys. Rev. B **95**, 195119 (2017)
83. F. Tang, Y. Ren, P. Wang, R. Zhong, J. Schneeloch, S.A. Yang, K. Yang, P.A. Lee, G. Gu, Z. Qiao, L. Zhang, Three-dimensional quantum Hall effect and metal–insulator transition in $ZrTe_5$. Nature **569**, 537–542 (2019)

84. X.-B. Li, W.-K. Huang, Y.-Y. Lv, K.-W. Zhang, C.-L. Yang, B.-B. Zhang, Y.B. Chen, S.-H. Yao, J. Zhou, M.-H. Lu, L. Sheng, S.-C. Li, J.-F. Jia, Q.-K. Xue, Y.-F. Chen, D.-Y. Xing, Experimental observation of topological edge states at the surface step edge of the topological insulator $ZrTe_5$. Phys. Rev. Lett. **116**, 176803 (2016)
85. Q. Ma, S.-Y. Xu, C.-K. Chan, C.-L. Zhang, G. Chang, Y. Lin, W. Xie, T. Palacios, H. Lin, S. Jia, P.A. Lee, P. Jarillo-Herrero, N. Gedik, Direct optical detection of Weyl fermion chirality in a topological semimetal. Nat. Phys. **13**, 842–847 (2017)
86. A.A. Soluyanov, D. Gresch, Z. Wang, Q.-S. Wu, M. Troyer, X. Dai, B. Andrei Bernevig, Type-II Weyl semimetals. Nature **527**, 495–498 (2015)
87. Y. Sun, S.-C. Wu, B. Yan, Topological surface states and Fermi arcs of the noncentrosymmetric Weyl semimetals TaAs, TaP, NbAs, and NbP. Phys. Rev. B **92**, 115428 (2015)
88. D. Grassano, O. Pulci, A.M. Conte, F. Bechstedt, Validity of Weyl fermion picture for transition metals monopnictides TaAs, TaP, NbAs, and NbP from *ab initio* studies. Nat. Sci. Rep. **8**, 3534 (2018)
89. S.-M. Huang, S.-Y. Xu, I. Belopolski, C.-C. Lee, G. Chang, B.K. Wang, N. Alidoust, G. Bian, M. Neupane, C. Zhang, S. Jia, A. Bansil, H. Lin, M.Z. Hasan, A Weyl Fermion semimetal with surface Fermi arcs in the transition metal monopnictide TaAs class. Nat. Comm. **6**, 7373 (2015)
90. S. Furuseth, K. Selte, A. Kjekshus, On the arsenides and antimonides of Tantalum. Acta Chem. Scand. **19**, 95 (1965)
91. L. Ye, T. Suzuki, C.R. Wicker, J.G. Checkelsky, Extreme magnetoresistance in magnetic rare-earth monopnictides. Phys. Rev. B **97**, 081108(R) (2018)
92. C.-G. Duan, R.F. Sabirianov, W.N. Mei, P.A. Dowben, S.S. Jaswal, E.Y. Tsymbal, Electronic, magnetic and transport properties of rare-earth monopnictides. J. Phys.: Condens. Matter 19, 315220 (2007)
93. M. Zeng, C. Fang, G. Chang, Y.-A. Chen, T. Hsieh, A. Bansil, H. Lin, L. Fu, Topological semimetals and topological insulators in rare earth monopnictides. arXiv:1504.03492v1
94. R. Lou, B.-B. Fu, Q.N. Xu, P.J. Guo, L.-Y. Kong, L.-K. Zeng, J.-Z. Ma, P. Richard, C. Fang, Y.-B. Huang, S.-S. Sun, Q. Wang, L. Wang, Y.-G. Shi, H.C. Lei, K. Liu, H.M. Weng, T. Qian, H. Ding, S.-C. Wang, Evidence of topological insulator state in the semimetal LaBi. Phys. Rev. B **95**, 115140 (2017)
95. J. Rossat-Mignod, P. Burlet, J. Villain, H. Bartholin, W. Tcheng-Si, D. Florence, O. Vogt, Phase diagram and magnetic structures of CeSb. Phys. Rev. B **16**, 440–461 (1997)
96. E. Tsymbal, Private communication
97. F.F. Tafti, Q.D. Gibson, S.K. Kushwaha, N. Haldolaarachchige, R.J. Cava, Resistivity plateau and extreme magnetoresistance in LaSb. Nat. Phys. **12**, 272–277 (2016)
98. S. Khalid, F.P. Sabino, A. Janotti, Topological phase transition in LaAs under pressure. Phys. Rev. B **98**, 220102(R) (2018)
99. N. Alidoust et al., A new form of (unexpected) Dirac fermions in the strongly-correlated cerium monopnictides. arXiv:1604.08571 [cond-mat.str-el] Table 1 in supplemental
100. X. Wan, A.M. Turner, A. Vishwanath, S.Y. Savrasov, Topological semimetal and Fermi-arc surface states in the electronic structure of pyrochlore iridates. Phys. Rev. B **83**, 205101 (2011)
101. A.B. Sushkov, J.B. Hofmann, G.S. Jenkins, J. Ishikawa, S. Nakatsuji, S. Das Sarma, H.D. Drew, Optical evidence for a Weyl semimetal state in pyrochlore $Eu_2Ir_2O_7$. Phys. Rev. B **92**, 241108(R) (2015)
102. N. Taira, M. Wakeshima, Y. Hinatsu, Magnetic properties of iridium pyrochlores $R_2Ir_2O_7$ R=Y, Sm, Eu and Lu. J. Phys.: Condens. Matter **13**, 5527 (2001)
103. R.A. McCauley, Structural characteristics of pyrochlore formation. J. Appl. Phys. **51**, 290–294 (1980)
104. H. Zhang, K. Haule, D. Vanderbilt, Metal-insulator transition and topological properties of pyrochlore iridates. Phys. Rev. Lett. **118**, 026404 (2017)
105. G.L. Stamokostas, G.A. Fiete, Mixing of t_{2g}-e_g orbitals in 4d and 5d transition metal oxides. Phys. Rev. B **97**, 085150 (2018)

106. H. Kumar, K.C. Kharkwal, K. Kumar, K. Asokan, A. Banerjee, A.K. Pramanik, Magnetic and transport properties of the pyrochlore iridates $(Y_{1-x}Pr_x)Ir_2O_7$: role of f-d exchange interaction and d-p orbital hybridization. Phys. Rev. B **101**, 064405 (2020)
107. J.P. Clancy, N. Chen, C.Y. Kim, W.F. Chen, K.W. Plumb, B.C. Jeon, T.W. Noh, Y.-J. Kim, Spin-orbit coupling in iridium-based 5d compounds probed by X-ray absorption spectroscopy. Phys. Rev. B **86**, 195131 (2012)
108. A.I. Liechtenstein, V.I. Anisimov, J. Zaane, Density functional theory: orbital ordering in Mott-Hubbard insulators. Phys. Rev. B **52**, R5467 (1995)
109. C. Berke, P. Michetti, C. Timm, Stability of the Weyl semimetal phase on the pyrochlore lattice. N. J. Phys. **20**, 043057 (2018)
110. V.K. Dwivedi, S. Mukhopadhyay, Influence of electronic structure parameters on the electrical transport and magnetic properties of $Y_{2-x}Bi_xIr_2O_7$ pyrochlore iridates. J. Appl. Phys. **126**, 165–112 (2019)
111. G. Chen, M. Hermele, Magnetic orders and topological phases from *f-d* exchange in pyrochlore iridates. Phys. Rev. B **86**, 235129 (2012)
112. J. Chakhalian, X. Liu, G.A. Fiete, Strongly correlated and topological states in [111] grown transition metal oxide thin films and heterostructures. APL Mater. **8**, 050904 (2020)
113. G.E. Topp, N. Tancogne-Dejean, A.F. Kemper, A. Rubio, M.A. Sentef, All-optical nonequilibrium pathway to stabilizing magnetic Weyl semimetals in pyrochlore iridates. Nat. Comm. **9**, 4452 (2018)
114. Y. Singh, P. Gegenwart, Antiferromagnetic Mott insulating state in single crystals of the honeycomb lattice material, Na_2IrO_3. Phys. Rev. B **82**, 064–412 (2010)
115. X. Liu, T. Berlijn, W.-G. Yin, W. Ku, A. Tsvelik, Y.-J. Kim, H. Gretarsson, Y. Singh, P. Gegenwart, J.P. Hill, Long-range magnetic ordering in Na_2IrO_3. Phys. Rev. B **83**, 220403 (2011)
116. C.H. Sohn, H.-S. Kim, T.F. Qi, D.W. Jeong, H.J. Park, H.K. Yoo, H.H. Kim, J.-Y. Kim, T.D. Kang, D.-Y. Cho, G. Cao, J. Yu, S.J. Moon, T.W. Noh, Mixing between $J_{\text{eff}} = 1/2$ and 3/2 orbitals in Na_2IrO_3: a spectroscopic and density functional calculation study. Phys. Rev. B **88**, 085125 (2013)
117. S. Boseggia, R. Springell, H.C. Walker, A.T. Boothroyd, D. Prabhakaran, D. Wermeille, L. Bouchenoire, S.P. Collins, D.F. McMorrow, Antiferromagnetic order and domains in $Sr_3Ir_2O_7$ probed by X-ray resonant scattering. Phys. Rev. B **85**, 184432 (2012)
118. B.J. Kim, H. Jin, S.J. Moon, J.-Y. Kim, B.-G. Park, C.S. Leem, J. Yu, T.W. Noh, C. Kim, S.-J. Oh, J.-H. Park, V. Durairaj, G. Cao, E. Rotenberg, Novel $J_{\text{eff}} = 1/2$ Mott State induced by relativistic spin-orbit coupling in Sr_2IrO_4. Phys. Rev. Lett. **101**, 076402 (2008)
119. F. Ye, S. Chi, B.C. Chakoumakos, J.A. Fernandez-Baca, T. Qi, G. Cao, Magnetic and crystal structures of Sr_2IrO_4: a neutron diffraction study. Phys. Rev. B **87**, 140406(R) (2013)
120. S. Fujiyama, K. Ohashi, H. Ohsumi, K. Sugimoto, T. Takayama, T. Komesu, M. Takata, T. Arima, H. Takagi, Weak antiferromagnetism of $J_{\text{eff}} = 12$ band in bilayer iridate $Sr_3Ir_2O_7$. Phys. Rev. B **86**, 174414 (2012)
121. P.D.C. King, T. Takayama, A. Tamai, E. Rozbicki, S. McKeown Walker, M. Shi, L. Patthey, R.G. Moore, D. Lu, K.M. Shen, H. Takagi, F. Baumberger, Spectroscopic indications of polaronic behavior of the strong spin-orbit insulator $Sr_3Ir_2O_7$. Phys. Rev. B **87**, 241106(R) (2013)
122. Z. Zhao, S. Wang, T.F. Qi, Q. Zeng, S. Hirai, P.P. Kong, L. Li, C. Park, S.J. Yuan, C.Q. Jin, G. Cao, W.L. Mao, Pressure induced second-order structural transition in $Sr_3Ir_2O_7$. J. Phys. Condens. Matter **26**, 215402 (2014)
123. M.K. Crawford, M.A. Subramanian, R.L. Harlow, J.A. Fernandez-Baca, Z.R. Wang, D.C. Johnston, Structural and magnetic studies of Sr_2IrO_4. Phys. Rev. B **49**, 9198–9201 (1994)
124. W. Al-Sawai, H. Lin, R.S. Markiewicz, L.A. Wray, Y. Xia, S.-Y. Xu, M.Z. Hasan, A. Bansil, Topological electronic structure in half-Heusler topological insulators. Phys. Rev. B **82**, 125208 (2010)
125. H. Lin, L.A. Wray, Y. Xia, S. Xu, S. Jia, R.J. Cava, A. Bansil, M. Zahid Hasan, Half-Heusler ternary compounds as new multifunctional experimental platforms for topological quantum phenomena. Nat. Mater **9**, 546–549 (2010)

126. S. Chadov, X. Qi, J. Kübler, G.H. Fecher, C. Felser, S.C. Zhang, Tunable multifunctional topological insulators in ternary Heusler compounds. Nat. Mat. **9**, 541–545 (2010)
127. H.C. Kandpal, C. Felser, R. Seshadri, Covalent bonding and the nature of band gaps in some half-Heusler compounds. J. Phys. D: Appl. Phys. **39**, 776–785 (2006). https://doi.org/10.1088/0022-3727/39/5/S02
128. C. Shekhar, N. Kumar, V. Grinenko, S. Singh, R. Sarkar, H. Luetkens, S.-C. Wu, Y. Zhang, A.C. Komarek, E. Kampert, Y. Skourski, J. Wosnitza, W. Schnelle, A. McCollam, U. Zeitler, J. Kübler, B. Yan, H.-H. Klauss, S.S.P. Parkin, C. Felser, Anomalous hall effect in Weyl semimetal half-Heusler compounds RPtBi (R=Gd and Nd). NAS **115**, 9140–9144 (2018)
129. D. Di Sante, P.K. Das, C. Bigi, Z. Ergönenc, N. Gürtler, J.A. Krieger, T. Schmitt, M.N. Ali, G. Rossi, R. Thomale, C. Franchini, S. Picozzi, J. Fujii, V.N. Strocov, G. Sangiovanni, I. Vobornik, R.J. Cava, G. Panaccione, Three-dimensional electronic structure of the type-II Weyl semimetal WTe_2. Phys. Rev. Lett. **119**, 026403 (2017)
130. A.J. Frenzel, C.C. Homes, Q.D. Gibson, Y.M. Shao, K.W. Post, A. Charnukha, R.J. Cava, D.N. Basov, Anisotropic electrodynamics of type-II Weyl semimetal candidate WTe_2. Phys. Rev. B **95**, 245140 (2017)
131. M.N. Ali, J. Xiong, S. Flynn, J. Tao, Q.D. Gibson, L.M. Schoop, T. Liang, N. Haldolaarachchige, M. Hirschberger, N.P. Ong, R.J. Cava, Large, non-saturating magnetoresistance in WTe_2. Nature **514**, 205–208 (2014)
132. L.S. Xie, L.M. Schoop, E.M. Seibel, Q.D. Gibson, W. Xie, R.J. Cava, A new form of Ca_3P_2 with a ring of Dirac nodes. APL Mat. **3**, 083602 (2015). https://doi.org/10.1063/1.4926545
133. Y.-H. Chan, C.-K. Chiu, M.Y. Chou, A.P. Schnyder, Ca_3P_2 and other topological semimetals with line nodes and drumhead surface states. Phys. Rev. B **93**, 205132 (2016)
134. Z. Wang, H. Weng, Q. Wu, X. Dai, Z. Fang, Three-dimensional Dirac semimetal and quantum transport in Cd_3As_2. Phys. Rev. B **88**, 125427 (2013)
135. M.N. Ali, Q. Gibson, S. Jeon, B.B. Zhou, A. Yazdani, R.J. Cava, The crystal and electronic structures of Cd_3As_2, the three-dimensional electronic analogue of graphene. Inorg. Chem. **53**, 4062–4067 (2014)
136. I. Crassee, R. Sankar, W.-L. Lee, A. Akrap, M. Orlita, 3D Dirac semimetal Cd_3As_2: a review of material properties. Phys. Rev. Matter. **2**, 120302 (2018)
137. E.O. Kane, Band structure of indium antimonide. J. Phys. Chem. Solids **1**, 249–261 (1957)
138. S.L. Chuang, C.S. Chang, k.p method for strained wurtzite semiconductors. Phys. Rev. B **54**, 2491–2504 (1996)
139. J. Bodnar, Band structure of Cd_3As_2 from Shubnikov—de Haas and de Haas—van Alphen effects, in *Proceedings of the III International Conference, Warsaw, 1977*, ed. by J. Raułuszkiewicz, M. Górska, E. Kaczmarek (Elsevier, New York, 1977), p. 311
140. L.C. Lew Yan Voon, M. Willatzen, *The k·p Method: Electronic Properties of Semiconductors* (Springer, Berlin, 2009). See p. 19 for the definition of parameters L, M, and P and 55–64 for a discussion of the four-band Kane model
141. T. Schumann, L. Galletti, D.A. Kealhofer, H. Kim, M. Goyal, S. Stemmer, Observation of the quantum hall effect in confined films of the three-dimensional Dirac semimetal Cd_3As_2. Phys. Rev. Lett. **120**, 016801 (2018)
142. Z. Wang, Y. Sun, X.-Q. Chen, C. Franchini, G. Xu, H. Weng, X. Dai, Z. Fang, Dirac semimetal and topological phase transitions in A_3Bi (A=Na, K, Rb). Phys. Rev. B **85**, 195320 (2012)
143. S.K. Kushwaha, J.W. Krizan, B.E. Feldman, A. Gyenis, M.T. Randeria, J. Xiong, S.-Y. Xu, N. Alidoust, I. Belopolski, Bulk crystal growth and electronic characterization of the 3D Dirac semimetal Na_3Bi. APL Mat. **3**, 041504 (2015)
144. K.-H. Ahn, K.-W. Lee, W.E. Pickett, Spin-orbit interaction driven collective electron-hole excitations in a noncentrosymmetric nodal loop Weyl semimetal. Phys Rev. B **92**, 115149 (2015)
145. B. Xu, Y.M. Dai, L.X. Zhao, K. Wang, R. Yang, W. Zhang, J.Y. Liu, H. Xiao, G.F. Chen, A.J. Taylor, D.A. Yarotski, R.P. Prasankumar, X.G. Qiu, Optical spectroscopy of the Weyl semimetal TaAs. Phys. Rev. B **93**, 121110(R) (2016)

146. D. Neubauer, A. Yaresko, W. Li, A. Löhle, R. Hübner, M.B. Schilling, C. Shekhar, C. Felser, M. Dressel, A.V. Pronin, Optical conductivity of the Weyl semimetal NbP. Phys. Rev. B **98**, 195203 (2018)
147. S. Polatkan, M.O. Goerbig, J. Wyzula, R. Kemmler, L.Z. Maulana, B.A. Piot, I. Crassee, A. Akrap, C. Shekhar, C. Felser, M. Dressel, A.V. Pronin, M. Orlita, Magneto-optics of a Weyl semimetal beyond the conical band approximation: case study of TaP. Phys. Rev. Lett. **124**, 176402 (2020)
148. Y.S. Kwon, M. Takeshigea, O. Nakamurab, T. Suzuki, T. Kasuya, On the common feature of the optical reflectivity in rare-earth monopnictides due to 5d electron screening. Phys. B **171**, 316–319 (1991)
149. K. Ueda, J. Fujioka, Y. Tokura, Variation of optical conductivity spectra in the course of bandwidth-controlled metal-insulator transitions in pyrochlore iridates. Phys. Rev. B **93**, 245120 (2016)
150. K. Matasuhira, M. Wakashima, R. Nakakanishi, T. Yamada, A. Nakamura, W. Kawano, S. Takagi, Y. Hinatsu, Metal-insulator transition in pyrochlore iridates $Ln_2Ir_2O_7$ (Ln=Nd, Sm, and Eu). J. Phys. Soc. Jpn. **76**, 043706 (2001)
151. J.P. Carbotte, Optical response of line node semimetal. J. Phys. Condens. Matter **29**, 045301 (2017)
152. R.W. Godby, P. García-Gonzälez, Density functional theories and self-energy approaches, in *A Primer in Density Functional Theory*, vol. 620, ed. by C. Fiolhais, F. Nogueira, M.A.L. Marques. Lecture Notes in Physics (Springer, Berlin, 2003), pp. 185–217
153. A.A. Burkov, Anomalous hall effect in Weyl metals. Phys. Rev. Lett. **113**, 187202 (2014)
154. Y. Sun, Y. Zhang, C. Felser, B. Yan, Strong intrinsic spin hall effect in the TaAs family of Weyl semimetals. Phys. Rev. Lett. **117**, 146403 (2016)
155. X. Huang, L. Zhao, Y. Long, P. Wang, D. Chen, Z. Yang, H. Liang, M. Xue, H. Weng, Z. Fang, X. Dai, G. Chen, Observation of the chiral-anomaly-induced negative magnetoresistance in 3D weyl semimetal TaAs. Phys. Rev. X **5**, 031023 (2015)
156. N.J. Ghimire, Y. Luo, M. Neupane, D.J. Williams, E.D. Bauer, F. Ronning, Magnetotransport of single crystalline NbAs. J. Phys. Condens. Matter **27**, 152201 (2015)
157. K.-Y. Yang, Y.-M. Lu, Y. Ran, Quantum Hall effects in a Weyl semimetal: possible application in pyrochlore iridates. Phys. Rev. B **84**, 075129 (2011)
158. S.M. Disseler, S.R. Giblin, C. Dhital, K.C. Lukas, S.D. Wilson, M.J. Graf, Magnetization and Hall effect studies on the pyrochlore iridate $Nd_2Ir_2O_7$. Phys. Rev. B **87**, 060403(R) (2013)
159. Y. Machida, S. Nakatsuji, S. Onoda, T. Tayama, T. Sakakibara, Time-reversal symmetry breaking and spontaneous Hall effect without magnetic dipole order. Nature **463**, 210–213 (2010)
160. S. Tang, C. Zhang, D. Wong, Z. Pedramrazi, H.-Z. Tsai, C. Jia, B. Moritz, M. Claassen, H. Ryu, S. Kahn, J. Jiang, H. Yan, M. Hashimoto, D. Lu, R.G. Moore, C.-C. Hwang, C. Hwang, Z. Hussain, Y. Chen, M.M. Ugeda, Z. Liu, X. Xie, T.P. Devereaux, M.F. Crommie, S.-K. Mo, Z.-X. Shen, Quantum spin Hall state in monolayer 1 T′-WTe_2. Nat. Phys. **13**, 683–688 (2017)
161. T.-R. Chang, P.-J. Chen, G. Bian, S.-M. Huang, H. Zheng, T. Neupert, R. Sankar, S.-Y. Xu, I. Belopolski, G. Chang, B.-K. Wang, F. Chou, A. Bansil, H.-T. Jeng, H. Lin, M. Zahid Hasan, Topological Dirac surface states and superconducting pairing correlations in $PbTaSe_2$. Phys. Rev. B **93**, 245130 (2016)
162. S.-Y. Guan, P.-J. Chen, M.-W. Chu, R. Sankar, F. Chou, H.-T. Jeng, C.-S. Chang, T.-M. Chuang, Superconducting topological surface states in the noncentrosymmetric bulk superconductor $PbTaSe_2$. Sci. Adv. **2**, e1600894 (2016)
163. T. Schumann, M. Goyal, D.A. Kealhofer, S. Stemmer, Negative magnetoresistance due to conductivity fluctuations in films of the topological semimetal Cd_3As_2. Phys. Rev. B **95**, 241113(R) (2017)
164. J. Xiong, S. Kushwaha, J. Krizan, T. Liang, R.J. Cava, N.P. Ong, Anomalous conductivity tensor in the Dirac semimetal Na_3Bi. EPL **114**, 27002 (2016)
165. J. Xiong, S.K. Kushwaha, T. Liang, J.W. Krizan, M. Hirschberger, W. Wang, R.J. Cava, N.P. Ong, Evidence for the chiral anomaly in the Dirac semimetal Na_3Bi. Science **350**, 413–416 (2015)

Appendix A
Mueller Matrix Spectroscopic Ellipsometry

The Mueller Matrix provides a complete characterization of the interaction of light with a material. The Mueller matrix is especially useful when characterizing optically anisotropic materials. The Mueller matrix transforms the Stokes vector that completely represents the intensity and polarization of the incident light into the Stokes vector for the scattered light. The Mueller Matrix has 16 elements or pieces of information at each wavelength instead of the usual 2 (ψ and Δ). As with ψ and Δ, Mueller matrix elements are typically determined over a range of wavelengths and Mueller matrix elements are sometimes referred to as spectra. Some of the key pioneers of Mueller Matrix spectroscopic ellipsometry are Collins and Drevillon [1, 2], and a very early publication by Hauge appeared in 1976 [3].

As discussed in Chap. 1, the 4 × 4 Jones matrix describes cross-polarized light scattering. The Mueller matrix provides a more complete characterization of light scattering that include depolarization effects and is very useful for highly optically anisotropic materials [4–6]. There are 16 sets of spectroscopic Mueller matrix element versus wavelength compared to the two measured in traditional ellipsometry.

Light can be fully represented by the Stokes vector which

$$S = \begin{pmatrix} I_p + I_s \\ I_p - I_s \\ I_{+45^\circ} - I_{-45^\circ} \\ I_R - I_L \end{pmatrix} = \begin{pmatrix} s_0 \\ s_1 \\ s_2 \\ s_3 \end{pmatrix} \tag{A.1}$$

where I_p and I_s refer to the intensity of the light polarized p or s, and I_{+45° and I_{-45° refer to the intensity of the light polarized at either + or -45°. I_R and I_L refer to the right and left circularly polarized components of the light. The Mueller Matrix transforms the Stokes vector from its incident value to its final value:

$$\begin{pmatrix} s_0 \\ s_1 \\ s_2 \\ s_3 \end{pmatrix}_{\text{out}} = \begin{pmatrix} M_{11} & M_{12} & M_{13} & M_{14} \\ M_{21} & M_{22} & M_{23} & M_{24} \\ M_{31} & M_{32} & M_{33} & M_{34} \\ M_{41} & M_{42} & M_{43} & M_{44} \end{pmatrix} \begin{pmatrix} s_0 \\ s_1 \\ s_2 \\ s_3 \end{pmatrix}_{\text{in}} \tag{A.2}$$

A. Diebold and T. Hofmann, *Optical and Electrical Properties of Nanoscale Materials*, Springer Series in Materials Science 318,
https://doi.org/10.1007/978-3-030-80323-0

The MM is usually normalized to the total reflectivity M_{11}. It is useful to relate the Mueller matrix to the ψ and Δ response of an isotropic material. For an isotropic material or thin film on a substrate, the Mueller Matrix is related to ψ and Δ as follows:

$$\begin{pmatrix} M_{11} & M_{12} & M_{13} & M_{14} \\ M_{21} & M_{22} & M_{23} & M_{24} \\ M_{31} & M_{32} & M_{33} & M_{34} \\ M_{41} & M_{42} & M_{43} & M_{44} \end{pmatrix} = \begin{pmatrix} 1 & -\cos 2\psi & 0 & 0 \\ -\cos 2\psi & 1 & 0 & 0 \\ 0 & 0 & \sin 2\psi \cos \Delta & \sin 2\psi \sin \Delta \\ 0 & 0 & -\sin 2\psi \sin \Delta & \sin 2\psi \cos \Delta \end{pmatrix}$$

$$= \begin{pmatrix} 1 & -N & 0 & 0 \\ -N & 1 & 0 & 0 \\ 0 & 0 & C & S \\ 0 & 0 & -S & C \end{pmatrix} \tag{A.3}$$

Equation (A.3) defines N, C, and S. The C and S terms are due to scattering into the same polarization direction. Next, the Mueller Matrix for any material is presented. Comparing (A.4) to (A.3), C refers to C_{pp} and S refers to S_{pp}.

The new terms are defined after (A.4). When the material is non-ideal because of edge roughness or the optical properties depend on the azimuthal direction (e.g., stress relaxation along a specific direction, then the fully non-symmetric Mueller Matrix becomes [4, 5]):

$$\begin{pmatrix} M_{11} & M_{12} & M_{13} & M_{14} \\ M_{21} & M_{22} & M_{23} & M_{24} \\ M_{31} & M_{32} & M_{33} & M_{34} \\ M_{41} & M_{42} & M_{43} & M_{44} \end{pmatrix} = \begin{pmatrix} 1 & -N-\alpha_{ps} & C_{sp}+\varsigma_1 & S_{sp}+\varsigma_2 \\ -N-\alpha_{sp} & 1-\alpha_{sp}-\alpha_{ps} & -C_{sp}+\varsigma_1 & -S_{sp}+\varsigma_2 \\ C_{ps}+\mathcal{E}_1 & -C_{ps}+\mathcal{E}_1 & C_{pp}+\beta_1 & S_{pp}+\beta_2 \\ -S_{ps}+\mathcal{E}_2 & S_{ps}+\mathcal{E}_2 & -S_{pp}+\beta_2 & C_{pp}-\beta_1 \end{pmatrix} \tag{A.4}$$

where $\frac{r_{ij}}{r_{ss}} = \tan \psi_{ij} e^{-i\Delta_{ij}}$ for i = p or s resulting in:

$$\alpha_{ij} = \frac{2\tan^2 \psi_{ij}}{D}$$

$$C_{ij} = \frac{2\tan \psi_{ij} \cos \Delta_{ij}}{D}$$

$$S_{ij} = \frac{2\tan \psi_{ij} \sin \Delta_{ij}}{D}$$

$$D = 1 + \tan^2 \psi_{pp} + \tan^2 \psi_{ps} + \tan^2 \psi_{sp}$$

$$N = \frac{1 - \tan^2 \psi_{pp} - \tan^2 \psi_{ps} - \tan^2 \psi_{sp}}{D}$$

$$\varsigma_i = \frac{D\left(C_{ps}^2 + S_{ps}^2 (-1)^{i+1 \bmod 2}\right)}{2}$$

$$\mathcal{E}_i = \frac{D\left(C_{sp}^2 + S_{sp}^2 (-1)^{i+1 \bmod 2}\right)}{2}$$

$$\beta_i = \frac{D\left(C_{sp} C_{ps} + S_{sp} S_{ps} (-1)^{i+1 \bmod 2}\right)}{2}$$

Arteaga described the relationship between optical materials and the Mueller Matrix [7, 8]. Some examples are shown in Table A.1.

Table A.1 Example Mueller matrix symmetries for uniaxial, biaxial orthorhombic, and biaxial monoclinic materials

Symmetry type	$M_{ij}^* = 0$ for transparent materials	Uniaxial	Biaxial orthorhombic	Biaxial monoclinic
A	$\begin{pmatrix} 1 & M_{12} & 0 & 0 \\ M_{12} & 1 & 0 & 0 \\ 0 & 0 & M_{33} & M_{34} \\ 0 & 0 & -M_{34} & M_{33} \end{pmatrix}$	• OA ⊥ surface • OA in POI • OA ⊥ POI	• One OA ⊥ surface other OA in POI ‖ to surface • One OA ⊥ POI other OA in POI any angle	• PA ⊥ POI
B	$\begin{pmatrix} 1 & M_{12} & M_{13}^* & M_{14} \\ M_{12} & M_{22} & M_{23}^* & M_{24} \\ -M_{13}^* & -M_{23}^* & M_{33} & M_{34}^* \\ M_{14} & M_{24} & -M_{34}^* & M_{44} \end{pmatrix}$			
C	$\begin{pmatrix} 1 & M_{12} & M_{13} & M_{14}^* \\ M_{12} & M_{22} & M_{23} & M_{24}^* \\ -M_{13} & -M_{23} & M_{33} & M_{34}^* \\ M_{14}^* & M_{24}^* & -M_{34}^* & M_{44} \end{pmatrix}$	• OA ‖ to surface any angle	• One OA in plane ⊥ surface other OA ‖ to surface any angle • One OA in plane ⊥ surface other OA at any angle	• PA ⊥ surface
D	$\begin{pmatrix} 1 & M_{12} & M_{13} & M_{14}^* \\ M_{12} & M_{22} & M_{23} & M_{24}^* \\ M_{13} & M_{23} & M_{33} & M_{34}^* \\ -M_{14}^* & -M_{24}^* & -M_{34}^* & M_{44} \end{pmatrix}$	• OA in plane ⊥ POI	• One OA ⊥ POI at any angle to POI and other OA in POI	• PA ‖. to surface in POI
E	$\begin{pmatrix} 1 & M_{12} & M_{13} & M_{14}^* \\ M_{21} & M_{22} & M_{23} & M_{24}^* \\ M_{31} & M_{23} & M_{33} & M_{34}^* \\ M_{41}^* & M_{42}^* & M_{43}^* & M_{44} \end{pmatrix}$	Other directions for OA	• Other orientations	• Other directions for PA

Arteaga's symmetry type notation is used here [7]. OA is the optical axis, PA is the principal axis, and POI is the plane of incidence of the reflected light. Symmetry type B is for isotropic and optically active materials. The two optical axes for biaxial orthorhombic materials are perpendicular to each other

References

1. R.W. Collins, J. Koh, Dual rotating-compensator multichannel ellipsometer: instrument design for real-time Mueller matrix spectroscopy of surfaces and films. J. Opt. Soc. Am. A **16**, 1997–2006 (1999)
2. E. Compain, S. Poirier, B. Drevillon, General and self-consistent method for the calibration of polarization modulators, polarimeters, and Mueller-matrix ellipsometers. Appl. Opt. **38**, 3490–3502 (1999)
3. P.S. Hauge, Automated Mueller matrix ellipsometry, Opt. Comm. **17**, 74–76 (1976)
4. M. Foldyna, A. De Martino, R. Ossikovski, E. Garcia-Caurel, C. Licitra, Characterization of grating structures by Mueller polarimetry in presence of strong depolarization due to finite spot size. Opt. Commun. **282**, 735–741 (2009)
5. J.W. Hovenier, Structure of a general pure Mueller matrix. Appl. Opt. **33**, 8318–8324 (1994)
6. K. Kim, L. Mandel, E. Wolf, Relationship between Jones and Mueller matrices for random media. J. Opt. Soc. Am. A **4**, 433 (1987)
7. O. Arteaga, Useful Mueller matrix symmetries for ellipsometry. Thin. Sol. Films **571**, 584–588 (2014)
8. O. Arteaga, E. Garcia-Caurel, R. Ossikovski, Anisotropy coefficients of a Mueller matrix. J. Opt. Soc. Am. A, Opt. Image Sci. Vis. **28**, 548–553 (2011)

Appendix B
Kramers–Kronig Relationships for the Complex Refractive Index and Dielectric Function

The real and imaginary parts of the complex refractive index are not independent. The Kramers–Kronig formulas relate the real and imaginary parts of the complex refractive index using the law of causality. Causality means that the effect cannot precede the cause. Based on the relationship between the complex refractive index and the complex dielectric function, this is also true for the real and imaginary parts of the dielectric function. The Kramers–Kronig integral relationships determine the real (imaginary) part of the complex refractive index from the principle part of a complex integral [1]:

$$n(\omega) = 1 + \frac{2}{\pi}\wp \int_0^\infty \frac{\omega' \kappa(\omega')}{\omega'^2 - \omega^2} d\omega'$$

$$\kappa(\omega) = -\frac{2}{\pi}\wp \int_0^\infty \frac{n(\omega')}{\omega'^2 - \omega^2} d\omega'$$

Thus if one knows one part of the complex refractive index, the other part can be determined. The same law of causality holds for the complex dielectric function:

$$\varepsilon_1(\omega) = 1 + \frac{2}{\pi}\wp \int_0^\infty \frac{\omega' \varepsilon_2(\omega')}{\omega'^2 - \omega^2} d\omega'$$

$$\varepsilon_2(\omega) = -\frac{2\omega}{\pi}\wp \int_0^\infty \frac{\varepsilon_1(\omega')}{\omega'^2 - \omega^2} d\omega' + \frac{\sigma_0}{\varepsilon_0 \omega}$$

Here, σ_0 is the dc conductivity. All of these integral extend to infinity.

One of the advantages of using parameterized functions such as the Lorentz, Cody–Lorentz, and Tauc–Lorentz optical models, which were discussed in Chap. 1, is that they are Kramers–Kronig consistent. Some software is able to determine the

A. Diebold and T. Hofmann, *Optical and Electrical Properties of Nanoscale Materials*, Springer Series in Materials Science 318,
https://doi.org/10.1007/978-3-030-80323-0

optical properties from ψ and Δ data. The Kramers–Kronig consistency of software determined optical properties must be checked. In some instances, it is possible to fit the resulting optical properties to a Kramers–Kronig consistent optical model and then compare values of the optical properties across the measured wavelength range. We note that a Cauchy model that only has real values for the complex refractive index is not Kramers–Kronig consistent. Cauchy models with an imaginary component have been introduced. This so-called Urbach absorption tail is fully Kramers–Kronig consistent [2].

References

1. D.G. Seiler, S. Zollner, A.C. Diebold, P.M. Amirtharaj, Optical properties of semiconductors, in *Handbook of Optics*, vol. IV, 3rd edn (McGraw Hill, New York, 2009) (Chapter 5)
2. J.N. Hilfiker, T. Tiwald, Dielectric function modeling, in *Spectroscopic Ellipsometry for Photovoltaics Volume 1: Fundamental Principles and Solar Cell Characterization*, ed. by H. Fujiwara, R.W. Collins (Springer, New York, 2018)

Appendix C
Topological Periodic Tables

The importance of spatial and non-spatial symmetries along with the electronic band structure in determining the potential for a crystalline material to exhibit topological properties has motivated the search of a Periodic Table of Topological Materials [1–21]. The use of non-spatial symmetry in determining the nature of quantum systems can be traced back to 1962 [22–24]. In 1996 and 1997, Altland and Zirnbauer [22, 23] expanded the three symmetry class system introduced by Dyson [24] in 1962 to a ten class symmetry system for random matrices. These symmetry classes did not include the influence of spatial symmetry. Since the symmetry indicators, which lead to the potential for topologically protected states, can arise from both non-spatial symmetries and the spatial (crystallographic) symmetry of the material, additional classification required the addition of spatial symmetry. The ground breaking work of Schnyder et al. [2] and that of Kitaev [3] provides a classification method for all insulators and superconductors based on a classification that included the Altland–Zirnbauer classification plus the system dimension, e.g., 1D, 2D, 3D, etc. The non-spatial symmetry classification system was further expanded to include the spatial symmetry of the crystal creating addition Topological Periodic Tables (for a review see [12]). The inclusion of point group and space group spatial symmetry has continued to evolve. Cornfeld and Chapman as well as Cornfeld and Carmeli provided a complete spatial and non-spatial symmetry based classification of gapped systems [18, 20, 21]. Initial efforts at classification focused on topological insulators which have small band gaps, at the location in the Brillouin zone where the electronic band structure. The next step was to classify other topological materials that are gapless. Thus, subsequent efforts were aimed at expanding topological classification to Weyl and Dirac semimetals as well as other gapless crystalline materials [5, 6, 12]. Much of this work is based on K Theory and Clifford Algebra which are beyond the scope of this book. One of the challenging aspects of this discussion is to distinguish between the classification of surface states, ground states, and symmetry indicators [18]. In the context of the literature associated with topological classification, the ground state refers to the thermodynamically stable phase of the material [19]. In this section, an introductory overview of topological classification and the Periodic Table of Topological Materials is presented. The reader is referred to [11,

A. Diebold and T. Hofmann, *Optical and Electrical Properties of Nanoscale Materials*,
Springer Series in Materials Science 318,
https://doi.org/10.1007/978-3-030-80323-0

12] for a more complete discussion of additional topics such as the classification of superconductors.

First, we review some definitions for the terms used in differential geometry and topology. $\mathbb{Z}$ is the group of integers, and $\mathbb{Z}_2$ is called a quotient group that classifies whether an integer is even or odd. The classification of a crystal with $\mathbb{Z}\times\mathbb{Z}$ protection can be stated as having $\mathbb{Z}^2$ classification, and $\mathbb{Z} \times \mathbb{Z} \times \mathbb{Z}$ can be written as $\mathbb{Z}^3$, etc. The same holds for $\mathbb{Z}_2$. $\mathbb{R}$ is the set of real numbers, and $\mathbb{R}^n = \mathbb{R} \times \mathbb{R} \times \mathbb{R} \times \cdots$ n times is known as an n real dimensional space.

It is useful to recall the discussion of topological classification started in Sect. 9.1. The topological index $\mathbb{Z}_2$ for a specific material in a specific phase or crystal structure has four values as discussed in Sect. 9.1 with the first index indicating whether or not the material is a strong topological insulator or not. The other three are weak indices as described in Sect. 9.1. We also discussed the Chern number which is an example of a $\mathbb{Z}$ classified invariant [17]. The $\mathbb{Z}_4$ invariant was introduced [11] to classify strong topological phases that are time reversal symmetry invariant and also have single glide symmetry [11, 14].

We now expand on our discussion of the topological classification of the potential for a crystal to show topologically protected properties. The initial step is to separate crystals with gapped electronic band structures from gapless ones. First we discuss the classification of crystals having band gaps ("gapped"). Time reversal symmetry, particle-hole (charge conjugation) symmetry, and chiral symmetry play a key role in categorizing the symmetry of the single particle Hamiltonian. These symmetry properties were discussed in Sect. 9.1. These three non-spatial symmetry properties are used to classify ten symmetry classes which are the Altland–Zirnbauer classes of random matrices [10, 13]. The classification is based on the result of operating on the wavefunctions twice with the time reversal symmetry, particle-hole (charge conjugation) symmetry, or chiral symmetry operators $\mathcal{T}$, $\mathcal{C}$, *and* $\mathcal{S}$. We note that different references use different symbols for the particle—hole and chiral operators. For spin ½ particles (fermions), $\mathcal{T}^2|n\rangle = -|n\rangle$, and for spin 1 particles $\mathcal{T}^2|n\rangle = +|n\rangle$. The same is true for $\mathcal{C}^2$. The chiral operator is $\mathcal{S} = \mathcal{T} \cdot \mathcal{C}$ so that time reversal and particle hole symmetry determine chiral symmetry for all classes except for class AIII. The ten Altland–Zirnbauer symmetry classes are listed in Table C.1. The topological index $\mathbb{Z}$ is for integer topological invariants, $2\mathbb{Z}$ is for even integers, and $\mathbb{Z}_2$ was discussed in the Sect. 9.1, and is discussed in most references used for this appendix, has only two values 0 or 1.

An important consideration when using the classification schemes is that the ground state (thermodynamically stable phase at zero Kelvin) of a system considers the electronic structure in the absence of a magnetic field. A magnetic field breaks time reversal symmetry. Consider a 2D electron gas in the presence of a magnetic field that is perpendicular to the 2DEG. The Lorentz force is linked to the direction of the velocity. For example, one can see in Fig. 6.1 in Chap. 6 where the observer is looking at layer of conducting materials with the electron traveling to the right and the magnetic field is normal to the surface. When an electron with wavevector of magnitude $|k|$ is traveling to the right, the electron is deflected away from the observer. Consider an electron with wavevector of magnitude $|k|$ traveling to the

Table C.1 The topological classification for non-spatial symmetry classes for crystals with band gaps based on the Altland–Zirnbauer is listed

Symmetry class	Time reversal symmetry	Particle hole symmetry	Chiral symmetry	$d = 0$	$d = 1$	$d = 2$	$d = 3$
A	No	No	No	$\mathbb{Z}$	0	$\mathbb{Z}$	0
AIII	No	No	Yes	0	$\mathbb{Z}$	0	$\mathbb{Z}$
AI	$\mathcal{T}^2\|n\rangle = +\|n\rangle$	No	No	$\mathbb{Z}$	0	0	0
BDI	$\mathcal{T}^2\|n\rangle = +\|n\rangle$	$\mathcal{C}^2\|n\rangle = +\|n\rangle$	Yes	$\mathbb{Z}_2$	$\mathbb{Z}$	0	0
D	No	$\mathcal{C}^2\|n\rangle = +\|n\rangle$	No	$\mathbb{Z}_2$	$\mathbb{Z}_2$	$\mathbb{Z}$	0
DIII	$\mathcal{T}^2\|n\rangle = -\|n\rangle$	$\mathcal{C}^2\|n\rangle = +\|n\rangle$	Yes	0	$\mathbb{Z}_2$	$\mathbb{Z}_2$	$\mathbb{Z}$
AII	$\mathcal{T}^2\|n\rangle = -\|n\rangle$	No	No	$2\mathbb{Z}$	0	$\mathbb{Z}_2$	$\mathbb{Z}_2$
CII	$\mathcal{T}^2\|n\rangle = -\|n\rangle$	$\mathcal{C}^2\|n\rangle = -\|n\rangle$	Yes	0	$2\mathbb{Z}$	0	$\mathbb{Z}_2$
C	No		No	0	0	$2\mathbb{Z}$	0
CI	$\mathcal{T}^2\|n\rangle = +\|n\rangle$	$\mathcal{C}^2\|n\rangle = -\|n\rangle$	Yes	0	0	0	$2\mathbb{Z}$

The non-spatial symmetries are time reversal, particle–hole, and chiral symmetry. The dimension d of the structure under consideration is restricted to 3 in this table. This table can be further expanded to classify a quantity that is the dimension minus the dimension of the defect [12]. Other tables report dimensions 0–8 (see Table C.1, [12])

right which is the time reversal state. This electron will be deflected toward the observer. This breaks time reversal symmetry.

The Chern number is zero for gapped crystal materials that have inversion and time reversal symmetry. That lead Fu et al. [25] along with Moore and Balents [8] to introduce the $\mathbb{Z}_2$ index for 3D topological insulators with inversion symmetry which has strong index υ_0 and the $\upsilon_1\upsilon_2\upsilon_3$ weak indices are all $\mathbb{Z}_2$. Here, the first index, υ_0, describes whether or not the insulator is a strong or weak TI, and is calculated from a product over all TRIM (time reversal invariant momenta): $(-1)^{\upsilon_0} = \prod \delta_i$ [25]. In general, the parity indicators δ_i are determined using the Pfaffian calculated at the TRIM points Λ_i: $\delta_i = \frac{Pf[w(\Lambda_i)]}{\sqrt{Det[w(\Lambda_i)]}}$ [25]. $w(\Lambda_i)$ is the unitary sewing matrix where the matrix elements come from the operation of the time reversal symmetry operator Θ on the wavefunctions of the filled bands. That can be simplified when time reversal and inversion symmetry are present. Then, the δ_i are related to the band parity $\xi_{2m}(\Lambda_i)$ at the TRIM through: $\delta_i(\Gamma) = \prod_{m=1}^{N} \xi_{2m}(\Lambda_i)$ for $2N$ occupied bands [1, 25]. Each weak invariant ν_j is determined using the four TRIM located at the three independent planes in the 3D Brillouin zone using $(-1)^{\upsilon_j} = \prod_{i=1}^{4} \delta_i$. The reader is referred to [8, 12, 25] for a more detailed discussion about the calculation of the strong and weak $\mathbb{Z}_2$ topological indices.

Table C.2 Symmetry Indicators for materials with band gaps (gapped materials) having the point group symmetry D_{3d} for space group #166 $(R\overline{3}m)$ for all the Altland–Zirnbauer classes

Symmetry class	Strong bulk topological invariants	Total anomalous surface states	Γ	L	F	Z
A	$\mathbb{Z}^3$	$\mathbb{Z}$	$\mathbb{Z}^6 \to 0$	$\mathbb{Z}^2 \to 0$	$\mathbb{Z} \to 0$	$\mathbb{Z}^3 \to \mathbb{Z}$
AIII	0	0	·	·	·	·
AI	$\mathbb{Z}^3$	$\mathbb{Z}$	$\mathbb{Z}^4 \to 0$	$\mathbb{Z} \to 0$	$\mathbb{Z} \to 0$	$\mathbb{Z}^3 \to \mathbb{Z}$
BDI	0	0	·	·	·	·
D	0	0	$\mathbb{Z}^2 \to 0$	$\mathbb{Z} \to 0$	·	·
DIII	0	0	·	·	·	·
AII	$\mathbb{Z}^3$	$\mathbb{Z} \times \mathbb{Z}_2^2$	$\mathbb{Z}^4 \to 0$	$\mathbb{Z} \to 0$	$\mathbb{Z} \to \mathbb{Z}_2$	$\mathbb{Z}^3 \to \mathbb{Z} \times \mathbb{Z}_2$
CII	$\mathbb{Z}_2^3$	$\mathbb{Z}_2^3$	$\mathbb{Z}_2^2 \to 0$	·	$\mathbb{Z}_2 \to \mathbb{Z}_2$	$\mathbb{Z}_2^3 \to \mathbb{Z}_2^2$
C	$\mathbb{Z}_2^3$	$\mathbb{Z}_2$	$\mathbb{Z}_2 \times \mathbb{Z}_2^2 \to 0$	$\mathbb{Z} \to 0$	$\mathbb{Z}_2 \to 0$	$\mathbb{Z}_2^3 \to \mathbb{Z}_2$
CI	0	0	·	·	·	·

The strong bulk topological invariants for a trigonal crystal having a point group of D_{3d} $(\overline{3}m)$ are listed in the second column [21]. The third column shows total topological invariants for anomalous surface states lists all strong, weak, and high order surface states for the point group symmetry D_{3d} for space group #166 $(R\overline{3}m)$ [20]. In columns $\Gamma, L, F, and M$, the complete topological classification of for both strong and weak states for topologically distinct ground states and for topological phases with anomalous surface states (AAS) is provided for the high symmetry time reversal invariant momentum located at specific points in the bulk Brillouin zone is shown for the $\Gamma, L, F, and M$ point is the Brillouin zone (see Fig. 9.8b) [21]. These classifications are shown as the bulk to ASS transition $K \to AAS$. The · is for a trivial classification. Comparison of the sum of the distinct ground states in columns Γ, L, F, *and* M with column 2 shows that not all topologically protected ground states will be classified as being strong

Next we discuss how spatial symmetry can be used to further classify the topology of a crystal. Spatial symmetry classification is based on both the local point group symmetry in a Brillouin zone and the space group symmetry. Cornfeld and Chapman determined the impact of point group symmetry building on the Altland–Zirnbauer classification [20]. The point group symmetry tables for monoclinic, tetragonal, trigonal, hexagonal, and cubic crystals can be found in [20]. In Table C.2, we present only one example where the symmetry indicators for the point group symmetry D_{3d} for space group #166 $(R\overline{3}m)$ are listed. This was followed by Cornfeld and Carmeli's classification based on space group symmetry [21].

It useful to illustrate that classification provides a means of identifying crystals that may have topologically protected states. $Bi_{1-x}Sb_x$ is considered to be the first material identified as a 3D TI. Here, we classify bismuth, antimony, and their alloy $Bi_{1-x}Sb_x$ using Table C.2 to demonstrate classification and the idea that although all three materials have the same classification, one of the materials does not have topologically protected states. Bismuth, antimony, and $Bi_{1-x}Sb_x$ have the rhombohedral A7 structure with spatial symmetry space group #166 $(R\overline{3}m)$ and point group

Table C.3 The parity invariants γ of bismuth, antimony, and their alloy $Bi_{1-x}Sb_x$ at the $\Gamma, L, T, and X$ points in the Brillouin zone [1]

	$\delta(\Gamma)$	$3L$ points with $\delta(L)$	$\delta(T)$	$3X$ points with $\delta(X)$	$(\upsilon_0, \upsilon_1\upsilon_2\upsilon_3)$
Bi	-1	-1	-1	-1	(0, 000)
Sb	-1	1	-1	-1	(1, 111)
$Bi_{1-x}Sb_x$	-1	1	-1	-1	(1, 111)

symmetry D_{3d} [1]. This crystal lattice has inversion symmetry, and the Brillouin zone for all three crystals is the same as that of Bi_2Se_3 which is shown in Chap. 9, Fig. 9.8. We note that the high symmetry locations in the Brillouin zone for Bi, Sb, and $Bi_{1-x}Sb_x$ are labeled differently throughout the literature than those used for the tetradymites as shown in Chap. 9, Fig. 9.8 and associated references in Chap. 9. For example Z in Chap. 9, Fig. 9.8 is labeled T in [1]. Also, F in Fig. 9.8, Chap. 9 is labeled X in [1]. For $0.9 < x < 0.18$, $Bi_{1-x}Sb_x$ is a direct-gap semiconductor with a gap on order of 30 meV at the L points [1]. The states at high symmetry points in the Brillouin zone have time reversal symmetry. First we determine which of these materials are predicted to be strong topological insulators. Therefore, it is useful to compare values of the $\mathbb{Z}_2$ index which we determine using the product of the band parity related δ_i at the eight TRIM points Λ_i with one at Γ and T and 3 each at the $L\, and\, X$ TRIM points for $Bi_{1-x}Sb_x$ using $(-1)^{\upsilon_0} = \prod_{i=1}^{8} \delta_i$. Here, the parity invariant at a TRIM is a product of the parity values $\xi_{2m}(\Lambda_i)$ for the Kramers pair at that TRIM Λ_i, and here we illustrate this using the TRIM at Γ for the $2N$ occupied bands $\delta_i(\Gamma) = \prod_{m=1}^{N} \xi_{2m}(\Lambda_i)$. Using the parity invariants δ_i at the eight TRIM points for Bismuth, antimony, and $Bi_{1-x}Sb_x$ as provided in Table C.3, one can see that the strong $\mathbb{Z}_2$ index $\upsilon_0 = 1$ for Sb and $Bi_{1-x}Sb_x$. The weak $\mathbb{Z}_2$ indices $\upsilon_1\upsilon_2\upsilon_3$ are determined from the $\delta_i(\Gamma)$ using $(-1)^{\upsilon_i} = \prod_{i=1}^{4} \delta_i$ [25]. The other weak $\mathbb{Z}_2$ indices $\upsilon_1\upsilon_2\upsilon_3$ are also 1 for Sb and $Bi_{1-x}Sb_x$.

Now we explore the topological classification of these materials. In referring to Table C.1 which provides classifications based only on the non-spatial symmetries, the classification of both AII and CII is $\mathbb{Z}_2$. Now we add the spatial symmetry classification for the space group #166 crystal structure crystal structure and focus on both Altland–Zirnbauer class CII. In referring to Table C.2 which provides classification based on both spatial and non-spatial symmetry, the space group #166 crystal structure has a total classification for strong class CII bulk invariants of $\mathbb{Z}_2^3$ and for the total strong and weak bulk class CII invariants a total of $\mathbb{Z}_2^6$ [18]. The total classification for weak and strong bulk states comes from combining the bulk states classifications from the high symmetry points in the Brillouin zone $\Gamma, L, F, and\, Z$. In addition, Table C.2 lists the classification of the strong surface states as $\mathbb{Z}_2^2$ [18]. The high symmetry points in the Brillouin zone $\Gamma, L, F, and\, Z$ have different invariant classifications because each one is protected by different spatial symmetries such as rotational or mirror symmetry. For example, there is three fold rotational symmetry for the point group D_{3d} which goes the z axis which passes through Z and Γ. There is mirror symmetry at numerous places in the lattice. It is important to note that a mirror reflection of a spin state changes the spin, for example +1/2 becomes –1/2. In referring to Table C.2, the strong bulk topological invariants are shown in column 2 as $\mathbb{Z}_2^3$. For CII the symmetry indicators at $\Gamma, L, F,$ *and* Z result in $\mathbb{Z}_2^6$ for the bulk ground states of which $\mathbb{Z}_2^3$ occur at points in the Brillouin zone which have additional protection due to spatial symmetry and thus are strong ground states. The distribution of the strong surface states for Altland–Zirnbauer class CII between the high symmetry points for in the Brillouin zone is also shown in Table C.2.

A second example of materials with a band gap is $ZrTe_5$ and $HfTe_5$. As described in Chap. 9, the bulk crystal is an orthorhombic layered material with the Cmcm (#63) (D_{2h}^{17}) space group. The bulk crystal has a band gap at or close to the Γ point in the Brillouin zone, and the exact location depends on the lattice constants used in the calculation [26]. The TRIM for bulk $ZrTe_5$ are Γ, X, Y, and M were used to calculate the Kane and Fu $\mathbb{Z}_2$ which has been reported as $(1, 001)$ [26] and $(1, 110)$ [27]. A single layer of $ZrTe_5$ or $HfTe_5$ has space group is $P_{mmn}(D_{2h}^{13})$ (#59) with an inversion center that is not located at the origin but at (¼, ¼). The inversion center is shown in Fig. 9.13(d). Bulk $HfTe_5$ is also expected to a strong topological insulator, and it exhibits a chiral anomaly in magnetotransport measurements [28, 29]. Single layer $ZrTe_5$ has an indirect band gap which is negative in the sense that energy of the lowest occupied conduction band energy lies below that of the highest occupied valence band energy [25]. The TRIM for single layer $ZrTe_5$ occurs at $(\pi, 0)$ and (π, π) in pairs with the opposite parity giving the Kane and Fu $\mathbb{Z}_2$ of 1 [26]. Reference [30], indicates that the total classification of Topological Insulators (strong bulk invariants) plus Topological Crystalline Insulators (which I interpret to be weak invariants) for Space Group # 59 and # 63 is $\mathbb{Z}^2 \times \mathbb{Z}_2^2$.

Now we turn to the classification of materials with no band gap such as Weyl and Dirac semimetals and nodal line materials. The topological classification for gapless crystals is done for crystal with stable Fermi surfaces. An example of an unstable Fermi surface is one where the Cooper pairs associated with superconducting states form thus opening a band gap. Again we start with the Altland–Zirnbauer classes based on only the non-spatial symmetries of time reversal, particle hole, and chiral symmetry. This is shown in Table C.4.

Graphene: Graphene is an interesting example of the challenges associated with topological classification. As described in Chaps. 6 and 7, graphene is a gapless material that has topologically protected states and quantized Hall conductance as shown in Chap. 6, Fig. 6.12. The Chern number of graphene has been calculated [31], and the relationship between the spin Chern number and the $\mathbb{Z}_2$ classification for topological insulators described [7, 32, 33]. The unusual Chern number observed for quantum Hall effect, [$\sigma_{xy} = \upsilon \frac{e^2}{h}$ *with* $\upsilon = \left(n + \frac{1}{2}\right)g_s$ with the degeneracy factor $g_s = 4$ due to spin and two Dirac cones] has been attributed to the presence of disorder in the graphene [31]. The presence of a magnetic field breaks time reversal symmetry, and Hall conductivity clearly demonstrates the violation of time reversal symmetry.

Table C.4 Classification of Gapless Materials

Symmetry class	Time reversal symmetry	Particle hole symmetry	Chiral symmetry	FS1 FS2	$p = 8$ $p = 2$	$p = 1$ $p = 3$	$p = 2$ $p = 4$	$p = 3$ $p = 5$
A	No	No	No		0	$\mathbb{Z}$	0	$\mathbb{Z}$
AIII	No	No	Yes		$\mathbb{Z}$	0	$\mathbb{Z}$	0
AI	+1	No	No		0	0^{a}	0	$2\mathbb{Z}$
BDI	+1	+1	Yes		$\mathbb{Z}$	0	0^{a}	0
D	No	+1	No		${\mathbb{Z}_2}^{b}$	$\mathbb{Z}$	0	0^{a}
DIII	−1	+1	Yes		${\mathbb{Z}_2}^{a,\,b}$	${\mathbb{Z}_2}^{b}$	$\mathbb{Z}$	0
AII	−1	No	No		0	${\mathbb{Z}_2}^{a,\,b}$	${\mathbb{Z}_2}^{b}$	$\mathbb{Z}$
CII	−1	−1	Yes		$2\mathbb{Z}$	0	${\mathbb{Z}_2}^{a,\,b}$	${\mathbb{Z}_2}^{b}$
C	No	−1	No		0	$2\mathbb{Z}$	0	${\mathbb{Z}_2}^{a,\,b}$
CI	+1	−1	Yes		0^{a}	0	$2\mathbb{Z}$	0

The topological classification for non-spatial symmetry classes for crystals with stable Fermi surfaces based on the Altland–Zirnbauer is listed. The non-spatial symmetries are time reversal, particle–hole, and chiral symmetry. The dimension p of the structure is related to the dimension d and the dimension of the Fermi surface where the Weyl points, Dirac points, or nodal lines occur, d_{FS}. For Weyl and Dirac points, $d_{FS} = 0$, and for nodal lines, $d_{FS} = 1$. The table is restricted to only a few of the many dimension reported in the literature. There are table entries for the dimension at FS1 and FS2 which refer to the dimension at high symmetry points for FS1 and away from high symmetry points for FS2. Other tables report all p dimensions 0–8 (see Table C.1, [12])

[a]Note that the designated entry for FS2 can have surface or bulk Fermi surfaces that are $\mathbb{Z}$ protected. The $\mathbb{Z}$ protection does not require time reversal or particle hole symmetry to be present

[b]The $\mathbb{Z}_2$ can't protected a stable Fermi surface away from high symmetry points [12]. Table adapted from [12]

Two different tight binding approaches have been used to make the Hamiltonian topologically non-trivial. Kane and Mele added spin orbit coupling to the Hamiltonian which opens a small bad gap [34], and other approaches have used a complex next nearest neighbor interaction [12] first introduced by Haldane [35, 37]. Spin orbit interactions open a gap at the Dirac points that is twice the spin orbit coupling [34]. This would mean that graphene would be classified as material with a bandgap (see Table C.1), and if the non-spatial symmetry is not altered by spin orbit interactions, graphene would be CII which has no topological classification for a 2D material (see Table C.1). *However*, the amount of spin orbit interaction in graphene is too small to change the properties of the nearest neighbor model [36]. This analysis does not consider spatial symmetry which can impart topological protection.

The addition of real (non-imaginary) next nearest neighbor interactions to the tight binding Hamiltonian does not change the time reversal and inversion symmetry of the Hamiltonian [34]. However, the addition of imaginary next nearest neighbor interactions breaks time reversal symmetry since the interaction depends on the direction of the hopping [12, 35, 36]. The next nearest neighbor hopping does not break time reversal symmetry, but there is no particle-hole symmetry, and since this approach does not introduce a band gap, the Altland–Zirnbauer classification is AI. When mirror symmetry is added, the topological classification becomes $M\mathbb{Z}$, and a mirror Chern number of $+1$ can be calculated [12]. We note that the details of the quantized Hall conductance data shown in Chap. 6, Fig. 6.12 are reproduced using the nearest neighbor tight binding model plus disorder [31].

Weyl Semimetals: In order to use the Altland–Zirnbauer classification, we first recall that either inversion symmetry or time reversal symmetry is broken in a Weyl semimetal. Here classify the inversion symmetry lacking Weyl semimetals TaAs, TaP, NbAs, and NbP where the 0D Weyl nodes are located away from high symmetry points [12]. These materials are time reversal symmetric initially giving them the AII classification [12]. However, the Weyl nodes are located away from high symmetry points and thus are not time reversal symmetric giving the Weyl Nodes Altland–Zirnbauer class A [12]. For $p = 3$, the class A the FS2 Weyl nodes are away from high symmetry points (FS2) and have $\mathbb{Z}$ protection. However, the Dirac surface states in the Fermi arc between the Weyl nodes have $\mathbb{Z}_2$ protection. We note that in general, Weyl nodes will have $\mathbb{Z}$ protection.

Note that Table C.4 would not be used for the classification of materials with magnetic ordering such as rare earth monopnictides which have $\mathbb{Z}_2$ protection.

Dirac Semimetals: In contrast to Weyl semimetals, Dirac semimetals preserve both inversion and time reversal symmetry. However, Dirac nodes that occur at momenta away from high symmetry points are not time reversal symmetric [12]. This results in a material that has an Altland–Zirnbauer classification of CII but the Dirac nodes having a different classification. Thus, it is useful to provide an example of the topological classification of a specific gapless material. One example is Na_3Bi which has the space group $P6_3/mmc$ (#194), and time reversal and inversion symmetry are present. The Altland–Zirnbauer classification for this space group is AII [38], but the Dirac nodal points ($p = 3$) are away from the high symmetry points in the Brillouin zone. Thus, the Dirac points for this material are not protected by time reversal symmetry. Using the FS2 designation and $p = 3$, a $\mathbb{Z}_2$ invariant can be defined for the crystal, but the Dirac points do not have the $\mathbb{Z}_2$ classification. Spatial symmetry stabilizes the topological protection for the Dirac nodal line for Na_3Bi [38].

Nodal Line Semimetals: For these materials, the Fermi surface, where the valence and conduction bands touch, is a line thus we use $d_{FS} = 1$, and $p = 2$ when classifying these materials. An example material is Ca_3P_2 which as discussed in Sect. 9.5.1.3 crystallizes in the hexagonal $P6_3/mcm$ (#193) space group. Ca_3P_2 structure has inversion and mirror symmetry and electronic band structure has time reversal symmetry. A Dirac nodal line occurs away from the high symmetry points in the Brillouin zone as shown in Fig. 9.2, and the line is on the Fermi surface [12]. The crystal structure is AZ class AII giving the crystal a $\mathbb{Z}_2$ classification, but the nodal line is away from the high symmetry points in the Brillouin zone. When spatial symmetry is also considered, the nodal line is protected by reflection and spin rotation symmetry.

References

1. J.C.Y. Teo, L. Fu, C.L. Kane, Surface states and topological invariants in three-dimensional topological insulators: application to $Bi_{1-x}Sb_x$. Phys. Rev. B 78, 045426 (2008)
2. A.P. Schnyder, S. Ryu, A. Furusaki, A.W.W. Ludwig, Classification of topological insulators and superconductors in three spatial dimensions. Phys. Rev. B **78**, 195125 (2008)
3. A. Kitaev, *Periodic Table for Topological Insulators and Superconductors* (2009). arXiv: 0901.2686v2 [cond-mat.mes-hall]
4. Z. Song, S.-J. Huang, Y. Qi, C. Fang, M. Hermele, Topological states from topological crystals. Sci. Adv. **5**, eaax2007 (2019)
5. M.G. Vergniory, L. Elcoro, C. Felser, N. Regnault, B. Andrei Bernevig, Z. Wang, A complete catalogue of high-quality topological materials. Nature **566**, 48–485 (2019)
6. B.-J. Yang, N. Nagaosa, Classification of stable three-dimensional Dirac semimetals with nontrivial topology. Nat. Comm. **5**, 4898 (2014)
7. P. Hořava, Stability of Fermi surfaces and *K* theory. Phys. Rev. Lett. **95**, 016405 (2005)
8. J.E. Moore, L. Balents, Topological invariants of time-reversal-invariant band structures. Phys. Rev. B **75**, 121306 (2007)
9. F. Zhang, C.L. Kane, E.J. Mele, Surface states of topological insulators. Phys. Rev. B **86**, 081303(R) (2012)
10. T. Morimoto, A. Furusaki, Topological classification with additional symmetries from Clifford algebras. Phys. Rev. B **88**, 125129 (2013)
11. K. Shiozaki, M. Sato, K. Gomi, Topology of nonsymmorphic crystalline insulators and superconductors. Phys. Rev. B **93**, 195413 (2016)
12. C.-K. Chiu, J.C.Y. Teo, Andreas P. Schnyder, S. Ryu, Classification of topological quantum matter with symmetries. Rev. Mod. Phys. **88**, 035005 (2016)
13. K. Shiozaki, M. Sato, K. Gomi, Topological crystalline materials: general formulation, module structure, and wallpaper groups. Phys. Rev. B **95**, 235425 (2017)
14. B.J. Wieder, B. Bradlyn, Z. Wang, J,. Cano, Y. Kim, H.-S.D. Kim, A.M. Rappe, C.L. Kane, B. Andrei Bernevig, Wallpaper fermions and the nonsymmorphic Dirac insulator. Science **361**, 246–251 (2018)
15. Z. Song, T. Zhang, Z. Fang, C. Fang, Quantitative mappings between symmetry and topology in solids. Nat. Comm. **9**, 3530 (2018)
16. E. Khalaf, H.C. Po, A. Vishwanath, H. Watanabe, Symmetry indicators and anomalous surface states of topological crystalline insulators. Phys. Rev. X **8**, 031070 (2018)
17. R.M. Kaufmann, D. Li, B. Wehefritz-Kaufmann, Notes on topological insulators. Rev. Math. Phys. **28**, 1630003 (2016)
18. E. Cornfeld, Private communication
19. E. Cobanera, Private communication
20. E. Cornfeld, A. Chapman, Classification of crystalline topological insulators and superconductors with point group symmetries. Phys. Rev. B **99**, 075105 (2019)
21. E. Cornfeld, S. Carmeli, *Tenfold Topology of Crystals*, arXiv:2009.04486v2 (2020)
22. A. Altland, M.R. Zirnbauer, Random matrix theory of a chaotic Andreev quantum dot. Phys. Rev. Lett. **76**, 3420–3423 (1996)
23. A. Altland, M.R. Zirnbauer, Nonstandard symmetry classes in mesoscopic normal-superconducting hybrid structures. Phys. Rev. B **55**, pp. 1142–1161 (1997)
24. F.J. Dyson, The threefold way. Algebraic structure of symmetry groups and ensembles in quantum mechanics. J. Math. Phys. **3**, 1199 (1962)
25. L. Fu, C.L. Kane, E.J. Mele, Topological insulators in three dimensions, Phys. Rev. Lett. **98**, 106803 (2007)
26. H. Weng, X. Dai, Z. Fang, Transition-metal pentatellurides $ZrTe_5$ and $HfTe_5$: a paradigm for large-gap quantum spin Hall insulators. Phys. Rev. X **4**, 011002 (2014)

27. G. Manzoni, L. Gragnaniello, G. Autès, T. Kuhn, A. Sterzi, F. Cilento, M. Zacchigna, V. Enenkel, I. Vobornik, L. Barba, F. Bisti, P. Bugnon, A. Magrez, V.N. Strocov, H. Berger, O.V. Yazyev, M. Fonin, F. Parmigiani, A. Crepaldi, Evidence for a strong topological insulator phase in $ZrTe_5$. Phys. Rev. Lett. **117**, 237601 (2016)
28. H. Wang, C.-K. Li, H. Liu, J. Yan, J. Wang, J. Liu, Z. Lin, Y. Li, Y. Wang, L. Li, D. Mandrus, X.C. Xie, J. Feng, J. Wang, Chiral anomaly and ultrahigh mobility in crystalline $HfTe_5$. Phys. Rev. B **93**, 165127 (2016)
29. N. Kumar, C. Shekhar, M. Wang, Y. Chen, H. Borrmann, C. Felser, Large out-of-plane and linear in-plane magnetoresistance in layered hafnium pentatelluride. Phys. Rev. B **95**, 155128 (2017)
30. Z. Song, S.-J. Huang, Y. Qi, C. Fang, M. Hermele, Topological states from topological crystals. Sci. Adv. **5**, eaax2007 (2019)
31. D.N. Sheng, L. Sheng, Z.Y. Weng, Quantum Hall effect in graphene: disorder effect and phase diagram. Phys. Rev. B **73**, 233406 (2006)
32. T. Fukui, Y. Hatsugai, Topological aspects of the quantum spin-Hall effect in graphene: Z_2 topological order and spin Chern number. Phys. Rev. B **75**, 121403 (2007)
33. C.L. Kane, Topological band theory and the Z_2 classification, in *Topological Insulators*, ed. by M. Franz, L. Molenkamp (Elsevier, Oxford, 2013), pp. 3–34 (Chapter 1)
34. C.L. Kane, E.J. Mele, Quantum spin hall effect in grapheme. Phys. Rev. Lett. **95**, 226801 (2005)
35. F.D.M. Haldane, Model for a quantum Hall effect without Landau levels: condensed-matter realization of the "parity anomaly". Phys. Rev. Lett. **61**, 2015 (1988)
36. T. Fukui, Y. Hatsugai, Topological aspects of the quantum spin-Hall effect in graphene: Z_2 topological order and spin Chern number. Phys. Rev. B **75**, 121403 (2007)
37. T.-W. Chen, Z.-R. Xiao, D.-W. Chiou, G.-Y. Guo, High Chern number quantum anomalous Hall phases in single-layer graphene with Haldane orbital coupling. Phys. Rev. B **84**, 165453 (2011)
38. C.-K. Chiu, A.P. Schnyder, Classification of crystalline topological semimetals with an application to Na_3Bi. J. Phys. Conf. Ser. **603**, 012002 (2015)

Index

A. Diebold and T. Hofmann, *Optical and Electrical Properties of Nanoscale Materials*,
Springer Series in Materials Science 318,
https://doi.org/10.1007/978-3-030-80323-0

D

E

F

L

M

N

O

P

Q

R